ANIMAL MODELS OF HUMAN PATHOLOGY

A Bibliography of a Quarter Century of Behavioral Research 1967–1992

Editors

J. Bruce Overmier, PhD
Center for Research in Learning, Perception & Cognition
University of Minnesota
Minneapolis, MN

Patricia D. Burke
Analyst
PsycINFO Coverage Development & Management Unit
American Psychological Association
Washington, DC

Bibliographies in Psychology No. 12

American Psychological Association

Library of Congress Cataloging-in-Publication Data

Animal models of human pathology: a bibliography of a quarter century of behavioral research, 1967-1992
/ J. Bruce Overmier. Patricia D. Burke, editors.
p. cm. -- (Bibliographies in psychology ; no. 12)
Includes bibliographical references and indexes.

ISBN 1-55798-184-1
1. Mental illness--Animal models--Bibliography. 2. Animal psychopathology--Bibliography. I. Overmier, J. Bruce. II. Burke, Patricia D.
III. Series
[DNLM: 1. Disease models, Animal--abstracts. 2. Pathology--abstracts. 3. Research--abstracts. ZQZ 20.5 A598 1967-92]
Z6665.7.A53A45 1992
[RC455.4.A54]
016.61689'027--dc20
DNLM/DLC CIP
for Library of Congress 92-49962

First Edition

Published by the
American Psychological Association
750 First Street, N.E.
Washington, DC 20002

Copies may be ordered from:
APA Order Department
P.O. Box 2710
Hyattsville, MD 20784-0710
Item Number 431-9140

Typesetter: PageCentre, Phoenix, AZ
Printer: Patterson Printing, Benton Harbor, MI
Production Coordinator: Deborah Segal

Printed in the United States of America

BIBLIOGRAPHIES IN PSYCHOLOGY SERIES

1. Black Males in the United States: An Annotated Bibliography from 1967 to 1987
2. AIDS: Abstracts of the Psychological and Behavioral Literature, 1983–1988
   Out of print. Replaced by second edition (see no. 6) July 1989.
3. Black Females in the United States: A Bibliography from 1967 to 1987
4. Alzheimer's Disease: Abstracts of the Psychological and Behavioral Literature
5. Environmental Toxins: Psychological, Behavioral, and Sociocultural Aspects, 1973–1989
6. AIDS: Abstracts of the Psychological and Behavioral Literature, 1983–1989, Second Edition
   Out of print. Replaced by third edition, July 1991.
   AIDS: Abstracts of the Psychological and Behavioral Literature, 1983–1991, Third Edition
7. Homelessness: Abstracts of the Psychological and Behavioral Literature, 1967–1990
8. Hispanics in the United States: Abstracts of the Psychological and Behavioral Literature, 1980–1989
9. Child Abuse: Abstracts of the Psychological and Behavioral Literature, Volume 1, 1967-1985
   Child Abuse: Abstracts of the Psychological and Behavioral Literature, Volume 2, 1986–1990
10. Peace: Abstracts of the Psychological and Behavioral Literature, 1967-1990
11. Asians in the United States: Abstracts of the Psychological and Behavioral Literature, 1967-1991
12. Animal Models of Human Pathology: A Bibliography of a Quarter Century of Behavioral Research, 1967-1992

ACKNOWLEDGMENTS

We gratefully acknowledge the assistance of the following PsycINFO staff persons: Sylvia Mitchell for guidance of this publication; Verna Walker, Gary Broyhill, Weldon Bagwell, Jody Kerby, James Whitfield, Anita Garvey, and Kathleen McEvoy for proofreading/editing the records; Donald Dailey and Karen Monroe for modifying computer programs to structure the output of the data; Marion Coates and Maureen Madison for gathering and preparing the book records; Elizabeth Simon for reviewing and modifying the photocomposition specifications; and James Whitfield for assistance in desktop publishing. Elizabeth Bulatao of the Office of Communications and Deborah Segal of APA Books have also made significant contributions.

CONTENTS

CONTENTS

Bibliographies in Psychology, No.12
Animal Models of Human Pathology:
A Quarter Century of Behavioral Research, 1967-1992
Published by the American Psychological Association, 1992

On the Nature of Animal Models of Human Behavioral Dysfunction

J. Bruce Overmier
Center for Research in Learning, Perception, and Cognition
University of Minnesota

The Dark Ages were so designated because in Europe there were substantial religious and social prohibitions against activities that might yield new information or new perspectives. These prohibitions applied to the use of animals for dissection as part of anatomical study, and medicine stagnated for nearly a millennium. Yet, although this prohibition on the study of the anatomy and physiology of animals passed with the coming of the Age of Enlightenment, the church still limited what we might infer about humans from the study of animals. This was captured formally in Descartes's seventeenth-century philosophical doctrine of mind–body dualism: The mental life of humans was set apart from that of animals. This metatheoretic belief system plagues us to this day (Dennett, 1991; Koestler, 1967), despite the seminal contributions of Charles Darwin and Herbert Spencer who argued the continuity of emotions and mind from animals to man and the potential value of a comparative psychology.

It was Pavlov, following Sechenov's dicta on the objective study of the mind, who was most instrumental in breaking the conceptual barriers to animal models in psychology, with his materialistic application of conditioned reflex methodology. Pavlov not only suggested that we might learn about the function of the mind from the objective study of behavior, but he was also the first to suggest that experimentally induced behavioral dysfunctions might indeed be informative to the study of human mental dysfunction as well. We are all familiar with the famous Shenger-Krestinova (1921) appetitive classical conditioning experiments with dogs involving a series of increasingly difficult discriminations between a circle and an ellipse in which the animals' behavior finally became so agitated and erratic that the dysfunction was designated *experimental neurosis*. Many are less familiar with the follow-up work in which Krasnogorski (1925) carried out similar experiments with children in an auditory discrimination task—and with somewhat similar results. Moreover, Pavlov and his associates found bromide salts to be an effective "therapy" for <u>both</u> the dogs and the children (see Babkin, 1938).

There followed a number of efforts to create what we call animal models of psychopathology. The works of two of Pavlov's American students are most readily recognized in this regard. There was Gantt's 12-year series of experiments with the neurotic dog, Nick (Gantt, 1944). And there was the work of Liddell, at the Cornell Behavior Farm, in which young goats and sheep were subject to defensive classical conditioning either in the presence or absence of their mothers; this work eventuated in Liddell's intriguing little book, *Emotional Hazards in Animals and Man* (1956).

The important observations made by Gantt and Lidell that maladaptive behavior patterns analogous to human neuroses could be conditioned—that is, in essence taught—inspired a number of others to explore the possibility of a new scientific study of psychopathology based in animal research. Two important exemplars are N. R. F. Maier's (1949) research with rats, in which being challenged with unsolvable problems caused behavioral fixations that he considered a form of *compulsivity,* and Masserman's (1943) brief punishment of cat's consummatory behavior that caused persistent avoidant behavior that he thought a form of *phobic neuroses,* which he showed were treatable by *environmental press*—a type of extinction procedure. The latter is significant for con-

I wish to thank Susan Mineka and Robert Murison for their constructive criticisms on an earlier draft of this chapter.

temporary clinical psychologists because it was Masserman's research that prompted Wolpe (1958) to his experiments with cats from which he derived the principles for *reciprocal inhibition* therapy so widely and effectively used today to treat phobias. Whether the experimental procedures were in fact a scientifically valid basis for the inferences Wolpe drew in regard to appropriate therapy may be argued at length elsewhere (e.g., Mineka, 1985). Nonetheless, the heuristic value of animal models is clearly established in contemporary psychology by this exemplar, and it also encouraged others to pursue animal models research in their study of psychopathology (which will be returned to) as well as other applied issues such as self-control (e.g., Mahoney & Bandura, 1972).

Animal models offered a promise to treat psychopathology, not as bizarre distortions of behavior, but rather as involving lawful processes whose principles and mechanisms we could come to understand scientifically—to move psychiatry into the modern age. These experimental models of neuroses, or *experimental neurosis*, were seen as critically important as a counter influence to earlier kinds of mythological and anecdotal speculations about the causal factors and to Freudian analysis in human psychopathology that led to treatments such as bleeding the evil humors away for depression, castration as a "treatment" for masturbation, and confinement in an orgone box for neuroses.

A substantial portion of the research efforts in all areas of psychology involves the use of models, and clinical psychology should not be any different if such models can give purchase on phenomena at issue. One sees reliance on models that are mechanical (e.g., cochlear models for audition), conceptual (layered network models of brain function), process-oriented (computer models of thinking), as well as biological (the invertebrate *aplysia* model of vertebrate learning), as only a few types of models (and exemplars among literally hundreds of models) in psychology (Bekesy,1960; Schmajuk & DiCarlo, 1992; Wagman, 1991; and Byrne & Berry, 1989, respectively as examples). Models are basic and powerful tools in science. The aeronautical scientist builds a miniature airplane for testing in a wind tunnel; the chemical scientist imagines electrons in shared planet-like orbits around atomic nuclei. These two types of models (one physical and one conceptual) aid in the discovery of useful principles for addressing real world problems such as optimal wing shape for speed or how smooth a shape molecules will take when bonded to the wing's surface.

Models are equally important in biological and behavioral sciences, and this accounts in part for why so much research has been and continues to be done with animals in psychology. Explicit contemporary uses of animal models in psychology are more accepted and less controversial in research on neural mechanisms of learning (Kandel, 1979), perceptual–cognitive mechanisms (Goldman-Rakic, 1987), the mysteries of memory dysfunction in aging populations (e.g., Kesner & Olton, 1990; Lister & Weingartner, 1991; Squire & Zola-Morgan, 1985), mechanisms of drug abuse by humans (Siegel, 1983), and psychopharmacology (Willner, 1991) than they are in addressing behavioral issues in psychopathology (Davey, 1981; Mineka & Zinbarg, 1991). And these models are fundamental components of the successes in those areas where they have been applied. So why is there so substantial a conceptual resistance to their use in understanding human psychopathology? Indeed, the question seems even more cogent given that the existing limited uses of such models have in fact led to development of new therapies for phobias (Stampfl & Levis, 1967; Wolpe, 1958) and depression (Klein & Seligman, 1976). I believe that the bases for this resistance are that (a) there is a lingering Cartesian dualism and (b) the structure and function of models—and animal models in particular—are not well understood. The former has been addressed by Dennett (1991) among others; the latter we address below.

Elements of models

The key word for understanding models is *analogy*. A model is not considered a claim of identity with that which is being modeled. Rather, a model is a convergent set of several kinds of analogies between the "target" real world phenomenon to be understood and the system that is being studied as a model for the target phenomenon. Two key kinds of analogies involved are (a) initial analogy and (b) formal analogy.

To understand the interplay of these different kinds of analogies in the modeling process, we need to note that any phenomenon we wish to model is not "just a thing" arising *deus ex machina*, but the consequence of causal relationships between levels of factors—perhaps unknown factors—in the real world. Similarly, then, any potentially useful model will involve a set of causally related factors. These causal chains of factors in both the target domain and the model domain may be several steps long. Models arise from the claims of correspondence between factors in the two domains. The two domains can have obvious similarities, as between a miniature airplane and a Boeing 747, or they can be dissimilar, as between the ball-and-stick arrays in molecular chemistry that represent the genetic substance DNA and the molecule itself.

Now let us consider the animal models case in psychology where there are both similarities and differences. One might note that some set of dysfunctional physiological and behavioral symptoms characterize patients with a given psychiatric disorder (e.g., refractory but elevated steroids, inability to cope with challenges, and distortions of memory); one might further note that animals exposed to some drug or to some experimental learning treatment exhibit behaviors that are similar to those behavioral symptoms of the patients. A hypothesis that the dysfunctional behavior of the animal and the dysfunctional behavior of the patient were similar in important ways would constitute an initial analogy in the modeling process. An additional hypothesis might be that the patient's dysfunctional physiological symptom is related to the animal's drug-induced physiological state; this would be a second initial analogy. The degree of descriptive similarity between the two sets of behaviors or between the two physiological states would constitute the degree of material or conceptual equivalences.

Now if a relation between the patient's physiology and the patient's behavior is hypothesized to parallel the empirical causal relation between the animal's physiology and its behavior, a formal analogy can be drawn between these two parallel, within-domain relations. It is this formal analogy--the hypothesized parallelism of causal relations in the two domains—that constitutes a functional model. Indeed, this is exactly how many models in psychology have arisen such as the amphetamine-based model of schizophrenia (Bell, 1965) or the cholinergic depletion model of Korsakov's syndrome (Overstreet & Russell, 1984); development of other models emphasized environmental and learning history as causal factors rather than drug history (e.g., the avoidance model of phobias, Mowrer, 1947; the learned-helplessness model of depression, Seligman, 1975; and the opponent process model of addictions, Solomon & Corbit, 1974). These are only five examples among many models.

The degree of descriptive similarity between the elements in an initial analogy constitutes the degree of material equivalence—sometimes confusingly referred to as "material analogy" (Overmier & Patterson, 1988). But material equivalence is often simply out of the question for some modeling efforts. In such cases, one may rely upon additional theoretical notions to place elements for an initial analogy into correspondence. This then is conceptual equivalence. Conceptual equivalences may also vary in degree depending on how many theoretical transformations are required to set the elements of the initial analogy into correspondence. Adrenal cortical output in a rat (corticosterone) and a human (cortisol) are not identical, but have a very high degree of material equivalence. On the other hand, a rat's pressing of a lever and a human's engaging in grocery shopping must be put into correspondence on the basis of a conceptual equivalence: Both are instrumental acts of food procurement. This reconceptualization seems straight forward enough; although some such shopping is certainly engaged in for reasons other than food-getting, the same is true for rats' lever pressing. However, placing inescapable electric shocks received by dogs into correspondence with failure by humans to solve anagrams requires several conceptual transformations: Minimally, one must first conceive of failure to be as aversive event as shock; then, these aversive events must be seen as functionally equivalent; and finally, the behavior of the dog in the presence of the shock must be conceived of as directed problem-solving behavior. Clearly, the conceptual distance here is greater than in the case of food procurement. Although there is often a preference for material equivalence or close conceptual equivalence (perhaps because it does not strain one's intellectual powers much, although it may be a substantial technical challenge), it is important to note the greater material or conceptual distance does not directly influence the validity or value of a model. Validity and value inheres in the degree to which the formal analogy provides correct information.

It should be clear now that an initial analogy alone is not a functional model; assuming that an initial analogy is a model is a common mistake. A true model involves both initial analogies and formal analogies, and the power of the modeling process is that one can use the known and explicated casual relations in one domain (typically the model domain, but it can work both ways [e.g., Blackman, 1983; Dorworth & Overmier, 1977]) as a guide for finding parallel relations in the second domain.

Some scientists (e.g., Abramson & Seligman, 1977, and McKinney, 1974) require that a large number of formal analogy parallels be proved, all involving substantial material equivalence, before a claim of a model be made. This requirement emphasizes the important representative functions of models (e.g., for development of drug therapy) to the exclusion of their heuristic and evidentiary functions important for learning about the causal and sustaining mechanisms of the dysfunction. Two very important points need to be made here: First, material analogy is not critical to the functional validity of a model; mathematical equations are often powerful models which when processed by computers generate knowledge about systems' behavior—even human decision systems—but without material equivalence. Material equivalence between elements is akin to *face validity*; although it offers promise, it ensures nothing. Second, if we wait until all the causal relations in each domain are fully and independently explicated before we set them into formal analogies, then the model yields little or no evidentiary power for new understanding.

The exemplar psychological models, already noted, all began with an initial analogy being made between behaviors—a so-called *symptomological match*. One can immediately see why this would be so: When two things have the same form, we often assume they are functionally similar, perhaps even homologous. But one might well argue that finding such similarities in symptoms between species is a very chancy process—possibly a misleading one, although subject to empirical testing within the modeling process. This is because each species brings its own evolved propensities and biological constraints on its behavior to the test arena. Different behaviors in different species could serve the same function although common behaviors could serve different functions. Thus, it is conceivable that two different species (e.g., humans and rats) might have opposite dysfunctional behavioral manifestations arising from the same underlying physiological or psychological state. For example, one might freeze, whereas the other might show agitation; yet these two forms of behavior could be part of a useful initial analogy despite their lack of material equivalence. Although, in actual practice, this has not proved a common problem, it does suggest an alternative approach to choosing one's initial analogy.

One might well choose to put into initial correspondence etiological factors rather than symptoms (e.g., see Cullen, 1974, for illustrations). This strategy is common in medical research and is based upon the assumption that etiology and therapy are necessarily linked because in some diseases the causes also sustain the disease process, as in the case of bacterial infection. In contrast, in psychology, we have found that the sustaining conditions for some behavioral processes are often different from those that led to their development. Korsakov's syndrome and phobias seem to be exact exemplars of this. On the other hand, posttraumatic stress disorder (PTSD) may prove to be a case requiring an approach based on an etiological initial analogy if one seeks an animal model of the consequences of traumatic stress, because the reported human symptoms tend to take the form of thought disruptions, flashbacks, and other symptoms that are not directly observable in animals (Basoglu & Mineka, in press; Foa, Zinbarg, & Rothbaum, in press; Pitman, 1988, 1989; Saporta & VanderKolk, in press).

It is perhaps also worth noting here that one area of application that assumes that etiology and therapy are not inextricably linked is operant behavior modification. This form of treatment also has as one of its metatheoretic assumptions the generality of the behavioral principles studied in the laboratory—and commonly with animals (Brown, Weinckowski, & Stolz, 1975; Davey, 1981;

Skinner, 1972). And, operant behavior modification has proved a successful basis for treatments of a wide range of behavioral dysfunctions (illustrated in Feldman & Broadhurst, 1976; Krasner & Ullman, 1965; and the articles published in the *Journal of Applied Behavior Analysis*). But the conceptual origins of behavior modification are to be found in Watson's early effort to "teach a phobia" to Little Albert (Watson & Raynor, 1920) and Mary Cover Jones's treatment of Peter for such "fears" (Jones, 1924). Although this work was with children, it was nonetheless an instance of modeling.

Final Considerations

The foregoing has tried to show that animal models have been and continue to be a ubiquitous component of psychological research into human dysfunction. The space provided here has allowed merely for an introduction to the structure and functions of models and the correlated considerations in the use of models. The interested reader is directed to key analytic discussions (Bond, 1984; Fox, 1971; Hanin & Usdin, 1977; Henn & McKinney, 1987; Hinde, 1976; McKinney, 1988; Overmier & Patterson, 1988; Russell, 1964; VonCranach, 1976; Willner, 1986), illustrative critical evaluations of particular models (Costello, 1978; Eysenck, 1979; Green, 1983; Kaufman, 1973; Katz, 1981; Maser & Seligman, 1977; Mineka, 1985; Willner, 1984), and interesting historical reviews (Abramson & Seligman, 1977; Babkin, 1938; Broadhurst, 1960; Cook, 1939; Mowrer, 1947; Richter, 1957; Wolpe, 1952; Zubin & Hunt, 1967). Review of these will put into better perspective the approximately 2,000 items in the following bibliography.

It is also appropriate to note that this modeling process is not always as clearly understood as it should be by either its proponents or its detractors. Unfortunately, some detractors also criticize contemporary animal modeling efforts based on an outdated grasp of the current state of knowledge about animal learning and behavior (Mineka, 1985, gives instances of this). Criticisms based on poor appreciation of the process or outdated information should not be allowed to dissuade either those wishing to use such models or those who provide the funds for such research.

This is not to say that the use of animal models is without its share of problems (e.g., Kornetsky, 1977; Rollin & Kesel, 1990). The model-building process is fraught with difficulties—conceptual, analogic translational, empirical, interpretive, and extensional—but then so is *all* research.

Behavioral dysfunctions and psychiatric disorders must be studied if we are to bring relief to the literally tens of millions of sufferers. And until we have some significant grasp on the processes involved, we are ethically deterred from some classes of research with human subjects and patient populations—in particular those experimenting with etiologies or with therapies that involve physiological changes not yet understood. This grasp is to be gained only through the use of animal models.

There are also ethical considerations in the use of animals for such research. We are obligated to ensure that we fully understand the modeling process and that the analogies in the model we are developing are sound and will pass the test of review by other scientists with a relevant knowledge base. Even such models may sometimes involve necessarily the induction of distress—after all, it appears that physical and emotional distress is involved in the etiology of many common forms of human dysfunctional behavior. When the protocol properly requires the induction of distress, we should not shirk from this if there is not an equally effective alternative. This is because the failing to do appropriate research also carries with it significant costs in future human suffering.

References

Abramson, L. Y. & Seligman, M. E. P. (1977). Modeling psychopathology in the laboratory: History and rationale. In J. D. Maser & M. E. P. Seligman (Eds.), *Psychopathology: Experimental model* (pp. 1–27). San Francisco: Freeman.

Babkin, B. P. (1938). Experimental neuroses in animals and their treatment with bromides. *Edinburgh Medical Journal, 45,* 605–619.

Basoglu, M., & Mineka, S. (in press). The role of uncontrollable and unpredictable stress in post-traumatic stress responses in torture survivors. In M. Basoglu (Ed.), *Torture and its consquences: Current treatment approaches*. London: Cambridge University Press.

Bekesy, G. von (1960). *Experiments in hearing*. New York: McGraw-Hill.

Bell, D. S. (1965). Comparison of amphetamine psychosis and schizophrenia. *British Journal of Psychiatry, 111,* 701–707.

Blackman, D. E. (1983). On cognitive theories of animal learning: Extrapolations from humans to animals? In G. C. L. Davey (Ed.), *Animal models of human behaviour* (pp. 37–50). Bristol: John Wiley.

Bond, N. (1984). Animal models in psychopathology: An introduction. In N. Bond (Ed.), *Animal models of psychopathology* (pp 1–21). Sydney: Academic Press.

Boulton, A. A., Baker, G. B., & Martin-Iverson, M. T. (Eds.). (1991). *Animal Models in Psychiatry*. Clifton, NJ: Humana Press.

Broadhurst, P.L. (1960). Abnormal animal behavior. In H. J. Eysenck (Ed.), *Handbook of abnormal psychology* (pp. 726–763). New York: Basic Books.

Brown, B. S., Weinckowski, L. A., & Stolz, S. B. (1975). *Behavior modification: Perspective on a current issue* (Publication No. ADM 75-202). Washington, DC: U. S. Department of Health Education & Welfare.

Byrne, J. H., & Berry, W. O. (Eds.). (1989). *Neural models of plasticity*. New York: Academic Press.

Cook, S. W. (1939). A survey of methods used to produce 'experimental neurosis.' *American Journal of Psychiatry, 95,* 1259–1276.

Costello, C. G. (1978). A critical review of Seligman's laboratory experiments on learned helplessness and depression in humans. *Journal of Abnormal Psychology, 87,* 21–31.

Cullen, J. H. (Ed.). (1974). *Experimental behaviour*. Dublin: Halsted Press.

Davey, G. (Ed.). (1981). *Applications of conditioning theory*. London: Methuen.

Dennett, D. C. (1991). *Consciousness explained.* New York: Little, Brown & Company.

Dorworth, T., & Overmier, J. B. (1977). On learned helplessness: The therapeutic effects of electroconvulsive shocks. *Physiological Psychology, 5,* 355–358.

Eysenck, H. J. (1979). The conditioning model of neurosis (plus commentaries). *Behavioral & Brain Sciences, 2,* 155–199.

Feldman, M. P., & Broadhurst, A. (Eds.). (1976). *Theoretical and experimental bases of the behaviour therapies*. London: John Wiley.

Foa, E., Zinbarg, R., & Rothbaum, B. (in press). Uncontrollability and unpredictability in post-traumatic stress disorder: An animal model. *Psychological Bulletin, 112,* 218–238.

Fox, M. W. (1971). Towards a comparative psychopathology. *Zeitschrift fur Tierpsychologie, 29,* 416–437.

Gantt, W. H. (1944). *Experimental basis of neurotic behavior*. New York: Hoebner.

Goldman-Rakic, P. S. (1987). Circuitry of primate prefrontal cortex and regulation of behavior by representational knowledge. In F. Plum (Ed.), *Handbook of physiology: The nervous system, Vol V: Higher cortical function* (pp. 373–417). Bethesda, MD: American Physological Society.

Green, S. (1983). Animal models of schizophrenia. In G.C.L. Davey (Ed.), *Animal models of human behaviour* (pp. 315–337). Chichester: John Wiley.

Hanin, I., & Usdin, E. (Eds.). (1977). *An animal model in psychiatry & neurology*. Oxford: Pergamon.

Henn, F. A., & McKinney, W. T. (1987). Animal models in psychiatry. In H. Y. Meltzer (Ed.), *Psychopharmacology: The third generation of progress.* (pp. 687–695). New York: Raven.

Hinde, R. A. (1976). The use of differences and similarities in comparative psychopathology. In G. Serban & A. Kling (Eds.), *Animal models in human psychobiology.* (pp. 187–202). New York: Plenum.

Jones, M. C. (1924). The elimination of children's fears. *Journal of Experimental Psychology, 7,* 382–390.

Kandel, E. R. (1979). Cellular insights into behavior and learning. *Harvey lectures, 73,* 19–92.

Kaufman, I. .C. (1973). Mother–infant separation in monkeys—an experimental model. In J.P. Scott & E. Senay (Ed.), *Separation and depression.* Washington, DC: AAAS.

Katz, R. J. (1981). Animal models of human depressive disorders. *Neuroscience & Biobehavioral Reviews, 5,* 231–246.

Kesner, R. P., & Olton, D. S. (1990). *Neurobiology of comparative cognition.* Hillsdale, NJ: Lawrence Erlbaum.

Klein, D. C., & Seligman, M. E. P. (1976). Reversal of performance deficits in learned helplessness and depression. *Journal of Abnormal Psychology, 85,* 11–26.

Koestler, A. (1967). *The ghost in the machine.* London: Hutchinson.

Kornetsky, C. (1977). Animal models: Promises and problems. In I. Hanin & E. Usdin (Eds.), *Animal models in psychiatry and neurology.* (pp. 1–7). Oxford: Pergamon.

Krasner, L., & Ullman, L. (Eds.). (1965). *Research in behavior modification.* New York: Holt, Rinehart, & Winston.

Krasnogorski, N. I. (1925). The conditioned reflexes and children's neuroses. *American Journal of Diseases of Children, 30,* 753–768.

Liddell, H. S.(1956). *Emotional hazards in animals and man.* Springfield, IL: Charles C. Thomas

Lister, R. G., & Weingartner, H. J. (1991). *Perspectives in cognitive neuroscience.* Oxford: Oxford University.

Mahoney, M. J. & Bandura, A. (1972). Self-reinforcement in pigeons. *Learning & Motivation, 3,* 293–303.

Maier, N. R .F. (1949). *Frustration: The study of behavior without a goal.* New York: McGraw Hill.

Maser, J. D., & Seligman, M. E. P. (Eds.). (1977). *Psychopathology: Experimental models.* San Francisco: Freeman.

Masserman, J. H. (1943). *Behavior and neurosis.* Chicago: University of Chicago.

McKinney, W. T. (1974). Animal models in psychiatry. *Perspectives in Biology and Medicine, 17,* 529–541.

McKinney, W. T. (1988). *Models of mental disorders: A new comparative psychiatry.* New York: Plenum Press.

Mineka, S. (1985). Animal models of anxiety based disorders: Their usefulness and limitations. In A. H. Tuma & J. D. Maser (Eds.), *Anxiety and anxiety disorders* (pp. 199–244). Hillsdale: Erlbaum.

Mineka, S., & Zinbarg, R. (1991). Animal models of psychopathology. In C. E. Walker (Ed.), *Clinical psychology: Historical and research foundations* (pp. 51–86) New York: Plenum Press.

Mowrer, O. H. (1947). On the dual nature of learning—A reinterpretation of "conditioning" and "problem solving." *Harvard Educational Review, 17,* 102–148.

Overmier, J. B., & Patterson, J. (1988). Animal models of human psychopathology. In P. Simon, P. Soubrie', & D. Wildlocher (Eds.), *Selected models of anxiety, depression, and psychosis.* (pp 1–35). Basel: Karger.

Overstreet, D. H., & Russell, R. W. (1984). Animal models of memory disorders. In A. A. Boulton et al. (Eds.), *Animal models in psychiatry: Neuromethods, 19* (pp. 315–368). Clifton, NJ: Humana Press.

Pitman, R. K. (1988). Post-traumatic stress disorder, conditioning and network theory. Psychiatric Annals, *18*, 182–189.

Pitman, R. K. (1989). Post-traumatic stress disorder, hormones, and memory. *Biological Psychiatry, 26*, 221–223.

Richter, C. P. (1957). On the phenomenon of sudden death in animals and man. *Psychosomatic Research, 19*, 191–198.

Rollin, B. E., & Kesel, M. L. (Eds.). (1990) *The experimental animal in biomedical research. Vol I.* Boca Raton: CRC Press.

Russell, R. W. (1964). Extrapolation from animals to man. In H. Steinberg (Ed.), *Animal behaviour and drug action* (pp. 410–418). London: JA Churchill.

Saporta, J. A., & VanderKolk, B. A. (in press). Psychobiological consequences of severe trauma. In M. Basoglu (Ed.), *Torture and its consequences: Current treatment approaches.* London: Cambridge University Press.

Schmajuk, N. A., & DiCarlo, J. J. (1992). Stimulus configuration, classical conditioning, and hippocampal function. *Psychological Review, 99*, 268–305.

Seligman, M. E. P. (1975). *Helplessness: On death, dying, and depression.* San Francisco: Freeman.

Shenger-Krestinova, N. R. (1921). Contributions to the question of differentiation of visual stimuli and the limits of differentiation by the visual analyser of the dog. *Bulletin of the Lesgaft Institute of Petrograd, 3*, 1–43.

Siegel, S. (1983). Classical conditioning, drug tolerance, and drug dependence. In Y. Israel, F. B. Glaser, H. Kalant, R. E. Popham, W. Schmidt, & R. G. Smart (Eds.), *Research advances in alcoholsim and drug problems (Vol 7).* New York: Plenum.

Skinner, B. F. (1972). What is psychotic behavior? *Cumulative Record.* New York: Appleton-Century-Crofts.

Solomon, R. L., & Corbit, J. D. (1974). An opponent process theory of motivation: I. The temporal dynamics of affect. *Psychological Review, 81*, 119–145.

Squire, L., & Zola-Morgan, S. (1985). The neuropsychology of memory: New links between humans and experimental animals. *Annals of the New York Academy of Sciences, 444*, 137–149.

Stampfl, T., & Levis, D. (1967). Essentials of implosive therapy: A learning theory based on psychodynamic behavioral therapy. *Journal of Abnormal Psychology, 28*, 496–503.

VonCranach, M. (1976). *Methods of inference from animal to human behaviour.* Chicago: Aldine.

Watson, J. B., & Raynor, R. (1920). Conditioned emotional reactions. *Journal of Experimental Psychology, 3*, 1–14.

Wagman, M. (1991). *Artificial intelligence and human cognition: A theoretical intercomparison of two realms of human intellect.* New York: Praeger.

Willner, P. (1984). The validity of animal models of depression. *Psychopharmacology, 83*, 1–16.

Willner, P. (1986). Validation criteria for animal models of human mental disorders: Learned helplessness as a paradigm case. *Progress in Neuro-psychopharmacological & Psychiatry, 10*, 677–690.

Willner, P. (1991). *Behavioural models in psychopharmacology.* London: Cambridge University Press.

Wolpe, J. (1952). Experimental neuroses as learned behavior. *British Journal of Psychology, 43*, 243–268.

Wolpe, J. (1958). *Psychotherapy by reciprocal inhibition.* Stanford, CA: Stanford University.

Zubin, J., & Hunt, H. F. (1967). *Comparative Psychopathology: Animal and Human.* New York: Grune & Stratten.

Section I. Selected References to Journal Articles on Animal Models of Human Pathology

This section contains a bibliography of annotated references focusing on the behavioral and psychological aspects of animal models of human pathology. References were retrieved from the PsycINFO database and are sorted alphabetically by first author within the major and minor classification categories used by *Psychological Abstracts* and the PsycINFO database.

General Psychology

History & Systems

1. Cancro, Robert. (1986). **Social and biologic psychiatry: Fusion or fission?** ***American Journal of Social Psychiatry,*** 6(2), 86–89.
Discusses the importance of external factors that are conceptualized as social in nature and the implication of the social environment for mental development and its attendant disorders. Animal models are used to demonstrate that organisms interact with their environments at molecular and at molar levels and to show that complex social events alter genetic expression in animals, including humans.

2. Gorenstein, Ethan E. & Newman, Joseph P. (1980). **Disinhibitory psychopathology: A new perspective and a model for research.** ***Psychological Review,*** 87(3), 301–315.
The syndrome produced by septal lesions in animals can serve as a functional research model of human disinhibitory psychopathology which appears to span several traditionally separate psychological categories—psychopathy, hysteria, hyperactivity, antisocial and impulsive personality, and alcoholism. It is proposed that these categories are separate manifestations of the same genetic diathesis and that the "septal syndrome" may constitute a valid model of behavioral aspects of this diathesis. A program of experimentation utilizing this animal model is outlined.

3. Overmier, J. Bruce. (1981). **Interference with coping: An animal model.** ***Academic Psychology Bulletin,*** 3(1), 105–118.
Animal models may be useful to both researchers and practitioners in revealing causes of behavioral dysfunction. Factors governing the heuristic value of such models are often misunderstood. The present paper comments on the use of animal models in general, with a brief illustration using the learned helplessness model.

4. Tryon, Warren W. (1976). **Models of behavior disorder: A formal analysis based on Woods's Taxonomy of Instrumental Conditioning.** ***American Psychologist,*** 31(7), 509–518.
Discusses the issue of formal analysis of operations in psychology. A total of 120 conditioning paradigms are derived from P. J. Woods's (1974) "Taxonomy of Instrumental Conditioning," of which 64 are predicted to produce behavior disorder and 32 are predicted to produce no behavior disorder. The remaining 24 are either redundant or "degenerate" paradigms. Eight paradigms producing behavior disorder are selected for review because they represent simple symmetrical formal relationships and have been more widely investigated. Among the phenomena covered are "superstitious" behavior, learned helplessness, experimental neurosis, anaclitic depression as a result of maternal separation, and physiological disturbances such as ulceration. In addition to systematizing animal models of behavior disorder, it is argued that respondent conditioning is a special case in which 2 operant paradigms are programmed simultaneously. The possibility of using the present analysis for behavioral diagnosis, treatment selection, and considering ethical issues is also discussed.

5. Weiner, Paul. (1977). **Applications of catastrophe theory in psychopathology.** ***Evolution Psychiatrique,*** 42(3–2), 955–974.
Discusses the development of the topographic model of catastrophe theory and the characteristics of the model that make it applicable to various forms of behavior. The use of the model is demonstrated as a means of analyzing the fight–flight behavior of dogs in relationship to dimensions of aggression (rage–fear). The same model is then applied to the analysis of manic-depressive states, epilepsy, and states of psychic disequilibrium generally. Modifications of the model describe the rapprochement or distancing phenomena in psychoanalysis and can assist therapeutic action in the face of patient resistance.

Psychometrics & Statistics & Methodology

Tests & Testing

6. Harrington, Gordon M. (1988). **Two forms of minority-group test bias as psychometric artifacts with an animal model (*Rattus norvegicus*).** ***Journal of Comparative Psychology,*** 102(4), 400–407.
Controversy abounds over attributing group differences on tests to nature, nurture, or test bias. Limitations of correlational sampling from natural populations necessitate experimental methods to resolve underlying issues. In classicial psychometrics test items are selected from a larger item pool through analysis of item responses in a sample of subjects. Rats of six inbred strains ($n = 366$) were tested in multiple mazes to provide a large item pool. Six populations were created, each with differing proportions of each strain. Items selected through independent item analyses within each population yielded six tests. An independent cross-validation sample ($n = 146$) provided scores on all six items. This sample was also tested in another set of maze problems defined as the criterion to be predicted. Strain means and intrastrain predictive validities for the six tests varied with strain representation in the population used for item selection ($p < .001$). Conventional item-selection procedures clearly produced two forms of minority test bias.

Neuropsychological Assessment

7. Welsh, Marilyn C. & Pennington, Bruce F. (1988). **Assessing frontal lobe functioning in children: Views from developmental psychology.** ***Developmental Neuropsychology,*** 4(3), 199–230.

Reviews evidence for the view that frontally mediated executive functions emerge in the 1st yr of life and continue to develop at least until puberty, if not beyond. It is contended that measures used to detect executive functions must be developmentally appropriate, and suggestions regarding viable executive function measures are offered. The contribution of animal models to an understanding of the rudimentary executive functions in infancy is discussed. Another behavioral domain, self-control, is proposed as a possible source of frontal assessment tools for very young children. Several cognitive tasks from developmental psychology are highlighted as potential frontal measures for school-age children.

Research Methods & Experimental Design

8. Baum, Morrie. (1986). **An animal model for agoraphobia using a safety-signal analysis.** ***Behaviour Research & Therapy,*** 24(1), 87–89.
Reviews avoidance-extinction responding in rats as a model for phobias and their treatment by exposure (flooding). A modified procedure involving the addition of Pavlovian safety signals is presented in line with S. Rachman's (1984) safety-signal perspective of agoraphobia. Proposed uses of the animal mode include investigations of whether gender or trauma affect safety-seeking behavior.

9. Crabbe, John C. (1984). **Pharmacogenetic strategies for studying alcohol dependence.** ***Alcohol,*** 1(3), 185–191.
Discusses the importance of genotypic differences in the determination of sensitivity to ethanol, tolerance development, and physical dependence susceptibility. It is now generally accepted by investigators studying the biochemical and physiological bases for alcoholism that genotype can influence all these different aspects of sensitivity to the effects of ethanol. Although there is convincing evidence that susceptibility to alcoholism is inherited in humans, researchers have no idea what it is that is inherited. By examining a family history for a particular individual, individuals at familial risk for developing problems with alcohol abuse can be identified. However, environmental as well as genetic factors are important in determining who does and who does not become an alcoholic. Thus, one critical need is for a genetic marker for alcoholism. Since the search for such markers in human research is both expensive and time-consuming, this has led to the use of animal models for alcoholism. Animal models are particularly helpful for genetic research since their genetics are well-understood and can be specifically tooled to the task at hand. The principal genetic methodologies that have been employed to study the human and animal pharmacogenetics of alcohol are illustrated, and future directions in this area are identified.

10. Hughes, Carroll W. & Preskorn, Sheldon H. (1981). **Consideration of physiologic mechanisms in animal models of "sudden death."** ***Omega: Journal of Death & Dying,*** 11(2), 113–118.
Contends that animal research of the "sudden death" phenomenon purporting to demonstrate causal psychological states has serious methodological problems. Neither models nor the definition of the term, "sudden death," are uniformly adopted; thus, the literature contains many conflicting reports. Further, much of the work has dealt with assumed psychological causation, which is nontestable in nonverbal animals and tends to obscure study of quantifiable behavioral and physiological mechanisms.

11. McKinney, William T. & Moran, Elaine C. (1981). **Animal models of schizophrenia.** ***American Journal of Psychiatry,*** 138(4), 478–483.
Reviews some general issues concerning the development and use of animal models of schizophrenia and presents a summary of the criteria necessary for validating models. Some of the major attempts at recreating animal models of schizophrenia are described, including drug and nondrug methods. The etiologic, phenomenologic, and treatment relevance of the various systems are discussed, and approaches are suggested that might produce improved animal models of schizophrenia.

12. Miller, Neal E. (1985). **The value of behavioral research on animals.** Meeting of the American Psychological Association (1984, Toronto, Canada). ***American Psychologist,*** 40(4), 423–440.
Presents facts documented by references to provide evidence regarding the value of behavioral research on animals. It is argued that attempts by radical animal activists to mislead humane people by repeatedly asserting such research is completely without any value and by other false statements are a disservice to animal welfare by deflecting funds from worthy activities. Some of the contributions of animal research have led to improvements in the welfare of animals. Animal research has also led to advances in psychotherapy, especially behavior therapy and behavioral medicine; rehabilitation of neuromuscular disorders; understanding and alleviating effects of stress and pain; discovery and testing of drugs for treatment of anxiety, psychosis, and Parkinson's disease; knowledge about mechanisms of drug addiction, relapse, and damage to the fetus; and understanding the mechanisms of some deficits of memory that occur with aging.

13. Neuringer, Allen. (1984). **Melioration and self-experimentation.** ***Journal of the Experimental Analysis of Behavior,*** 42(3), 397–406.
Contends that operant researchers rarely use the arena of applied psychology to motivate or to judge their research and that absence of tests by application weakens the field of basic operant research. Early in their development, the physical and biological sciences emphasized meliorative aspects of research, with improvement of human life as a major goal. It is argued that if basic operant researchers analogously invoked a melioration criterion, the operant field might avoid its tendency toward ingrowth and instead generate a broadly influential science. Operant researchers could incorporate melioration by (a) creating animal models to study applied problems; (b) confronting questions raised by applied analysts and testing hypotheses in applied settings; or (c) performing self-experiments—that is, using experimental methods and behavioral techniques to study and change the experimenter's behavior.

14. Riley, Edward P. & Meyer, Linda S. (1984). **Considerations for the design, implementation, and interpretation of animal models of fetal alcohol effects.** ***Neurobehavioral Toxicology & Teratology,*** 6(2), 97–101.

Discusses methodological issues related to the use of animal models in research on the effects of prenatal exposure to alcohol on morphology, biochemistry, and behavior. The selection of a test animal is considered in terms of the species' comparability to humans, the cost of obtaining an adequate sample size, and necessary control procedures. Various methods of ethanol administration are also considered, including oral intubation, liquid diet, and ip injection. Appropriate methods of data analysis are examined with particular reference to the issue of whether the individual or the litter should be the unit of analysis.

15. Tullis, Katherine V.; Sargent, William Q.; Simpson, John R. & Beard, James D. (1977). **An animal model for the measurement of acute tolerance to ethanol.** ***Life Sciences,*** 20(5), 875–882.
Tested an apparatus for the measurement of acute tolerance to ethanol in small animals. 44 male Sprague-Dawley rats were trained on the apparatus to leap to a descending platform to avoid being shocked. After an ip injection of 2 g/kg ethanol, Ss were tested repeatedly on the apparatus, and the plasma ethanol concentration was measured after each trial. Results demonstrate that (a) the jumping ability of the Ss was significantly more impaired during the ascending portion of the plasma ethanol curve than during the descending portion and (b) the improvement in jumping ability during the descending portion of the curve was not dependent on a lowered plasma ethanol concentration. In a 2nd experiment, the possibility of practice effects was eliminated by measuring the jumping ability and plasma ethanol concentration in one group of Ss on the ascending portion of the plasma ethanol curve and in another group on the descending portion of the curve. A significant improvement in jumping ability was again observed during the descending portion of the curve, even though the plasma ethanol concentrations of the 2 groups were comparable. The development of acute tolerance to ethanol is thus demonstrated in both experiments.

16. Valdes Miyar, Manuel & Flores i Formenti, Tomas. (1981). **Concerning animal experimentation in psychology and psychopathology.** ***Revista de Psicología General y Aplicada,*** 36(2), 203–212.
Maintains that the progressive use of animal models in psychology and psychopathology compels an evaluation of results as well as an epistemological reconsideration. The present authors consider the major animal models within the experimental frame and briefly comment on common criticisms. (English abstract).

17. Williams, Rick A.; Boothe, Ronald G.; Kiorpes, Lynne & Teller, Davida Y. (1981). **Oblique effects in normally reared monkeys (*Macaca nemestrina*): Meridional variations in contrast sensitivity measured with operant techniques.** ***Vision Research,*** 21(8), 1253–1266.
Describes a newly completed operant methodology for the assessment of spatial vision in pigtail macaque monkeys. Automated techniques for the generation, calibration, and presentation of sinusoidal grating stimuli, and for control of the operant experiment, are described. Contrast sensitivity functions were obtained in 4 Ss for gratings in vertical, oblique, and horizontal orientations. Data demonstrate that the monkey visual system, like that of humans, shows variations of contrast sensitivity with grating orientation at high spatial frequencies. One S showed a classical oblique effect; i.e., similar sensitivity for vertical and horizontal gratings and a lower sensitivity for oblique gratings. The other 3 Ss showed contrast sensitivity differences between horizontal and vertical gratings. The similarity of monkey and human contrast sensitivity variations and the implications for use of the macaque monkey as an animal model for human vision are discussed.

18. Wolthuis, Otto L. (1991). **Some animal models and their probability of extrapolation to man.** Symposium on Animal-To-Human Extrapolation (1990, San Antonio, Texas). ***Neuroscience & Biobehavioral Reviews,*** 15(1), 25–34.
Presents examples of experimental models subdivided into models that provide results with a high or a moderate probability for qualitative extrapolation to man. Models with a high predictive value for man produce results that can be directly verified in man or human organs, such as in testing for radiation damage, for fast-acting compounds such as alkylating agents, and for neuromuscular transmission. In contrast, models that have a moderate predictive value rely on similarities or analogies of signs and symptoms between man and the animal model. Examples include models for skin penetration, skin damage, and some models for neurotoxicity.

Human Experimental Psychology

19. Prokasy, William F. (1973). **A two-phase model account of aversive classical conditioning performance in humans and rabbits.** ***Learning & Motivation,*** 4(3), 247–258.
Applied a 2-phase model to previously reported results of classical conditioning procedures with 138 undergraduates and 101 rabbits. During Phase 1, response probability remained constant. During Phase 2, response probability generally increased, although 1 operator was sufficient for a majority of Ss and 2 operators were required for a minority. The latter Ss exhibited increases in responding after a CR trial and decreases after a non-CR trial. The pattern of parameters was similar for man and rabbits. In man, increases in UCS intensity resulted in a decrease in the duration of Phase 1 and an increase in the limit of the operators for those Ss requiring more than a single operator to describe performance during Phase 2. The value of employing a model to describe the data and understand the effects of independent variable manipulation is discussed.

Learning & Memory

20. Maier, Steven F. & Seligman, Martin E. (1976). **Learned helplessness: Theory and evidence.** ***Journal of Experimental Psychology: General,*** 105(1), 3–46.
Reviews the literature which examined the effects of exposing organisms to aversive events which they cannot control. Motivational, cognitive, and emotional effects of uncontrollability are examined. It is hypothesized that when events are uncontrollable the organism learns that its behavior and outcomes are independent, and this learning produces the motivational, cognitive, and emotional effects of uncontrollability. Research which supports this learned helplessness hypothesis is described along with alternative hypotheses which have been offered as explanations of the learned helplessness effect. The application of this hypothesis to rats and man is examined.

21. Plevová, Jarmila. (1977). **System approach in experimental psychopharmacology and memory.** ***Studia Psychologica,*** 19(3), 261–262.
Discusses the contributions to psychopharmacology of A. V. Valdman and a team of Leningrad pharmacologists. Based on P. K. Anokhin's (1973) universal schema of the functional system as a dynamic organization of processes oriented to the organism's adaptation to its environment, the findings of the Valdman team, who study animal models of emotional and motivational reactions, are applicable to humans and to the study of memory functioning and disorders.

22. Riccio, David C.; Richardson, Rick & Ebner, Debbie L. (1984). **Memory retrieval deficits based upon altered contextual cues: A paradox.** ***Psychological Bulletin,*** 96(1), 152–165.
It has long been recognized that memory retrieval depends on the similarity of the stimulus context at testing to the context present at encoding. This contextual-cues model appears to draw support from a variety of studies in animal and human research indicating that performance is disrupted by manipulations that reduce the correspondence in background stimuli between acquisition and testing. Although this approach has much in common with a stimulus generalization decrement interpretation of performance change, little attention has been paid to data from stimulus control research showing that generalization gradients typically flatten over time. A literature review indicates that loss of stimulus control with delayed testing occurs for background contexts as well as for conditioned or discriminative stimuli and supports the interpretation that stimulus attributes are forgotten. These findings pose a paradox: The functionally increased interchangeability of stimuli after a retention interval makes it unlikely that disruption of memory can be attributed to subtle shifts in context under the nominally identical conditions of a retention test.

23. Riccio, David C. & Richardson, Rick. (1984). **The status of memory following experimentally induced amnesias: Gone, but not forgotten.** Invited Symposium Midwestern Psychological Association Meetings: Long-term memories: How durable, and how enduring? (1984, Chicago, IL). ***Physiological Psychology,*** 12(2), 59–72.
Considers data that demonstrate that postacquisition traumatic insult can produce profound memory losses and suggests that these retrograde amnesias are not attributable to disruption or failure of the storage process. It is suggested that memory loss is linked to the lack of appropriate cues for retrieval. The storage-disruption hypothesis became problematic as findings of spontaneous recovery and rapid reversal of retrograde amnesia became common. The storage-failure interpretation was challenged by findings of familiarization effects and delayed onset of retrograde amnesia. Studies of retrograde amnesia for old memory, reactivation, exogenous hormones and memory recovery, and reversal of amnesia through reexposure to the amnestic agent support the view that memory losses are often the result of retrieval failure. An interpretation of amnesia in terms of the interactive nature of encoding and retrieval makes it understandable why the severity of amnesia is characteristically time dependent. Two important issues related to these findings need to be explored: the permanence of a recovered memory and the "content" of the recovered amnestic memory.

24. Walk, Richard D. & Schwartz, Michael L. (1982). **Birdsong learning and intersensory processing.** ***Bulletin of the Psychonomic Society,*** 19(2), 101–104.
Two experiments were performed in which 124 undergraduates learned to attach names to birdsongs. In Exp I, Ss who were instructed to generate their own visual codes were far superior to those not given any instructions except those of learning the birdsongs. In the 2nd experiment, both those given a model code for half of the birdsongs and those who made their own visual codes were superior to controls without visual codes. The studies demonstrated the way in which learning in the auditory modality can benefit from visual symbols, and there are implications for the study of "higher-order invariances" or relations between the modalities of vision and audition.

Motivation & Emotion

25. Abramson, Lyn Y.; Seligman, Martin E. & Teasdale, John D. (1978). **Learned helplessness in humans: Critique and reformulation.** ***Journal of Abnormal Psychology,*** 87(1), 49–74.
Criticizes and reformulates the learned helplessness hypothesis. It is considered that the old hypothesis, when applied to learned helplessness in humans, has 2 major problems: (a) It does not distinguish between cases in which outcomes are uncontrollable for all people and cases in which they are uncontrollable only for some people (universal vs personal helplessness), and (b) it does not explain when helplessness is general and when specific, or when chronic and when acute. A reformulation based on a revision of attribution theory is proposed to resolve these inadequacies. According to the reformulation, once people perceive noncontingency, they attribute their helplessness to a cause. This cause can be stable or unstable, global or specific, and internal or external. The attribution chosen influences whether expectation of future helplessness will be chronic or acute, broad or narrow, and whether helplessness will lower self-esteem or not. The implications of this reformulation of human helplessness for the learned helplessness model of depression are outlined.

Consciousness States

26. Demaret, A. (1983). **Ethology and hypnosis.** ***Perspectives Psychiatriques,*** 21(2)[91], 97–102.
L. Chertok (1969, 1974) has suggested that hypnosis is a 4th organic state, equivalent to wakefulness, sleep, and dreaming. The present author argues that if this view is correct, hypnosis should be observable in both humans and animals. However, the experts are far from reaching any agreement on the similarity of conditions in human hypnosis and aspects of animal behavior designated as "animal hypnosis." An ethological model of hypnosis is presented that takes into account criticisms aimed at previous animal models. The possible adaptive function of the hypnotic state in both humans and animals is discussed.

27. Demaret, D. (1974). **Toward an ethological theory of hypnosis.** ***Feuillets Psychiatriques de Liege,*** 7(3), 332–334.

Describes how "animal hypnosis," or the tendency of animals to become immobilized in the face of danger, corresponds to the state of catelepsy in human hypnotic states. Examples from studies of birds are presented to demonstrate how an ethological model of the "tendency to follow," or imprinting, is similar to tendencies in human infants and to the human state of hypnotic sleepwalking.

Animal Experimental & Comparative Psychology

28. Abramson, Lyn Y. (1976). **Relevance of animal therapy learning models to behavioral psychology.** ***B.A.B.P. Bulletin,*** 4(1), 1–7.
Discusses the use of animals in modeling human behavioral disorders in the laboratory as analogues of behavior therapy and psychopathology, with possibilities for clinical application. Animal experimental models for phobias, learned helplessness, and neurosis have generated hypotheses about causes, therapy, and prevention in humans.

29. Ader, Robert. (1967). **Emotional reactivity and susceptibility to gastric erosions.** ***Psychological Reports,*** 20(3, Pt 2), 1188–1190.
Contradictory results in the literature led to a reanalysis of previously collected data which failed to provide evidence for a relationship between behaviors in an open-field or a reaction-to-handling test and immobilization-produced gastric erosions.

30. Baysinger, C. M.; Plubell, P. E. & Harlow, H. F. (1973). **A variable-temperature surrogate mother for studying attachment in infant monkeys.** ***Behavior Research Methods & Instrumentation,*** 5(3), 269–272.
Describes an apparatus which facilitates manipulation of the infant–surrogate attachment bond. Data showing significant behavioral changes in ventral contact and locomotion of 4 rhesus monkey infants as a function of depressed surrogate temperature are presented. The value of this technique in the production of psychopathology is indicated by a dramatic and progressive increase in disturbance behaviors during a 9-wk test period. Implications for the use of a variable-temperature surrogate in studying animal models of psychopathology are noted.

31. Bell, R. W.; Hendry, G. H. & Miller, C. E. (1967). **Prenatal maternal conditioned fear and subsequent ulcerproneness in the rat.** ***Psychonomic Science,*** 9(5), 269–270.
Increased ulceration under immobilization stress was observed in rats that had been subjected to prenatal maternal fear conditioning coincident with the development of the fetal gut. Presentation of the CS (buzzer) or UCS (shock) alone did not affect ulceration, nor did the conditioning procedure if it occurred after the fetal gut development was complete.

32. Binik, Yitzchak M. et al. (1979). **Sudden swimming deaths: Cardiac function, experimental anoxia, and learned helplessness.** ***Psychophysiology,*** 16(4), 381–391.
Examined C. P. Richter's (1957) hypothesized mechanism of sudden swimming deaths—parasympathetic overstimulation mediated by helplessness—in 3 experiments with 82 male Holtzman rats. EKG recordings of sudden swimming deaths (SSD) indicated a consistent pattern of cardiac function including severe bradycardias and arrhythmias followed by atrioventricular block leading ultimately to asystole. Ss that survived the swimming stress showed a transient bradycardia followed by a return to baseline heart rate levels. Swimming behavior and cardiac function were closely related. Anoxia is suggested as the mechanism of death, and data indicating the involvement of the diving reflex in Ss are presented. Helplessness induced by pretreatment with inescapable shock did not affect SSD. Data are consistent with a parasympathetic overstimulation hypothesis but disconfirm mediation by helplessness. The implications for the generality of learned helplessness and for the study of psychosocial factors in SSD are discussed.

33. Bruner, Carlos A. (1988). **The perception of personal causality from the viewpoint of the modern theory of behavior.** ***Revista Mexicana de Psicología,*** 5(2), 167–172.
Describes an animal (pigeon) model for studying the development of perception of personal causality. The concept of perception is discussed in relation to behaviorist objections to studying internal states. Implications of preliminary locus of control data from the animal model are discussed.

34. Carmona, Alfredo; Miller, Neal E. & Demierre, Terrie. (1974). **Instrumental learning of gastric vascular tonicity responses.** ***Psychosomatic Medicine,*** 36(2), 156–163.
Trained 10 Sprague-Dawley male albino rats with a chronically implanted device that allowed photoelectric plethysmographic measures to be made of the stomach wall while paralyzed with dextrotubocurarine and being artifically respirated. One group was reinforced by avoidance of and/or escape from electric shocks to the tail, whenever increases in the transmission of light through the stomach wall occurred during a CS; another group was reinforced for decreases. Reliable changes in the rewarded direction were learned. Results are interpreted as evidence for the instrumental learning of a gastric response which, in the light of control experiments, very probably was a vasomotor one producing changes in the amount of blood in the stomach wall.

35. Chamove, A. S.; Rosenblum, L. A. & Harlow, H. F. (1973). **Monkeys (Macaca mulatta) raised only with peers: A pilot study.** ***Animal Behaviour,*** 21(2), 316–325.
Compared 4 infant rhesus monkeys raised in a group (4-TT) and 6 raised in pairs (2-TT) with 8 Ss raised on mother surrogates and 20 raised with real mothers. When tested with peers early in life 4-TT and 2-TT Ss showed less play, hostility, and sex, and the 2-TT Ss exhibited a preponderance of social cling. When tested as adults the 2-TT and 4-TT Ss were below controls on measures of play, above controls on social proximity, hostility, and withdrawal, and the 2-TT Ss showed inadequate sexual adjustment. Data suggest that behaviors normally associated with affectional ties can become so extreme as to inhibit normal social development.

36. Cheal, MaryLou. (1987). **Adult development: Plasticity of stable behavior.** ***Experimental Aging Research,*** 13(1–2), 29–37.
62 gerbils observed from 6 to 18 mo of age displayed a highly significant individual stability of such behaviors as locomotor activity, rearing on the hind legs, latency to jump down from a platform, and marking with the ventral gland. In Ss tested monthly, there was less locomotor activity and rearing on the hind legs than in Ss tested only once. Ss with a strong tendency to have seizures had more locomotor activity, more rearing, fewer marking, and shorter latency to jump down from a platform than those who rarely seized.

When Ss had a seizure prior to testing, their activity scores and number of rearings were greater than on days when they did not have a seizure. Females ($n = 31$) crossed more lines, had more seizures, and reared more, while males marked more and were slower to jump down from the platform.

37. Coile, D. Caroline & Miller, Neal E. (1984). **How radical animal activists try to mislead humane people.** ***American Psychologist,*** 39(6), 700–701.
Investigated charges of animal cruelty by analyzing 608 articles published from 1979 through 1983 in 2 American Psychological Association journals that report animal studies. Only 10% of studies used shock, no Ss were starved to death, and no studies used periods of water or food deprivation over 48 hrs.

38. Crawley, Jacqueline N. (1984). **Evaluation of a proposed hamster separation model of depression.** ***Psychiatry Research,*** 11(1), 35–47.
Phodopus sungorus, the Siberian dwarf hamster, exhibits a number of reproducible and quantifiable behavioral changes when the members of a male–female pair bond are separated. Three pairs of Ss were maintained in cages for a minimum of 3 wks; behaviors observed included food consumption, nest building, nest occupation, self-grooming, and responses to gentle handling. Ss were also weighed twice weekly. To measure exploratory and social behavior, 2 challenge paradigms were administered. Separation consisted of transferring each S to a new home cage, after which the procedures were repeated. Results indicate that a significant increase in body weight, decrease in social interaction, and decrease in exploratory behaviors occurred predominantly in separated males. Some, but not all, of the behavioral effects of separation were reversed by the tricyclic antidepressant imipramine. It is suggested that the separation of pair-bonded Siberian dwarf hamsters may provide a new animal model for depression, incorporating the practical advantages of a rodent model with the conceptual advantages of a naturalistic life-event precipitant.

39. Crawley, Jacqueline N. (1984). **Preliminary report of a new rodent separation model of depression.** ***Progress in Neuro-Psychopharmacology & Biological Psychiatry,*** 8(3), 447–457.
Quantified behavioral changes of Siberian dwarf hamsters (*Phodopus sungorus*) when members of a male–female pair bond were separated and evaluated the potential of the separation syndrome as a new animal model for depression. Groups of 3 juvenile male and 3 juvenile female Ss were allowed to form nesting pairs; each S was later placed with a new hamster of the opposite sex for 5-min periods during which they were monitored by computer. Clorgyline (1 mg/kg, sc) was administered once daily to all separated Ss; in a group of control Ss, whole-brain MAO was assayed. Results indicate that separation induced increases in body weight in males; both males and females showed significant decreases in social encounters with an unfamiliar conspecific. Clorgyline reduced whole brain MAO to approximately one-third of control values. Imipramine (10 mg/kg sc for 2 wks) reversed some of the effects of separation in male Ss. The ratio of 5-hydroxylindoleacetic acid to serotonin was significantly reduced in the cerebral cortex, diencephalon, and mesencephalon of separated males.

40. Dicara, Leo V. & Miller, Neal E. (1968). **Changes in heart rate instrumentally learned by curarized rats as avoidance responses.** ***Journal of Comparative & Physiological Psychology,*** 65(1), 8–12.
Two groups of curarized rats learned to increase or decrease, respectively, their heart rates in order to escape or avoid mild electric shocks. Responses in the appropriate direction were greater during the stimulus preceding shock than during control intervals between shock; they changed in the opposite direction, toward the initial pretraining level, during the different stimulus preceding nonshock. Electromyograms indicated complete paralysis of the gastrocnemius muscle throughout training and for a period of at least 1 hr thereafter.

41. Dicara, Leo V. & Miller, Neal E. (1968). **Instrumental learning of vasomotor responses by rats: Learning to respond differentially in the two ears.** ***Science,*** 159(3822), 1485–1486.
12 curarized and artificially respirated male Sprague-Dawley rats were rewarded by electrical stimulation of the brain for changes in the balance of vasomotor activity between the 2 ears. Ss learned vasomotor responses in 1 ear that were independent of those in the other ear, in either forepaw, or in the tail, or of changes in heart rate or temperature. In addition to implications for learning theory and psychosomatic medicine, the results indicate a greater specificity of action in the sympathetic nervous system than is usually attributed to it.

42. Drugan, Robert C.; Skolnick, Phil; Paul, Steven M. & Crawley, Jacqueline N. (1989). **A pretest procedure reliably predicts performance in two animal models of inescapable stress.** ***Pharmacology, Biochemistry & Behavior,*** 33(3), 649–654.
Results of 4 experiments with male rats revealed a significant correlation between Ss that displayed learned helplessness on an inescapable tailshock test and those that displayed learned helplessness on a 2nd shuttle-escape test performed either 2 or 4 wks later. A significant correlation was found between Ss that learned this task on the 1st test and those that learned 2 or 4 wks later. In the forced-swimming-induced behavioral despair test, a significant correlation was observed for floating time for Ss on the 1st test and on the 2nd test either 2 or 4 wks later. The lack of cross-predictability strongly suggests that the learned helplessness and the behavioral despair models may be mediated by different neurochemical mechanisms.

43. Dubin, William J. & Levis, Donald J. (1973). **Influence of similarity of components of a serial conditioned stimulus on conditioned fear in rats.** ***Journal of Comparative & Physiological Psychology,*** 85(2), 304–312.
Reports 2 experiments in which 80 male and 96 female Blue Spruce rats were presented a serial CS procedure consisting of 2 components: S1 followed by S2 (S1/S2). In both experiments the stimulus similarity of S1 to S2 in terms of tonal frequency was systematically manipulated. Exp I tested the effects of similarity of components in a standard shuttle-box avoidance situation. Exp II employed a conditioned emotional response paradigm measuring the suppression of consummatory licking. Results suggest that the amount of fear elicited by S1 is a direct function of the stimulus similarity of S1 to S2, and support a generalization interpretation of fear transference.

44. Elias, Merrill F. & Schlager, Gunther. (1974). **Discrimination learning in mice genetically selected for high and low blood pressure: Initial findings and methodological implications.** ***Physiology & Behavior,*** 13(2), 261–267.
Mice bred from an 8-way cross of 8 inbred strains were selected for high or low blood pressure and compared with randomly mated controls on discrimination learning performance. The high blood pressure Ss performed more poorly than the low blood pressure Ss. Subsequent comparisons of high blood pressure Ss from a different foundation stock and extreme blood pressure groups from an F2 distribution, formed from the cross of F1 (High × Low) hybrids, indicated that these differences were not due to a causal relationship between blood pressure and performance, linked gene effects, or pleiotropic effects of the same genes. Implications of these findings for the development of animal models for hypertension and behavior studies are discussed.

45. Gluck, John P.; Harlow, Harry F. & Schiltz, Kenneth A. (1973). **Differential effect of early enrichment and deprivation on learning in the rhesus monkey (Macaca mulatta).** ***Journal of Comparative & Physiological Psychology,*** 84(3), 598–604.
Conducted an experiment with 24 rhesus monkeys to extend research findings concerning the effects of early experience on nonhuman primate learning ability. Enriched Ss tested in their home living environment performed more proficiently than Ss separated from their living environments and tested in an adjoining room. Further, Ss reared in enriched environments were superior to partially isolated controls on the complex oddity tasks but not on 2-choice discrimination or delayed-response problems.

46. Goeckner, Daniel J.; Greenough, William T. & Maier, Steven F. (1974). **Escape learning deficit after overcrowded rearing in rats: Test of a helplessness hypothesis.** ***Bulletin of the Psychonomic Society,*** 3(1B), 54–56.
In a previous study, rats reared in overcrowded environments exhibited characteristics similar to those of the learned helplessness phenomenon described following exposure to inescapable shock. In 2 replications, a total of 112 Long-Evans hooded rats were reared alone or in groups of 4 or 32 for 50 days after weaning. Ss were then tested in a shuttlebox, for which 2 successive crossings were required to avoid or escape shock. 10 Ss from each population group received inescapable shock prior to shuttlebox training, while another 10 were identically handled but not shocked. Ss given inescapable shock failed to perform the shuttlebox task, regardless of the size of the population with which they were reared. Ss not preshocked performed the shuttle task, although those reared in groups of 32 required several trials to reach the level of the smaller groups. Results indicate that impaired learning following overcrowded rearing does not result from helplessness-like behavior.

47. Gunderson, Virginia M.; Grant-Webster, Kimberly S. & Fagan, Joseph F. (1987). **Visual recognition memory in high- and low-risk infant pigtailed macaques (*Macaca nemestrina*).** ***Developmental Psychology,*** 23(5), 671–675.
Pigtailed macaque infants were administered a series of visual recognition problems adapted from a standardized test developed for use with human infants. The subjects were classified as either low risk or high risk. The low-risk animals were normal, whereas the high-risk animals had developmental problems (e.g., hypoxia, failure-to-thrive) that sometimes are correlated with cognitive deficits later in life in humans. The test consisted of a series of problems in which two identical abstract black-and-white patterns were presented for a study period, followed by a two-part test trial in which the previously exposed pattern was paired with a novel one. Looking time to each target was recorded. The low-risk group easily differentiated novel from previously seen targets. The high-risk group gave no evidence of recognition. The results have implications for an animal model to examine factors contributing to intellectual deficits in human infants.

48. Gunderson, Virginia M.; Grant-Webster, Kimberly S. & Sackett, Gene P. (1989). **Deficits in visual recognition in low birth weight infant pigtailed monkeys (*Macaca nemestrina*).** ***Child Development,*** 60(1), 119–127.
15 low birth weight and 15 normal birth weight monkey infants were administered an adaptation of a standardized test of visual recognition memory, originally developed for human infants. Ss were given a series of problems in which 2 identical black-and-white patterns were presented for a familiarization period. The previously exposed pattern was then paired with a novel one, and looking time to each pattern was recorded. The normal birth weight Ss directed a significant amount of their visual attention to the novel stimuli, thus demonstrating recognition abilities. The performance of the low birth weight Ss remained at chance. Findings have implications for an animal model to examine factors contributing to poor cognitive outcome in low birth weight human infants.

49. Gunderson, Virginia M. & Sackett, Gene P. (1984). **Development of pattern recognition in infant pigtailed macaques (*Macaca nemestrina*).** ***Developmental Psychology,*** 20(3), 418–426.
Examined the development of pattern recognition in 31 infant pigtailed macaques using the familiarization–novelty technique. Ss were familiarized with 2 identical black and white patterns and tested on the familiar pattern paired with a novel one. Cross-sectional data revealed that a novelty preference occurred with increasing age. Younger Ss (mean age 178 days postconception or 1 postnatal week) did not show a reliable visual preference for either the novel or the familiar patterns. Infants with a mean age of 203.2 days postconception (about 4 postnatal weeks) fixated novel patterns significantly longer than familiar ones. Data suggest that by 200 days postconception, infant macaques are able to remember some aspects of previously exposed stimuli and will perform consistently on a familiarization–novelty task. Results are discussed in relation to the development of human infant pattern-recognition abilities. Pigtailed macaques provide an excellent model for the investigation of human infant recognition memory, 1 wk of maturation in the pigtailed infant being equivalent to 1 mo in the human.

50. Harlow, Harry F. & Novak, Melinda A. (1973). **Psychopathological perspectives.** ***Perspectives in Biology & Medicine,*** 16(3), 461–478.
Describes research at the University of Wisconsin Primate Laboratory on the induction and treatment of psychopathological states in subhuman simians. Advantages of nonhuman S populations, criteria for mimicking human syndromes, and success to date in achieving animal models for human psychiatric disorders are discussed. Also discussed are effects on primate behavior of (a) privation (no prior experience) of social contact, (b) deprivation (following some experience) of social contact, (c) traumatization by intense

fear, and (d) separation from mother or infant friends. These treatments result in behavioral syndromes briefly described as severe personality deterioration, reversible personality deterioration, bizarre behavior including stereotypy, and depression.

51. Harlow, Harry F. & Suomi, Stephen J. (1974). **Induced depression in monkeys.** ***Behavioral Biology,*** 12(3), 273–296.
Presents a review of recent studies conducted at the Wisconsin Primate Laboratory that have been directed toward the experimental production and cure of human-type psychopathologies in rhesus monkeys, with the primary emphasis being on depression. Data are presented that (a) identify different procedures used to produce depression (e.g., maternal separation, peer or agemate separation, vertical chamber-induced depression, and isolation); (b) determine the susceptibility of any given monkey to these procedures; and (c) outline methods and modes of therapy. Results suggest that profound depressions can be produced in monkeys relatively easily by a variety of techniques and that these induced depressions either closely resemble human depression or have such similarity that closely correlated human and animal depressive patterns can be obtained with refined techniques.

52. Held, Joe R. (1983). **Appropriate animal models.** ***Annals of the New York Academy of Sciences,*** 406, 13–19.
Asserts that the appropriate animal model for scientific research on human diseases should be relevant to the disease under study, available to multiple investigators, exportable, polytocous if the disease under study is genetic, large enough for multiple biopsies of samples, able to fit into available animal facilities, easily handled by investigators, available in multiple species, and alive long enough to be usable. Other criteria for selecting animal models are discussed; these include the animal's genetic homo- or heterogeneity and unique anatomical, physiological, or behavioral attributes.

53. Hinde, R. A. (1972). **Mother–infant separation in rhesus monkeys.** ***Journal of Psychosomatic Research,*** 16(4), 227–228.
Removed the mothers of 20–32 wk old rhesus infants to a distant room for 6–13 days. The symptoms shown by the infants during separation (distress calling and reduced locomotor activity) continued in a reduced form for some time after reunion but had largely disappeared after a few weeks in most infants. Tests 6 mo and 2 yrs after the separation experience revealed marked differences between infants that had and had not had such a separation experience in their responses to strange objects or mildly frustrating situations. Findings indicate that brief separation can have a direct or indirect long-lasting effect on infant behavior.

54. Hughes, Carroll W.; Stein, Elliot A. & Lynch, James J. (1978). **Hopelessness-induced sudden death in rats: Anthropomorphism for experimentally induced drownings?** ***Journal of Nervous & Mental Disease,*** 166(6), 387–401.
Five experiments examined reports of hopelessness-induced sudden death in wild and laboratory *Rattus norvegicus* in the Richter swimming cylinder. Data support alternative behavioral and physiological interpretations of this phenomenon. Deaths in wild and domestic Ss were dependent upon the following conditions: (a) the jet pressure of water shooting into the swimming cylinder and depth of the water, both of which handicapped swimming; (b) kinesthetic cues of the vibrissae; (c) genetic variations in swimming ability; (d) stock differences in emotionality and differences resulting from rearing and handling conditions; (e) body size and sex; and (f) differences in behavior leading to susceptibility to anoxia-induced drownings. Sudden death occurred in those Ss either unable to keep their nostrils above the water surface or those who spent a great deal of time diving below the surface and exploring the bottom of the cylinder. Both were more susceptible to anoxia and drowning. Bradycardia previously associated with these hopelessness-induced deaths proved to be the dive or oxygen-conserving reflex and only occurred when the S was below the surface.

55. Julian, O. A. (1967). **A comparative psychosomatic and pharmacological study, in man and animal, within homeopathic medicine.** ***Annales de Therapeutique Psychiatrique,*** 3, 305–309.
Homeopathic dilutions from a subtropical plant from the genus nepenthes were used on 16 male and 5 female healthy Ss, divided into 2 groups in which 1 group received the solution and the other placebos. Analysis of symptoms observed indicated the possible use of nepenthes for treating neurotic anxiety, stomach ulcers, menstrual disorders, frigidity, and eczema. A water injectable solution of nepenthes extract was used on 20 rats, followed by psychopharmacological testing, with results suggesting its possible clinical treatment of growth, metabolic, and memory disorders. It is believed that this study illustrates the abundance of information obtainable with humans in contrast to the modest results with animals, but it is concluded that the comparative approach yields additional and necessary data to the therapist.

56. Levis, Donald J. (1971). **One-trial-a-day avoidance learning.** ***Behavior Research Methods & Instrumentation,*** 3(2), 65–67.
Compared a 1-trial-a-day discrete-trial avoidance conditioning procedure run for 30 consecutive days with a massed-trial procedure where Ss received 30 consecutive trials within 1 day. 20 naive male Blue Spruce rats in 2 equal groups were equated for time each S spent in the nonshock compartment of the apparatus, number of times the transport box carrying Ss was lifted out of the nonshock compartment, and the number of times each S was handled. The main difference between groups was the intersession trial length of 24 hrs for the 1-trial-a-day Ss. Learning was rapid for both groups. The groups did not differ reliably on 6 acquisition indices. The methodological advantages of the 1-trial-a-day procedure and its theoretical importance are discussed.

57. Levis, Donald J. (1970). **Serial CS presentation and shuttlebox avoidance conditioning: A further look at the tendency to delay responding.** ***Psychonomic Science,*** 20(3), 145–147.
Exp I, with 72 male hooded rats, compared the effects of nonserial and serial CS presentations, while Exp II with 90 Ss, compared nonserial and serial compound CS presentations on the acquisition of Ss' avoidance responses in a shuttlebox situation. The avoidance-response latencies of Ss presented the serial CS conditions tended to be delayed until after the onset of the last stimulus segment introduced into the sequence. This was not the case for the nonserial Ss who tended to avoid shortly after the onset of the CS–UCS interval. The delay in responding noted for the serial CS Ss occurred when either 2 or 3 stimuli comprised the sequence

and when either serial or serial compound presentation was employed. Serial compound conditions increased avoidance responding, but the serial condition did not. Other effects of CS complexity on avoidance responding were also reported.

58. Levis, Donald J. (1971). **CS complexity and intertrial responding in shuttlebox avoidance conditioning.** ***American Journal of Psychology,*** 84(4), 555–564.
As CS complexity (2 levels serial, 2 nonserial) increased, the intertrial responding of 200 naive male Sprague-Dawley rats decreased. Learner Ss did much more intertrial responding than did nonlearners, and responding correlated positively with but had no direct effect on the absolute level of conditioned avoidance responding, increasing over trials and then tapering off. The general findings, replicated with Iowa hooded rats, imply that intertrial responses were generalized avoidance responses contingent upon CS–UCS pairing and subject to CS discriminability.

59. Levis, Donald J. & Boyd, Thomas L. (1973). **Effects of shock intensity on avoidance responding in a shuttlebox to serial CS procedures.** ***Bulletin of the Psychonomic Society,*** 1(5-A), 304–306.
Used 96 male Blue Spruce rats to investigate whether a serial CS delayed-responding effect could be altered by manipulating the shock intensity. 2 serial procedures (S1/S2, S1/S1S2) were tested at 3 shock levels (.5, 1.0, and 2.0 mA). Shock level was not found to have a reliable effect on any of the avoidance response indexes analyzed. An important difference, however, did emerge between the 2 serial CS procedures tested. The theoretical implications of the finding are discussed.

60. Levis, Donald J.; Bouska, Sally A.; Eron, Joseph B. & McIlhon, Michael D. (1970). **Serial CS presentation and one-way avoidance conditioning: A noticeable lack of the delay in responding.** ***Psychonomic Science,*** 20(3), 147–149.
Compared 7 different CS conditions on the acquisition and extinction of avoidance responses in a 1-way situation in 96 male hooded rats. 3 of the groups received nonserial CS manipulations comprising 1, 2, or 3 components; 2 received straight serial CS presentations, comprising either 2 or 3 components; and 2 received serial compound CS conditions comprising either 2 or 3 components. The groups did not differ reliably on 5 different acquisition indices. The extinction data did produce reliable differences between combined serial and nonserial CS conditions with the serial CS conditions producing greater resistance to extinction. Reliable differences were not found between serial and serial compound conditions or between number of CS components. The tendency for the avoidance response to be delayed for the serial CS conditions which is characteristic of shuttlebox data was noticeably absent. The majority of responses for both serial and nonserial CS conditions occurred close to the onset of the CS–UCS interval. The differences between 1-way and shuttlebox avoidance situations is discussed.

61. Levis, Donald J. & Dubin, William J. (1973). **Some parameters affecting shuttle-box avoidance responding with rats receiving serially presented conditioned stimuli.** ***Journal of Comparative & Physiological Psychology,*** 82(2), 328–344.
Gave 48 male Blue Spruce rats shuttle-box training with serial or nonserial CS procedures. The serial CS condition (S1/S1S2), which involved a single stimulus for the 1st ½ of a 16-sec CS–UCS interval and 2 stimuli for the latter ½, produced shorter avoidance latencies and more avoidance responses when compared with a serial condition (S1/S2), in which the latter ½ of the interval involved only 1 stimulus. Both serial conditions resulted in longer avoidance latencies when compared to nonserial conditions. Exp II with 144 Ss demonstrated that the above latency differences could be eliminated with shorter CS–UCS intervals, and Exp III with 72 Ss suggested that avoidance-latency differences obtained at longer CS–UCS intervals were independent of the CS duration ratio between serial components. These and other findings were predicted from a generalization-decrement hypothesis.

62. Levis, Donald J. & Stampfl, Thomas G. (1972). **Effects of serial CS presentation on shuttlebox avoidance responding.** ***Learning & Motivation,*** 3(1), 73–90.
Studied shuttlebox avoidance responding in 400 male albino rats as a function of various CS conditions. In Exp I the CS condition consisted of (a) a single stimulus (S1), (b) a 2-component compound (S1S2), (c) a 2-component serial (S1/S2) in which the 1st stimulus terminated at the midpoint of the CS–UCS interval and the onset of the 2nd stimulus immediately followed the termination of S1, or (d) a 2-component serial compound (S1/S1S2) which consisted of S1 for the 1st 1/2 of the CS–UCS interval and of S1 + S2 for the latter ½ of the CS–UCS interval. The serial conditions (S1/S2, S1/S1S2) reliably produced more avoidance responses than the nonserial CS conditions (S1, S1S2). The modal response latency of the nonserial conditions was approximately 2.5 sec, measured from CS onset, and the modal response latency in the serial groups was 10.5 sec, or 2.5 sec after the onset of the 2nd stimulus in the chain. Exps II and III provided data which indicate that increased avoidance responding obtained with serial CS presentations was not due to the increased number of stimulus onsets and terminations for these groups or to the shorter CS durations resulting from the division of the serial conditions into 2 distinctive segments. Various theoretical implications of these findings are discussed.

63. Lewis, Jonathan K.; McKinney, William T.; Yong, Laurens D. & Kraemer, Gary W. (1976). **Mother–infant separation in rhesus monkeys as a model of human depression: A reconsideration.** ***Archives of General Psychiatry,*** 33(6), 699–705.
Separated 19 5.9–8.5 yr old Ss from their mothers in 5 different studies. While in 2 of the studies data indicated behavioral responses roughly parallel to J. Bowlbey's (1961) protest-despair response to maternal separations, data across all 5 studies were sufficiently variable to bring this technique into serious question as a reliable and predictable animal model for neurobiologic and rehabilitative studies.

64. Maier, Steven F. (1970). **Failure to escape traumatic electric shock: Incompatible skeletal-motor responses or learned helplessness?** ***Learning & Motivation,*** 1(2), 157–169.
Trained 10 experimentally naive mongrel dogs, the passive differential reinforcement of other behavior (DRO) group, to escape electric shock in a Pavlov harness by inhibiting the head movements normally elicited by that electric shock. 10 other Ss, the yoked group, received in the harness inescapable electric shock equivalent to those taken by the passive DRO group. A 3rd group of 10 Ss, the naive control group, received no experience in the harness. All Ss subsequently received escape-avoidance training in a shuttlebox. The passive DRO group learned to escape-avoid in the shuttlebox

more slowly than did controls, but eventually learned. In contrast, ½ of the yoked Ss did not learn to escape. The relevance of these results for a theory of learned helplessness is discussed.

65. Maier, Steven F. (1984). **Learned helplessness and animal models of depression.** ***Progress in Neuro-Psychopharmacology & Biological Psychiatry,*** 8(3), 435–446.
Reviews learned helplessness effects—effects caused by the uncontrollability of events that are beyond the organism's control rather than by the events per se. At a behavioral level, uncontrollable aversive events result in associative, motivational, and emotional deficits. At a neurochemical level, uncontrollable but not controllable aversive events have been reported to lead to disturbances in cholinergic, noradrenergic, dopaminergic, serotonergic, and GABAergic systems. Interpretive difficulties in this literature are discussed. The controllability—uncontrollability of aversive events has a role in producing stress-induced analgesia and the activation of endogenous opiate systems. These relationships are reviewed. It is proposed that the learning that aversive events cannot be controlled activates an opiate system. The research reviewed is related to depression, and the general issue of animal models of depression is discussed. It is concluded that no experimental paradigm can be a model of depression in some general sense, but can only model a particular aspect. Learned helplessness may model stress and coping.

66. Maier, Steven F.; Allaway, Thomas A. & Gleitman, Henry. (1967). **Proactive inhibition in rats after prior partial reversal: A critique of the spontaneous recovery hypothesis.** ***Psychonomic Science,*** 9(1), 63–64.
48 rats were taught a simultaneous visual discrimination, S1+ vs S–. Subsequently, they learned a 2nd discrimination, involving an old and a new stimulus. 24 Ss now learned S3+ vs S–. The others learned S2+ vs S–. Retention of the 2nd discrimination was tested 1 or 32 days after training. The 2 different partial reversals resulted in proactive inhibition to the same degree. It is argued that this result is inconsistent with the spontaneous recovery interpretation of proactive inhibition.

67. Maier, Steven F.; Anderson, Christine & Lieberman, David A. (1972). **Influence of control of shock on subsequent shock-elicited aggression.** ***Journal of Comparative & Physiological Psychology,*** 81(1), 94–100.
Shocks which cannot be controlled by an organism have been shown to interfere with subsequent escape-avoidance training more than do equivalent shocks which can be controlled. Two experiments extended the generality of this phenomenon by examining the effects of the escapability of shock on subsequent shock-elicited aggression. Exp I (with 48 male Sprague-Dawley rats) found that prior exposure to inescapable shock reduced the frequency of shock-induced fighting, while escapable shock did not produce such a reduction. The theory that yoked-control procedures can capitalize on individual differences and produce a systematic difference between groups was ruled out as an explanation of the data of Exp I by the results of Exp II, conducted with 22 similar Ss.

68. Maier, Steven F.; Albin, Richard W. & Testa, Thomas J. (1973). **Failure to learn to escape in rats previously exposed to inescapable shock depends on nature of escape response.** ***Journal of Comparative & Physiological Psychology,*** 85(3), 581–592.
Results of previous studies show that dogs exposed to inescapable shocks in a Pavlov harness subsequently fail to learn to escape shock in a shuttle box. The present 6 experiments attempted to replicate this finding with 182 male Sprague-Dawley rats. In agreement with many previous investigations, Exp I found that Ss exposed to inescapable shock did not fail to learn to escape in a shuttle box. Exp II, III, and IV varied the number, intensity, and temporal interval between inescapable shocks and did not find failure to learn in the shuttle box. An analysis of responding in the shuttle box revealed that Ss shuttled rapidly from the very 1st trial, whereas dogs acquire shuttling more gradually. Exp V and VI revealed that Ss exposed to inescapable shock failed to learn to escape when the escape response was one that was acquired more gradually. Exp V utilized a double crossing of the shuttle box as the escape response and Exp VI utilized a wheel-turn response.

69. Maier, Steven F. & Gleitman, Henry. (1967). **Proactive interference in rats.** ***Psychonomic Science,*** 7(1), 25–26.
Four groups of rats were taught a visual simultaneous discrimination with a noncorrection procedure. Two of the groups had previously been trained to make the reverse discrimination, the other 2 had no prior training. Two groups were given a relearning test 1 day after training, while 2 others were tested 32 days after training. There was no retention loss unless prior interference had been provided. Thus, proactive interference was demonstrated.

70. Maier, Steven F. & Testa, Thomas J. (1975). **Failure to learn to escape by rats previously exposed to inescapable shock is partly produced by associative interference.** ***Journal of Comparative & Physiological Psychology,*** 88(2), 554–564.
Two experiments demonstrated that the effects of prior exposure to inescapable shock on the subsequent acquisition of an escape response in rats is determined by the nature of the contingency that exists between responding and shock termination during the escape learning task, and not by the amount of effort required to make the response or the amount of shock that the S is forced to receive during each trial. Exp I, using 48 male Simonsen rats, showed that inescapably shocked Ss did not learn to escape shock in a shuttle box if 2 crossings of the shuttle box were required (fixed ratio [FR] -2) to terminate shock, but did learn this FR-2 response if a brief interruption of shock occurs after the 1st crossing of the FR-2. Exp II with 72 Ss showed that inescapably shocked Ss learned a single-crossing escape response as rapidly as did controls, but were severely retarded if a brief delay in shock termination was arranged to follow the response. Results are discussed in terms of the learned helplessness hypothesis, which assumes that prior exposure to inescapable shock results in associative interference.

71. Mason, William A. (1976). **Environmental models and mental modes: Representational processes in the great apes and man.** ***American Psychologist,*** 31(4), 284–294.
Argues that psychologists are interested principally in those behaviors that imply some type of functional image or representation of the environment. When man's categories of experience are imposed on the great apes, solid evidence of correspondence between their construction of reality and our

own is obtained. Their sensory capacities are similar, and they spontaneously classify their experiences with objects in ways that resemble those of man. These analytic dispositions are complemented by the ability to synthesize heterogeneous attributes as different properties of the same "object." It is clear that the *unwelten* of ape and man are similar in many respects. Although it is true that the contrast between ape and man seems greatest in the ability for creative reconstruction of experience—to display foresight, to imagine, to plan ahead—even here the differences may be more in degree than in kind.

72. Mason, William A. & Berkson, Gershon. (1975). **Effects of maternal mobility on the development of rocking and other behaviors in rhesus monkeys: A study with artificial mothers.** ***Developmental Psychobiology,*** 8(3), 197–211.
Studied 19 infant rhesus monkeys to test the hypothesis that a mobile artificial mother would establish the critical condition for preventing the development of habitual body-rocking. Ss were maternally separated at birth and assigned to 2 groups. Both groups were placed with surrogates, identical in construction except that for 1 group the surrogate was in motion for 9.5 hrs each day, and for the other group the surrogate was stationary. All but 1 of the 10 Ss raised with stationary artificial mothers developed rocking as an habitual pattern whereas none of the 9 Ss raised with mobile mothers did so. Data also suggest that emotional responsiveness was reduced in Ss raised with mobile mothers, compared to those raised with stationary devices.

73. Mason, William A. & Lott, Dale F. (1976). **Ethology and comparative psychology.** ***Annual Review of Psychology,*** 27(12), 9–154.
Reviews recent developments in the study of animal behavior and describes an emerging new synthesis increasingly influenced by biological thought. The synthetic theory of organic evolution is seen as the major unifying theme in the current perspective. Major sections are devoted to the problems of comparisons, social behavior, and learning. It is concluded that the synthesis between ethology and comparative psychology is essentially complete.

74. McKinney, William T. (1974). **Primate social isolation: Psychiatric implications.** ***Archives of General Psychiatry,*** 31(3), 422–426.
A review of the literature shows that social isolation of rhesus monkeys for the 1st 6–12 mo of life produces severe and persistent behavioral effects including social withdrawal, rocking, huddling, self-clasping, stereotyped behaviors, and inappropriate heterosexual and maternal behaviors as adults. The mechanisms by which these effects are produced are uncertain and require additional investigations. The social isolation syndrome has been likened to several human psychopathological states, but exact labeling of it in human terms is premature at present. It is suggested that the syndrome be viewed in terms of its heuristic value as a model system for further clarifying the interactions among early rearing conditions, their possible neurobiological consequences, and subsequent social behaviors.

75. McKinney, William T.; Suomi, Stephen J. & Harlow, Harry F. (1971). **Depression in primates.** ***American Journal of Psychiatry,*** 127(10), 1313–1320.
Presents the results of a number of experiments designed to produce depressive behavior in young rhesus monkeys. Ss were observed in their home cages and/or playroom daily, and rated on a scale which included various behavioral categories. The reaction of the infant monkey occurred in 2 stages; an initial stage of protest, followed by a stage of despair and withdrawal. The studies are part of a research program aimed at creating an animal model of depression that should make it possible to study the effects of manipulation of the social and biological variables that are thought to be important in human depression.

76. Mikhail, A. A. (1969). **Relationship of conditioned anxiety to stomach ulceration and acidity in rats.** ***Journal of Comparative & Physiological Psychology,*** 68(4), 623–626.
Tested the effects of conditioned anxiety on gastric ulceration (Exp I) and acidity (Exp II). Ss were 24 male rats and 16 male Sprague-Dawley rats, respectively. Contrary to the view that conditioned anxiety promotes ulcer formation, both ulceration and acidity of conditioned groups of rats were not greater than those of controls. Findings are considered in the framework of the physiological influences of the sympathetic system on gastric functions. It is concluded that the inhibitory effects of sympathetic activity on gastric secretions which are usually associated with acute anxiety do not provide support for a positive relationship between conditioned anxiety and gastric ulceration.

77. Miller, Barbara V. & Levis, Donald J. (1971). **The effects of long-term auditory exposure upon the behavioral preference of rats for auditory stimuli.** ***Developmental Psychology,*** 5(1), 178.
Conducted 2 experiments with a total of 150 rats who were reared from birth in constant tone and nontone environments and subsequently tested with tone and white noise stimuli. Tone-reared Ss indicated no preference for tone, absence of tone, or white noise. Non-tone-reared Ss preferred the tone to white noise. Tone-reared Ss were significantly heavier and less emotional in open-field testing. Results (a) pose difficulty for the hypothesis that exposure to a stimulus increases its attractiveness, and (b) supports the hypothesis "that long-term exposure to tone retards learning of a stressful but not a nonstressful task."

78. Miller, Neal C.; DiCara, Leo V. & Banuazizi, A. (1968). **Instrumental learning of glandular and visceral responses.** ***Conditional Reflex,*** 3(2), 129.
Reports experiments on the instrumental learning of various autonomically controlled glandular and visceral responses with 100 artificially respirated rats whose skeletal muscles were completely paralyzed by curare. Results showed that increases in heart rate were produced when increases were rewarded by electrical stimulation of the medial forebrain bundle, and similar decreases were produced when they were rewarded. The occurrence of intestinal contractions and differences in breathing rates were also noted.

79. Miller, Neal E. (1983). **Understanding the use of animals in behavioral research: Some critical issues.** ***Annals of the New York Academy of Sciences,*** 406, 113–118.
Defends the use of animals in behavioral research. It is contended that animals were caused to suffer in some of the research that led unpredictably to the drugs that greatly reduced the intense suffering of many patients in the wards of mental hospitals. If there had been a requirement that the suffering of all animals in this type of research be justified

by any clear-cut prospect for the relief of any specific human suffering, many patients would still be without any hope of recovery. Experimental studies of behavioral stress are another example of necessary suffering. It was found that when mice were forced to live under stressful conditions, they gradually acquired high blood pressures—a finding that made significant contributions to the understanding of behavioral measures to prevent and/or treat high blood pressure.

80. Miller, Neal E. & Banuazizi, Ali. (1968). **Instrumental learning by curarized rats of a specific visceral response, intestinal or cardiac.** ***Journal of Comparative & Physiological Psychology,*** 65(1), 1–7.
When deeply curarized rats, maintained on artificial respiration, were rewarded by electrical stimulation of the medial forebrain bundle for relaxation of the large intestine, spontaneous intestinal contraction decreased; but if subsequently rewarded for intestinal contraction, it increased to above base-line level. When rewards stopped, the learned response extinguished. Deeply curarized rats rewarded for increased or decreased intestinal contraction showed progressive changes in the appropriate direction but heart rate did not change; conversely, rats rewarded for high or low rates of heartbeat, respectively, learned to change rates appropriately but showed no changes in intestinal contraction. Thus, instrumental learning of 2 visceral responses can be specific to the rewarded response, a fact that rules out several alternative interpretations.

81. Miller, Neal E. & Carmona, Alfredo. (1967). **Modification of a visceral response, salivation in thirsty dogs, by instrumental training with water reward.** ***Journal of Comparative & Physiological Psychology,*** 63(1), 1–6.
Thirsty dogs rewarded by water for bursts of spontaneous salivation showed progressive increases in salivation, while other dogs, rewarded for brief periods without salivation, showed progressive decreases. No obvious motor responses were involved, but dogs rewarded for decreasing salivation appeared more drowsy than those rewarded for increasing it. Implications for learning theory and psychosomatic medicine are mentioned.

82. Miller, Neal E. & Dicara, Leo. (1967). **Instrumental learning of heart rate changes in curarized rats: Shaping, and specificity to discriminative stimulus.** ***Journal of Comparative & Physiological Psychology,*** 63(1), 12–19.
Artificially respirated rats with skeletal muscles completely paralyzed by curare were rewarded by electrical stimulation of the medial forebrain bundle for either increasing or decreasing their heart rates. After achieving the easy criterion of a small change, they were required to meet progressively more difficult criteria for reward. Different groups learned increases or decreases, respectively, of 20%; 21 of 23 rats showed highly reliable changes. The electrocardiogram indicated that decreased rates involved vagal inhibition. Rats learned to respond discriminatively to the stimuli signaling that cardiac changes would be rewarded.

83. Mineka, Susan; Keir, Richard & Price, Veda. (1980). **Fear of snakes in wild- and laboratory-reared rhesus monkeys (*Macaca mulatta*).** ***Animal Learning & Behavior,*** 8(4), 653–663.
Exp I compared the responses of 10 laboratory-reared and 10 wild-reared rhesus monkeys to a real snake and to a range of snake-like objects. Most wild-reared Ss showed considerable fear of the real, toy, and model snakes, whereas most lab-reared Ss showed only mild responses. Fear was indexed by unwillingness to approach food on the far side of the snake and by behavioral disturbance. Exp II examined the effectiveness of 7 flooding sessions in reducing snake fear in 8 wild-reared Ss. Mean latency to reach for food, trials to criterion (4 consecutive short latency responses), and total exposure time to criterion declined significantly across flooding sessions. Behavioral disturbance declined within but not across sessions. Results of a final behavioral test reveal that substantial long-lasting changes had occurred in only 3 of the 8 Ss. Results are discussed in the context of dissociation between different indices of fear.

84. Novak, Melinda A. & Harlow, H. F. (1975). **Social recovery of monkeys isolated for the first year of life: I. Rehabilitation and therapy.** ***Developmental Psychology,*** 11(4), 453–465.
Previous research demonstrated that 12 mo of total social isolation initiated at birth produced severe and seemingly permanent social deficits in rhesus monkeys. Such monkeys exhibited self-clasping, self-mouthing, and other stereotypic, self-directed responses. Recent research has indicated that 6-mo-isolated monkeys could develop social behaviors if exposed to younger, socially unsophisticated "therapist" monkeys. In the present experiment, 4 12-mo isolate-reared monkeys developed appropriate species-typical behavior through the use of adaptation, self-pacing of visual input, and exposure to 4 younger therapist monkeys. Adaptation enabled the isolate Ss to become familiar with their post-isolation environment, while self-pacing facilitated their watching the therapist Ss' social interactions. The isolates showed a marked decrease in self-directed behaviors following extensive intimate contact with the therapists. Species-typical behaviors significantly increased during this period, so that the isolate behavioral repertoire did not differ substantially from the therapist behavioral repertoire by the end of the therapy period. Results clearly fail to support a critical period for socialization in the rhesus monkey, and an alternative environment-specific learning hypothesis is proposed.

85. Ogawa, Nobuya & Hara, Chiaki. (1979). **Characteristics of behavioral changes induced by long-term appetitive competition in rats.** ***Japanese Journal of Experimental Social Psychology,*** 19(1), 41–47.
Abnormal behavior induced by long-term competition for food between 5 pairs of male Wistar rats (each pair in operant chamber) is discussed with regard to its application to the animal model of psychopathology and with reference to D. O. Hebb's (1947) criteria of neurosis. Pairs of Ss that had learned individually to leverpress under an FR schedule were put into an operant chamber with 2 levers and 1 feeder. Ss were successively exposed to 3 conditions and showed various behavioral patterns on the basis of the dominance hierarchy as follows: (a) With the competition for food under the free-feeding condition, only the subordinate Ss showed abnormal behavior (i.e., heterophagia). (b) Under the restricted feeding condition, the subordinate S's abnormality increased, and concomitantly, circadian

rhythms of various activities shifted to feeding time except for aggressive activity. (c) With a return to individual housing and free feeding, only the subordinate S showed abnormal behavior.

86. Overmier, J. Bruce & Seligman, Martin E. (1967). **Effects of inescapable shock upon subsequent escape and avoidance responding.** ***Journal of Comparative & Physiological Psychology,*** 63(1), 28–33.
Exposure of dogs to inescapable shocks under a variety of conditions reliably interfered with subsequent instrumental escape-avoidance responding in a new situation. Use of a higher level of shock during instrumental avoidance training did not attenuate interference; this was taken as evidence against an explanation based upon adaptation to shock. Ss curarized during their exposure to inescapable shocks also showed proactive interference with escape-avoidance responding, indicating that interference is not due to acquisition, during the period of exposure to inescapable shocks, of inappropriate, competing instrumental responses. Magnitude of interference was found to dissipate rapidly in time, leaving an apparently normal S after only 48 hrs.

87. Paré, William P. (1967). **The role of gross locomotor restriction in the acquisition of motor and cardiac constitutional discrimination in the dog.** ***Conditional Reflex,*** 2(4), 277–284.
Four littermate dogs were used in a counterbalanced design involving 2 treatment conditions (free and restrained) and 2 discrimination learning problems. Each learning problem included 8 daily training sessions with 10 CS+ tones and 10 CS– tones presented randomly. The UCS was a shock to the left foreleg. Degree of foot flexion and percentage change in heart rate in response to CS tones were recorded. Ss subjected initially to the free condition manifested a superior and more rapid motor and cardiac discrimination than Ss initially restrained. The free condition was not as facilitatory when presented to Ss who 1st had been exposed to the restraint condition. Results are discussed in terms of activation and negative transfer.

88. Reite, Martin, et al. (1974). **Normal physiological patterns and physiological-behavioral correlations in unrestrained monkey infants.** ***Physiology & Behavior,*** 12(6), 1021–1033.
Recorded normal physiological data simultaneously with behavior in 4 unrestrained M. nemestrina monkey infants, ages 24, 24, 26, and 35 wks, living with their mothers in their social group. Circadian body temperature rhythms, waking EEG rhythms, sleep patterns, and heart rate measurements were obtained and were correlated with behavior. Sleep stage values and body temperature circadian rhythms showed relatively little inter-S variability. Heart rate and EEG patterns were more individually characteristic of a given S. Using both physiological and behavioral data collected in this manner, a bio-behavioral developmental profile for the monkey infant can be constructed. By comparing the S's position with respect to normative physiological and behavioral developmental indices for its chronological age, Ss showing bio-behavioral developmental retardation or acceleration in 1 or several areas can be identified and the underlying mechanisms investigated. It is believed that this capability will greatly enhance the value of the infant monkey as an animal model system of value to psychiatry and the behavioral sciences.

89. Richelle, Marc. (1991). **Animal models of human behavior, yesterday and today.** ***Schweizerische Zeitschrift für Psychologie,*** 50(3), 198–207.
Discusses the development and use of animal models for the study of comparative psychology and ethology. Emphasis is on the use of these models to study human psychology. The contributions of A. Rey to animal psychology, the relation between animal and human ethology, characteristics of laboratory animals, animal cognition, and clinical implications of animal models are considered. (English abstract).

90. Rosellini, Robert A.; Binik, Yitzchak M. & Seligman, Martin E. (1976). **Sudden death in the laboratory rat.** ***Psychosomatic Medicine,*** 38(1), 55–58.
In 2 experiments with male laboratory rats, vulnerability to sudden death was produced in the Ss by manipulating their developmental history. Ss who were reared in isolation died suddenly when placed in a stressful swimming situation. Handling of these singly-housed Ss from 25 to 100 days of age potentiated the phenomenon. However, Ss who were group housed did not die even when they had been previously handled.

91. Rosellini, Robert A. & Seligman, Martin E. (1975). **Frustration and learned helplessness.** ***Journal of Experimental Psychology: Animal Behavior Processes,*** 104(2), 149–157.
Conducted 2 experiments with a total of 64 male Holtzman rats to examine the transfer of learned helplessness from one aversive motivator (shock) to another (frustration). In Exp I, Ss were trained to approach food in a runway and concomitantly exposed to either escapable, inescapable, or no shock in a different situation. Extinction was conducted in the runway, and subsequently Ss were tested for hurdle-jump escape from the frustrating goal box. Inescapably shocked Ss failed to learn to hurdle-jump, whereas escapably or nonshocked Ss learned the frustration escape response. Exp II replicated the basic findings of Exp I and showed transfer of learned helplessness from shock to frustration when no running response had been first acquired in the runway.

92. Rowan, Andrew N. (1988). **Animal anxiety and animal suffering.** International Symposium on Animal Bio-Ethics and Applied Ethology: Bio-ethics/87 (1987, Montreal, Canada). ***Applied Animal Behaviour Science,*** 20(1–2), 135–142.
Reviews literature that examines some of the philosophical and biological facets of what constitutes anxiety and suffering in animals. The concept of animal suffering is explored, and symptoms of animal anxiety states are described. The relationship of new drugs that relieve anxiety in humans to human and animal anxiety is discussed. The author concludes that benzodiazepine data from humans and animals do help to clarify the issue of sentience and some key animal welfare issues.

93. Royce, Joseph R. & Poley, Wayne. (1975). **Invariance of factors of mouse emotionality with changed experimental conditions.** ***Multivariate Behavioral Research,*** 10(4), 479–487.
Used data from 1 unpublished and 2 published studies to compare the factors of autonomic balance, motor discharge, acrophobia, territoriality, tunneling-1, and tunneling-2. Each study included 19 measures from 5 tests of emotionality: open field, straightaway, pole, cell, and hole-in-wall. In the 1st study, the measures taken were part of a large test battery. In the 2nd study, a reduced test battery included only the 19 measures. In the 3rd study, this reduced battery

was used again, but Ss were injected with psychoactive drugs prior to testing. Ss in the 1st study were tested as part of a 6 by 6 diallel table. In the 2nd and 3rd studies, 2 emotionally contrasted strains (SWR and SJL) were tested. Each population was refactored by alpha factoring with varimax, followed by promax rotations. Factors obtained were compared by quantitative means using S-index and rc coefficients of factor matching. Although support was obtained for the invariance of all 6 factors, results indicate invariance as being strongest for motor discharge and acrophobia and weakest for tunneling-1 and tunneling-2 factors.

94. Royce, Joseph R. & Poley, Wayne. (1976). **Acrophobia factor scores as a function of pole height and habituation.** ***Multivariate Behavioral Research,*** 11(2), 189–194.
Tested 15 male and 15 female C57BL/ALB mice for acrophobia via the pole test. Ss were randomly distributed into 3 groups of 6-, 12-, or 24-in (15.24- 30.48-, or 60.96-cm) high poles. Both factor scores and individual pole measures were analyzed by analysis of variance, with main effects for Sex, Pole Height, and Days of Testing. Results provide experimental support for the interpretation of this factor as acrophobia (i.e., the greater the pole height, the higher the factor score). The reduction of factor scores over days (habituation) is interpreted as an arousal-based response.

95. Sackett, Gene P. (1984). **A nonhuman primate model of risk for deviant development.** ***American Journal of Mental Deficiency,*** 88(5), 469–476.
Tested a nonhuman primate model for studying genetic, physiological, and psychosocial processes producing delayed development. The model focuses on parental risk for poor reproductive outcomes using pigtailed monkeys (*Macaca nemestrina*). Data are available for over 100 deliveries and over 80 surviving offspring. In initial experiments, Ss were offspring from high- and low-risk breeders who were nursery-reared with no postnatal contact with their parents. Results show that both maternal and paternal factors were associated with delayed development in systems ranging from tooth eruption to concept-learning ability.

96. Sawrey, William L. & Sawrey, James M. (1968). **UCS effects on ulceration following fear conditioning.** ***Psychonomic Science,*** 10(3), 85–86.
Six groups of rats were given fear conditioning trials while immobilized. Conditioning was at 3 levels of UCS intensity (.5, 1.5, and 2.5 ma) and 2 lengths of duration (.25 and .75 sec). Ulceration rate increased as a function of increasing UCS intensity but not UCS duration.

97. Scheuer, Cynthia & Sutton, Cary O. (1973). **Discriminative vs motivational interpretations of avoidance extinction: Extensions to learned helplessness.** ***Animal Learning & Behavior,*** 1(3), 193–197.
Assessed resistance to extinction of discriminated barpress avoidance in 15 male albino rats through the use of 3 procedures. Each procedure served to break the response-reinforcement contingency: classical extinction (CE), operant extinction (OE), and a variable-ratio shock schedule (VR). Greatest-resistance to extinction was found for the VR group, followed by OE and then by CE Ss, supporting a discriminative rather than a motivational analysis. Reacquisition rates following extinction suggested evidence of learned helplessness in some Ss exposed to noncontingent CS-UCS presentations.

98. Seligman, Martin E. (1969). **For helplessness: Can we immunize the weak?** ***Psychology Today,*** 3(1), 42–44.
Dogs that receive inescapable electric shock do not later learn to escape shock even when it is possible. However, dogs that learn to escape shock who then experience inescapable shock, continue to escape the shock whenever possible. The pretraining with escapable shock seemed similar to an inoculation, giving animals immunity from becoming helpless. Parallels to human behavior suggest that it may be possible to prepare people so that they do not react to failure with total helplessness.

99. Seligman, Martin E. (1972). **Learned helplessness.** ***Annual Review of Medicine,*** 23, 207–412.
Reviews animal studies concerned with the impact of traumatic events over which the organism has no control. Findings on basic behavioral effects, results of hypothesis-testing on etiology, and work on possible cures, are included. The relationship of learned helplessness in animals to maladaptive behaviors in man is examined.

100. Seligman, Martin E. & Beagley, Gwyneth. (1975). **Learned helplessness in the rat.** ***Journal of Comparative & Physiological Psychology,*** 88(2), 534–541.
Four experiments, using a total of 159 male albino Sprague-Dawley rats, attempted to produce behavior in the rat parallel to the behavior characteristic of learned helplessness in the dog. When Ss received escapable, inescapable, or no shock and were later tested in jump-up escape, both inescapable and no-shock controls failed to escape. When bar-pressing, rather than jumping up, was used as the tested escape response, fixed ratio (FR) 3 was interfered with by inescapable shock, but not lesser ratios. With FR-3, the no-shock control escaped well. Interference with escape was a function of the inescapability of shock and not shock per se: Ss that were "put through" and learned a prior jump-up escape did not become passive, but their yoked, inescapable partners did. It is concluded that rats, as well as dogs, fail to escape shock as a function of prior inescapability, exhibiting learned helplessness.

101. Seligman, Martin E. & Groves, Dennis P. (1970). **Nontransient learned helplessness.** ***Psychonomic Science,*** 19(3), 191–192.
18 cage-raised beagle dogs and 15 mongrel dogs of unknown history who received repeated, spaced exposure to inescapable electric shock in a Pavlovian hammock failed to escape shock in a shuttlebox 1 wk. later, while 1 session of inescapable shock produced only transient interference. Beagles were more susceptible to interference produced by inescapable shock than were mongrels. Results are compatible with learned helplessness and contradict the hypothesis that failure to escape shock is produced by transient stress.

102. Seligman, Martin E. & Maier, Steven F. (1967). **Failure to escape traumatic shock.** ***Journal of Experimental Psychology,*** 74(1), 1–9.
Dogs that had 1st learned to panel press in a harness in order to escape shock subsequently showed normal acquisition of escape/avoidance behavior in a shuttle box. In contrast, yoked, inescapable shock in the harness produced profound interference with subsequent escape responding in the shuttle box. Initial experience with escape in the shuttle box led to enhanced panel pressing during inescapable shock in the harness and prevented interference with later responding in the shuttle box. Inescapable shock in the harness and

failure to escape in the shuttle box produced interference with escape responding after a 7-day rest. These results are interpreted as supporting a learned "helplessness" explanation of interference with escape responding: Ss failed to escape shock in the shuttle box following inescapable shock in the harness because they had learned that shock termination was independent of responding.

103. Seligman, Martin E.; Rosellini, Robert A. & Kozak, Michael J. (1975). **Learned helplessness in the rat: Time course, immunization, and reversibility.** ***Journal of Comparative & Physiological Psychology,*** 88(2), 542–547.
Previous research has shown that rats, like dogs, fail to escape following exposure to inescapable shock. Three experiments were conducted with a total of 121 male Sprague-Dawley rats to further explore parallels between rat and dog helplessness. The failure to escape did not dissipate in time; Ss failed to escape 5 min, 1 hr, 4 hrs, 24 hrs, and 1 wk after receiving inescapable shock. Ss that first learned to jump up to escape were not retarded later at barpressing to escape following inescapable shock. Failure to escape could be broken up by forcibly exposing the S to an escape contingency. Therefore, the effects of inescapable shock in the rat parallel learned helplessness effects in the dog.

104. Senay, E. C. (1966). **Toward an animal model of depression: A study of separation behavior in dogs.** ***Journal of Psychiatric Research,*** 4(1), 65–71.
Six dogs were raised in an experimental setting with only 1 consistent object. The results tentatively indicate that models of separation and depression can be constructed in experimental animals.

105. Senf, Gerald M. & Miller, Neal E. (1967). **Evidence for positive induction in discrimination learning.** ***Journal of Comparative & Physiological Psychology,*** 64(1), 121–127.
The principle of positive induction states that the occurrence of the negative stimulus of a discrimination during extinction serves to strengthen the CR to the positive stimulus, thus slowing extinction. Rats were used as Ss in 3 experiments (the 1st using a classical conditioning procedure, the 2nd and 3rd instrumental learning situations) with controls for (1) possible stimulus-generalization decrements, (2) rate of trials during extinction, and (3) possible disinhibiting effects of presenting any stimulus between trials. The negative stimulus interspersed between nonreinforced positive stimulus trials in an extinction series significantly retarded the extinction of the CR. It is believed that the fact of positive induction requires a modification in traditional theories of discrimination learning.

106. Shipley, Robert H.; Mock, Lou A. & Levis, Donald J. (1971). **Effects of several response prevention procedures on activity, avoidance responding, and conditioned fear in rats.** ***Journal of Comparative & Physiological Psychology,*** 77(2), 256–270.
Conducted 2 experiments with 75 and 45 male Blue Spruce hooded rats to evaluate various response prevention procedures of extinction when CS exposure/trial was held constant. Preventing the occurrence of a response resulted in a low activity level for Ss during prevention trials and rapid extinction of the conditioned avoidance response (CAR). S's activity level and CAR extinction also appeared to be related to instrumental response contingencies and where the iti was spent. Fear extinction occurred during response prevention and appeared to occur in direct proportion to the total amount of CS exposure, irrespective of S's activity level during the CS exposure, number of trials, or the instrumental contingencies involved.

107. Suomi, Stephen J.; Collins, Mary L. & Harlow, Harry F. (1973). **Effects of permanent separation from mother on infant monkeys.** ***Developmental Psychology,*** 9(3), 376–384.
12 rhesus monkeys reared from birth with their mothers in a common pen were separated permanently from their mothers at 60, 90, or 120 days of age. One-half of the Ss in each age group were housed individually following separation; the other half were housed in pairs. Although Ss did not differ appreciably in levels of behaviors scored prior to separation, their behaviors both in the week following separation and at 6 mo of age varied according to age at separation and subsequent housing condition manipulations. All Ss showed agitation immediately following separation. Ss separated at 90 days of age showed quantitatively more severe immediate reaction to separation but by 6 mo of age did not differ appreciably from Ss separated at other ages. Ss housed singly following separation showed significantly higher levels of disturbance and self-clasping behavior and lower levels of locomotion during the week following separation than Ss housed in pairs. These differences were exaggerated at 6 mo of age.

108. Suomi, Stephen J.; Eisele, Carol D.; Grady, Sharon A. & Harlow, Harry F. (1975). **Depressive behavior in adult monkeys following separation from family environment.** ***Journal of Abnormal Psychology,*** 84(5), 576–578.
Studied the reactions of 10 nuclear family-reared young adult rhesus monkeys to separation from their families. Ss housed with friends during family separation were relatively unaffected by the separation, as were Ss housed with both friends and strangers. However, Ss individually housed following family removal exhibited depressive-like behaviors previously observed only in infant monkeys separated from mother and/or peers.

109. Suomi, Stephen J. & Harlow, Harry F. (1975). **Effects of differential removal from group on social development of rhesus monkeys.** ***Journal of Child Psychology & Psychiatry & Allied Disciplines,*** 16(2), 149–164.
16 rhesus monkey (Macaca mulatta) infants were reared with surrogates in groups of 4 for the 1st 4 mo of life, then were individually removed from their group for 28 days during the succeeding 6-mo period. Members of 2 groups were placed in vertical chambers during their period of group separation, while members of a 3rd group were housed in single cages during their time of separation. A control group of 4 Ss remained intact throughout the study. Results indicate that following removal from group, Ss showed lower levels of locomotion and play and higher levels of clinging and self-clasping than nonseparated controls. Differences from controls were exaggerated in Ss confined in vertical chambers during separation. All separation groups were socially inferior to the control group by the end of the study.

110. Suomi, Stephen J. & Harlow, Harry F. (1975). **Early experiences and induced psychopathology in Rhesus monkeys.** ***Revista Latinoamericana de Psicologia,*** 7(2), 205–229.

Reviews the main research conducted at the University of Wisconsin Primate Laboratory during the past 20 yrs on the effects of early experience on later social behavior in Rhesus monkeys.

111. Testa, Thomas J.; Juraska, Janice M. & Maier, Steven F. (1974). **Prior exposure to inescapable electric shock in rats affects extinction behavior after the successful acquisition of an escape response.** ***Learning & Motivation,*** 5(3), 380–392.
Describes 2 experiments with a total of 91 male Sprague-Dawley rats. In Exp I Ss exposed to 64 inescapable electric shocks in a restrainer or merely restrained were later given either 0, 5, 15, or 30 escape-avoidance training trials with a 2-way shuttlebox procedure that does not lead to interference with escape acquisition due to prior exposure to inescapable shock. After escape training all Ss were given an escape-avoidance extinction procedure in which shock was inescapable. Ss which had received exposure to shock responded less often and with longer latencies in extinction than did the restrained Ss. Exp II demonstrated that this effect was caused by the inescapability of the initial shock treatment. Results are explained in terms of (a) associative interference which minimized the effect of shuttlebox escape training for the preshocked Ss and (b) a stronger tendency to recognize the presence of an inescapable shock situation during extinction for the preshocked Ss.

112. Uyeno, Edward T. & Newton, Harold. (1972). **Learning ability and retention performance of a weanling squirrel monkey (Saimiri sciureus).** ***Primates,*** 13(4), 339–346.
Administered adaptations of the visual discrimination, delayed-response, and conditioned-avoidance response tests to a 33-wk-old male weanling squirrel monkey. Data show that the S learned to perform all the tests successfully. Findings indicate that the tractable squirrel monkey is a potentially satisfactory animal model for a comparative and developmental study of learning ability and retention performance during the early stages of life.

113. Varnell, J. Neil; McDaniel, Max H. & McCullough, James P. (1975). **Effect of flooding and order of cue presentation on extinction of a serial conditioned avoidance response.** ***Psychological Reports,*** 36(2), 623–629.
Investigated the effect of flooding and the order of serial-cue presentation on the extinction of a conditioned avoidance response using 40 female albino rats. Flooding significantly facilitated the extinction of the conditioned avoidance response. Partial reversal of the order in which the serial cues were presented to the Ss during flooding produced no significant differential effects. A brief analogy is drawn between the serial-cue conditioned avoidance learning model and the human phobic response. Results support a current therapeutic technique which treats human phobic clients with a flooding procedure.

114. Veraart, Claude; Crémieux, Jacques & Wanet-Defalque, Marie-Chantal. (1992). **Use of an ultrasonic echolocation prosthesis by early visually deprived cats.** ***Behavioral Neuroscience,*** 106(1), 203–216.
Designed an animal model of sensory substitution in the case of blindness. Six kittens were binocularly enucleated; as adults, they were fitted with an ultrasonic echolocation prosthesis. This device provided the animals with auditory signals that coded distance and direction of obstacles. Animals were trained by operant conditioning to use the prosthesis in various behavioral situations. Results showed that visually deprived animals tried to solve the task using natural information and that they only used artificial information provided by the prosthesis when they were unable to succeed with natural cues. Under these conditions, it was asserted that in a jumping test these animals evaluated depth by means of the prosthesis; in a locomotion task in a maze, it was also demonstrated that they could use the prosthesis for avoiding obstacles.

115. Weiss, Jay M. (1967). **Effects of coping behavior on development of gastrointestinal lesions in rats.** ***Proceedings of the 75th Annual Convention of the American Psychological Association,*** 2, 135–136.
Rats that could perform a coping response to avoid or escape electric shock developed less severe gastrointestinal lesions than yoked Ss that received the same shocks but could not perform such a coping response. Since the amount of shock that matched avoidance and yoked Ss received was exactly the same (shock was delivered to both Ss through fixed tail electrodes wired in series), differences resulting between them are attributable to psychological factors produced because 1 S could "cope with" the stress situation while the other S was "helpless." These results agreed with those of other studies in the same series.

116. Weiss, Jay M. (1968). **Effects of coping responses on stress.** ***Journal of Comparative & Physiological Psychology,*** 65(2), 251–260.
Male, Sprague-Dawley, 90-day-old albino rats that could perform a coping response to avoid or escape electric shock developed less severe physiological symptoms of stress than yoked Ss that received the same shocks but could not perform such a coping response. Yoked Ss showed a greater decrease in body weight and more extensive gastric lesions than avoidance and nonshock Ss. Since avoidance and yoked Ss received the same amount of shock through fixed tail electrodes wired in series, differences between groups are attributable to psychological factors produced because 1 S could "cope with" the stress situation whereas the other S could not.

117. Weiss, Jay M.; Krieckhaus, E. E. & Conte, Richard. (1968). **Effects of fear conditioning on subsequent avoidance behavior and movement.** ***Journal of Comparative & Physiological Psychology,*** 65(3, Pt 1), 413–421.
In 2 experiments with male albino Sprague-Dawley rats, 8 fear-conditioning trials, given to rats before shuttle avoidance training, markedly impaired avoidance learning through as many as 250 training trials. Measurement of movement showed that fear conditioning produced extensive freezing, which interfered with avoidance learning. Two additional experiments showed that the avoidance decrement could be eliminated by teaching Ss the correct response prior to fear conditioning, or could be overcome by giving fear extinction to Ss performing poorly, procedures that in both cases reduced freezing. Freezing was considered to be a UCR to strong fear, which has implications for the inverted U-shaped curve relating performance to motivation and for 2-factor learning theory.

118. Willner, Paul. (1986). **Validation criteria for animal models of human mental disorders: Learned helplessness as a paradigm case.** ***Progress in Neuro-Psychopharmacology & Biological Psychiatry,*** 10(6), 677–690.

Proposes 3 sets of criteria for assessing animal models of human psychopathology: predictive validity (performance in the test predicts performance in the condition being modeled); face validity (phenomenological similarity); and construct validity (theoretical rationale). Problems inherent in each procedure are discussed, and their application to the learned helplessness model of depression is examined. It is suggested that while the model has good predictive validity, important questions about face validity remain unanswered, and construct validity has not yet been established. Distinctions between animal models and related experimental procedures are also discussed.

119. Zamble, Edward; Mitchell, John B. & Findlay, Helen. (1986). **Pavlovian conditioning of sexual arousal: Parametric and background manipulations.** ***Journal of Experimental Psychology: Animal Behavior Processes,*** 12(4), 403–411.
Recent studies have shown that a Pavlovian conditioned stimulus (CS) for unconsummated sexual arousal can increase the rate of copulation in the rat. The present report includes 5 experiments examining the effects of parametric manipulations on the conditioned arousal response in 216 male and more than 136 ovariectomized female Long-Evans hooded rats. Results show that 6–9 trials are necessary for reliable conditioning, but extinction is somewhat slower than acquisition. The function for the CS/unconditioned stimulus (UCS) interval is quadratic, with a minimum of several minutes required for effective conditioning. In the 1st 3 experiments, it appeared that background cues were conditioned as well as the designated CSs, and this was tested explicitly in the last 2 studies. In one, the effect of background cues was shown by training and testing in different situations; in the 2nd, background cues were shown to be subject to latent inhibition. Results demonstrate the influence of Pavlovian learning in sexual behavior and help to provide the basis for an animal model of the acquisition of deviant sexual arousal in humans.

120. Zolman, James F. & McDougall, Sanders A. (1983). **Young precocial birds: Animal models for developmental neurobehavioural research.** ***Bird Behaviour,*** 5(1), 31–58.
Reviews the general advantages and limitations of the young precocial bird, particularly the domestic chick, as an animal model for research on the neurobiological mechanisms involved in the ontogeny of behavior. Most neurophysiological studies have focused on widespread changes in cerebral electrical activity during embryogenesis and the hatching period. More information is needed about brain anatomy and physiology in the 1st wk after hatching to study the neurobiological correlates of emerging behavioral patterns in the young chick. There is evidence for multiple neurotransmitter involvement in emerging behavioral patterns of the young chick during ontogeny. Comparative drug-behavior profiles are necessary to determine the relationships among age, brain neurotransmitters, and behavioral patterns. The detail and precision in the temporal structure of memory consolidation has been established using drug manipulations. Neurobiological studies of learning have identified parallels in the biochemical sequelae induced in the chick's brain by imprinting and taste-aversion learning.

Learning & Motivation

121. Alloy, Lauren B. & Bersh, Philip J. (1979). **Partial control and learned helplessness in rats: Control over shock intensity prevents interference with subsequent escape.** ***Animal Learning & Behavior,*** 7(2), 157–164.
Studied 36 male Sprague-Dawley rats to examine the effect on subsequent escape acquisition of control over shock intensity in the absence of control over other shock characteristics. Pretreatment involved random shocks of 1.6 and .75 mA at a density of approximately 10/min. The experimental group could avoid the higher shock intensity if they leverpressed at least once every 15 sec. Yoked and no-shock Ss completed the triadic design. Experimental and yoked Ss received all scheduled shocks. Triads were later tested for FR 2 shuttlebox escape at either the .75 mA (low) or 1.6 mA (high) intensity. During testing, avoidance Ss performed as well as no-shock Ss at the low intensity and escaped even more rapidly at the high intensity. Yoked Ss showed interference at both intensities, with interference very marked, including many failures to escape, at the low intensity. Findings indicate that control over shock intensity alone is sufficient to prevent learned helplessness and suggest that control over any salient characteristic of shock may be sufficient for immunization.

122. Altenor, Aidan; Kay, Edwin & Richter, Martin. (1977). **The generality of learned helplessness in the rat.** ***Learning & Motivation,*** 8(1), 54–61.
Two experiments were simultaneously conducted in which 2 different groups of 40 Sprague-Dawley rats each were exposed to 1 of 2 different stressors. In both experiments half the Ss were pretreated with shock, half with underwater exposure. For each pretreatment stressor, half the Ss were allowed to escape, the other half were not. The experiments differed in the test task used. Approximately 24 hrs after pretreatment, half the Ss from each pretreatment group received 20 water-escape trials in an underwater maze, the other half received 20 shock-escape trials in a 2-way shuttle box. Ss in each of the inescapable pretreatment conditions were slower to escape in the subsequent shock-escape and water-escape tasks when compared with Ss in the corresponding escapable pretreatment condition. The "learned helplessness" effect appeared to be no smaller when aversive stimuli were changed between pretreatment and test than when they remained the same.

123. Altenor, Aidan; Volpicelli, Joseph R. & Seligman, Martin E. (1979). **Debilitated shock escape is produced by both short- and long-duration inescapable shock: Learned helplessness vs. learned inactivity.** ***Bulletin of the Psychonomic Society,*** 14(5), 337–339.
In an experiment with 50 male Sprague-Dawley rats, inescapable shocks of short (.5 sec) and long (5 sec) duration interfered with subsequent shock escape. In addition, there were no differences between groups that received the pretreatment shocks and testing in the same or different apparatuses. These results are consistent with the learned helplessness account but conflict with recent learned inactivity accounts for the interference produced by inescapable shocks.

124. Anisman, Hymie; DeCatanzaro, Denys & Remington, Gary. (1978). **Escape performance following exposure to inescapable shock: Deficits in motor response maintenance.** ***Journal of Experimental Psychology: Animal Behavior Processes,*** 4(3), 197–218.
13 experiments employing 607 Swiss-Webster mice investigated shock-elicited activity in a circular field and escape performance in a shuttlebox following exposure to either escapable or inescapable shock. Upon shock inception in the circular field, Ss exhibited a 2–3 sec period of constant or increasing motor excitation followed by a decline in motor activity toward or below preshock levels. Prior exposure to inescapable shock decreased the magnitude of initial excitation and increased the rate at which locomotor excitation declined. Inescapable shock did not detectably affect escape performance if escape was possible immediately upon shock onset. If escape was briefly delayed 4–6 sec until the time at which a marked decline in the shock-elicited excitation would be expected, marked deficits of escape performance were seen. Treatments that attenuated the reduction in activity produced by inescapable shock (e.g., shock interruption during test) mitigated the escape deficits. It was demonstrated that duration and intensity of inescapable shock influence later escape behavior in a manner that is highly correlated with the motor changes induced by the treatments seen in the circular field. It is concluded that the escape interference induced by inescapable shock may be interpreted in terms of a decreased tendency for shock to sustain vigorous motor activity for protracted periods.

125. Anisman, Hymie; Irwin, Jill; Beauchamp, Christine & Zacharko, Robert M. (1983). **Cross-stressor immunization against the behavioral deficits introduced by uncontrollable shock.** ***Behavioral Neuroscience,*** 97(3), 452–461.
In 4 experiments with 204 male CD-1 mice, exposure to inescapable shock disrupted performance in both shock- (SE) and water-escape (WE) tasks. These deficits were prevented in Ss that were previously trained in the same task. However, an asymmetrical immunization effect was seen in a cross-stressor paradigm. Whereas deficits of WE performance engendered by inescapable shock were prevented by prior SE training, the deficits of SE were not eliminated by prior WE training. Evidently, the immunization effect occurs when initial training and subsequent testing are conducted in the same task or when the initial training and uncontrollable stress session involve the same aversive stimulus. Norepinephrine (NE) determinations revealed that reductions of NE introduced by inescapable shock were unaffected by prior SE training and were enhanced by prior exposure to the stress of water immersion. Thus, although the performance deficit introduced by inescapable shock may be related to variations of NE, the immunization effect probably was unrelated to alterations of NE. Data provisionally suggest that the immunization stems from 2 independent factors: Initially training Ss in an active escape task may (a) disrupt subsequent learning that the inescapable stress actually is uncontrollable and (b) limit the influence of the motor deficits introduced by uncontrollable shock on subsequent escape performance.

126. Baker, A. G. (1976). **Learned irrelevance and learned helplessness: Rats learn that stimuli, reinforcers, and responses are uncorrelated.** ***Journal of Experimental Psychology: Animal Behavior Processes,*** 2(2), 130–141.
In 4 experiments with 120 male hooded Long-Evans rats, preexposure to uncorrelated presentations of noises and shocks retarded the acquisition of conditioned emotional response (CER) and signaled punishment suppression when this preexposure was carried out while Ss were not permitted to respond. When preexposure was carried out while Ss were responding, preexposure to uncorrelated noises and shocks interfered with signaled punishment suppression on the 1st test day, and both preexposure to uncorrelated noises and shocks and preexposure to the shocks caused similar interference on the remaining test days. Signaled punishment suppression was interfered with more when preexposure to uncorrelated noises and shocks was carried out while Ss were responding, but CER suppression was similarly affected whether Ss were permitted to continue responding or not. Results suggest that rats may learn that both noises and shocks and responses and shocks are uncorrelated, and these 2 types of learning may interfere with learning CER suppression and the punishment contingency, respectively.

127. Baum, Morrie. (1988). **Contributions of animal studies of response prevention (flooding) to human exposure therapy.** ***Psychological Reports,*** 63(2), 421–422.
Describes the behavioral technique of response prevention (e.g., flooding, exposure therapy, implosive therapy), which involves thwarting avoidance responses and having organisms confront anxiety evoking situations. Contributions of animal studies of avoidance-extinction to therapy of human phobic disorders include the delineation of the importance of treatment duration and the facilitative effects of distraction during treatment. Animal studies showing that certain pharmaceutical and hormonal agents increase treatment efficiency are reviewed, and improvements in human therapy are suggested.

128. Beatty, William W. (1979). **Failure to observe learned helplessness in rats exposed to inescapable footshock.** ***Bulletin of the Psychonomic Society,*** 13(4), 272–273.
Conducted 2 experiments with 71 Holtzman rats to determine whether exposure to inescapable footshock would impair acquisition of difficult (FR 3 or higher) instrumental escape responses by rats. In Exp I, Ss received 250 trials of inescapable or escapable shock (mean duration: 7.7 sec/trial). In Exp II, Ss were exposed to durations of 5, 10, or 20 sec/trial of inescapable footshock. No evidence of an interference effect (helplessness) was observed, despite the use of conditions that should have detected this effect.

129. Bersh, Philip J.; Whitehouse, Wayne G.; Blustein, Joshua E. & Alloy, Lauren B. (1986). **Interaction of Pavlovian conditioning with a zero operant contingency: Chronic exposure to signaled inescapable shock maintains learned helplessness effects.** ***Journal of Experimental Psychology: Animal Behavior Processes,*** 12(3), 277–290.
Four studies, with 372 male Holtzman rats, examined the effect of Pavlovian contingencies and a zero operant contingency (i.e., uncontrollability) on subsequent shock-escape acquisition in the shuttle box using triads consisting of escapable-shock (ES), yoked inescapable-shock (IS), and no-shock (NS) rats. After exposure to 50 signals and shocks per session for 9 sessions, interference with shuttlebox escape acquisition for IS Ss was a monotonically increasing function of the percentage of signal–shock pairings during training (Exp I), with 50% pairings producing little or no impairment. Without regard to signaling, ES Ss performed as well as NS Ss. Exp II demonstrated that training and test con-

ditions led to substantial and equal impairment in IS Ss preexposed for 1 session to 100 or 50% signal–shock pairings or to unsignaled shocks. In Exp III, chronic exposure to 100% signaled ISs resulted in impairment only if the signal (light) was present during the shuttlebox test. The continuous presence of the signal during the test contrasted with its discrete (5-sec) presentation during training and suggested that an antagonistic physiological reaction rather than a specific competing motor response had been conditioned. Exp IV provided evidence for possible conditioned opioid mediation. Findings suggest that chronic exposure to uncontrollable shocks maintains the impairment produced by acute exposure only if the shocks are adequately signaled.

130. Binik, Yitzchak & Seligman, Martin E. (1979). **Sudden swimming deaths.** ***American Psychologist,*** 34(3), 270–273.
Suggests that C. W. Hughes and J. J. Lynch's (1978) review of data on the phenomenon of sudden swimming deaths is in need of clarification. The review's dismissal of psychosocial influences is said to be unfounded, and the assertion that helplessness is an essentially untestable concept in nonverbal animals is said to be uninformed.

131. Black, Abraham H. (1977). **Comments on "Learned helplessness: Theory and evidence." by Maier and Seligman.** ***Journal of Experimental Psychology: General,*** 106(1), 41–43.
S. F. Maier and M. E. Seligman (1976) argued against response competition explanations of the effects of preshock on subsequent escape and avoidance learning. One of their main points was that competing responses interfere only with performance of subsequently trained responses, whereas learned helplessness accounts for effects on both performance and learning. The present author contends that competing responses could interfere with both learning and performance.

132. Boyd, Thomas L. & Levis, Donald J. (1976). **The effects of single-component extinction of a three-component serial CS on resistance to extinction of the conditioned avoidance response.** ***Learning & Motivation,*** 7(4), 517–531.
Examined the differential effects of extinguishing separate components of the CS complex upon responding to the complete CS complex during extinction. In Phase 1 of the study, 72 male albino rats were classically conditioned to a 3-component serial CS complex followed by shock. Each S was then given avoidance training in a 1-way apparatus to a criterion of 1 successful avoidance. In Phase 2, Ss were divided into 4 groups, with 3 groups receiving nonreinforced exposure for 25 trials to 1 of the components of the serial CS complex; a 4th group was exposed for the same period of time to the apparatus cues. In Phase 3, the total stimulus complex was reintroduced in its original order, and Ss were tested until extinction of the instrumental response was reached. Results are consistent with the hypothesis that a fear gradient exists in extinction and decreases in magnitude as the distance from the point of UCS onset increases. Data are believed to have implications for response-prevention procedures of extinction as employed at the human level in the context of therapeutic treatment.

133. Boyd, Thomas L. & Levis, Donald J. (1979). **The interactive effects of shuttlebox situational cues and shock intensity.** ***American Journal of Psychology,*** 92(1), 125–132.
Studied the effects of 3 levels of shock intensity (0.5, 1.0, and 2.0 mA) and 2 levels of situational-cue change (grid-floor to solid-floor and grid-floor to grid-floor) on shuttle-avoidance performance in 80 male Blue-Spruce hooded rats. Shock-intensity effects interacted with changes in situational cues even though topography of the avoidance responses (shuttling) was similar for all Ss. When the response requirement was from grid-floor to solid-floor, avoidance responding either improved with increases in shock intensity or remained unchanged. When the response requirement was from grid-floor to grid-floor, the typical finding of an inverse relationship between performance and intensity effects was obtained. The former stimulus condition also resulted in superior avoidance performance when compared to the latter condition. It is argued that the data are consistent with a stimulus change or reinforcement interpretation of shock-intensity effects.

134. Brown, Gary E. & Dixon, Paul A. (1983). **Learned helplessness in the gerbil?** ***Journal of Comparative Psychology,*** 97(1), 90–92.
27 male Mongolian gerbils were assigned to escapable, inescapable, and control groups and subjected to 2 consecutive days of jump-up escape followed 24 hrs later by testing in a barpress escape task. Reliable differences occurred between escapable and inescapable Ss and between inescapable and control Ss. Findings provide evidence of learned helplessness in Mongolian gerbils.

135. Brown, Gary E.; Smith, Paul J. & Peters, R. Brian. (1985). **Effect of escapable versus inescapable shock on avoidance behavior in the goldfish (*Carassius auratus*).** ***Psychological Reports,*** 57(3, Pt 2), 1027–1030.
Following exposure to either escapable, inescapable, or no shock, goldfish were tested on an avoidance task. Differences in latency consistent with M. Seligman and G. Beagley's (1975) hypothesis that helplessness is learned were present only on the 1st block of 5 trials. While reliable differences in the number of trials to the 1st avoidance response supported an interpretation as learned helplessness, differences in the total number of avoidance responses did not. It is concluded that for goldfish, inescapable preshock produces an interference with subsequent avoidance behavior, but this interference effect does not meet Seligman and Beagley's criteria for demonstrating that helplessness is learned, nor is the effect based on the fish's learning to be inactive during the pretest.

136. Buchanan, Shirley L. & Ginn, Sheryl R. (1988). **Classically conditioned cardiac responses in "old" and "young" Fischer 344 rats.** ***Psychology & Aging,*** 3(1), 51–58.
Male and female Fischer 344 rats, 12 or 26–28 months of age, received two sessions of Pavlovian heart rate conditioning, and were compared with same-sex and same-age controls receiving unpaired presentations of the tone conditional stimulus (CS) and the shock unconditional stimulus (US). Older rats of both sexes demonstrated slower acquisition of the heart rate (HR), conditioned response (CR), and smaller magnitude changes than did the younger animals. Control experiments in 6-, 12-, 24-, and 30-month-old animals indicated that these differences were not due to an impaired sensitivity to the CS or US in the older animals. Results are discussed in terms of their implications for use of this animal model in investigations of age-related deficits in associative learning.

137. Burdette, David R.; Krantz, David S. & Amsel, Abram. (1975). **Effects of inescapable shock in the rat: Learned helplessness or response competition.** ***Bulletin of the Psychonomic Society,*** 6(1), 96–98.
Attempted a new procedure for alleviating the effects of inescapable shock (preshock) on subsequent learning. Trials were introduced in a straight alley between preshock and escape/avoidance learning designed to countercondition an approach response to fear cues in 24 male Holtzman albino rats. The success of the procedure in reducing the interference effect supports a response-competition interpretation of the effect. However, performance of 8 controls that received only preshock led to consideration of other explanations of the results.

138. Calef, Richard S. et al. (1984). **Acquisition of running in the straight alley following experience with response-independent food.** ***Bulletin of the Psychonomic Society,*** 22(1), 67–69.
During Phase 1 of the present study, 24 male Sprague-Dawley albino rats received contingent food reinforcement, noncontingent reward, or no treatment in an operant chamber. In Phase 2, acquisition of a runway task was more rapid in the noncontingent-reward group than in the contingent-reward and no-treatment groups. Findings conflict with the learned helplessness interpretation of the effects of receiving response-independent reward, which suggests that uncontrollability should produce a generalized reduction in motivation. Results are interpreted in terms of A. Amsel's (1958, 1972) conditioned-frustration and general persistence theories, which suggests that Ss receiving response-independent food should have acquired the runway task faster than the controls because of the generalization of learning to persist in the presence of a disruptive stimulus.

139. Calef, Richard S.; Choban, Michael C.; Dickson, Marcus W.; Newman, Paul D. et al. (1989). **The effects of noncontingent reinforcement on the behavior of a previously learned running response.** ***Bulletin of the Psychonomic Society,*** 27(3), 263–266.
During Phase 1, all rats received a delay of food reward following a traversal of a straight alley. During Phase 2, Ss received contingent (CN), noncontingent (NCN), or no food reward (NR) in an operant chamber. During the 1st day of Phase 3 (running in the straight alley), no differences in speeds occurred between groups receiving CN and NCN food. However, during the 3rd day of Phase 3, Group NCN ran significantly slower than Groups CN and NR, suggesting that noncontingent reinforcement does not interfere with the retention of prior learning, but may impair the further learning of a response partially learned prior to receiving uncontrollability (response-independent rewards).

140. Calef, Richard S.; Choban, Michael C.; Shaver, Jim P.; Dye, Jack D. et al. (1986). **The effects of inescapable shock on the retention of a previously learned response in an appetitive situation with delay of reinforcement.** ***Bulletin of the Psychonomic Society,*** 24(3), 213–216.
Investigated the effects of uncontrollable shock on retention and examined the mechanisms contributing to learned helplessness in an experiment with 33 male albino Sprague-Dawley rats. During Phase 1 of the experiment, all Ss received a delay of (food) reward in a straight alley. During Phase 2, Ss received escapable, inescapable, or no footshock in an operant chamber. Results, which support a motivational interpretation of learned helplessness, show that only inescapable shock reduced speed in a response previously attained (running in the straight alley) in Phase 3.

141. Caspy, Tamir & Lubow, R. E. (1981). **Generality of US preexposure effects: Transfer from food to shock or shock to food with and without the same response requirements.** ***Animal Learning & Behavior,*** 9(4), 524–532.
Four experiments with 152 male BALB-c mice investigated the generality of the learned helplessness phenomenon by employing preexposure to aversive stimuli (shock) and appetitive stimuli (food). Ss were preexposed to contingent, noncontingent, or no stimuli (except for Exp II) in a Skinner box. During the test, Ss preexposed to shock were tested with food, and those preexposed to food were tested with shock. The test was conducted in a similar situation, a Skinner box, or a different situation, a runway. Performance decrements were evident when Ss preexposed to a noncontingent stimulus were compared with Ss preexposed to contingent stimuli. The differences between the contingent and the noncontingent groups were significant, as were the differences between the contingent and the nonpreexposed groups (except Exp I). The effects cut across the different types of stimuli, situations, and response requirements of the preexposure and test phases.

142. Childress, Anna R. & Thomas, Earl. (1979). **A comparison of central aversive stimulation and peripheral shocks in the production of learned helplessness.** ***Physiological Psychology,*** 7(2), 131–134.
30 male Sprague-Dawley rats were given helplessness training with either inescapable peripheral shocks or inescapable, similarly aversive, central stimulation in the dorsal midbrain. The effects of helplessness treatment were assessed in an FR 3 leverpress escape task administered 1 day later. Helplessness treatment with peripheral shocks produced severe deficits in later escape to peripheral shock, but not to dorsal midbrain stimulation. Helplessness treatment with dorsal midbrain stimulation produced no deficits in later escape to either dorsal midbrain stimulation or peripheral shocks. These data pose problems for a strict cognitive model of the helplessness syndrome.

143. Chiszar, David & Carpen, Karlana. (1979). **Natural selection and "sudden death."** ***American Psychologist,*** 34(3), 274–275.
C. W. Hughes and J. J. Lynch's (1978) article supports arguments relating survival time of rats in the Richter swimming apparatus to ontogenetic, physiological, and situational variables. Experiences with the sudden death of zoo animals, particularly ungulates, are discussed with reference to this argument.

144. Clark, Lincoln D. & Gay, Patricia E. (1980). **Behavioral defeat in squirrel monkeys: An experimental model of reactive depression.** ***Psychological Reports,*** 47(3, Pt 2), 1175–1184.
In 3 experiments with 10 adult male squirrel monkeys, the combination of a progressive FR schedule of reinforcement and the opportunity to "escape" from this schedule produced a syndrome with prima facie similarity to human reactive depression. This condition was created by a progressive increase in work demanded per reward and a corresponding reduction in density of reinforcement. The syndrome was characterized by progressive reduction of posi-

tively reinforced behavior, withdrawal from the environment, task ambivalence, and signs of emotional stress. These behaviors were ameliorated by environmental change that reduced the experientially produced stress and were dramatically reversed by the antianxiety agent, chlordiazepoxide (3, 6, and 9 mg/kg, im).

145. Cook, Michael & Mineka, Susan. (1987). **Second-order conditioning and overshadowing in the observational conditioning of fear in monkeys.** ***Behaviour Research & Therapy,*** 25(5), 349–364.
Explored the extent to which 2nd-order conditioning and overshadowing occur in the context of observational conditioning (OC). In Exp I, rhesus monkeys that had previously acquired a fear of snakes through OC underwent 6 sessions of 2nd-order conditioning in which a black-striped box, the 2nd-order conditioned stimulus (CS), was paired with snake stimuli, the 1st-order CS. Results indicate that small but significant amounts of fear were conditioned to the 2nd-order CS. In Exp II, a modified overshadowing paradigm was used to determine whether a fear-relevant CS (e.g., a snake) would overshadow a fear-irrelevant CS (e.g., a flower). After 6 sessions of OC, Ss acquired a fear of snakes but not a fear of flowers.

146. Cook, Michael; Mineka, Susan & Trumble, Dennis. (1987). **The role of response-produced and exteroceptive feedback in the attenuation of fear over the course of avoidance learning.** ***Journal of Experimental Psychology: Animal Behavior Processes,*** 13(3), 239–249.
Possible mechanisms mediating fear attenuation over prolonged avoidance learning were examined. In Replication 1, two groups of rats (masters) received 50 or 200 trials of signaled avoidance training. Six groups were yoked to each master group: Three were strictly yoked, and three were yoked only for reinforced conditioned stimulus (CS) presentations (yoked fear conditioning). Of the six groups, two (one strictly yoked and one yoked fear conditioning) received exteroceptive feedback contingent upon the reponses of the masters, two received random/noncontingent feedback, and two received no feedback. Fear of the conditioned stimulus (CS), indexed by freezing during the CS, was lowest in the 200-trial masters and in the two 200-trial groups receiving contingent feedback. In Replication 2, which was procedurally identical to Replication 1 except that the master groups received contingent exteroceptive feedback, fear was lowest in the same three groups. These results support the conclusion that the response-produced feedback that an avoidance response provides is responsible for the fear attenuation seen in well-trained avoidance responders. Several hypotheses concerning the effects of feedback in mediating fear attenuation are examined.

147. Cook, Michael & Mineka, Susan. (1990). **Selective associations in the observational conditioning of fear in rhesus monkeys.** ***Journal of Experimental Psychology: Animal Behavior Processes,*** 16(4), 372–389.
Three experiments explored the issue of selective associations in the observational conditioning of fear. Experiment 1 results indicated that observer rhesus monkeys acquired a fear of snakes through watching videotapes of model monkeys behaving fearfully with snakes. In Experiment 2, observers watched edited videotapes that showed models reacting either fearfully to toy snakes and nonfearfully to artificial flowers (SN+/FL–) or vice versa (FL+/SN–). SN+/FL– observers acquired a fear of snakes but not of flowers; FL+/SN– observers did not acquire a fear of either stimulus. In Experiment 3, monkeys solved complex appetitive discriminative (PAN) problems at comparable rates regardless of whether the discriminative stimuli were the videotaped snake or the flower stimuli used in Experiment 2. Thus, monkeys appear to selectively associate snakes with fear.

148. Cook, Michael & Mineka, Susan. (1989). **Observational conditioning of fear to fear-relevant versus fear-irrelevant stimuli in rhesus monkeys.** ***Journal of Abnormal Psychology,*** 98(4), 448–459.
Two experiments examined whether superior observational conditioning of fear occurs in observer rhesus monkeys that watch model monkeys exhibit an intense fear of fear-relevant, as compared with fear-irrelevant, stimuli. In both experiments, videotapes of model monkeys behaving fearfully were spliced so that it appeared that the models were reacting fearfully either to fear-relevant stimuli (toy snakes or a toy crocodile), or to fear-irrelevant stimuli (flowers or a toy rabbit). Observer groups watched one of four kinds of videotapes for 12 sessions. Results indicated that observers acquired a fear of fear-relevant stimuli (toy snakes and toy crocodile), but not of fear-irrelevant stimuli (flowers and toy rabbit). Implications of the present results for the preparedness theory of phobias are discussed.

149. Cook, Michael; Mineka, Susan; Wolkenstein, Bonnie & Laitsch, Karen. (1985). **Observational conditioning of snake fear in unrelated rhesus monkeys.** ***Journal of Abnormal Psychology,*** 94(4), 591–610.
Extended the findings of the present 2nd author et al (1984) in 2 experiments, using 4 rhesus monkeys as models and 18 rhesus monkeys as observers. In Exp I, 2 wild-reared Ss with a strong fear of snakes served as models, and 10 laboratory-reared Ss with no initial snake fear, who were acquainted with but not related to their models, served as observers. The observers showed asymptotic levels of fear in another context (the Sackett Self-Selection Circus) after only 8 min of watching their models behave fearfully in the presence of snake stimuli in the Wisconsin general test apparatus. In Exp II, 2 observers from Exp I who had acquired snake fear vicariously served as models for 8 other unrelated, and for the most part unacquainted, laboratory-reared Ss. Results are similar to those for Exp I, except that the level of acquired or maintained fear was slightly lower in Exp II. The differences in level of fear in the 2 experiments are discussed in relation to the possible effects of the model's age, dominance status, rearing history, and level of fear and to possible mechanisms underlying observational conditioning.

150. Cotton, M. M. et al. (1982). **Learned helplessness in shuttlebox-avoidance behavior.** ***Psychological Reports,*** 51(1), 215–221.
18 male Wistar rats were given experience in a Skinner box before being trained in a 2-way shuttlebox-avoidance task. Ss experiencing controllable shock (avoidable or escapable) performed significantly better than those that had experienced no shock or unavoidable shock in the Skinner box. Results are discussed in terms of the learned helplessness hypothesis proposed by S. F. Maier et al (1969).

151. Davis, Joel L.; Pico, Richard M. & Flood, James F. (1987). **Differences in learning between hyperprolinemic mice and their congenic controls.** ***Behavioral & Neural Biology,*** 48(1), 128–137.

Examined learning differences in PRO/Re-bb (genetically hyperprolinemic) mice and PRO/Re-aa (congenic nonhyperprolinemic controls). Results indicate that hyperprolinemic Ss, with a defect in proline metabolism, were impaired in acquiring T-maze and shuttlebox footshock avoidance behavior. Radial maze performance was poor in both groups. This may have been due to observed acrophobia and lack of exploratory behavior. Findings suggest that high-brain proline in conjunction with other amino acid changes account for the learning deficits in these mice. It is suggested that the lack of difference between groups in passive avoidance reflects a sensory impairment rather than difficulty in learning the task.

152. DeCola, Joseph P.; Rosellini, Robert A. & Warren, Donald A. (1988). **A dissociation of the effects of control and prediction.** ***Learning & Motivation,*** 19(3), 269–282.
Assessed the ability of a feedback stimulus presented during inescapable shock exposure to mimic the effects of control over shock offset, using 37 male rats. It was found that Ss exposed to inescapable shock with a feedback stimulus and Ss that received escapable shock demonstrated equivalent levels of fear conditioned to the shock context; these 2 groups showed less fear than Ss exposed to inescapable shock without feedback. This pattern of group differences was observed also in a shuttlebox escape test but not in an appetitive noncontingent test. In the latter test, Ss previously exposed to inescapable shock with and without feedback demonstrated equivalent response patterns that were different from those of the escape group.

153. DeCola, Joseph P. & Rosellini, Robert A. (1990). **Unpredictable/uncontrollable stress proactively interferes with appetitive Pavlovian conditioning.** ***Learning & Motivation,*** 21(2), 137–152.
Prior research on the proactive effects of exposure to uncontrollable aversive events has demonstrated interference with the formation of stimulus–stimulus (S–S) associations. The present study investigated whether such interference could be observed on an appetitive Pavlovian discrimination (PD) task. After stress exposure, rats showed equivalent acquisition of an excitatory association to a tone consistently paired with a food unconditioned stimulus/stimuli (UCS) during the 1st phase of conditioning. During the appetitive PD phase, the group exposed to uncontrollable and unpredictable shock showed retarded acquisition of the PD. However, the group exposed to uncontrollable stress accompanied by a feedback stimulus did not show interference. The transfer observed may result from learning about S–S independence during the original stress exposure.

154. Domjan, Michael. (1987). **Animal learning comes of age.** ***American Psychologist,*** 42(6), 556–564.
The current status of the field of animal learning is described in historical context. Three themes that have provided major impetus for the study of animal learning are identified: comparative cognition, animal models of human behavior, and functional neurology. The historical roots of these themes are described, followed by examples of contemporary research relevant to each theme. Important recent changes in conceptualization of basic conditioning phenomena are also described. The review suggests that the field of animal learning continues to contribute in unique and important ways to the understanding of behavior.

155. Drugan, Robert C. & Maier, Steven F. (1982). **The nature of the activity deficit produced by inescapable shock.** ***Animal Learning & Behavior,*** 10(3), 401–406.
Two experiments with 32 male albino rats investigated the nature and etiology of the reduced activity in the presence of shock produced by prior exposure to inescapable shock. Previous experiments have demonstrated this deficit in the presence of gridshock. However, gridshock hurts less if movement across grids is reduced. It is unclear whether the inescapable-shock-produced deficit represents facilitation of learning to reduce movement across the grids in order to alleviate pain or is an "unconditioned" reduction in movement in response to shock. Exp I examined the effects of inescapable shock on subsequent movement during shock delivered via fixed tail electrodes to freely moving Ss. Inescapably shocked Ss still moved less in response to shock than did escapably shocked and restrained control Ss. Exp II examined the possibility that this deficit occurred because unconditioned movement in response to shock during pretreatment diminished after a few seconds, the reduction being adventitiously reinforced by shortly ensuing shock termination. Activity during inescapable shock was closely monitored by ultrasonic motion detection. Although activity did decrease across trial blocks, the required within-trial patterns did not occur. Shock-elicited activity did not diminish after a few seconds of shock but remained unchanged across the 5-sec shock presentations.

156. Drugan, Robert C.; Moye, Thomas B. & Maier, Steven F. (1982). **Opioid and nonopioid forms of stress-induced analgesia: Some environmental determinants and characteristics.** ***Behavioral & Neural Biology,*** 35(3), 251–264.
J. W. Grave et al (1981) found that exposure to a series of inescapable shocks produces both an early nonopioid analgesia and a late-appearing opioid analgesia. The nonopioid analgesia was observed following 20 shocks, while the opioid analgesia was aroused after 60 and 80 shocks. Three experiments are reported with 72 male albino Holtzman rats in which the environmental conditions necessary to produce and maintain the 2 forms of analgesia were examined. It was revealed in Exp I that an extended number of continued shocks was necessary to arouse the late-appearing opioid analgesia. Neither remaining in the stress situation following an initial series of shocks nor receiving fewer shocks spread across the entire session was sufficient to produce the opioid analgesia. In Exp II, it was found that only 5 5-sec shocks were sufficient to produce the nonopiate analgesia. In Exp III, the dissipation of the nonopioid analgesia aroused by 5 shocks was examined. The analgesia decayed to control levels within 10 min of shock termination. However, if the shocks were continued, the analgesia was still well above control levels 20 min following initial arousal. A large dose of an opiate antagonist failed to attenuate this analgesia.

157. Falk, John L. & Tang, Maisy. (1988). **What schedule-induced polydipsia can tell us about alcoholism.** ***Alcoholism: Clinical & Experimental Research,*** 12(5), 577–585.
Presents an animal model of chronic and excessive voluntary (unforced) alcohol ingestion in which, by drinking, animals produce repeated, substantial elevations in blood ethanol concentration and develop physical dependence. The conditions inducing the ethanol overindulgence can generate a variety of behavioral excesses that place alcoholism in a context of environmentally determined malfunctions, subject

to therapeutic change by altering situational parameters. Efficacious experiments utilizing therapeutic and preventive strategies are described that may serve as suggestions for corresponding human alcoholism intervention strategies.

158. Ferrándiz, Pilar & Pardo, Antonio. (1990). **Immunization to learned helplessness in appetitive noncontingent contexts.** ***Animal Learning & Behavior,*** 18(3), 252–256.
Examined immunization against learned helplessness in 36 dogs. The experiment consisted of 5 phases: (1) appetitive contingent training, (2) immunization training, (3) inescapable noise training, (4) recovery, and (5) an appetitive noncontingent test. The immunization effect was assessed by measuring the acquisition of an appetitive response when food was not contingent upon responding. The immunization effect was observed in a noncontingent appetitive context. The effects of escapable noises that ensure immunization against the motivational deficit and predictable noises that immunize against the associative deficit seem to be additive.

159. Follick, Michael J. (1981). **Aggression during learned-helplessness pretraining as a coping response.** ***Psychological Reports,*** 48(2), 471–485.
The effects of the opportunity to aggress during pretraining on learned helplessness was examined in 56 male albino Sprague-Dawley rats. While Ss individually given inescapable shock showed deficits in subsequent chain-pull escape, the performance of Ss given inescapable shock in pairs did not differ from that of Ss exposed to escapable shock or no shock. Results are contrary to the prediction of the learned-helplessness hypothesis and are consistent with the notion that shock-induced aggression serves an adaptive function.

160. Garber, Judy; Fencil-Morse, Ellen; Rosellini, Robert A. & Seligman, Martin E. (1979). **Abnormal fixations and learned helplessness: Inescapable shock as a weanling impairs adult discrimination learning in rats.** ***Behaviour Research & Therapy,*** 17(3), 197–206.
Weanling rats received escapable, yoked inescapable, or no electric shock. Tested as adults, the inescapable groups were poor at appetitive discrimination learning in either a parallel arm maze (Exp I, 24 male Holtzman rats) or a Lashley jumping stand (Exp II, 24 male Sprague-Dawley rats). They were slower to respond, slower to reach criterion, and made fewer correct responses and more nonresponses than the escapable or no shock groups. It is suggested that Ss retain learned helplessness from weaning to adulthood, that such learning generalizes widely, and that learned helplessness may explain the abnormal fixation results of N. R. Maier (1949).

161. Gibson, E. Leigh & Booth, David A. (1989). **Dependence of carbohydrate-conditioned flavor preference on internal state in rats.** ***Learning & Motivation,*** 20(1), 36–47.
24 rats rapidly acquired preferences for a flavor incorporated in meals of a dilute carbohydrate diet, whether or not they were trained following consumption of a substantial volume of nonnutritive fluid. However, this carbohydrate-conditioned flavor preference was elicited only when the Ss were tested in the same gastrointestinal distension condition in which they had been trained (i.e., either with or without the nonnutritive preload). This is evidence that ingestion can be controlled by associatively conditioned combinations of dietary cues with internal cues or that preferences can be conditioned to depend on physiological states. This compound conditioning of physiological and dietary stimuli provides a mechanism for tuning the motivation to eat to nutritional needs and supplies.

162. Glazer, Howard I. & Weiss, Jay M. (1976). **Long-term and transitory interference effects.** ***Journal of Experimental Psychology: Animal Behavior Processes,*** 2(3), 191–201.
If animals receive inescapable electric shocks, their subsequent avoidance-escape learning is poor. This phenomenon, which can be called the interference effect, was studied in 4 experiments with a total of 133 male albino Holtzman rats. Exp I demonstrated that, depending on the parameters of the inescapable shock used, there exists a transitory effect and a separable, more permanent, long-term interference effect. Exps II and III investigated the long-term effect, showing that it (a) required inescapable shocks of at least 5 sec in order to develop and (b) was still evident 1 wk after such shock. It is suggested that, whereas the transient effect is produced by a short-lived neurochemical change, the long-term effect is mediated by a learned response. Consistent with this differentiation, Exp IV showed that the interference effect measured 30 min after inescapable shock did not occur when Ss had been repeatedly exposed to the type of inescapable shock that produced the transitory effect, whereas the interference effect measured 72 hrs after shock became more pronounced when Ss had been repeatedly exposed to the type of inescapable shock that produced the long-term deficit. Aspects of the data suggest that learned helplessness is not the basis of the long-term interference phenomenon.

163. Glazer, Howard I. & Weiss, Jay M. (1976). **Long-term interference effect: An alternative to learned helplessness.** ***Journal of Experimental Psychology: Animal Behavior Processes,*** 2(3), 202–213.
Three experiments, with a total of 112 male albino Holtzman rats, explored whether inescapable shock of long duration and moderate intensity (LoShk) produces an avoidance-escape deficit (interference effect) by causing animals to learn to respond less actively or by causing them to learn to be "helpless." Exp I shows that if Ss given LoShk were subsequently tested in an avoidance-escape "nosing" response that required little motor activity, they learned and performed better than no-shock controls. Exp II verified that the same LoShk treatment that led to this better performance would indeed interfere with subsequent avoidance-escape acquisition in those test situations that have previously been used to demonstrate the interference effect, that is, the 3-response leverpress, 2-response shuttle, and single-response barrier jump. In Exp III, using triplets in the classical avoidance-escape, yoked, and control animal paradigm, it was shown that yoked Ss performed more poorly than either avoidance-escape or control Ss on a subsequent 3-response leverpress task but performed better on the nosing avoidance-escape task. Results are compatible with the idea that animals acquire the tendency to be inactive as a result of exposure to long-duration, moderate-intensity, inescapable shock. Results are incompatible with the idea that animals learn "helplessness."

164. Goodkin, Franklin. (1976). **Rats learn the relationship between responding and environmental events: An expansion of the learned helplessness hypothesis.** ***Learning & Motivation,*** 7(3), 382–393.

Previous research has shown that preexposure to inescapable shock interferes with subsequent acquisition of escape responding, while pretraining with escapable shock facilitates subsequent acquisition of a different escape response. It has also been demonstrated that interference and facilitation persist when the aversive event is changed between the 2 phases of training. The present experiment extended these findings, showing generalized learning from an appetitive to an aversive situation. Six groups of 10 male Long-Evans rats each received the following treatment in the presence of discriminative stimuli: One group was trained to nosepress for food, a 2nd to chain-pull for food, and a 3rd to chain-pull to escape or avoid shock. Two groups received either signalled free food or inescapable shock, and a naive control group received no pretreatment. All groups were then tested in a nosepress escape-avoidance situation. The 3 groups with prior response training acquired responding most rapidly, and at the same rate. The controls acquired responding slowly, and the 2 groups with response-independent histories did not acquire responding during the 5 days of training.

165. Hamm, Robert J.; Knisely, Janet S. & Dixon, C. Edward. (1983). **An animal model of age changes in short-term memory: The DRL schedule.** ***Experimental Aging Research,*** 9(1), 23–25.
Trained 8 mature (6-mo-old) and 7 aged (24-mo-old) male Sprague-Dawley rats for 20 30-min sessions on a DRL 6-sec schedule. Responses were recorded according to their interresponse times (0–2 sec, 2–4 sec, 4–6 sec, 6–8 sec, 8–10 sec, and more than 10 sec). Results indicate that the aged Ss initially had difficulty in inhibiting short interresponse time responses, but with extended training this performance deficit was overcome and both age groups exhibited characteristic and effective patterns of responding. These data suggest that aged rats suffer from a temporary response bias and not a deficit in short-term memory. The majority of errors made by Ss were perseveration errors.

166. Hannum, Robert D.; Rosellini, Robert A. & Seligman, Martin E. (1976). **Learned helplessness in the rat: Retention and immunization.** ***Developmental Psychology,*** 12(5), 449–454.
Conducted 3 experiments with 128 male albino Sprague-Dawley and Holtzman rats to determine (a) whether experience with uncontrollable trauma shortly after weaning interfered with an adaptive responding as an adult and (b) if early experience with controllable trauma protected adults against the helplessness-inducing effects of uncontrollable trauma received as an adult. Inescapable shock given to weanling rats produced large deficits in adult escape behavior. Therefore, helplessness learned as a weanling was retained in later life and interfered with adaptive instrumental responding. Experience with escapable shock while a weanling immunized the animal against the deficits produced by inescapable shock received as an adult. The implications of these findings for animal models of human depression are discussed.

167. Hughes, Carroll W. & Lynch, James J. (1979). **Sudden swimming deaths: No longer hopelessness in rats?** ***American Psychologist,*** 34(3), 273–274.
Reiterates C. W. Hughes and J. J. Lynch's (1978) position that the psychological state of nonverbal animals cannot be assessed, in response to comments by Y. Binik and M. E. Seligman (1979) on Hughes and Lynch's review of studies of sudden swimming death as a measure of learned helplessness.

168. Hughes, Carroll W. & Lynch, James J. (1978). **A reconsideration of psychological precursors of sudden death in infrahuman animals.** ***American Psychologist,*** 33(5), 419–429.
Psychological notions of helplessness–hopelessness have been invoked as hypotheses to account for many cases of sudden death in animals and man. At issue in this review is the validity of what is essentially an untestable concept in infrahuman animals—the role the psychological state of helplessness plays in animal sudden death. Recent research using the classic animal model of sudden death is integrated with the equally important issue of whether or not the term "sudden death" is a valid construct for all situations. It is suggested that caution should be used in extrapolating from infrahuman research that links aversive emotional states in humans with abnormal behavior and psychosomatic pathology.

169. Ingram, Donald K. (1985). **Analysis of age-related impairments in learning and memory in rodent models.** ***Annals of the New York Academy of Sciences,*** 444, 312–331.
Discusses methodological issues in the analysis of age-related impairments in learning and memory in rodent models, including sampling, confounding variation, motivational questions, sensory issues, and response biases and strategies. It is argued that the study of aging is not merely the comparison of presumably young and aged groups, but is rather the analysis of a complex biological phenomenon that requires specific methodological considerations and rigorous investigation to fully comprehend its impact on memory processes and to fully assess the utility of animal models. The issues discussed have their counterpart in human gerontological research. Their influence has the potential to alter the direction of research, similar to the impact of benchmark issues such as interpretation of cross-sectional vs longitudinal studies in the literature on humans.

170. Jackson, Raymond L.; Alexander, James H. & Maier, Steven F. (1980). **Learned helplessness, inactivity, and associative deficits: Effects of inescapable shock on response choice escape learning.** ***Journal of Experimental Psychology: Animal Behavior Processes,*** 6(1), 1–20.
Explored whether exposure to inescapable shock produces a deficit in the organism's propensity to associate its behavior with shock termination. Four experiments with a total of 139 male Holtzman rats examined the effects of inescapable shock on the acquisition of Y-maze escape, in which escape is accomplished by choosing the correct response from 2 alternatives rather than by simple locomotion. By itself, reduced activity should not produce inaccurate choices, only slow choices. Exp I found that inescapable shock produced slow learning of the correct choice, even though active choices occurred on every trial. The speed and accuracy of choice were not correlated. Exp II showed that the choice escape learning deficit was produced by the inescapability of the shocks. Exp III demonstrated that the choice accuracy of inescapably shocked Ss was not improved by increases in

Y-maze shock intensity, even though speed of responding was increased. Exp IV showed that the effects of inescapable shock on Y-maze acquisition did not dissipate across a 1-wk period.

171. Jackson, Raymond L.; Maier, Steven F. & Rapaport, Peter M. (1978). **Exposure to inescapable shock produces both activity and associative deficits in the rat.** ***Learning & Motivation,*** 9(1), 69–98.
Five experiments with a total of 152 male Holtzman rats investigated interference with shuttle-box escape learning following exposure to inescapable shock, which is often difficult to obtain in rats. Exp IA demonstrated that the magnitude of the interference effect was systematically related to shock intensity during shuttle-box testing. At .6 mA, a robust effect was obtained, whereas at .8 mA and 1.0, little or no deficit in the escape performance of inescapably shocked Ss was observed. Exp IB demonstrated that the deficit observed in Exp IA depended on whether or not Ss could control shock offset. Exp II suggested that preshock may suppress activity and that higher shock levels may overcome this deficit. Exp III tested this as the sole cause of the escape deficit by requiring an escape response which exceeded the level of activity readily elicited by a 1.0-mA shock in both restrained and preshocked Ss. In such a task, preshocked Ss performed more poorly than did restrained controls. These results are consistent with the possibility that inescapable shock may, in addition to reducing activity, produce an associative deficit. Exp IV more clearly demonstrated that inescapable shock produces deficits in performance that cannot be explained by activity deficits and which appear to be associative in nature. It was shown that inescapable shock interfered with the acquisition of signaled punishment suppression but not conditioned emotional response suppression. Theoretical implications are discussed.

172. Jackson, Raymond L.; Maier, Steven F. & Coon, Deborah J. (1979). **Long-term analgesic effects of inescapable shock and learned helplessness.** ***Science,*** 206(4414), 91–93.
Although exposure to inescapable shock induced analgesia in 16 rats, the analgesia was not manifest 24 hrs later unless Ss were reexposed to shock. Long-term analgesic effects depended on the controllability of the original shock and not on shock exposure per se. Implications for learned helplessness and stress-induced analgesia are discussed.

173. Jackson, Raymond L. & Minor, Thomas R. (1988). **Effects of signaling inescapable shock on subsequent escape learning: Implications for theories of coping and "learned helplessness."** ***Journal of Experimental Psychology: Animal Behavior Processes,*** 14(4), 390–400.
The present experiments reveal that shuttle-escape performance deficits are eliminated when exteroceptive cues are paired with inescapable shock. Experiment 1 indicated that, as in instrumental control, a signal following inescapable shock eliminated later escape performance deficits. Subsequent experiments revealed that both forward and backward pairings between signals and inescapable shock attenuated performance deficits. However, the data also suggest that the impact of these temporal relations may be modulated by qualitative aspects of the cues because the effects of these relations depended upon whether an increase or decrease in illumination (Experiment 2) or a compound auditory cue (Experiment 4) was used. Preliminary evidence suggests that the ability of illumination cues to block escape learning deficits may be related to their ability to reduce contextual fear (Experiment 3). The implications of these data for conceptions of instrumental control and the role of fear in the etiology of effects of inescapable shock exposure are discussed.

174. Job, R. F. (1988). **Interference and facilitation produced by noncontingent reinforcement in the appetitive situation.** ***Animal Learning & Behavior,*** 16(4), 451–460.
The results of experiments with 128 male rats on learned helplessness in the appetitive situation have varied from facilitation to debilitating effects produced by exposure to uncontrollable food. The conditions under which the interference effect (debilitation) may occur were examined in the 1st 3 experiments, employing the triadic design. Results suggest that the effect occurs when (1) Ss are preexposed to the manipulandum to be used in the test stage and (2) the manipulandum employed during pretreatment is absent during the test stage. Under the reverse conditions and partial reinforcement of the response-contingent Ss during pretreatment, the test performance of Ss exposed to uncontrollability was facilitated. Exp 4 confirmed the occurrence of the interference effect under the suggested conditions.

175. Job, R. F. (1987). **Learned helplessness in chickens.** ***Animal Learning & Behavior,*** 15(3), 347–350.
Evidence of the learned helplessness effect was obtained in 24–48 hr old domestic chickens. 24 hrs after exposure to escapable, inescapable, or no shock, Ss were tested on a 1-way shuttle task with shock-offset reinforcement. The inescapable-shock group showed retarded learning compared with other groups. Data are difficult to account for in terms of C. G. Costello's (1973) application of the systematic bias in the triadic design.

176. Job, R. F. (1987). **Learned helplessness in an appetitive discrete-trial T-maze discrimination test.** ***Animal Learning & Behavior,*** 15(3), 342–346.
Four groups of rats were exposed to response-contingent, yoked noncontingent, or en-masse food deliveries in a Skinner box or to no pretreatment. All groups were subsequently tested for transfer of the learned-helplessness effect to an appetitive discrete-trial T-maze discrimination. The yoked group showed retarded discrimination learning compared with the response-contingent and naive control groups. The nonsignificant difference between the yoked group and the en-masse group may reflect the effect of limited exposure to uncontrollability in the en-masse group. The groups did not differ in terms of the speed of maze traversal, suggesting that the learned-helplessness effect observed in discrimination was not due to a competing response.

177. Kesner, Raymond P. (1990). **New approaches to the study of comparative cognition.** ***National Institute on Drug Abuse: Research Monograph Series,*** 97, 22–36.
The rat's memory capacity and performance render it an excellent model of human memory. Rats display serial position, serial anticipation learning, temporal coding, and repetition lag functions as well as use of retrospective and prospective codes that are nearly equivalent to that observed for humans. There are comparable memory-deficit patterns between brain-damaged rats and humans. The rat can serve as an animal model to evaluate the efficacy of pharmacological treatments or brain damage on memory.

178. Kirk, Raymond C. & Blampied, Neville M. (1986). **Transituational immunization against the interference effect (learned helplessness) by prior passive and active escape.** ***Psychological Record,*** 36(2), 203–214.
Following 1 of 6 pretreatments, 42 male Wistar rats were taught to escape shock in a shuttlebox by completing 2 crossings per trial (FR 2) for 25 trials. Compared with untreated controls, the group receiving prior inescapable shock alone showed learned helplessness. Both previous passive and active escape training immunized against learned helplessness, with passive escape training being the most effective. Restraint alone also produced some interference with shuttle escape, unaffected by immunization pretreatments. Activity measures during inescapable shock showed that passive-escape-trained Ss were most active, while active-escape-trained and control Ss showed the same lower level of activity.

179. Klosterhalfen, Wolfgang & Klosterhalfen, Sibylle. (1983). **A critical analysis of the animal experiments cited in support of learned helplessness.** ***Psychologische Beiträge,*** 25(3–4), 436–458.
In 5 reviews (e.g., L. B. Alloy and M. E. Seligman, 1979; S. F. Maier and R. L. Jackson, 1979), the major proponents of the theory of learned helplessness have cited 49 publications reporting animal experiments that they consider to support the theory. 92 potentially supportive experiments from these publications are listed, and whether these studies in fact support the theory's basic propositions is questioned. Even if the inactivity accounts are ignored as interpretational alternatives, comparably few experiments lend support to the predicted associative and motivational deficits. Most experiments fail to do so because they are not pertinent to the theory, they do not show the predicted differences, or their results can be interpreted more parsimoniously. It is shown that the recommended triadic design cannot isolate the effects of controllability and uncontrollability of shock from each other or from the effects of shock per se. It is therefore concluded that the theory of learned helplessness is still in need of conclusive animal experiments. (French & German abstracts).

180. Kumar, K. B. & Karanth, K. Sudhakar. (1991). **Enhanced retrieval of unpleasant memory in helpless rats.** ***Biological Psychiatry,*** 30(5), 493–501.
Examined the effect of learned helplessness on the retrieval of unpleasant memory in male rats. Ss initially exposed to a single unpleasant event in a passive avoidance task were subjected, respectively, to inescapable, escapable, or no shock stress exposure. A retention test conducted 48 hrs following stress exposure showed an enhanced performance for the passive avoidance task in Ss subjected to inescapable shock stress. This improved performance was not observed in escapable or no shock stress groups. The finding that learned helplessness primed the retrieval of unpleasant memory in rats is comparable to the qualitative shift that is seen in the retrieval process in clinical depression, in negative mood induction procedures, and in induced helplessness in humans.

181. Lashley, Robin L. & Rosellini, Robert A. (1986). **Conditioning of odors in compound with taste is a function of factors other than potentiation.** ***Bulletin of the Psychonomic Society,*** 24(2), 159–162.
Conducted 2 experiments with 229 male Holtzman rats in which Ss ingested odor-alone and/or odor-taste solutions and were subsequently made ill by LiCl injections. Following poisoning, aversions to the odor stimuli were assessed using a 2-bottle choice test. The results failed to provide clear evidence of odor-taste potentiation for all the stimuli employed, regardless of the nature of the odor conditioned stimulus/stimuli (CS) administered and the use of both within- and between-group analyses. Findings suggest that previous reports of odor-taste potentiation may be somewhat tenuous and that odor saliency and order of conditioning may play more important roles than potentiation does in the development of odor aversion.

182. Lashley, Robin L. & Rosellini, Robert A. (1987). **Associative control of schedule-induced polydipsia.** ***Psychological Record,*** 37(4), 553–561.
Discusses the contribution of Pavlovian conditioned signals for pellet delivery in experiments with rats (positive conditioned stimulus [CS]) and reviews investigations of the role of signals for nonreinforcement (negative conditioned stimulus [CS]). It is concluded that both lines of research highlight the importance of associative factors in the development and maintenance of adjunctive drinking.

183. Lawry, J. A. et al. (1978). **Interference with avoidance behavior as a function of qualitative properties of inescapable shocks.** ***Animal Learning & Behavior,*** 6(2), 147–154.
Combined temporal form (continuous vs pulsating) and shock source (alternating current vs direct current) factorially to produce 4 shock treatments. The effects of inescapable presentations of these stimuli on subsequent avoidance response acquisition were measured in 36 adult dogs (Exp I) and in 36 male Sprague-Dawley albino rats (Exp II) and revealed an interaction of shock variables. Initially, all groups that received alternating current (AC) shock showed impaired performance for the pulsating and continuous shock conditions; groups that received direct current (DC) continuous shock were also impaired, while those that received DC pulsating shock were not. While this pattern of interference persisted for dogs, it was transient in rats, with only the AC continuous-shock group continuing to be impaired. Mean avoidance performances were positively related to mean activity levels during inescapable shocks for the DC shock groups but not for the AC shock groups.

184. Lee, Robert K. & Maier, Steven F. (1988). **Inescapable shock and attention to internal versus external cues in a water discrimination escape task.** ***Journal of Experimental Psychology: Animal Behavior Processes,*** 14(3), 302–310.
In these experiments we examined discrimination learning in a water escape task following exposure to escapable, yoked inescapable, or no electric shock. Inescapable shock did not have an effect on swim speeds in any of the experiments. Inescapable shock interfered with the acquisition of a position (left–right) discrimination when an irrelevant brightness cue (black and white stimuli) was present. However, inescapable shock did not affect the acquisition of the position discrimination when the irrelevant brightness cue was removed. Inescapably shocked Subjects showed *facilitated* learning relative to escapably shocked and nonshocked subjects when the brightness cue was included as a *relevant* cue. These data may resolve discrepancies between studies that did, and did not, find inescapable shock to interfere with the acquisition of discriminations. Moreover, they

point to attentional processes as one locus of the cognitive changes produced by inescapable shock and suggest the exposure to inescapable shock biases attention away from "internal" response-related cues toward "external" cues.

185. Levis, Donald J. & Boyd, Thomas L. (1979). **Symptom maintenance: An infrahuman analysis and extension of the conservation of anxiety principle.** ***Journal of Abnormal Psychology,*** 88(2), 107–120.
Focuses on the paradoxical issue of why nonadaptive behavior or symptom maintenance persists over extended periods of time in the absence of any apparent UCS. Recent critics have challenged accounts of this phenomenon that rely on the classical 2-factor avoidance interpretation formulated from the principles developed in the infrahuman conditioning laboratory. They argue that human symptoms persist over extended periods of time, yet infrahuman researchers have only infrequently conditioned avoidance behavior that has resulted in extreme resistance to extinction. Furthermore, laboratory data suggest that classical conditioned fear behavior extinguishes rapidly following CS exposure. In those few cases in which persistent avoidance behavior was noted, there has been a failure to document the presence of fear as the elicitor of such behavior. A model of symptom formation and maintenance is outlined that extends the conservation of anxiety hypothesis to incorporate the concept of CS complexity and sequencing of cues. Two infrahuman (Ss were 60 male rats) avoidance studies are presented that focus on the critical issues raised at this level of analysis. The data are supportive of the model proposed. The principles underlying the model are believed to be operating at the human level and responsible for symptom maintenance.

186. Levis, Donald J.; Dubin, William J. & Holzman, Arnold D. (1978). **Effects of component training and subsequent sequencing of stimuli on shuttlebox avoidance responding of rats.** ***Animal Learning & Behavior,*** 6(3), 335–340.
The serial presentation of 2 different CSs, with each stimulus having an 8-sec duration ($S1_8/S2_8$), has consistently resulted in most of the shuttlebox avoidance responses being recorded to the S2 component. Exp I attempted to attenuate this serial CS, delayed-response effect by conditioning the separate components of a serial CS prior to ordering them sequentially. 10 component-training trials were administered, with 216 male Blue Spruce rats receiving CS–UCS pairing to S1 only, S2 only, or to both S1 and S2 presented on separate trials. Two CS durations (8 or 16 sec) during this phase also were compared. Ss were then given 100 avoidance test trials using the standard serial procedure. The 10 best avoidance responders in each group were selected for analysis. Shorter avoidance latencies were obtained only for Ss receiving component conditioning to S1. CS duration was not a factor in establishing the shorter latencies. Component conditioning to S2 resulted in increasing the total avoidances. Exp II increased the number of component-training trials and the generality of the findings by using a different strain of rats and by extending the testing phase of the study so that all Ss (24 male albino Sprague-Dawley rats) could be included in the analysis. Comparable results were obtained.

187. MacLennan, A. John; Jackson, Raymond L. & Maier, Steven F. (1980). **Conditioned analgesia in the rat.** ***Bulletin of the Psychonomic Society,*** 15(6), 387–390.
It has been suggested by recent studies that the analgesic reaction to electric shock can be conditioned. However, these studies either lacked shocked controls or used an indirect measure of analgesia (freezing). In the present investigation, 76 albino rats were exposed an equal number of times to 2 distinct environmental contexts. Ss were shocked in one context and reexposed to the same context before test, shocked in one context and reexposed to the nonshock context before test, or not shocked at all and reexposed to 1 of the 2 contexts. Immediately following reexposure, the pain reactivity of Ss was assessed by a hot plate (Exp I) and a tail-flick apparatus (Exp II). It was found that Ss reexposed to the context in which they had been shocked were significantly more analgesic than Ss in the other 2 groups (which did not differ). Results confirm that it is possible to condition shock-induced analgesia in the rat.

188. Maier, Steven F. (1977). **Competing motor responses: A reply to Black.** ***Journal of Experimental Psychology: General,*** 106(1), 44–46.
A. H. Black (1977) argued that S. F. Maier and M. E. Seligman (1976) incorrectly interpreted competing motor response explanations of the learned helplessness effect. Whereas Maier and Seligman assumed that the putative competing response interferes with only the performance of escape responses, Black argued that the existence of an association between one response and shock could interfere with the formation of an association between the escape response and shock termination. Here, Maier argues that no article that has proposed a competing motor response explanation of the learned helplessness effect has alluded to a mechanism similar to the one discussed by Black and that it is not clear how such a process would be possible.

189. Maier, Steven F. et al. (1979). **The time course of learned helplessness, inactivity, and nociceptive deficits in rats.** ***Learning & Motivation,*** 10(4), 467–487.
Explored the effects of inescapable shock on subsequent shuttlebox escape learning in 4 experiments with 136 male Holtzman albino rats. In Exp I shuttle escape deficits dissipated within 48 hrs after treatment with inescapable shock. Exp II showed that exposure to inescapable shock suppressed unlearned activity in the shuttlebox and that this activity deficit recovered within 48 hrs. Exp IIA demonstrated that this shuttlebox crossing decrement was at least partly attributable to the inescapability of the shocks. The hypothesis that activity and shuttle escape learning deficits are subserved by the effects of inescapable shock on pain sensitivity was confirmed in Exp III: Ss were less sensitive to painful stimulation 24 hrs after inescapable shock, and this analgesic tendency dissipated within 48 hrs after pretreatment.

190. Maier, Steven F. & Jackson, Raymond L. (1977). **The nature of the initial coping response and the learned helplessness effect.** ***Animal Learning & Behavior,*** 5(4), 407–414.
The availability of an effective coping response has been shown to attenuate the deleterious behavioral and physiological consequences of inescapable electric shock. In the current study with 40 Holtzman male albino rats, 2 groups of Ss could escape tailshock by turning a wheel. When short-latency responses that appeared to be elicited by shock onset were permitted to terminate shock, Ss subsequently failed to learn to escape in a shuttle box and did not differ from Ss that received an equivalent amount of inescapable shock;

however, when a relatively long-latency response was required and short-latency responses were not allowed to affect shock, Ss subsequently readily learned to escape in the shuttlebox. Implications of results for explanations of the manner in which prior exposure to shock influences subsequent escape learning are discussed.

191. Maier, Steven F. & Watkins, Linda R. (1991). **Conditioned and unconditioned stress-induced analgesia: Stimulus preexposure and stimulus change.** ***Animal Learning & Behavior,*** 19(4), 295–304.
Two experiments, with 174 male rats, explored the associative nature of the analgesia that follows exposure to electric tailshocks. Preexposure to the environment in which shock later occurred had no effect on analgesia soon after shock but eliminated later analgesia. The initial postshock analgesia was unaffected by removing the S from the shock environment to a different environment, but the later reaction was prevented by such a change. Also, returning the S to the shock environment after confinement in a nonshock environment rearoused an analgesic reaction. This rearousal did not occur if the S had first been confined to the shock environment without shock. Data suggest that shock produces analgesia as an unconditioned reaction.

192. Marcus, Emilie A. & Carew, Thomas J. (1990). **Ontogenetic analysis of learning in a simple system.** Conference of the National Institute of Mental Health et al: The development and neural bases of higher cognitive functions (1989, Philadelphia, Pennsylvania). ***Annals of the New York Academy of Sciences,*** 608, 128–149.
Discusses experiments in which an invertebrate simple system strategy was combined with a developmental approach to examine at a mechanistic level how different forms of learning emerge and are assembled through ontogeny. The defensive gill syphon withdrawal reflex of the marine mollusc *Aplysia* was used for a developmental analysis of learning and its underlying cellular and biochemical mechanisms. Data are presented on experiments that examined the developmental emergence of 3 forms of nonassociative learning: habituation, dishabituation, and sensitization. A cellular analog of sensitization is examined, along with the cellular and subcellular events that mediate the developmental expression of sensitization.

193. Mauk, Michael D. & Pavur, Edward J. (1979). **Interconsequence generality of learned helplessness.** ***Bulletin of the Psychonomic Society,*** 14(6), 421–423.
Tested the interconsequence generality of the learned helplessness (LH) phenomenon. 36 adult female Sprague-Dawley rats received escapable, inescapable, or no shock in a shuttlebox. Intact triads were then randomly assigned to 2 groups. The 1st group was required to learn to escape shock via barpressing; the 2nd was required to learn a 6-unit maze for food reward. The shock-escape triads demonstrated the standard LH effect, with inescapable Ss inferior to escapable and unshocked Ss. However, Ss tested in the appetitive situation did not exhibit LH, rather, they exhibited a trauma-like effect, with both escapable and inescapable Ss inferior to no-shock control Ss on early trials. These findings demonstrate a limit to the generality of LH. It is suggested that LH follows the rules of stimulus and response generalization.

194. McGonigle, Brendan O. & Chalmers, Margaret. (1977). **Are monkeys logical?** ***Nature,*** 267(5613), 694–696.
Used a modification of P. E. Bryant and T. Trabasso's (1971) procedure of assessing formal reasoning processes in children to assess such capabilities in 8 adult squirrel monkeys. Ss were trained to judge the relative weights of nonadjacent members of a set of differently colored cylindrical containers in a series of reinforced and nonreinforced trials. A comparison of results obtained from monkeys with those obtained by Bryant and Trabasso from 4-yr-old children show a choice profile consistent with the notion that monkeys' choices are transitive and are congruent in almost every detail with the child data. The interpretation that in order to solve problems of the type devised in this experiment, Ss must coordinate 2 pieces of information is contrasted with the idea that transitive choices result from single binary decison making. Additional data from a test of the "reduced" transitive effect demanded by the binary model demonstrate its predictiveness for overall choice profiles in a 3-item comparison.

195. McSweeney, Frances K.; Melville, Cam L. & Higa, Jennifer. (1988). **Positive behavioral contrast across food and alcohol reinforcers.** Special Issue: Behavior analysis and biological factors. ***Journal of the Experimental Analysis of Behavior,*** 50(3), 469–481.
Examined behavioral contrast during concurrent and multiple schedules that provided food and alcohol reinforcers, using a total of 19 rats in 5 experiments. Concurrent-schedule contrast occurred in responses reinforced by food when alcohol reinforcers were removed and in responses reinforced by alcohol when food was removed. Multiple-schedule contrast appeared for food when alcohol reinforcers were removed but not for alcohol when food was removed. Results suggest that behavioral contrast may occur across qualitatively different reinforcers and that multiple-schedule contrast may be more difficult to produce than concurrent-schedule contrast. Results have implications for a model of alcohol consumption.

196. Miceli, Dom; Marfaing-Jallat, Pierrette & le Magnen, Jacques. (1980). **Ethanol aversion induced by parenterally administered ethanol acting both as CS and UCS.** ***Physiological Psychology,*** 8(4), 433–436.
Examined the effects of different doses of ethanol (.0–4.0 g/kg) injected ip during 3 consecutive days on subsequent voluntary consumption of the drug using 100 male Wistar rats (Exp I). Ethanol treatment induced a significant suppression of ethanol drinking. Five injections of naloxone (10 mg/kg) combined with ethanol (1.75 g/kg) performed on alternate days significantly enhanced aversion (Exp II, 10 Ss). Findings indicate that ethanol administered via the ip route generated both negative reinforcement (UCS) and the associated oral ethanol flavor (CS) for aversion conditioning. Implications for obtaining an animal model of behavioral dependence toward ethanol are discussed.

197. Miller, Stephanie; Mineka, Susan & Cook, Michael. (1982). **Comparison of various flooding procedures in reducing fear and in extinguishing jump-up avoidance responding.** ***Animal Learning & Behavior,*** 10(3), 390–400.
Two experiments with 136 male albino Fischer rats examined the effectiveness of 3 variations in flooding techniques on hastening extinction of a jump-up avoidance response (Exp I) and on reducing fear (Exp II), as assessed by the multivariate fear-assessment techniques of D. P. Corriveau and N. F. Smith (1978). Traditional flooding involved blocking Ss' response by making the safety ledge unavailable;

barrier flooding involved inserting a Plexiglas barrier in front of the ledge to make it inaccessible and moving the wall periodically; no-barrier flooding involved allowing Ss to jump onto the ledge periodically but, if they did so, immediately dumping them back onto the grids. In both experiments, all 3 flooding treatments were more effective than a home cage treatment, although the no-barrier procedure was significantly more effective than the other 2 procedures. In addition, activity measures revealed interesting and significant group differences in the patterns of activity shown during treatment.

198. Mineka, Susan. (1976). **Effects of flooding an irrelevant response on the extinction of avoidance responses.** ***Journal of Experimental Psychology: Animal Behavior Processes,*** 2(2), 142–153.
Conducted 3 experiments with 90 male albino Fischer rats. Exp I shows that flooding a jump-up avoidance response can hasten the extinction of a shuttlebox avoidance response learned to a very different conditioned stimulus (CS). Exp II demonstrates that an irrelevant fear extinction experience can also hasten the extinction of a shuttlebox avoidance response. This supports the hypothesis that Pavlovian generalization of extinction of conditioned fear across CS modalities mediates the irrelevant flooding effect. Exp III shows that flooding a shuttlebox avoidance response does not hasten the extinction of a jump-up avoidance response learned to a different CS. Possible reasons for the asymmetry of the irrelevant flooding effect are discussed.

199. Mineka, Susan. (1979). **The role of fear in theories of avoidance learning, flooding, and extinction.** ***Psychological Bulletin,*** 86(5), 985–1010.
Summarizes the major lines of evidence that demonstrate a dissociation or desynchrony between measures of fear and avoidance responding. The evidence bearing on the role of fear in theories of avoidance learning and extinction is reviewed and critically evaluated. In addition, research is discussed regarding the determinants of fear over the course of avoidance acquisition, flooding, and extinction. Particular emphasis is placed on discussing the extent to which fear extinction is necessary and/or sufficient for avoidance response extinction with conventional extinction procedures and with response prevention techniques.

200. Mineka, Susan. (1978). **The effects of overtraining on flooding of jump-up and shuttlebox avoidance responses.** ***Behaviour Research & Therapy,*** 16(5), 335–344.
Reports 2 experiments that tested the effects of overtraining (2, 4, or 6 days) on the efficacy of flooding (response prevention) in hastening the extinction of jump-up and 2-way shuttlebox avoidance responses; 96 male Fischer rats were Ss. In the jump-up box, overtraining reduced the effectiveness of flooding in 2 ways: Ss trained for 6 days and given a flooding treatment were more resistant to extinction than Ss trained for only 2 or 4 days and given a flooding treatment. Ss trained for 6 days and given a control treatment were less resistant to extinction than controls trained for 2 or 4 days. In the shuttlbox, overtraining neither reduced the effectiveness of flooding in hastening avoidance response extinction nor reduced the resistance to extinction of control Ss. Implications of the effects of overtraining on jump-up and 2-way shuttlebox responses are discussed.

201. Mineka, Susan; Cook, Michael & Miller, Stephanie. (1984). **Fear conditioned with escapable and inescapable shock: Effects of a feedback stimulus.** ***Journal of Experimental Psychology: Animal Behavior Processes,*** 10(3), 307–323.
Four experiments, with 140 male Fischer rats, compared the level of fear conditioned with escapable and inescapable shock. In Exps I and II, master Ss that had received 50 unsignaled escapable shocks were less afraid of the situation where the shock had occurred than were yoked Ss that had received inescapable shocks. Comparable results were found in Exps III and IV, which used freezing as an index of fear of a discrete CS that had been paired with shock. Control per se was not necessary to produce the low level of fear seen in the master Ss. Yoked groups receiving a feedback signal at the time the master made an escape response showed a low level of fear that was comparable to that of the masters and significantly less than that seen in the yoked Ss without feedback. In addition, there were strong suggestions that control and feedback exert their effects through the same or highly similar mechanisms. Possible explanations for how control and the exteroceptive feedback signal produce this effect are discussed.

202. Mineka, Susan & Cook, Michael. (1986). **Immunization against the observational conditioning of snake fear in rhesus monkeys.** ***Journal of Abnormal Psychology,*** 95(4), 307–318.
Examined the effects of extensive prior exposure to snakes on subsequent observational conditioning of snake fear in 24 rhesus monkeys (*Macaca mulatta*). Three groups of Ss were given 1 of 3 kinds of pretreatment: (1) An immunization group spent 6 sessions watching a nonfearful monkey behave nonfearfully with snakes; (2) a latent inhibition group spent 6 sessions by themselves behaving nonfearfully with snakes with total exposure time to snakes equal to that for the immunization group; and (3) a pseudoimmunization group spent 6 sessions of observational conditioning in which they watched fearful monkeys behave fearfully with snakes. When subsequently tested for acquisition of snake fear, the pseudoimmunization and latent inhibition groups showed significant acquisition, but 6 out of 8 Ss in the immunization group did not. Thus, it seems that for a majority of Ss, prior exposure to a nonfearful model behaving nonfearfully with snakes can effectively immunize against the subsequent effects of exposure to fearful models behaving fearfully with snakes.

203. Mineka, Susan; Davidson, Mark; Cook, Michael & Keir, Richard. (1984). **Observational conditioning of snake fear in rhesus monkeys.** ***Journal of Abnormal Psychology,*** 93(4), 355–372.
Hypothesized that observational conditioning is involved in the origins of many human and nonhuman primates' fears and phobias. In Exp I, a new index of snake fear in 7 19–28 yr old wild-reared rhesus monkeys and 9 laboratory-reared offspring (aged 8 mo to 6 yrs) was tested. Results show the measure was useful and demonstrated that young Ss raised by parents who had a fear of snakes did not acquire the fear in the absence of any specific experience with snakes. In Exp II, using 5 of the wild-reared Ss and 6 of the laboratory-reared Ss from Exp I, 5 of 6 offspring acquired an intense and persistent fear of snakes as a result of observing their

wild-reared parents behave fearfully in the presence of real, toy, or model snakes for a short period of time. The fear was not context specific and showed no significant signs of diminution at 3-mo follow-up.

204. Mineka, Susan & Gino, Antonio. (1979). **Dissociative effects of different types and amounts of nonreinforced CS exposure on avoidance extinction and the CER.** ***Learning & Motivation,*** 10(2), 141–160.
Two experiments with 108 male albino Sprague-Dawley rats examined the effectiveness of 2 amounts of flooding or response-prevention on hastening avoidance response extinction and on reducing CS-produced suppression of barpressing for food. In Exp I, 20 and 30 flooding trials were both effective in hastening the extinction of a well-learned shuttlebox avoidance response. In Exp II, Ss trained under conditions comparable to those in Exp I were tested following flooding for the conditioned emotional response (CER) in a different apparatus. Results indicate that 30, but not 20, flooding trials were sufficient to reduce the CER. Comparisons with control groups show that although 30 flooding trials did reduce the CER, the same total duration of nonreinforced CS exposure in the form of avoidance extinction trials did not. Thus the context in which CS exposure occurs may affect the dynamics of extinction of the CER. The experiments are discussed in the broader context of dissociation of various indices (e.g., autonomic, subjective) of fear in humans.

205. Mineka, Susan & Gino, Antonio. (1980). **Dissociation between conditioned emotional response and extended avoidance performance.** ***Learning & Motivation,*** 11(4), 476–502.
Three experiments with 112 Fisher male rats examined the dissociation between the strength of a shuttlebox avoidance response (AR) and 1 index of fear of the avoidance CS. Avoidance response strength was indexed by resistance to extinction of AR and by changes in response latency; fear of the CS was indexed by the conditioned emotional response (CER) technique. Results show that Ss trained to a criterion of 27 consecutive avoidance responses (CARs) showed response strength comparable or superior to Ss trained to a criterion of 9 CARs. Ss trained to 27 CARs showed less suppression of barpressing during the avoidance CS (less CER) than Ss trained to 9 CARs. When extinguished in the shuttlebox to a moderate criterion before CER testing, Ss trained to 9 CARs showed some loss of CER, whereas Ss trained to 27 CARs showed no loss of CER. Although Ss taking 2 days to reach a criterion of 27 CARs showed somewhat greater resistance to extinction of AR than Ss reaching criterion in 1 day, this variable had no apparent effect on the CER.

206. Mineka, Susan & Keir, Richard. (1983). **The effects of flooding on reducing snake fear in rhesus monkeys: 6-month follow-up and further flooding.** ***Behaviour Research & Therapy,*** 21(5), 527–535.
Administered 2 6-mo follow-up tests to 7 wild-reared rhesus monkeys to assess for spontaneous recovery of snake fear that had been somewhat reduced following 7 flooding sessions 6 mo earlier. Both tests revealed essentially complete spontaneous recovery of fear. In addition, all 7 Ss received 4 further mixed flooding sessions that involved exposure to real, toy, and model snakes, and the 4 most fearful Ss also received 3 more hours of exposure to the real snake alone. A final behavioral test following these additional flooding sessions revealed a pattern of changes very similar to that observed after the original 7 sessions 6 mo earlier. In particular, there were some significant changes in the behavioral avoidance component of the fear but no changes in the behavioral disturbance component of the fear. Results are discussed in the context of earlier studies by S. Mineka et al (1980) that purported to demonstrate that snake fear is easy to abolish. It is concluded that these earlier studies erred by not having tests for spontaneous recovery and by only testing for changes in the behavioral avoidance component of fear. Possible reasons for the failure to produce significant changes in the behavioral disturbance component of fear are discussed.

207. Minor, Thomas R. (1990). **Conditioned fear and neophobia following inescapable shock.** ***Animal Learning & Behavior,*** 18(2), 212–226.
Lick-suppression tests were used in 6 experiments to assess the transsituational transfer of fear in the learned helplessness paradigm in inescapably shocked male rats. A situational odor was strongly associated with shock pretreatments and mediated the transfer of conditioned fear during testing. Fear of the pretreatment odor was greater following inescapable shock than after escapable shock or restraint. Neophobia was enhanced as a 2nd, nonassociative reaction to inescapable shock. The pretreatment odor elicited fear only when tested in a novel context. Data provide evidence for odor-mediated transfer of helplessness. Conditioned fear and neophobia are discussed in relation to anxiety interpretations of the phenomenon.

208. Minor, Thomas R.; Jackson, Raymond L. & Maier, Steven F. (1984). **Effects of task-irrelevant cues and reinforcement delay on choice-escape learning following inescapable shock: Evidence for a deficit in selective attention.** ***Journal of Experimental Psychology: Animal Behavior Processes,*** 10(4), 543–556.
Five experiments with 152 male albino rats examined the conditions under which choice-escape learning in an automated **Y**-maze is impaired by pretreatment with inescapable shock. Ss exposed to inescapable shock made more errors and responded more slowly than did controls only when shock termination was delayed and task-irrelevant cues were present during choice-escape training. Findings are discussed in terms of information processing and neurochemical consequences of exposure to inescapable shock. It is suggested that the concurrent manipulation of irrelevant cues and reinforcement delay may have masked the choice contingency and increased task difficulty. Other possible explanations of the results include lessened attention to proprioceptive stimuli, differences in arousal levels between conditions, and depletion of norepinephrine in the locus coeruleus following inescapable shock.

209. Minor, Thomas R.; Trauner, Michael A.; Lee, Chi-yuarn & Dess, Nancy K. (1990). **Modeling signal features of escape response: Effects of cessation conditioning in "learned helplessness" paradigm.** ***Journal of Experimental Psychology: Animal Behavior Processes,*** 16(2), 123–136.
Six experiments examined the effects of signaling the termination of inescapable shock (cessation conditioning) or shock-free periods (backward conditioning) on later escape deficits in the learned helplessness paradigm, using rats (Sprague-Dawley and Bantin–Kingman). A cessation signal prevented later performance deficits when highly variable inescapable shock durations were used during pretreatment.

The inclusion of short minimum intertrial intervals during pretreatment did not alter the benefits of cessation conditioning but eliminated the protection afforded by a safety signal. The beneficial effects of both cessation and backward signals were eliminated when a single stimulus signaled shock termination and a shock-free period. Finally, a combination of cessation and backward signals was found to be most effective in immunizing against the effects of subsequent unsignaled, inescapable shock on later escape performance. These data suggest that cessation conditioning may be crucial to the prophylactic action of an escape response.

210. Moran, Peter W. & Lewis-Smith, Marion. (1979). **Learned helplessness and response difficulty.** ***Bulletin of the Psychonomic Society,*** 13(4), 250–252.
To determine the effects of preshock (PS) and sex on subsequent escape learning, 32 male and 32 female Sprague-Dawley rats were tested on barpress escape responding. Ss were randomly assigned to an inescapable PS treatment or to the no-shock control treatment (C). These groups were further randomly assigned to 4 groups of 4 according to a required frequency of the barpress escape response: FR 2, FR 3, FR 4, and FR 5. Results show marked learning deficits in the PS groups at intermediate task-difficulty levels (FR 3 and FR 4). Female C Ss tended to have slightly lower mean escape latencies than male controls. There were no sex differences among the PS Ss. Thus, helplessness was demonstrated at intermediate task levels (FR 3 and FR 4), but not at extreme levels (FR 2 and FR 5).

211. Mullins, G. P. & Winefield, A. H. (1977). **Immunization and helplessness phenomena in the rat in a nonaversive situation.** ***Animal Learning & Behavior,*** 5(3), 281–284.
In a study with 72 male Wistar rats, performance on a visual discrimination task was found to be impaired following experience with an insoluble problem. The experiment was designed to preclude the possibility that the effect could be attributed to the development of an incompatible response. Prior experience with a soluble problem significantly reduced the deleterious effects of the insoluble problem. Results are interpreted as evidence of learned helplessness and behavioral immunization in a context where no aversive stimulation was employed.

212. Mullins, G. P. & Winefield, Anthony H. (1978). **Helplessness in the rat: Interfering response or motivational/cognitive deficit?** ***Perceptual & Motor Skills,*** 47(3, Pt 2), 1059–1068.
An experiment with 64 male Wistar rats showed that experience with an insoluble problem interfered with subsequent visual discrimination learning. Prior experience with a soluble problem significantly reduced the deleterious effects of the insoluble problem, but this "immunization" did not benefit Ss that were able to develop an incompatible position response in the insoluble problem. Implications of these and other recent results for the theory of learned helplessness are discussed.

213. Mumby, Dave G.; Pinel, John P. & Wood, Emma R. (1990). **Nonrecurring-items delayed nonmatching-to-sample in rats: A new paradigm for testing nonspatial working memory.** ***Psychobiology,*** 18(3), 321–326.
Rats were trained on a nonrecurring-items delayed nonmatching-to-sample task, using a newly designed apparatus and a training protocol for nonspatial working memory. On each trial, a sample object was briefly presented, withdrawn, and reintroduced with a new object; choosing the novel object resulted in a food reward. New stimuli were used on each trial. With a 4-sec delay between the sample and choice, Ss performed with 90% accuracy in fewer than 250 trials. When the delay was increased to 15, 60, 120, and 600 sec, the Ss scored approximately 91%, 81%, 77%, and 57%, respectively. Thus rats can perform a nonspatial working-memory task like those used in monkey models of amnesia. This paradigm may be useful in modeling brain-damage-produced amnesia in rats.

214. Murison, Robert; Isaksen, Eva & Ursin, Holger. (1981). **"Coping" and gastric ulceration in rats after prolonged active avoidance performance.** ***Physiology & Behavior,*** 27(2), 345–348.
Rats performing an avoidance task involving conflict with massed trials exhibit severe gastric ulceration, and high post session levels of adrenocortical activity. At the same time, rats performing a 2-way active avoidance task, which also involves conflict, with spaced trials are known to exhibit "coping," i.e., deactivation of the adrenocortical system. In the present experiment, 30 male Moll-Wistar rats were made to perform a 2-way active avoidance task with massed trials for 20 hrs. Ss exhibited more severe gastric ulceration than did shock-yoked or nonshock controls, but no differences were found between groups on adrenocortical activity. "Coping" as measured here was independent of gastric ulceration.

215. Nakai, Ikuo & Matsuyama, Yoshinori. (1979). **Effects of inescapable shock upon subsequent fixed ratio-2 shuttle response.** ***Japanese Journal of Psychology,*** 50(2), 97–101.
Groups of 8 rats were given 64, 48, 32, 16, or 0 inescapable preshocks and then were trained on an escape-avoidance task in a shuttle box for 5 FR-1 trials followed by 25 FR-2 trials. Findings support the hypothesis that the rate of learning would be reduced because of learned helplessness. (English summary).

216. Nash, Susan M.; Martinez, Sheena L.; Dudeck, Michael M. & Davis, Stephen F. (1983). **Learned helplessness in goldfish under conditions of low shock intensity.** ***Journal of General Psychology,*** 108(1), 97–101.
10 goldfish that had received inescapable shock presentations during a 1st experimental phase made significantly fewer responses and had significantly longer response latencies during a subsequent avoidance-learning phase than did 10 Ss that had not received the inescapable shocks. Results indicate that the experimental Ss learned that their behavior and its consequences were independent and confirm the existence of the learned helplessness phenomenon in goldfish.

217. Nation, Jack R. & Matheny, Jimmy L. (1980). **Instrumental escape responding after passive avoidance training: Support for an incompatible response account of learned helplessness.** ***American Journal of Psychology,*** 93(2), 299–308.
Two experiments (48 male Sprague-Dawley albino rats) examined the relative degree of interference (helplessness) occasioned by pretreatment experiences with either passive-escape (PE) training or noncontingent, inescapable shock. In Exp I PE Ss learned to terminate shock by not running and subsequently were tested in a shuttlebox that required an instrumental running response for termination of shock. Ss yoked to PE Ss during the initial training stage experienced inescapable shock prior to shuttlebox testing. An additional group of Ss experienced no pretreatment but was required to

learn that shuttling (running) terminated shock during the test phase. Exp II included the same 3 experimental conditions except that a more intense shock was used during shuttlebox testing. Results indicate that interference effects occurred uniformly for PE and yoked Ss with a moderately aversive test stimulus (Exp I). With a highly aversive test stimulus, interference effects were observed following PE training but were absent following pretreatment exposure to inescapable shock (Exp II). Implications for interference theories (e.g., the learned helplessness hypothesis and the incompatible motor response hypotheses) are discussed.

218. Nealis, Perry M.; Harlow, Harry F. & Suomi, Stephen J. (1977). **The effects of stimulus movement on discrimination learning by rhesus monkeys.** ***Bulletin of the Psychonomic Society,*** 10(3), 161–164.
Conducted 2 experiments with test-sophisticated adult rhesus monkeys to determine effects of differentiating and nondifferentiating movement cues on 2-choice discrimination learning. In Exp I, 12 Ss were given initial learning problems followed by intradimensional shift problems with movement- and color-differentiated stimulus cues. Original learning progressed most rapidly with movement cues, but subsequent task-shift performance was comparable for movement- and color-cue problems. There was no evidence that learning with either cue dimension transferred to subsequent problems involving the other cue dimension. In Exp II, 14 Ss were given 2-choice discrimination problems in which discriminanda differed in color only. Problems were presented under 2 conditions in which (a) both discriminanda moved synchronously during the choice period, or (b) both discriminanda remained stationary. Ss learned more rapidly when discriminanda were moving. It is concluded that motion is a dimension of comparable saliency to color, although the psychophysical properties need further clarification. It is hypothesized that stimulus movement enhances discrimination learning through facilitation of the appropriate allocation of attention.

219. Oakes, William F.; Rosenblum, Jan L. & Fox, Paul E. (1982). **"Manna from heaven": The effect of noncontingent appetitive reinforcers on learning in rats.** ***Bulletin of the Psychonomic Society,*** 19(2), 123–126.
Three groups of female Sprague-Dawley rats received food pellets in Phase 1. Members of the contingent group received the pellets contingent on their nose-poke responses. Ss in the noncontingent group, yoked to members of the contingent group, received pellets whenever their yoked-contingent S nose-poked, but independent of the noncontingent S's behavior. Members of the control group received the same numbers of pellets in their home cages 50 min later. In Phase 2, the 3 groups learned a leverpress response with food reinforcement. Two of the 3 dependent measures showed a detrimental effect of the 1st-phase experience on leverpressing performance for the noncontingent group, compared with the contingent and control groups, which did not differ from each other. It is concluded that an appetitive learned-helplessness effect was demonstrated.

220. Overmier, J. Bruce & Wielkiewicz, Richard M. (1983). **On unpredictability as a causal factor in "learned helplessness."** ***Learning & Motivation,*** 14(3), 324–337.
In 2 replications, 2 groups of 9 mongrel dogs were exposed to a series of uncontrollable electric shocks. For Group 1, the shocks were preceded by a tone; for Group 2, the shocks were randomly related to the tones and were unpredictable. In both replications, a 3rd group of Ss was exposed to only the series of tones (CS-only) initially then was exposed only to the series of shocks. All Ss were required to learn a discriminative choice escape and/or avoidance task in which the required response was to lift the correct paw in the presence of each of 2 visual stimuli to escape or avoid the shocks. Results show that Ss preexposed to random tones and shocks were least successful in learning the task relative to those Ss that experienced either predicted shocks, only the tones, or only the shocks. These latter groups did not differ from each other in learning. It is suggested that proactive interference with choice behavior following random-tone CSs and shocks was attributable to a learned irrelevance generalized with respect to the CSs.

221. Overstreet, David H.; Rezvani, Amir H. & Janowsky, David S. (1990). **Impaired active avoidance responding in rats selectively bred for increased cholinergic function.** ***Physiology & Behavior,*** 47(4), 787–788.
The Flinders Sensitive Line of rats (selectively bred for increased cholinergic function) performed poorly in a tone-cued, 2-way active avoidance task in comparison with the control Flinders Resistant Line, suggesting a genetic animal model of depression.

222. Plonsky, Mark & Rosellini, Robert A. (1986). **The effects of a pretrained excitatory stimulus on schedule-induced polydipsia in the rat.** ***Psychological Record,*** 36(3), 387–397.
Tested 2 predictions derived from the application of Pavlovian conditioning principles to the schedule-induced polydipsia (SIP) phenomena, using 51 male albino rats. The 1st prediction was that a pretrained excitatory signal for pellet delivery would facilitate the development of SIP. A 2nd prediction was that phenomena that are typically observed in Pavlovian conditioning experiments would also be observed in the SIP paradigm. Results indicate that Ss given excitatory pretraining to a tone failed to develop SIP more rapidly than controls when this tone was used to signal pellet deliveries during SIP training. However, learned irrelevance, latent inhibition, and the unconditioned stimulus/stimuli (UCS) preexposure effect were demonstrated. Data generally support the predictions derived from the Pavlovian conditioning view of SIP.

223. Plonsky, Mark; Warren, Donald A. & Rosellini, Robert A. (1984). **The effects of inescapable shock on appetitive motivation.** ***Bulletin of the Psychonomic Society,*** 22(3), 229–231.
Exposure to inescapable shock is known to proactively interfere with the acquisition of instrumental responses to escape shock, as well as to obtain food. These effects have been termed *learned helplessness.* The present experiment used 23 male albino rats, to investigate the possibility that the learned helplessness effect observed in an appetitive context may, at least in part, be due to the motivational effects of inescapable shock. The schedule-induced polydipsia paradigm, which is known to be sensitive to both deprivation level and incentive motivation, assessed the effects of inescapable shock on appetitive motivation. Despite the fact that learned helplessness was demonstrated in a shuttle escape task, no effect of inescapable shock was observed on polydipsia. Thus, the reinforcer generality of helplessness appears not to be due to shock-induced motivational factors.

224. Polaino-Lorente, Aquilino & Vázquez Valverde, C. (1981). **Learned helplessness: An experimental model—critical review.** ***Psiquis: Revista de Psiquiatría, Psicología y Psicosomática,*** 2(5), 26–40.
Discusses the learned helplessness phenomenon in animals and reviews some of the major theories about the deficits that follow an uncontrollable aversive event. It is suggested that the best explanation of learned helplessness is the learned helplessness theory, which is characterized by its cognitive approach. However, a number of methodological criticisms of this theory are presented.

225. Prescott, Louisa; Buchanan, S. L. & Powell, D. A. (1989). **Leg flexion conditioning in the rat: Its advantages and disadvantages as a model system of age-related changes in associative learning.** ***Neurobiology of Aging,*** 10(1), 59–65.
12- and 28-mo-old rats received 5 Pavlovian conditioning sessions in which the conditioned stimulus/stimuli (CS) was a tone and the unconditioned stimulus/stimuli (UCS) was a footshock. Right foreleg flexion was measured as the conditioned response (CR). Other Ss received a random sequence of unpaired tones and footshock and served as pseudoconditioning control groups. Interstimulus intervals (ISIs) of 1.5 and 3.5 sec were studied. The longer ISI resulted in higher rates of responding in both the conditioning and pseudoconditioning groups. Except for young males, all Ss showed significantly higher levels of responding in the conditioning groups. Females showed faster acquisition and higher levels of responding than males. Old males were slower to reach a criterion of 5 successive CRs than either young males or young or old females.

226. Rankin, Catherine H.; Beck, Christine D. & Chiba, Catherine M. (1990). ***Caenorhabditis elegans*: A new model system for the study of learning and memory.** ***Behavioural Brain Research,*** 37(1), 89–92.
Because extensive information on the neuroanatomy, development, and genetics of the nematode *Caenorhabditis elegans* (CE) makes it an ideal candidate model system for the analysis of the mechanisms underlying learning and memory, nonassociative learning in CE was investigated by observing changes in reversal reflex response amplitude to a mechanical vibratory stimulus. Results show that CE was capable of short-term habituation, dishabituation, and sensitization, as well as long-term retention of habituation training lasting for at least 24 hrs. Findings support the use of CE in developmental, genetic, and physiological analyses of learning and memory.

227. Riccio, David C.; MacArdy, Elayne A. & Kissinger, Steven C. (1991). **Associative processes in adaptation to repeated cold exposure in rats.** ***Behavioral Neuroscience,*** 105(4), 599–602.
Learning processes have been implicated in drug tolerance, but the role of associative mechanisms in adaptation to stressors has not previously been determined. Rats that received daily brief cold exposures demonstrated adaptation to the cold as measured by an attenuation of hypothermia. Tolerance to the cold was disrupted by changing the context in which the S experienced the cold. These findings provide evidence of associative processes in adaptation to cold exposure and illustrate that these processes are not limited to drug tolerance.

228. Riley, Anthony L.; Lotter, Elizabeth C. & Kulkosky, Paul J. (1979). **The effects of conditioned taste aversions on the acquisition and maintenance of schedule-induced polydipsia.** ***Animal Learning & Behavior,*** 7(1), 3–12.
A series of 4 experiments showed that conditioned taste aversions produced a moderate, but transient, suppression of schedule-induced polydipsia in 66 female Long-Evans rats. This suppression was greater and longer lasting when Ss were offered a choice between water and the previously poisoned solution on the polydipsia baseline. A final experiment demonstrated that taste aversions were more effective in suppressing schedule-induced consumption when superimposed on a developing schedule-induced drinking baseline as opposed to a stable pattern of schedule-induced drinking. It is suggested that schedule-induced polydipsia is insensitive to conditioned taste aversions. This conclusion is discussed in terms of schedule-induced alcohol consumption and its potential as an animal model of alcoholism.

229. Rosadini, Guido; Cupello, Aroldo; Ferrillo, Franco & Sannita, Walter G. (1981). **Quantitative EEG and neurochemical aspects of memory and learning.** ***Acta Neurologica Scandinavica,*** 64(Suppl 89), 109–118.
Studied different, established, or putative animal models of learning by electrophysiological (EEG) and neurochemical methods. The training of Wistar rats to new behavioral patterns resulted in the stimulation of total RNA synthesis rate in specific brain structures, as well as in modifications of the EEG organization of the hippocampus. This 2-fold approach to the assessment of modification in learning-involved brain structures was extended to the study of mirror focus and kindling to verify the suitability of these phenomena as experimental models of learning. Reduced uridine incorporation and the proportion of poly (A)-associated RNA were found in mirror focus in comparison with control regions. These variations seem related to brain damage rather than congruent with learning-related modifications.

230. Rosellin, Robert A.; Woodruff, Guy & Gamzu, Elkan. (1982). **The role of response-reinforcer contiguity in maintenance of autoshaped key-pecking.** ***Behaviour Analysis Letters,*** 2(1), 21–29.
Studied the maintenance of autoshaped keypecking in the context of a 2-component multiple schedule of reinforcement, in which either a differential or a nondifferential probability of reinforcement was obtained in the presence of the 2-component stimuli across successive phases of the experiment. 15 Silver King pigeons trained to eat from a hopper served as Ss. Reinforcement was delivered at a constant rate independent of responding by a variant of a linear interresponse time schedule with a variable delay component. Ss were then divided into 3 groups: all reinforcers delivered immediately after response, 50% reinforcers delivered after a delay, or all reinforcers delivered after a delay. Keypecking was acquired and maintained in all groups under differential conditions, and declined abruptly under nondifferential conditions. Responding persisted at appreciable levels under nondifferential conditions for Ss with relatively good response–reinforcer contiguity, but not for Ss with relatively poor contiguity. Results support the view that operant and Pavlovian factors interact directly in determining maintenance of autoshaped responding.

231. Rosellini, Robert A. (1978). **Inescapable shock interferes with the acquisition of an appetitive operant.** ***Animal Learning & Behavior,*** 6(2), 155–159.

Reports on the reinforcer generality of the interference effect resulting from exposure to inescapable shock in 2 experiments with 48 male Holtzman rats. In Exp I, Ss that received inescapable shock showed weak interference with the acquisition of an appetitive operant compared to Ss exposed either to escapable or no shock. In Exp II, the response-reinforcer contingency was degraded by introducing a 1-sec delay of reinforcement on the appetitive task. Inescapable shock produced much stronger interference with the acquisition of the operant response than in Exp I. The results demonstrate reinforcer generality of the debilitating effects produced by inescapable shock.

232. Rosellini, Robert A. et al. (1984). **Uncontrollable shock proactively increases sensitivity to response-reinforcer independence in rats.** ***Journal of Experimental Psychology: Animal Behavior Processes,*** 10(3), 346–359.
Learned helplessness theory predicts that animals exposed to inescapable shock acquire an expectancy of response-reinforcer independence, which proactively interferes with learning of response-reinforcer dependence. The theory also predicts that this expectancy can increase sensitivity to subsequent instances of response-reinforcer independence. Two experiments, with 42 male Holtzman rats, tested the latter prediction in a paradigm that minimized the confounding effects of shock-induced activity deficits. In Exp I, Ss were trained to respond for food, then given either escapable, inescapable, or no shock. Subsequently, they received 2 sessions of response-contingent food followed by sessions of noncontingent food deliveries. During this phase, inescapably shocked Ss decreased responding faster than did controls. Exp II replicated this finding with a different schedule of food delivery and a procedure that more directly minimized the possibility that the outcome was due to either direct or indirect shock-induced activity changes. Results support the prediction that uncontrollable aversive events can increase an animal's sensitivity to noncontingent response-reinforcer relationships.

233. Rosellini, Robert A.; DeCola, Joseph P. & Shapiro, Neil R. (1982). **Cross-motivational effects of inescapable shock are associative in nature.** ***Journal of Experimental Psychology: Animal Behavior Processes,*** 8(4), 376–388.
An appetitive choice discrimination test was used to assess the relative contribution of activity and associative effects of inescapable shock (IS) in a cross-motivational paradigm. A 2-response nosepoke test was used following IS treatment. In Exp I, male Holtzman rats demonstrated separate associative and activity effects of IS. Ss exposed to IS made more incorrect responses than controls and were lower in activity. Exp II demonstrated that these effects resulted from the uncontrollability of the shock, not from shock exposure per se. In Exp III, residual effects of IS were investigated by exposing Ss to discrimination reversals. On these tests, shocked Ss showed performance inferior to nonshocked controls, a result indicating that the effects of IS were not completely reversed by experience with contingent reward in the discrimination task. Results suggest that associative factors play a more important role than activity reduction in mediating the effects of IS, at least when these are measured in an appetitive context.

234. Rosellini, Robert A. & DeCola, Joseph P. (1981). **Inescapable shock interferes with the acquisition of a low-activity response in an appetitive context.** ***Animal Learning & Behavior,*** 9(4), 487–490.
Investigated the effects of exposure to inescapable shock on 24 male Holtzman rats' acquisition of a low-activity appetitive response using a trial procedure. Inescapable shock was found to interfere with the acquisition of a nose-poke response to obtain food as compared with Ss exposed to either escapable shock or no shock. General activity levels were measured separately during the trial and the intertrial interval during the appetitive test. Inescapably shocked Ss were less active during the trial component than were either escapably shocked or nonshocked Ss. However, no differential levels of activity were observed during the intertrial component of the appetitive test. The relevance of these findings for the learned helplessness and learned inactivity hypotheses are discussed.

235. Rosellini, Robert A. & Seligman, Martin E. (1976). **Failure to escape shock following repeated exposure to inescapable shock.** ***Bulletin of the Psychonomic Society,*** 7(3), 251–253.
J. M. Weiss et al (1975) reported that repeated sessions of uncontrollable stressors failed to produce an escape learning deficit in rats. This finding is unexpected in view of the often reported learned helplessness findings where a deficit in escape learning is found following a single session of uncontrollable shock. It is suggested that the discrepancy between these 2 sets of results may be due to procedural differences. The present experiment determined whether the usual parameters for producing learned helplessness in the rat produce escape deficits after repeated sessions. Results from 24 male Holtzman rats demonstrate an escape learning deficit, as expected, after repeated exposure to inescapable shock. The phenomenon reported by Weiss et al appears to be different in kind from learned helplessness.

236. Rosellini, Robert A. & Seligman, Martin E. (1978). **Role of shock intensity in the learned helplessness paradigm.** ***Animal Learning & Behavior,*** 6(2), 143–146.
Manipulated factorially 4 levels of shock intensity during exposure to inescapable shock and 3 levels of intensity during the test for interference in an experiment with 96 male Holtzman rats. Interference occurred at each training shock intensity when training and test shocks were similar. Interference was not obtained when training intensity was high but testing intensity low or medium or when training intensity was low or medium and test intensity was high. These findings pose problems for learned helplessness, learned inactivity, competing motor response, and catecholamine depletion hypotheses of the interference effect in the rat.

237. Rosellini, Robert A.; Warren, Donald A. & DeCola, Joseph P. (1987). **Predictability and controllability: Differential effects upon contextual fear.** ***Learning & Motivation,*** 18(4), 392–420.
Investigated under several parametric conditions whether controllability and predictability similarly affect contextual fear. 109 male rats served as Ss for all 3 experiments. In Exp I, control over shock termination reduced contextual fear at an earlier point in training than prediction of shock absence. Exp II demonstrated an effect of controllability under conditions in which the feedback effect is precluded. Exp III examined the possibility that the group differences observed in the above experiments could be due to a potential difference in the conditionability of the response-produced stimulation and the external feedback stimulus. Re-

sults suggest that the effects of controllability may not be reducible to those of predictability. Furthermore, they have important implications for theoretical proposals concerning the effect of feedback on contextual fear.

238. Rush, Douglas K.; Mineka, Susan & Suomi, Stephen J. (1983). **Therapy for helpless monkeys.** ***Behaviour Research & Therapy,*** 21(3), 297–301.
During the course of pilot studies and 2 formal experiments examining the learned-helplessness phenomenon in rhesus monkeys, 5 Ss failed to escape in a shuttlebox following earlier experience with aversive stimulation in primate-restraining chairs. The authors detail a therapeutic manipulation designed to reverse these Ss' maladaptive behavior in the shuttlebox. Introduction of a different fear stimulus (a net previously used to restrain the Ss) was found to be effective in inducing shuttlebox escape and avoidance learning. Implications for an understanding of the learned-helplessness phenomenon and their relevance to therapy for human depression are discussed.

239. Rush, Douglas K.; Mineka, Susan & Suomi, Stephen J. (1982). **The effects of control and lack of control on active and passive avoidance in rhesus monkeys.** ***Behaviour Research & Therapy,*** 20(2), 135–152.
Most studies on the effects of lack of control have utilized test tasks in which active responding is required, and generally they have found impaired learning. Those few studies that have required passive responding in the test task generally have found facilitation of learning. The present 2 experiments examined the effects of lack of control in both active and passive avoidance tasks in a primate species (*Macaca mulatta*) not previously used in this research area. In Exp I, with 14 Ss, although the group without control (IE) tended to be inferior at active and superior at passive avoidance in comparison to the group with control (E), there were no significant differences. In Exp II (16 Ss), which used a difficult discrimination task in which Ss were required to learn when and when not to respond actively to avoid aversive stimulation, greater group differences were found. Two monkeys from Group IE failed to escape in active avoidance acquisition and, as a whole, Group IE was somewhat slower to respond than Group E. At passive avoidance, however, Group IE was superior to Group E and, as a consequence, more efficiently solved the discrimination problem. Implications for interpreting the effects of lack of control as deficits are discussed.

240. Samuels, Owen B.; DeCola, Joseph P. & Rosellini, Robert A. (1981). **Effects of inescapable shock on low-activity escape/avoidance responding in rats.** ***Bulletin of the Psychonomic Society,*** 17(4), 203–205.
Exposure to inescapable shock proactively interferes with the acquisition of shuttle or barpress escape responding. The learned inactivity hypothesis proposes that this deficit results primarily from the active nature of the test escape response. In support of this position, it has been demonstrated that inescapable shock not only fails to interfere with but also actually facilitates the acquisition of a low-activity escape/avoidance response. However, in this demonstration, it is unclear whether the response was emitted by the animal or elicited by the test procedure employed. The present experiment (35 male Holtzman albino rats) examined the effects of inescapable shock on the acquisition of a low-activity response in a context that precluded the possibility of recording elicited responses as escape/avoidance responses. No evidence was found of facilitated acquisition of the low-activity response as a result of inescapable shock. This finding, in conjunction with others, questions the data base for the learned inactivity hypothesis.

241. Smith, Jane E. & Levis, Donald J. (1991). **Is fear present following sustained asymptotic avoidance responding?** ***Behavioural Processes,*** 24(1), 37–47.
Provided a direct test between a "fear" and an "expectancy" interpretation of avoidance in extinction. The test was suggested by M. E. Seligman and J. C. Johnston (1973). 72 rats were given discrete-trial avoidance training (AVT) in a modified shuttlebox apparatus until they reached a criterion of either 5, 25, or 50 consecutive avoidance responses with a latency of less than 4 sec. Asymptotic responding was reached quickly and continued to be maintained until AVT was discontinued. A transfer test using a lick suppression paradigm followed AVT. Data clearly support the theory that "fear" needs to be present for avoidance responding to occur. All Ss displayed good suppression to the conditioned stimulus/stimuli (CS).

242. Spector, Alan C.; Breslin, Paul & Grill, Harvey J. (1988). **Taste reactivity as a dependent measure of the rapid formation of conditioned taste aversion: A tool for the neural analysis of taste-visceral associations.** ***Behavioral Neuroscience,*** 102(6), 942–952.
Investigated the ability of animals to form taste aversions following neural manipulations. In Exp 1, 10 rats received intraoral infusions of sucrose every 5 min starting immediately after the injection of LiCl. 12 controls were injected with NaCl. Oromotor and somatic taste reactivity behaviors were videotaped and analyzed. Lithium-injected Ss decreased their ingestive taste reactivity over time; aversive behavior increased. Controls maintained high levels of ingestive responding and demonstrated virtually no aversive behavior following sodium injection. Ss were tested several days later for a conditioned taste aversion (CTA). Rats previously injected with lithium demonstrated significantly more aversive behavior than controls. Exp 3 revealed that when similarly treated rats were tested for a CTA while in a lithium-induced state, difference in the ingestive behavior was observed. In Exp 2, naive rats were injected with NaCl or LiCl but did not receive their 1st sucrose infusion for 20 min. Ss also received infusions at 25 and 30 min postinjection. There were no differences in the task reactivity behavior displayed. Rats dramatically changed their oromotor responses to sucrose during the period following LiCl administration, provided the infusions started immediately after injection, a change attributable to associative processes.

243. Starr, Mark D. & Mineka, Susan. (1977). **Determinants of fear over the course of avoidance learning.** ***Learning & Motivation,*** 8(3), 332–350.
Most theorists have explained attenuation of fear over the course of avoidance learning by assuming that fear extinguishes with repeated nonreinforced avoidance trials. Exp I, with 72 Fisher-344 male rats, replicates the finding that rats trained to a criterion of 27 consecutive avoidance responses (CARs) show less fear during the CS than rats trained to a criterion of 3 or 9 CARs. This attenuation of fear cannot, however, be accounted for by simple Pavlovian fear extinction, because yoked partners receiving the exact same pattern of CSs and UCSs did not show this attenuation and did not differ from yoked partners receiving only reinforced CS presentations. Exp II, with 24 Ss, found that feedback from

the master avoidance learner's response is sufficient to produce this attenuation in yoked Ss; "control" per se is not necessary. Several possible explanations are discussed regarding the mechanism underlying this role of feedback in diminishing fear of the CS in the avoidance learning context.

244. Straub, Richard O.; Singer, Jerome E. & Grunberg, Neil E. (1986). **Toward an animal model of Type A behavior.** ***Health Psychology,*** 5(1), 71–85.
Attempted to differentiate behaviors of Mongolian gerbils analogous to Type A (coronary-prone) and Type B (non-coronary-prone) human behavior. Preliminary classification of 20 Ss was based on performance on differential reinforcement of low rates 20-sec and 60-sec schedules. To retain their preliminary classification, Type A and Type B Ss were required to be dominant and subordinate, respectively, in matches with Ss of opposite behavioral classification. Ss that exhibited Type A timing won significantly more dominance matches than did Type B Ss. Incidence rates of Type A and Type B behavior in the 2 selectively bred generations were significantly greater than frequencies in the original stock.

245. Thomka, Michael L. & Rosellini, Robert A. (1975). **Frustration and the production of schedule-induced polydipsia.** ***Animal Learning & Behavior,*** 3(4), 380–384.
Conducted 2 experiments with a total of 42 male Holtzman rats to test the hypothesis that frustration mediates the production of schedule-induced polydipsia. In Exp I, a group in which reward was reduced from 6 to 2 pellets of food in an operant chamber was found to increase water intake compared to a group maintained at 2 pellets reward. In Exp II, Ss trained to approach food on a partial reinforcement schedule in a runway subsequently showed lower levels of water intake in the operant test for polydipsia than Ss given continuous reinforcement during runway training. Results are interpreted as supporting a frustration hypothesis of schedule-induced polydipsia and are discussed within the context of persistence theory.

246. Tsuda, Akira. (1981). **Influence of preshock conditions on subsequent** ***Annual of Animal Psychology,*** 30(2), 115–126.
Tested whether shock-induced aggression would reduce the development of learned helplessness in 36 rats. Ss in the Fight group were placed in a shock chamber in pairs, and electric shock (1 mA for 5 sec, preceded by a tone) was administered through sc electrodes for 64 trials. Ss in the Alone group were shocked alone in the chamber, and those in the Restraint group did not receive any shock. 24 hrs later all were trained on a discriminative escape-avoidance task in a **Y**-maze. A tone signaled the start of a trial, a light indicated the correct arm of the maze, and 5 sec later footshock (1 mA) was given. The Alone group showed signs of learned helplessness. The Fight group showed as long latencies as the Alone group but as many correct choices of the arm as the Restraint group. Aggressive responses induced by inescapable shock may have served as a coping behavior and reduced learned helplessness. (English abstract).

247. Tsuda, Akira & Hirai, Hisashi. (1975). **Effects of the amount of required coping response tasks on gastrointestinal lesions in rats.** ***Japanese Psychological Research,*** 17(3), 119–132.
To study the effects of the coping response on gastrointestinal lesions in 108 male Wistar albino rats, the values of FR schedules were changed for the coping response task in the free operant avoidance situation over 24 hrs. In FR-1 or FR-2, experimental Ss which could control electric shock to avoid or escape by pushing a flapper only once or twice developed less severe gastrointestinal lesions and body weight loss than yoked Ss which received the same amount of shocks but could not perform any coping responses. In FR-5 or FR-8, however, experimental Ss developed more severe stress pathology than the yoked Ss or controls. These reversal effects show that the interaction between controllability of experimental environment and the amount of required coping response plays an important role. These findings support J. M. Weiss's (1971) ulcer-prediction model.

248. Volpicelli, Joseph R.; Ulm, Ronald R.; Altenor, Aidan & Seligman, Martin E. (1983). **Learned mastery in the rat.** ***Learning & Motivation,*** 14(2), 204–222.
Investigated the bidirectional effects of prior experience with both control or lack of control over shock on subsequent shock-motivated activity and escape learning in 108 male Sprague-Dawley rats. Ss were tested with inescapable shock rather than escapable shock. Naive Ss initially shuttled frequently during shock but decreased activity as testing continued. Pretraining with inescapable shock reduced shuttle responding throughout testing. Ss that first learned to leverpress to escape shock continued unabated shuttling through 200 trials of 10-sec duration of inescapable shocks (Exp I). During 2 uninterrupted 1,000-sec duration inescapable shocks (Exp II), escape Ss continued to leverpress through the 2,000 sec of shock. In Exp III, escapable shock facilitated and inescapable shock hindered later learning when the escape contingency was degraded by a 3-sec delay of shock termination. Exps IV and V demonstrated that (1) this associative facilitation effect was not simply due to an increase in active responding by escape Ss and (2) no associative facilitation was observed if the contingency was not initially degraded by a 3-sec delay. Results indicate bidirectional effects of control on aversively motivated behavior in animals. In addition to typical helplessness effects, a mastery phenomenon was observed that was the opposite of helplessness.

249. Volpicelli, Joseph R.; Ulm, Ronald R. & Altenor, Aidan. (1984). **Feedback during exposure to inescapable shocks and subsequent shock-escape performance.** ***Learning & Motivation,*** 15(3), 279–286.
Two experiments with 84 male Sprague-Dawley rats assessed the ability of a feedback stimulus during helplessness training to reduce the performance deficits common to inescapable shock. In each experiment, 4 groups of Ss were exposed to either escapable shock, inescapable shock with a feedback stimulus following shock termination, inescapable shock with no feedback stimulus, or no shock. The feedback stimulus eliminated the interference effects of inescapable shock when tested with an FR 3 leverpress escape task (Exp I) or on an FR 1 task with a 3-sec delay between the response and shock termination (Exp II). It is concluded that stress-induced biochemical changes may mediate the interference effects seen in inescapably shocked rats.

250. Wade, Stephen E. & Maier, Steven F. (1986). **Effects of individual housing and stressor exposure upon the acquisition of watermaze escape.** ***Learning & Motivation,*** 17(3), 287–310.
In 4 experiments with 152 male Sprague-Dawley rats, Ss housed in groups showed fast and reproducible acquisition of watermaze tasks; Ss housed individually showed a gradual decrease in acquisition performance of both spatial and nonspatial tasks. This deficit was detectable after 2 or 3 wks of isolation and continued to increase in severity thereafter. The deficit was reversible by a brief period of group housing, by episodic exposure to loud noise, and by episodic restraint. The efficacy of these corrective treatments declined when isolation was prolonged beyond 3 wks. Findings are interpreted as indicating an abnormality in stress responsiveness in the individually housed Ss.

251. Wagner, H. Ryan; Hall, Thomas L. & Cote, Ila L. (1977). **The applicability of inescapable shock as a source of animal depression.** ***Journal of General Psychology,*** 96(2), 313–318.
27 male albino rats were given initial experience with escapable shock, equivalent amounts of inescapable shock, or no shock. Measures were obtained in the ensuing 15 hrs on food intake, water intake, number of cage crossings, and weight change for all groups. Ss were then tested on an escape task. Inescapably shocked Ss showed significant decreases in food and water consumption in comparison to both nonshocked and escapably shocked control Ss. Weight gains were significantly decreased by exposure to shock, irrespective of the availability of a coping response. Consistency of these findings with proposals suggesting that exposure to inescapable shock leads to a state of animal depression (learned helplessness) is discussed and compared to alternative stress explanations.

252. Wall, Lida G.; Ferraro, John A. & Dunn, Derek E. (1981). **Temporal integration in the chinchilla.** ***Journal of Auditory Research,*** 21(1), 29–39.
Extended ranges of frequencies and signal durations were used to define the temporal integration function of 5 chinchillas and to examine the effect of stimulus frequency on critical duration. Behavioral thresholds using pure-tone stimuli were determined by shock-avoidance audiometry, and Ss were conditioned to respond to pure tones of varying duration. The stimulus sound pressure level (SPL) required for threshold response decreased as the stimulus duration increased. Frequency did not have a significant effect on threshold SPL, and no interaction was found between frequency and duration. Critical duration appeared to be 100–200 msec. Similarities between these data and those obtained with human Ss suggest that the chinchilla can serve as a model for human listeners in extended studies involving temporal integration and short-duration stimuli, which may be an effective tool in identifying the presence of specific sensorineural lesions.

253. Warren, Donald A.; Rosellini, Robert A.; Plonsky, Mark & DeCola, Joseph P. (1985). **Learned helplessness and immunization: Sensitivity to response–reinforcer independence in immunized rats.** ***Journal of Experimental Psychology: Animal Behavior Processes,*** 11(4), 576–590.
Exps I and II, with 62 male Holtzman rats, examined the learned helplessness and immunization effects using a test in which appetitive responding was extinguished by delivering noncontingent reinforcers. Contrary to learned helplessness theory, "immunized" Ss showed performance virtually identical to that of Ss exposed only to inescapable shock and different from that of nonshock controls. Exp II suggested that the helplessness effect and the lack of immunization were not due to direct response suppression resulting from shock. In Exp III, in which the immunization effect was assessed in 28 Ss by measuring the acquisition of a response to obtain food when there was a positive response–reinforcer contingency, immunization was observed. Results cannot be explained on the basis of proactive interference and instead suggest that Ss exposed to the immunization procedure acquired an expectancy of response–reinforcer independence during inescapable shock. Thus, immunization effects may reflect the differential expression of expectancies rather than their differential acquisition as learned helplessness theory postulates.

254. Warren, Donald A.; Rosellini, Robert A. & Plonsky, Mark. (1985). **Regularity of inescapable shock duration affects behavioral topography, but not shuttle escape performance.** ***Psychological Record,*** 35(2), 227–238.
Investigated the effects of exposure to fixed and variable duration inescapable shock on the subsequent acquisition of a shuttle escape response in 24 male Holtzman rats. A 2nd purpose of the experiment was to examine closely the behavior of Ss during exposure to inescapable shock in an attempt to assess directly the behavioral changes occurring during this experience. Competing response accounts of the learned helplessness effect propose the mechanism of adventitious reinforcement during inescapable shock for the learning of low activity responses. Exposure to inescapable fixed duration shock should therefore be more effective than variable duration shock of the same average value in producing subsequent escape deficits. However, the present experiment demonstrated equivalent shuttle escape deficits resulting from the 2 types of shock exposure. In addition, behavioral observations made during inescapable shock indicated that those behaviors immediately preceding shock offset during the 1st several trials were not more likely to develop than other behaviors that were virtually absent early in the session. This pattern of results does not support current competing response hypotheses but is in agreement with learned helplessness theory.

255. Weisker, Sandra M. & Barkley, Marylynn. (1991). **Female genotype influences the behavioral performance of mice selected for reproductive traits.** ***Physiology & Behavior,*** 50(4), 807–813.
Behavioral performance of mice that differ in regularity of estrous cycle and litter size was studied after female exposure to a male of the same or a different strain. Emotional reactivity was measured using the pole, straightaway, and open field tests. Factor interpretations of emotionality included motor discharge, autonomic imbalance, and acrophobia. Mice characterized by regular estrous cycles and large litters were more explorative and emotionally reactive with respect to motor discharge and autonomic imbalance. In contrast, mice with less regular estrous cycles and small litter size were more acrophobic. These strain differences in behavioral performance were influenced by the genotype of the female rather than the cohabitating male.

256. Wieland, Scott; Boren, James L.; Consroe, Paul F. & Martin, Arnold. (1986). **Stock differences in the susceptibility of rats to learned helplessness training.** ***Life Sciences,*** 39(10), 937–944.

Tested male rats from 8 different stocks for susceptibility to learned helplessness (LH) training. Results show that Lewis, Brown Norway, Fischer F-344, and Sasco Holtzman rats were virtually nonsusceptible to LH training. Harlan SD and Buffalo rats evidenced intermediate susceptibilities of 28% and 33%, respectively. Wistar Kyoto and Charles River Holtzman rats were the most susceptible at 53% and 55%, respectively. It is suggested that suppliers and researchers should choose Ss from a susceptible stock.

257. Williams, Jon L. (1982). **Influence of shock controllability by dominant rats on subsequent attack and defensive behaviors toward colony intruders.** ***Animal Learning & Behavior,*** 10(3), 305–313.
30 colonies, each consisting of a female and 2 male adult albino Holtzman rats, remained intact for an 8-wk period. Naive conspecific intruders were then introduced into each colony for a 10-min test for 5 consecutive days. Videotapes of the tests were scored for aggressive and defensive behaviors. In every colony, aggression was greatest for a single alpha male. Alpha Ss were randomly given 1 of 3 treatments: wheel-turn escape training, inescapable yoked shock, or restraint without shock. Alpha Ss were returned to their colonies, and an intruder test was given 26 hrs later. Significant decreases in aggressive responses and increases in defensive behaviors occurred in the alpha yoked group but not in the other alpha groups. The nonalpha colony partners of the alpha yoked Ss showed the opposite changes following treatment. A final intruder test 72 hrs later revealed that the deficits in aggression of the alpha yoked group were still present but that the behaviors of most of the other groups were beginning to return to their pretreatment levels. Findings are discussed in terms of the concept of learned helplessness and alternative theoretical explanations.

258. Williams, Jon L. & Maier, Steven F. (1977). **Transituational immunization and therapy of learned helplessness in the rat.** ***Journal of Experimental Psychology: Animal Behavior Processes,*** 3(3), 240–252.
Two experiments used a total of 100 male albino Sprague-Dawley rats. In Exp I, Ss that had wheel-turn escape training prior to inescapable shock in a restraining tube showed enhanced FR-2 shuttlebox acquisition relative to Ss not given wheel training. This "immunization" effect was not found to be simply the result of positive transfer between the 2 responses. Exp II showed that inescapably shocked Ss, which were given escape training with a chained-response sequence of leverpressing and jumping onto a platform, were not retarded in acquiring the shuttling response. The "therapy" task resulted in negative transfer in shuttling for Ss that were restrained but not given prior inescapable shock. The major finding of these experiments is that immunization and therapy effects occur when such procedures involve responses and settings that are different from the ones used in later testing. Implications of the breadth of transfer obtained with the immunization and therapy procedures employed in the present experiments are discussed in terms of the various interpretations of the learned helplessness phenomenon.

259. Wilson, W. Jeffrey & Butcher, Larry L. (1980). **A potential shock-reducing contingency in the backshock technique: Implications for learned helplessness.** ***Animal Learning & Behavior,*** 8(3), 435–440.
Delivered ECS to 15 female albino Sprague-Dawley rats through an sc implanted back electrode. Exp I evaluated the relationship between number of paws grounded and total power dissipated in the rat. In Exp II, the threshold of shock-induced vocalization, a putative index of aversiveness, was positively correlated with the number of paws grounded. Findings suggest that when the backshock technique is used, the aversiveness of shock potentially can be modified by the posture adopted by the experimental animal. Caution should be exercised in attributing deficits in escape behavior following inescapable shock administered with back electrodes to learned helplessness.

260. Woodruff-Pak, Diana S.; Logan, Christine G. & Thompson, Richard F. (1990). **Neurobiological substrates of classical conditioning across the life span.** Conference of the National Institute of Mental Health et al: The development and neural bases of higher cognitive functions (1989, Philadelphia, Pennsylvania). ***Annals of the New York Academy of Sciences,*** 608, 150–178.
Eyeblink conditioning in the rabbit shows promise as a model system for the study of the neurobiology of learning and memory over the lifespan. The eyeblink classical conditioning (ECC) paradigm is described, and a summary of the lifespan data on ECC in rabbits and humans is presented. The roles of the cerebellum and the hippocampus in ECC are considered, and a cerebellar model of learning and memory is outlined. Data from the animal model lead to the prediction that Alzheimer's disease patients, having hippocampal dysfunction, would show poorer acquisition of ECC than normal adults.

261. Young, A. Grant & Speier, A. H. (1982). **Learned helplessness in the rat: An elusive phenomenon.** ***Journal of General Psychology,*** 107(1), 75–83.
Tested the learned helplessness hypothesis, which states that during inescapable shock the organism learns that shock termination is independent of the organism's behavior, and this learning interferes with the ability to associate behavior and shock termination in a new escape situation. In Exp I, 48 Wistar rats received no, low, or high inescapable footshock followed by low or high escapable shock. In Exp II, with 80 Ss, the length of the shock varied. In Exp III, with 32 Ss, intensity and duration of shock varied. Results of all experiments show no interference effects whether test shock was the same, less intense, or more intense than treatment shock.

262. Zhuravlyova, N. G. (1982). **Disturbances of conditioned activity in hypokinesia rats and normalizing effect of motor loads.** ***Zhurnal Vysshei Nervnoi Deyatel'nosti,*** 32(4), 642–650.
Examined partial limitation of motor activity (A) in rats as a model of human hypokinesis, and moderate motor loads (B) as a means of hypokinesis prevention. A and B facilitated elaboration of chains of motor alimentary CRs. Reversal of the 2-link chain of reflexes into a 3-link one and that into a chain CR was impeded under Condition A and facilitated in Condition B. Combination of A and B normalized both kinds of reversals. It is concluded that internal inhibition was weakened in A and enhanced in B, which provided for normalization of conditioned activity.

Social & Instinctive Behavior

263. Alonso, S. J.; Arevalo, R.; Afonso, D. & Rodríguez Díaz, Manuel. (1991). **Effects of maternal stress during pregnancy on forced swimming test behavior of the offspring.** ***Physiology & Behavior,*** 50(3), 511–517.
Three studies with male and female rats revealed that (1) there were sex differences for depression in 2 different animal models (swimming-induced immobility and natatory tests), (2) there were sex differences in open-field behavior, (3) prenatal maternal restraint decreased sex differences for depression but did not affect sex differences in open-field behavior, and (4) prenatal maternal restraint affected female but not male behavior in the 2 depression tests. Results suggest that (1) sex differences reported in animal models of depression are under the control of gonadal steroids during prenatal brain development and (2) stress during early phases of development increases the risk for depression in adulthood.

264. Anisko, Joseph J.; Suer, Sharon F.; McClintock, Martha K. & Adler, Norman T. (1978). **Relation between 22-kHz ultrasonic signals and sociosexual behavior in rats.** ***Journal of Comparative & Physiological Psychology,*** 92(5), 821–829.
The copulatory performance of male rats, tested in a large seminaturalistic environment, was assessed to determine the relation between 22-kHz ultrasonic vocalizations and a range of sociosexual behaviors. The 7 sexually experienced Charles River male rats were tested until sexual exhaustion. Ultrasonic signals were shown to occur in a wider range of sociosexual circumstances than previously reported; for example, the calls occurred in particular social circumstances during the preejaculatory period as well as during the postejaculatory interval. There was no consistent evidence that the emission of this call during the postejaculatory period consistently functions to keep the female away from the male. The nature and occurrence of postejaculatory ultrasonic signals showed increasing variability in successive ejaculatory series. The results of this and previous studies are interpreted within a semiotic theory of communication. The 22-kHz call is described as a message that makes available the information that the sender is in a socially depressed and withdrawn state.

265. Blanchard, Robert J. & Blanchard, D. Caroline. (1989). **Attack and defense in rodents as ethoexperimental models for the study of emotion.** ***Progress in Neuro-Psychopharmacology & Biological Psychiatry,*** 13(Suppl), S3–S14.
Suggests ethological studies of the reactions of wild and laboratory rats to conspecifics and to predators as an alternate approach to the study of aggression and fear. The organization of these behaviors is described, and their relevance to the study of human emotion is considered. Behaviorally and functionally, risk assessment shows similarity to the apprehensive expectation and vigilance and scanning components of generalized anxiety reactions. There are parallels between lower mammal offense and human angry aggression in terms of eliciting stimuli, and a variety of experiential factors including inhibition by fear/pain, and reinforcement effects. Animal models may be valuable in the study of relatively primitive and conservative neurobehavioral systems, including many emotional states.

266. Blanchard, Robert J. & Blanchard, D. Caroline. (1987). **An ethoexperimental approach to the study of fear.** Symposium of the International Society for Research on Aggression (1986, Chicago, Illinois). ***Psychological Record,*** 37(3), 305–316.
Reviews research that complements natural observations with laboratory studies of animal behaviors to better control the factors influencing the relationships between fear responses and precipitating events. The discussion covers work on threatening stimuli, species-typical defensive behavior, organization of defense against predators, organization of conspecific defense, effects of domestication on defensive behavior, interaction of pain and learning, mechanisms of the conditioning of defense to painful stimuli, and interaction of defensive behavior and avoidance conditioning. It is suggested that the ethoexperimental approach is particularly adapted to drawing inferences from animal studies to human behavior.

267. Chamove, Arnold S. & Harlow, Harry F. (1975). **Cross-species affinity in three macaques.** ***Journal of Behavioural Science,*** 2(4), 131–136.
46 macaques (24 rhesus, 4 stumptailed, and 18 pigtail) were choice-tested for species preference subsequent to periods of infant social experience with members of 1 or more of the 3 species. Evidence was found for attachment to the 1st (3–9 mo) species but not to later (9–15 mo) species whether the same species but different animals or different species. Monkeys appear to have some innate preference for their own species, at least as objects of certain behaviors.

268. Crawley, Jacqueline N.; Sutton, Mary E. & Pickar, David. (1985). **Animal models of self-destructive behavior and suicide.** Special Issue: Self-destructive behavior. ***Psychiatric Clinics of North America,*** 8(2), 299–310.
Explores naturalistic and laboratory-induced animal behaviors that may have conceptual or neurochemical similarities with self-destructive behaviors and suicide. Analogies to human behaviors associated with inborn errors of metabolism or severe mental retardation are excluded, and focus is placed on principles and examples of animal behaviors that are self-destructive and that may ultimately result in injury or death.

269. Crews, David. (1988). **The problem with gender.** Special Issue: Sexual differentiation and gender-related behaviors. ***Psychobiology,*** 16(4), 321–334.
Contends that gender becomes nebulous when used as a concept to guide how sexuality is studied and perceived. To some, gender is defined in terms of behavior, whereas others regard gender as indicative of chromosome constitution. However, gender is an anthropocentric psychological construct that has little meaning when dealing with nonhuman organisms. Two animal model systems in which behavior or chromosomes cannot be used to define gender are discussed (the partheno-genetic whiptail lizard and the leopard gecko). It is suggested that (1) sexuality is not a unitary phenomenon, but rather a multifaceted composite of different sexes that may or may not be concordant; and (2) the components of sexuality be considered in terms of complementary mechanisms and outcomes.

270. Cyrulnik, Boris & Leroy, Roger. (1977). **An ethology of the family.** ***Annales Médico-Psychologiques,*** 2(1), 15–42.

Summarizes the descriptive and experimental literature on animal models of family organization, and suggests several implications for understanding human behavior. Assuming a genetic substrate for all behavior, the steps by which complex group interactions (e.g., affiliation, mating, care of the young) arise from a genetically programmed base are analyzed from a comparative and phylogenetic point of view. A range of family structures are described, especially those observed among primates, for insight into the growth of social behaviors. The plasticity of familial organizations is great, but not unlimited. Biological, economic, and (for humans) political pressures all have an identifiable share in determining the final complex social organization needed to continue the species successfully. The systematic study of such evidence is called the "ethology of the family," a discipline which may be expected to improve our understanding of deranged human family constellations and our ability to intervene positively in them. (English summary).

271. Ellison, Gaylord D. (1981). **A novel animal model of alcohol consumption based on the development of extremes of ethanol preference in colony-housed but not isolated rats.** ***Behavioral & Neural Biology,*** 31(3), 324–330.
Studied the fluid preference of 66 male Long-Evans hooded rats housed in an enriched social colony environment for several months with ad-lib access to water and 10% ethanol solutions. Marked individual differences were found. When subsequently tested in separate cages, some Ss showed a marked preference for 10% ethanol and consumed little water, while others exhibited opposite fluid preferences. Upon return to the colony, the extreme ethanol consumers were observed at the ethanol spout more than nonconsumers. Since this extreme variability in alcohol vs water preference does not develop in rats housed in individual cages, a new animal model is therefore presented for voluntary ethanol intake based on social factors.

272. Frances, Henriette; Lienard, Catherine; Fermanian, Jacques & Lecrubier, Yves. (1989). **Isolation-induced social behavioral deficit: A proposed model of hyperreactivity with a behavioral inhibition.** ***Pharmacology, Biochemistry & Behavior,*** 32(3), 637–642.
The behavior of mice isolated for 7–9 days was compared to that of mice reared in groups. When individually observed the isolated Ss attempted to escape slightly but significantly more often than the grouped Ss. When a pair of Ss (one isolated + one grouped) were tested together, the number of escape attempts of the isolated Ss was half of that of the grouped Ss: This phenomenon was named the isolation-induced social behavioral deficit. These opposed behaviors may mean the same thing: a hyperreactivity to the novelty. In a variety of new situations under the beaker (presence of a lifeless object, of grouped Ss or of an isolated S), the isolated Ss were more reactive than the grouped Ss. In conclusion, the social behavioral deficit test may be seen as a model of hyperreactivity with a behavioral inhibition.

273. Gould, Edwin & Bres, Mimi. (1986). **Regurgitation in gorillas: Possible model for human eating disorders (rumination/bulimia).** ***Journal of Developmental & Behavioral Pediatrics,*** 7(5), 314–319.
Compared regurgitation and reingestion behavior in gorillas with data on 2 human disorders: rumination and bulimia. Through questionnaires, zoo visits, and phone interviews concerning 117 gorillas at 17 zoos, it was found that 76 of 91 captive gorillas that were more than 5 yrs old regurgitated and reingested. Comparisons were made based on ontogeny, context, motor pattern, and intervention. Results indicate that there were more similarities between regurgitation and reingestion and rumination than between regurgitation and reingestion and bulimia. Regurgitation and reingestion resembled bulimia in parent–infant separation, lack of eating control, methods of induction, and some aspects of motor pattern. Regurgitation and reingestion resembled rumination in disrupted mother–infant communication, context of the behavior (enjoyment of the taste of the regurgitant), several aspects of motor pattern, and treatment (increased food volume).

274. Harlow, Harry. (1978). **Affectivity.** ***Psychologia Wychowawcza,*** 21(1), 13–23.
Discusses 5 distinct kinds of emotionality observed during research with young rhesus monkeys. The view that sexuality is primary in emotional development is rejected. The significance of body contact in mother–child relations is considered. (English & Russian summaries).

275. Hinde, R. A. & McGinnis, Lynda. (1977). **Some factors influencing the effects of temporary mother-infant separation: Some experiments with rhesus monkeys.** ***Psychological Medicine,*** 7(2), 197–212.
Reviews and compares the findings of 4 experiments conducted by Y. Spencer-Booth and R. A. Hinde (1971) on the effects of mother–infant separation in rhesus monkeys. Those experiments involved 4 groups in which (a) mothers were removed for 13 days leaving the infant in the social group, (b) infants were removed, (c) mothers and infants were removed and separated, and (d) mothers and infants were removed but not separated. The nature of the separation experience had a profound effect on the infant's response. Infants left in a familiar environment while their mothers were removed showed marked but brief "protest" and then profound "despair," while infants removed to a strange cage showed more prolonged "protest." A major factor determining the effects of the separation experience in the weeks following reunion was the degree to which the mother–infant relationship had been disturbed by it. The multiplicity of factors affecting the outcome of a separation experience are discussed.

276. Hinde, Robert A. (1991). **Relationships, attachment, and culture: A tribute to John Bowlby.** Special Issue: The effects of relationships on relationships. ***Infant Mental Health Journal,*** 12(3), 154–163.
As a tribute to John Bowlby (published 1960–1980), some of the work that he stimulated and facilitated is reviewed. In rhesus monkeys, dyadic relationships and group structure are crucial to behavior of individuals. A few days' separation between mother and infant can produce long-term effects, but the outcome depends on a large number of factors. The need to maintain a proper balance between a research focus on the individual, the relationship, and the family or group is stressed. Bowlby's use of comparative data shed light on aspects of infant and child behavior and was crucial in the development of attachment theory. The dangers of equating what is natural with what is best are stressed: Cultural desiderata interact with the biological desiderata on which natural selection operated in the environment of evolutionary adaptedness. (French, Spanish & Japanese abstracts).

277. Hinde, Robert A. (1985). **Was "the expression of the emotions" a misleading phrase?** ***Animal Behaviour,*** 33(3), 985–992.
Suggests that the title of Darwin's (1872) book, *The Expression of the Emotions in Man and Animals,* implies that emotional behavior is a direct expression of a relatively simple internal state. Data from both animals and humans indicate that neither emotional states nor emotional behavior can be considered independently of the factors that give rise to them and may involve complex affective–cognitive interactions. Such expression of emotion often involves processes of negotiation between 2 or more individuals. It is suggested that an attempt to grasp the nettle of the difficulty of defining emotion may (1) aid understanding of the processes involved and (2) facilitate liaison between studies of signal movements in animals and human emotions.

278. Hofer, Myron A. (1987). **Early social relationships: A psychobiologist's view.** ***Child Development,*** 58(3), 633–647.
Outlines a strategy for studying early social relationships in relatively simple animal model systems. Some of the differences in approach between neuroscientists and human developmentalists in dealing with the same research problems are illustrated, using studies on early social attachment in 2-wk-old rats and on their acute and long-term responses to separation from their mothers. A mechanism by which the mother regulates physiological and behavioral systems of the infant is suggested by findings that infant hyperactivity in response to separation can be moderated using artificial thermal, olfactory, and tactile maternal cues. Long-term and acute effects of separation are hypothesized to be 2 distinct processes. Studies on the organization of nursing that reveal an unexpected degree of synchrony and reciprocity are discussed.

279. LeFevre, Judith & McClintock, Martha K. (1991). **Isolation accelerates reproductive senescence and alters its predictors in female rats.** ***Hormones & Behavior,*** 25(2), 258–272.
Found that the process of reproductive senescence in female rats was altered by their social environment during adulthood. The incidence of constant estrus (CE), which marks the end of estrous cyclicity, was nearly twice as high in females living in isolation as it was in females living in groups. The characteristics of the estrous cycle when Ss were young predicted whether or not they would enter CE during reproductive senescence. However, these characteristics were different for isolated Ss than for Ss living in groups. In isolated Ss, entry into CE was predicted by a pattern of regular cycles followed by irregular cycles, a lordosis reflex of consistently high intensity during irregular cycles, and an absence of spontaneous pseudopregnancies. In Ss living in groups, increased estrogenization during irregular cycles predicted entry into CE. In both environments, the onset of acyclic lordosis reflex predicted the timing of CE.

280. Levine, Seymour; Franklin, Deborah & Gonzalez, Carol A. (1984). **Influence of social variables on the biobehavioral response to separation in rhesus monkey infants.** ***Child Development,*** 55(4), 1386–1393.
Five 4-mo-old rhesus monkeys were removed from their social group under 3 counterbalanced conditions: (1) infant totally isolated from its familiar physical and social environment and placed alone in a cage for 4 days; (2) "mother in," in which the infant was housed in a single cage in front of its social group with mother remaining in the group; and (3) "mother out," in which the mother was removed from the group during separation. Infant behavior was recorded, and, at selected times during separation, a blood sample was obtained for analysis of plasma cortisol. Infants rarely vocalized when totally isolated, but showed high vocalization and movement in the presence of their social groups. Vocalization was transiently higher in the mother-out than in the mother-in condition. Infants never showed signs of depression, and plasma cortisol response did not differentiate between groups. Ss showed significantly high levels of plasma cortisol 3 hrs following separation. Data indicate that the responses of the infant following separation are attempts to produce effective coping responses. The concepts of "protest" and "despair" are discussed as they relate to behavioral differences observed following different separation paradigms.

281. Meyer, Jerrold S.; Novak, Melinda A.; Bowman, Robert E. & Harlow, Harry F. (1975). **Behavioral and hormonal effects of attachment object separation in surrogate-peer-reared and mother-reared infant rhesus monkeys.** ***Developmental Psychobiology,*** 8(5), 425–435.
Four mother-reared and 4 surrogate-peer-reared monkeys were separated from their respective attachment objects at 6 mo of age and tested for the following 9 wks to determine their home-cage behavior and their pituitary-adrenocortical responses to stress. Both groups displayed a strong immediate behavioral response to separation which was characterized by increased vocalization, increased locomotion, and decreased self-play. The surrogate-peer-reared Ss showed a subsequent recovery in their levels of self-play, whereas the mother-reared Ss developed stereotypic behavior patterns. The groups displayed similar plasma cortisol responses to weekly sessions in an apparatus equipped with animated toy "monsters." Mother-reared but not surrogate-peer-reared Ss also manifested elevated cortisol levels when an S in an adjacent cage was captured and removed for stress testing. Mother-reared Ss thus responded in a stronger and more prolonged manner to the loss of their attachment object than surrogate-peer-reared Ss. Results suggest that infant rhesus monkeys form stronger attachments to monkey mothers than to inanimate surrogate mothers.

282. Mineka, Susan; Gunnar, Megan & Champoux, Maribeth. (1986). **Control and early socioemotional development: Infant rhesus monkeys reared in controllable versus uncontrollable environments.** ***Child Development,*** 57(5), 1241–1256.
Studied the effects of controllable vs uncontrollable appetitive stimulation on socioemotional development in 20 infant rhesus monkeys reared in 5 peer groups of 4 Ss each. In 2 groups (Masters), Ss had access to operant manipulanda that permitted them to control the delivery of food, water, and treats. In 2 other groups (Yoked), Ss received access to these reinforcers noncontingently. In a 5th group (Standard Rearing control), Ss were reared in a standard laboratory cage within access to the manipulanda and the variety of reinforcers available in the Master and Yoked groups. Ss were introduced to these rearing environments during the 2nd mo of life. During the 2nd half of the 1st yr, tests of socioemotional behavior commenced. Results indicate that Master Ss displayed less fear, as measured by reactions to a mechanical toy robot, and exhibited more exploratory behavior, as measured by responses to a standard primate playroom, than did Yoked Ss. Master and Yoked Ss did not differ on measures of response to social separation from peers, unless opportu-

nities to make active coping attempts were provided, in which case Master Ss appeared to adapt or cope better. On most measures, Yoked Ss did not differ from Standard Rearing controls. It is suggested that results can be attributed to the effects of experience with increased control over appetitive environmental events rather than to the effects of prolonged exposure to noncontingent or uncontrollable appetitive stimulation.

283. Mineka, Susan & Suomi, Stephen J. (1978). **Social separation in monkeys.** ***Psychological Bulletin,*** 85(6), 1376–1400.
Reviews phenomena associated with social separation from attachment objects in nonhuman primates. A biphasic protest–despair reaction to social separation is often seen in monkeys, as in human children. However, upon reunion there is generally a temporary increase in attachment behaviors rather than a temporary phase of detachment, as has been reported in the human literature. Gross factors such as age and sex do not appear to influence the responses to separation or reunion substantially. Rather, behavioral repertoires prior to separation and the nature of the separation and reunion environments appear to be more important determinants of the severity of separation reactions. These findings are consistent with the human literature. Possible long-term consequences of early separations are also discussed. Four theoretical treatments of separation phenomena are presented and evaluated: J. Bowlby's attachment-object-loss theory, I. C. Kaufman's conservation–withdrawal theory, M. E. P. Seligman's learned helplessness theory, and R. L. Solomon and J. D. Corbit's opponent-process theory.

284. Mineka, Susan; Suomi, Stephen J. & DeLizio, Roberta. (1981). **Multiple separations in adolescent monkeys: An opponent-process interpretation.** ***Journal of Experimental Psychology: General,*** 110(1), 56–85.
Examined the effects of a series of 8 short-term separations from life-long partners on 29 adolescent rhesus monkeys, some of whom had previous separation histories. In 3 experiments, groups of Ss were observed during a 6-wk baseline period, during a series of 8 wkly cycles of 4 days of separation and 3 days of reunion, and during a 6-wk recovery period. Ss included 7 3½-yr-old peer-reared Ss, 4 groups of 4 1½-yr-old peer-reared Ss (Exp II), and 6 3–4 yr old nuclear-family-reared Ss (Exp III). The changes seen within reunion periods, across repeated reunions, across repeated separations, and from baseline to recovery were all generally consistent with predictions from R. L. Soloman and J. D. Corbit's (1974) opponent-process theory of affective dynamics. Results indicate that all groups of Ss showed decreases in normal socially directed activity across repeated reunions and concurrent increases in depressive or agitated/depressive behavior across repeated separations. In addition, all groups showed increases over baseline levels in some socially directed activity on Day 1 of reunion, followed by declines in some or all of the activities by Day 3 of reunion.

285. Rapaport, Peter M. & Maier, Steven F. (1978). **Inescapable shock and food-competition dominance in rats.** ***Animal Learning & Behavior,*** 6(2), 160–165.
Demonstrated a deficit in a nonlearning task in which no aversive stimulus occurs. In Exp I, using 16 male Holtzman albino rats, inescapable shock lowered Ss' dominance in a food-competition situation relative to restrained controls. In Exp II (16 Ss), inescapable shock lowered Ss' dominance in the same food-competition situation relative to a group that received the equivalent amount of escapable shock, demonstrating that the inescapability of the shock caused at least part of the decrement observed in Exp I. In Exp III (16 Ss), inescapable shock did not cause a significant difference in food consumed or running time when the Ss were tested alone, showing it unlikely that the dominance effects were caused by decreased hunger or reduced running following inescapable shock.

286. Reese, William G. (1979). **A dog model for human psychopathology.** ***American Journal of Psychiatry,*** 136(9), 1168–1172.
Reviews studies of the development and scientific exploitation of 2 true-breeding strains of pointer dogs, one of which is basically normal and one of which is nervous, particularly around people. Basic studies, which generally contrast the nervous dogs with normal dogs, include studies of inheritance, early experience, conditioning, psychophysiology, neuropharmacology, and neurochemistry. It is suggested that the nervous line is an animal model of human psychopathology and probably of cardiac pathology.

287. Reite, Martin L. (1987). **Infant abuse and neglect: Lessons from the primate laboratory.** Special Issue: Child abuse and neglect. ***Child Abuse & Neglect,*** 11(3), 347–355.
Reviews several areas in which research on nonhuman primates contributes to an understanding of child abuse and neglect in human children. One advantage of primate studies is that the experimental method can be used to examine the short- and long-term effects of well-defined and circumscribed alterations in early experience and the manner in which they can affect later behavioral and physiological development. Four studies with monkeys are described in which short social separation experiences in infancy were associated with evidence of persistent changes in certain aspects of social behavioral functioning and immunological functioning, up to 6 yrs later, when the previously separated Ss were in late adolescence or early adulthood. (French abstract).

288. Rosellini, Robert A. & Widman, David R. (1989). **Prior exposure to stress reduces the diversity of exploratory behavior of novel objects in the rat (*Rattus norvegicus*).** ***Journal of Comparative Psychology,*** 103(4), 339–346.
Although a large literature has documented the varied effects of stress on an organism, relatively little attention has been devoted to investigating stress effects in ecologically relevant situations. Our experiment was conducted to assess the effects of stress on rats' (*Rattus norvegicus*) exploration of novel objects in a seminaturalistic and familiar environment that is relatively free of the constraints that have been placed on rats in prior investigations of stress. The results show that prior exposure to stress decreased the rats' diversity of exploration but did not affect general activity in comparison with animals not exposed to stress. We propose that the effect of stress on the qualitative aspects of exploratory behavior may be due to effects on organisms' sensitivity to predation.

289. Rosenblatt, Jay S. (1978). **Evolutionary background of human maternal behavior: Animal models.** ***Birth & the Family Journal,*** 5(4), 195–199.

Reviews animal studies with relevance to maternal–infant behavioral synchrony. Physiological and behavioral changes as well as forced separations during "critical" periods are considered in terms of their implications for enhancement or inhibition of bonding.

290. Rosenblum, Leonard A. & Paully, Gayle S. (1987). **Primate models of separation-induced depression.** ***Psychiatric Clinics of North America,*** 10(3), 437–447.
Argues that nonhuman primate response to separation, in some cases, meets the criteria for symptoms of clinical depression. A variety of social, environmental, and genetic factors can influence the separation response in terms of intensity and specific types of symptomatology. Experimental manipulations have revealed the importance of these and other variables in mediating depressive reactions. Also discussed are the effects of catecholamine depleting drugs and therapeutic agents on separation-induced anxiety, which attempt to determine underlying mechanisms of depression and how neurochemical factors may interact synergistically with environmental and psychosocial factors in determining severity and nature of depressive reactions to separation.

291. Ruppenthal, Gerald C.; Walker, Coleen G. & Sackett, Gene P. (1991). **Rearing infant monkeys (*Macaca nemestrina*) in pairs produces deficient social development compared with rearing in single cages.** ***American Journal of Primatology,*** 25(2), 103–113.
Compared the social development of pigtailed macaque infants raised in male and female pairs with that of Ss raised in individual cages. All Ss received 30 min of daily socialization in a playroom. Ss paired from postnatal 3 wks through 4 mo developed a playroom behavioral repertoire consisting largely of mutual clinging, fear, and social withdrawal. This was especially true of females. Unlike the singly caged Ss, pair-reared Ss did not successfully adapt to living in a large social group at 8–10 mo of age. In this situation, pair-reared Ss were subordinate and spent almost all of their time huddling on the pen floor. Rearing macaque infants in pairs apparently produces a behavioral repertoire that is maladaptive with respect to social development.

292. Skolnick, Neil J.; Ackerman, Sigurd H.; Hofer, Myron A. & Weiner, Herbert. (1980). **Vertical transmission of acquired ulcer susceptibility in the rat.** ***Science,*** 208(4448), 1161–1163.
Reports that premature separation of rat pups from their dams greatly increases their susceptibility to restraint-induced gastric erosions. When prematurely separated female rats grow to adulthood and mate with stock males, their normally reared F_1 progeny also have increased susceptibility to restraint-induced erosions. Cross-fostering studies show that prenatal factors in maternal behavior rather than postnatal factors transmit this susceptibility to the F_1 progeny.

293. Stevenson-Hinde, J. & Simpson, M. J. (1981). **Mothers' characteristics, interactions, and infants' characteristics.** ***Child Development,*** 52(4), 1246–1254.
Observed 25 infant–mother pairs of rhesus monkeys when infants were 8, 16, and 52 wks old. Stable characteristics of mothers in terms of Confident (C) and Excitable (E) scores were significantly positively correlated with the respective scores of their daughters but not their sons. With sons, mothers' scores were significantly negatively correlated with sons' scores. Correlations of the maternal scores with earlier mother–infant interactions suggest how mothers' characteristics could have influenced infant characteristics. In terms of attachment theory, C mothers appeared to provide a "secure base" for their daughters, but rejected their sons when very young. E mothers appeared to behave inappropriately to both sons and daughters, producing infants who may have been "insecurely attached."

294. Stevenson-Hinde, J.; Zunz, M. & Stillwell-Barnes, R. (1980). **Behaviour of one-year-old rhesus monkeys in a strange situation.** ***Animal Behaviour,*** 28(1), 266–277.
When 25 1-yr-olds and their mothers were removed from the colony to a strange situation, infants' activity was initially low compared with behavior 6 days later. Although absolute values did not differ between males and females, correlations did. That is, 3 days after return to the colony, correlations with baseline colony behavior were significantly positive for females, while males had 1 significantly negative correlation. Males had more significant correlations than females when behavior in the strange situation was correlated with colony scores. Of the 6 individuals that had adverse early experience, 5 showed extreme behavior. The exception was the son of a female with a very low "excitable" score. In fact for all males, mothers' colony scores were significantly correlated with sons' behavior in the strange situation.

295. Suomi, Stephen J. (1991). **Adolescent depression and depressive symptoms: Insights from longitudinal studies with rhesus monkeys.** Special Issue: The emergence of depressive symptoms during adolescence. ***Journal of Youth & Adolescence,*** 20(2), 273–287.
Presents data suggesting that insights regarding adolescent depressive phenomena in humans can be provided through systematic studies of nonhuman primates such as rhesus monkeys. Species-normative patterns of social changes that emerge as rhesus monkeys pass through puberty are described. Developmental changes in depressive-like behavioral and physiological response to separation shown by monkeys as they become adolescents are outlined. Issues of developmental continuity and risk factors for depressive symptomatology are discussed, and the issue of sex differences that emerge in adolescence is considered.

296. Suomi, Stephen J.; Collins, Mary L.; Harlow, Harry F. & Ruppenthal, Gerald C. (1976). **Effects of maternal and peer separations on young monkeys.** ***Journal of Child Psychology & Psychiatry & Allied Disciplines,*** 17(2), 101–112.
Eight rhesus monkeys reared with mothers and peers for the 1st 6 mo of life were subjected to 3 consecutive 2-wk periods of social separations, progressively increasing the degree of social isolation for the infants. During each period of social separation—mesh permitting physical contact with mothers, physical but not visual separation from mothers but unlimited access to peers, and finally physical separation from both mothers and peers—clear patterns of protest and despair were recorded. Reactions to mesh separations were mild relative to the latter 2 separations, both equally debilitating. Following reunion, separated Ss returned to behavior patterns similar to those of nonseparated controls. Later in the study mothers were removed permanently from the homecages of all Ss, and all Ss reacted with protest and cessation of play activity. However, there was little evidence of despair and previously separated and controls reacted in similar form despite their different early histories.

297. Timmermans, Paul J.; Roder, Else L. & Hunting, Paul. (1986). **The effect of absence of the mother on the acquisition of phobic behaviour in cynomolgus monkeys (*Macaca fascicularis*).** ***Behaviour Research & Therapy,*** 24(1), 67–72.
Examined the phobic behavior of 6 monkeys reared in a peer group with surrogate mothers in comparison to the behavior of 6 monkeys reared by their parents for the 1st yr. Monkeys reared in a peer group with surrogate mothers avoided a large unfamiliar object provided with food that was presented to them at the age of 1 yr. Peers reared by their parents approached the same object and took food from it. This difference in behavior appeared to persist for years, although no aversive experience whatsoever had been conditioned to the object. In adulthood, the difference between the 2 groups was absolute; the mother-reared monkeys all approached the object, whereas the monkeys reared by surrogate mothers all avoided the object. The behavioral difference was tested systematically, and an explanation of its origin is presented in terms of G. W. Bronson's (1968) theory concerning the phasic development of fear. The phenomenon under study shows a striking resemblance to human phobic behavior and offers a new animal analog for the study of the acquisition and therapy of human phobias.

298. Troisi, Alfonso; Aureli, Filippo; Piovesan, Paola & D'Amato, Francesca R. (1989). **Severity of early separation and later abusive mothering in monkeys: What is the pathogenic threshold?** ***Journal of Child Psychology & Psychiatry & Allied Disciplines,*** 30(2), 277–284.
Presents a nonexperimentally induced case of infant abuse by a macaque mother living in a stable social group. The abusive mother had been abandoned right after birth by her biological mother, adopted, and reared by an adoptive mother who provided adequate maternal care. The abusive mother alternated violent abuse and attentive maternal care and was shown to have a very possessive relationship with her infant. This study raises questions about etiology of primate infant abuse and supports the view that maternal anxiety may play a role in precipitating abuse.

299. Wiepkema, P. R.; Van Hellemond, K. K.; Roessingh, P. & Romberg, H. (1987). **Behaviour and abomasal damage in individual veal calves.** ***Applied Animal Behaviour Science,*** 18(3–4), 257–268.
Analyzed the relationship between abomasal (digestive stomach) damage and the individual behavioral patterns of crated veal calves fed with milk replacer only. No relationship was found between abomasal damage and anxiety levels (measured by startle response). Two stereotypies were observed at age 12–20 wks. Biting/licking decreased during this period and tongue-playing increased. Ss that developed tongue-playing had no ulcers or scars, while those that did not develop tongue-playing all showed damage.

300. Winslow, James T. & Insel, Thomas R. (1991). **Infant rat separation is a sensitive test for novel anxiolytics.** ***Progress in Neuro-Psychopharmacology & Biological Psychiatry,*** 15(6), 745–757.
Proposes the rat pup ultrasonic vocalization as a new paradigm for the study of anxiolytics (ANXs) since this test has been shown to detect a wide range of ANXs and may be useful for investigating the ontogeny and physiology of anxiety. The ultrasonic calls that rat pups emit during brief episodes of social separation have been variously described as distress calls and may be related to the separation cries expressed by the young of many mammalian species. These calls are modulated by environmental stimuli such as ambient temperature and olfactory and tactile stimuli associated with the nest. Calls are also sensitive to a variety of purported ANX and anxiogenic drugs, including the benzodiazepines, serotonin agonists, and ligands at the N-methyl-D-aspartate/glycine receptor complex.

Physiological Psychology & Neuroscience

301. Abercrombie, Elizabeth D.; Bonatz, Alfred E. & Zigmond, Michael J. (1990). **Effects of L-DOPA on extracellular dopamine in striatum of normal and 6-hydroxydopamine-treated rats.** ***Brain Research,*** 525(1), 36–44.
Administration of L-3,4-dihydroxyphenylalanine (L-DOPA) in an animal model of Parkinson's disease, involving striatum of adult male rats treated with the catecholaminergic neurotoxin 6-hydroxydopamine (6-OHDA), led to increases in striatal extracellular dopamine (DA) level, as measured with in vivo microdialysis. These increases were greater than elevations observed in male rats with intact striatum. This phenomenon is attributed to the formation of DA from L-DOPA in residual DA terminals and in nondopaminergic striatal elements coupled with a decrease in the capacity for efficient DA inactivation due to the loss of high-affinity DA uptake sites. Such a phenomenon may produce oxidative stress in the parkinsonian striatum that contributes to the loss of therapeutic efficacy of L-DOPA with long-term use.

302. Ackerman, Sigurd H.; Hofer, Myron A. & Weiner, Herbert. (1975). **Age at maternal separation and gastric erosion susceptibility in the rat.** ***Psychosomatic Medicine,*** 37(2), 180–184.
Studied the development of susceptibility to immobilization-induced gastric erosions in 54 Wistar rats previously separated from their mothers at 15, 21, or 25 days of age. Early separation (Day 15) produced Ss whose maximum susceptibility occurred at a much younger age, generated a susceptibility curve over life that was the inverse of the curve for Ss separated later, and led to severe gastrointestinal hemorrhage as a common and distinguishing complication in younger Ss. The pathogenesis of erosion formation in early separated Ss may be unique in that, for that group only, the food deprivation component of the immobilization paradigm, when presented alone, also produced erosion of the glandular stomach, with hemorrhage.

303. Ackerman, Sigurd H. & Shindledecker, Richard D. (1987). **Chronobiologic factors in experimental stress ulcer.** Special Issue: Chronobiology and ulcerogenesis. ***Chronobiology International,*** 4(1), 3–9.
Reviews the English language literature from 1960 to 1986 concerning aspects of chronobiologic factors in stress ulcer. Both the *Index Medicus* and the files of Medlar and Medline were consulted, and personal contacts were made with several leaders in the field of chronobiology. The literature focuses on circadian rhythms (rhythms related to light–dark cycles, activity, and body temperature).

304. Adkins-Regan, Elizabeth. (1988). **Sex hormones and sexual orientation in animals.** Special Issue: Sexual differentiation and gender-related behaviors. ***Psychobiology,*** 16(4), 335–347.

Examines whether sex hormone exposure during prenatal or early postnatal development determines the adult sexual orientation (sexual preferences) of animals. In rats, hamsters, ferrets, pigs, zebra finches, and possibly dogs, either early castration of males or early testosterone treatment of females or both have been shown to change or reverse sexual orientation. The effects of early sex hormone exposure on copulatory behavior, studies of prenatal hormones and sexual orientation in humans, different animal models for research on sexual orientation, and the desirability of a broad psychobiological investigation of sexual orientation in animals are discussed.

305. Ahlers, Stephen T. & Riccio, David C. (1987). **Anterograde amnesia induced by hyperthermia in rats.** ***Behavioral Neuroscience,*** 101(3), 333–340.
Anterograde amnesia (AA), forgetting of events that occur following a traumatic episode, has recently been demonstrated by using hypothermia as the amnestic agent. However, no data currently exist to indicate if hyperthermia might affect memory processing in a similar manner. Exps 1 and 2 demonstrated that increasing the colonic body temperature of the rat to 3–4°C or more above normal during avoidance training produced a significant retention loss when the test occurred 24 hrs after training. In Exp 3, AA resulting from an elevation in temperature was reversed by reheating "amnestic" Ss just prior to the 24-hr test. By rapidly reversing hyperthermia immediately after the training trial with a cooling procedure, Exp 4 demonstrated that hyperthermia-induced AA was not the result of retrograde influences of the heating treatment. Results are discussed in terms of possible retention deficits which could conceivably follow environmental heat stress or fever hyperthermia resulting from bacterial infection.

306. Alonso, S.; Navarro, E.; Castellano, M. A.; García, C. et al. (1991). **Animal models of depression.** ***Psiquis: Revista de Psiquiatría, Psicología y Psicosomática,*** 12(3), 35–50.
Describes behavioral and biochemical animal models and tests used to study depression. Behavioral models include, among others, separation, social isolation, and learned helplessness. Biochemical tests include those using serotonin, dopamine, and noradrenaline. Implications of these tests for depression in humans are also discussed. (English abstract).

307. Alonso, S. Josefina; Castellano, Miguel A.; Afonso, D. & Rodriguez, Manuel. (1991). **Sex differences in behavioral despair: Relationships between behavioral despair and open field activity.** ***Physiology & Behavior,*** 49(1), 69–72.
Examined sex differences in depression using 2 animal models. Both the R. D. Porsolt et al (1977) test and the L. A. Hilakivi and I. Hilakivi (1987) forced swimming test have shown that the duration of immobility is higher in male than in female rats. Sexual differences in the animal models of depression are probably unrelated to general activity differences because there is no significant correlation between activity in both tests. However, the correlation between the 2 models of depression used reached significance. Immobility levels in the Porsolt test were similiar in the different stages of the estrous cycle.

308. Altenor, Aidan & Kay, Edwin J. (1980). **The effects of postweaning rearing conditions on the response to stressful tasks in the rat.** ***Physiological Psychology,*** 8(1), 88–92.
Examined the effects of individual- and group-housing conditions in 3 experiments. In Exp I, individually housed (IH) Ss were more susceptible to sudden death by drowning than were group-housed Ss, and daily handling increased this susceptibility. In Exp II, the test was one typically used in "learned helplessness" experiments: escape from shock. IH Ss took longer to escape from shock than did Ss housed in groups. In Exp III, IH and group-housed Ss were handled and given either no further treatment of ECS or a change in housing conditions before being tested for escape from shock. IH Ss throughout were slower to escape shock than were each of the other 5 groups. Findings suggest that there are both behavioral and physiological similarities between the operations of individual housing and exposure to inescapable shock. Explanations in terms of learned helplessness and noradrenergic depletion are discussed.

309. Anderson, Brenda J. et al. (1985). **Prenatal exposure to aluminum or stress: II. Behavioral and performance effects.** ***Bulletin of the Psychonomic Society,*** 23(6), 524–526.
Presents results of shock-elicited aggression and learned helplessness testing of 129 offspring of Holtzman rats exposed to stress, aluminum, or a control condition during gestation. Compared to controls, aluminum- and stress-exposed Ss displayed significantly more aggressive responses. However, aluminum-exposed Ss spent significantly less time in contact with the target rod during aggression testing and had significantly longer latencies than did stress-exposed Ss during the escape-training phase of the learned helplessness study. Results indicate that the prenatal treatments produced lasting behavioral effects, including the disruption of a response inhibition/direction mechanism in the aluminum-exposed Ss.

310. Anderson, G. Harvey; Leprohon, Carol; Chambers, John W. & Coscina, Donald V. (1979). **Intact regulation of protein intake during the development of hypothalamic or genetic obesity in rats.** ***Physiology & Behavior,*** 23(4), 751–755.
Results of 2 experiments show that both medial hypothalamic lesioned female Wistar rats and spontaneously hyperphagic Zucker rats became obese by consuming more of a low protein diet and less of a high protein diet. Results imply that the stimulus for hyperphagia in both animal models of obesity is some physiological and/or behavioral error in energy regulation but not protein regulation.

311. Andrade, Luiz A.; Lima, José G.; Tufik, Sérgio; Bertolucci, Paulo H. et al. (1987). **REM sleep deprivation in an experimental model of Parkinson's disease.** ***Arquivos de Neuro-Psiquiatria,*** 45(3), 217–223.
Male rats were bilaterally lesioned in the nigrostriatal pathway through a stereotaxically directed electrical current. Seven days postsurgery, the Ss were REM sleep deprived (SD) for 72 hrs and immediately after the end of this period were observed in an open field for ambulation, rearing, grooming, and latency. In comparison with nondeprived Ss there was a significant increase in ambulation and rearing, a response that appeared again after a 2nd REM SD period on Day 21 postsurgery. Findings demonstrating improvement of 2 parameters of an experimental model of Parkinson's disease suggest that SD may be useful in this condition. (Portuguese abstract).

312. Appleton, D. B. & DeVivo, D. C. (1974). **An animal model for the ketogenic diet: Electroconvulsive threshold and biochemical alterations consequent upon a high-fat diet.** ***Epilepsia,*** 15(2), 211–227.
Placed adult male albino Charles River rats on a high-fat diet. After 10 days the electrical stimulus necessary to produce a minimal convulsion began to rise, reaching a maximum at about 20 days. The convulsive threshold did not change in Ss on a high-carbohydrate diet. When the diet was changed from high-fat to high-carbohydrate, the threshold reverted rapidly to prestudy levels. Biochemical studies revealed the high-fat diet to be ketogenic as the blood concentrations of dextro-b-hydroxybutyrate and acetoacetate were elevated. Serum chloride, triglycerides, esterified fatty acids, and total lipids were also elevated, but the serum cholesterol was lowered. In brain, the only changes were elevations of dextro-b-hydroxybutyrate and sodium and lowering of the acetoacetyl CoA transferase activity. It is suggested that ketosis alters cerebral metabolism of glucose leading to an elevation in the electroconvulsive threshold. (French, German & Spanish summaries).

313. Armario, Antonio; Restrepo, Carlos & López-Calderon, Asunción. (1988). **Effect of a chronic stress model of depression on basal and acute stress levels of LH and prolactin in adult male rats.** ***Biological Psychiatry,*** 24(4), 447–450.
Chronic stress did not alter basal levels of luteinizing hormone (LH) or prolactin (PRL) in adult male rats relative to unstressed controls. However, slightly reduced LH and PRL levels were observed in response to a novel acute stressor in chronically stressed Ss. Data indicate that a relatively mild chronic stress model that induces some depressivelike behavioral changes also induces abnormalities in anterior pituitary gland secretion that are mediated in the central nervous system (CNS).

314. Bankiewicz, K. S.; Oldfield, E. H.; Plunkett, R. J.; Schuette, W. H. et al. (1991). **Apparent unilateral visual neglect in MPTP-hemiparkinsonian monkeys is due to delayed initiation of motion.** ***Brain Research,*** 541(1), 98–102.
Made 8 monkeys hemiparkinsonian by infusing a solution of 1-methyl-4-phenyl-1,2,3,6-tetrahydropyridine (MPTP) into one carotid artery. Ss appeared to ignore food presented from the contralateral side, and initial observations suggested neglect of visual stimuli presented as fruit treats in the half-field contralateral to MPTP treatment. Further evaluation in which fruit treats were left in the "neglected" visual field indicated that this apparent neglect, unlike neglect attending cortical lesions, was rather a marked delay in initiating movements (unilateral hypokinesia). These observations may explain apparent subcortical neglect and are consistent with the role of nigrostriatal dopaminergic neurones in movement regulation. This is a useful animal model in which difficulties in initiation of movement (hypokinesia), a cardinal symptom of Parkinson's disease, can be studied separately from other deficits in motor performance.

315. Barbaree, H. E. & Harding, R. K. (1973). **Free-operant avoidance behavior and gastric ulceration in rats.** ***Physiology & Behavior,*** 11(2), 269–271.
14 Fischer 344 young adult female rats trained to perform a free-operant wheel-turn avoidance response had gastric lesions but their yoked controls did not. The study constitutes a replication of J. V. Brady's "Executive" ulcer experiment but with random assignment of animals to conditions and demonstrates that a free-operant avoidance schedule is ulcerogenic.

316. Barnes, Deborah M. (1986). **Tight money squeezes out animal models.** ***Science,*** 232(4748), 309–311.
Discusses funding restrictions that have resulted in an abandonment of animal strains used in biomedical research. Particular problems are noted with respect to the use of animal models based on cats, dogs, and primates (which have also been targeted by animal-rights groups) used to investigate normal and pathological neurobiological functioning in humans.

317. Baum, Morrie. (1986). **An animal model for situational panic attacks.** ***Behaviour Research & Therapy,*** 24(5), 509–512.
Presents an animal model of human panic attacks that is based on the abortive avoidance activity (AAA) displayed by rats undergoing flooding. A shortcoming of the model is noted in the difficulty involved in assessing the physiological correlates of AAA in rats during flooding procedures. The use of drugs, behavioral distractions, and intracranial stimulation to suppress AAA is discussed.

318. Bernardi, Mara; Genedani, Susanna; Tagliavini, Simonetta & Bertolini, Alfio. (1989). **Effect of castration and testosterone in experimental models of depression in mice.** ***Behavioral Neuroscience,*** 103(5), 1148–1150.
In the behavioral despair (forced swimming) test and in the tail-suspension test, long-term (30–32 days) castration significantly increased the duration of immobility in mice. Testosterone propionate (1 or 10 mg/kg/day^{-1} sc for 4 days), although not affecting the duration of immobility in sham-operated mice, reduced the duration of immobility in castrated mice to within normal limits. Desipramine (20 mg/kg ip) decreased the duration of immobility both in sham-operated and in castrated animals. These results indicate that castration favors an inactive behavior and that testosterone, although having no "antidepressant" effect per se, is necessary for the male animal to cope normally with adverse environmental situations.

319. Bernardi, Mara; Vergoni, A. V.; Sandrini, M.; Tagliavini, Simonetta et al. (1989). **Influence of ovariectomy, estradiol and progesterone on the behavior of mice in an experimental model of depression.** ***Physiology & Behavior,*** 45(5), 1067–1068.
In the tail suspension test (an animal model of depression) the duration of immobility during the 6 min of observation was 56.84 ± 6.54 sec in sham-ovariectomized mice and 113.11 ± 7.86 sec 30–32 days after ovariectomy. Estradiol (10, 100 or 1,000 µg/kg) and progesterone (50, 1,000 or 10,000 µg/kg), sc injected daily 4 times before the test, restored the duration of immobility in ovariectomized mice to normal, while having no effect on sham-operated animals. Desipramine (20 mg/kg ip 1 hr before testing) significantly reduced the duration of immobility both in ovariectomized and in sham-operated mice. Data indicate that ovarian sex

hormones, while having no antidepressant, desipramine-like, effect on the behavior of intact adult female mice, have such an effect in ovariectomized mice, and enable the animal to cope in a normal way with adverse environmental situations.

320. Bignami, Giorgio. (1989). **Experimental models in the animal and clinical models in psychiatry.** ***Confrontations Psychiatriques,*** 30, 179–199.
Reviews requirements and problems that may be encountered in developing experimental animal models and human clinical models of mental illness. The ideal model and different levels of this model are described, including the phenomenologic, etiologic, pathogenic, and heuristic levels. Different definitions used to classify animal and human models are also presented, including the model as a body of knowledge, as an animal species that can replace humans as research objects, as an analysis of form and function, and as an assemblage of symptoms. The use of analogies, homologies, correlations, and treatment effects in developing models and the apparent, construct, and predictive validity of models are also discussed.

321. Billewicz-Stankiewicz, Jaroslaw; Gorny, Dionizy & Zajaczkowska, Malgorzata. (1972). **Secretory and vascular changes in the rat stomach after compulsory swimming and vagotomy.** ***Acta Physiologica Polonica,*** 23(2), 249–253.
Produced gastric ulcers in 111 white male rats by vagotomy; 34 Ss were also given compulsory swimming. Disconnection of the vagus nerves inhibited gastric secretion but did not prevent development of ulcers in the gastric mucosa. Compulsory swimming intensified gastric secretion. Decentralization of the stomach also contributed to the appearance of vascular changes in the form of subserous extravasations which intensified after swimming and also appeared in the submucous layer. It is believed that the vascular changes appearing after severance of the vagus nerves resulted from trophic changes in gastric tissues caused by the contraction of capillaries.

322. Blanchard, D. Caroline & Blanchard, Robert J. (1988). **Ethoexperimental approaches to the biology of emotion.** ***Annual Review of Psychology,*** 39, 43–68.
Reviews research literature (1872–1987) on the relationship of certain biological systems to aggression, anxiety, and fear. The applicability of an ethoexperimental approach in using lower mammals as models for research on human behavior is highlighted. Research areas examined include (1) natural patterns of offense manipulated and measured in laboratory settings and (2) the biological bases of defense and offense. The research suggests that careful attention to specific behavior patterns is essential to an understanding of the relationship between biological and behavioral systems.

323. Błaszczyk, Janusz W. & Dobrzecka, Czesława. (1989). **Alteration in the pattern of locomotion following a partial movement restraint in puppies.** ***Acta Neurobiologiae Experimentalis,*** 49(1), 39–46.
Studied the pattern of locomotion following partial movement restraint in 5 mongrel puppies. Ss were forced to pace during the restraint period and exhibited significant, time-dependent gait alterations after removal of the restraint. Ss gradually recovered to normal trotting, but the time to recovery depended on the duration of movement restraint. One S forced to practice pacing for only 2 mo switched almost instantaneously to normal trotting, whereas longer periods of selective movement restraint caused longer-lasting gait alterations. Results are discussed as a potential animal model of the plasticity of the motor system.

324. Blundell, J. E. (1987). **Nutritional manipulations for altering food intake: Towards a causal model of experimental obesity.** ***Annals of the New York Academy of Sciences,*** 499, 144–155.
Reviews dietary manipulations in experimental animal models to help define variables associated with the development and maintenance of obesity. Research is presented regarding dietary contents and feeding paradigms. A causal model for the development and maintenance of obesity is proposed, which distinguishes between vulnerability, protection, provoking, and amplification factors. The model proposes that obesity arises from combinations of factors rather than from unique causes. A similar conceptualization for human obesity may help in the development of therapeutic strategies.

325. Boles, John & Russell, Roger W. (1970). **Relations between the electrogastrogram and gastric ulceration during exposure to stress.** ***Psychophysiology,*** 6(4), 404–410.
Studied relations between development of gastric ulceration and changes in motility of the stomach, as recorded by the electrogastrogram (EGG), under conditions where 30 naive male Cox rats were exposed to physical (immobilization) and chemical (reserpine) stressors and to the 2 in combination. Following 36 hrs of the various treatments, all restraint plus reserpine Ss were ulcerated; 60% of the restraint-only Ss had lesions; and, only 3 of 10 no-restraint Ss showed gastric damage. The same rank order was obtained for the mean number of lesions/group. There were no significant differences among groups in total acidity of the stomach at autopsy. The restraint-only groups showed a trend toward increased gastric motility during the treatment period, while the restraint plus reserpine groups showed a progressive decrease in motility, the divergence leading to statistically significant differences as exposure to the different stressors continued. Results do not support the hypothesis that the experimentally-produced lesions are related to an increase in gastric acidity; they do support the hypothesis that the ulceration is related to changes in gastric motility, more severe lesions appearing in Ss whose motility decreased during the treatment period.

326. Booth, David A. & Miller, Neal E. (1969). **Lateral hypothalamus mediated effects of a food signal on blood glucose concentration.** ***Physiology & Behavior,*** 4(6), 1003–1009.
Found a distinctive biphasic hyperglycemia, similar to that observed after electrical stimulation of the lateral hypothalamus in the region of the feeding system, in 22 hungry Holtzman and Charles River male albino rats after a 3-min presentation of an unreinforced visual discrimination stimulus (DS) during which Ss pressed a lever for food reinforcement at variable intervals, as they had been trained under food deprivation. A monophasic hyperglycemia was seen after presentation of the same visual stimulus to undeprived Ss trained to escape or avoid foot shock by pressing the lever during DS presentation. No appreciable blood glucose variations were seen with a water DS under thirst or with the visual stimulus before training. Food deprivation was necessary to the oscillatory glucose response to the food DS. The response was eliminated by injections of novocaine

bilaterally through chronic intrahypothalamic cannulae which were also capable of eliciting eating when used to guide noradrenaline injections. Results provide further evidence that the feeding system mediates direct metabolic effects as well as behavioral control.

327. Borsini, Franco; Lecci, Alessandro; Volterra, Giovanna & Meli, Alberto. (1989). **A model to measure anticipatory anxiety in mice?** ***Psychopharmacology,*** 98(2), 207–211.
Among male mice from the same cage, Ss removed last had a higher body temperature than Ss removed first. This phenomenon (1) persisted 2 and 24 hrs later; (2) was present regardless of the number of Ss in each cage; and (3) could be observed by reversing the order of removal from the cage. When fewer Ss were allocated to a cage, the percentage of hyperthermic Ss increased. The rise in rectal temperature of Ss removed last from a less crowded cage was prevented by po diazepam and nitrazepam and was observed in a greater percentage of Ss following sc yohimbine treatment. This model may be used in the study of the psychological basis of anxiety, especially the alarm reaction due to emotional stimuli.

328. Bourgeois, Marc. (1987). **Are there animal models of suicidal behavior?** XVIIIth Meeting of the Group for the Study and Prevention of Suicide: Suicide and institution (1986, Bordeaux, France). ***Psychologie Medicale,*** 19(5), 739–740.
Presents several animal models of suicidal behavior available to ethological research (e.g., unfavorable dispersions due to demographic pressures or human encroachment, "altruistic" sacrifice to protect the group, grief over loss of a beloved owner). Laboratory models of learned helplessness (with rats, dogs, and primates), "anaclitic depression" in monkeys, and biological anomalies can induce suicidal behaviors. (English abstract).

329. Boyd, Suellyn C.; Caul, William F. & Bowen, Bruce K. (1977). **Use of cold-restraint to examine psychological factors in gastric ulceration.** ***Physiology & Behavior,*** 18(5), 865–870.
Five experiments examined the influence of prior classical conditioning and predictability of shock on stress-induced gastric ulceration produced by cold-restraint in a total of 400 Holtzman and 48 Sprague-Dawley male rats. Optimal durations of cold-restraint treatment were first determined. Secondly, the ulcerogenic effect of reserpine and the ulcer-inhibiting influence of atropine were demonstrated using cold-restraint. Next, it was found that neither prior classical conditioning nor the psychological manipulation of predictability of shock affected gastric ulceration produced by cold-restraint. The final study suggested the hypothesis that hypothermia produced by cold-restraint inhibits the influence of the psychological factors on stress ulceration. Cold-restraint is viewed as an inappropriate technique to use in the further examination of the influence of specific psychological factors on stress-induced ulceration.

330. Boyd, Suellyn C.; Sasame, Henry A. & Boyd, Michael R. (1981). **Effects of cold-restraint stress on rat gastric and hepatic glutathione: A potential determinant of response to chemical carcinogens.** ***Physiology & Behavior,*** 27(2), 377–379.
Demonstrated stressor-induced decreases in glutathione (GSH) concentrations in the liver and the gastric mucosa of male Sprague-Dawley rats. Either food deprivation and/or the time of day the stressor was administered could affect the extent of stressor-induced decreases in GSH. The system described may be useful for further exploration of potential stressor-induced alterations in susceptibility to chemical carcinogens and/or other toxins whose expression of biological activity may be dependent on GSH.

331. Brown, Gary E. & Stroup, Kevin. (1988). **Learned helplessness in the cockroach (*Periplaneta americana*).** ***Behavioral & Neural Biology,*** 50(2), 246–250.
For 3 consecutive days cockroaches were exposed to either escapable, inescapable, or no shock in an escape task. When tested 24 hrs later in a shuttlebox escape task, reliable differences were found between escapable or inescapable Ss and between inescapable and control Ss in both escape latencies and the number of failures to escape.

332. Buchanan, Denton C. & Caul, William F. (1974). **Gastric ulceration in rats induced by self-imposed immobilization or physical restraint.** ***Physiology & Behavior,*** 13(4), 583–588.
Assessed gastric ulceration in 12 male Holtzman rats after they were placed in a self-immobilization condition in which the unrestrained S could avoid footshock by remaining motionless. 12 Ss received equivalent but unavoidable footshock which was not contingent on movement within the apparatus. A 3rd group of 12 received no shock and served as a deprivation control. The 1st 2 groups showed a similar degree of ulceration which was greater than that observed in the deprivation control Ss. Two methods of physical restraint were also yoked in series to the shock circuit of the self-imposed condition to assess the influence of different forms of restraint. In 2 groups of 12 Ss, restraint via the legs plus shock condition produced a greater degree of ulceration than did whole body restraint plus shock. Heart rate and body weight changes were evaluated for each of the 5 conditions.

333. Calvino, Bernard; Crepon-Bernard, Marie-Odile & le Bars, Daniel. (1987). **Parallel clinical and behavioural studies of adjuvant-induced arthritis in the rat: Possible relationship with "chronic pain."** ***Behavioural Brain Research,*** 24(1), 11–29.
Studied the course of adjuvant-induced arthritis over an 11-wk postinoculation period in 198 6-wk-old male Sprague-Dawley rats using clinical observations (e.g., body weight, diameters of radiocarpal and tibiotarsal joints, radiological analysis of forepaws) and behavioral observations (e.g., mobility, exploring, rearing). Findings show that the acute phase of the disease (Weeks 2–4) was characterized by the onset of clinical symptoms and dramatic changes in behavior, all of which could be related to the occurrence of pain; signs of pain were still present, albeit weaker, in the post-acute phase (Weeks 5–8).

334. Canver, Charles C. (1985). **Chlorpromazine in experimental gastric ulcers induced by restraint and cold stress.** ***New York State Journal of Medicine,*** 85(3), 101–102.
In three random groups of male Swiss albino mice ($N = 35$) arbitrarily given chlorpromazine (CPZ) in low (2.5 mg/100 g), medium (5 mg/100 g), and high im doses (10 mg/100 g) 30 min prior to cold stress, CPZ inhibited the development of gastric ulcers. CPZ induced characteristic cataleptic im-

mobility in higher doses that resulted in lowered vulnerability to the restraint and cold stress. In low, medium, and high doses, CPZ increased titratable gastric acidity. It is suggested that this secretory influence during stress is predominant in the mechanism of protective action of this drug.

335. Capell, Howard; Ginsberg, Ronald & Webster, C. D. (1972). **Amphetamine and conditioned "anxiety."** ***British Journal of Pharmacology,*** 45(3), 525–531.
16 male Wistar rats bar pressed for milk reward at a steady rate, but this baseline responding was suppressed in the presence of an auditory stimulus associated with electric shock (conditioned suppression). Effects of (+)-amphetamine sulphate on this conditioned suppression were studied in 2 experiments. At doses of .5, 1.0, or 2.0 mg/kg, it reduced the baseline rate of responding and also reduced the conditioned suppression. Both these effects were dose related. In a further experiment, effects of 1.0 mg/kg on 2 levels of conditioned suppression were studied. Regardless of its degree, (+)-amphetamine attenuated suppression. Results are compared to previous research which found that amphetamine increased baseline responding and exaggerated conditioned suppression. It is concluded that the conditioned suppression procedure should be used with caution as an animal model of anxiety in psychopharmacological investigations.

336. Capitanio, John P. & Lerche, Nicholas W. (1991). **Psychosocial factors and disease progression in simian AIDS: A preliminary report.** ***AIDS,*** 5(9), 1103–1106.
Studied 22 rhesus macaques (aged 9–42 mo) who had been experimentally infected with simian immunodeficiency virus (SIV), an immunodeficiency syndrome that closely parallels human immunodeficiency virus (HIV) infection. After statistically controlling for variables that might have an obvious impact on the latencies to display hematological and physical signs of simian acquired immune deficiency syndrome (AIDS), psychosocial variables pertaining to rearing conditions, separation history, and cage-move history explained significant variance in outcome measures following SIV infection. In particular, separation from familiar companions, such as the primary attachment figure and/or others, prior to inoculation resulted in shorter latencies to leukopenia and lymphopenia, as well as an increased likelihood of weight loss > 10%. Data suggest that such psychosocial stressors may exacerbate the pathogenic effect of SIV.

337. Caputo, Daniel V., et al. (1968). **Housing modification as a variable in fasting-induced ulcerogenesis.** ***Journal of Psychosomatic Research,*** 12(2), 129–135.
Assessed the effect of a modification in housing conditions on susceptibility to gastric ulcerogenesis in mice in 2 experiments. A significantly greater incidence of lesion development was found for those Ss that had experienced housing modification. Food consumption was influenced by the altered housing conditions while weight and activity levels were not. Results are discussed in the context of activation theory. It is concluded that, in concurrence with other suggestions, there may be different "varieties" of arousal.

338. Carey, Robert J. (1990). **Dopamine receptors mediate drug-induced but not Pavlovian conditioned contralateral rotation in the unilateral 6-OHDA animal model.** ***Brain Research,*** 515(1–2), 292–298.
Following Pavlovian conditioning treatment sessions with apomorphine (APO), rats receiving this paired treatment showed substantial contralateral rotation (COR) when placed without drugs into the test environment previously paired to the APO injection. Ss in an unpaired control treatment showed only ipsilateral rotation (IPR). Subsequent tests with SCH 23390 or haloperidol partially suppressed the APO-induced COR response; combined treatment with the D_1-D_2 antagonists completely suppressed this response. The same COR response in the paired APO treatment group was not attenuated by dopamine receptor blockage. In both paired and unpaired groups, the spontaneous IPR response was completely blocked.

339. Castañeda, Edward; Becker, Jill B. & Robinson, Terry E. (1988). **The long-term effects of repeated amphetamine treatment *in vivo* on amphetamine, KCl and electrical stimulation evoked striatal dopamine release *in vitro.*** ***Life Sciences,*** 42(24), 2447–2456.
Results of the present study, using *in vitro* striatum sections from ovariectomized rats, support the hypothesis that an enhancement in striatal dopamine release may be at least partly responsible for the behavioral sensitization seen in amphetamine (AM)-pretreated animals and perhaps even the hypersensitivity to the psychotogenic effects of AM seen in former AM addicts.

340. Caul, William F.; Buchanan, Denton C. & Hays, Robert C. (1972). **Effects of unpredictability of shock on incidence of gastric lesions and heart rate in immobilized rats.** ***Physiology & Behavior,*** 8(4), 669–672.
Assessed degree of gastric ulceration and heart rate over a 19-hr immobilization period during which 5 groups of 12 male Holtzman rats received (a) CS presentations, (b) CS presentations 50% of which were paired with a UCS, (c) CS presentations all of which were paired with the UCS, (d) 50% CS–UCS pairings with the number of UCS presentations equal to Group c, or (e) unrelated presentations of the CS and UCS. The unpredictable condition of Group e resulted in significantly greater ulceration than seen in Group a with the other groups falling between these extremes. While Groups b, c, and d showed reliable and equivalent bradycardic responses to the CS, the pre-CS heart rate measure failed to discriminate among groups.

341. Chance, William T.; von Meyenfeldt, Maarten F. & Fischer, Josef E. (1983). **Changes in brain amines associated with cancer anorexia.** ***Neuroscience & Biobehavioral Reviews,*** 7(4), 471–479.
Analysis of indole amine metabolism within acute (Walker 256 carcinosarcoma) and chronic (methycholanthrene-induced sarcoma) animal models of cancer anorexia demonstrated elevated levels of plasma free tryptophan, whole brain tryptophan, serotonin, and 5-hydroxyindoleacetic acid in anorectic tumor-bearing female Sprague-Dawley rats. Whole brain levels of catecholamines were not changed within either tumor line; regional CNS determination of tryptophan metabolism in Ss with Walker 256 tumors revealed elevated tryptophan in the hypothalamus, corpus striatum, mesencephalon, diencephalon, cerebellum, and cortex; increased serotonin in the diencephalon and cerebellum; and elevated 5-hydroxyindoleacetic acid in the diencephalon, hippocampus, pons-medulla, cerebellum, and cortex. Findings suggest that the increased serotonin metabolism observed in tumor-bearing rats is involved in the anorexia of cancer.

342. Cheal, MaryLou. (1986). **The gerbil: A unique model for research on aging.** ***Experimental Aging Research,*** 12(1), 3–21.
Suggests that the Mongolian gerbil (*Meriones unguicultatus*) be used as a model for aging research because of its unique physiological attributes, ease of handling, and the current availability of data. Factors that demonstrate the gerbils' suitability in fulfilling practical and scientific considerations important in determining a model for aging research are listed. Behavioral processes considered include agonistic behavior, vigilance and attention, taste and hearing, learning, and environmental enrichment. Several unique physiological attributes of gerbils are described. Based on these attributes and on review of research in gerbils, it is suggested that gerbils can serve as animal models for behavioral and biological processes and for normative and pathological aspects of aging.

343. Cho, C. H.; Fong, L. Y.; Ma, P. C. & Ogle, C. W. (1987). **Zinc deficiency: Its role in gastric secretion and stress-induced gastric ulceration in rats.** ***Pharmacology, Biochemistry & Behavior,*** 26(2), 293–297.
Studied the effects of zinc deficiency on gastric secretion and on cold-restraint stress-induced ulceration in the stomachs of weanling male Sprague-Dawley rats. Administration of graded zinc deficient diets for 5 wks significantly depressed the serum zinc concentration and decreased body weight gain. These diets significantly increased the gastric secretory volume, acid, and pepsin. Zinc deficiency produced or aggravated the formation of glandular ulceration in the absence or presence of stress, respectively. It is concluded that zinc deficiency adversely affected the rats by reducing the body weight gain and producing ulceration that was probably mast cell-mediated. On the other hand, it increased gastric secretory functions.

344. Clark, William W. (1991). **Recent studies of temporary threshold shift (TTS) and permanent threshold shift (PTS) in animals.** ***Journal of the Acoustical Society of America,*** 90(1), 155–163.
Reviews laboratory studies published since 1966 on TTS and PTS and hearing loss induced in animals as a result of exposure to noise. Focus is on studies that measured hearing directly (i.e., by behavioral methods). The review is organized around (1) the question of whether there is an appropriate animal model for studies of noise-induced hearing loss, (2) the types of effects on hearing produced by continuous exposure to noise, and (3) the role of periodic rest on hearing loss from noise exposure.

345. Coenen, A. M.; Drinkenburg, W. H.; Peeters, B. W.; Vossen, J. M. et al. (1991). **Absence epilepsy and the level of vigilance in rats of the WAG/Rij strain.** ***Neuroscience & Biobehavioral Reviews,*** 15(2), 259–263.
Examined the occurrence of spike-wave discharges (SWDs) during various spontaneous or experimentally induced levels of vigilance in WAG/Rij rats. Exp 1 examined the circadian rhythmicity of absence discharges, and Exp 2 attempted to determine the vigilance level (i.e., sleep-wake state) at which discharges occurred most frequently. Three other experiments examined whether 3 methods of modulating vigilance affected the number of SWDs. Findings indicate that SWDs preferentially occur when the level of vigilance is close to the level of transitions from sleep to waking. Alertness present during active wakefulness or induced by REM sleep deprivation, a learning task, or photic stimulation, leads to reduced SWDs. Findings validate the WAG/Rij strain as a model for generalized absence epilepsy.

346. Coscina, Donald V.; Goodman, Jeff; Godse, Damodar D. & Stancer, Harvey C. (1975). **Taming effects of handling on 6-hydroxydopamine induced rage.** ***Pharmacology, Biochemistry & Behavior,*** 3(3), 525–528.
Experimental results in male Wistar albino rats demonstrate the importance of handling to the expression of 6-hydroxydopamine-induced rage and emphasize the importance of controlling for handling as a variable which can significantly affect the assessment of rage by behavioral criteria in this animal model of hyperemotionality.

347. Coveney, Joseph R. & Sparber, Sheldon B. (1990). **Delayed effects of amphetamine or phencyclidine: Interaction of food deprivation, stress and dose.** ***Pharmacology, Biochemistry & Behavior,*** 36(3), 443–449.
Describes 4 experiments with 120 male rats. Tritium-labelled phencyclidine hydrocholoride (PCP; 12 mg/kg) was injected sc into Ss maintained at 85% of their initial free-feeding weights. Application of an acute stressor resulted in redistribution of tissue stores of PCP. The direction of the redistributions was to fat. After 6 daily injections of PCP (2 or 4 mg/kg, sc), exploratory behavior of the 4 mg/kg dose group was abruptly altered at 6 days of food deprivation. PCP (4 or 8 mg/kg, sc) or dextroamphetamine sulfate (DAM; 3.2 or 6.4 mg/kg, sc) was injected into Ss for 6 days and food deprivation followed afterward for 9 consecutive days. Exploratory behavior was altered in PCP-treated Ss (at the 4 mg/kg dose level) when Ss reached about 70% of initial weights. Behavior of DAM-treated Ss (at 3.2 mg/kg) suggests that interactions between food deprivation (stress) and lipophilic drugs of abuse occur when body fat stores reach a threshold.

348. Curzon, Gerald. (1989). **5-hydroxytryptamine and corticosterone in an animal model of depression.** 11th Annual Meeting of the Canadian College of Neuropsychopharmacology (1988, Montreal, Canada). ***Progress in Neuro-Psychopharmacology & Biological Psychiatry,*** 13(3–4), 305–310.
Describes a rat model for depression that implies that high corticoid responses, low 5-hydroxytryptamine (5-HT) functional activity at certain sites, and female sex oppose adaptation and predispose to depression. The model responds appropriately to chronic antidepressant pretreatment. The 5-HT_{1A} agonists (8-OH-DPAT, buspirone, ipsapirone, gepirone) may have antidepressant activity. Both behavioral and neurochemical evidence indicates that the adaptive effects of these agonists on the depression model are associated with desensitization of somatodendritic 5-HT_{1A} autoreceptors.

349. Dai, S. & Ogle, C. W. (1974). **Gastric ulcers induced by acid accumulation and by stress in pylorus-occluded rats.** ***European Journal of Pharmacology,*** 26(1), 15–21.
Exposure to stress for 2 hrs produced a high incidence of lesions in the glandular part of the stomach. Stress ulceration was prevented only by atropine. Gastric ulcer, produced during a 5-hr period following pyloric occlusion, was located only in the rumenal part of the stomach. Both atropine and antacids prevented its development, indicating causation mainly by accumulation of gastric juice.

350. Dai, Soter & Ogle, Clive W. (1975). **Effects of stress and of autonomic blockers on gastric mucosal microcirculation in rats.** ***European Journal of Pharmacology,*** 30(1), 86–92.
Studied changes in gastric mucosal microcirculation in Wistar rats by using the method of intra-aortic injection of India ink, followed by microdissection of the mucosa. Acute stress, induced by restraint and exposure to cold for 2 hrs, caused marked and significant vasodilatation in the gastric mucosa. This vasodilatation was prevented by pretreatment with atropine or chlorpromazine, but not by a- or b-adrenoceptor blocking agents. Phentolamine caused significant vasoconstriction in the gastric mucosa of nonstressed Ss, but when Ss were stressed phentolamine induced a greater vasodilatation than was obtained with stress alone. These observations provide added support for the hypothesis that stress induces vagal overactivity, probably of central origin. The resulting strong contractions of the gastric wall and compression of the intramural vessels are probably responsible for degeneration of the mucosal cells leading to the formation of stress-induced ulcers in the rat.

351. Dark, Kathleen; Peeke, Harman V.; Ellman, George & Salfi, Mary. (1987). **Behaviorally conditioned histamine release: Prior stress and conditionability and extinction of the response.** ***Annals of the New York Academy of Sciences,*** 496, 578–582.
In an effort to develop an animal model to study the role of classical conditioning in allergic disease, 2 experiments were conducted with male guinea pigs to explore the role that mild stress plays in the conditioning of the histamine response and the extinction of this response. When handled (stressed) and nonhandled Ss were sensitized to bovine serum albumin and trained with a classical discrimination conditioning design, handling stress administered weeks before the conditioning procedures predisposed Ss to learn an association between an antigen and a neutral odor. The conditioned histamine response could be extinguished.

352. Davidson, T. L.; Flynn, Francis W. & Grill, Harvey J. (1988). **Comparison of the interoceptive sensory consequences of CCK, LiCl, and satiety in rats.** ***Behavioral Neuroscience,*** 102(1), 134–140.
In 3 experiments we assessed the degree to which ad lib feeding, injection of cholecystokinin (CCK), and injection of lithium chloride (LiCl) produce states with similar sensory consequences. In each experiment, two groups of rats were trained to use cues arising from food deprivation and satiation of discriminative signals for shock. One group was shocked when deprived but not when nondeprived. The other group received the reversed discrimination. Testing began when incidence of freezing was greater under the shocked deprivation than under the nonshocked deprivation condition. In Exp 1, the rats were tested under 24-hr food deprivation after injections of CCK, LiCl, and saline (in counterbalanced order). The effects of CCK on freezing did not differ from those of saline, whereas both CCK and LiCl had effects that were different from ad lib feeding. This pattern of results was also obtained when deprivation level during training and testing was reduced to 8 hrs (Exp 1A) and also when rats received small amounts of food in conjunction with CCK (Exp 2). The intubation of a high-calorie stomach load (Exp 1A) produced a response profile like that observed after free feeding. Freezing after LiCl treatment differed from that observed after free feeding and from that found after injection of CCK.

353. Davies, Christina, et al. (1974). **Lithium and a-methyl-p-tyrosine prevent "manic" activity in rodents.** ***Psychopharmacologia,*** 36(3), 263–274.
Examined the effects of lithium and a-methyl-p-tyrosine (AMT) in a series of experiments with 214 female Porton strain mice and 89 female hooded rats. Lithium in moderate doses seemed to have little effect on the normal activities of laboratory rodents, but did prevent some kinds of hyperactivity, especially repetitive movements involving the whole body. High levels of such activities were induced by appropriately combining drug administrations, Ss' previous experience, and the kind of environment in which they were tested. Pretreatment with lithium and also with AMT blocked the hyperactivity without affecting controls. The possibility of using such artificially unbalanced states as animal models of manic-depressive disorders is discussed.

354. Desan, Paul H.; Silbert, Lee H. & Maier, Steven F. (1988). **Long-term effects of inescapable stress on daily running activity and antagonism by desipramine.** ***Pharmacology, Biochemistry & Behavior,*** 30(1), 21–29.
Investigated the effects of inescapable shock on daily activity in a familiar home cage/running wheel environment. Rats lived in the wheel environment for 44–85 days before treatment. Results show that inescapable shock produced only a transient reduction of water intake and body weight, but daily running was depressed for 14–42 days (the maximum period studied) depending on the conditions. This long-term effect on activity occurred despite the fact that shock was administered in an environment different from the S's home running wheel environment. The activity reduction was reversed by desipramine in a dose-dependent fashion. Indeed, the activity of inescapably shocked Ss treated with the optimum dose of desipramine exceeded that of controls undergoing neither stress nor drug treatment.

355. Desiderato, Otello; MacKinnon, John R. & Hissom, Helene. (1974). **Development of gastric ulcers in rats following stress termination.** ***Journal of Comparative & Physiological Psychology,*** 87(2), 208–214.
Observed gastric ulcer formation in 123 female albino Charles River unrestrained rats sacrificed at varying intervals following the end of a single, 6-hr shock-stress session. Significant ulcer production was not found unless Ss experienced a minimum of 2 hrs poststress rest prior to sacrifice. Findings with appropriate control groups implicate the sudden reversal from stressful to "safe" (home cage) conditions, rather than delay per se, as the major ulcerogenic factor.

356. Desiderato, Otello & Testa, Marcia. (1976). **Shock-stress, gastric secretion and habituation in the chronic gastric fistula rat.** ***Physiology & Behavior,*** 16(1), 67–73.
In 2 experiments with 16 experimental and 8 yoked-control male Long-Evans rats, unrestrained Ss exposed to shock-stress showed an abrupt reduction in both the volume and acidity of gastric secretions, followed by a rapid poststress recovery. The gastric response pattern did not distinguish between Ss receiving avoidable or unavoidable shock. Marked habituation of the acid inhibition response occurred within 4 stress sessions.

357. Dess, Nancy K.; Raizer, Jeffrey; Chapman, Clinton D. & Garcia, John. (1988). **Stressors in the learned helplessness paradigm: Effects on body weight and conditioned taste aversion in rats.** 17th Annual Meeting of the Society for Neuroscience: Appetite, thirst and related disorders (1987, San Antonio, Texas). ***Physiology & Behavior,*** 44(4–5), 483–490.
In Exp 1, 44 male rats drank saccharin or a control solution, followed by 100 inescapable shocks or simple restraint. Ss were weighed daily and were tested for saccharin aversion 2 days after the stress session. Shocked Ss gained less weight than restrained controls. Saccharin aversion was apparent only among Ss that had consumed saccharin before the stress session. In Exp 2, 72 Ss drank saccharin solution, followed by shock, restraint, or no treatment. Half of each group was injected with saline; the other half was injected with lithium chloride. Shock reduced body weight relative to restraint or no treatment and produced taste aversion among saline-treated Ss. However, shock attenuated the aversion produced by lithium chloride, as did restraint. Results suggest a role for stress in anorexia and weight loss associated with depression and may have implications for theories of learning and learned helplessness.

358. Di Cara, L. & Miller, N. E. (1968). **Instrumental learning of systolic blood pressure responses by curarized rats: Dissociation of cardiac and vascular changes. *Psychosomatic Medicine,*** 30(5, Pt 1), 489–494.
28 male Sprague-Dawley rats were divided into experimental and yoked control groups to determine whether conditioned increases and decreases of systolic blood pressure could be obtained in curarized rats, using as a reward escape from and/or avoidance of, an electric shock to the tail. Blood pressure was measured by means of a catheter in the abdominal aorta. "Over-all group increases and decreases of 22.3% and 19.2% respectively, were obtained for the experimental groups, and were significantly different from the changes in the control groups. All experimental Ss without exception changed their blood pressure in the rewarded direction. Analyses of heart rate and temperature did not reveal any significant changes in either group. It is possible to instrumentally condition changes in blood pressure."

359. Di Cara, Leo V. & Weiss, Jay M. (1969). **Effect of heart-rate learning under curare on subsequent noncurarized avoidance learning. *Journal of Comparative & Physiological Psychology,*** 69(2), 368–374.
Male Sprague-Dawley rats which learned to decrease heart rate under curare in order to avoid electric shock showed good subsequent learning of a free-moving skeletal-avoidance response in a modified shuttle device, while Ss which learned to increase heart rate under curare showed poor subsequent avoidance and escape learning. Similar poor avoidance was shown by naive Ss trained with strong electric shock. Results suggest that instrumental learning of heart rate may alter emotional responses. An alternative explanation for poor avoidance performance based on learned "helplessness" is ruled out.

360. DiCara, Leo V. & Miller, Neal E. (1969). **Transfer of instrumentally learned heart-rate changes from curarized to noncurarized state: Implications for a mediational hypothesis. *Journal of Comparative & Physiological Psychology,*** 68(2, Pt 1), 159–162.
Two groups each consisting of 8 Sprague-Dawley curarized rats learned to increase or decrease, respectively, their heart rates to escape and/or avoid mild electric shock. Two weeks later, a statistically significant increase of 5% and decrease of 16% transferred to the noncurarized state. Immediately subsequent retraining without curare produced additional significant increases and decreases, bringing the overall changes to 11 and 22%, respectively. Initial significant differences between the 2 groups in respiration and activity decreased until they were far from statistically reliable. It is difficult to account for the pattern of these results by the hypothesis that all of the heart-rate changes in these rats were mediated by learned changes in central commands for activity or breathing.

361. DiCara, Leo V. & Miller, Neal E. (1969). **Heart-rate learning in the noncurarized state, transfer to the curarized state, and subsequent retraining in the noncurarized state. *Physiology & Behavior,*** 4(4), 621–624.
Trained 2 groups of 7 male Sprague-Dawley freely-moving rats to increase, or decrease, heart rate in order to avoid and/or escape electric shock. This training produced a statistically reliable difference between groups, which persisted when the Ss were later tested and trained while paralyzed with curare and maintained on artificial respiration. The difference persisted during subsequent nonreinforced tests without curare and was increased by additional reinforced training in the normal, free-moving state. During the initial session of noncurarized training, the group rewarded for increases in heart rate showed more activity than the 1 rewarded for decreases, but there were no reliable differences in respiration rate. At the end of the 2nd session of noncurarized training, the difference between the groups did not approach statistical reliability in activity ($t = .7$), respiration rate ($t = .2$), or the variability of that rate ($t = .6$), but was highly reliable in heart-rate ($t = 7.2$, $df = 12, p < .001$).

362. Dickson, C. T. & Vanderwolf, C. H. (1990). **Animal models of human amnesia and dementia: Hippocampal and amygdala ablation compared with serotonergic and cholinergic blockade in the rat. *Behavioural Brain Research,*** 41(3), 215–227.
94 male hooded rats were prepared with: (1) bilateral surgical lesions of the hippocampus and amygdala; (2) pharmacological blockade of central cholinergic and serotonergic function by systemic injections of scopolamine and *p*-chlorophenylalanine; and (3) neurotoxic lesions of the rostrally projecting serotonergic nuclei in the brainstem using intracerebral injections of 5,7-dihydroxytryptamine, later combined with scopolamine. The behavioral tests used were: an open field test, a swim-to-platform test, and a Lashley III maze. In all 3 tests, Ss with either the neurotoxin lesions plus scopolamine or *p*-chlorophenylalanine plus scopolamine treatment showed greater impairments in comparison with controls than did the combined lesion group. Simultaneous blockade of central serotonergic and cholinergic transmission had a greater effect on some aspects of the organization of behavior than large surgical lesions of the hippocampus and amygdala.

363. DiMascio, Alberto. (1973). **The effects of benzodiazepines on aggression: Reduced or increased? *Psychopharmacologia,*** 30(2), 95–102.

Reviews research on the effects of various benzodiazepines on animal models of aggression. The differences in response noted after single dose or after chronic drug administrations are stressed. Implications for predicting responses to these drugs in humans are discussed.

364. Dobbing, J. (1971). **Undernutrition and the developing brain: The use of animal models to elucidate the human problem.** ***Psychiatria, Neurologia, Neurochirurgia,*** 74(6), 433–442.
Experimental studies of the long-term effects of undernutrition on the developing brain have shown results which emphasize the paramount importance of the timing of the undernutrition in relation to the brain "growth spurt." In humans this period is extensive, occupying the last ½–⅓ of gestation and the 1st 18 mo to 2 yrs of postnatal life. This implies that permanent physical deficits in the brain could result from fetal growth retardation in the later months of pregnancy, and it underlines the importance of good postnatal growth in the prematurely born. The 1st 2 yrs of postnatal life probably present a further hazard to the developing brain if malnutrition or any other growth-retarding influence is allowed to persist. Alternatively the postnatal period may also present an opportunity in humans for compensatory brain growth, provided good nutrition or the correction of other deleterious influences is diligently pursued at this early stage. These remarks are based on the double assumption that (a) the changes which can be permanently produced in developing animal brain are functionally significant; and (b) they can be extrapolated to man by carefully matching comparable stages of brain growth, especially the complex of interrelated events known collectively as the brain "growth spurt."

365. Donohoe, Thomas P.; Kennett, Guy A. & Curzon, Gerald. (1987). **Immobilisation stress-induced anorexia is not due to gastric ulceration.** ***Life Sciences,*** 40(5), 467–472.
Studied the relationship between postimmobilization anorexia and associated gastric changes in Sprague-Dawley rats in 2 experiments. 32 male rats were used in Exp I, and 32 female rats were used in Exp II. The findings indicate that anorexia induced by immobilization stress is not due merely to gastric ulceration. It is suggested that the increased 5-hydroxytryptamine (5-HT) metabolism caused by immobilization stress may be responsible for anorexia and weight loss.

366. Dougherty, John & Pickens, Roy. (1973). **Fixed-interval schedules of intravenous cocaine presentation in rats.** ***Journal of the Experimental Analysis of Behavior,*** 20(1), 111–118.
Examined FI schedules of presentation of cocaine as a function of injection dose (.32–.64 mg/kg, iv) and interval duration (200–400 sec) in 2 male Sprague-Dawley rats. Cocaine was found to exert a dose-related temporal control over the initiation of responding that was unaffected by the FI contingency. FI pause duration was linearly related to injection dose and was the same duration as the interresponse time found on continuous reinforcement schedules of cocaine presentation. The FI pause remained constant with changes in interval duration. Characteristic FI patterns of responding were observed. However, overall response rates were inversely related to injection dose and directly related to interval duration. Running response rates varied unsystematically with both variables. Findings are at variance with results typically found in studies of FI food and electric shock presentation.

367. Drago, Filippo; Continella, Giuseppe; Conforto, Gaetano & Scapagnini, Umberto. (1985). **Prolactin inhibits the development of stress-induced ulcers in the rat.** ***Life Sciences,*** 36(2), 191–197.
Hyperprolactinemia, as induced by pituitary homografts, was accompanied by an inhibition of development of gastric ulcers following cold-plus-restraint stress in male Wistar rats. Intracisternal administration of prolactin (3 μg) and intraperitoneal (ip) domperidone (1 mg/kg), but not haloperidol (1 mg/kg), also inhibited the development of stress-induced ulcers. The prostaglandin synthesis inhibitor indomethacin (5 mg/kg, ip) increased the incidence of gastric ulcers in hyperprolactinemic Ss subjected to stress. Data suggest that the cytoprotective effect of prolactin on development of gastric ulcers in stressed animals may involve both central (i.e., dopamine transmission) and peripheral (i.e., prostaglandin synthesis) mechanisms.

368. Driscoll, Peter; Martin, J. R.; Kugler, P. & Baettig, K. (1983). **Environmental and genetic effects on food-deprivation induced stomach lesions in male rats.** ***Physiology & Behavior,*** 31(2), 225–228.
When Roman high- and low-avoidance (RHA and RLA, respectively) rats ($N = 224$) were individually housed in plastic cages with sawdust bedding and food-deprived (FD) for 4–5 days, it was found that FD RHA Ss had more lesions than their unfasted controls and more lesions than FD RLA Ss. Also, FD RHA Ss that were housed in the same room as the controls, as well as FD RHA Ss that were housed in a separate room with a strong food odor present, had more lesions than RD RHA Ss housed in the same separate room when there was no food odor and when none of the Ss present had access to food. When FD RHA and FD RLA Ss were individually housed in metal cages with grid floors, however, a general increase in lesion scores resulted and differences between the 2 rat lines disappeared, as did differences among the room conditions.

369. Drugan, R. C.; McIntyre, Todd . D.; Alpern, Herbert P. & Maier, S. F. (1985). **Coping and seizure susceptibility: Control over shock protects against bicuculline-induced seizures.** ***Brain Research,*** 342(1), 9–17.
96 Holtzman-derived male albino rats were either given 80 escapable shocks, yoked inescapable shocks, restraint, or given no treatment. Two hours later, all Ss received intraperitoneal injection of bicuculline (4, 6, or 8 mg/kg) and were immediately tested for latency to initial myoclonic jerk and clonus. The latency to clonic convulsion was dramatically affected by prior shock treatment, and the direction of this change depended upon the escapability/inescapability of the shock. Ss that were given escapable shock showed a delay of onset to seizure, while Ss inescapably shocked demonstrated a decreased latency to clonus in comparison to restrained and naive controls. It was also demonstrated with 24 Ss that if the Ss were tested immediately following a stress experience, both the 80 escapable and inescapable shock conditions protected against bicuculline-induced seizures in comparison to the control condition. Exp II with 32 Ss confirmed a previous finding that less stress (i.e., 20 inescapable shocks) protects against seizures when the Ss are challenged with bicuculline either immediately or 2 hrs later. It is suggested that control over stress may facilitate the

transmission of gamma-aminobutyric acid (GABA), and this may be the mechanism whereby coping protects against the behavioral and physiological disruption produced by exposure to a stressor.

370. Eastgate, Sheila M.; Wright, James J. & Werry, John S. (1978). **Behavioural effects of methylphenidate in 6-hydroxydopamine-treated neonatal rats.** ***Psychopharmacology,*** 58(2), 157–159.
Neonatal Wistar rats treated at 7 days of age with 6-hydroxydopamine (25 µl, intracisternally) showed normal levels of activity during maturation, but less hyperactivity than normals did when additionally treated with methylphenidate hydrochloride (8.6 mg/kg, ip) between 14 and 22 days of age. Comparison of these results with those of other workers suggests that several experimental variables must be controlled precisely if reproducible results analogous to the disturbed behavior of children with minimal brain dysfunction are to be obtained.

371. Edwards, Emmeline; Harkins, Kelly; Wright, G. & Henn, F. A. (1991). **5-HT$_{1b}$ receptors in an animal model of depression.** ***Neuropharmacology,*** 30(1), 101–105.
Exposure to a 0.8 mA course of uncontrollable shocks differentiated male rats into 2 distinct groups defined in terms of their performance in a shock escape paradigm. Learned helpless (LH) rats do not learn to escape a controllable shock while nonhelpless rats learn this response as quickly as naive control rats do. Experiments were designed to study 5-hydroxytryptamine (5-HT) receptors in these 3 groups of rats. The major finding concerned postsynaptic 5-HT receptor effects in the cortex, hippocampus, septum, and hypothalamus of LH rats. These included an up-regulation of 5-HT$_{1b}$ receptors in the cortex, hippocampus, and septum in LH rats. In contrast, 5-HT$_{1b}$ receptors in the hypothalamus of LH rats were down-regulated. Results implicate serotonergic mechanisms in the behavioral deficit caused by uncontrollable shock with a limbic hypothalamic circuit serving as a center for adaptation to stress.

372. Ellis, Fred W. & Pick, James R. (1973). **Animal models of ethanol dependency.** ***Annals of the New York Academy of Sciences,*** 215, 215–217.
Administered 2–4 g/kg of ethanol twice daily to rhesus monkeys and beagle dogs. Severe reactions of a withdrawal syndrome were observed for 2–3 wks in monkeys and 4–6 wks in dogs. Ethanol administration interrupted withdrawal reactions at any stage.

373. Elmes, David G.; Jarrard, Leonard E. & Swart, Peter D. (1975). **Helplessness in hippocampectomized rats: Response perseveration?** ***Physiological Psychology,*** 3(1), 51–55.
In a test for learned helplessness, 20 hippocampectomized albino rats, 19 cortical controls, and 20 unoperated controls were given shock escape training after being subjected to a series of noncontingent shocks. During these preshock treatments, the same amount of gross body activity was elicited in all 3 groups. Compared to Ss that did not receive the preshock treatments, all the preshocked Ss performed poorly in shock escape; hippocampals, cortical controls, and normals all exhibited learned helplessness. The fact that hippocampectomy resulted in the most marked learned helplessness is contrary to the response-perseveration interpretation of hippocampal function.

374. Essman, Walter B.; Essman, Shirley G. & Golod, Mark I. (1971). **Metabolic contributions to gastric ulcerogenesis in mice.** ***Physiology & Behavior,*** 7(4), 509–516.
Subjected 360 male cf-1 mice to immobilization and/or prior varying intervals of food deprivation. Stomachs were excised and examined for pathology, degree of acidity, and differences in gastric amine content and metabolism. Immobilized Ss showed a greater incidence of gastric lesions than nonimmobilized Ss, and immobilization also contributed to changes in gastric 5-hydroxytryptamine (5-HT) concentration. Ulcerated Ss, in general, had higher gastric 5-HT levels and elevated gastric tissue ph. Inhibition of a major metabolic degredative pathway for 5-HT in gastric tissue resulted in a paradoxical effect not found in either brain or heart; changes in the rate of gastric 5-HT concentration change after enzyme inhibition was related to immobilization, duration of fasting, and ulcer incidence. A biochemical mechanism is suggested whereby the etiology of stress-induced lesions may be explained.

375. Fine, Alan. (1986). **Transplantation in the central nervous system.** ***Scientific American,*** 255(2), 52–58B.
Contends that grafts of embryonic brain tissue can be anatomically and functionally incorporated into the adult central nervous system (CNS) and suggests that this treatment can be useful in treating Parkinson's and Alzheimer's diseases in which parts of the CNS degenerate. Experimentation with rat brains has shown that grafts improved dopamine input to damaged nerve cells or improved Ss' abilities to run mazes. It is proposed that denervation supersensitivity accounts for some of the ability of embryonic neuronal transplants to reverse behavioral abnormalities in animal models of degenerative neurological diseases.

376. Fischette, Christine T.; Biegon, Anat & McEwen, Bruce S. (1984). **Sex steroid modulation of the serotonin behavioral syndrome.** ***Life Sciences,*** 35(11), 1197–1206.
Investigated the hormonal basis of the serotonin behavioral syndrome in drug response in Sprague-Dawley and King-Holtzman rats. Endocrine influences were tested through gonadectomy, hormone replacement, and neonatal hormone treatment, and the animal model for genetic androgen insensitivity, the pseudohermaphroditic rat, was used. Results indicate that the sex difference observed in frequency of rats exhibiting the serotonin behavioral syndrome induced by pargyline/1-tryptophan depended on hormonal state. Castration eliminated the sex difference in drug response in adult and prepubertal males, whereas ovariectomy had little effect. Dihydrotestosterone administration to males (10–30 days) reinstated the sex difference but had little effect in females. Testicular feminized mutants (Tfm/y), deficient in androgen receptors, responded like females. Estrogen administration had no effect in either sex. Manipulation of the hormonal environment on Postnatal Days 0–7 (blockade of aromatization in males or estradiol administration to females) had no effect on the expression of the sex difference when the Ss were tested as adults. It is concluded that androgens acting via androgen receptors mediate this subsensitivity of male rats to the drug challenge. Results indicate that sex and hormonal environment are important variables in determining the experimental and perhaps clinical responses to drugs.

377. Flemmer, Duane D.; Dilsaver, Steven C. & Peck, Jason A. (1991). **Exposure to constant darkness enhances the thermic response of the rat to a muscarinic agonist.** ***Pharmacology, Biochemistry & Behavior,*** 38(1), 227–230.

Continual exposure to bright light for 7 days or during discrete portions of the photoperiod blunts the thermic response to a muscarinic agonist (oxotremorine) in the rat, while exposure to either 24 hrs per day of bright light or darkness tends to produce free running, suggesting that the reduced responsiveness to oxotremorine may result from the induction of free running. It was hypothesized that constant exposure to darkness would, contrary to the free-running hypothesis, enhance the thermic response to oxotremorine. 12 male rats exposed to constant darkness for 7 days exhibited supersensitivity to oxotremorine 5 days after return to standard light/dark cycle in the vivarium. This argues against the hypothesis that the induction of free running enhances sensitivity to the thermic effects of oxotremorine.

378. Flemmer, Duane D. & Dilsaver, Steven C. (1989). **Chronic restraint stress does not sensitize a muscarinic mechanism.** ***Pharmacology, Biochemistry & Behavior,*** 34(1), 207–208.
12 male rats were subjected to daily restraint stress for 2 hrs for 7 days. Ss received methylscopolamine nitrate (a peripherally antimuscarinic agent) prior to an injection of oxotremorine. While chronic prolonged restraint activated the hypothalamic-pituitary-adrenal axis, this stressor did not, unlike forced swim stress and footshock, enhance sensitivity to the hypothermic effects of a muscarinic agonist.

379. Freund, Gerhard. (1973). **Alcohol, barbiturate, and bromide withdrawal syndromes in mice.** ***Annals of the New York Academy of Sciences,*** 215, 224–234.
Describes several recently developed animal models to test various pathogenic hypotheses and empirical therapeutic procedures in the field of alcohol withdrawal research.

380. Fukushima, Masataka et al. (1981). **Circadian pattern of stress response to affective cues of foot shock.** ***Physiology & Behavior,*** 27(5), 915–920.
Two experiments studied formation of gastric lesions, changes in body weight, and changes in serum level of endogenous chemical substances in response to stressful stimuli in male adult CF1 mice. To examine the influences of stressful events on bodily responses, the stimuli were introduced by using a communication box. The "sender" received physical stimuli of electric footshocks, and the "responder" received affective information such as visual, auditory, and olfactory cues from sender with no exposure to footshock. The responder exhibited significantly more severe gastric lesions than the sender. The degree of susceptibility to the formation of gastric lesions and weight reduction increased when the stimuli were administered during the dark cycle. Results suggest that a circadian rhythm can be detected in susceptibility to gastric lesions in response to indirect stimulus cues.

381. Fuller, Ray W. & Yen, Terence T. (1987). **The place of animal models and animal experimentation in the study of food intake regulation and obesity in humans.** ***Annals of the New York Academy of Sciences,*** 499, 167–178.
Identifies commonly used animal models (AMs) for studying obesity and discusses similarities and differences between AMs of obesity and obesity in humans with respect to genetics, control of food intake, intermediary metabolism, and therapeutic approaches (i.e., drug interventions) to obesity.

382. Gamallo, A.; Trancho, G. J. & Alario, P. (1988). **Behavioral and physiologic effects of early nutrition and social factors in the rat.** ***Physiology & Behavior,*** 44(3), 307–311.
Two experimental methods were used to provoke caloric restriction during suckling in 2 different rat groups: Low Growth (LG) and High Growth (HG). In one method, the groups also differed in a social factor (litter size). Growth differences and high levels of social competition were found among pups of the crowded group with LG compared with the HG group with small litters. Both methods resulted in growth differences between respective groups from the 1st wk of suckling that persisted 40 days after weaning. LG Ss had higher defecation scores with lower activity in the open-field test and higher susceptibility to restraint ulcers and adrenal hypertrophy than HG Ss in litters of equal size. However, early stimulation from social competition among pups in larger different litters in the crowded LG group counteracted nutritional factor effects.

383. German, Dwight. (1971). **Advantages and disadvantages of sub-human animal models for human neuropsychology.** ***Biological Psychology Bulletin,*** 1(1), 24–28.
Briefly considers several areas of research including: agnosias, localization of brain functions, and species specific differences.

384. Giordano, Magda; Ford, Lisa M.; Shipley, Michael T. & Sanberg, Paul R. (1990). **Neural grafts and pharmacological intervention in a model of Huntington's disease.** 19th Annual Meeting of the Society for Neuroscience: Neural basis of behavior: Animal models of human conditions (1989, Phoenix, Arizona). ***Brain Research Bulletin,*** 25(3), 453–465.
Examined 2 potential therapies for Huntington's disease, using animal models. Exp 1 explored the effect of MK801 pretreatment on quinolinic acid (QA)-induced striatal degeneration and behavioral deficits. Findings indicate that MK801 prevented QA-induced neuropathological changes in the striatum; this anatomical projection was correlated with the absence of deficits in the cataleptic response to haloperidol. Exp 2 tested the ability of 3 types of fetal grafts to reverse behavioral deficits induced by kainic acid (KA) lesions. Striatal grafts attenuated KA-induced deficits in motor coordination, haloperidol-induced catalepsy, and amphetamine-induced locomotor activity. Tectal grafts had a partially beneficial effect. Early cortical grafts reversed the exaggerated response to amphetamine observed after KA lesions.

385. Giral, Philippe; Martin, Patrick; Soubrié, Philippe & Simon, Pierre. (1988). **Reversal of helpless behavior in rats by putative 5-HT_{1A} agonists.** ***Biological Psychiatry,*** 23(3), 237–242.
Assessed the effects of 5-hydroxytryptamine 1A (5-HT_{1A}) agonists on escape deficits in rats produced by inescapable shock. Ss were first exposed to 60 inescapable shocks and 48 hrs later were subjected to daily 15-min shuttle-box sessions. Twice daily ip injection of the 5-HT_{1A} agonists buspirone, gepirone, 8-OH dipropylaminotetralin, and ipsapirone, eliminated escape failures induced by shock. Results indicate that an antidepressant-like effect on helpless behavior can be obtained with drugs assumed to stimulate 5-HT_1 receptors. The role of serotonergic mechanisms in mediating helpless behavior is discussed.

386. Glavin, Gary B. et al. (1983). **Regional rat brain noradrenaline turnover in response to restraint stress.** ***Pharmacology, Biochemistry & Behavior,*** 19(2), 287–290.
Male Wistar rats were starved for 12 hrs and then subjected to either 2 hrs of restraint in a wire mesh "envelope" at room temperature, 2 hrs of supine restraint in a specially constructed harness at room temperature, or no restraint. Eight brain regions were examined for noradrenaline (NA) level and the level of its major metabolite 3-methoxy-4-hydroxy phenylethylene glycol sulfate (MHPG-SO/4). Plasma corticosterone and gastric ulcer incidence were also measured. All restrained Ss displayed marked elevations in MHPG-SO/4 levels in most brain regions. Several brain regions in restrained Ss also showed a reduction in NA levels. All restrained Ss showed elevated plasma corticosterone levels and evidence of gastric lesions. In general, supine restraint produced greater alterations in regional brain NA turnover, greater evidence of ulcer disease, and higher plasma corticosterone levels than wire mesh restraint. Data suggest that acute but intense stress in the form of restraint causes markedly altered brain NA activity—a possible neurochemical mechanism underlying the phenomenon of stress-induced disease.

387. Glavin, Gary B. (1984). **Prenatal maternal stress: Differential effects upon male and female offspring responses to restraint stress as adults.** ***Pavlovian Journal of Biological Science,*** 19(3), 157–159.
22 pregnant primiparous Holtzman rats were subjected to 4 days of light restraint stress on postconception Days 7 through 10, coincident with the development of the fetal gastrointestinal system. 20 male and 20 female offspring from prenatally stressed (PNS) and nonstressed (NPS) Ss were then subjected to 2 hrs of supine cold-restraint as adults. 80% of NPS offspring developed gastric lesions, while 47.5% of offspring of PNS Ss displayed lesions. A significant sex-stress interaction was detected, indicating that male offspring from PNS Ss displayed less severe gastric lesions in response to restraint stress as adults than did male offspring from NPS Ss. Female offspring from both conditions showed similar levels of stress-induced lesions. Results suggest an adaptational rather than a predispositional effect of prenatal maternal stress.

388. Glavin, Gary B. (1982). **Adaptation effects on activity–stress ulcers in rats.** ***Pavlovian Journal of Biological Science,*** 17(1), 42–44.
120 male albino Wistar rats were given either 7 or 14 days of experience with either restricted feeding (2 hrs/day) or activity (24 hrs/day of access to activity wheels with ad lib feeding). They were then subjected to the activity–stress ulcer procedure involving 1 hr/day of feeding and continuous access to running wheels. Neither experience with restricted feeding nor with activity wheels attenuated gastric ulceration, indicating that adaptation did not occur. Ss with either restricted feeding experience or activity experience died faster and exhibited more frequent and more severe stomach damage than did nonexperienced controls.

389. Glavin, Gary B. & Mikhail, Anis A. (1976). **Stress and ulcer etiology in the rat.** ***Physiology & Behavior,*** 16(2), 135–139.
Varied shock predictability and dietary physical property in an unavoidable shock situation for 48, 72, or 96 hrs using 168 male Wistar rats. Unpredictable shock did not markedly affect rumenal or corpus pH nor did it produce corpus ulcers. Ss in the liquid diet condition developed significantly more ulceration and had significantly lower pH in the rumen than Ss in the solid diet groups. A strong ulceration and pH was found in the rumen. Under all experimental conditions, the rumenal portion of the stomach was consistently more acidic than the corpus. Although increased acidity is common to ulcer development in both portions of the rat stomach, results indicate that a uniform intra-gastric acidity increase may not result from exposure to a particular stressor.

390. Glavin, Gary B. & Pare, William P. (1985). **Early weaning predisposes rats to exacerbated activity-stress ulcer formation.** ***Physiology & Behavior,*** 34(6), 907–909.
Investigated whether early weaning of 24 Holtzman rats would produce susceptibility only to restraint ulcers that involve a gastric acid etiology, or whether a more general stress ulcer susceptibility would be produced by this procedure. 24 Ss weaned early (Day 15), 14 Ss weaned normally (Day 21), and 16 Ss weaned late (Day 27) were placed into an activity-stress paradigm. Early weaned Ss died at a faster rate and exhibited a significantly greater cumulative ulcer length than did normally or late weaned Ss, although ulcer incidence did not differ between the groups. Results suggest that early maternal deprivation may generally predispose rats to stress-induced gastrointestinal disease, whether or not such disease has an acid etiology.

391. Glavin, Gary B. & Vincent, George P. (1979). **Species differences in restraint-induced gastric ulcers.** ***Bulletin of the Psychonomic Society,*** 14(5), 351–352.
30 each of male Wistar rats, LHC/LoK hamsters, gerbils, HPB Swiss-Webster mice, and HOR guinea pigs were starved for 12 hrs and then subjected to 3 hrs of cold restraint (4–6° C) in either the prone or the supine position. Significant differences occurred between restraint positions and between species. All species exhibited glandular stomach ulceration in response to both forms of stress.

392. Gliner, Jeffrey A. (1972). **Predictable vs. unpredictable shock: Preference behavior and stomach ulceration.** ***Physiology & Behavior,*** 9(5), 693–698.
Gave 90 male Sprague-Dawley albino rats a choice for either predictable or unpredictable shock or only unpredictable shock. Ss in the 1st group chose predictable shock and developed fewer stomach ulcers than those who could not make this choice. A yoked control procedure was utilized in a tilt shuttlecage apparatus with 2 levels of shock intensity and 4 treatments of predictability. Predictable shock was preferred by both low and high intensity shock groups but was slower to develop at the high intensity shock level. Unpredictable shock resulted in a high frequency of stomach ulcers at both shock intensities, whereas significantly fewer Ss developed ulcers when they had received predictable shock at the low intensity shock level. Results are discussed with respect to the effect of the safety signal hypothesis in rats.

393. Gloor, P. & Testa, G. (1974). **Generalized penicillin epilepsy in the cat: Effects of intracarotid and intravertebral pentylenetetrazol and amobarbital injections.** ***Electroencephalography & Clinical Neurophysiology,*** 36(5), 499–515.

Presents evidence from studies with 21 cats that generalized penicillin epilepsy in the cat produced by large intramuscular injections of penicillin is characterized by clinical and EEG features resembling human myoclonic petit mal. It was concluded that the origin of the convulsive discharges in this animal model is most likely cortical. (French summary).

394. Goesling, Wendell J.; Buchholz, Allan R. & Carriera, Charlotte J. (1974). **Conditioned immobility and ulcer development in rats.** ***Journal of General Psychology,*** 91(2), 231–236.
Investigated the effects of conditioned immobility on the development of gastrointestinal pathology in 30 naive male hooded rats. Ss placed in a Passive Avoidance situation where shock escape or avoidance was made contingent upon reduction of skeletal activity were compared to yoked Ss which received the same duration and frequency of shocks, but were not able to avoid or escape the aversive stimulus by their behavior. The 15 experimental Ss made significantly fewer responses and developed less severe gastric ulcers than did the 15 yoked Ss. The data are consistent with J. Weiss's (1971) theory which proposes that Ss which have control over a stressor should ulcerate less than their yoked controls. It is suggested that the present results demonstrate that passive avoidance is as strong a potentiator of ulcerations as J. Weiss's (1971) active avoidance procedure.

395. Golda, V. & Petr, R. (1986). **Higher susceptibility to stomach lesions and long lasting retention of shock-induced motor depression in genetically hypertensive male rats.** 20th Interdisciplinary Conference of Experimental & Clinical Research of Higher Nervous Activity (1984, Olomouc, Czechoslovakia). ***Activitas Nervosa Superior,*** 28(2), 152–153.
Exposed genetically hypertensive and normotensive rats of both sexes to 96 hrs of isolation and starvation. Hypertensives, namely males, showed more stomach pathology and lower plasma corticosterone than did normotensives. Thus, hypertensives appear to show elevated physiological, as well as behavioral (i.e., depression, hyponeophagia), reactivity to stress.

396. Golda, V. & Petr, R. (1986). **Behaviour of genetically hypertensive rats in an animal model of depression and in an animal model of anxiety.** 20th Interdisciplinary Conference of Experimental and Clinical Research of Higher Nervous Activity (1984, Olomouc, Czechoslovakia). ***Activitas Nervosa Superior,*** 28(4), 274–275.
A retarded extinction of motor depression and maximal latency in the hyponeophagia test was found in genetically hypertensive Wistar male rats; hypertensive Wistar females displayed a marked lengthening of hyponeophagia. It is suggested that different mechanisms underlie models of depression and anxiety.

397. Goldenberg, Marvin M. (1973). **Study of cold + restraint stress gastric lesions in spontaneously hypertensive, Wistar and Sprague-Dawley rats.** ***Life Sciences,*** 12(11, Pt. 1), 519–527.
Assigned 136 spontaneously hypertensive (Wistar-derived), Wistar, and Sprague-Dawley (SD) rats to cold + restraint stress. Additional Ss served as controls. The incidence of gastric lesions (percent of Ss with lesions) was significantly greater in the SD than in the other 2 types of rats. The increased susceptibility of the SD strain to lesion formation was tentatively attributed to differences in hereditary factors and also to less variation in sensitivity within this strain. The incidence of gastric lesions was significantly higher in the Spontaneously Hypertensive (SH) than in the normotensive Wistar Ss. Evidence from the literature suggests that the SH rat synthesizes more catecholamines in the adrenal glands than the normotensive Wistar rat, and thus under stress these amines may contribute to gastric lesion formation by virtue of their mucosal vascular constriction and gastric ischemia potential. The SD and the SH rat were estimated to have the same average lesions/stomach values-significantly higher than in the Wistar rats-following cold + restraint stress.

398. Goldstein, Dora B. (1973). **Quantitative study of alcohol withdrawal signs in mice.** ***Annals of the New York Academy of Sciences,*** 215, 218–223.
Presents an animal model for alcohol physical dependence in mice, consisting of an intoxication phase lasting a few days and a withdrawal period in which the intensity of the withdrawal reaction was measured.

399. Goldstein, Robert & Wozniak, David F. (1979). **Effect of age, food deprivation and stress on gastric erosions in the rat.** ***Physiology & Behavior,*** 23(6), 1011–1015.
Reports 2 experiments with 80 male Charles River CD rats. In Exp I, 40- and 570-day-old Ss exhibited a significant degree of glandular gastric erosion due to the stress and to the deprivation. Both of these were attributable to the single condition wherein deprivation and stress were combined. No effect involving age was significant. In Exp II, 22-day-old weanlings exposed to the same conditions evinced a significant glandular erosion effect of the deprivation, but neither the stress nor the interaction effects were significant. It is concluded that 3 hrs of cold restraint in the sated mature rat has a relatively minor effect on the stomach. 48 hrs of food deprivation, also ineffective by itself in such animals, renders the stomach vulnerable to the effects of stress. In the weanling, in contrast, deprivation alone can cause glandular erosion, but its potentiating effect is lacking. With respect to ruminal lesions, it is hypothesized that time without food rather than initial body weight or nutritional deficit is the critical variable.

400. Greenberg, Danielle & Ackerman, Sigurd H. (1984). **Genetically obese (*ob/ob*) mice are predisposed to gastric stress ulcers.** ***Behavioral Neuroscience,*** 98(3), 435–440.
In Exp I, 20 adult male genetically obese (*ob/ob*) mice and 20 lean littermate controls were food deprived and subsequently physically restrained at normal room temperatures. Obese Ss became hypothermic and developed gastric stress ulcers. Lean Ss maintained normal body temperatures and did not form gastric ulcers. In Exp II, 5 male obese and 4 lean littermates were used to test the effects of noradrenaline (NA) during restraint, and 5 obese and 5 lean mice were used to test the effects of NA alone. It was expected that in lean, but not in obese, Ss that NA would induce an increase in O_2 consumption beyond that induced by initial restraint. O_2 consumption was measured during food deprivation and restraint. Obese and lean Ss had parallel metabolic responses, with obese Ss using significantly less O_2 at all times. The predisposition to formation of gastric ulcers is a new phenotypic expression of the *ob/ob* genotype. The pathogenesis of this susceptibility appears to be related to a genetic disturbance in heat production.

401. Greenwood, M. R. et al. (1974). **Food motivated behavior in genetically obese and hypothalamic-hyperphagic rats and mice. *Physiology & Behavior,*** 13(5), 687–692.
Studied the food-associated behavior of 4 genetically obese Zucker rats (fa/fa) and 4 yellow obese mice (aAy) using classic operant procedures. When obese Zucker rats and yellow mice were compared to their lean littermates ($n = 8$) and lean littermates made obese by electrolytic lesions and chemical lesions ($n = 8$), respectively, the naturally obese animals did not display the behavioral patterns associated with rodents made obese by hypothalamic damage. These experiments point out the necessity for careful selection of animal models in studying behaviors associated with regulatory disturbances of normal food intake.

402. Guile, Michael N. & McCutcheon, N. Bruce. (1984). **Effects of naltrexone and signaling inescapable electric shock on nociception and gastric lesions in rats. *Behavioral Neuroscience,*** 98(4), 695–702.
Two experiments, with 144 male Long-Evans hooded rats, examined the antinociceptive effects of signaled shock and its physiological underpinnings. In Exp I, Ss were exposed to 1 of 3 shock conditions: no shock, unsignaled shock, and signaled (by a 10-sec, 1,000-Hz tone) shock. In each condition, Ss were tested hourly in the absence of tones for nociception, with vocalization to shock used as the behavioral measure. Ss receiving signaled shocks had stomach ulcer scores intermediate between those of no-shock and unsignaled shock Ss. Signaled-shock Ss also displayed a pronounced vocalization antinociception effect. This suggested that signaled shock may be less aversive. Exp II investigated a possible role of endogenous opiate peptides in these effects. Ss received hourly injections of either the opiate antagonist naltrexone (7 mg/kg, ip) or saline. There were no significant effects of naltrexone on either stomach pathology or nociception scores. The same effects of signaled shock were obtained as in Exp I. It is concluded that the role of endogenous opiates in the effects of signaled shock seen here is minimal.

403. Guile, Michael N. & McCutcheon, N. Bruce. (1980). **Prepared responses and gastric lesions in rats. *Physiological Psychology,*** 8(4), 480–482.
Examined the importance of prepared responses (M. E. Seligman and J. L. Hager, 1972) in the induction of stomach lesions by tailshock restraint in male Long-Evans rats in 2 experiments. In Exp I with 22 Ss, the prevention of a species-specific defense reaction by bodily restraint led to increased levels of stomach ulceration as compared with controls. In Exp II with 18 Ss, giving highly restrained Ss the opportunity to engage in gnawing, a prepared response for the rat, resulted in a significant reduction in gastric pathology. The data were related to J. M. Weiss' (1968, 1971, and 1977) 2-factor model of coping behavior and ulceration.

404. Hackmann, Eva; Wirz-Justice, Anna & Lichtsteiner, M. (1973). **The uptake of dopamine and serotonin in rat brain during progesterone decline. *Psychopharmacologia,*** 32(2), 183–191.
Investigated a possible relevant animal model for the hormone-triggered emotional disturbances of premenstruum, postpartum, and the menopause. The common characteristic situation of progesterone decline was simulated in ovariectomized Wistar rats, studied 12, 24, 36, and 48 hrs after the last of 2 wk-long 12-hourly series of progesterone injections. In vitro uptake of 3H-dopamine and 3H-serotonin in slices from cortex, preoptic hypothalamus, thalamus, midbrain, and striatum was measured. Serotonin uptake was reduced in progesterone-treated slices from the preoptic-hypothalamus 24 hrs, in the thalamus 36 hrs, after the last injection, whereas no change in dopamine uptake in any region was noted. Dopamine uptake increased with time after the last injection (a possible effect of the particular stress situation of chronic injection) and was also significantly higher at 8 hrs than at 20 hrs in cortex, midbrain, and striatum. Serotonin uptake did not vary with the time of day.

405. Hagan, M. M. & Moss, D. E. (1991). **An animal model of bulimia nervosa: Opioid sensitivity to fasting episodes. *Pharmacology, Biochemistry & Behavior,*** 39(2), 421–422.
Deprived and maintained female rats at 75–80% of body weight at 3 different times during development. Following recovery to normal weight, food intake was measured with and without butorphanol tartrate (BUTR), a kappa-sigma agonist, 8 mg/kg sc. Ss with a history of deprivation showed an increase in postrecovery feeding when they were tested at normal body weight and not food deprived. BUTR prolonged food intake in the 3-hr eating test only in Ss with a developmental history of food restriction. A developmental history of fasting in eating disorders may trigger changes in opiate systems that result in atypical feeding behavior in the adult.

406. Hämäläinen, Heikki. (1983). **Psychophysical and neurophysiological studies on logia. *Psykologia,*** 18(5), 336–340.
Applied the 3 approaches of psychophysics, animal models, and microneurography in an experiment to determine the principles and neural mechanisms of sensations aroused with short tactile pulses and bursts of vibration of varying amplitude and frequency applied to 2 functionally different areas, dorsal and glabrous palm, of the human hand. The glabrous palm was more sensitive than the dorsal palm when determined with both pulses and vibration, and lower detection thresholds were measured on both skin areas by increasing the frequency of the pulse or vibratory stimulus. However, higher stimulus intensities were needed on both skin areas to elicit qualitatively distinct sensations of touch or vibration. The peripheral neural mechanisms transmitting cutaneous information from the 2 skin areas differed considerably. There were differences in both receptor populations and in physiological properties of some receptor types in these skin areas. The peripheral nervous system of the cat was found to be valid for use as a model for neural mechanisms of human tactile and vibrotactile sensations. The basic difference between the 2 species was the nonexistence of differentiated, slowly adapting receptor types in the cat paw, reflecting the highly qualified use of the human glabrous palm and fingers as manipulative tools.

407. Hamamura, Yoshihisa & Kobayashi, Juichi. (1986). **Some social factors relevant to the stress-reducing effect of fighting in rats. *Japanese Psychological Research,*** 28(2), 87–93.
Conducted an experiment with male Wistar albino rats, using a 2-factorial design. One factor concerned the nature of reactive feedback that came from each S, and another concerned the nature of social contact (visual and/or bodily). Results were as follows: (1) The Ss that were shocked together and fought with each other tended to show less severe gastric lesions than both of those Ss that received the same

amount of shock either solitarily or under the condition that they were permitted to attack their target Ss exposed to no shock. (2) The stress-reducing effect of fighting was observed only when Ss were allowed to make bodily contact with their target Ss. Results indicate that the mere release of aggression is not sufficient to reduce the degree of gastric lesions, but some social elements play an important role in the stress-reducing effect of fighting.

408. Hara, Chiaki; Manabe, Kazue & Ogawa, Nobuya. (1981). **Influence of activity-stress on thymus, spleen and adrenal weights of rats: Possibility for an immunodeficiency model.** ***Physiology & Behavior,*** 27(2), 243–248.
64 Wistar rats were housed in laboratory or activity-wheel cages and fed either 1 or 24 hrs/day. The incidence of ulcers was higher in Ss fed 1 hr at night than for those fed 1 hr in the daytime though their mortality rate was similar. Thymus and spleen weights of Ss exposed to activity-stress (AS) decreased, while adrenal weights increased. Victims always revealed pulmonary infection and the lack of immunologically competent cells. Results suggest that rats exposed to AS reveal, not only ulceration, but also immunosuppression. Stress factors and utilization of the AS rat for an immunodeficiency model are discussed.

409. Hara, Chiaki & Ogawa, Nobuya. (1983). **Influence of maturation on ulcer-development and immunodeficiency induced by activity-stress in rats.** ***Physiology & Behavior,*** 30(5), 757–761.
Young adult rats housed in wheel-activity cages and fed only 1 hr daily developed stomach ulcers and immunodeficiency. In the present study, 80 6- or 10-wk-old male Wistar rats were stressed 0, 3, 5, or 7 days to separate these effects and to examine activity-stress as an animal model of the human stress ulcer. Ulceration was found in younger Ss in all 3 stress conditions, while immunodeficiency was observed in Ss stressed 5 or 7 days. Ulceration occurred in older Ss after 5 or 7 days of stress, and immunodeficiency only in 7-day stressed Ss. Both age groups showed thymus and spleen atrophy following stress. Results suggest that immature rats are more susceptible to stress than are young mature rats and that mature rats may provide a better experimental model for the examination of the development of stress ulcers.

410. Harlow, Henry J.; Darnell, Diana K. & Phillips, John A. (1982). **Pinealectomy in ground squirrels: Effect on behavioral and physiological responses to heat stress.** ***Physiology & Behavior,*** 28(3), 501–504.
Seven Richardson's ground squirrels, 1 yr after pinealectomy, showed altered behavioral and physiological responses to heat stress when compared to intact controls. Pinealectomized Ss barpressed more often for a cool temperature reward in a hot environment. When deprived of behavioral control of the hot environment, pinealectomized Ss increased their oxygen consumption, had a higher body temperature, and displayed signs of greater thermal stress, including death, compared to intact Ss. When the intact Ss were pinealectomized and the experiments repeated, differences in response to heat stress were not as great as with the 1-yr pinealectomized group. An explanation of the pineal gland's influence on central and peripheral control of evaporative water loss and peripheral blood circulation is offered, and it is suggested that the pineal gland may exert a subtle influence on heat transfer mechanisms and adaptations to thermal stress.

411. Hawkins, James et al. (1981). **Swimming immobility and rat REM deprivation: A pilot study on time-delay effects.** ***Bulletin of the Psychonomic Society,*** 18(4), 215–217.
Hypothesized that the swimming activity level of rats would increase as a result of REM deprivation (REMD). Three groups of 40-day-old male Sprague-Dawley rats were divided among a control group, REMD on a large platform, and REMD on a small platform. On each of 7 days, 1 S from each group was tested for swimming immobility (a posture that is adopted after vigorous activity) 3 hrs, 30 min, and with no delay after being removed from the REMD bucket. Control and large platform Ss showed increasing time in immobility across trials. Small platform Ss did not increase time in immobility until they had consolidated sleep. The experiment also solves a time-delay problem in the ongoing sequential validation of an animal model concerned with emotionality and adaptation.

412. Hayes, Ronald; Clifton, Guy & Kreutzer, Jeffrey S. (1989). **A laboratory model for testing methods of improving cognitive outcome following traumatic brain injury.** ***Journal of Head Trauma Rehabilitation,*** 4(3), 9–19.
Discusses investigations, using a fluid percussion model of traumatic brain injury (TBI), of the effects of experimental TBI on neurological outcome measures in the cat and rodent. Different neurochemical mechanisms may mediate reversible unconsciousness and persistent neurological deficits following TBI. Data on vestibular motion suggest that specific pharmacological interventions may reduce cognitive deficits following mild to moderate TBI in humans.

413. Hellhammer, D. H. et al. (1984). **Learned helplessness: Effects on brain monoamines and the pituitary-gonadal axis.** ***Pharmacology, Biochemistry & Behavior,*** 21(4), 481–485.
Determined the effect of the learned helplessness paradigm, a model of depression, on biogenic amines in 8 brain regions and on the serum levels of luteinizing hormone, corticosterone, and testosterone in 15 male Wistar rats. Ss that were exposed to uncontrollable and unpredictable shocks (HY Ss) had hormone levels similar to those in appropriate control Ss. However, HY Ss had higher levels of 5-hydroxyindoleacetic acid in the pons/medulla oblongata and lower levels of serotonin (5-HT) in the cortex than Ss that could escape the shocks (HE Ss). Striatal levels of norepinephrine (NE) were higher in HY Ss when compared to HE Ss and nonshocked controls. Shock treatment per se resulted in lower NE levels in the hippocampus. Data implicate the serotonergic and noradrenergic systems as possible mediators of the learned helplessness phenomenon but do not support the view that this behavior is associated with impaired pituitary-gonadal function.

414. Henke, Peter G. (1984). **The bed nucleus of the stria terminalis and immobilization-stress: Unit activity, escape behaviour, and gastric pathology in rats.** ***Behavioural Brain Research,*** 11(1), 35–45.
Conducted 3 experiments with 45 male Wistar rats to investigate (1) the effects of immobilization on the multiple-unit activity in the bed nucleus of the stria terminalis (BNST), (2) whether or not such units could be conditioned to respond to an auditory stimulus that was paired with the stress-treatment, (3) whether or not the presentation of the stimulus also produced behavioral escape, and (4) the effects of bilateral lesions in the BNST on such escape responses and development of gastric stress-pathology. Results show

that Ss escaped from the auditory stimulus in behavioral tests, and bilateral lesions in the bed nucleus reduced the latencies of escape responses. The lesion also increased the severity of restraint-induced mucosal erosions. The latter effect was most pronounced when the damage was in the lateral portion of the bed nucleus. It is concluded that the BNST is part of a coping system that responds when the organism is placed in a stressful situation.

415. Henke, Peter G. (1982). **The telencephalic limbic system and experimental gastric pathology: A review.** ***Neuroscience & Biobehavioral Reviews,*** 6(3), 381–390.
The effects of lesions and stimulations of the telencephalic limbic system on experimental gastric ulcers and erosions in animals are reviewed. It is concluded that the centromedial amygdala and the anterior cingulate gyrus are facilitatory structures, whereas the medial prefrontal cortex, posterior cingulate cortex, entorhinal cortex, hippocampus, and posterolateral amygdala are inhibitory areas during stressful experiences (e.g., immobilization). Both the centromedial amygdala and the anterior cingulate gyrus may be part of an "ancillary" pain system, mediating the affective components of aversive experiences. The inhibitory structures, on the other hand, are assumed to be part of a "preventive" mechanism initiated by the selective nuclear binding of glucocorticoids under stress.

416. Henke, Peter G. (1988). **Electrophysiological activity in the central nucleus of the amygdala, emotionality and stress ulcers in rats.** ***Behavioral Neuroscience,*** 102(1), 77–83.
Multiple-unit activity in the central nucleus of the amygdala was continuously recorded during 4 hr of restraint stress in rats. Five different activity profiles were found. Two types were associated with stress ulceration: one with increased stomach pathology, and the other with decreased stomach pathology. The same unit profiles were also differentially related to the emotionality characteristics of Wistar-derived rats, as well as to those of the genetically selected lines of Roman high- and low-avoidance rats. The type of profile that had been associated with increased pathology was generally seen in the Roman low-avoidance rats and in the Wistar rats that had been judged to be more emotional, that is, defecated before five "rearings" had occurred in an open-field test. The other unit profile was significantly more frequent in the Roman high-avoidance animals and the less emotional Wistar rats. Low-level electrical stimulation of both types of units produced stomach erosions in all cases. It was concluded that the unit activity in the central nucleus of the amygdala reflects certain emotionality characteristics of rats and also their susceptibility to stress ulcers.

417. Henke, Peter G. (1988). **Recent studies of the central nucleus of the amygdala and stress ulcers.** ***Neuroscience & Biobehavioral Reviews,*** 12(2), 143–150.
Reviews studies that indicate that the multiple-unit activity of the central nucleus of the amygdala differentiates stress-susceptible from stress-resistant rats, highly emotional from less emotional rats, and genetically-selected Roman high-avoidance and low-avoidance rats. Kindling of this region increases the susceptibility to stress ulcer formation. It is suggested that the amygdala codes the stressfulness of aversive inputs, the central nucleus being the point of output to areas controlling visceral responses to such information.

418. Henn, Fritz A.; Edwards, Emmeline & Anderson, David. (1986). **Receptor regulation as a function of experience.** ***National Institute on Drug Abuse: Research Monograph Series,*** 74, 107–116.
Investigated mechanisms by which exposure to uncontrollable shock (UN) produces a subsequent escape deficit. Sprague-Dawley rats that responded to UN with no deficit in subsequent shock-escape tests and Ss that responded with a profound learning deficit were studied. Neurobiochemical differences between the groups are identified. Clinically effective tricyclic and 2nd generation antidepressants and monoamine oxidase (MAO) inhibitors reversed the learning deficit and the up-regulation of beta-adrenergic receptors found in response-deficient Ss. Findings are discussed in terms of a modified version of M. E. Seligman's (1975) learned helplessness theory and implications for drug abuse research.

419. Herner, Dorothy & Caul, William F. (1972). **Restraint induced ulceration in rats during estrus and diestrus.** ***Physiology & Behavior,*** 8(4), 777–779.
Assessed the degree of ulceration found in 36 male, estrous female, and diestrous female Wistar rats following immobilization stress. After 19 hr of restraint, diestrous females showed greater ulceration than males while the extent of ulceration for estrous females fell between. Neither weight loss during restraint nor measures of heart rate were consistently related to the development of ulcers.

420. Hicks, Robert A. & Sawrey, James M. (1978). **REM sleep deprivation and stress susceptibility in rats.** ***Psychological Record,*** 28(2), 187–191.
Measured with the ANOVA the ulceration rate in 54 45-day-old Sprague-Dawley rats as a function of REM sleep deprivation and combinations of 2 stressors, fear conditioning and immobilization, as a test of the hypothesis that deprivation of REM sleep acts to produce stress-related change by increasing the salience of stressful environmental stimuli. Results support this hypothesis.

421. Ho, Ing K.; Yamamoto, Ikuo & Loh, Horace H. (1975). **A model for the rapid development of dispositional and functional tolerance to barbiturates.** ***European Journal of Pharmacology,*** 30(2), 164–171.
The subcutaneous implantation of a 75-mg pentobarbital pellet in the back of a conscious male ICR mouse resulted in a much more rapid development of tolerance to barbiturates than that produced in Ss receiving daily intraperitoneal injections of 75 mg/kg sodium pentobarbital. Acceleration in tolerance development by pellet implantation was evidenced by a decrease in sleeping time after the challenge with either sodium pentobarbital or sodium barbital. The degree of hepatic microsomal drug enzyme induction after pellet implantation was significantly higher than that produced by the injection technique. Further studies demonstrated that the threshold for pentylenetetrazol-induced seizures was significantly reduced compared to that of the sodium pentobarbital daily injected and control groups. These studies provide an animal model for studying the mechanism of barbiturate tolerance and dependence.

422. Horita, A.; Carino, M. A.; Zabawska, J. & Lai, H. (1989). **TRH analog MK-771 reverses neurochemical and learning deficits in medial septal-lesioned rats.** ***Peptides,*** 10(1), 121–124.

Microinjection of ibotenic acid into the medial septum of rats decreased choline acetyltransferase and high-affinity choline uptake (HACU) activities in the hippocampus. Learning of a spatial memory task in a radial-arm maze was also retarded. Administration of MK-771, a stable analog of thyrotropin-releasing hormone (TRH), restored HACU to normal levels. Daily treatment with MK-771 prior to maze running also restored learning ability. MK-771 did not enhance HACU or maze performance in sham-lesioned Ss. It is suggested that MK-771 and other TRH analogs may help improve memory deficits produced by cholinergic insufficiency in Alzheimer's disease.

423. Hornbuckle, Phyllis A. & Isaac, W. (1969). **Activation level and the production of gastric ulceration in the rat.** ***Psychosomatic Medicine,*** 31(3), 247–250.
"Sensory conditions and frontal cortical ablations thought to increase activation levels were found to produce a significant increase in gastric ulceration as a result of restraint."

424. Houser, Vincent P. & Pare, William P. (1973). **Measurement of analgesia using a spatial preference test in the rat.** ***Physiology & Behavior,*** 10(3), 535–538.
Used a rectangular tilt cage with 30 male Dublin DR rats to define the aversive threshold to grid shock as that intensity avoided 75% of the time. Each shock intensity (30, 60, 90, 120, and 150 mA) was presented for 10 min during the 50-min daily session. Various narcotic (codeine sulfate, meperidine hydrochloride) and narcotic antagonist (pentazocine, cyclazocine) analgesics, as well as a sedative hypnotic (sodium pentobarbital), were assayed using this procedure. Both the narcotic and narcotic antagonist analgesics reliably elevated the aversive threshold in a dose-dependent manner. Results cannot be accounted for in terms of drug-induced sedation. Sodium pentobarbital, in agreement with the clinical literature, demonstrated poor analgesic properties and was able to elevate reliably the aversive threshold only in doses that severely hampered the execution of the escape response. This technique appears to be specifically sensitive to the analgesic properties of drugs. As such, it should be useful as a screening procedure in the development of new analgesics and for the investigation of possible mechanisms underlying drug addiction.

425. Houser, Vincent P. & Pare, William P. (1972). **A method for determining the aversive threshold in the rat using repeated measures: Tests with morphine sulfate.** ***Behavior Research Methods & Instrumentation,*** 4(3), 135–137.
Describes a procedure which provides an objective automated technique for the psychophysical assessment of the aversive threshold and allows animal Ss to serve as their own controls. Tests of sensitivity to drug effects, as well as comparisons between drugs, are also possible. Using a rectangular tilt cage, the aversive threshold of 6 male Dublin DR rats to grid shock was defined as the intensity of shock avoided 75% of the time. Each shock intensity (30, 60, 90, 120, and 150 m amp) was presented for 5 min on 1 side of the cage and then switched to the other for 5 min, forcing S to sample each shock intensity. A 2nd experiment compared the analgesic effects of morphine sulfate to saline. Ss were tested for 3 days at each morphine dose level (2, 4, 8, and 16 mg/kg), alternated with 3 days of testing under saline. Significant differences were detected between saline and morphine at 4, 8, and 16 mg/kg.

426. Houser, Vincent P. & Pare, William P. (1974). **Long-term conditioned fear modification in the dog as measured by changes in urinary 11-hydroxycorticosteroids, heart rate and behavior.** ***Pavlovian Journal of Biological Science,*** 9(2), 85–96.
Recorded heart rate (HR), keypressing, and urinary 11-hydroxycorticosteroids (11-OH-CS) while dogs were subjected to aversive conditioning schedules over a 6-mo period. The schedules consisted of Sidman avoidance, followed by a Sidman schedule which paired unavoidable shocks with offset of 7 discrete conditioned stimuli. The Sidman avoidance schedule always resulted in an increase in 11-OH-CS. Neither HR nor 11-OH-CS were correlated with rate of operant keypressing. Differences in the dependent variable reflected the different topological characteristics of the Ss studied. Results demonstrate (a) causal independence of physiological and behavioral responses conditioned to the same stimulus complex and (b) that long-term HR increases can be maintained if the experimental situation is manipulated to maintain the fear eliciting characteristics of the conditioning situation.

427. Houser, Vincent P. & Pare, William P. (1974). **Anticholinergics: Their effects on fear-motivated behavior, urinary 11-hydroxycorticosteroids, urinary volume, and heart rate in the dog.** ***Psychological Reports,*** 34(1), 183–197.
Administered scopolamine hydrobromide (.5, .7 mg) and scopolamine methylbromide (.5, .7 mg) to 2 mongrel female dogs while they were subjected to a Sidman nondiscriminated avoidance schedule which contained 7 conditioned stimuli-unavoidable shock pairings. Both anticholinergics significantly elevated urinary 11-hydroxycorticosteroids and heart rate, while only the central acting agent, scopolamine hydrobromide, affected behavior. These results suggest that the behavioral effects of scopolamine hydrobromide are not mediated through its effects on the adrenal-pituitary system. Response rates under scopolamine hydrobromide were substantially reduced leading to increased shock rates, especially during the CS segments of this schedule. These behavioral results are interpreted to suggest that cognitive (possibly memory) functions were altered in response to scopolamine administration.

428. Hughes, Carroll W. et al. (1984). **Cerebral blood flow and cerebrovascular permeability in an inescapable shock (learned helplessness) animal model of depression.** ***Pharmacology, Biochemistry & Behavior,*** 21(6), 891–894.
Tested the effects of an animal model of depression (inescapable shock [IS]) on (a) escape behavior; (b) regional brain levels of norepinephrine (NE), 5-HT, and dopamine; and (c) the response of the cerebromicrovasculature to metabolic demand as mimicked by manipulation of arterial CO_2 content, using 32 male Sprague-Dawley rats. IS treatment ($n = 12$) resulted in increased escape latency and lowered levels of NE and 5-HT in the locus coeruleus but not in terminal fields in distant regions. This treatment also did not alter cerebral blood flow or capillary permeability in distant regions when compared to 15 untreated or 5 sham-treated controls.

429. Hunt, Tony; Poulos, Constantine X. & Cappell, Howard. (1988). **Benzodiazepine-induced hyperphagia: A test of the hunger-mimetic model.** ***Pharmacology, Biochemistry & Behavior,*** 30(2), 515–518.

28 male rats were either food deprived (FD) or maintained on ad lib food but injected with chlordiazepoxide (CDP) 30 min prior to food presentation. Results indicate that in single-bottle tests, while FD Ss exhibited a greater augmentation of eating when given the high-palatability food, CDP Ss exhibited an indiscriminate elevation of eating across both foods. On 2-bottle choice tests, FD Ss exhibited an enhanced preference for the high-palatability food, whereas the CDP Ss did not change from baseline food preference. Results fail to support the hunger-mimetic model of benzodiazepine-induced hyperphagia.

430. Huttunen, Matti O. (1971). **Persistent alteration of turnover of brain noradrenaline in the offspring of rats subjected to stress during pregnancy.** ***Nature,*** 230(5288), 53–55.
Describes "a neurochemical examination of the progeny of stressed pregnant rats, which could be a preliminary step in the construction of an animal model for human mental illnesses." Sprague-Dawley rats were stressed from the 15th–19th day of pregnancy with footshocks (5 min/day). Control mothers were kept in similar cages but received no shock. When the offspring were 40–45 days old they were injected with radioactive noradrenaline, were decapitated, and brains and spinal cords were disected. An increased rate of the disappearance of noradrenaline in telencephalon-diencephalon was found. The possibility of other causal factors (e.g., postnatal effects) is considered. It is suggested that the guinea-pig might better serve as a model for human studies. However, it is concluded that it is "too early to speculate about the theoretical implications of our observations for the etiology of human mental illnesses."

431. Imaizumi, K.; Kudo, Y.; Shiosaka, S.; Lee, Y. et al. (1991). **Specific cholinergic destruction in the basal magnocellular nucleus and impaired passive avoidance behavior of rodents.** ***Brain Research,*** 551(1–2), 36–43.
Examined passive avoidance learning behavior in male mice injected with a nerve growth factor-diphtheria toxin conjugate, to further establish the value of an animal model of Alzheimer's disease. In the model, the cholinergic neurons of the basal forebrain were selectively destroyed while the noncholinergic system in this region and the other cholinergic neurons of the brain remained unaffected. Memory loss was still detected, indicating that the basal cholinergic system is closely involved in the retention of memory.

432. Irle, Eva. (1983). **Physiological psychology in the Federal Republic of Germany: A situation report.** ***Psychologische Rundschau,*** 34(3), 125–133.
Analyzes 29 international neuroscience journals to compare the extent and type of neuro-scientific research in the Federal Republic of Germany (FRG) with those of other countries. It was found that psychological departments in the US, Canada, and Great Britain contributed more neuroscientific research than FRG psychological departments. The former countries' research centered on animal models: research in the FRG focused on human models. Professional education in physiological psychology was of a lower standard in the FRG than in the US. (English abstract).

433. Irving, George W. (1991). **A perspective on the selection of experimental models.** Symposium on Animal-To-Human Extrapolation (1990, San Antonio, Texas). ***Neuroscience & Biobehavioral Reviews,*** 15(1), 15–20.
The use of models of biological systems, including animal models, can be considered in 4 categories: theoretical, in vitro, nonmammalian, and mammalian. Each category has advantages and limitations in describing the dynamic milieu of events that characterize human biologic response. Although individual models can be good predictors, multiple models are better than single models; the most critical drawback is lack of human information for comparison. Other important criteria for the use of animal models include the use of scientific teams as well as combinations of models, the use of institutional review committees, and better communication to the general public of the risks and benefits involved.

434. Jinnah, Hyder A.; Gage, Fred H. & Friedmann, Theodore. (1990). **Animal models of Lesch-Nyhan syndrome.** 19th Annual Meeting of the Society for Neuroscience: Neural basis of behavior: Animal models of human conditions (1989, Phoenix, Arizona). ***Brain Research Bulletin,*** 25(3), 467–475.
In humans, deficiency of the enzyme hypoxanthine-guanine phosphoribosyltransferase (HPRT) is associated with Lesch-Nyhan syndrome, a disorder that includes severe neurobehavioral abnormalities. Several animal models developed to examine the neurobiologic substrates of this disorder have suggested a role for abnormal function in purine/dopamine neurotransmission, but the relationship between HPRT-deficiency and these abnormalities remains unknown. Researchers have recently produced HPRT-deficient mice that appear to have similar, though more subtle, changes in brain dopamine function. These strains may be useful in elucidating the relationship between HPRT-deficiency and the neurological deficits observed in patients with this disorder.

435. Kaltwasser, Maria T. (1991). **Acoustic startle induced ultrasonic vocalization in the rat: A novel animal model of anxiety?** ***Behavioural Brain Research,*** 43(2), 133–137.
Ultrasonic vocalization was induced by either high intensity acoustic stimuli or by electric footshock in male rats. High intensity acoustic stimuli elicited a startle response, while electric footshocks provoked an immediate withdrawal of the feet often accompanied by a pain reaction. Flunitrazepam (.5 mg/kg), diazepam (5 mg/kg), and ipsapirone (5 mg/kg) reduced the vocalization induced by both averse stimuli. Maprotiline (10–25 mg/kg) enhanced the vocalization. The anxiogenic compound FG 7142 (10 mg/kg) had no effect. The acoustic startle-induced vocalization paradigm like the electric footshock-induced vocalization paradigm may provide a simple and reliable tool in the study of anxiety. The advantage of an acoustic pulse as the averse stimulus is discussed.

436. Karpiak, S. E.; Tagliavia, A. & Wakade, C. G. (1989). **Animal models for the study of drugs in ischemic stroke.** ***Annual Review of Pharmacology & Toxicology,*** 29, 403–414.
Describes animal stroke models that meet the criterion of ischemic (i.e., arterial occlusions), involving occlusions of the vascular supply to the central nervous system (CNS). The models include the elimination or reduction of CNS blood supply to (1) the entire cerebral hemispheres (global ischemia), (2) focal regions of the brain, and (3) multiple loci in the brain. Characteristics of the models are detailed by including surgical procedures, pathology, resulting dysfunctions, and their relevance to the human condition.

437. Kennett, Guy A.; Chaouloff, Francis; Marcou, Margaret & Curzon, Gerald. (1986). **Female rats are more vulnerable than males in an animal model of depression: The possible role of serotonin.** ***Brain Research,*** 382(2), 416–421.
Compared the effects of repeated restraint stress on male and female Sprague-Dawley rats. A single 2-hr restraint stress reduced locomotion and increased defecation of male rats placed in an open field 24 hrs later. After daily 2-hr restraints for 5 days, these effects were no longer observed. This adaptation was associated with enhanced sensitivity to the serotonin agonist 5-methoxy-N, N-dimethyltryptamine (5MD). Females were less affected by a single restraint but failed to adapt to the repeated stress procedure and did not exhibit enhanced sensitivity to 5MD. Females killed 24 hrs after the final restraint period had decreased brain regional 5-hydroxyindoleacetic acid concentrations, particularly in the frontal cortex. Similar differences could be involved in the higher incidence of depressive illness in women.

438. Kerr, Frederick W. & Wilson, Peter R. (1978). **Pain.** ***Annual Review of Neuroscience,*** 1, 83–102.
Reviews recent contributions toward understanding pain mechanisms, including stimulus-produced analgesia, identification of the opiate receptor, elucidation of the mechanisms and sites of action of morphine analgesia, and the discovery and characterization of an entirely new endogenous analgesic system. The literature cited deals primarily with animal research, but the application of these findings to human investigations and to medical procedures (such as dorsal rhizotomy and cordotomy) is also considered. The lack of an acceptable animal model for human chronic pain is noted. The authors theorize that pain is mediated via a specific system that is particularly sensitive to noxious stimuli. However, pain control as an ultimate research goal is still elusive.

439. Killam, K. F.; Killam, E. K. & Naquet, R. (1967). **An animal model of light sensitive epilepsy.** ***Electroencephalography & Clinical Neurophysiology,*** 22(6), 497–513.
Studied the EEG and motor patterns of 40 baboons to 25/sec intermittent light stimulation (ILS) which indicated 3 types of stimulus-bound paroxysmal responses, individually determined. In certain Ss, self-sustained epileptiform discharges, divisible into 4 different patterns, followed termination of ILS. The majority of the paroxysmal responses were elicited at 20–30/sec with 25/sec about optimal. Night-long EEG and polygraphic recordings of the least sensitive Ss showed no slow sleep as muscle activity lessened and REM sleep was reduced. The 2 most sensitive Ss showed spontaneous paroxysmal activity resembling spike and wave discharges during their nocturnal sleep and no identifiable REM sleep. Evoked neocortical responses to 1-sec ILS, usually studied as summed potentials in groups of 30, differed with eyes open and closed, and between Ss. Ss showed low thresholds to pentylenetetrazol-induced seizures in the presence of ILS. Anticonvulsant activity of phenobarbital, trimethadione, and diazepam was demonstrated against ILS-induced epileptiform seizures in 2 Ss displaying stable, daily severe paroxysmal responses to ILS. (French summary).

440. Kiss, J.; Schlumpf, M. & Balázs, R. (1989). **Selective retardation of the development of the basal forebrain cholinergic and pontine catecholaminergic nuclei in the brain of trisomy 16 mouse, an animal model of Down's syndrome.** ***Developmental Brain Research,*** 50(2), 251–264.
Examined specific neuronal systems in fetal murine trisomy 16 (TS16) mice brains (considered to be a model of Down's syndrome [DS]), focusing on the morphological detection of cholinergic, catecholaminergic, and serotoninergic nerve cells. These neuronal systems were chosen because in Alzheimer's disease (AD) they are characteristically affected, and DS adults from the 4th decade exhibit neuropathological and neurochemical alterations comparable to those in AD. In addition to a reduction in brain size and cortical thickness, a severe reduction throughout the brain in the density of muscarinic receptors was observed. TS16 results in selective retardation of some neuronal systems, which may lead to a perturbation of brain development. Systems whose development was retarded are those which in DS adults exhibit pronounced deficits of cells.

441. Kokkinidis, Larry & Zacharko, Robert M. (1980). **Response sensitization and depression following long-term amphetamine treatment in a self-stimulation paradigm.** ***Psychopharmacology,*** 68(1), 73–76.
The effects of long-term dextroamphetamine (DAM) treatment (7.5 mg/kg 2 times/day for 5 days, ip) were evaluated on responding supported by self-stimulation of the substantia nigra. Charles River CD rats repeatedly treated with DAM and tested with a dose of the drug that ordinarily has no behavioral effect (0.3 mg/kg) showed higher response rates than Ss repeatedly treated with saline and tested with the same dose of DAM. In contrast, a depression in responding was observed among Ss that received long-term DAM and were tested with saline. These results cannot be explained by the intrusion of drug-induced competitive behaviors such as locomotor activity and stereotypy. Results are attributed to changes in dopamine neurotransmission following prolonged exposure to DAM and are discussed in terms of an animal model for amphetamine psychosis and postamphetamine depression in humans.

442. Koo, M. W.; Cho, C. H. & Ogle, C. W. (1989). **Does acidosis contribute to stress-induced ulceration in rat stomachs?** ***Pharmacology, Biochemistry & Behavior,*** 33(3), 563–566.
Exposed female rats to either cold restraint stress or normal housing conditions. Cold restraint stress for 2 hrs did not affect the blood lactate level; however, it produced respiratory acidosis, as reflected by the depressed respiratory rate which was associated with increased CO_2 tension and a lowered blood pH. Severe hemorrhagic ulceration was found in the glandular mucosa. The effects of stress on blood pH and the stomach were reversed by iv infusion of $NaHCO_3$. Infusion of HCl iv decreased the blood pH and HCO_3 level and produced gastric ulceration. Respiratory acidosis could be involved in stress ulceration. The metabolic acidosis evoked by HCl also induced gastric damage, but the effect was much less.

443. Kornetsky, Conan & Eliasson, Mona. (1969). **Reticular stimulation and chlorpromazine: An animal model for schizophrenic overarousal.** ***Science,*** 165(3899), 1273–1275.
Developed a model to test the hypothesis that certain schizophrenic patients are in a state of continual central excitation and that improvement in these patients after treatment with chlorpromazine is a result of the action of the drug in reducing this excitation. Six male albino holtzman rats were electrically stimulated in the mesencephalic reticular formation while performing a simple attention task.

Stimulation or treatment with chlorpromazine impaired the performance of the Ss; however, the 2 treatments together resulted in performance indistinguishable from that seen after injections of saline alone.

444. Kraemer, Gary W. & Clarke, A. Susan. (1990). **The behavioral neurobiology of self-injurious behavior in rhesus monkeys.** ***Progress in Neuro-Psychopharmacology & Biological Psychiatry,*** 14(Suppl), S141–S168.
Describes a neurobiological "cause" of self-injurious behavior (SIB) in animals that is at least analogous, and preferably homologous, to the probable cause(s) of SIB in humans. There are 3 lines of evidence that nonhuman primate SIB is linked to malfunctions in the norepinephrine and serotonin (5-hydroxytryptamine [5-HT]) neurotransmitter systems. The activity of these systems appears to be altered by psychosocial deprivation. The functional relationship between the 2 systems appears to be altered or absent in socially deprived monkeys. Pharmacologic agents that act on these systems alter SIB in monkeys. Preliminary data from socially deprived monkeys are consistent in major respects with evidence linking altered serotonin systems to SIB and suicidal motivation in humans who also probably suffer from social deprivation.

445. Krasnegor, Norman A. (1987). **Developmental psychobiology research: A health scientist administrator's perspective.** ***Developmental Psychobiology,*** 20(6), 641–644.
Surveys trends in developmental psychobiology research, based on the author's experiences as chief of the Human Learning and Behavior Branch of the National Institute of Child Health and Human Development (NICHD). The NICHD's emphasis on interdisciplinary research is noted. Trends discussed include the focus on perinatal development and the shift from animal models to studies of human neonates.

446. Kshama, Devi; Hrishikeshavan, H. J.; Shanbhogue, R. & Munonyedi, U. S. (1990). **Modulation of baseline behavior in rats by putative serotonergic agents in three ethoexperimental paradigms.** ***Behavioral & Neural Biology,*** 54(3), 234–253.
Examined the effects on behavior of alteration in serotonin (5-hydroxytryptamine [5-HT]) neurotransmission following treatment with site-specific neuropharmacological probes. Three animal models of anxiety, the hole-board, elevated plus maze, and bright/dark arena, were used to test 226 inbred male rats. Uniform results were obtained for each probe in all anxiety models, suggesting that the battery of anxiety tests chosen is reliable and sensitive to detect unknown pharmacological responses. Stimulation of serotonergic neurotransmission appeared to heighten normal anxiety, whereas its blockade released normal behavioral inhibition. The results establish the validity of the 3 paradigms in evaluating the involvement of multiple neurotransmitter receptors in the control of behavior of rodents under natural circumstances.

447. Kumar, R. & Stolerman, I. P. (1973). **Morphine dependent behaviour in rats: Some clinical implications.** ***Psychological Medicine,*** 3(2), 225–237.
Reviews the literature and uses published reports to describe an animal model of morphine dependence. In this model, rats learn to overcome their aversion to solutions of morphine and eventually drink such solutions in preference to water on choice trials. Selected aspects of the acquisition, maintenance, and elimination of this type of morphine dependent behavior are studied and the findings discussed in terms of their relevance to man.

448. Lambert, Kelly G. & Peacock, L. J. (1989). **Feeding regime affects activity-stress ulcer production.** ***Physiology & Behavior,*** 46(4), 743–746.
Divided 64 male rats into either an active (ACT) group housed in activity wheels or a control (CTL) group housed in stationary laboratory cages. Both ACT and CTL groups were further divided into groups receiving 1, 2, 3, or 4 meals daily for a total feeding time of 1 hr. CTLs were food-yoked to ACT Ss. ACT Ss fed 1 meal daily developed significantly more ulceration, lost more weight, and consumed less food and water than other groups. The number of daily meals had no effect on the amount of activity. No CTLs developed ulcers, although they received the same amount of food. Frequent feedings mitigated gastric peptic ulcer formation in rats placed in the activity-stress ulcer paradigm.

449. Lane, John D. et al. (1982). **Amino acid neurotransmitter utilization in discrete rat brain regions is correlated with conditioned emotional response.** ***Pharmacology, Biochemistry & Behavior,*** 16(2), 329–340.
Male F-344 conditioned emotional response (CER) rats suppressed responding and exhibited anxious behavior after presentation of the CS, while shock and tone only control groups, or CER Ss that received diazepam (2.5–50 mg/kg) prior to testing, did not suppress. Few changes were observed in the content of amino acids, suggesting that the behavioral manipulations were acting within normal physiological limits. Numerous changes were observed in the utilization of the amino acid neurotransmitters. The effects of a history of shock presentation were persistent and resulted in a decreased turnover of the amino acids in many areas. CER conditioning-emotion produced an increase in the turnover of aspartate and glutamate in many structures, while changes in GABA turnover were generally limited to decreases in limbic areas. If CER represents an animal model of anxiety, these observations may suggest roles for neurons that utilize amino acids in mediating or responding to emotional components of the paradigm.

450. Langlais, Philip J.; Mair, Robert G.; Anderson, Clint D. & McEntee, William J. (1987). **Monoamines and metabolites in cortex and subcortical structures: Normal regional distribution and the effects of thiamine deficiency in the rat.** ***Brain Research,*** 421(1–2), 140–149.
Examined concentrations of monoamines and metabolites, and estimates of turnover rate in the brains of rats fed a pyrithiamine and thiamine deficient diet (PTD) and normally fed controls. In 17 behaviorally tested PTD Ss, a significant reduction in norepinephrine (NE) was observed in the entorhinal cortex. Diminished NE was also observed in entorhinal, hippocampal, septal, and olfactory areas of 8 nonbehaviorally tested PTD Ss. Serotonin and 5-hydroxyindoleacetic acid were increased in the midbrain-thalamus of both groups of PTD Ss. Findings are discussed in terms of thiamine deficiency as an animal model of Korsakoff's disease. In 12 controls, the entorhinal cortex contained the highest levels of NE and 5-hydroxytryptamine (5-HT), while dopamine was highest in the somatosensory cortex.

451. Lanum, Jackie et al. (1984). **Effects of restraint on open-field activity, shock avoidance learning, and gastric lesions in the rat.** ***Animal Learning & Behavior,*** 12(2), 195–201.
Suggests that many studies of learned helplessness in rats confound the stress of restraint with inescapable shock. In the present experiment, 108 Sprague-Dawley rats were held immobile for 0, 2, 8, 14, or 18 hrs. Behavioral deficits were observed in an open-field activity maze and in 2-way shuttlebox avoidance acquisition. In the activity maze, a sex × restraint interaction was observed for latency to leave the center square, ambulation, and frequency-of-center-square crossing. Males were slower to leave the center square, had fewer ambulations, and crossed the center square less frequently than females. These effects were potentiated by restraint. Males reared significantly less than females, and restrained Ss reared significantly less than non-restrained Ss. Restraint also significantly increased the frequency of grooming. On the avoidance tasks, a significant restraint × trial block interaction indicated slower learning for restrained Ss. The severity of the decrements increased with restraint duration. The presence of stomach lesions was positively correlated with stress duration and the severity of the behavioral decrement. Data indicate that restraint produces a variety of behavioral changes, which may result in interpretive difficulties for helplessness studies that confound restraint and shock.

452. Lawler, James E. & Cox, Ronald H. (1985). **The borderline hypertensive rat (BHR): A new model for the study of environmental factors in the development of hypertension.** ***Pavlovian Journal of Biological Science,*** 20(3), 101–115.
Results of studies with the offspring of male Wistar-Kyoto and female spontaneously hypertensive rats show that (1) these Ss did not develop spontaneous hypertension but displayed a more sensitive cardiovascular response to environmental stressors than rats with normotensive parents; (2) Ss developed spontaneous borderline hypertension; and (3) when subjected to shock-shock conflict, these BHRs developed permanent hypertension that failed to abate even after a 10-wk, shock-free recovery period. The hypertension was accompanied by elevated heart weight/body weight ratios and by significant cardiac pathology. Findings also indicate that BHRs became hypertensive when fed a high-sodium diet and that BHRs subjected to a shock stressor were protected against stress-induced hypertension if they exercised daily. The potential of this model for studies of the mechanisms by which environmental variables produce permanent hypertension is discussed.

453. Lehman, Constance D.; Rodin, Judith; McEwen, Bruce & Brinton, Roberta. (1991). **Impact of environmental stress on the expression of insulin-dependent diabetes mellitus.** ***Behavioral Neuroscience,*** 105(2), 241–245.
To investigate the influence of environmental factors on inherited tendencies, the impact of chronic environmental stress on the expression of a genetically determined autoimmune disease was explored in the bio-breeding (BB) rat, which is an animal model for human autoimmune insulin-dependent diabetes mellitus. Animals assigned at random to the experimental group received a triad of stressors designed to model chronic moderate stress over a 14-wk period. Animals from 25 to 130 days of age were weighed and tested for glycosuria twice weekly. Weekly blood sampling was performed on all animals. Diabetes was diagnosed on the basis of weight loss, 2+ glycosuria, and blood glucose levels of 250+ mg/dl. We found that in the BB rat chronic stress significantly increased the incidence of the phenotypic expression of the gene for Type I diabetes. 80% of the male stress and 70% of the female stress animals developed diabetes, compared with 50% in both control groups. Stressed males developed manifest diabetes at the same time as their matched controls, whereas stressed females had significantly delayed onset in relation to controls.

454. le Moal, Michel & Jouvent, Roland. (1991). **Psychopathology and experimental models.** ***Psychologie Française,*** 36(3), 211–220.
Discusses the development and use of psychopathology models based on animal models. Different types of experimental models, difficulties in establishing analogies between animal behavior and psychiatric disorders, and criteria for using experimental models are considered. Several types of experimental and clinical models are described, including physiopathological models, syndromatic or symptomatic models, and transnosographic models. (English abstract).

455. le Roch, Kaisu; Riche, Danielle & Sara, Susan J. (1987). **Persistence of habituation deficits after neurological recovery from severe thiamine deprivation.** ***Behavioural Brain Research,*** 26(1), 37–46.
Examined rats that were fed a thiamine-deficient diet for 4 wks and injected daily with pyrithiamine during the last 2 wks of the diet. This regime induced severe neurological anomalies such as ataxia, loss of righting reflex and visual place reflex, and finally full tonic–clonic seizures, reminiscent of the clinical Wernicke-Korsakoff syndrome. Injection of thiamine reversed these symptoms within 1 or 2 hrs as seen in Wernicke patients. Six weeks later these Ss showed a deficit in habituation of exploratory behavior and of the auditory orienting response. Histological data indicate a heterogeneous pattern of damage to the brainstem, including mammillary bodies and several thalamic nuclei, reminiscent of that seen in Korsakoff patients.

456. Lester, David & Freed, Earl X. (1973). **Criteria for an animal model of alcoholism.** ***Pharmacology, Biochemistry, & Behavior,*** 1(1), 103–107.
Argues that a correspondence between the various components of human alcoholism and their animal analogue has not yet been achieved; in some part, this failure resides with experimental attempts which obtain only surface equivalencies and which lack an underlying motivational structure. Seven criteria for an animal model are proposed including the oral ingestion of alcohol without food deprivation, substantial ingestion of alcohol with competing fluids available, drinking directed to the intoxicating effect of alcohol, the performance of work to obtain alcohol, the maintenance of intoxication over a long period and, finally, the production of physical dependence and, on withdrawal, the abstinence syndrome.

457. Levine, R. J. & Senay, E. C. (1970). **Studies on the role of acid in the pathogenesis of experimental stress ulcers.** ***Psychosomatic Medicine,*** 32(1), 61–65.
To examine directly the relationships between gastric acidity and stress ulcer formation, rats were stressed for 2 hr. by restraint in a cold environment after some of them were treated by intragastric administration of basaljel. "The results indicate that the role of histamine in the pathogenesis of stress ulcers in rats is probably mediated through stimula-

tion of gastric acid secretion and that basaljel afforded significant protection against stress ulcers without preventing the sharp increase in histidine decarboxylase activity in response to stress."

458. Levis, Donald J. & Caldwell, Donald F. (1971). **The effects of a low dosage of mescaline and 3,4-dimethoxyphenylamine under two levels of aversive stimulation.** ***Biological Psychiatry,*** 3(3), 251–257.
Reports an experiment in which male rats received either mescaline sulfate, 3,4-dimethoxyphenylethylamine (dmpea), isotonic sodium chloride, or no injection ($n = 40$ per group). Half the Ss in each group received .6-ma shocks and the other ½ 1-ma shocks in a signaled pole climbing escape-avoidance task. Injections preceded 150 conditioning trials by 10 min About ½ the Ss in each group did not learn the task (10 consecutive avoidances = criterion). For those Ss who did learn, at high shock levels both mescaline sulfate and dmpea receivers responded faster than controls, and both groups of drugged Ss made longer strings of avoidances than controls at low shock levels, although the reverse was true at high shock levels. Data are interpreted as showing that mescaline and dmpea facilitate behavior during low drive states and inhibit it when drive is high.

459. Loch, R. K.; Rafales, L. S.; Michaelson, I. A. & Bornschein, R. L. (1978). **The role of undernutrition in animal models of hyperactivity.** ***Life Sciences,*** 22(22), 1963–1970.
Studied the influence of neonatal growth retardation on subsequent spontaneous activity and activity following dextroamphetamine (10 mg/kg, ip) in CD-1 mice. Different growth rates were obtained by raising mice in litters of either 8 or 16 pups per lactating dam. The testing protocol was specifically designed to duplicate a procedure used to assess the influence of neonatal lead exposure on locomotor activity. At 35–37 days of age Ss were individually tested for general locomotor activity and drug response. Developmental growth retardation influenced their pattern of habituation to the test apparatus and their locomotor response to amphetamine. It is concluded that growth retardation may partially account for behavioral effects previously attributed to the neurotoxic effects of viruses, 6-hydroxydopamine or inorganic lead.

460. Loskota, W. J.; Lomax, P. & Rich, S. T. (1974). **The gerbil as a model for the study of the epilepsies: Seizure patterns and ontogenesis.** ***Epilepsia,*** 15(1), 109–119.
Describes a new strain of selectively bred seizure-sensitive Mongolian gerbils. These WJL/UC gerbils exhibit stereotyped seizures in response to increased stimulus input that can be rated on a 7-point scale of severity. Integrated motor activity during seizures and latency and duration of seizures are related to severity. Susceptibility to seizures starts at 9 wks and reaches 97% at 6 mo. The severity increases from a mean of 1.2 at 2 mo to 4.1 at 6 mo of age. Death in seizure is rare. These attributes make the WJL/UC seizure-sensitive Mongolian gerbil suitable for testing as an animal model for study of the epilepsies. (French, German, & Spanish summaries).

461. Luparello, T. J. (1966). **Restraint and hypothalamic lesions in the production of gastroduodenal erosions in the guinea pig.** ***Journal of Psychosomatic Research,*** 10(3), 251–254.
Using 6 groups of guinea pigs the production of ulcers in the guinea pig by restraint and hypothalamic lesions was investigated. It was found that gastroduodenal erosions can be induced in the species by restraint for 24 hrs and by bilateral electrolytic lesions in the anterior and posterior hypothalamus.

462. Luther, Irene G.; Heistad, G. T. & Sparber, S. B. (1969). **Influence of pregnancy upon gastric ulcers induced by restraint.** ***Psychosomatic Medicine,*** 31(1), 45–56.
Using a modification of Senay and Levine's technique of restraint in a cold environment gastric ulcers were induced in 31 pregnant and 31 control Sprague-Dawley female rats to explore the problem of whether pregnancy protects against ulcer induction. "Under these experimental conditions, pregnancy did not generate ulcer protection and, in fact, increased ulcer incidence was found at the end (20th day) of pregnancy.... The general statement that pregnancy has a protective effect against ulceration should be restricted to the specific experimental conditions under which it has been demonstrated."

463. Luther, Irene G.; Heistad, G. T. & Sparber, S. B. (1969). **Effect of ovariectomy and of estrogen administration upon gastric ulceration induced by cold-restraint.** ***Psychosomatic Medicine,*** 31(5), 389–392.
Gastric ulcers were induced in 66 ovariectomized and 14 intact rats using a technique of restraint in a cold environment. Under these conditions ovariectomy strikingly reduced sensitivity to ulcer induction while estrogen administration showed no effect in either intact or ovariectomized animals.

464. MacLennan, A. John & Maier, Steven F. (1983). **Coping and the stress-induced potentiation of stimulant stereotypy in the rat.** ***Science,*** 219(4588), 1091–1093.
Administered controllable or uncontrollable shocks to rats, followed by ip injections of amphetamine or cocaine. In both cases, Ss that received uncontrollable shocks were more sensitive to the drugs than those that received controllable shocks. Findings have implications for the role of stress and coping in amphetamine and cocaine psychoses, endogenous psychoses, and some forms of schizophrenia.

465. Mah, Chris; Suissa, Albert & Anisman, Hymie. (1980). **Dissociation of antinociception and escape deficits induced by stress in mice.** ***Journal of Comparative & Physiological Psychology,*** 94(6), 1160–1171.
Six experiments (327 Swiss-Webster mice) assessed the conditions under which stress would induce antinociception in a subsequent hot-plate test. Both footshock and tail shock produced the antinociception. This effect was apparent with as little as a single shock trial. The magnitude of the antinociception was maximal following 15 shock presentations and was largely reduced after 60 shocks. In contrast to the results of R. L. Jackson et al (1979), whether stress was escapable was not a necessary feature needed to produce the antinociception. Moreover, the magnitude of the antinociception induced by stress was not enhanced in mice that had previously been exposed to stress. Finally, morphine (10.0, 20.0, and 30.0 mg/kg, ip) produced a pronounced antinociception but did not appreciably influence escape performance in a shuttle task in which performance was disrupted by inescapable shock. It is suggested that the antinociception and shuttle-escape deficits induced by uncontrollable shock are independent of one another.

466. Maier, Steven F. et al. (1980). **Opiate antagonists and long-term analgesic reaction induced by inescapable shock in rats.** ***Journal of Comparative & Physiological Psychology,*** 94(6), 1172–1183.
Five experiments with 240 male albino rats examined the influence of opiate antagonists (naltrexone; 1–14 mg/kg, ip) on both the short-term analgesic reaction resulting 30 min after exposure to inescapable shock and the long-term analgesic reaction resulting after reexposure to shock 24 hrs after inescapable shock exposure. Exp I showed that the long-term analgesic reaction could be reduced by administration of naltrexone prior to exposure to inescapable tail shock. Exp II showed that the reduction in the long-term analgesic reaction produced by naltrexone was dose-dependent. Exp III showed that the long-term analgesic reaction could also be reduced by administration of naltrexone prior to reexposure to shock. Exp IV showed that the long-term analgesic reaction could be reduced by administration of a large dose of naloxone prior to reexposure to shock. Exp V showed that the short-term analgesic reaction was reduced by naltrexone administered prior to inescapable shock. Implications for the biochemical substrates of both learned helplessness and stress-induced analgesia are discussed.

467. Maier, Steven F. (1989). **Determinants of the nature of environmentally induced hypoalgesia.** ***Behavioral Neuroscience,*** 103(1), 131–143.
Factors that determine whether naltrexone and pentobarbital anesthesia block the hypoalgesia produced by electric shock were investigated. One shock was followed by an initial hypoalgesia that was reversed by naltrexone but was not affected by pentobarbital. When testing was conducted in the shock context, early hypoalgesia was followed by a hypoalgesia that was still reversed by naltrexone but was eliminated by pentobarbital anesthesia. This second hypoalgesia did not occur when testing was conducted after removing the animal from the shock apparatus. Five shocks were also followed by 2 separable responses. The first response was not reduced by pentobarbital, but it was not blocked by naltrexone. This initial response was not prevented by removal from the shock context, but the 2nd response did not occur after removal. The pattern after 80 shocks was quite different. Here the hypoalgesic response did not appear to change in character across the 10 min of testing or with removal from the shock apparatus. The response persisted for 10 min in the shock context or after removal and was blocked by both naltrexone and pentobarbital throughout its entire duration.

468. Mair, R. G.; Anderson, C. D.; Langlais, P. J. & McEntee, W. J. (1985). **Thiamine deficiency depletes cortical norepinephrine and impairs learning processes in the rat.** ***Brain Research,*** 360(1–2), 273–284.
Examined behavioral deficits in 23 male Long-Evans rats after recovery from a bout of thiamine deficiency to determine whether thiamine deficiency causes Wernicke-Korsakoff syndrome. Impairments were observed for a spatial delayed-alternation task that had been learned prior to experimental treatment. Experimental Ss were impaired in their ability to acquire 2 novel tasks, active and passive shock avoidance, after recovery from the acute effects of thiamine deficiency. Data demonstrate that a bout of thiamine deficiency can produce persistent deficits in brain norepinephrine and concomitant decrements in behavioral measures of learning and memory. Results are consistent with evidence that noradrenergic deficits contribute to the amnesic symptoms of Korsakoff's psychosis.

469. Manning, Frederick J. et al. (1978). **Microscopic examination of the activity-stress ulcer in the rat.** ***Physiology & Behavior,*** 21(2), 269–274.
Gastric lesions were produced in Wistar-derived albino rats by allowing continuous access to a running wheel, but only 1 hr of food access each day. A reliable pattern of changes in activity, food intake, and body weight, culminating in severe debilitation, resulted. Ss were sacrificed when food intake fell below 2 g. The glandular stomachs of these Ss showed extensive pathology, ranging from mucosal hemorrhage, with or without focal necrosis, through acute ulcers characterized by discontinuity of the muscularis mucosa. With 1 exception, all experimental Ss showed this entire range of pathology, and in addition, 33% had at least 1 lesion showing signs of regeneration. Focal erosions predominated, but all experimental Ss had at least 1 ulcer penetrating the muscularis mucosa. No control had a lesion of any sort.

470. Manuck, Stephen B.; Kaplan, Jay R. & Clarkson, Thomas B. (1983). **Social instability and coronary artery atherosclerosis in cynomolgus monkeys.** ***Neuroscience & Biobehavioral Reviews,*** 7(4), 485–491.
Two experiments were conducted to facilitate development of an appropriate animal model for examining the atherogenic effects of psychosocial variables: In Exp I, an experimental stressor—involving repeated reorganization of 2 socially housed groups of 15 adult, male cynomolgus monkeys (*Macaca fasicularis*) fed a moderately atherogenic diet—resulted in increased coronary artery atherosclerosis only among Ss that retained dominant social status. In Exp II, which employed the same experimental procedures among Ss fed a low cholesterol/low saturated fat diet, periodic social group reorganization similarly led to development of greater atherosclerosis in the coronary arteries.

471. Marchand, Serge; Trudeau, Nathalie; Bushnell, M. Catherine & Duncan, Gary H. (1989). **A primate model for the study of tonic pain, pain tolerance and diffuse noxious inhibitory controls.** ***Brain Research,*** 487(2), 388–391.
Used a primate model of cold pressor pain in which the animal itself initiated all painful stimuli and sustained the painful stimulus with an active response, thereby avoiding the inadvertent administration of intolerable pain. Results with a female monkey suggest that the test can be used successfully in primates as a pain tolerance measure. A comparison with subjective reports of cold pain for 2 humans in a no-reward experiment indicate the major influence of motivational level on tolerance time. Implications regarding analgesic manipulations are noted.

472. Martin, P.; Soubrie, P. & Simon, P. (1986). **Shuttle-box deficits induced by inescapable shocks in rats: Reversal by the beta-adrenoreceptor stimulants clenbuterol and salbutamol.** ***Pharmacology, Biochemistry & Behavior,*** 24(2), 177–181.
Studied the antidepressant effects of clenbuterol and salbutamol on male Wistar rats subjected to helplessness training. Ss were exposed to inescapable shock pretreatment (60 shocks, 15-sec duration) and 48 hrs later, shuttle-box training (30 trials/day) was initiated to evaluate escape and avoidance deficits. Ss pretreated with inescapable shocks

exhibited escape and avoidance deficits when tested for subsequent responding in a shuttle-box. Deficits were particularly marked at the 3rd training session. Daily injections of clenbuterol (0.5 and 0.75 mg/kg) and salbutamol (16 and 24 mg/kg) prevented escape deficits as did daily injections of classical antidepressants such as desipramine and clomipramine (16 and 24 mg/kg/day). These data extend previous results bearing on the similarity of action of beta receptor stimulants and tricyclic antidepressants and further support the notion of a close relationship between noradrenergic function, more especially beta-adrenoreceptors, and helpless behavior.

473. Martin, P.; Soubrié, P. & Simon, P. (1986). **Noradrenergic and opioid mediation of tricyclic-induced reversal of escape deficits caused by inescapable shock pretreatment in rats.** ***Psychopharmacology,*** 90(1), 90–94.
Male Wistar rats were first exposed to 60 inescapable shocks and 48 hrs later were subjected to daily shuttlebox sessions during 3 consecutive days. Twice-daily injections of desipramine or clomipramine prevented escape deficits. Penbutolol, prazosin, and naloxone given once a day dose-dependently attenuated the beneficial effect of tricyclic antidepressants (TADs) in reducing the number of escape failures in Ss exposed to shock pretreatment. In agreement with data obtained in the forced-swimming model, these findings support the notion that activation of noradrenergic and opioid receptors is an important factor in the mediation of the effects of TADs in animal models of depression.

474. McCarty, Richard. (1983). **Stress, behavior and experimental hypertension.** ***Neuroscience & Biobehavioral Reviews,*** 7(4), 493–502.
Presents a review in which the advantages and limitations of studying animal models of essential hypertension are evaluated. Emphasis is placed on the relationship between stressful stimulation and behavioral and physiological responsiveness in 2 animal models of essential hypertension. Previous studies have examined sympathetic nervous system activity and behaviors of rats under basal conditions and following acute or chronic exposure to stressful stimulation. These findings indicate that the spontaneously hypertensive (SHR) strain is excessively responsive behaviorally and physiologically to a variety of stressful stimuli when compared to a normotensive control strain. However, behavioral and physiological responses of genetically hypertensive and normotensive rats do not differ following acute exposure to stress. Thus, the hyperreactivity of SHR rats to stressful stimulation is not necessarily related to the development of hypertension, but may be a valuable marker of the predisposition to develop high blood pressure in the SHR strain. An experimental approach is outlined for examining the causal relationship between a genetically determined physiological or behavioral marker and the development of hypertension.

475. McClearn, G. E.; Wilson, James R.; Petersen, Darrel R. & Allen, D. L. (1982). **Selective breeding in mice for severity of the ethanol withdrawal syndrome.** ***Substance & Alcohol Actions/Misuse,*** 3(3), 135–143.
Discusses the advantages of using multivariate indices to define pharmacological concepts, using 200 genetically heterogeneous HS/Ibg mice selectively bred for a multivariate index of the ethanol withdrawal syndrome. The 7 variables included in the index are seizure score, body temperature, hole-in-wall crossings, hole-in-wall rearings, hole-in-wall seizure scores, vertical screen crossings, and total ethanol consumption (in g/kg). Because 5 of the 7 variables showed a significant sex difference, separate principal component analyses were performed to establish selection indices appropriate for each sex. Within-family mating was employed to minimize inbreeding, and replicate high, low, and control lines were included. It is concluded that these selectively bred lines should eventually provide a useful animal model for studying physiological and neurological mechanisms that may be involved in alcohol withdrawal syndrome.

476. McClintock, Martha K. & Adler, Norman T. (1978). **Induction of persistent estrus by airborne chemical communication among female rats.** ***Hormones & Behavior,*** 11(3), 414–418.
Female Sprague-Dawley rats housed in a constantly lighted environment to induce estrus were used as an odor source to test whether they could facilitate estrus in other female rats via airborne chemical communication. Results show that airborne chemical communication can mediate the induction of persistent estrus in rats.

477. McCutcheon, N. Bruce; Rosellini, Robert A. & Bandel, Steven. (1991). **Controllability of stressors and rewarding brain stimulation: Effect on the rate-intensity function.** ***Physiology & Behavior,*** 50(1), 161–166.
Examined rate-intensity functions for rewarding brain stimulation in male rats 24 and 48 hrs following stressor exposure at 2 electrodes, 1 on each side of the brain, aimed at the posterior, lateral hypothalamus. Stimulation of 1 electrode always caused stronger activity than equivalent stimulation of the other. If the rats had been stressed with uncontrollable shock the preceding day, their responding was significantly depressed to stimulation at the weaker electrode site but not the stronger electrode site. However, if the rat had control over the stressor, then rate-intensity functions were unchanged. Repeating the test for rate-intensity function 48 hrs after exposure to the stressor showed no effect of prior stress on self-stimulation. It is suggested that some effects of exposure to uncontrollable stressors on acquisition and maintenance of responding for rewards may be due to attenuation of reward strength.

478. McFarland, Dennis J. & Hotchin, John. (1980). **Early behavioral abnormalities in mice due to scrapie virus encephalopathy.** ***Biological Psychiatry,*** 15(1), 37–44.
The behavioral effects of scrapie virus infection, a slow degenerative disease of the CNS, were examined in Nya: NYLAR mice. Testing was conducted at 50, 100, and 150 days postinfection and included open-field behavior, Y-maze alternation and activity, and 2-way shuttle-box avoidance. The behavioral pathology was found to be task-specific rather than of a global nature. Furthermore, the effects observed could be classified as either early or late components of the disease. The relevance of this animal model to human presenile dementias is discussed.

479. McGaugh, James L. (1983). **Preserving the presence of the past: Hormonal influences on memory storage.** ***American Psychologist,*** 38(2), 161–174.
Reviews current research on memory storage, particularly that pertaining to the modulating influences of peripheral epinephrine and of amygdala stimulation on memory. It is known that memory in animals and humans is influenced by treatments such as drugs and electrical stimulation of the brain administered shortly after training. The fact that newly acquired memories are so readily influenced suggests that

memory storage processes may be modulated by endogenous physiological systems activated by experience. This suggestion is supported by findings that retention is influenced by posttraining administration of hormones of the adrenal medulla as well as other hormones released by training experiences. It is suggested that the effects of amygdala stimulation on memory are due to influences mediated by the stria terminalis. Further, findings indicating that the effect of electrical stimulation on memory is altered by treatments affecting peripheral epinephrine suggest that endogenous modulation of memory may involve cooperative influences of peripheral hormones and brain systems. Endogenous modulating systems seem to provide a basis for selecting experiences for storage.

480. McGaugh, James L. (1976). **Neurobiology and the future of education.** ***School Review,*** 85(1), 166–175.
Refers to animal studies showing that stimulant drugs and electrical stimulation apparently act under certain conditions to improve learning, rather than merely facilitating its acquisition by maintaining alertness or attention. In animals, changes in hormonal states affect learning and may be the medium by which drugs and electrical stimulation work. Human responses are probably similar to those of animals; in the future, it may be possible to develop drug and/or hormone treatments to temporarily or permanently improve human memory. Such treatments could be important in the control of learning and memory disorders but might be unwise as aids to learning in persons without these disorders. The effects of drugs on complex cognitive processes are unknown. Use of drugs might divert attention from the problems of why the learning of certain individuals or groups is poor.

481. McKenzie, G. M.; Gordon, R. J. & Viik, K. (1972). **Some biochemical and behavioural correlates of a possible animal model of human hyperkinetic syndromes.** ***Brain Research,*** 47, 439–456.
Analyzed brain perfusates in the cat for release of (3H) catechols and behavioral studies of rats after unilateral application of D-tubocurarine in the nucleus caudatus putamen. The latter treatment produced choreiform activity characterized by involuntary movements of the contralateral forelimb, contralateral rotation of the head, facial grimacing, and teeth chattering. These effects were blocked by intraperitoneal doses of either haloperidol or chlorpromazine, and intrastriatal injections of neostigmine. It is suggested that the choreiform activity in the rat could be due to an inappropriate increased release of catecholamines in the striatum as evidenced by the increased release of basic (3H) catechols from the caudate nucleus of the cat. It is noted that the limited pharmacological and biochemical data in human hyperkinetic syndromes correlate well with the data derived from this investigation and suggest that the D-tubocurarine-induced choreiform activity in the rat may represent at least the beginning of a workable animal model of human hyperkinetic disorders.

482. Mefford, Ivan N. et al. (1983). **Narcolepsy: Biogenic amine deficits in an animal model.** ***Science,*** 220(4597), 629–632.
Conducted a study on the brains of 5 Doberman pinschers with genetically transmitted narcolepsy and 7 control Dobermans. It was found that concentrations of biogenic amine metabolites in discrete brain areas differed significantly between narcoleptic Ss and controls. Narcoleptic Ss consistently exhibited lower utilization of dopamine and higher intraneuronal degradation of dopamine but no uniform decrease in serotonin utilization.

483. Mello, Nancy K. (1973). **A review of methods to induce alcohol addiction in animals.** ***Pharmacology, Biochemistry, & Behavior,*** 1(1), 89–101.
Notes that, since alcoholism has been shown to be a form of addiction, there has been increasing attention to the development of an animal model of alcoholism. Within the past 5 yrs, several techniques have been developed to produce physical dependence upon alcohol in a number of species. This review presents a summary and evaluation of the various approaches used and a discussion of the relative merits of the behavioral (self-administration) and pharmacological (forced administration) models. Several questions are raised concerning the relationship between physical dependence and subsequent self-administration of alcohol. It is concluded that although it has been possible to produce physical dependence upon alcohol in animals, the critical determinants of the addictive process are still unknown.

484. Meyer, A.-E. (1989). **On the psychology of separation: Animal experiments in psychosomatic medicine.** ***Psychotherapie Psychosomatik Medizinische Psychologie,*** 39(3–4), 106–109.
Discusses the psychosomatic effects of early separation of rats and rhesus monkeys from their mothers, based on findings in the literature. Sleep and temperature disorders, cardiac disorders, and increased gastric ulcer risk are considered. Applications to human psychosomatic medicine are discussed.

485. Mezinskis, J.; Gliner, J. & Shemberg, K. (1971). **Somatic response as a function of no signal, random signal, or signaled shock with variable or constant durations of shock.** ***Psychonomic Science,*** 25(5), 271–272.
Studied gastric ulceration in a total of 72 male Holtzman albino rats exposed to predictable shock (signaled) or unpredictable shock (either random signals and shock or shock alone). These shock programs were delivered under conditions of either fixed or variable shock durations. Results are compatible with previous work, suggesting that predictable shocks are less ulcerogenic than unpredictable shocks. Results also suggest that this relationship can be obtained when unpredictability is defined as signals and shocks at random or as shocks with no signals. Predictable shock was less ulcerogenic than unpredictable shock irrespective of whether the shock durations were fixed or variable.

486. Mikhail, A. A. (1969). **Genetic predisposition to stomach ulceration in emotionally reactive strains of rats.** ***Psychonomic Science,*** 15(5), 245–247.
Applied a simplified method of inducing ulceration by stress to 20 male Maudsley reactive (MR), 20 male Maudsley nonreactive (MNR), and 20 male and female rats of both strains. Contrary to a previous report, the severity of stomach ulceration as assessed by a refined technique of stomach examination was found to be greater in the MR Ss than in the MNR Ss. It was argued that the physiological mechanism underlying the higher emotional elimination of the MR strain reflects greater parasympathetic discharge of the sacral division under certain stressful conditions and that the parasympathetic hyperactivity may account for the genetic predisposition of the MR strain to stress ulceration.

487. Mikhail, A. A. (1973). **Stress and ulceration in the glandular and nonglandular portions of the rat's stomach.** ***Journal of Comparative & Physiological Psychology,*** 85(3), 636–642.
Previous studies indicate that conditioned fear and conflict produce ulceration in the nonglandular portion of the rat's stomach (rumen), while immobilization produces ulceration in the glandular portion (corpus). The present 3 experiments were conducted with Maudsley Reactive, Maudsley Nonreactive, and male Sprague-Dawley rats (N = 66). Exp I and II tested the course of ulceration in the corpus and rumen under stress conditions. Ulceration was induced in the corpus by partial restraint. Ulcerated Ss were then exposed to conditioned fear for 72 hrs (Exp I) and 48 hrs (Exp II) and were compared with controls. Exp I showed that glandular lesions healed while nonglandular lesions developed in food-deprived Ss. Nonglandular ulcers did not appear during the 48 hrs of Exp II. Exp III showed a positive relationship between rumenal ulceration and food deprivation. Rumenal ulceration induced by conflict and conditioned fear procedures mainly seems to reflect the suppression of food intake by these procedures.

488. Mikhail, A. A. (1972). **The effects of conditioned anxiety on the recovery from experimental ulceration.** ***Journal of Psychosomatic Research,*** 16(2), 115–122.
Compared the recovery from gastric ulcers in 20 male MNR rats divided randomly into an experimental group and a control group and tested periodically during 87 hrs in a series of 3 experiments. It was found that the recovery was not delayed in experimental Ss exposed to continuous shock stress (Exp I) or intermittent stress (Exp II), relative to controls. It is argued that experimental production of ulcers in rats occurred via reduced food intake and that this literature has doubtful relevance to the problem of peptic ulcer in man.

489. Mikhail, Anis A.; Kamaya, Valerie A. & Glavin, Gary B. (1978). **Stress and experimental ulcer: Critique of psychological literature.** ***Canadian Psychological Review,*** 19(4), 296–303.
Although several authors have claimed that the psychological procedures of conflict, conditioned fear, and avoidance could be used to develop ulcers in rats, the present authors' evaluation of these procedures suggests that the physical factors of shock and food deprivation were the primary agents responsible for the observed lesions. It is concluded that a convincing demonstration that psychological factors per se are sufficient to induce ulcers and the pathophysiological changes necessary for their formation is lacking at present in the experimental literature. (French summary).

490. Miller, Alan D. & Wilson, Victor J. (1983). **Vestibular-induced vomiting after vestibulocerebellar lesions.** ***Brain, Behavior & Evolution,*** 23(1–2), 26–31.
Developed an animal model for motion sickness using sinusoidal electrical polarization of the labyrinths of decerebrate cats to recreate natural vestibular stimulation. 14 adult cats served as Ss. 10 Ss received large lesions of the posterior cerebellar vermis. After decerebration, at least 2 hrs usually elapsed before the start of electrical stimulation of the labyrinth. Results show that the model produced vomiting and related activity resembling that seen in motion sickness. The symptoms included panting, salivation, swallowing, and retching as well as vomiting. In contrast to a previous proposal by S. C. Wang and H. I. Chinn (1956), it is suggested that a transcerebellar pathway from the vestibular apparatus through the nodulus and uvula to the vomiting center is not essential for vestibular-induced vomiting and, by analogy, for the occurrence of many symptoms of motion sickness.

491. Mills, David E. & Ward, Ron P. (1986). **Attenuation of stress-induced hypertension by exercise independent of training effects: An animal model.** ***Journal of Behavioral Medicine,*** 9(6), 599–605.
Examined whether exercise, in the absence of physical training, could alter development of hypertension during chronic exposure to a psychosocial stressor in 14 adult male Sprague-Dawley rats. Ss were exposed to social stress for 7 days, following 5 wks of acclimation to social isolation. Half of the Ss had access to exercise in a running wheel during the stress period, while the other half did not. Blood pressure increased significantly to hypertensive levels on Days 4 and 7 in the group denied access to exercise but was unchanged in the exercise group. Degree of attenuation of stress-induced hypertension was unrelated to amount of running activity; there were no differences in body weight, heart rate, or organ weight between groups. Exercise appeared to act specifically via diversional (coping) mechanisms to buffer the response of the body to stress.

492. Minor, Thomas R.; Pelleymounter, Mary A. & Maier, Steven F. (1988). **Uncontrollable shock, forebrain norepinephrine, and stimulus selection during choice-escape learning.** ***Psychobiology,*** 16(2), 135–145.
Examined the role of stress-induced depletion of forebrain norepinephrine (NE) in choice-accuracy deficit, using 112 male albino rats. Ss were administered 0, 40, or 120 inescapable tailshocks prior to a maze learning task, and NE levels were assessed (Exps I and II). A relationship was observed between shock pretreatment, magnitude of the deficit in later choice escape, and depletion of forebrain NE upon reexposure to a few shocks 24 hrs after pretreatment. Exp III established that NE depletion is sufficient to impair choice performance. Bilateral 6-hydroxydopamine (6-OHDA) lesions of the ascending dorsal tegmental bundle mimicked the effects of inescapable shock. Data support the notion that the ascending dorsal tegmental bundle is involved in attentionlike processing and suggest that deficits in stimulus selection following inescapable shock may result from stress-induced depletion of forebrain NE.

493. Molina, V. A.; Volosin, M.; Cancela, L.; Keller, E. et al. (1990). **Effect of chronic variable stress on monoamine receptors: Influence of imipramine administration.** ***Pharmacology, Biochemistry & Behavior,*** 35(2), 335–340.
Adult male rats were exposed to a series of unpredictable stressors with or without concurrent administration of imipramine. One day after the last stress event of the chronic regime, binding of cortical beta-adrenoreceptors and the behavioral serotonin, 5-hydroxytryptamine (5-HT), syndrome induced by 5-methoxy-N,N,dimethyltryptamine were determined in all the experimental groups. Stressed Ss showed an up-regulation of cortical beta-adrenergic sites; similar values to controls were observed when stressed Ss were administered imipramine. Regarding the behavioral 5-HT syndrome, comparable behavioral scores were observed between controls and chronically stressed Ss. The probable facilitation of behavioral deficits induced by this scheme of chronic stress and the recovery following concurrent administration of imipramine are discussed.

494. Mollmann, H. & Egbers, H. J. (1972). **Comparative study on the influence of psychotropic agents on stress-induced lesions of gastric mucuso.** ***Arzneimittel-Forschung,*** 22(4), 743–751.
Investigated inhibitory effects of fluphenazine, pentobarbital, a mixture of the 2, and a formulation of chlordiazepoxide and amitriptyline (given orally to 970 female Wistar albino rats for 2–10 days) with regard to psychogenic gastric ulcer induced by intensive immobilization. Effectiveness was estimated by size and frequency of ulcerations with results indicating: (a) fluphenazine (.012 mg/day) caused steady decrease in degree of ulceration, (b) pentobarbital (1.2 or 2.4 mg/day) reduced the mean degree of affection, (c) fluphenazine (.012 mg/day) with pentobarbital (1.2 mg/day) gave best results, and (d) only a highly out-of-proportion dosage of chlordiazepoxide and amitriptyline had any effect. These results make it probable that the psychogenic origin of ulcer is associated with certain brain centers as well as their neuronal junctions. It is assumed that direct site of action of psychotropic agents in exerting their effect is through A- , B- and C-receptors, which affect metabolism of biogenic amines in certain brain centers. (English summary).

495. Moot, Seward A.; Cebulla, Ralph P. & Crabtree, J. Michael. (1970). **Instrumental control and ulceration in rats.** ***Journal of Comparative & Physiological Psychology,*** 71(3), 405–410.
Trained 2 groups of 22 male Sprague-Dawley rats each to bar press for food on a variable interval schedule. During testing, each food reinforcement was paired with electric shock, and the groups were differentiated on the basis of the degree of instrumental control over this aversive stimulus. Significantly fewer Ss which were able to terminate the shock by means of an instrumental escape response developed stomach ulcers than did Ss with unescapable shocks. It is concluded that the degree of instrumental control the organism is able to exert over the aversive stimulus is an important determinant of the psychological severity of a conflict situation, and of ulceration.

496. Morgan, Laura L.; Schneiderman, Neil & Nyhan, William L. (1970). **Theophylline: Induction of self-biting in rabbits.** ***Psychonomic Science,*** 19(1), 37–38.
Presents a tentative animal model for studying the self-mutilating behavior seen in children with an abnormality of purine metabolsim. Each of 4 groups of 10 New Zealand male albino rabbits was given a quarter-normal diet and daily injections of saline or 46, 61.5, or 92 mg/kg of anhydrous theophylline (1,3-dimethylxanthine). It was found that: (a) the greatest number of self-biters were found in the 61.5 mg/kg group, (b) mortality was directly related to drug dosage, and (c) the onset of biting occurred earliest with the highest dosage.

497. Mower, George D.; Burchfiel, James L. & Duffy, Frank H. (1982). **Animal models of strabismic amblyopia: Physiological studies of visual cortex and the lateral geniculate nucleus.** ***Developmental Brain Research,*** 5(3), 311–327.
Receptive field properties of visual cortical and lateral geniculate cells were studied in 4 models of amblyopia in the cat: monocular deprivation (MD), surgical esotropia (SC), optically induced concomitant strabismus (CS), and optically induced incomitant strabismus (IS). Comparison observations were made in normal cats. Recordings in visual cortex indicated a reduction in responsiveness to the treated eye in MD and IS Ss. SC and CS Ss showed mainly a loss of binocular cells. Recordings in the lateral geniculate nucleus indicated reduction in spatial resolving capacity of X-cells driven by the treated eye in MD, SC, and IS Ss. The magnitude of this effect was comparable in all of these preparations. CSs showed comparable spatial resolving capacities in X-cells driven by either eye. Y-cells were unaffected in any preparation except MD. These results indicate that (1) interocular differences in spatial patterns without form deprivation are sufficient to produce a loss of responsiveness to 1 eye in visual cortex, (2) incomitant disparities are necessary to produce the physiological correlates of amblyopia in cats, and (3) deficits in spatial resolution in geniculate neurons are comparable in magnitude in various amblyopic preparations.

498. Mower, George D. & Duffy, Frank H. (1983). **Animal models of strabismic amblyopia: Comparative behavioral studies.** ***Behavioural Brain Research,*** 7(2), 239–251.
Assessed visual acuity and visuo-motor behavior in various models of experimental amblyopia in 10 cats. Three models of strabismic amblyopia were studied: (1) surgical esotropia by sectioning 1 lateral rectus muscle, (2) comitant optical strabismus by rearing Ss with goggles that placed a stationary wedge prism before 1 eye, and (3) incomitant optical strabismus by rearing Ss with goggles that placed a rotatable wedge prism before 1 eye. These Ss were compared with 5 normal and monocularly deprived Ss. Clear amblyopic deficits were found in monocularly deprived, esotropic, and rotating prism Ss. The amblyopic deficits were graded among these preparations, which were the most severe in monocularly deprived Ss and least severe in esotropic Ss. The degree of behavioral amblyopia in these preparations was correlated with the extent of physiological abnormalities in visual cortex and lateral geniculate nucleus. Variable optical strabismus produced clear deficits in 1 eye, both behaviorally and physiologically, without impaired ocular motility.

499. Moye, Thomas B.; Hyson, Richard L.; Grau, James W. & Maier, Steven F. (1983). **Immunization of opioid analgesia: Effects of prior escapable shock on subsequent shock-induced and morphine-induced antinociception.** ***Learning & Motivation,*** 14(2), 238–251.
In Exp I, using 30 male Holtzman rats, experience with escapable footshock 4 hrs prior to a session of 80 inescapable tailshocks prevented the occurrence of an analgesic response normally observed immediately following the tailshock. Exp II, using 48 Ss, examined the effects of prior escapable shock on the long-term analgesia reaction that occurs with brief exposure to shock 20 hrs after morphine administration. Ss were given escapable shock, inescapable shock, or no shock 4 hrs prior to a morphine injection. 20 hrs following the injection, all Ss received 5 brief footshocks and were then immediately given tailflick analgesia tests. Ss that received inescapable shock or no shock prior to the morphine injection displayed a significant analgesic response. However, Ss that received escapable shock prior to morphine were not analgesic following brief exposure to shock. Results suggest that escapable shock directly influences the activation of opioid analgesia systems.

500. Murison, Robert C. (1983). **Time course of plasma corticosterone under immobilisation stress in rats.** ***IRCS Medical Science: Psychology & Psychiatry,*** 11(1–2), 20–21.

50 male Sprague-Dawley rats were food-derived for 16 hrs, lightly anesthetized with ether, and returned to their home cages or placed in an immobilization apparatus for 20 min to 23 hrs 20 min. Plasma corticosterone levels were higher in the immobilized group than in the control group, and levels declined over time in both groups. Immobilization also led to an increase in gastric ulceration. Results indicate that immobilization stress is associated with sustained activation and suggest that adrenocortical activity may only be used as an index of stress in acute situations.

501. Murison, Robert et al. (1982). **Sleep deprivation procedure produces stomach lesions in rats.** ***Physiology & Behavior,*** 29(4), 693–694.
18 Sprague-Dawley male rats were deprived of food for 40 hrs and subjected to sleep deprivation (SD) by the "flower pot" or handling intervention procedures during the final 24 hrs. Five of 6 Ss in the flower pot group exhibited stomach erosions. Only 1 S in the handling group and no Ss in a food-deprived control group showed such erosions. The flower pot group also had higher levels of corticosterone than the food-deprived control group. Findings indicate the need for more serious consideration of the effects of the stress of procedures for SD.

502. Murison, Robert C. & Isaksen, E. (1982). **Gastric ulceration and adrenocortical activity after inescapable and escapable pre-shock in rats.** ***Scandinavian Journal of Psychology,*** 1, 133–137.
Investigated the effects of preshock treatment in male Sprague-Dawley rats on adrenocortical activity and susceptibility to later restraint-induced ulceration. Ss exposed to inescapable shock exhibited sensitization of the adrenocortical response, increased adrenocortical reactivity to open field testing, and a trend for more severe ulceration after restraint. Ss given escapable or yoked preshock trials showed no sensitization of the adrenocortical response. Escape Ss exhibited less gastric ulceration and a reduced adrenocortical response after restraint.

503. Murison, Robert; Overmier, J. Bruce & Glavin, Gary B. (1989). **Stress-rest cyclicity in the pathogenesis of restraint-induced stress gastric ulcers in rats.** ***Physiology & Behavior,*** 45(4), 809–813.
Explored the influence of stress-rest cycles on gastric ulceration (UC) after 2 forms of immobilization stress in rats. In Exp 1, a single 180-min exposure to cold supine restraint produced more extensive UC than did a series of 6 30-min stress periods interspersed with 30-min rest periods in the home cage. In Exp 2, Ss were subjected to either a single 150-min stress period or to a 30- or 150-min priming exposure followed by a 2nd 150-min exposure. There was no evidence for protective effects of priming against UC. Findings point to the role of cycles in determining the extent of stress pathology. Such data must be accounted for in any description of the mechanisms of stress-related UC.

504. Murison, Robert; Overmier, J. Bruce & Skoglund, Even J. (1986). **Serial stressors: Prior exposure to a stressor modulates its later effectiveness on gastric ulceration and corticosterone release.** ***Behavioral & Neural Biology,*** 45(2), 185–195.
Explored the effects of prior exposure (PE) of 36 male Møll-Wistar rats to water-restraint stress on gastric ulceration and corticosterone responses to a final stressor. The number of PEs to water-restraint challenge was varied from 1 to 4, and the durations of the PEs varied from 30 min to 2 hrs. The duration of final stress challenge was 75 min. The 2 indices of the stress response to the final stressor challenge yielded different results. For neither index was the duration of PE a significant factor in modulating the response to the final challenge. Ss receiving 1 PE exhibited less gastric ulceration than Ss receiving no PE or 4 PEs. Ss receiving 4 PEs exhibited a lower adrenocortical response to the final stressor than did Ss receiving no or only 1 PE. Results indicate the complexity of proactive effects of earlier stress experience on responses to a final stressor challenge.

505. Murison, Robert & Overmier, J. Bruce. (1986). **Interactions amongst factors which influence severity of gastric ulceration in rats.** ***Physiology & Behavior,*** 36(6), 1093–1097.
58 male Sprague-Dawley rats were preexposed to either signaled or unsignaled shock (using learned helplessness parameters) or no shock. They were later subjected to 2 hrs restraint-in-water (immersion) stress or appropriate handling control procedures. Half of the Ss were sacrificed immediately on removal from the water-restraint, and half were sacrificed after a 2-hr recovery period in their home cages. Analysis of the severity and number of glandular stomach lesions indicated that Ss subjected to the preshock exhibited greater ulceration than unshocked Ss as long as no poststress rest period was allowed. The effect of poststress rest was masked by experience with preshock. It is suggested that this reciprocal modulation of treatment influences may help explain discrepancies in the literature on the effects of these 2 modulating variables.

506. Murison, Robert; Overmier, J. Bruce; Hellhammer, Dirk H. & Carmona, Manuel. (1989). **Hypothalamo-pituitary-adrenal manipulations and stress ulcerations in rats.** ***Psychoneuroendocrinology,*** 14(5), 331–338.
Both gastric ulceration and activation of the hypothalamo-pituitary adrenal axis are considered integral to the stress response, and a causal relationship between the two has been suggested. In the present study, corticosterone secretion in rats was either stimulated with corticotropin-releasing factor (CRF) or lowered with metyrapone during a known ulcerogenic stress. Reduction of circulating corticosterone during the stress had no effect on ulceration severity compared to saline-treated stressed controls. Treatment with CRF in stressed Ss reduced ulceration severity. The mechanism of this protective effect remains unclear. Findings do not support a simple causal relationship between adrenocortical activity and gastric ulceration.

507. Murphy, Helen M. & Wideman, Cyrilla H. (1983). **Self-starvation and activity stress in Brattleboro rats.** ***Physiological Psychology,*** 11(3), 209–213.
Examined the susceptibility of 40 male DI Brattleboro rats (which lack the hormone vasopressin) and 40 male Long-Evans rats (controls) to self-starvation and activity stress, as evidenced by ulcer formation. Four conditions were studied: (1) ad lib access to food and water in an individual home cage, (2) ad lib access to food and water in an activity-wheel cage, (3) ad lib access to water and 1-hr access to food in an individual home cage, and (4) ad lib access to water and 1-hr access to food each day in an activity-wheel cage. In the 1st 2 conditions, neither Brattleboro nor the control Ss developed ulcers. In the 3rd condition, some Ss in each group developed pits in the glandular portion of the stomach. Significant differences appeared in the 4th condition: Brattleboro Ss developed more stomach pathology in both

the nonglandular and glandular portions of the stomach than controls. Results are shown not to be due to variations in running behavior. Data indicate that the lack of vasopressin or physiological results of this deficit render Brattleboro rats more sensitive to self-starvation and activity stress than control animals.

508. Muscat, Richard; Towell, Anthony & Willner, Paul. (1988). **Changes in dopamine autoreceptor sensitivity in an animal model of depression.** ***Psychopharmacology,*** 94(4), 545–550.
Male rats exposed for 6 wks to a variety of mild unpredictable stressors showed reduced consumption of a preferred sucrose solution. The deficit was apparent after 1 wk of stress and was maintained for at least 2 wks after termination of the stress regime. Sucrose preference was unaffected by 2 wks of treatment with the tricyclic antidepressant DMI but returned to normal after 3 wks of DMI treatment. Subsensitivity to the anorexic effect of a low dose of apomorphine was seen in vehicle-treated stressed animals, and in unstressed animals following withdrawal from DMI. In both cases, the changes resulted from a failure of apomorphine to reduce eating time. It is concluded that dopamine autoreceptor desensitization probably does not contribute to clinical improvement following chronic antidepressant treatment.

509. Natelson, Benjamin H. et al. (1984). **An analysis of some sensitizing agents in the pathogenesis of stress-induced gastric erosive disease.** ***Pavlovian Journal of Biological Science,*** 19(4), 195–199.
Two experiments investigated the effect of gender, prior stress, and immobilization-induced hypothermia on gastric disease in catheterized and noncatheterized Sprague-Dawley rats. Exp I showed that prior surgical implantation of a venous catheter sensitized Ss to cold-immobilization stress; 3 or 6 catheterized females succumbed during the stress. Uncatheterized males had higher temperatures than Ss in the other 3 groups. No relation was found between catheter patency and magnitude of hypothermia. The degree of gastric disease paralleled the core temperature findings in that uncatheterized males had significantly fewer gastric erosions. There was a robust gender effect, with uncatheterized females showing more hypothermia and more gastric disease than uncatheterized males. Exp II evaluated whether anesthesia or wearing a protective spring was responsible in part for the sensitization. Results show that the gender difference was less, although females consistently averaged lower core temperatures after stress than did males. Females that were prepared with the protective spring apparatus developed more gastric disease than female controls or similarly treated males. Findings suggest that gender is not critical by itself but instead plays a role in the pathogenesis of stress-induced disease.

510. Natelson, Benjamin H.; Cagin, Norman A.; Donner, Kenneth & Hamilton, Bruce E. (1978). **Psychosomatic digitalis-toxic arrhythmias in guinea pigs.** ***Life Sciences,*** 22(24), 2245–2250.
Adult male guinea pigs were exposed daily to 6 1-min tone/light signals, each 5 min apart. For half of the Ss, the signal was followed by rump shock 40% of the time (signal/shock group); for the remaining Ss, shock was never delivered (signal/no-shock group). Ss in both groups were given weekly injections of ouabain, a fast-acting digitalis glycoside. For the signal/shock group, significantly more potentially lethal arrhythmias began during the signal than in the minutes preceding it. Also, a significantly higher percentage of arrhythmias began during the signal periods for the signal/shock group when compared to the signal/no-shock group. Findings indicate that psychogenic factors may contribute to the development of digitalis-toxic arrhythmias.

511. Natelson, Benjamin H.; Hoffman, Scott L. & McKee, Cynthia N. (1979). **Duodenal pathology in rats following cold-restraint stress.** ***Physiology & Behavior,*** 23(5), 963–966.
39 female Sprague-Dawley rats developed duodenal abnormalities as a monotonic function of stress (fasting, exposure to cold, and restraint). The existence of gastric and duodenal pathology in stressed rats mirrors that found in stressed humans and thus gives credence to the notion that the rat is a useful animal for study of acute, stress-induced gastro-duodenal disease.

512. Natelson, Benjamin H.; Janocko, Laura & Jacoby, Jacob H. (1981). **An interaction between dietary tryptophan and stress in exacerbating gastric disease.** ***Physiology & Behavior,*** 26(2), 197–200.
Female Sprague-Dawley rats ($N = 28$) fed a tryptophan (TR) deficient diet composed of corn developed significantly more gastric erosive disease when subjected to prolonged immobilization than similarly stressed Ss fed a TR enriched diet. No gastric disease developed in controls that were not immobilized. No significant correlations were found between treatment groups and levels of indoleamines in brain and gut. However, serum TR was significantly lowered only in the group of stressed Ss fed a TR deficient diet. Results suggest the possibility that humans eating preponderantly a corn-based diet, deficient in TR, may be at increased risk of having stress ulcer disease of the gut.

513. Natelson, Benjamin H.; Tapp, Walter N.; Drastal, Susan; Suarez, Ronald et al. (1991). **Hamsters with coronary vasospasm are at increased risk from stress.** ***Psychosomatic Medicine,*** 53(3), 322–331.
In 2 experiments, 129 cardiomyopathic hamsters (CMH) whose ages differed by about 3 mo were placed under stress from cold immobilization using 2 durations of cold exposure. The younger of the 2 groups of stressed hamsters was in the vasospastic phase of the disease. No difference in mortality occurred with the more intense stressor. Significantly fewer of the older CMHs succumbed to the less intense stressor. Examination of the hearts in the experiment where mortality rate was the same for both groups revealed evidence of cardiac dilation, indicative of heart failure, only in the older CMHs following stress. Results indicate that the process of coronary vasospasm should be viewed as an independent risk factor in determining the consequences of stress.

514. Navarro, José I. et al. (1984). **Influence of non-behavioral factors on learned helplessness.** ***Revista de Análisis del Comportamiento,*** 2(3), 297–311.
Conducted 4 experiments with Sprague-Dawley rats to evaluate the differential effects of experimentally induced stress (via immobilization and injections) and exposure to uncontrollable shocks on operant learning, the sex-dependent effects of uncontrollable shocks, and ponderal and gastric mucus variations occurring as a result of uncontrollable shock. Results indicate that while conditions of uncontrollability produced significant increases in number of trials

required to acquire operant responding, separate analyses according to sex yielded significant effects for male but not for female Ss. Experimentally induced stress by either procedure did not significantly affect operant acquisition and, in contrast to other studies, did not cause significant differences between experimental and control groups in gastric mucus.

515. Nazaretyan, R. A. (1969). **The influence of benzofuran derivatives on several functions of the stomach in the normal state and in experimental ulceration.** ***Ek'sperimental ev Klinikakan Bzhshkowt'yan Handes,*** 9(2), 18–26.
Rats and rabbits (gastric ulceration caused by 10-min compression of the pyloroduodenal region) and 8 dogs (by injection of various pharmacological agents) were Ss in a study establishing the high antiulcerative activity of diiodmethylate N-ethyl-N-benzofurfuryl N1,N1-dimethylethylenediamine in comparison with well-known ganglioblockers. Depending on the administered dose and the initial functional state, the preparation manifested a stimulatory (small doses) or inhibitory (large doses) influence on the motor, secretory, and excretory function of the stomach. Under conditions of chronic experimental ulceration in dogs, the preparation hastened the restoration of disturbed functions of the stomach and promoted a more rapid healing of dystrophic lesions of the mucous membrane of the stomach. The action of the preparation was more apparent in the region of lesser curvature where the disturbances of the functions of the stomach and the trophic state of its wall are more pronounced.

516. Nishimura, Hiroshi; Tsuda, Akira; Oguchi, Masanobi; Ida, Yoshishige et al. (1988). **Is immobility of rats in the forced swim test "behavioral despair?"** ***Physiology & Behavior,*** 42(1), 93–95.
Examined the relationship between sinking and immobility, which has been reported to reflect behavioral despair, in a forced-swimming test with rats classified into sinking and nonsinking groups, according to the appearance of sinking behavior. Sinking Ss' immobility times during the 1st 15 min of testing were shorter than those of nonsinking Ss. Results suggest that sinking is a sign of emotional behavior such as fear or anxiety. Discriminant analysis showed that immobility during the 1st 15 min predicted sinking. It is concluded that immobility in nonsinking rats reflects a lower level of emotional reaction than that of immobile rats that sink.

517. Northway, Margaret G.; Morris, Mariana; Geisinger, Kim R. & MacLean, David B. (1989). **Effects of a gastric implant on body weight and gastrointestinal hormones in cafeteria diet obese rats.** ***Physiology & Behavior,*** 45(2), 331–335.
In order to evaluate the effectiveness of a gastric implant in an animal model of dietary obesity, silicone implants were inserted into the stomachs of 40 male rats maintained on a chow or "cafeteria" diet (CD). At the time of implantation, the CD Ss weighed 14% more than chow fed controls. Overweight CD Ss lost weight in response to the gastric implant, whereas controls did not. Both implant groups had significant increases in stomach weights in contrast to sham implant groups, but the increase was much less in the CD Ss. The fasting plasma levels of the gastrointestinal hormones, gastrin and pancreatic polypeptide, and oxytocin (a marker of vagal afferent function) were measured by radioimmunoassay. CD sham or implanted Ss had significantly higher fasting levels of plasma oxytocin and gastrin, and significantly lower plasma levels of pancreatic polypeptide than the chow fed groups.

518. Öhman, Arne. (1986). **Face the beast and fear the face: Animal and social fears as prototypes for evolutionary analyses of emotion.** 25th Annual Meeting of the Society for Psychophysiological Research (1985, Houston, Texas). ***Psychophysiology,*** 23(2), 123–145.
Applies a functional-evolutionary perspective to fear in the context of encounters with animals and threatening humans. It is argued that animal fear originates in a predatory defense system whose function is to allow animals to avoid and escape predators. Social fears are viewed as originating in a dominance/submissiveness system. Signs of dominance paired with aversive outcomes provide for learning fear to specific individuals. Data supportive of this conceptualization are reviewed. To explain the mechanism behind the causal relationships suggested in the evolutionary analysis, an information-processing model is presented and empirically tested. It is argued that responses to evolutionary fear-relevant stimuli can elicit the physiological concomitants of fear after only a very quick unconsciousness or preattentive stimulus analysis.

519. Olmstead, Charles E. (1987). **Neurological and neurobehavioral development of the mutant "twitcher" mouse.** ***Behavioural Brain Research,*** 25(2), 143–153.
Assessed the neurological and locomotor development of the mutant twitcher mouse, an enzymatically authentic model of globoid cell (Krabbe) leukodystrophy (GCL), using a neurological developmental test battery and a set of behavioral measures. Neurological development was slowed in homozygous affected (*twi/twi*) Ss, and there were subtle differences between heterozygous carriers and normals. *Twi/twi* mice reached all functional milestones except grasp. There was a rapid deterioration of motor indices but not sensory markers after age 20 days. Data are discussed in terms of the relationship between human GCL and the animal model.

520. Olton, David S. (1990). **Dementia: Animal models of the cognitive impairments following damage to the basal forebrain cholinergic system.** 19th Annual Meeting of the Society for Neuroscience: Neural basis of behavior: Animal models of human conditions (1989, Phoenix, Arizona). ***Brain Research Bulletin,*** 25(3), 499–502.
Reviews research on cognitive impairments resulting from selective damage to neurons in the basal forebrain cholinergic system (BFCS). Lesions of the nucleus basalis magnocellularis and the medial septal area reproduce the behavioral symptoms following lesions of their respective target sites, the frontal cortex, and the hippocampus. Impairments of recent memory are one of the most striking symptoms in Alzheimer's disease patients at the beginning of the disease. Lesions of the BFCS induce similar impairments. Comparisons of the effects of the lesions produced by different neurotoxins have raised questions about the role of cholinergic and noncholinergic neurotransmitter systems in the basal forebrain. Implications for the cholinergic hypothesis of mnemonic functions are discussed.

521. Ordy, J. M.; Thomas, G. J.; Volpe, B. T.; Dunlap, William P. et al. (1988). **An animal model of human-type memory loss based on aging, lesion, forebrain ischemia, and drug studies with the rat.** Special Issue: Experimental models of age-related memory dysfunction and neurodegeneration. ***Neurobiology of Aging,*** 9(5–6), 667–683.
Compared the effects of (1) age; (2) basal forebrain, medial septal, and amygdala lesions; (3) 4-vessel occlusion (4-VO), forebrain ischemia; and (4) physostigmine, scopolamine, arecoline, piracetam, and clonidine on memory and performance of young, middle-aged, and old male rats. Aging significantly impaired working memory and performance. Memory of septal and basal forebrain, but not of amygdala lesioned Ss was significantly impaired without effects on performance. Transient, 4-VO forebrain ischemia produced significant memory impairment, without effects on performance, and highly selective CA1 cell loss in the hippocampus. Physostigmine enhanced working memory in middle-aged and old Ss. Scopolamine impaired memory in young, middle-aged, and old Ss. Physostigmine reversed the scopolamine impairments of working memory. Arecoline enhanced memory in old Ss without effects on performance.

522. Overmier, J. Bruce; Murison, Robert & Ursin, Holger. (1986). **The ulcerogenic effect of a rest period after exposure to water-restraint stress in rats.** ***Behavioral & Neural Biology,*** 46(3), 372–382.
In Exp I with 32 male Sprague-Dawley rats, Ss restrained in water for 2 hrs and then allowed a 2-hr rest period exhibited significantly greater gastric ulceration than Ss sacrificed immediately after the 2-hr restraint-in-water stress period. In Exp II, with 20 Ss, Ss challenged with restraint in water for 1.25 hrs and then allowed a 1.25-hr rest period before sacrifice exhibited greater ulceration than Ss maintained in the restraint-in-water stress for the full 2½ hrs. Corticosterone levels for these 2 treatment groups were similar at the time of sacrifice.

523. Overmier, J. Bruce; Murison, Robert; Ursin, Holger & Skoglund, Even J. (1987). **Quality of poststressor rest influences the ulcerative process.** ***Behavioral Neuroscience,*** 101(2), 246–253.
Gastric ulceration in rats is exacerbated by allowing a so-called recovery period after exposure to an ulcerogenic stressor. One hypothesis, which has support from pharmacological studies, argues that this effect is brought about by a rebound of parasympathetic activation. We tested this parasympathetic rebound hypothesis by presenting animals with a fear-inducing (sympathetic-activating) conditioned stimulus (CS) after 2 hr of water-restraint stress. Contrary to the hypothesis, presentation of such a CS increased severity of ulceration compared with those animals that did not receive the CS after restraint stress and control animals. These ulceration data favor instead a sustained activation hypothesis for ulceration, whereby presentation of the CS effectively prolonged the length of time during which animals were under stress, thus enhancing the degree of ulceration. Measurement of plasma corticosterone however indicated a negative correlation between adrenocortical activity and degree of gastric ulceration, contrary to that expected by a sustained activation hypothesis.

524. Overmier, J. Bruce; Murison, Robert; Skoglund, Even J. & Ursin, Holger. (1985). **Safety signals can mimic responses in reducing the ulcerogenic effects of prior shock.** ***Physiological Psychology,*** 13(4), 243–247.
Tested the hypothesis that providing animals with safety signals during a Pavlovian conditioning session would also provide a proactive protection against restraint ulceration similar to that provided by escape responses. 27 male Sprague-Dawley rats were subjected to 5 daily sessions of 20 shocks before they were subjected to a single 23-hr restraint-stress procedure. Ss given safety signals during the conditioning sessions developed less ulceration than those subjected to random tone-shock pairings and those that were not shocked. This complements other reports of the similar properties shared by escape conditioning and safety-signal (backward) conditioning. Postrestraint corticosterone levels were higher in Ss provided earlier with safety signals.

525. Pallarés i Año, Marc; Nadal, Roser & Ferré, Nuria. (1991). **Alcohol and animal behavior.** Special Issue: Alcohol and alcoholism. ***Avances en Psicología Clínica Latinoamericana,*** 9, 33–87.
Reviews research findings and methods in the field of animal models of human alcohol consumption. The main animal models are: schedule-induced polydipsia, sucrose fading procedure, deprivation methods, genetic models, social models, negative reinforcement (forced consumption), and liquid diets (forced consumption). Topics discussed include the following: free access paradigm, operant self-administration, conditioned taste preference, conditioned place preference, open field behavior, grooming, exploratory behavior, aggression, and emotional behavior. (English abstract).

526. Panksepp, Jaak & Cox, James F. (1986). **An overdue burial for the serotonin theory of anxiety.** ***Behavioral & Brain Sciences,*** 9(2), 340–341.
Supports P. Soubrié's (1986) argument against the serotonin (SE) theory of anxiety based on studies using the animal model of fear/anxiety, in which the escape behavior was facilitated rather than retarded by SE depletion. It is suggested that the role of gamma-aminobutyric acid in initiating behavioral effects that have been ascribed to benzodiazepines and SE systems deserves closer attention.

527. Papp, Mariusz; Willner, Paul & Muscat, Richard. (1991). **An animal model of anhedonia: Attenuation of sucrose consumption and place preference conditioning by chronic unpredictable mild stress.** ***Psychopharmacology,*** 104(2), 255–259.
Male rats were subjected to chronic unpredictable stress for 4 wks. During Wks 3 and 4, Ss received 4 training trials in which rewards were presented in a distinctive environment, and 4 further nonrewarded trials in a different environment. The rewards used in different experiments were food pellets, diluted and concentrated sucrose solutions, and dl-amphetamine sulphate. Nonstressed Ss showed an increase in preference for the environment associated with reward; in stressed Ss, these effects were abolished or greatly attenuated. Chronic unpredictable mild stress, which may be comparable in intensity to the difficulties people encounter in their daily lives, appears to cause a generalized decrease in sensitivity to rewards.

528. Pappas, Bruce A.; DiCara, Leo V. & Miller, Neal E. (1970). **Learning of blood pressure responses in the noncurarized rat: Transfer to the curarized state.** ***Physiology & Behavior,*** 5(9), 1029–1032.

Trained 9 free-moving Sprague-Dawley rats to increase, and 8 to decrease, systolic blood pressure in order to avoid and/or escape electric shock. The initial difference in blood pressure between the 2 groups was small and did not approach significance. The training produced changes in the rewarded directions, with the difference between the changes being reliable and producing an overall difference of 9.7 mm Hg. No corresponding changes were found for heart rate or gross skeletal activity. The blood pressure responses failed to transfer to the curarized state. However, subsequent training in the curarized state produced a difference between the increase and decrease groups of 21.2 mm Hg. There was some indication that the blood pressure responses in the curarized and noncurarized training sessions were positively correlated.

529. Pappas, Bruce A.; DiCara, Leo V. & Miller, Neal E. (1972). **Acute sympathectomy by 6-hydroxydopamine in the adult rat: Effects on cardiovascular conditioning and fear retention.** ***Journal of Comparative & Physiological Psychology,*** 79(2), 230–236.
Reports results of an experiment with 38 Charles River male albino rats. Curarized Ss, chemically sympathectomized by 6-hydroxydopamine (6-OH), had lower base-line systolic blood pressures than vehicle-injected controls, but heart rates were equivalent. Maximum blood pressure increase in the vehicle group to unsignaled shock occurred about 2 sec after shock onset. In the 6-OH group the increase was smaller, reaching maximum 9 sec after shock. Heart-rate increases to shock were also attenuated by sympathectomy. After 41 paired or random CS and shock presentations, only the paired vehicle group showed a biphasic pressure increase to the CS. Paired and random 6-OH groups showed equivalent long-latency responses to the CS. No consistent group-conditioned heart-rate responses were observed. Vehicle and 6-OH groups did not differ on retention of fear of the CS when tested by a CER procedure 5 days after curarization.

530. Paré, William P. (1977). **Body temperature and the activity-stress ulcer in the rat.** ***Physiology & Behavior,*** 18(2), 219–223.
Conducted 2 experiments, using a total of 90 Sprague-Dawley male rats at the outset. Ss that developed lesions after exposure to the activity-stress procedure showed a drop in core body temperature. Body temperature and running activity are inversely related, thereby suggesting that excessive running is an attempt at behavioral thermoregulation. The restricted feeding schedule, however, did not provide sufficient energy to support the high running activity. Ss placed in a cold environment (7'C) and fed only 1 hr daily also developed stomach lesions but not as extensive as Ss housed in activity wheel cages and also fed 1 hr daily. Results suggest that the activity-stress procedure is a high energy demand situation and that subsequent stomach lesions represent a disturbance in energy metabolism.

531. Paré, William P. (1986). **Prior stress and susceptibility to stress ulcer.** ***Physiology & Behavior,*** 36(6), 1155–1159.
216 male Sprague-Dawley rats were exposed to daily sessions of either restraint plus cold, forced exercise, or grid shock for 1 mo and then subjected to the ulcerogenic effects of either supine restraint plus cold, or activity-stress. Chronic stress generally protected against stress ulcer, but the protection depended on the degree of similarity between the pretreatment stressor and the ulcerogenic procedure. Pretreatment with restraint plus cold provided protection that applied to supine restraint plus cold and generalized to activity-stress. Footshock pretreatment significantly exacerbated activity-stress ulcer.

532. Paré, William P. & Glavin, Gary B. (1986). **Restraint stress in biomedical research: A review.** ***Neuroscience & Biobehavioral Reviews,*** 10(3), 339–370.
Summarizes the methods for, the parameters of, and known drug effects on, restraint-induced pathology. It is suggested that this technique is a useful one for the examination of both central and peripheral mechanisms of stress-related disorders, as well as for studying drug effects upon these disorders. 11 tables of information on restraint techniques and effects are presented.

533. Paré, William P.; Natelson, Benjamin H.; Vincent, George P. & Isom, Kile E. (1980). **A clinical evaluation of rats dying in the activity-stress ulcer paradigm.** ***Physiology & Behavior,*** 25(3), 417–420.
Male Sprague-Dawley rats fed 1 hr daily and housed in running-wheel activity cages exhibited excessive running and developed stomach ulcers as compared to food control, body weight control, and home cage control Ss. In addition to the observed gastric disease, experimental Ss had increased bilirubins, decreased glycogen, and decreased serum proteins, suggesting that hepatic disease played a role in the lethal consequence of exposing Ss to the activity-stress procedure. The decreases in liver glycogen and serum glucose suggested that the terminal problem was related to incipient exhaustion of metabolic substrates.

534. Paré, William P. & Schimmel, Gregg T. (1986). **Stress ulcer in normotensive and spontaneously hypertensive rats.** ***Physiology & Behavior,*** 36(4), 699–705.
Studied stress ulcer susceptibility in 30 spontaneously hypertensive rats (SHRs) and 30 normotensive Wistar-Kyoto rats (WKYs), subjected to 2 ulcerogenic procedures—restraint plus cold and activity-stress. SHRs were more active and judged less fearful in the open-field test. Changes in adrenal and thymus weights and in core body temperature did not differentiate between SHRs and WKYs in the cold-restraint procedure. A significant adrenal hypertrophy was observed for SHRs in the activity-stress procedure. WKYs were more susceptible to stress ulcer in both the cold-restraint and the activity-stress procedures. Results illustrate the importance of genetic and constitutional variables in the development of ulcer disease. It is suggested that SHR and WKY rats represent a valuable tool for investigating how genetic factors, as well as behavioral traits, contribute to ulcer disease.

535. Paré, William P.; Vincent, George P. & Hsu, C. K. (1989). **The effect of different stressors, before and after inoculation, on growth of a mammary tumor in the rat.** ***Research Communications in Psychology, Psychiatry & Behavior,*** 14(2), 157–163.
Examined the effects of 3 stressors (restraint, cold, or forced swimming) and the timing of the introduction of the stressors before, after, or before and after, on the introduction of an R-3230 AC mammary tumor in 110 female rats. Stress did not significantly influence tumor onset, longevity or metastasis score. Body weight was retarded in more rats exposed to stress after inoculation. Stress inhibited tumor growth and was most effective if applied after inoculation.

536. Paré, William P. & Valdsaar, Elo. (1985). **The effects of housing and preshock on activity-stress ulcer.** ***Physiological Psychology,*** 13(1), 33–36.
48 male Long-Evans rats were either group-housed 4/cage or singly for 6 wks. Half of the subgroups were subsequently exposed to a daily 3-min 1.25-mA grid shock for 5 days; the other half were not shocked. All of the Ss were then housed in running-wheel activity cages and, following a 4-day free-feeding period, were fed only 1 hr daily for 20 days. Group-housed Ss had more ulcers, but the previously reported protective effect of preshock was not observed.

537. Paré, William P.; Vincent, George P.; Isom, Kile E. & Reeves, Jesse M. (1978). **Sex differences and incidence of activity-stress ulcers in the rat.** ***Psychological Reports,*** 43(2), 591–594.
Data from a study of 30 female and 30 male Sprague-Dawley rats show that female Ss housed in running-wheel activity cages and fed 1 hr daily ran significantly more than similarly housed and fed male Ss. Male Ss survived an average of 10 days, whereas mean survival time for female Ss was 7.3 days. Experimental activity Ss developed glandular stomach ulcers. Pair-fed controls housed in cages without activity wheels were ulcer-free. Among experimental activity Ss, there were no sex differences with respect to the number or severity of ulcers.

538. Paré, William P.; Vincent, George P. & Natelson, Benjamin H. (1985). **Daily feeding schedule and housing on incidence of activity-stress ulcer.** ***Physiology & Behavior,*** 34(3), 423–429.
Evaluated the roles of learning and duration of food availability in the pathogenesis of the disease state known as activity-stress (AS) ulcer, using 168 male Sprague-Dawley rats in 2 experiments. It was hypothesized that Ss with longer prestress experience with restricted feeding would be less susceptible to the lethal outcome of AS. In Exp I, the number and size of ulcers resulting from exposure to an AS procedure were inversely related to the length of the prestress experience with either 1- or 2-hr daily feeding schedules. In Exp II, Ss housed in group cages during the prestress period were more vulnerable to the ulcerogenic effects of the AS procedure. A pellet- or powdered-food treatment condition failed to provide significant group differences. Vulnerability to stress-ulcer is discussed in terms of the disparity of the environmental conditions between the acclimation period and the AS period.

539. Paré, William P. & Vincent, George P. (1984). **Early stress experience and activity-stress ulcer in the rat.** ***Research Communications in Psychology, Psychiatry & Behavior,*** 9(3), 325–334.
160 male Sprague-Dawley rats were randomly assigned to 1 of 4 treatment (prestress) conditions—exposure to cold (-20° C), forced swimming, restraint in a hardware cloth tube, and a no-stress control condition—and were subsequently exposed to an activity-stress-ulcer (ASU) procedure to investigate the effect of early stress experience on the incidence of stomach ulcers in Ss. Ss were subjected to their prestress condition for 5 days/wk for 4 wks, and, at the beginning of the 5th wk, Ss were exposed to the ASU procedure. Results indicate that ulcer severity in Ss prestressed with forced swimming or cold did not differ from control Ss. However, the number and severity of ASUs were significantly reduced in Ss prestressed by restraint.

540. Pare, William P. (1971). **Six-hour escape-avoidance work shift and production of stomach ulcers.** ***Journal of Comparative & Physiological Psychology,*** 74(3), 459–466.
Studied 16 male Long-Evans rats in each of 6 experiments in which they could make either an escape or an avoidance response to grid shock which was presented every 20 sec. Escape-avoidance sessions lasted 6 hrs and alternated with a 6-hr rest period for 21 days. Response rate, body weight, food and water intake, and diurnal consummatory pattern were recorded daily. Adrenal weight and ulcer incidence were noted at the termination of each experiment. Experimental Ss did not differ significantly from yoked controls on the dependent variables recorded. Only a small percentage (7.2%) of experimental "executive" Ss developed ulcers.

541. Pare, William P. (1972). **Conflict duration, feeding schedule, and strain differences in conflict-induced gastric ulcers.** ***Physiology & Behavior,*** 8(2), 165–171.
Conducted 4 experiments on the effects of chronic conflict using a total of 570 Long-Evans male rats. Ss exposed to 7 days of conflict developed more ulcers as compared to 4, 10, and 21 days of conflict. Ulcer incidence did not differ between 50- and 120-day old Ss, but ulcer incidence was significantly reduced if a 1-hr conflict-free feeding period was scheduled every day instead of every other day. Strain comparisons with Wistar, Long-Evans, Sprague-Dawley, and Holtzman rats showed that Wistar Ss were more resistant and Long-Evans Ss more susceptible to conflict-induced ulcers.

542. Pare, William P. (1972). **Conditioning and avoidance responding effects on gastric secretion in the rat with chronic fistula.** ***Journal of Comparative & Physiological Psychology,*** 80(1), 150–162.
Conducted 3 experiments with a total of 42 male Long-Evans rats. Base-line data was obtained by collecting gastric juice from Ss with chronic gastric fistula (CSF). Results show the hourly volume of gastric secretion decreased over the 23-hr period, whereas free and total acid increased. When Ss were given a 2-hr shock-stress period, secretion was inhibited during this period and higher volumes were obtained during 2-hr pre- and poststress periods. The same gastric secretory inhibition occurred during conditioning test trials in which shock was omitted. When Ss were trained on a Sidman avoidance task, hourly avoidance work sessions were characterized by a decrease in volume of gastric secretion and an increase in total acid. Results are discussed in terms of the psychological etiology of gastrointestinal lesions.

543. Pare, William P. (1974). **Feeding environment and the activity-stress ulcer.** ***Bulletin of the Psychonomic Society,*** 4(6), 546–548.
Male Sprague-Dawley rats were housed in standard activity cages and fed 1 hr each day. 18 Ss were fed in their activity cages and 12 Ss in the colony home cages in which they resided before being moved to the activity wheel. The latter operation was designed to eliminate novelty stress inherent in the activity wheel environment which would suppress feeding. However, home cage feeding Ss ate less than did the 12 controls and had as many ulcers as activity Ss fed in the activity wheel. Novelty stress did not contribute significantly to the development of the activity-stress ulcer.

544. Paré, William P. & Houser, Vincent P. (1973). **Activity and food-restriction effects on gastric glandular lesions in the rat: The activity-stress ulcer.** ***Bulletin of the Psychonomic Society,*** 2(4), 213–214.
Housed 18 60-day-old and 18 110-day-old Sprague-Dawley rats in standard activity cages and fed all Ss 1 hr a day for 21 days. 10 60-day-old and 5 110-day-old Ss died before the end of the experiment. All Ss that died had extensive lesions in glandular portions of the stomach. Controls housed in standard cages and also fed 1 hr each day, did not die. Experimental Ss that died were more active than experimental Ss that survived, and ate less than survivors and normal controls. This experimental technique is proposed as a new animal model for studying gastrointestinal pathology.

545. Paré, William P. & Isom, Kile E. (1975). **Gastric secretion as a function of acute and chronic stress in the gastric fistula rat.** ***Journal of Comparative & Physiological Psychology,*** 88(1), 431–435.
Conducted 2 experiments using a total of 54 male Long-Evans rats chronically implanted with gastric cannulas. In Exp I Ss exposed to signaled and unsignaled grid shock secreted more gastric acid after shock stress (chronic stress) for 8 days compared to the 1st 12 hrs of shock stress (acute stress). However, Exp II indicated that the higher gastric acid values under chronic stress were not significantly greater than prestress baseline values. Results are interpreted to reflect an inhibition of gastric acid secretion as a function of acute stress. During chronic stress this inhibition was followed by an habituation of gastric secretory processes which was observed as a return of secretion volume to baseline levels.

546. Paré, William P. & Livingston, Andrew. (1975). **Shock predictability and plasma gastrin in the rat.** ***Bulletin of the Psychonomic Society,*** 5(4), 289–291.
18 male Sprague-Dawley rats with chronic gastric cannulas were immobilized in restraint cages and were subjected either to predictable shock, unpredictable shock, or simply restrained. All Ss were immobilized only for 4 hrs. During the 2nd and 3rd hrs, Ss in the predictable shock and unpredictable shock groups received their respective treatments. Gastric secretion was collected after each hr, and plasma gastrin estimates were obtained after the 1st and 3rd hrs. Shock stress produced a significant inhibition of total acid output for the 2 shock treatment groups. However, statistically significant plasma gastrin differences were not observed either between treatment groups or between pre- and poststress measures. The data support contemporary clinical observations that gastrin activity is greater in patients with duodenal ulcers, and not significantly different from control values in patients with gastric ulcers.

547. Paré, William P. & Temple, Lester J. (1973). **Food deprivation, shock stress and stomach lesions in the rat.** ***Physiology & Behavior,*** 11(3), 371–375.
Assigned 140 male Sprague-Dawley rats to groups deprived of food for 1–12 days. Rumenal ulcers were observed in some Ss after 3 days of fasting and in all Ss fasted for 5 or more days. In Exp II with 120 Ss, Ss were fasted or fasted plus subjected to grid shock stress for 1–5 days. The addition of shock stress to the fasting schedule did not increase ulcer incidence.

548. Paul, Steven M. (1988). **Anxiety and depression: A common neurobiological substrate?** Symposia: Consequences of anxiety (1988, Montreal, Canada). ***Journal of Clinical Psychiatry,*** 49(Suppl), 13–16.
Clinical studies suggest that anxiety is not only accompanied by depression but may be an expected precursor in the development of some forms of depression. Genetic and epidemiological data indicate that some forms of anxiety and depression may represent different phenotypic manifestations of the same genetic predisposition resulting from varying environmental conditions. Recent findings on the learned helplessness syndrome show that the administration to rats of anxiogenic inverse agonists of the benzodiazepine gamma-aminobutyric acid (GABA) receptor complex produces the same behavioral syndrome evoked by inescapable stress. Pretreating animals with benzodiazepine anxiolytics prevents learned helplessness after exposure to inescapable stress. Data suggest a common neurobiological substrate for some forms of anxiety and depression.

549. Peachey, John E. & Stancer, Harvey C. (1973). **An animal model for psychopharmacological research with relevance to psychiatry.** ***Canadian Psychiatric Association Journal,*** 18(2), 139–146.
Presents the concept that animal models for psychopharmacological research should take into account clinical methods if they are to yield information which is relevant to psychiatry. It is suggested that animal species with appropriate behavioral characteristics should be selected, drugs should be administered chronically, and groups of animals should be used if social interaction is to be observed. Some of the relevant pharmacological, behavioral, and environmental factors are presented and illustrated with examples taken from a study on groups of squirrel monkeys who had received parachlorophenylalanine or alpha-methylparatyrosine to lower specific brain amines. It is noted that social interaction may affect the expression of the drug effect. (French summary).

550. Phillips, James D., Jr. & Boone, Daniel C. (1968). **Effects of adrenalin supplement on the production of stress induced ulcers in adrenal sympathectomized male rats.** ***Proceedings of the 76th Annual Convention of the American Psychological Association,*** 3, 261–262.

551. Pieper, W. A.; Skeen, Marianne J.; McClure, Harold M. & Bourne, Peter G. (1972). **The chimpanzee as an animal model for investigating alcoholism.** ***Science,*** 176(4030), 71–73.
Describes a program in which 6 young chimpanzees (Pan troglodytes) were induced to accept ethanol, in quantities sufficient to produce symptoms of withdrawal when ethanal was subsequently discontinued after 6–10 wks of chronic oral intake. Mild to severe symptoms of physical dependence, including grand mal seizures, were observed when ethanol was abruptly withdrawn. The rate of disappearance of ethanol in blood increased during periods of chronic ingestion, an indication of developing metabolic tolerance. Results suggest that the young chimpanzee may be a suitable model for experimental studies of alcoholism.

552. Poley, Wayne & Royce, J. R. (1973). **Behavior genetic analysis of mouse emotionality: II. Stability of factors across genotypes.** ***Animal Learning & Behavior,*** 1(2), 116–120.

Mated 3 emotionally divergent strains of mice (SWR/J, A/HeJ, and SJL/J) to produce 3 populations: a pure strain population of 90 Ss, an F1 population of 120 Ss, and an F2 population of 96 Ss. All Ss were tested on a battery of measures of emotionality. Each population was factored separately by principal components factoring with varimax and promax rotations. Factorial invariance was assessed quantitatively by congruence coefficients. Eight of the resultant factors were found to be replicable across populations. These were Autonomic Balance, Motor Discharge, Territorial Marking, Acrophobia, Tunneling-1, Tunneling-2, Underwater Swimming, and Audiogenic Reactivity. Of the 8 factors, Motor Discharge was most distinct in each population and Tunneling-1 was least distinct.

553. Price, Kenneth P. (1972). **Predictable and unpredictable shock: Their pathological effects on restrained and unrestrained rats.** ***Psychological Reports,*** 30(2), 419–426.
Examined the effects on 76 male Sprague-Dawley albino rats of predictable and unpredictable shock under 2 conditions: unrestrained Ss were shocked intermittently for 12/24 hrs on a 6-hrs on, 6-hrs off schedule; restrained Ss were shocked intermittently for 12/24 hrs on a 6-hrs on, 6-hrs off schedule. All Ss were shocked via tail electrode so that unrestrained Ss were completely unable to avoid "unavoidable" shock. Under Condition I, Ss receiving unpredictable shock lost more weight than Ss receiving predictable shock. Under Condition II, more Ss in the unpredictable-shock group developed ulcers than Ss in the predictable-shock group. Findings partially confirm those of J. Weiss (1968, 1970) and extended results to unrestrained Ss.

554. Prince, Christopher R.; Collins, Carolyn & Anisman, Hymie. (1986). **Stressor-provoked response patterns in a swim task: Modification by diazepam.** ***Pharmacology, Biochemistry & Behavior,*** 24(2), 323–328.
Studied the effect of treatment with a benzodiazepine on the response invigoration and stimulus perseveration engendered by inescapable shock in experiments with 174 naive, male CD-1 mice. In a forced-swim task Ss initially exhibited vigorous responding followed by a rapid decay of active swimming. Following exposure to inescapable shock, the response invigoration was enhanced, as was the tendency to remain in the illuminated region of the arena. Low doses of diazepam (0.5 and 1.0 mg/kg) prior to testing eliminated the response invigoration, as well as the response of approaching the illuminated region. It is proposed that the behaviors evident after uncontrollable shock are related to a transient increase of anxiety or vigilance. It is suggested that several time-dependent behavioral variations associated with inescapable shock may be related to alterations of anxiety.

555. Pritzel, Monika & Markowitsch, Hans J. (1985). **Animal experiments: A commentary from the viewpoint of the physiological psychologist.** ***Psychologische Rundschau,*** 36(1), 16–25.
Presents arguments for and against research using animals, stressing the value of animal research in addressing physiological and psychological human problems. While opponents of animal research assert that the results of such research are not transferable to human behavior or the human nervous system, it is argued that these studies are needed to develop treatments for brain-injury patients and to investigate brain functions that guide human behavior. The role of animal models in studying interactions between brain and behavior in humans is also emphasized. (English abstract).

556. Raab, Achim; Kojer, Gerti & Oswald, Regine. (1982). **Social status and coping with conflict: Features of the mesolimbic dopaminergic system.** ***Aggressive Behavior,*** 8(2), 212–216.
Findings from 3 studies with male tree shrews show that (a) a decrease in tyrosine hydroxylase activity in the septal area, as a response to conflict, decreased the Ss' coping abilities; (b) a chemical impairment of the mesolimbic system by destruction of the dopaminergic neurons in the A10 region by 6-hydroxydopamine resulted in performance decrements in a passive-avoidance task; and (c) conflict induced an impairment of the mesolimbic dopaminergic system in chronically subordinate Ss. Despite this impairment, however, subordinates were able to adapt to new conditions. Implications for models of depression are noted.

557. Rabinovich, M. Ya. (1986). **Experimental modeling of cellular mechanisms of some psychopathologic syndromes.** ***Zhurnal Vysshei Nervnoi Deyatel'nosti,*** 36(2), 242–251.
Discusses experimental modeling of syndromes of pathologic brain production (delirium, hallucinations, mental automatisms) in animals. Emphasis is on the present author's research involving the chronic intoxication of alert rabbits with d,l-amphetamine and single administration of haloperidol to analyze the "antipsychotic" action of haloperidol on the cellular analog of amphetamine stereotypy.

558. Rea, M. A. & Hellhammer, D. H. (1984). **Activity wheel stress: Changes in brain norepinephrine turnover and the occurrence of gastric lesions.** 7th World Congress of the International College of Psychosomatic Medicine: Psychosomatic research and practice (1983, Hamburg, West Germany). ***Psychotherapy & Psychosomatics,*** 42(1–4), 218–223.
Investigated the effects of activity wheel stress (AW) on brain regional norepinephrine (NE) and 3-methoxy-4-hydroxyphenylglycol (MHPG) content and on the occurrence of gastric lesions in adult male Wistar rats. Multiple gastric lesions were present in the stomachs of all Ss exposed to AW. No gastric lesions were observed in any of the food consumption (FC) or untreated control (UC) Ss. In AW Rats, NE levels were significantly different from UC and/or FC Ss in the hypothalamus, striatum, and hippocampus. MHPG levels in AW Ss were significantly elevated in the hypothalamus, thalamus, neocortex, midbrain, pons medulla, and cerebellum, indicating increased NE turnover in these brain regions. Findings are discussed in terms of a possible role for brain NE in the mediation of activity stress-induced gastric lesions.

559. Resnick, O. & Morgane, Peter J. (1984). **Generational effects of protein malnutrition in the rat.** ***Developmental Brain Research,*** 15(2), 219–227.
Studied the generational effects of protein malnutrition in Sprague-Dawley rats fed either low (8% casein) or normal (25% casein) amounts of protein. Data indicate that a mild protein restriction in the 1st generation became a more severe protein restriction in the 2nd generation. This is based on weight gains of the dams during pregnancy, the mean number of pups per litter, the mean pup body and brain weight at birth, growth curves, levels of brain tryptophan, 5-HT, and 5-hydroxyindoleacetic acid from birth to

weaning, and the levels of certain plasma constituents, especially nonesterified fatty acids. This paradigm is proposed as an animal model for some types of chronic undernutrition in socioeconomically underprivileged human populations.

560. Resnick, Oscar; Morgane, Peter J.; Hasson, Rachelle & Miller, Maravene. (1982). **Overt and hidden forms of chronic malnutrition in the rat and their relevance to man.** ***Neuroscience & Biobehavioral Reviews,*** 6(1), 55–75.
Examined the physiological weight changes in 61 rat dams and their offsprings as sequelae of overt or hidden chronic protein malnutrition. In the overt model, produced by feeding Ss a low protein diet starting 5 wks prior to conception and continuing through lactation, Ss showed significant weight losses at all ages. These Ss had small offspring, and the inadequate milk production resulted in their pups displaying almost total failure of growth and peripheral imbalances. In contrast, the hidden form of malnutrition produced by feeding Ss a somewhat higher protein diet caused no marked weight losses during their pregnancy compared to the normal dams. Although the pups had the same birth weight indexes as the normal offspring, previous data indicate that the pups with hidden malnutrition show many metabolic imbalances at birth that are indicators of severe gestational malnutrition in humans. Additional nutrition later on was unable to rehabilitate most of the prenatally determined biochemical alterations affecting the pups but, additionally, this form of malnutrition would remain undetected if weight indexes alone were used as assessors of normalcy. This model can be used to study human children who often show a "normal" weight at birth, but actually suffer from hidden malnutrition that is later manifested in mental disabilities.

561. Richardson, J. Steven. (1991). **Animal models of depression reflect changing views on the essence and etiology of depressive disorders in humans.** Special Issue: Perspectives in Canadian neuro-psychopharmacology: Proceedings of the 13th Annual Canadian College of Neuro-psychopharmacology. ***Progress in Neuro-Psychopharmacology & Biological Psychiatry,*** 15(2), 199–204.
The hypothesis that depression is caused by stress led to the development of animal models of depression based on behavioral abnormalities induced in animals exposed to prolonged or intense stress. Selective therapeutic efficacy of antidepressant drugs (ADs) suggested the hypothesis that depression is caused by a neurochemical abnormality that is alleviated by chronic exposure to ADs. This hypothesis has been refined to state that depressive disorders are caused by genetically based neurochemical dysregulation of neural activity in the limbic system. Development of a genetic strain of animals that shows comparable behavioral and neurovegetative deficits and similar selective drug responses to those seen in patients with depressive disorder would contribute to the understanding of depression and the neurobiology of the limbic system.

562. Ridley, R. M. & Baker, H. F. (1991). **A critical evaluation of monkey models of amnesia and dementia.** ***Brain Research Reviews,*** 16(1), 15–37.
Reviews models of amnesia and dementia in monkeys and examines their validity. Memory tasks available for use with monkeys are described, and the extent to which these tasks assess different facets of memory according to present theories is discussed. Demonstrating episodic memory in monkeys is problematic, and the term recognition memory has been used too loosely. It is particularly difficult to dissociate episodic memory for stimulus events from the use of semantic memory for the rule of the task. Classification of tasks into those that assess memory for stimulus-reward associations and those that tax stimulus–response associations, including spatial and conditional responding, may be useful in clarifying the contribution of the temporal lobes and the cholinergic system to memory.

563. Robinson, Susan E.; Martin, Rebecca M.; Davis, Todd R.; Gyenes, Carol A. et al. (1990). **The effect of acetylcholine depletion on behavior following traumatic brain injury.** ***Brain Research,*** 509(1), 41–46.
Examined the role of acetylcholine (ACh) in behavioral deficits (BDs) following traumatic brain injury (TBI), using male rats injected with ACh depleting agents (A-4 and A-5 forms of hemicholinium) or saline. In Exp 1, Ss were assessed 1 hr after TBI for recovery time of reflexes and responses (e.g., tail flexion), gross vestibulomotor function, locomotor activity, and fine motor coordination. Brain tissue from these Ss was analyzed in Exp 2. Pretreatment with A-4 or A-5 attenuated BDs following TBI, supporting involvement of a cholinergic mechanism in these deficits. Effects of TBI and A-5 treatment on ACh levels in the cerebrospinal fluid (CSF) suggest that TBI may allow plasma constituents to gain access to the central nervous system (CNS).

564. Rockman, G. E.; Borowski, T. B. & Glavin, G. B. (1986). **The effects of environmental enrichment on voluntary ethanol consumption and stress ulcer formation in rats.** ***Alcohol,*** 3(5), 299–302.
43 male Wistar rats were reared in an enriched environment for 90 days. Following an initial 36-day ethanol exposure period, voluntary ethanol preference was assessed. At the conclusion of the ethanol test session, Ss were exposed to restraint stress for 3 hrs. Results indicate that exposure to an enriched environment produced increased voluntary ethanol consumption as compared to nonenriched controls. Furthermore, Ss reared in the enriched environment demonstrated reduced gastric ulcer severity in response to restraint stress. Ethanol per se did not affect ulcer formation.

565. Rockman, G. E. & Glavin, G. B. (1986). **Activity stress effects on voluntary ethanol consumption, mortality and ulcer development in rats.** ***Pharmacology, Biochemistry & Behavior,*** 24(4), 869–873.
Investigated the relationship between activity stress, alcohol consumption, and ulcer proliferation with 86 male Wistar rats. Ethanol-consuming Ss were initially divided into low-, medium-, or high-ethanol-preferring groups on the basis of daily ethanol intake. Following a habituation period in activity cages, Ss were fed for 1 hr per day. Access to both water and ethanol remained ad lib. Yoked control home cage Ss were fed the same amount of food consumed by their wheel-housed partners. This procedure continued until wheel-housed Ss died, at which time they and their yoked home cage control partners were examined for ulcers. Results indicate that, in contrast to the yoked controls, only the high-ethanol-preferring Ss reduced their ethanol consumption. Although no differences were apparent in ulcer frequency or severity, Ss exposed to ethanol had a lower ulcer incidence and mortality rate than non-ethanol-exposed Ss.

566. Rodríguez Echandía, E. L.; Gonzalez, A. S.; Cabrera, R. & Fracchia, L. N. (1988). **A further analysis of behavioral and endocrine effects of unpredictable chronic stress.** ***Physiology & Behavior,*** 43(6), 789–795.
Analyzed behavioral and endocrine effects in male and female rats treated daily with unpredictable emotional stressors (ES-groups) or unpredictable physical stressors (PS-groups) and in male and female undisturbed controls over a 14-day period. Ss were then submitted to 3 behavioral tests at 24-hr intervals. Exp 1 showed that when Ss were tested in an enriched environment, both total motor activity and exploration of the novel object were impaired by the PS treatment. This suggests the occurrence of motivational deficit. Females were more resistant to the behavioral effects of PS treatment. In Exp 2, the PS treatment caused constant diestrus in females, whereas the ES treatment caused minor estrous cycle disturbances. Results indicate that the chronic administration of unpredictable chronic stress is a useful animal model of human depression.

567. Royce, J. R.; Poley, Wayne & Yeudall, L. T. (1973). **Behavior-genetic analysis of mouse emotionality: I. Factor analysis.** ***Journal of Comparative & Physiological Psychology,*** 83(1), 36–47.
Obtained 775 pure-strain and F1 mice from a 6×6 diallel mating plan. Each S completed a battery of 12 tests of emotionality, requiring 1 mo for completion. The subsequent 42 measures were factor analyzed by alpha factoring with varimax and promax rotations. 15 factors with eigenvalues > 1.0 were found. 10 of these factors were interpreted as different facets of emotionality. Another factor was identified as weight, and 4 were underdetermined. The 10 emotionality factors were interpreted as motor discharge, acrophobia, underwater swimming, tunneling 1, audiogenic reactivity, food motivation, autonomic balance, territorial marking, activity level, and tunneling 2. Similarities with other major factor-analytic studies of emotionality are discussed.

568. Russell, Roger W. (1991). **Essential roles for animal models in understanding human toxicities.** Symposium on Animal-To-Human Extrapolation (1990, San Antonio, Texas). ***Neuroscience & Biobehavioral Reviews,*** 15(1), 7–11.
Asserts that there are no viable alternatives to the use of animal models in toxicology and that virtually every major advance in the biomedical sciences and technologies stems in whole or in part from research performed with animals. Analogies may involve modeling at levels of bio-organization from molecules to the behavior of the total, integrated organism: Toxicity of exposure to a chemical varies with the structure(s) or function(s) modeled. Trade-offs between biological limitations and social cost–benefits place great importance on the validities of extrapolations from the animal data. Animal models play essential roles in the biomedical sciences and technologies provided that due precautions are taken in creating models and in generalizing from them.

569. Ryan, Susan M. & Maier, Steven F. (1988). **The estrous cycle and estrogen modulate stress-induced analgesia.** ***Behavioral Neuroscience,*** 102(3), 371–380.
In this article we investigate the impact of estrous cycle, ovariectomy, and estrogen replacement on both opioid and nonopioid stress-induced analgesia. Stage of estrous strongly influenced analgesia. Diestrus females exhibited the typical male pattern produced by the analgesia inducing procedures used—strong nonopioid analgesia following 10–20 tailshocks, and strong opioid analgesia following 80–100 taskshocks. In these experiments the nonopioid analgesia was slightly attenuated during estrus, but the opioid analgesia was markedly reduced. The role of estrogen in producing these changes was studied with estrogen replacement in ovariectomized subjects. Ovariectomy only slightly altered nonopioid analgesia but eliminated opioid analgesia, which suggests that some estrogen might be necessary to maintain the integrity of the system(s) underlying opioid analgesia. Estrogen administration restored opioid analgesia, but further estrogen suppressed opioid analgesia, duplicating the estrus pattern. It did not suppress nonopioid analgesia. Opioid analgesia was enhanced 102 hr after estrogen replacement, thus duplicating the diestrus pattern. Estrogen thus appears to be responsible for the impact of estrous cycle on opioid but not on nonopioid analgesia. These results suggest that ovarian hormones may modulate the impact of stressors on endogenous pain inhibition and other stress-responsive systems.

570. Sagvolden, Terje. (1991). **Attention deficit disorder: The problems may be secondary to changed learning processes.** ***Tidsskrift for Norsk Psykologforening,*** 28(8), 672–678.
Discusses new data from experiments on attention deficit disorder using hypertensive and hyperactive rat strains. Diagnostic problems, the contribution of animal research, changes in reactivity to reinforcers, behavior therapy for rat hyperactivity, the development of attention problems, and the effects of psychomotor stimulants (especially methylphenidate) are considered. (English abstract).

571. Salim, Aws. S. (1987). **Stress, the adrenergic hypothalamovagal pathway, and chronic gastric ulceration.** ***Journal of Psychosomatic Research,*** 31(1), 85–90.
Tested the hypothesis that stress activates the hypothalamus, which stimulates the adrenergic hypothalamovagal pathway, producing acute gastric mucosal injury, using groups of 20 Sprague-Dawley rats. Stimulation of the pathway with a single intraperitoneal injection of reserpine (5 mg/kg) every 24 hrs for 5 days produced chronic gastric ulceration in 80% of Ss, demonstrating the relationship between stress and chronic gastric ulceration.

572. Salim, Aws S. (1987). **Stress, the adrenergic hypothalamovagal pathway, and the aetiology of chronic duodenal ulceration.** ***Journal of Psychosomatic Research,*** 31(2), 231–237.
In 48 Sprague-Dawley rats stressed pharmacologically by administration of intraperitoneal reserpine (0.1 mg/kg), the stimulated pathway released gastric 5-hydroxytryptamine (5-HT) that in turn significantly increased gastric acid secretion. Administration of intramuscular reserpine (0.1 mg/kg) every 24 hrs for 6 wks significantly increased the basal acid output of the rat stomach and produced duodenal ulceration in 83.3% of Ss. Vagotomy completely protected the rat stomach against these effects. Results are discussed in relation to the effects of stressful life events on the occurrence of ulcerative disease in humans.

573. Sanberg, Paul R. & Emerich, Dwaine F. (1990). **Neural basis of behavior: Animal models of human conditions.** 19th Annual Meeting of the Society for Neuroscience: Neural basis of behavior: Animal models of human conditions (1989, Phoenix, Arizona). ***Brain Research Bulletin,*** 25(3), 447–451.

Presents research on animal models of Huntington's disease (HD) and Tourette syndrome (TS). These models can elucidate the underlying neurobiological mechanisms of human disease states and can suggest innovative therapies. The finding that kainic acid-lesioned rats show marked changes in body weight and regulatory behavior suggests that underlying striatal degeneration may also be related to weight loss seen in HD. Body weight has proven useful in diagnosis and in designing interventions. Animal studies have also suggested that nicotine could potentiate the effects of haloperidol. If this is the case, then nicotine might also be useful for augmenting the therapeutic effects of neuroleptics in patients with extrapyramidal movement disorders, such as TS.

574. Sanberg, Paul R.; Johnson, David A.; Moran, Timothy H. & Coyle, Joseph T. (1984). **Investigating locomotion abnormalities in animal models of extrapyramidal disorders: A commentary.** ***Physiological Psychology,*** 12(1), 48–50.
Notes that diseases of the extrapyramidal system usually result in disorders of movement and that the location of the pathology within this system is critical in the expression of the type of movement disorder (e.g., Huntington's disease, Parkinson's disease, and Tourette's syndrome). Research has also shown that alterations in the functions of the basal ganglia are evident in many movement disorders, although the role of this important group of telencephalic nuclei in behavior is little understood. The present authors, therefore, review research in this area to analyze in detail the pattern of abnormal locomotion found in neonatal and adult rats with various types of striatal lesions. This information may be a useful addition to knowledge about the role of the striatum in movement (and related) disorders and may lead to the development of reliable animal models for human extrapyramidal disorders.

575. Sarter, Martin & Van der Linde, Angelika. (1987). **Vitamin E deprivation in rats: Some behavioral and histochemical observations.** ***Neurobiology of Aging,*** 8(4), 297–307.
Conducted 5 experiments with 30 rats deprived of vitamin E from age 4 wks and 30 rats fed a control diet. Vitamin E deprivation did not alter locomotor and exploratory activity as measured by the use of a hole-board. Deprived Ss were significantly impaired in relearning a 6-arm radial maze following a break in training, but mastered a reverse configuration of the tunnel maze as well as controls. Deprived Ss showed weaker effects than controls when treated with scopolamine. The highest density of neuronal lipofuscin accumulation resulting from the deprivation was found in the hippocampal areas. Since lipofuscin accumulation represents a prominent alteration of the aging brain, the usefulness of vitamin E deprivation as an animal model for accelerated aging is discussed.

576. Schapiro, H.; McDougal, H. D.; Albert, I. & Boone, D. H. (1978). **The effect of visual deprivation on gastrointestinal ulceration.** ***Physiology & Behavior,*** 21(5), 705–709.
Visual deprivation reduced the incidence of perforated ulcers and increased the survival time of dogs subjected to the Exalto-Mann-Williamson ulcer and reduced both the time of onset and the degree of ulceration in rats subjected either to the Shay ulcer or to the restraint-cold ulcer procedure.

577. Schmidt, K. M.; Kangas, J. A. & Solomon, G. F. (1971). **The effects of ethanol on the development of gastric ulceration in the rat.** ***Journal of Psychosomatic Research,*** 15(1), 55–61.
Investigated (a) the effects of a moderate dose of alcohol (10% solution per 100 g of body weight) and a strong dose of alcohol (20% solution per 100 g of body weight) on ulceration of the stomach in the albino rat, and (b) the extent of ulceration at different times throughout the duration of a 10-day stress period. Results suggest that stress due to dietary disturbance was a greater factor in ulcer production than the approach-avoidance conflict or the alcohol treatment. The administered alcohol solutions did not either increase the production of ulcers or interfere with their healing.

578. Schouten, M. Joris & Bruinvels, Jacques. (1986). **Endogenously formed norharman (β-carboline) in platelet rich plasma obtained from porphyric rats.** ***Pharmacology, Biochemistry & Behavior,*** 24(5), 1219–1223.
Porphyria was induced in male Wistar rats starved for 24 hrs by an injection of 400 mg/kg allylisopropylacetamide. Intraperitoneal injection of 2 mmol/kg serine 24 hrs later was used as an animal model for an acute psychosis, by measuring catalepsy scores 30 min after serine injection. An increase in the concentration of norharman (NH) in platelet rich plasma, ranging from 0.57 nmoles/l in controls to 1.88 nmoles/l in serine-treated porphyric Ss, was found. It is concluded that an elevated conversion of serine into glycine via serine hydroxymethyltransferase may be responsible for the enhanced NH biosynthesis.

579. Schuster, Richard H.; Rachlin, Howard; Rom, Meni & Berger, Barry D. (1982). **An animal model of dyadic social interaction: Influence of isolation, competition, and shock-induced aggression.** ***Aggressive Behavior,*** 8(2), 116–121.
Describes a method of analyzing dyadic interactions by creating a situation for social interaction with reinforcement and manipulating social or individual experiences inserted at specified times during learning and later performance. Examples of data obtained with the procedure, in an experiment in which a pair of laboratory rats had to coordinate their movements in space by attending to each other's behavior, are presented. Findings show that 2 hypothetically stressful interactions, competition and shock-induced aggression, led to contrasting social effects depending on whether the interacting dyad had previously learned a positively reinforced coordination.

580. Schwille, P. O.; Schellerer, W.; Reitzenstein, M. & Hermanek, P. (1974). **Hyperglucagonemia, hypocalcemia and diminished gastric blood flow: Evidence for an etiological role in stress ulcer of rat.** ***Experientia,*** 30(7), 824–826.
Studied the effect of pancreatic glucagon (pGl) on the formation of mucosal lesions and ulcers in male SPF Wistar rats. Ss were divided into groups of intact, sham-operated, and adrenalectomized animals. Half of each were stressed by the restraint technique. Local oxygen pressure was measured by polarography and pGl by radioimmunoasssay. Results indicate that stress induced mucosal lesions and that the pGl level in stressed Ss is elevated. The highest pGl occurs in adrenalectomized Ss during stress. (German summary).

581. Sechzer, Jeri A.; Faro, Maria D. & Windle, William F. (1973). **Studies of monkeys asphyxiated at birth: Implications for minimal cerebral dysfunction.** ***Seminars in Psychiatry,*** 5(1), 19–34.
A review of previous studies of specific neuropathology of asphyxia neonaturum reveals functional deficits in monkeys subjected to asphyxiation for varying amounts of time up to 17 min. Although these deficits persist for some time, compensation eventually occurs. In the present series of studies, adaptive behaviors (visual depth perception, visual placing, and independent locomotion) and acquired behaviors were examined in neonatally asphyxiated and nonasphyxiated rhesus monkeys ($N = 16$). Brain damage was found after only 7 min of asphyxiation. Although adaptive behaviors of infant Ss asphyxiated for 15 min were recovered, acquired behaviors (e.g., memory and learning) were severely impaired in adult Ss asphyxiated at birth. Comparison of results with studies of human Ss reveals similar deficits in asphyxiated infant monkeys and children with minimal brain dysfunction. It is suggested that much could be understood about minimal brain dysfunction in children through the use of an animal model.

582. Seligman, Martin E. & Meyer, Bruce. (1970). **Chronic fear and ulcers in rats as a function of the unpredictability of safety.** ***Journal of Comparative & Physiological Psychology,*** 73(2), 202–207.
A variable number of unpredictable electric shocks presented to 20 male Sprague-Dawley albino rats bar pressing for food produced substantial nontransient suppression across 70 sessions. With a fixed number of otherwise unpredictable shocks in each session, Ss recovered by pressing after the last shock, using its occurrence as a safety signal. When signals predicted a variable number of shocks Ss bar pressed in the absence of the CS and not in its presence. Pressing recovered with milder predictable shock and more slowly with mild unpredictable shock. Inhibition of delay was found in predictable shock groups. Fear, measured by suppression, correlated with gastrointestinal ulceration ($p = .74$). Findings confirmed a safety-signal explanation of the effects of unpredictable shock.

583. Seltzer, Ze'ev; Cohn, Sergiu; Ginzburg, Ruth & Beilin, BenZion. (1991). **Modulation of neuropathic pain behavior in rats by spinal disinhibition and NMDA receptor blockade of injury discharge.** ***Pain,*** 45(1), 69–75.
Showed in male rats that local anesthetic blockade of the discharge emitted by damaged sensory peripheral nerve fibers suppressed autotomy (a behavioral model of neuropathic pain). Mimicking prolonged injury discharge with electrical stimulation increased autotomy. Data support the hypothesis that injury discharge plays a role in the triggering of neuropathic pain. The mechanism of triggering autotomy was investigated using intrathecal injection of agents affecting glutamatergic transmission. A single injection at the lumbar enlargement of the NMDA receptor blockers MK-801 and 5-aminophosphonovalerate prior to neurectomy significantly suppressed autotomy. Blocking glycinergic inhibition prior to neurectomy with a single strychnine injection enhanced autotomy. The expression of autotomy in rats, and by inference neuropathic pain in humans, appears to be affected by injury discharge, possibly mediated by long-lasting, NMDA receptor-related, spinal disinhibition.

584. Shull, Ronald N. & Holloway, Frank A. (1985). **Effects of caffeine and L-PIA on rats with selective damage of the hippocampal system.** ***Pharmacology, Biochemistry & Behavior,*** 22(3), 449–459.
Observed the effects of both caffeine and the adenosine analog levo-N6-phenyliso-propyladenosine (L-PIA) on "abnormal" behavior generated by 12 naive male Sprague-Dawley rats with selective damage of various areas of the hippocampal system. Ss with electrolytic lesions of either the medial septum or hippocampus or with colchicine-induced lesions of the dentate granule cells produced significantly higher amounts of leverpressing compared to controls on both VI50-sec and VI50-sec reinforcement schedules. Caffeine (3.2, 10, and 32 mg/kg) produced a dose-related decrease in operant responding for all lesioned Ss while having little effect on the responding of controls. L-PIA (0.01, 0.05, and 0.10 mg/kg) produced similar but more variable effects in some of the groups tested and did not alter the rate-reducing effects of caffeine (32 mg/g) when given concurrently (0.05 mg/kg). Caffeine administration also appeared to coincide with a long-term decrease in response rates that continued after cessation of its administration. It is postulated that colchicine-induced damage of dentate granule cells might be a viable animal model of some forms of hyperactivity.

585. Silbergeld, Ellen K. & Goldberg, Alan M. (1974). **Lead-induced behavioral dysfunction: An animal model of hyperactivity.** ***Experimental Neurology,*** 42(1), 146–157.
Developed an animal model of lead poisoning in which suckling CD-1 mice were exposed to lead acetate from birth indirectly through their mothers and then directly after weaning. For the 1st 60 days, no deaths of offspring occurred due to lead but growth and development were significantly retarded. Activity was measured between 40 and 60 days of age. Treated Ss were more than 3 times as active as age-matched controls. Treated and control Ss were given drugs used in the treatment and diagnosis of minimal brain dysfunction hyperactivity in children: dextro- and levoamphetamine, methylphenidate, phenobarbital, and chloral hydrate. Lead-treated hyperactive Ss responded paradoxically to all drugs except chloral hydrate: the amphetamines and methylphenidate suppressed hyperactivity, while phenobarbital increased levels of motor activity. Chloral hydrate was an effective sedative. Implications for the study of the central effects of lead poisoning and for the relationship between lead poisoning and minimal brain dysfunction hyperactivity are discussed.

586. Simpson, C. Wayne et al. (1975). **Stress-induced ulceration in adrenalectomized and normal rats.** ***Bulletin of the Psychonomic Society,*** 6(2), 189–191.
Examined the influence of plasma corticosterone on stress-induced ulceration in restrained rats. Ss were 30 adult male Sprague-Dawley rats. Gastric pathology and plasma corticosterone levels were examined in normal (UNOP) and adrenalectomized (ADX) Ss under 2 conditions of shock predictability. The severity of the gastric ulceration in unoperated Ss given noncontingent presentations of a tone and an electric shock was greater than in Ss receiving shocks contingent upon CS presentation. When ulcer pathology was evaluated for ADX Ss, no differences were found between the predictable and the unpredictable shock groups. Results are dis-

cussed in terms of the necessary and sufficient conditions for stomach ulcer formation as a function of shock predictability and restraint stress and the necessity to investigate alternative hypotheses.

587. Simson, Peter E. & Weiss, Jay M. (1989). **Peripheral, but not local or intracerebroventricular, administration of benzodiazepines attenuates evoked activity of locus coeruleus neurons.** ***Brain Research,*** 490(2), 236–242.
Demonstrated a profound inhibitory effect of benzodiazepines on sensory-evoked activity of the locus coeruleus (LC) in male rats. This is consistent with the notion that LC neurons are activated during anxiety and that this activation is characterized by heightened responsiveness of LC neurons to stimulation. Benzodiazepines attenuate the sensory input to the LC neurons.

588. Simson, Peter E. & Weiss, Jay M. (1988). **Altered activity of the locus coeruleus in an animal model of depression.** ***Neuropsychopharmacology,*** 1(4), 287–295.
Conducted an experiment with male rats in which one group was subjected to an uncontrollable stressor and another group was not. In Ss showing depression of active behavior, locus coeruleus (LC) neurons were not inhibited as they are normally; in particular, electrophysiologic recording showed LC neurons were hyperresponsive to excitatory input. Moreover, the degree to which LC neurons of individual Ss were hyperresponsive correlated positively with the degree to which active behavior was depressed. Recently, hyperresponsiveness of LC neurons has been found to occur when the inhibitory influence of α_2-receptors in LC firing is blocked. Results show that activity of the LC is altered in behavioral depression in an animal model and suggest that abnormalities in this system and its regulatory elements, such as α_2-receptors, may be present in some types of clinical depression.

589. Sines, J. O. (1979). **Non-pharmacological and non-surgical resistance to stress ulcers in temperamentally and physiologically susceptible rats.** ***Journal of Psychosomatic Research,*** 23(1), 77–82.
It has been demonstrated that animals' physiological responses to a variety of stressful environmental conditions can be modified or lessened by repeated exposure to brief periods of those stressors. The current 2 studies investigated whether resistance to certain stressors can be induced in animals that are either temperamentally or physiologically vulnerable or predisposed to develop stomach lesions under stressful conditions. In Study 1, 40 university-bred male rats were assigned to either a high- or low-activity group on the basis of a 2-min activity trial. For both groups, gradually increasing preexposure to stressors of enforced running or restraint helped to protect them against stomach lesions known to develop after prolonged exposure to the stressor. Preexposure to one stressor, however, did not generalize to the other. In Study 2, 58 male hooded rats of a stress-ulcer-susceptible strain were assigned to 1 of 3 groups: no-stress controls, preexposure to enforced running, or preexposure to restraint. Results indicate that even these Ss were significantly protected from lesions by preexposure to the same stressor. In this experiment, marginal cross-stress protection was also found.

590. Sines, J. O. & McDonald, D. G. (1968). **Heritability of stress-ulcer susceptibility in rats.** ***Psychosomatic Medicine,*** 30(4), 390–394.
Two parental populations of Sprague-Dawley rats and animals of the 6th generation of a stress-ulcer-susceptible strain were used to produce F1, F2, and B1 and B2 backcross generations to provide estimates of the degree to which stress-ulcer susceptibility in the rat is genetically determined; whether that characteristic is a dominant, intermediate, or recessive trait; and how many genic units may be necessary to account for the observed trait variability. Analyses of the means and variance in lesion severity that developed during 12 hrs of restraint indicated that over ½ the observed trait variance is due to genetic factors and susceptibility to restraint-ulcer is an incompletely dominant characteristic.

591. Sivam, S. P. (1989). **D_1 dopamine receptor-mediated substance P depletion in the striatonigral neurons of rats subjected to neonatal dopaminergic denervation: Implications for self-injurious behavior.** ***Brain Research,*** 500(1–2), 119–130.
Examined the influences of dopamine (DA) receptor stimulation on enkephalin (Met^5-enkephalin [ME]) and tachykinin (substance P [SP]) systems of basal ganglia of rats, lesioned as neonates with 6-hydroxydopamine (6-OHDA). It was proposed that the neonatal 6-OHDA-lesioned rat could serve as a model for the DA deficiency and self-injurious behavior (SIB) observed in the childhood neurological disorder, Lesch-Nyhan syndrome. Neonatal 6-OHDA treatment at 3 days of age reduced DA and caused an increase in ME and a decrease in SP content in the striatum and substantia nigra when tested as adults. Administration of L-dihydroxyphenylalanine to lesioned Ss induced SIB, increased DA and dihydroxyphenylacetic acid levels, and produced a greater decrease in SP levels in the striatum and substantia nigra than was observed with lesion alone.

592. Slangen J. L. & Miller, N. E. (1969). **Pharmacological tests for the function of hypothalamic norepinephrine in eating behavior.** ***Physiology & Behavior,*** 4(4), 543–552.
Injected drugs that affect peripheral adrenergic neurotransmission via implanted cannulas into the perifornical region near the posterior part of the anterior hypothalamus of satiated rats, to test the hypothesis that in this part of the hypothalamus particularly norepinephrine (NE) functions as a modulator of the neural system involved in regulating eating behavior. Results strongly support the hypothesis.

593. Smith, Stanley G.; Werner, Toreen E. & Davis, W. Marvin. (1975). **Intravenous drug self-administration in rats: Substitution of ethyl alcohol for morphine.** ***Psychological Record,*** 25(1), 17–20.
Following a period in which the opportunity was given for morphine sulfate self-administration by 7 adult male Wistar rats, the effects of substituting an ethyl alcohol solution for morphine were examined. Ss which had acquired the morphine self-administration behavior readily switched to ethyl alcohol, while Ss that did not acquire morphine self-administration did not respond for ethyl alcohol. Results indicate that this paradigm is a suitable animal model for the analysis of interrelationships between states of dependence toward opiates and alcohol.

594. Snoddy, Andrew M.; Heckathorn, Dale & Tessel, Richard E. (1985). **Cold-restraint stress and urinary endogenous β-phenylethylamine excretion in rats.** ***Pharmacology, Biochemistry & Behavior,*** 22(3), 497–500.

Stress applied to humans increases the urinary excretion of the endogenous amphetaminelike substance beta-phenylethylamine (BPEA), a potentially common mediator of amphetamine and stress effects. The present study was conducted to determine if cold-restraint stress in the rat could represent an animal model for stress-induced changes in BPEA disposition in humans. Ss were male Sprague-Dawley rats. The stressor markedly elevated the urinary excretion of endogenous BPEA in a manner that was not attributable to changes in urinary pH, glomerular filtration rate, or in food consumption. In addition, a large diurnal variation in BPEA excretion was noted. The data suggest that the variables responsible for stress-induced alterations in endogenous BPEA disposition in humans and rats are generally similar. However, they also indicate that in rats, in contrast to humans, BPEA disposition is subject to diurnal changes.

595. Soubrie, P.; Martin, P.; el Mestikawy, S.; Thiebot, M. H. et al. (1986). **The lesion of serotonergic neurons does not prevent antidepressant-induced reversal of escape failures produced by inescapable shocks in rats.** ***Pharmacology, Biochemistry & Behavior,*** 25(1), 1–6.
Investigated the hypothesis implicating brain serotonergic neurons in the induction of learned helplessness (escape deficit) and its reversal by antidepressants in Wistar AF rats. After desipramine (25 mg/kg)-pretreatment, Ss were either sham-operated or infused with 5,7-dihydroxytryptamine (5,7-DHT) into the midbrain raphe area. Three weeks later, experimental Ss were exposed to 60 randomized inescapable shocks and subjected to daily shuttlebox sessions. Shock pretreatment groups were given twice daily injections of clomipramine (32 mg/kg/day), desipramine (24 mg/kg), imipramine (32 mg/kg), nialamide (32 mg/kg), or saline; tryptophan hydroxylase activity was assayed in the cerebral cortex, the hippocampus, and the striatum. Findings question the hypothesized role of serotonergic neurons in helpless behavior and its reversal by antidepressants.

596. Soubrié, Philippe; Martin, P.; el Mestikawy, S. & Hamon, M. (1987). **Delayed behavioral response to antidepressant drugs following selective damage to the hippocampal noradrenergic innervation in rats.** ***Brain Research,*** 437(2), 323–331.
Investigated in male rats the role of the noradrenergic innervation of the hippocampus in the reversal by antidepressant drugs of escape failures caused by previous exposure to inescapable shocks (learned helplessness design). The data suggest that the noradrenergic innervation of the hippocampus and perhaps the cerebral cortex might be a crucial neuronal target in the antidepressant action of imipramine-like drugs and monoamine oxidase (MAO) inhibitors in animals.

597. Soubrié, Philippe; Martin, Patrick; Massol, Jacques & Gaudel, Gilbert. (1989). **Attenuation of response to antidepressants in animals induced by reduction in food intake.** ***Psychiatry Research,*** 27(2), 149–159.
Investigated whether restriction of food intake concomitant to antidepressant challenge would influence the behavioral actions of these drugs in the learned helplessness model of depression. Rats previously exposed to inescapable footshock were administered antidepressants, and reversal of apomorphine-induced hypothermia and clenbuterol-induced reduction in locomotor activity were examined in mice following administration of antidepressants. Findings indicate that food deprivation concomitant to antidepressant challenge attenuated the ability of these drugs to eliminate escape failure after exposure to inescapable shock. Clenbuterol also lost its ability to reverse escape failure following food deprivation. Nutritional status could be a factor in delayed therapeutic response to antidepressants or drug-resistant depression.

598. Sparber, Sheldon B.; Bollweg, George L. & Messing, Rita B. (1991). **Food deprivation enhances both autoshaping and autoshaping impairment by a latent inhibition procedure.** ***Behavioural Processes,*** 23(1), 59–74.
35 young, male rats were deprived to 75%, 80%, 85%, 90%, and 95% of ad lib weight and were subjected to an autoshaping procedure in which a 6-sec delay was interposed between lever retraction and food pellet delivery. 40 male rats were deprived to 80 or 90% of ad lib weight prior to testing in a latent inhibition variation of the same autoshaping procedure. Greater food deprivation was associated both with fast acquisition of autoshaped lever responding and with more reliable failure to increase lever responding in the latent inhibition paradigm. Thus, increasing food deprivation was associated with enhanced acquisition regardless of whether the required performance was an increase or a failure to increase the same behavior, indicating a specific effect on learning.

599. Stanton, Mark E.; Patterson, Jeffrey M. & Levine, Seymour. (1985). **Social influences on conditioned cortisol secretion in the squirrel monkey.** ***Psychoneuroendocrinology,*** 10(2), 125–134.
Examined the phenomenon of "social buffering," the apparent capacity of group membership to reduce the adrenocortical response to stress, in 12 male squirrel monkeys by means of a between-within experimental design. Ss were assigned to 2 groups: group paired, who received pairing of a conditional stimulus (CS) with footshock, and group control, who received CS presentations without shock. All Ss were then tested with 10 presentations of the CS without shock under 3 social-housing conditions, in 4 successive phases of the experiment: individual, dyad, group, and individual housing. Neither group showed a cortisol response to the CS prior to training. Following training, CS-evoked elevations of cortisol were found only in group paired and only in the individual housing conditions. Results extend the previous finding that the presence of conspecifics can ameliorate a neuroendocrine response to psychological stressors in squirrel monkeys.

600. Stark, L. G.; Edmonds, H. L. & Keesling, P. (1974). **Penicillin-induced epileptogenic foci: I. Time course and the anticonvulsant effects of diphenylhydantoin and diazepam.** ***Neuropharmacology,*** 13(4), 261–267.
Diphenylhydantoin and diazepam exerted anticonvulsant effects which were dose related and of short duration in 41 adult cats with penicillin-induced seizures. Results suggest that the evaluation of new drugs for their potential anticonvulsant effects must include careful attention to details of the time course of abnormal activity and to the timing of drug administration when using this animal model of epilepsy.

601. Steinert, Harriett R.; Holtzman, Stephen G. & Jewett, Robert E. (1973). **Some agonistic actions of the morphine antagonist levallorphan on behavior and brain monoamines in the rat.** ***Psychopharmacologia,*** 31(1), 35–48.

Investigated the effects of levallorphan, a narcotic-antagonist analgesic, on locomotor activity, operant behavior (continuous avoidance schedule), and brain monoamine content in 24 male CFE rats. Levallorphan produced an increase in locomotor activity and in the rate of avoidance responding. Brain norepinephrine was significantly decreased 1 hr after 256 mg/kg of levallorphan. Brain dopamine (DA) levels were lowered by 64 and 256 mg/kg of levallorphan. There was no effect on brain serotonin levels. The stimulant effects of levallorphan on operant behavior were blocked by simultaneous administration of naloxone. A clear antagonism of the effects of levallorphan on locomotor activity by naloxone could be demonstrated for low doses of levallorphan but not for doses above 16 mg/kg. Naloxone also failed to prevent the depletion of brain catecholamines produced by levallorphan. Naloxone alone had no consistent effect on either of the behaviors under observation or on brain monoamine content. Findings indicate that levallorphan is a stimulant of behavior in the rat and that the stimulant action is mediated by at least 2 mechanisms: 1 which is blocked by naloxone and 1 which is not. It is suggested that the rat is a possible animal model in which to study the agonistic properties of certain narcotic-antagonist analgesics on behavior.

602. Stevens, Robin. (1989). **Estradiol benzoate potentiates the effects of body-restraint in suppressing food intake and reducing body weight.** ***Physiology & Behavior,*** 45(1), 1–5.
Food intakes and body weights were recorded daily over a 15-day period for 40 ovariectomized rats. During Days 6–10, Ss were injected with estradiol benzoate at 1 of 4 concentrations, and half the Ss in each group were stressed by being restrained for 30 min/day. During the treatment period there was a dose response inhibitory effect of estradiol on intake and body weight that was greater in confined than in unrestrained Ss. This synergistic relationship between small doses of estrogen and stress in producing anorexic effects provides an animal model of anorexia nervosa; a possible neuroendocrine mechanism that could explain the effect is discussed.

603. Stiglick, Alexander & Woodworth, Ian. (1984). **Increase in ethanol consumption in rats due to caloric deficit.** ***Alcohol,*** 1(5), 413–415.
Notes that advocates of the caloric hypothesis of ethanol (ETOH) consumption by laboratory animals maintain that ETOH's reinforcing properties are derived from its caloric value. However, evidence has accumulated suggesting that animals consume ETOH for its pharmacological effect. Solution to this question is important for an animal model of alcoholism. In the present study, 60 male food-restricted and food-satiated Long-Evans rats were given a choice between ETOH (8, 16, or 32%) and water for 22 hrs/day over 14 days. On all days and at all concentrations, intakes of ETOH were significantly higher in the food-restricted Ss. Doses consumed by these Ss were highest when 32% ETOH was used, with a mean daily intake of 6.83 g/kg. Preference scores, calculated as the percent of total fluid intake as ETOH, were also much higher in the food-restricted Ss. These findings demonstrate that the caloric value of ETOH may be an important factor in ETOH self-administration, but they do not rule out the possible importance of pharmacological effects.

604. Stockert, Marta; Serra, Jorge & de Robertis, Eduardo. (1988). **Effect of olfactory bulbectomy and chronic amitryptiline treatment in rats: ^{3}H-imipramine binding and behavioral analysis by swimming and open field tests.** ***Pharmacology, Biochemistry & Behavior,*** 29(4), 681–686.
Investigated an animal model of depression in rats, based on bulbectomy, followed by chronic treatment with amitryptiline. In the synaptosomal membranes of the cerebral cortex plus hippocampus, the number of binding sites for ^{3}H-imipramine increased significantly when bulbectomy was associated with the antidepressant. In the bulbectomized Ss, the tendency was toward a decrease in binding. The behavioral parameters analyzed by the swimming with a water wheel and the open field test revealed a series of differences in the groups of Ss, with respect to handling, bulbectomy, and antidepressant treatment. Handling resulted in an increase in swimming time in controls, while bulbectomy reduced this parameter.

605. Stricker, Edward M. & Miller, Neal E. (1968). **Saline preference and body fluid analyses in rats after intrahypothalamic injections of carbachol.** ***Physiology & Behavior,*** 3(3), 471–475.
Supported the hypothesis that drinking induced by intrahypothalamic injections of carbachol directly activates a thirst motivational system. In Exp I (Ss were 14 Sprague-Dawley adult male rats) the palatability and osmotic properties of the drinking fluids were varied by presenting different concentrations of NaCl solution in 1- and 2-bottle drinking tests. Drinking induced by carbachol was found to be closely related to the concentration of the drinking fluid, being augmented when a palatable dilute saline solution was available but decreasing when the unpalatable concentrated solutions were presented instead. Fluid intakes were similar in this regard to the well known preference-aversion curves for NaCl solutions produced by Ss made thirsty by water deprivation. These findings indicate that the carbachol injections, like normal thirst, produce a motivated behavior and not reflexive drinking. In Exp II using 9 Ss from Exp I, body fluid and tissue analyses revealed that there were no changes in plasma volume or osmolarity associated with the carbachol injections. These findings indicate that the drinking which follows the cholinergic stimulations are due to the direct activation of central neural structures controlling water balance and are not secondary to systemic alterations of body fluids.

606. Strupp, Barbara J.; Levitsky, David A. & Blumstein, Lisa. (1984). **PKU, learning, and models of mental retardation.** ***Developmental Psychobiology,*** 17(2), 109–120.
Induced experimental phenylketonuria (PKU) in 20 male Wistar rats by daily injections of alpha-methylphenlalanine and phenylalanine on postnatal Days 3–31. Beginning at 8 wks of age, Ss were tested for observational learning and latent learning (2 tests of advantageous learning). Results show that Ss who received PKU treatment early in life showed significant learning deficits in both tests as compared to a control group of Ss from the same litters. It is suggested that tests of advantageous learning are sensitive to the kinds of biological insults that cause mental retardation in humans. A model making the distinction between advantageous and essential learning and suggesting that human retardation is manifested in impairment of advantageous learning is proposed.

607. Takeda, Toshio; Hosokawa, Masanori & Higuchi, Keiichi. (1991). **Senescence-Accelerated Mouse (SAM): A novel murine model of accelerated senescence.** ***Journal of the American Geriatrics Society,*** 39(9), 911–919.
Describes an animal model for accelerated senescence (AS) in mice, focusing on the circumstances related to development, the characteristics of aging, and pathologic phenotypes (PPhs). The genetic background of this model is also discussed. Starting with several pairs of AKR/J strain of mice, brother and sister mating were done to maintain this inbred strain, selective breeding was applied based on the data of the grading score of senescence, life span, and PPhs. PPhs that may distinguish each of the SAM strains include senile amyloidosis, senile osteoporosis, and age-related deficits in learning and memory. SAM-P strains, with AS and age-dependent pathologies without experimental manipulation, and SAM-R strains, with physiologic senescence or normal aging, appear to be the sole animal model available for research on AS and aging. Most age-dependent geriatric disorders seen in humans are included in the SAM model.

608. Teicher, Martin H.; Kootz, Hedevig L.; Shaywitz, Bennett A. & Cohen, Donald J. (1981). **Differential effects of maternal and sibling presence on hyperactivity of 6-hydroxydopamine-treated developing rats.** ***Journal of Comparative & Physiological Psychology,*** 95(1), 134–145.
Assessed the influence of test environment on the expression of hyperactivity produced by neonatal 6-hydroxydopamine (6-OHDA) administration in Sprague-Dawley rat pups at 15, 19, and 22 days of age in 3 experiments. The 6-OHDA Ss and an equal number of controls were tested in isolation, mixed groupings of 2 treated and 2 control pups, mixed groupings with their anesthetized mothers, mixed groupings with an anesthetized sibling, and homogeneous groupings of all treated and all control siblings. Social factors had differential effects on activity, particularly at 19 days of age: the 6-OHDA Ss were hyperactive relative to controls in isolation, and both treated and control pups were equally active in the mixed grouping and were hyperactive relative to control isolation levels. The addition of an anesthetized mother sharply attenuated the activity of both types of pups. It is concluded that social factors strongly influenced the behavior of rat pups with whole brain dopamine depletion produced by 6-OHDA and, within the confines of this animal model of hyperactivity, exerted greater attenuating effects on their activity than previously observed with stimulant medication.

609. Thieme, R. E.; Dijkstra, H. & Stoof, J. C. (1980). **An evaluation of the young dopamine-lesioned rat as an animal model for minimal brain dysfunction (MBD).** ***Psychopharmacology,*** 67(2), 165–169.
Analyzed the behavioral response of Wistar rats (3–4 wks old) with lesioning of the central dopaminergic (DA) system to a novel environment. Ss were administered apomorphine (.2 mg/kg, ip), dextroamphetamine sulfate (.75 mg/kg, ip) and haloperidol (.25 mg/kg, ip). Treatment did not produce any "therapeutic" effect on the 3 main symptoms of minimal brain dysfunction (MBD)—hyperactivity, learning disabilities, and attention deficits. It is concluded that the young DA-lesioned rat is not an appropriate animal model for MBD.

610. Thierry, B.; Stéru, L.; Simon, P. & Porsolt, R. D. (1986). **The tail suspension test: Ethical considerations.** ***Psychopharmacology,*** 90(2), 284–285.
Presents an analysis of hemodynamic, behavioral, physiological and pharmacological factors in suggesting that the tail suspension test (TST) of L. Stéru et al (1985) is considerably less stressful to experimental animals than the traditional behavioral despair test. The ethical consideration of minimizing the animal's discomfort in seeking new antidepressant agents is a factor in making the TST a useful behavioral test.

611. Trullas, Ramon; Jackson, Barrington & Skolnick, Phil. (1989). **Genetic differences in a tail suspension test for evaluating antidepressant activity.** ***Psychopharmacology,*** 99(2), 287–288.
Marked differences in tail suspension-induced immobility were observed among 9 inbred mouse strains. These 9 strains could be ranked in 4 distinct groups based on immobility. Balb/cJ and DBA/2J Ss displayed the highest and the lowest immobility times, respectively. While significant differences in open field activity were also observed among strains, these differences were unrelated to immobility times in a tail suspension test. Findings suggest that performance in this proposed animal model of depression is under specific genetic control. This may provide a useful tool to study neurochemical and neuroendocrine correlates of depression and antidepressant action.

612. Tsuda, Akira & Hirai, Hisashi. (1976). **Psychological stress and development of gastric lesions in animals.** ***Japanese Psychological Review,*** 19(2), 116–139.
Discusses how certain psychological conditions are related to the development of gastric lesions: (a) In shock predictability, the interaction between regions and predictability of stressor plays an important role, which seems to explain the inconsistent effects of shock predictability that have been reported by some investigators. (b) A major determinant of gastric ulceration is the degree of psychological conflict engendered by the introduction of the aversive stimuli. The conflict technique has been used to investigate social variables related to ulcer susceptibility in rats. (c) The rapid production of gastric lesions in rats by immobilizing them through physical restraint offers a method for the study of fundamental variables influencing the incidence of gastrointestinal lesions. It is concluded that the consequences of the same physical stressor can be markedly altered by purely psychological factors such as predictability, conflict, and immobilization. These animal experiments confirm the clinical observation that psychological variables play a role in the production of peptic lesions in humans.

613. Tsuda, Akira; Tanaka, Masatoshi; Nishikawa, Tadashi & Hirai, Hisashi. (1983). **Effects of coping behavior on gastric lesions in rats as a function of the complexity of coping tasks.** ***Physiology & Behavior,*** 30(5), 805–808.
In an experiment with 132 male Wistar rats, experimental Ss in an FR-2 coping task condition which could avoid and/or escape shock by emitting a disk-pulling operant response developed less stomach ulceration than did yoked "helpless" Ss that had no control over shock. In a VR-5 coping task condition, however, experimental Ss developed more lesions than did the yoked Ss. Neither a VR-2 nor an FR-5 experimental group was significantly different from its yoked group. Ulceration in a nonshock control group was negligible compared to experimental and yoked Ss in each of the 4 coping task conditions. Results indicate that the level of a complexity or difficulty of coping response tasks required

has a detrimental effect on ulcerogenesis for "coping" experimental rats. The effectiveness of a coping behavior covaries with the nature or ease of the coping tasks in a stressful situation.

614. Tsuda, Akira et al. (1981). **Effects of divided feeding on activity-stress ulcer and the thymus weight in the rat.** ***Physiology & Behavior,*** 27(2), 349–353.
80 male Wistar rats housed in running-wheel activity cages and fed 1 or 2 hrs daily exhibited excessive running and subsequently died revealing large stomach ulcers, reduced absolute thymus weight, and an increase in relative weight of adrenal glands. However, 2 0.5-hr or 2 1-hr daily feedings did significantly reduce ulcer incidence. Controls for the 4 feeding schedules did not die, were ulcer free, and did not exhibit the changes in thymus and adrenal weight observed in experimental Ss. Results suggest that the divided daily feeding schedule ameliorates the ulcerogenic and immune processes in activity-stress rats.

615. Tsuda, Akira et al. (1982). **Priming effects of activity-stress ulcer in rats.** ***Physiology & Behavior,*** 29(4), 733–736.
Determined the effects of habituating rats to the running wheel prior to the restricted feeding schedule in the activity–stress ulcer paradigm and clarified the role of habituation processes on activity–stress related symptoms. 14 male Wistar rats that had been habituated with ad lib feeding to the running-wheel cage environment for 3 days prior to the restricted feeding phase developed significantly more gastric glandular ulcers and exhibited greater levels of running activity when compared to 7 Ss that had been given no habituation experience. Control Ss housed in standard laboratory cages but that received the same restricted feeding regimen developed significantly less stress pathology. Since allowing Ss access to the running-wheel during habituation resulted in enhanced stress pathology, this manipulation is referred to as a "priming effect." A possible explanation for such a "priming effect" is discussed in terms of procedures that may increase running wheel activity and decrease survival time during the restricted feeding phase of the activity–stress ulcer procedure.

616. Tsuda, Akira et al. (1982). **Influence of feeding situation on stomach ulcers and organ weights in rats in the activity-stress ulcer paradigm.** ***Physiology & Behavior,*** 28(2), 349–352.
20 male Wistar rats, housed in running-wheel activity cages except for 1 hr/day and fed in their home cages, revealed more stomach ulceration, a higher level of brain 3-methoxy-4-hydroxyphenyl-eneglycol sulfate and larger weight changes in the thymus, spleen, and adrenal gland, compared to 20 Ss housed in running-wheel activity cages and fed 1 hr daily in those same cages who, in turn, showed more stress pathology than did 20 controls housed in standard home cages but receiving the same feeding schedule. Results suggest that feeding activity-stressed Ss in their home cages might aggravate the development of stomach ulcers coincident with organ weight changes and the enhancement of noradrenaline turnover in the brain.

617. Tsuda, Akira et al. (1983). **Effects of unpredictability versus loss of predictability of shock on gastric lesions in rats.** ***Physiological Psychology,*** 11(4), 287–290.
The relative importance of unpredictability vs loss of predictability of electric tailshock affecting stress pathology was assessed using an index of the severity of gastric lesions in 64 male Sprague-Dawley rats. Results emphasize the deleterious effects of shock unpredictability; the effects of changes in predictability of shock on gastric erosions were not clearly supported.

618. Tsuda, Akira et al. (1982). **Marked enhancement of noradrenaline turnover in extensive brain regions after activity-stress in rats.** ***Physiology & Behavior,*** 29(2), 337–341.
32 male Wistar rats were housed in running-wheel activity cages and fed for only 1 hr/day. This activity-stress procedure elevated levels of the major metabolite of noradrenaline (NA), 3-methoxy-4-hydroxyphenylethyleneglycol sulfate, in 8 brain regions; a reduction of NA occurred in several regions. These Ss also exhibited excessive running and developed severe gastric glandular ulcers. Ss fed ad lib and housed in activity cages and Ss housed in individual cages and given either 1-hr daily or ad lib feeding showed neither significant changes in brain NA metabolism nor gastric ulcers. Results suggest that the interaction of a restricted feeding regimen and an increased running wheel activity caused marked enhancement of NA turnover in several brain regions, which is one of the neurochemical mechanisms underlying the physiological and behavioral changes produced by the activity-stress paradigm.

619. Ueki, Akinori & Miyoshi, Koho. (1991). **Reversal of learning impairment in ventral globus pallidus-lesioned rats by combination of continuous intracerebroventricular choline infusion and oral cholinergic drug administration.** ***Brain Research,*** 547(1), 99–109.
Found that the combination of oral 9-amino-1,2,3,4-tetrahydroacridine hydrochloride (THA) and 9-amino-2,3,5,6,7,8-hexahydro-1-H-cyclopental[*b*] quinoline monohydrate hydrochloride (NIK-247) administration and icv choline infusion elicited good acquisition of passive avoidance learning and produced a significant increase of choline and acetylcholine in the cerebral cortex of the bilateral ventral globus pallidus-lesioned male rat. Continuous icv choline infusion may intensify the ameliorating effect of THA or NIK-247 on learning disturbance. THA and NIK-247 may be beneficial in the treatment of Alzheimer's disease.

620. Ushijima, Itsuko; Mizuki, Yasushi; Hara, Takahide; Kudo, Ryoji et al. (1986). **The role of adenosinergic, GABAergic and benzodiazepine systems in hyperemotionality and ulcer formation in stressed rats.** ***Psychopharmacology,*** 89(4), 472–476.
In male Wistar rats, hyperemotionality (e.g., struggling, vocalization) evoked immediately after immobilization stress was attenuated by diazepam, adenosine, or adenosine plus diazepam. Conversely, pretreatment with these drugs produced rapid and potent exacerbation of gastric lesions observed after 12 hrs of stress. The potent adenosine A_1-receptor agonist N^6-cyclohexyl adenosine markedly inhibited the distress-evoked hyperemotional behaviors and potentiated the ulceration. Gamma-aminobutyric acid (GABA), muscimol, and aminooxyacetic acid attenuated both stress-induced hyperemotionality and ulceration.

621. VandeWoude, Susan; Richt, Juergen A.; Zink, Mary C.; Rott, Rudolf et al. (1990). **A Borna virus cDNA encoding a protein recognized by antibodies in humans with behavioral diseases.** ***Science,*** 250(4985), 1278–1281.
Identified the genome of Borna disease virus (BDV), which causes a rare neurological disease in horses and sheep, by constructing a subtractive complementary DNA expression library. The library was constructed with polyadenylate-selected RNA from a BDV-infected MDCK cell line. A clone was isolated that specifically hybridized to RNA isolated from BDV-infected brain tissue and cell lines. In vitro transcription and translation of the clone resulted in synthesis of the 14- and 24-kilodalton BDV-specific proteins. This BDV-specific clone may isolate the other BDV-specific nucleic acids and identify the virus responsible for Borna disease. Also, the significance of BDV or a related virus as a human pathogen can now be more directly examined.

622. Van Haaren, Frans; Van Hest, Annemieke & Van de Poll, Nanne E. (1988). **Differences between the sexes?!** ***Psycholoog,*** 23(1), 12–15.
Discusses the mechanisms by which gonadal hormones may affect brain development and nonreproductive behavior. A simple animal model of aggressive behavior is used to illustrate the behavioral effects of hormonal organizational and activational influences, and the development of the sexually dimorphic nucleus of the preoptic area is discussed. Similarities between effects observed in animal models and humans are considered. (English abstract).

623. Vincent, George P.; Paré, William P.; Prenatt, Jamie E. & Glavin, Gary B. (1984). **Aggression, body temperature, and stress ulcer.** ***Physiology & Behavior,*** 32(2), 265–268.
88 female Long-Evans rats were exposed to supine restraint plus cold for 3 hrs and then placed in bite or no-bite groups. Bite Ss were visually stimulated by 8-cm nylon laboratory brushes that passed over (within 2 cm) the S's nose. The bite Ss developed fewer gastric lesions as compared to the no-bite Ss, which were similarly restrained but did not have access to the aggressive biting response. A 2nd study, wherein Ss were exposed to 2 restraint sessions and received either bite/bite, bite/no-bite, no-bite/bite, or no-bite/no-bite treatments, replicated the results obtained from the 1st experiment. Core body temperature measures revealed that Ss with access to the biting response were more successful in maintaining body temperature. It is suggested that the protective effect of aggression may be due to the reduction in restraint hypothermia and not necessarily to the affective qualities of the aggressive response per se.

624. Vincent, George P. & Paré, William P. (1982). **Post-stress development and healing of supine-restraint induced stomach lesions in the rat.** ***Physiology & Behavior,*** 29(4), 721–725.
In Exp 1, 60 female Sprague-Dawley rats were subjected to 3 hrs of supine restraint and sacrificed either immediately, 30, 60, 90, 120, or 180 min following restraint. Ss sacrificed 90 min after restraint revealed significantly more stomach lesions as compared to other treatment conditions. The healing rate for supine-restraint ulcers was observed in Exp II (63 Ss), and comparisons with conventional restraint procedures, as reported in other publications, suggest a slower healing rate for lesions induced with supine restraint. Exp III, with 50 Long-Evans rats, indicated that cimetidine significantly accelerated the rate of healing for supine-restraint lesions.

625. Vincent, George P.; Paré, William P. & Glavin, Gary B. (1980). **The effects of food deprivation on restraint induced gastric lesions in the rat.** ***Physiology & Behavior,*** 25(5), 727–730.
Results of 2 experiments with 60 female Long-Evans rats suggest that the rat stomach is differentially susceptible to restraint-induced lesion formation. Shorter fasting intervals were related to lesions in the corpus of the stomach, while protracted periods of fasting intensified rumenal pathology.

626. Wade, George N. (1983). **Dietary obesity in golden hamsters: Reversibility and effects of sex and photoperiod.** ***Physiology & Behavior,*** 30(1), 131–137.
Previous findings indicate golden hamsters fed a high-fat diet (HFD) do not overeat but become obese because of decreases in energy expenditure (EE). This decrease in actual EE is accompanied by increases in thermogenic capacity and brown adipose tissue mass, protein content, and DNA content. The present 3 experiments with 82 golden hamsters examined this phenomenon. Exp I demonstrated that this form of dietary obesity was largely reversible simply by returning Ss to a high-carbohydrate chow diet. However, the obesity that developed solely because of decreased EE was reversed primarily by decreased energy intake. In this respect, fat-fed hamsters resemble tube-fed rats. Exp II revealed that the effects of HFD were as robust in female as in male Ss. Exp III examined the interactions between diet and photoperiod. Short days (10 hrs light/24 hrs) had almost no effect on male Ss fed Purina chow. However, nearly all of the effects of the HFD (i.e., increases in body weight gain, feed efficiency, carcass energy content, percent ingested energy stored in the carcass, carcass lipid content, brown adipose tissue protein, and brown adipose tissue DNA) were exaggerated in Ss housed in short days. It is suggested that HFD-induced increases in metabolic efficiency and thermogenic capacity may be of value in readying hamsters for winter. As winter approaches, decreasing day length might synergize with changes in diet quality to promote beneficial changes in energy metabolism. Fat-fed hamsters could be a useful animal model of some types of human obesity.

627. Wald, Elliott D.; MacKinnon, John R. & Desiderato, Otello. (1973). **Production of gastric ulcers in the unrestrained rat.** ***Physiology & Behavior,*** 10(4), 825–827.
Describes a simple and effective procedure for inducing gastric lesions in the unrestrained rat. A significant feature of the present technique is that gastrointestinal lesions are observed following a relatively brief (6-hr) exposure to the shock-stress condition. Lesions are found to occur in the glandular portion of the stomach, and initial evidence implicates conflict as the ulcerogenic agent.

628. Warren, Donald A. & Rosellini, Robert A. (1988). **Effects of Librium and shock controllability upon nociception and contextual fear.** ***Pharmacology, Biochemistry & Behavior,*** 30(1), 209–214.
Investigated with rats the possibility that chlordiazepoxide (CDP) would attenuate the impact of shock in a manner similar to that of providing control over shock. As shown by others, CDP administered prior to shock treatment blocked the long-term analgesic response, as did the provision of control during shock. Furthermore, whereas Ss given controllable shock subsequently exhibited less fear of the shock context than did yoked Ss, CDP treatment prior to uncontrollable shock did not appreciably reduce the contextual

fear subsequently shown. Results suggest that under some conditions, controllability attenuates the impact of stress by mechanisms other than those shared by benzodiazepine treatment.

629. Watson, Ronald R. (1989). **Murine models for Acquired Immune Deficiency Syndrome.** ***Life Sciences,*** 44(3), iii–xv.
Reviews modeling studies in which mice have been infected with murine retroviruses that cause functional changes similar to those produced by acquired immune deficiency syndrome (AIDS), as well as studies in which human immunodeficiency virus (HIV) has been incorporated into the DNA of mice.

630. Wayner, M. J.; Greenberg, I.; Carey, R. J. & Nolley, D. (1971). **Ethanol drinking elicited during electrical stimulation of the lateral hypothalamus.** ***Physiology & Behavior,*** 7(5), 793–795.
Presented intoxicating amounts of aversive ethyl alcohol solutions to 1 hooded and 3 Sprague-Dawley albino rats during electrical stimulation of the lateral hypothalamus in a standard test chamber. Results indicate that the ingestion of ethanol was produced rapidly and daily for as long as 25 days without any obvious deleterious effects. The development of an animal model for the study of alcoholism is discussed.

631. Weinberg, Joanne & Emerman, Joanne T. (1989). **Effects of psychosocial stressors on mouse mammary tumor growth.** ***Brain, Behavior & Immunity,*** 3(3), 234–246.
Developed a sensitive and replicable animal model developed to examine the effects of social housing condition and exposure to novel environments on mammary tumor growth (TG) rate. Two experiments used the transplantable androgen-responsive Shionogi mouse mammary carcinoma SC115 classified by N. Bruchovsky and P. S. Rennie (1978). 91 male mice and 123 male mice served as Ss in Exps 1 and 2, respectively. Being reared individually and remaining individually housed or being reared in a sibling group and then being singly housed following tumor cell injection markedly increased TG compared with TG in Ss remaining in their sibling rearing groups. Being reared individually and then being moved to a larger group consisting of 4 to 5 nonsiblings following tumor cell injection markedly reduced TG.

632. Weinstein, Howard & Driscoll, Janis W. (1972). **Immobilization-produced gastric pathology in wild rats (*Rattus norvegicus*).** ***Physiology & Behavior,*** 9(1), 39–41.
Deprived 7 groups of 6 wild rats of food and immobilized them in wire mesh cocoons for varying periods of time. Following immobilization, they were examined for gastric ulcers. Minimum immobilization necessary to produce ulcers, given that preimmobilization food deprivation was sufficiently long appeared to be from 12–24 hr. Results are compared to those obtained by other investigators using laboratory rats.

633. Weiss, Jay M. (1968). **Effects of predictable and unpredictable shock on development of gastrointestinal lesions in rats.** ***Proceedings of the 76th Annual Convention of the American Psychological Association,*** 3, 263–264.

634. Weiss, Jay M. (1970). **Somatic effects of predictable and unpredictable shock.** ***Psychosomatic Medicine,*** 32(4), 397–408.
12 male albino Sprague-Dawley rats were used in 4 experiments to examine the effects of stressor predictability on a variety of stress responses, such as stomach ulceration, plasma corticosterone concentration, and body weight changes. Ss that received electric shocks unpredictably showed greater somatic stress reactions and more stress-induced pathology than Ss that received the same shocks but could predict their occurrence by a signal. Ss in the unpredictable and predictable shock conditions received shock simultaneously through fixed body electrodes wired in series, so that shock was always of exactly the same intensity and duration for the 2 groups.

635. Weiss, Jay M. (1971). **Effects of punishing the coping response (conflict) on stress pathology in rats.** ***Journal of Comparative & Physiological Psychology,*** 77(1), 14–21.
Extended previous findings that when rats avoid and/or escape electric shock that is preceded by a warning signal, little gastric ulceration normally develops. In an experiment with 72 male albino rats, severe gastric ulceration developed when Ss were given a brief punishment shock each time they performed the avoidance-escape response, and thus were exposed to a conflict situation. Yoked "helpless" Ss did not develop as much ulceration as avoidance-escape Ss, showing that coping behavior can, under certain circumstances, be more ulcerogenic than helplessness. Results support a proposed theory which relates ulceration to certain psychological (or behavioral) variables. It is concluded that this theory subsumes conflict. An explanation is derived for why conflict situations are particularly pathogenic.

636. Weiss, Jay M. (1971). **Effects of coping behavior with and without a feedback signal on stress pathology in rats.** ***Journal of Comparative & Physiological Psychology,*** 77(1), 22–30.
Extended previous findings that when rats avoid and/or escape electric shock that is not preceded by a warning signal, considerable gastric ulceration normally develops. In an experiment with 96 male albino Sprague-Dawley rats, very little ulceration developed under these conditions when Ss were given a brief feedback signal after each avoidance-escape response. Ss showed only slightly more ulceration than nonshock controls and much less ulceration than either Ss that could also avoid and escape shock but had no feedback signal or yoked "helpless" Ss. These results, showing that excellent response feedback will greatly reduce or even eliminate ulceration, support a proposed theory, which also accounts for the "executive" monkey phenomenon. The explanation for this atypical effect is presented.

637. Weiss, Jay M. (1971). **Effects of coping behavior in different warning signal conditions on stress pathology in rats.** ***Journal of Comparative & Physiological Psychology,*** 77(1), 1–13.
Studied the effect of an electric shock that was preceded by either a warning signal, a series of signals forming an "external clock," or no signal at all on 180 male albino Sprague-Dawley rats. In all conditions, Ss which could avoid and/or escape shock developed less ulceration than did yoked "helpless" Ss that received exactly the same shock (through fixed electrodes wired in series) but had no control over shock. Presence or absence of a warning signal did, however, have an effect: a discrete warning signal reduced the ulcer-

ation of both Ss having control over shock and of yoked helpless Ss. A theory is proposed to explain how psychological factors determine the development of gastric ulceration in stress situations.

638. Weiss, Jay M. (1972). **Psychological factors in stress and disease.** ***Scientific American,*** 226(6), 104–113.
Studied ulceration effects on rats shocked under various conditions of avoidance. Ss able to avoid and escape shock showed less gastric ulceration than yoked, helpless partners. Helpless Ss shocked without a warning signal had more ulceration than Ss given a warning signal. The difference between these results and those reported by J. V. Brady (1958) in experiments on monkeys is accounted for in terms of a theory which states that ulceration is directly related to number of "coping" responses and inversely related to amount of relevant feedback. Supporting evidence is described.

639. Weiss, Jay M.; Glazer, Howard I. & Pohorecky, Larissa A. (1974). **Neurotransmitters and helplessness: A chemical bridge to depression?** ***Psychology Today,*** 8(7), 58–62.
Studied the effect of stress on neural transmitters and the effect of neural transmitters on an animal's ability to escape noxious stimuli. One experience with acute stress reduces the ability of the animal to escape noxious stimuli for the next 24 hrs. A series of studies which manipulated the amount and type of stress before the escape task, the presence of neural transmitters, and the time between the stress and the escape task, show that learned helplessness can be better understood as a chemical phenomenon than as a reaction to learning.

640. Weiss, Jay M.; Stone, Eric A. & Harrell, Nell. (1970). **Coping behavior and brain norepinephrine level in rats.** ***Journal of Comparative & Physiological Psychology,*** 72(1), 153–160.
One group of male albino Sprague-Dawley rats avoided or escaped electric shock while yoked Ss received the same shocks but could not cope. 2 avoidance-escape situations were used, 1 with high avoidance and few shocks and 1 with mainly escape responding and many shocks. Brain norepinephrine (NE) was elevated approximately 10% in Ss that could avoid or escape shock. Yoked Ss that were not able to avoid or escape did not show this elevation. When few shocks were received, yoked Ss' levels were not significantly different from controls, but when many shocks were received, their levels were depleted. Results provide evidence for the importance of brain NE in avoidance behavior, and suggest an explanation for avoidance deficits previously attributed to learned helplessness.

641. Weiss, Jay M.; Sundar, Syam K.; Becker, Kyra J. & Cierpial, Mark A. (1989). **Behavioral and neural influences on cellular immune responses: Effects of stress and interleukin-1.** Symposium: Interrelations between depression, the immune system, and the endocrine system (1988, Charleston, South Carolina). ***Journal of Clinical Psychiatry,*** 50(Suppl), 43–53.
A series of experiments with animals examined effects of stressful conditions on several cellular immune responses and the physiological mechanisms underlying these effects. Initial studies showed that stressful conditions can profoundly suppress immune responses of blood and splenic lymphocytes, including T-cell mitogenesis, natural killer cell activity, production of interleukin-2 (IL-2) and interferon, and IL-2 receptor expression. Subsequent studies found that (1) multiple physiological pathways mediate stress-induced suppression of these responses; (2) stress-induced suppression of these responses is partly produced by a peptide with molecular weight greater than 10 kilodaltons; (3) stressful conditions of moderate intensity can enhance cellular immune responses; and (4) extremely small quantities of interleukin-1 acting in the brain bring about suppression of cellular immune responses very rapidly.

642. Weiss, Jay M., et al. (1969). **Pituitary-adrenal influences on fear responding.** ***Science,*** 163(3863), 197–199.
Found that in a passive avoidance situation, hypophysectomized male albino rats show less fear than normal Ss, whereas adrenalectomized Ss show greater fear than normals. It is suggested that results occurred because hypophysectomized Ss lack ACTH, which increases arousal or emotionality, whereas adrenalectomized Ss lack certain adrenal steroids, which inhibit excitatory effects. Results indicate that ACTH, and certain adrenal steroids have opposite effects in regulating fear-motivated behavior.

643. Welch, K. M.; Spira, P. J.; Knowles, L. & Lance, J. W. (1974). **Simultaneous measurement of internal and external carotid blood flow in the monkey: An approach to the study of migraine mechanisms.** ***Neurology,*** 24(5), 450–457.
Internal and external carotid blood flow was simultaneously measured in monkeys. The experimental technique used appears to be an improved approach to the study of migraine mechanisms in the animal model. Intracarotid serotonin was shown by this method to constrict both the internal and external carotid arteries. This effect was prevented by intracarotid methysergide.

644. Welker, Robert L.; Garber, Judy & Brooks, Francine. (1977). **Stress as a function of irregular feeding of food deprived rats.** ***Physiology & Behavior,*** 18(4), 639–645.
In 2 experiments with 98 male Sprague-Dawley albino rats, Ss exposed to irregular meal durations and irregular intermeal intervals during protracted food deprivation ate less food, lost more weight, and exhibited more glandular-stomach lesions and deaths than did Ss receiving regular periods of feeding and fasting. Complete privation of food produced lesions primarily confined to the rumen of the stomach. Severity of deprivation per se, as defined by amount of weight lost, did not appear to account for the greater incidence in glandular-stomach lesions and death produced by protracted exposure to irregular as opposed to regular feeding. It is suggested that the irregular feeding schedules produce their deleterious effects via the unpredictability of feeding and fasting periods.

645. West, A. Preston. (1990). **Neurobehavioral studies of forced swimming: The role of learning and memory in the forced swim test.** ***Progress in Neuro-Psychopharmacology & Biological Psychiatry,*** 14(6), 863–877.
Reviews behavioral studies of forced swimming (FS), or behavioral despair, and compares them with certain behavioral effects of exposure to inescapable shock (IS), or learned helplessness. Exposure to IS impairs subsequent coping responses; however, detailed behavioral studies of FS indicate that immobility during FS is not a failure of coping but instead reflects a relatively successful coping strategy that employs energy conserving behaviors. Certain neurobiological studies of FS are reinterpreted in light of the behavioral

evidence that immobility during FS reflects effects of learning and memory rather than effects of despair or depression. Thus, the development of immobile behavior in the water cylinder no longer seems to be a valid model of depression.

646. Whitney, Glayde. (1970). **Timidity and fearfulness of laboratory mice: An illustration of problems in animal temperament.** ***Behavior Genetics,*** 1(1), 77–85.
Tested 2 inbred strains of mice (c57bl/1bi and jk/bi) and their derived generations ($n = 236$) for home cage emergence and open field behavior under 2 levels of environmental stimulation. The least timid (fast emerging) genotype was found to most fearful (high defecation in open field), whereas the most timid strain was least fearful. In addition, exposure to a loud noise during testing consistently resulted in a decrease in emergence latency and an increase in open field defecation (i.e., environmental stimulation sufficient to decrease timidity increased fearfulness). This apparent paradox illustrates a major problem in interspecific behavioral comparisons: a priori analogic reasoning from human theory to animal model, without regard for the meaning of constructs in the behavioral organization and evolutionary adaptation of the species studied, often results in rigorous investigation of operationally defined behavioral constructs devoid of meaning.

647. Wideman, Cyrilla H. & Murphy, Helen M. (1986). **The pathological effects of limited feeding in vasopressin-deficient animals.** ***Bulletin of the Psychonomic Society,*** 24(3), 225–228.
30 male Brattleboro (vasopressin-deficient) rats and 30 male Long-Evans rats were subjected to 1 of the following treatments: (1) ad-lib access to food each day, (2) 23 hrs of food deprivation each day, or (3) 22 hrs of food deprivation each day. Both groups of food-deprived, vasopressin-deficient Ss survived for a significantly shorter period of time, had a greater decrease in body weight, and developed significantly more ulcers than other Ss.

648. Wideman, Cyrilla H. & Murphy, Helen M. (1983). **The effects of restraint and restraint plus intermittent shock on ulcer formation in Brattleboro rats.** ***Physiological Psychology,*** 11(1), 78–80.
To examine the susceptibility of Brattleboro rats to stress as evidenced by ulcer formation, 40 Brattleboro and 40 Long-Evans male rats were examined under 4 conditions: (1) unrestrained, nondeprived; (2) unrestrained, 29-hr food deprived; (3) simple restraint, 29-hr food deprived; and (4) restraint plus intermittent shock, 29-hr food deprived. In the 2 unrestrained conditions, neither group developed ulcers. In the simple restraint and restraint plus intermittent shock conditions, Brattleboro rats developed more ulcers in the glandular portion of the stomach. It is concluded that Brattleboro rats show an enhanced reaction to stress as evidenced by ulcer formation.

649. Wideman, Cyrilla H. & Murphy, Helen M. (1985). **Effects of vasopressin deficiency, age, and stress on stomach ulcer induction in rats.** Fifth Annual Winter Neuropeptide Conference (1984, Breckenridge, Colorado). ***Peptides,*** 6(Suppl 1), 63–67.
Studied, in 2 experiments, susceptibility to ulceration induced by restraint, restraint plus intermittent shock, and activity stress in a total of 160 male Long-Evans and 160 male Brattleboro 6- and 18-wk-old rats. Older Ss developed more glandular ulcers than younger Ss with Brattleboro Ss having significantly greater ulceration than Long-Evans Ss in both conditions. With activity stress, younger Ss developed significantly more glandular ulcers than older Ss, whereas older Ss developed significantly more nonglandular ulcers than younger Ss. In both instances, the ulceration was significantly greater in Brattleboro Ss than in Long-Evans Ss. There were significantly high correlations among running behavior, survival time, and the development of glandular ulcers in younger Ss exposed to activity stress. The presence of vasopressin, as well as the age of the S and the nature of the stress, influenced the type and degree of stomach pathology induced.

650. Wiener, Sandra G.; Bayart, Francoise; Faull, Kym F. & Levine, Seymour. (1990). **Behavioral and physiological responses to maternal separation in squirrel monkeys (*Saimiri sciureus*).** ***Behavioral Neuroscience,*** 104(1), 108–115.
This study extends an examination of the behavioral and pituitary-adrenal responses of infant squirrel monkeys (*Saimiri sciureus*) separated from their mothers under different environmental conditions to another physiological system by measuring the metabolites of the central monoamines found in the cerebrospinal fluid (CSF). This study included spectrographic examination of the vocalizations emitted by the infant during separation. Infants were separated from their mothers for 24 hrs under 3 conditions: *Home,* infant remained in his home cage after removal of mother; *adjacent,* infant was placed in a cage adjacent to its mother; and *total,* infant was totally isolated. The behavioral results indicated that the number of calls emitted differed with condition (adjacent > total > home), and the peak frequency of the calls and number of multiple calls was greatest in the total condition. Plasma cortisol elevations after separation differentiated the conditions of separation (total > adjacent > home > base). The elevations in the CSF catecholamine metabolites (3-methoxy-4-hydroxyphenylglycol and homovanillic acid) were also sensitive to the conditions of separation (total > adjacent > base). These results are discussed in the context of coping theory.

651. Williams, Jon L. (1984). **Influence of postpartum shock controllability on subsequent maternal behavior in rats.** ***Animal Learning & Behavior,*** 12(2), 209–216.
Responses of 30 Holtzman mother rats were observed 24 hrs before and 24 and 72 hrs after exposure to 1 of 3 8-day postpartum treatments: shock escape training, yoked inescapable shock, or restrained with no shock. In contrast to the other 2 groups, the dams given inescapable shock showed slower speed to approach the nest, shorter durations of being on the nest, and lower frequency and shorter total duration of oral contact with their pups. These Ss also retrieved their pups less frequently, but this measure, as well as the frequency of leaving the nest, did not result in significant differences between groups. Since the traditional interpretations of the learned-helplessness effect are not entirely able to account for these findings, the observed uncontrollable-stress-produced changes in maternal behavior are examined from an ethological perspective.

652. Williams, Jon L.; Drugan, Robert C. & Maier, Steven F. (1984). **Exposure to uncontrollable stress alters withdrawal from morphine.** ***Behavioral Neuroscience,*** 98(5), 836–846.

Examined the influence of the controllability/uncontrollability of shock as a stressor on the severity of subsequent morphine withdrawal in 2 experiments with 84 male Holtzman rats. In Exp I (36 Ss), Ss that received 2 daily sessions of 80 yoked-inescapable shocks, in contrast to those given 80 escapable shocks or restrained without shock, showed an enhanced series of correlated withdrawal behaviors (i.e., mouthing, teeth chattering, head/body shakes) 24 hrs later when injected with morphine sulfate (5 mg/kg) followed by a naloxone HCl (5 mg/kg) challenge. In Exp II (48 Ss), this finding was replicated with escape-yoked-restrained Ss given saline injections during the pretreatment phase, but the impact that inescapable shock had on later precipitated withdrawal was completely blocked when Ss were administered naltrexone HCl (14 mg/kg) before each shock session. Findings are discussed in terms of the capability of inescapable shock to activate an endogenous opiate system, thereby leading to a sensitization of release or receptor processes that could protentiate later morphine withdrawal.

653. Willner, Paul; Towell, Anthony; Sampson, D.; Sophokleous, S. et al. (1987). **Reduction of sucrose preference by chronic unpredictable mild stress, and its restoration by a tricyclic antidepressant.** ***Psychopharmacology,*** 93(3), 358–364.
Rats exposed for 2 wks to mild unpredictable stressors showed a reduced consumption of and preference for saccharin or sucrose solutions that persisted for 2 wks after termination of the stress regime. Sucrose preference returned to normal after 2–4 wks of treatment with the tricyclic antidepressant desmethylimipramine (DMI). Blood corticosterone and glucose levels were reduced by DMI but not by stress. Results are discussed in terms of an animal model of endogenous depression.

654. Wilson, James R. et al. (1984). **Ethanol dependence in mice: Direct and correlated responses to ten generations of selective breeding.** ***Behavior Genetics,*** 14(3), 235–256.
Reports on research designed to produce an animal model of alcohol dependence; genetic selection in mice for severity of the alcohol withdrawal syndrome is underway. 10 generations of selection have been completed, and the lines selected for severe or mild expression of the withdrawal syndrome following a 9-day ethanol treatment period were found to be significantly different from each other. The index of selection was the unrotated 1st principal-component score derived from the intercorrelations among 7 measures of severity of ethanol withdrawal. Across the generations, Ss from all 6 closed mating populations increased their mean consumption of alcohol considerably, which may have been the result of natural selection operating concurrently with artificial selection. When Ss from Generation 9 were used to test for possible correlated responses to selection, there was an indication that Ss selected for mild expression of withdrawal symptoms were also less sensitive to an acute dose of ethanol.

655. Wirz-Justice, Anna. (1974). **Possible circadian and seasonal rhythmicity in an in vitro model: Monoamine uptake in rat brain slices.** ***Experientia,*** 30(11), 1240–1241.
Investigated a factor possibly causing periods of increased susceptibility characteristic of manic-depressive illness. Alterations in sleep–waking patterns, corticosteroid rhythms, and seasonal increases in depressions are associated with the metabolic rhythmicity of the amines. (German summary).

656. Wise, Roy A. (1989). **Opiate reward: Sites and substrates.** Special Issue: The neural basis of reward and reinforcement: A conference in honour of Peter M. Milner. ***Neuroscience & Biobehavioral Reviews,*** 13(2–3), 129–133.
Opiates appear to have rewarding actions at more than 1 locus in the brain. Studies of the effects of dopaminergic lesions and dopamine receptor blockade in rats indicate that iv heroin self-administration depends importantly on a dopaminergic substrate (e.g., H. O. Pettit et al; 1984). Mapping of effective injection sites for morphine-conditioned place preference establishes 1 site of rewarding action near the dopamine cell bodies of the ventral tegmental area (VTA). Opioid injections into the nucleus accumbens (NAS) also facilitate brain stimulation reward and serve as rewards in their own right. The VTA and NAS are firmly established as sites of opiate rewarding actions.

657. Wolffgramm, J. (1991). **An ethopharmacological approach to the development of drug addiction.** Ethopharmacology Conference: Advances in ethopharmacology (1991, Lísek, Czechoslovakia). ***Neuroscience & Biobehavioral Reviews,*** 15(4), 515–519.
In a rat model of alcoholism, different stages of the development toward a drug addiction can be discriminated. During the phase of "controlled" intake, drug consumption is reversibly modified by the social situation (housing conditions) and the individual's social role (in particular its dominance rank). During the next few months, the consumption of ethanol rises without a concomitant loss of its behavioral effects. After an abstinence period, the rats maintain a high preference for alcohol that cannot be suppressed by adulteration with (unpleasantly tasting) quinine. Ethanol-taking behavior is no longer modified by external stimuli or by dominance rank. This irreversible state is called behavioral dependence. It is drug-specific, not related to physical dependence. In behaviorally dependent rats, very low doses tranquilize and higher ones stimulate.

658. Woolf, Clifford J. (1983). **Evidence for a central component of post-injury pain hypersensitivity.** ***Nature,*** 306(5944), 686–688.
An animal model was developed of changes occurring in the threshold and responsiveness of the flexor reflex following injury to investigate the peripheral mechanism that is thought to be responsible for postinjury hypersensitivity. Decerebration was performed under pentobarbitol anesthesia on 8 rats, and electrophysiological experiments examining the hypersensitivity of the flexion reflex after thermal injury to the ipsilateral foot were performed on 28 adult Wistar rats decerebrated under Althesin anesthesia. It is noted that noxious skin stimuli, sufficiently intense to produce tissue injury, characteristically generate prolonged poststimulus sensory disturbances that include continuing pain, increased sensitivity to noxious stimuli, and pain following innocuous stimuli. This could result from a reduction in thresholds of skin nociceptors (sensitization) or an increase in the excitability of the CNS so that normal inputs evoke exaggerated responses. Electrophysiological analysis of the injury-induced increase in excitability of the flexion reflex showed that it resulted from changes in the activities of the spinal cord; therefore, long-term consequences of noxious stimuli result from central and peripheral changes.

659. Wozniak, David F. & Goldstein, Robert. (1980). **Effect of deprivation duration and prefeeding on gastric stress erosions in the rat.** ***Physiology & Behavior,*** 24(2), 231–235.

Conducted 2 experiments with 120 Wistar rats in which Ss were food deprived for 9–144 hrs and then subjected to cold-restraint. Comparison of each of these groups with a stressed, but nondeprived, control group revealed no significant differences in gastric erosions. It was suggested that many of the controls, though not deprived, may not have had full stomachs during the stress, and this may have masked the true potentiating effects of deprivation. In a 2nd experiment, Ss were divided into 2 groups and deprived, respectively, for 18 and 48 hrs. These groups were subdivided, and for 1 hr preceding stress, half were given food. This prefeeding significantly reduced glandular erosion scores, and there was a significant inverse relationship within the prefed groups between the amount of weight gained over the prefeeding hour and the subsequent erosion level. It is concluded that the potentiation of stress erosions by food deprivation occurs by virtue of the empty stomach associated with the deprivation rather than other concomitant changes. It is also concluded that prolonged deprivation (144 hrs) produces minor glandular effects but extensive damage to the rumen.

660. Yumatov, E. A.; Pevtsova, E. I. & Mesentseva, L. N. (1988). **Physiologically adequate experimental model of aggression and emotional stress.** ***Zhurnal Vysshei Nervnoi Deyatel'nosti,*** 38(2), 350–354.
Elaborated a simple experimental model of aggression and emotional stress by catching Ss' tails on the wall of their cage. Ss were 67 male outbred rats. Manifestations of aggressive vs passive-defensive behavior and vocalizations were analyzed. 12 Ss were in an aggressive state for 5 hrs of tail-catching (T-C); 23 were subjected to 4 days of T-C (5 hrs/day), during which aggressive-defensive behavior was observed; 10 Ss were subjected to this condition but did not manifest aggressive behavior; and 22 Ss served as controls. The thymus, adrenal glands, and hypothalamus of each S were removed for study. (English abstract).

661. Zahorik, Donna M.; Maier, Steven F. & Pies, Ronald W. (1974). **Preferences for tastes paired with recovery from thiamine deficiency in rats: Appetitive conditioning or learned safety?** ***Journal of Comparative & Physiological Psychology,*** 87(6), 1083–1091.
Conducted a study of 80 male Sprague-Dawley rats to investigate earlier findings that rats develop learned preferences for flavors paired with recovery from vitamin deficiencies. Results show that thiamine deficient Ss preferred flavors paired with recovery from deficiency to other familiar flavors, suggesting that part of the preference for flavors paired with recovery was the result of appetitive conditioning. Data are discussed in relation to "learned safety," specific hungers, illness-induced neophobia, and other phenomena in the taste-aversion literature.

Genetics

662. Bailey, William H. & Weiss, Jay M. (1979). **Evaluation of a "memory deficit" in vasopressin-deficient rats.** ***Brain Research,*** 162(1), 174–178.
Brattleboro rats that are homozygous for hereditary hypothalamic diabetes insipidus (DI) and totally deficient in vasopressin are reported to be unable to demonstrate retention of a passive avoidance (PA) response 24 hrs or more after training. Heterozygous (HE) Brattleboro rats that do not have DI and only exhibit a mild vasopressin deficiency retain a PA response for as long as 120 hrs after training. These findings have been interpreted as support for a role for vasopressin in memory processes. The present study of male and female DI and HE Brattleboro rats indicated (a) that DI rats do not exhibit an absolute retention deficit as previously reported and (b) that behavioral differences between DI and HE rats may not necessarily reflect the direct involvement of vasopressin in memory, since the absence of vasopressin in DI rats was not associated with a total impairment of PA performance. Comparisons of pre- and post-shock latencies indicated that quite good PA was displayed by DI rats on tests given 1 and 2 days after shock. It is suggested that a secondary elevation of oxytocin (which, when given to normal rats, has been reported to attenuate PA performance and facilitate extinction of active avoidance responding) rather than a deficiency of vasopressin could be responsible for the poor PA performance of DI compared to HE rats.

663. Bareggi, S. R.; Becker, R. E.; Ginsburg, B. E. & Genovese, E. (1979). **Neurochemical investigation of an endogenous model of the hyperkinetic syndrome in a hybrid dog.** ***Life Sciences,*** 24(6), 481–488.
A telomian-beagle hybrid has been recently proposed as a possible model for the hyperkinetic syndrome. They resemble hyperkinetic children in their poor ability to respond with appropriate behavior in an inhibitory training test. Two groups of hybrids could be differentiated, one of whose behavior improved with amphetamine (responders) while the other's did not (nonresponders). In the present study, the levels of homovanillic acid (HVA), 5-hydroxyindoleacetic acid (HIAA), and 3-methoxy-4-hydrophenylglycol sulfate were measured in cerebrospinal fluid (CSF), and the levels of noradrenaline (NA), dopamine (DA), HVA, dihydroxyphenylacetic acid, and HIAA were assayed in brain tissues from different regions, taken under basal conditions from beagles and telomian-beagle hybrids. Responder hybrids had lower levels of NA, DA, HVA in brain, and low HVA in CSF. Therefore, they can be distinguished biochemically as well as behaviorally from nonresponder hybrids and beagles and may prove to be useful as models for study of the mechanism and the therapy of this syndrome.

664. Cabib, Simona; Puglisi-Allegra, Stefano & Oliverio, Alberto. (1985). **A genetic analysis of stereotypy in the mouse: Dopaminergic plasticity following chronic stress.** ***Behavioral & Neural Biology,*** 44(2), 239–248.
Studied the effects of chronic stress on climbing behavior induced by apomorphine hydrochloride (0.1, 0.25, 0.5 mg/kg), using 192 male DBA/2, C57BL/6, and hybrid mice in 2 experiments. After repeated stressful experiences, DBA Ss showed an increase in apomorphine-induced climbing while C57 Ss showed a clear-cut decrease of this behavior. Genetic analysis involving F1 and F2 hybrids and the backcross populations (F1 × C57; F1 × DBA) indicated complete dominance of the C57 genotype and a significant genotype by environment interaction. Findings are discussed in terms of dopaminergic plasticity and of the heuristic value of this animal model in relation to disturbed behaviors triggered by stressful experiences.

665. Corson, Samuel A. & Corson, Elizabeth O. (1979). **Interaction of genetic and psychosocial factors in stress-reaction patterns: A systems approach to the investigation of stress-coping mechanisms.** ***Psychotherapy & Psychosomatics,*** 31(1-4), 161–171.

Stress has been referred to as "the nonspecific response of the body to any demand made upon it." Results from various breeds of dogs (e.g., beagle, German shepherd, and cocker spaniel) indicate that (a) there are significant constitutional differences in the types of stress reactions exhibited by different breeds, (b) inability to achieve an adaptive consummatory response or to develop a sense of control over stressful situations may lead in susceptible (low adaptation) Ss to the development of maladaptive distress reactions, evidenced by persistent psychovisceral turmoil; (c) such maladaptive distress reactions represent a physiologic substrate of anxiety and frustration; (d) exposure of the low adaptation Ss to similar stressors, but under conditions where Ss can develop avoidance responses, inhibited the psychovisceral disturbances, suggesting that it is the inability to develop control over psychosocially aversive situations that is primarily responsible for psychophysiologic disorders.

666. Coyle, Joseph T.; Oster-Granite, Mary L.; Reeves, Roger H. & Gearhart, John D. (1988). **Down syndrome, Alzheimer's disease and the trisomy 16 mouse.** ***Trends in Neurosciences,*** 11(9), 390–394.
Recent findings have implicated genes on human chromosome 21 as important in the pathophysiology of Alzheimer's disease (AD). These include the high incidence of the pathological features characteristic of AD in individuals with Down's syndrome (DS) and the localization of both a familial AD gene and the gene encoding amyloid precursor protein on chromosome 21. It is shown that substantial genetic homology exists between human chromosome 21 and mouse chromosome 16, including the gene encoding the amyloid precursor protein. Mice that are trisomic for chromosome 16 offer a genetic model for studies relevant to DS that may also clarify molecular mechanisms in AD.

667. Drewek, K. J. & Broadhurst, P. L. (1981). **A simplified triple-test cross analysis of alcohol preference in the rat.** ***Behavior Genetics,*** 11(5), 517–531.
Reanalyses of 1st-degree biometrical genetic data from previous studies (e.g., G. E. McClearn and D. A. Rodgers, 1961; G. Whitney et al, 1970) of alcohol preference in the mouse revealed little consistency beyond a basic additive genetic component. A simplified triple-test cross in the rat investigated the genetic architecture of alcohol preference for a 10% (w/v) alcohol solution or water. An initial survey of 8 selected and inbred strains identified high- and low-scoring strains, the Maudsley Nonreactive (MNR) and Irish (ACI), respectively, which were crossed as tester lines to 6 strains (the Roman high- and low-avoidance, Tryon maze-bright and maze-dull, MNR, and ACI) to produce the required set of largely F_1 families. The additive-dominance model proved adequate for males, and directional dominance for low alcohol preference was found on all 3 measures: alcohol intake, alcohol preference ratio, and alcohol calorie contribution ratio. For females, the model was adequate only for alcohol preference ratio, which showed ambidirectional dominance. The relevance of such genetic architecture to an animal model of alcoholism and the evolution of alcohol drinking in the rat is discussed.

668. Forster, Michael J. & Lal, Harbans. (1990). **Animal models of age-related dementia: Neurobehavioral dysfunctions in autoimmune mice.** 19th Annual Meeting of the Society for Neuroscience: Neural basis of behavior: Animal models of human conditions (1989, Phoenix, Arizona). ***Brain Research Bulletin,*** 25(3), 503–516.
Reviews research findings from studies of genetically defined mice with spontaneous age-related behavioral deficits and attempts to relate these findings to current understanding of the etiology of aging-associated cognitive decline and Alzheimer's disease. These studies have focused on the potential role of autoimmunity and other immunopathological changes in aging-associated dementia (AAD). When compared with normal genotypes, mutant mice with accelerated autoimmunity show learning and memory impairments at earlier chronological ages. The deficits of autoimmune and normal Ss are qualitatively similar. The behavioral deficits of normal aged and autoimmune mice are sensitive to similar pharmacologic interventions. Intervention strategies targeting the immune system might be useful in treating or preventing AAD.

669. Jinnah, H. A.; Gage, F. H. & Friedmann, T. (1991). **Amphetamine-induced behavioral phenotype in a hypoxanthine–guanine phosphoribosyltransferase-deficient mouse model of Lesch-Nyhan syndrome.** ***Behavioral Neuroscience,*** 105(6), 1004–1012.
In humans, congenital deficiency of the enzyme hypoxanthine–guanine phosphoribosyltransferase (HPRT) results in a disorder known as the Lesch-Nyhan syndrome. Patients with this disorder exhibit a prominent neurobehavioral phenotype that results in part from dysfunction of catecholaminergic systems in the striatum. HPRT-deficient mice produced as animal models for this syndrome curiously exhibit no spontaneous neurobehavioral abnormalities. However, the present study demonstrates that HPRT-deficient mice are more sensitive than their HPRT-normal littermates to the ability of amphetamine to stimulate locomotor or stereotypic behaviors. This behavioral supersensitivity to amphetamine indicates the existence of an underlying subclinical abnormality of catecholaminergic systems in the brains of HPRT-deficient mice, analogous to findings in human Lesch-Nyhan patients.

670. Marrazzi, Mary A.; Mullings-Britton, Janet; Stack, Lori; Powers, Robert J. et al. (1990). **Atypical endogenous opioid systems in mice in relation to an auto-addiction opioid model of anorexia nervosa.** ***Life Sciences,*** 47(16), 1427–1435.
The atypical opioid system in female mice may be representative of that in humans with anorexia nervosa (AN) and may account for a biological predisposition to AN. Different mouse strains were compared with reference to this AN model. Three patterns were found in different mouse strains: AN with hyperactivity in BALB/C and C57BL/6J Ss, AN without hyperactivity in DBA/J Ss, and a biphasic curve with hyperphagia at low doses and AN and hyperactivity at higher doses in CF-1 Ss. These atypical opioid systems may reflect a spectrum of biological predispositions to AN.

671. McClearn, Gerald E. (1981). **Animal models of genetic factors in alcoholism.** ***Advances in Substance Abuse,*** 2, 185–217.
Maintains that psychosocial models have been unable to account completely for causes of alcoholism and that animal models may clarify some of the biological factors of this addiction. However, various strains of animals can alter the results of such experiments since inbred strains, recombinant inbred strains, selectively bred lines, and systematically heterogeneous stocks have their unique advantages and disadvantages. Genes are often used as the manipulated variables in this kind of research. When genes are independent variables, the genotype is manipulated through knowledge of

the relationship between genotype and phenotype. For example, laboratory rats can be inbred so that by the 20th generation they are, for all practical purposes, genetically uniform. In other experiments, rats are selectively bred for a specific trait. Control of relevant variables may involve elimination of alleles, elimination of differential influence, or randomization of variables. An illustrative review of the literature on genes and alcohol is provided to demonstrate a genetic influence on several alcohol-related domains. Studies involving inbred strains and their derived generations have experimented with alcohol consumption, the effects of administered alcohol, and genotype control. Selectively bred lines of rats have also been used to study consumption, as well as alcohol sensitivity and dependence.

672. McGuffin, Peter & Buckland, Paul. (1991). **Major genes, minor genes and molecular neurobiology of mental illness: A comment on "Quantitative trait loci and psychopharmacology" by Plomin, McClearn and Gora-Maslak.** ***Journal of Psychopharmacology,*** 5(1), 18–22.
Argues that R. Plomin et al (1991) are skeptical about the fashion of applying linkage strategies in the search for behaviors with a monogenic basis. However, Plomin et al are optimistic that other approaches can help define the factors that contribute to "multifactorial" phenotypes in the study of the genetics of behavior. The authors are interested in the molecular genetics of schizophrenia and major affective illness. Linkage studies in psychiatry are based on the theoretical position that any trait where there is major gene transmission can be mapped using linkage methods. Association studies may, however, be capable of detecting minor susceptibility loci. The alternative proposal of Plomin et al for animal models where recombinant inbred strains can be used to look for quantitative trait loci associations is discussed.

673. Miyamoto, Masaomi; Kiyota, Yoshihiro; Yamazaki, Naoki; Nagaoka, Akinobu et al. (1986). **Age-related changes in learning and memory in the senescence-accelerated mouse (SAM).** ***Physiology & Behavior,*** 38(3), 399–406.
Studied age-related changes in learning ability in senescence-accelerated mice (SAM) reared under specific pathogen-free (SPF) conditions. SAM-P/8/Ta (SAM-P/8, senescence-prone substrain) showed an age-associated increase in spontaneous motor activity compared with SAM-R/1/Ta (SAM-R/1, senescence-resistant substrain) in a novel environment when the activity was measured in the light period. Results of other tests (e.g., a T-maze task) indicate that SAM-P/8 showed age-related deterioration of ability in learning and memory. SAM-P/8 may be useful as an experimental animal model for senile memory impairment in humans.

674. Overstreet, David H. (1991). **Commentary: A behavioral, psychopharmacological, and neurochemical update on the Flinders Sensitive Line rat, a potential genetic animal model of depression.** ***Behavior Genetics,*** 21(1), 67–74.
Recent studies reinforce the proposal that the Flinders Sensitive Line (FSL) rat is a useful genetic animal model of depression. The muscarinic supersensitivity exhibited by FSL rats is genetically influenced, as it is in humans, and develops early (human data in this area is limited). There is no evidence that FSL rats are any more anxious than Flinders Resistant Line rats, or that anxiety in humans is associated with muscarinic supersensitivity. The possibility that other neurotransmitter systems might have been altered is a potential deficiency in the model.

675. Rauch, Terry M. (1979). **The additive effects of two mutant genes on otolith formation in mice: An animal model to assess otolith function.** ***Journal of Auditory Research,*** 19(4), 259–265.
Four strains of mice were bred: (1) a control strain, heterozygous for both pallid (*pa*) and tilted head (*th*); (2) a pallid strain, homozygous for *pa*; (3) a tilted head strain, homozygous for *th*; and (4) a double mutant strain, homozygous for both *pa* and *th.* Findings suggest significant differences in mean otolith scores between all possible pairs of strains. As expected, the controls had normal otoliths and the *pa/th* strain the most severe otolith defects; there was a significant directional asymmetry in mean otolith scores.

676. Reeves, Roger H.; Gearhart, John D. & Littlefield, J. W. (1986). **Genetic basis for a mouse model of Down syndrome.** Special Issue: The neurobiologic consequences of autosomal trisomy in mice and men. ***Brain Research Bulletin,*** 16(6), 803–814.
The trisomy 16 (Ts16) mouse has been proposed as a model for Down's syndrome (DS) in humans, based on genetic homology between mouse chromosome 16 (MMU 16) and human chromosome 21 (HSA 21). Translocations of HSA 21 resulting in trisomy for only a portion of the genetic information contained on this chromosome can result in a DS phenotype. Techniques for localizing genes on chromosomes have been used to identify the portion of MMU 16 that corresponds to the DS region of HSA 21. This region appears to be highly conserved between mouse and human, providing further support for a mouse model of DS.

677. Royce, J. R.; Holmes, T. M. & Poley, Wayne. (1975). **Behavior genetic analysis of mouse emotionality: III. The diallel analysis.** ***Behavior Genetics,*** 5(4), 351–372.
Obtained a total of 775 pure-strain and F1 mice from a 6 by 6 diallel mating plan. Previous factor analysis of 42 measures of emotionality identified 14 behavioral factors, 10 of which were interpretable. B. I. Hayman's (1954) analysis of variance and analysis of diallel crosses were applied to each of the factors. Findings indicate that the mode of inheritance for emotionality factors is polygenic and in the direction of complete dominance. However, the mode of inheritance of highly complex behavior such as emotionality depends on the factor in question. For example, the breakdown of dominance effects by factor was as follows: partial dominance—motor discharge, food motivation, tunneling-2, and activity level (males); complete dominance—audiogenic reactivity, underwater swimming (males), and activity level (females); overdominance—acrophobia, territorial marking (males). Additional findings include directional dominance for underwater swimming and audiogenic reactivity.

678. Schaefer, Carl F.; Brackett, Daniel J.; Wilson, Michael F. & Gunn, C. G. (1978). **Lifelong hyperarousal in the spontaneously hypertensive rat indicated by operant behavior.** ***Pavlovian Journal of Biological Science,*** 13(4), 217–225.
Used instrumental conditioning techniques to obtain objective evidence of differences in behavioral arousal between spontaneously hypertensive rats (SHRs) and the normotensive ancestral Wistar Kyoto (WKY) strain. Subjective emotionality ratings previously indicated that the genetically hypertensive rats were more active and aggressive than their normotensive cousins. In a lengthy series of operant conditioning sessions using a small number of adult female SHRs and WKY Ss, hyperarousal in the SHRs was confirmed by their significantly higher response outputs on

either response- or time-contingent schedules of reinforcement. Conditioned emotionality tests also suggested hyperarousal and aggressiveness in the SHRs, since the fear CS suppressed barpressing in the SHRs much less than in the WKY. Further experiments with young prehypertensive SHRs provided the same evidence of hyperresponsivity in the SHRs compared to the WKY strain. Furthermore, the young SHRs failed to develop hypertension by the end of the study (14 wks of age), while their nonconditioned SHR cousins had become clearly hypertensive by the same age. This suggests that factors related to the conditioning methods modified the development of high blood pressure in this animal model of essential hypertension.

679. Shanks, Nola & Anisman, Hymie. (1988). **Stressor-provoked behavioral changes in six strains of mice.** ***Behavioral Neuroscience,*** 102(6), 894–905.
Behavioral changes induced by inescapable shock were examined in 6 strains of mice. Exposure to shock provoked time-dependent disturbances of shuttle escape performance. In some strains the shock treatment did not affect escape performance, whereas in others profound performance deficits were evident. The inescapable shock treatment induced strain-dependent alterations of performance in a forced-swim task. In most instances the shock treatment initially provoked invigorated responding, but in other strains the shock had no effect or depressed active responding. Y-maze spontaneous alteration performance was not affected by the shock treatment, although a strain-dependent increase of perseverative responses was evident. The occurrence of a stressor-induced deficit in 1 task in a particular strain was not predictive of behavioral alterations in a 2nd task. Data are discussed with respect to animal models of depression and genetic differences associated with response to stressors.

680. Steel, Karen P. & Bock, Gregory R. (1980). **The nature of inherited deafness in *deafness* mice.** ***Nature,*** 288(5787), 159–161.
Deafness mice, a mutant strain in which mice homozygous for the recessive *deafness* gene are unresponsive to sound, but unlike other mutants, show no behavioral abnormality, were studied to determine the nature of their inherited deafness. Electrophysiological (cochlear microphonic recordings) and anatomical data suggest that *deafness* mice are profoundly deaf during all stages of cochlear development, including the early stages when most hair cells are intact. It is suggested that the absence of microphonics in these mice can be explained by abnormal opening of the ion channels across the apical surfaces of the hair cells, although some of the endocochlear potential data raise questions about the generally accepted mechanism of endocochlear potential generation. The usefulness of *deafness* mutants as an animal model of profound hereditary deafness in humans is discussed.

681. Wimer, R. E. & Wimer, C. C. (1985). **Animal behavior genetics: A search for the biological foundations of behavior.** ***Annual Review of Psychology,*** 36, 171–218.
Discusses the basic concepts of behavior genetics and special genetic techniques such as recombinant inbred strains, chimeras and mosaics, and biometrical genetics. General topics such as ethology and ecology, social behavior, and psychopharmacology are considered, and animal models are examined for human disease and behaviors such as alcoholism, aging, autoimmune disorders, and emotionality. The application of behavior genetics to the analysis of normal behavior is discussed with respect to learning and to locomotor activity and exploratory behavior. Sexual and sex-related behaviors, such as aggression and nest-building, are examined in relation to studies of fruit flies and of mice and rats. Research on neurobiological foundations of learning is also reviewed.

682. Wolf, G. (1985). **Concerning the psychobiology of emotions.** ***Zeitschrift für Psychologie,*** 193(4), 385–396.
Discusses the application of animal models in psychobiological research on the mechanisms of human emotionality. Although the capacity of animals for subjective experience cannot be tested with rigorous scientific methods, neurophysiological, neurochemical, neuropharmacological, and vegetative parameters of emotional and motivational conditions, as well as phylogenetic considerations, argue in favor of emotionality in animals, at least in higher vertebrates. The basic neuronal substrate of human and animal emotionality is the result of phylogenesis and ontogenesis (i.e., it is genetically determined in a species-specific and individual-specific manner). However, during manifestation of those genetic predispositions, environmental factors play an important role. The significance of these environmental factors needs to be assessed more precisely, especially with regard to their potential for enhancing educational strategies. Future investigations of emotionality will have to pay more attention to psychobiological conceptions. (English & Russian abstracts).

683. Wood, W. Gibson; Elias, Merrill F. & Pentz, Clyde A. (1978). **Ethanol consumption in genetically selected hypertensive and hypotensive mice.** ***Journal of Studies on Alcohol,*** 39(5), 820–827.
Conducted 3 experiments with male mice from stocks with high and low blood pressure to investigate whether G. Schlager's (1974) results with mice can be a useful animal model for the study of cardiovascular disease and alcohol consumption. Results show that hypertensive mice drank more alcohol than did hypotensive animals over a wide range of concentrations. The intake was not related to blood pressure phenotype in a casual manner, nor was age a factor in the results.

684. Wysocki, Charles J.; Whitney, Glayde & Tucker, Don. (1977). **Specific anosmia in the laboratory mouse.** ***Behavior Genetics,*** 7(2), 171–188.
As an approach to a general theory of olfaction, different specific anosmia phenotypes characterized by different profiles of odorant sensitivities have been proposed for humans. In the present 4 experiments, 329 male inbred mice were tested for relative odorant sensitivity using a conditioned aversion technique and odors classified as primary or complex for humans. C57BL/6J and C57BL/10J Ss appeared to be less sensitive to the primary odorant isovaleric acid than were males of 7 other inbred strains (A/J, AKR/J, BALB/cJ, C3HeB/FeJ, DBA/2J, SJL/J, and SWR/J). In comparisons of C57BL/6J and AKR/J strains, the relative insensitivity of C57 to isovaleric acid did not generalize to the musklike primary odor of pentadecalactone or to the complex odor of amyl acetate. It is suggested that the C57BL/6J genotype may provide an animal model of a specific anosmia as characterized among humans.

685. Yagi, Hideo; Katoh, Seika; Akiguchi, Ichiro & Takeda, Toshio. (1988). **Age-related deterioration of ability of acquisition in memory and learning in senescence accelerated mouse: SAM-P/8 as an animal model of disturbances in recent memory.** ***Brain Research,*** 474(1), 86–93.
Investigated the memory, learning, and behavior of senescence accelerated mice (SAM-P/8) and senescence resistant mice (SAM-R/1), using passive avoidance response (PAR), T-maze, and open field (OF) testing. SAM-P/8 showed an age-related deterioration in PAR memory and learning. Findings show that this age-related memory and learning deficit was linked to a deterioration in the ability of acquisition and was not due to impairment in the ability of retention and hyperactivity, as observed in the OF test. In the alternation T-maze tests, SAM-P/8 showed as high a rate of alternations as did the SAM-R/1; in the T-maze avoidance tests, SAM-P/8 also showed as intact a memory ability as seen in the SAM-R/1, despite a memory deficit in the PAR. Thus, SAM-P/8 may prove useful for researching memory deficits in senile humans.

686. Zuckerman, Marvin. (1984). **Sensation seeking: A comparative approach to a human trait.** ***Behavioral & Brain Sciences,*** 7(3), 413–471.
Applies a comparative method to the human trait of sensation seeking (SS), noting that this method involves the comparison of SS dimensions with likely animal models in terms of genetic determination and common biological correlates. SS has been defined in previous research on the human level by questionnaires, reports of experience, and behavioral observation; on the animal level, SS has been defined by general activity, behavior in novel situations, and naturalistic behavior in animal colonies. Genetic determination has been found for human SS, and marked strain differences in rodents have been observed in open-field behavior, which may be related to basic differences in brain neurochemistry. Agonistic and sociable behaviors in animals and humans and SS in humans have been related to common biological correlates (i.e., gonadal hormones, monoamine oxidase, and augmenting of cortical evoked potential). Norepinephrine and enzymes in its production have been related to human SS. A model is presented that relates mood, behavioral activity, sociability, and clinical states to activity of the central catecholamine neurotransmitters and to neuroregulators and other transmitters that act in opposite ways on behavior to stabilize activity in arousal systems. Open-peer commentary on the present article by 26 authors is provided, and the present author's reply to these comments is presented.

Neuropsychology & Neurology

687. Ackerman, Sigurd H.; Hofer, Myron A. & Weiner, Herbert. (1978). **Early maternal separation increases gastric ulcer risk in rats by producing a latent thermoregulatory disturbance.** ***Science,*** 201(4353), 373–376.
Several experiments compared the effects of restraint on rats separated early and rats reared normally. Results show that Ss that were separated early from their mothers, at postnatal Day 15, became hypothermic when subjected to physical restraint on postnatal Day 30. Restraint of separated Ss also elicited an unusually high incidence of gastric erosions and insomnia and an increase in quiet wakefulness. When hypothermia during restraint was prevented, neither the erosions nor the behavioral responses occurred. Ss separated at the customary age (i.e., postnatal Day 22) did not become hypothermic during restraint, and the restraint of such Ss was not associated with either gastric erosion or insomnia.

688. Anderson, C. D.; Mair, Robert G.; Langlais, P. J. & McEntee, William J. (1986). **Learning impairments after 6-OHDA treatment: A comparison with the effects of thiamine deficiency.** ***Behavioural Brain Research,*** 21(1), 21–27.
Cortical norepinephrine, dopamine, and 3,4-dihydroxyphenylacetic acid were reduced by injection of 6-hydroxydopamine (6-OHDA) jointly into the cisterna magna and the dorsal noradrenergic bundle of male Long-Evans rats. On subsequent behavioral testing, deficits were observed for spatial-delayed alternation learning, but not for active or passive avoidance. Treatment with clonidine resulted in a significant improvement in spatial-delayed alternation for experimental compared with control Ss. Results are similar to previous observations following a bout of thiamine deficiency, in which cortical catecholamines were depleted in animals that had exhibited deficits for spatial-delayed alternation learning. It is argued that the cortical catecholamine deficits observed in post-thiamine-deficient animals are sufficient to account for the delayed alternation deficits observed in this animal model of Korsakoff's psychosis.

689. Araki, Hiroaki; Nojiri, Makiko; Kawashima, Kazuaki; Kimura, Masaaki et al. (1986). **Behavioral, electroencephalographic and histopathological studies on mongolian gerbils with occluded common carotid arteries.** ***Physiology & Behavior,*** 38(1), 89–94.
In male mongolian gerbils with ischemia induced by a bilateral occlusion of the carotid arteries, severe impairment of memory was apparent when the training session of a passive avoidance test was carried out 2 or 14 days after the ischemia. Amplitude of the hippocampal theta waves decreased, and Nissl's degradation was apparent in the CA1 neurons in the hippocampus. These neurons disappeared completely after 14 days. Similar hippocampal damage may be related to memory defects in humans after stroke.

690. Ball, Melvyn J. et al. (1983). **Paucity of morphological changes in the brains of ageing beagle dogs: Further evidence that Alzheimer lesions are unique for primate central nervous system.** ***Neurobiology of Aging,*** 4(2), 127–131.
Studied 12 regions of grey matter from the brains of 25 Beagle dogs, varying from 1 to over 16 yrs in age, by serially sectioning and sequentially scanning with a semi-automated sampling stage microscope, in a morphometric search for neuritic plaques, neurofibrillary tangles, and evidence of nerve cell loss. Examination of 227,776 light microscopic fields failed to reveal any senile plaques or neurofibrillary tangles. The neuronal densities, which ranged from 473 to 37,014 nucleolated neurons/mm^3, showed no significant relationship with aging. It is suggested that neuronal lesions of Alzheimer type may be more typical of the human CNS; and physiological evidence for regionally reduced glucose metabolic rate in this animal model may require other structural alterations for its explanation.

691. Barbeau, André; Dallaire, Lise & Poirier, Judes. (1986). **"Animals and experimentation: An evaluation of animal models of Alzheimer's and Parkinson's disease": Commentary.** ***Integrative Psychiatry,*** 4(2), 75–78.

Discusses J. H. Kordower and D. M. Gash's (1986) thesis that nonhuman primates constitute the best available model for the study of Alzheimer's disease (AD) and Parkinson's disease (PD). Although it is agreed that any animal model is only an approximation of the human disease being studied, it is argued that the advantages presented by nonhuman primates for pathologic and behavioral paradigms are less evident when the biochemistry, neuropharmacology, or pathophysiology of AD and PD are being investigated. Studies of PD are used to substantiate the premise that the choice of animal species or in-vitro system depends entirely on the questions being asked.

692. Barnes, Deborah M. (1987). **Neural models yield data on learning.** ***Science,*** 236(4809), 1628–1629.
Discusses research on neural models of plasticity that is revealing new information about nervous system changes that may underlie learning. Studies on learning in *Aplysia,* hippocampal research on long-term potentiation, and the development of a new theoretical model of changes in neuronal activity during learning are considered.

693. Bartus, Raymond T. (1988). **The need for common perspectives in the development and use of animal models for age-related cognitive and neurodegenerative disorders.** Special Issue: Experimental models of age-related memory dysfunction and neurodegeneration. ***Neurobiology of Aging,*** 9(5–6), 445–451.
Attempts to promote a more effective dialog among investigators with different approaches to studying the development and use of animal models of age-related neurodegeneration and memory loss by identifying and discussing issues that produce unnecessary confusion in the common pursuit of using animals to understand complex human neurodegenerative diseases. Some perspectives are suggested that may facilitate discussion and comparison of different animal models, independent of the paradigms and species used.

694. Bartus, Raymond T. (1986). **"Animals and experimentation: An evaluation of animal models of Alzheimer's and Parkinson's disease": Commentary.** ***Integrative Psychiatry,*** 4(2), 74–75.
Discusses J. H. Kordower and D. M. Gash's (1986) paper on the development of animal models for Alzheimer's and Parkinson's disease. The various perspectives presented on animal models for these diseases are addressed.

695. Beagley, Gwyneth H. & Beagley, Walter K. (1978). **Alleviation of learned helplessness following septal lesions in rats.** ***Physiological Psychology,*** 6(2), 241–244.
Subjected 10 male albino Sprague-Dawley rats to inescapable shock and subsequently found them to be helpless, unable to escape shock in an FR3 leverpress escape test. 10 Ss which received equal amounts of escapable shock were unimpaired in their ability to escape, performing as well as 10 controls given no pretest shock. Half of the Ss in each training group were given bilateral lesions in the septal area, and all Ss were retested. The lesions produced a large improvement in escape responding in the helpless group, but not in either of the nonhelpless groups. The results of this study suggest that septal lesions eliminate response inhibition caused by learned helplessness.

696. Becker, Jill B.; Curran, Eileen J. & Freed, William J. (1990). **Adrenal medulla graft induced recovery of function in an animal model of Parkinson's disease: Possible mechanisms of action.** ***Canadian Journal of Psychology,*** 44(2), 293–310.
Reviews evidence suggesting at least 3 fundamental processes that may underlie the functional effects of adrenal medulla grafts. These include (1) increased permeability of the blood-brain barrier to dopamine (DA), (2) increased serum DA concentrations, and (3) changes in extracellular concentrations of DA and dihydroxyphenylacetic acid in the host brain resulting in a restoration of symmetry in the striatal DA system. Five hypotheses are proposed to explain the behavioral effects of adrenal medulla grafts in light of these processes. (French abstract).

697. Beninger, Richard J. (1983). **The role of dopamine in locomotor activity and learning.** ***Brain Research Reviews,*** 6(2), 173–196.
Reviews findings of behavioral studies that provide clues to the function of dopamine (DA). Changes in overall level of activity of DA neurons appear to produce parallel changes in locomotor activity. Additionally, DA neurons seem to mediate in part the effects of biologically significant (reinforcing) stimuli on learning. Normal DA functioning appears to be required for the establishment and maintenance of incentive learning in naive animals. Previous incentive learning in trained animals can influence behavior for a time even when the function of DA neurons is disrupted; however, with continued testing in the absence of normal DA functioning, previously established conditioned incentive stimuli cease to influence behavior. From these observations and recent physiological, anatomical, and biochemical studies of DA systems, it is suggested that the biological substrate of DA-mediated incentive learning is a heterosynaptic facilitation of muscarinic cholinergic synapses. It is concluded that this model has important clinical implications since it has been suggested that DA hyperfunctioning underlies the development of schizophrenia.

698. Beninger, Richard J.; Wirsching, B. A.; Jhamandas, K. & Boegman, R. J. (1989). **Animal studies of brain acetylcholine and memory.** Symposium: Memory and aging (1988, Lausanne, Switzerland). ***Archives of Gerontology & Geriatrics,*** 1, 71–89.
Discusses the effects of manipulations of cholinergic neurotransmission on memory in delayed matching and nonmatching, delayed alternation, and radial maze tasks that were learned prior to testing, focusing on animal studies of the role of brain acetylcholine in memory. Performance decreased with increasing delays between sample and test stimuli. In studies with rats and monkeys, scopolamine produced an impairment that was delay dependent. In the radial maze, working memory was more impaired than reference memory following systemic anticholinergics or neurotoxic destruction of the nucleus basalis magnocellularis (NBM); physostigmine reversed these effects. It is suggested that degeneration of the NBM and associated memory impairment seen in aging and Alzheimer's disease may be related to a change in the ratio of endogenous tryptophan metabolites.

699. Berridge, Kent C.; Venier, Isabel L. & Robinson, Terry E. (1989). **Taste reactivity analysis of 6-hydroxydopamine-induced aphagia: Implications for arousal and anhedonia hypotheses of dopamine function.** ***Behavioral Neuroscience,*** 103(1), 36–45.
Deficits in feeding and drinking that result from 6-hydroxydopamine (6-OHDA) lesions of the mesostriatal dopamine system are often explained using either sensorimotor arousal or anhedonia hypotheses. Sensorimotor arousal hypotheses posit that dopamine systems facilitate the capacity of sensory stimuli to activate any motor output. The anhedonia hypothesis suggests that dopamine systems amplify the hedonic impact of positive reinforcers. Natural palatability-dependent ingestive and aversive actions, which are emitted by rats to tastes, provide a sensitive test that can discriminate between these hypotheses: A reduction of sensorimotor arousal should diminish the ability of tastes to elicit any actions; anhedonia should shift the balance between positive and aversive actions. To directly compare these hypotheses, taste reactivity was examined in rats made aphagic by intranigral 6-OHDA injections. Results did not support either of these predictions: Taste reactivity was essentially unchanged. The persistence of normal taste reactivity argues against both an anhedonia and a global sensorimotor arousal interpretation and provides further evidence that the capacity for hedonics can be neurologically dissociated from motivated appetitive behavior.

700. Björklund, Anders & Gage, Fred H. (1985). **Neural grafting in animal models of neurodegenerative diseases.** Institute for Child Development Research Conference: Hope for a new neurology (1984, New York, New York). ***Annals of the New York Academy of Sciences,*** 457, 53–81.
Reviews the results obtained by implantation of dopaminergic and cholinergic neurons in animals with lesions of the nigrostriatal or septohippocampal systems (i.e., in conditions that can be said to represent analogous models of Parkinson's and Alzheimer's diseases). It is argued that neural grafting in aged rats may offer new opportunities to study both the factors underlying the age-related decrements in motor and cognitive function and the possibilities of improving the performance of selected brain systems in the aged brain.

701. Boehme, Richard E. et al. (1984). **Narcolepsy: Cholinergic receptor changes in an animal model.** ***Life Sciences,*** 34(19), 1825–1828.
In an inbred colony of narcoleptic doberman pinschers analyzed for muscarinic receptor levels in 19 discrete brain regions, receptors were generally elevated in the brain stem and reduced in forebrain areas in comparison with age-matched controls. Findings suggest that cholinoceptive neurons in this region are hypersensitive and may be involved in the initiation of cataplexy and other aspects of the narcolepsy syndrome.

702. Bond, Nigel W.; Walton, Judie & Pruss, James. (1989). **Restoration of memory following septo-hippocampal grafts: A possible treatment for Alzheimer's disease.** Special Issue: The psychobiology of health. ***Biological Psychology,*** 28(1), 67–87.
Examines the role of cholinergic dysfunction in the memory deficits associated with Alzheimer's disease, the effects of hippocampal lesions on memory in infra-human animals, and the anatomy of the hippocampus. Methodological aspects of neural grafting are then examined. A review of the tasks employed to determine functional recovery following septo-hippocampal grafts suggests that although recovery is evident, its nature is unclear. An experiment is described that suggests that grafts from embryonic septum bring about recovery of working memory in rats. Different bases of the recovery of function are discussed, including the role of the graft in eliciting release of trophic factors from the host brain; the graft providing a pool of neurotransmitters; and the graft replacing the damaged circuitry of the host. Grafting may provide a viable treatment for Alzheimer's disease.

703. Bondarenko, T. T. (1988). **The sensitivity of amygdaloid neurons' dopamine receptors in rats with different alcohol motivation.** ***Fiziologicheskii Zhurnal SSSR im I.M. Sechenova,*** 74(10), 1373–1376.
Studied the relationship between sensitivity of dopamine receptors of amygdaloid neurons and alcohol consumption. Ss were 50 male mongrel white rats (immature). Chronic alcoholism was induced in experimental Ss in 6 mo. Sensitivity of chronic alcoholic Ss, other Ss preferring alcohol, and Ss preferring water was determined with microiontophoresis. (English abstract).

704. Buchanan, Shirley L. & Powell, D. A. (1988). **Age-related changes in associative learning: Studies in rabbits and rats.** Special Issue: Experimental models of age-related memory dysfunction and neurodegeneration. ***Neurobiology of Aging,*** 9(5–6), 523–534.
Describes 2 experimental models for studying age-related changes in associative learning: One involves classical (Pavlovian) conditioning of eyeblink and heart rate in the rabbit, and the other involves Pavlovian leg flexion and heart rate conditioning in the rat. Advantages and disadvantages of each model are discussed. Results with both models suggest differential effects of aging on acquisition of autonomic and somatomotor responses, underlining the utility of assessing multiple response systems to adequately characterize age-related changes in learning and memory.

705. Carman, Laurie S.; Gage, Fred H. & Shults, Clifford W. (1991). **Partial lesion of the substantia nigra: Relation between extent of lesion and rotational behavior.** ***Brain Research,*** 553(2), 275–283.
Attempted to create a partial lesion model of Parkinson's disease (PD) in adult female rats that would mimic the pattern of cell loss in humans in early stages of PD and permit examination of experimental manipulations that promote sprouting of axons of the surviving dopaminergic cells in the midbrain. Ss with unilateral 6-hydroxydopamine (6-OHDA) lesions of the substantia nigra pars compacta (SNPC) were tested weekly for rotational asymmetry after sc administration of apomorphine (APM) or amphetamine. Analysis of anatomical and behavioral data revealed a strong correlation between number of remaining tyrosine-hydroxylase-immunoreactive cells in the SNPC and the number of rotations induced by APM. APM appears to be a valuable tool for reliably estimating the number of remaining dopaminergic cells in rats with partial lesions of the SNPC.

706. Cassidy, John W. (1990). **Neurochemical substrates of aggression: Toward a model for improved intervention: I.** ***Journal of Head Trauma Rehabilitation,*** 5(2), 83–86.

The study of the neurobiologic foundations of aggression has been hampered by uncertainty regarding the correlation between animal models and human mechanisms. The absence of a conceptual model has hindered the rational use of pharmacologic agents in the treatment of the aggressive brain-injured patient. The development of a provisional model of the neurochemical mechanisms that underlie aggression is discussed. Focus is on alterations that are currently amenable to pharmacologic manipulation.

707. Chappell, E. T. & LeVere, T. E. (1988). **Recovery of function after brain damage: The chronic consequence of large neocortical injuries.** ***Behavioral Neuroscience,*** 102(5), 778–783.
In the present experiment we addressed the common clinical finding that subsequent to recovery of function, there is often a lingering chronic dysfunction associated with extensive neocortical injury. We have confirmed this observation in the laboratory setting, and the data is compatible with the theoretical position that brain injury induces a shift in dominance of functional neural systems that normally control behavior. Although the present data do not suggest how this shift in dominance may be reversed, it does, nonetheless, demonstrate that the persistent chronic dysfunctions associated with neocortical injury may be effectively moderated within certain environmental situations.

708. Cotman, Carl W. & Lynch, Gary S. (1989). **The neurobiology of learning and memory.** Special Issue: Neurobiology of cognition. ***Cognition,*** 33(1–2), 201–241.
Reviews approaches that have resulted in progress in understanding memory formation and considers the potential of these advances for treatment of clinical problems. Current progress and emerging concepts derived from a simple model system approach using animal models is emphasized. (French abstract).

709. Coyle, Joseph T.; Schwarcz, R.; Bennett, J. P. & Campochiaro, P. (1977). **Clinical, neuropathologic and pharmacologic aspects of Huntington's disease: Correlates with a new animal model.** ***Progress in Neuro-Psychopharmacology,*** 1(1–2), 13–30.
A review of the literature indicates that movement dysfunctions in Huntington's disease (HD), a neurologic disorder inherited in an autosomal dominant fashion characterized by dementia and a movement disorder, are due to degeneration of neurons intrinsic to the striatum; this results in a functional imbalance between the intact dopaminergic input and the paucity of cholinergic and gamma-aminobutyric-acid-ergic neurons in the striatum. Similar neurochemical and histologic alterations in the nigro-striatal axis can be produced in rats by stereotaxic injection of kainic acid, a rigid analog of glutamate, into the corpus striatum. It is suggested that the animal model of HD resulting from the striatal kainate lesions offers opportunities for better understanding the pathophysiology of HD as well as for testing pharmacologic agents that may correct neurotransmitter imbalances. The kainate-induced lesion suggests testable hypotheses concerning the fundamental defect in HD.

710. Darlington, Cynthia L. & Smith, Paul F. (1991). **Behavioural recovery from peripheral vestibular lesions as a model of recovery from brain damage.** ***New Zealand Journal of Psychology,*** 20(1), 25–32.
Describes an animal model of lesion-induced central nervous system (CNS) plasticity, known as vestibular compensation, which results from the destruction of the vestibular receptor cells in the inner ear. Unilateral inner ear lesions of this sort cause dramatic eye movement and postural deficits in humans and other animals, many of which disappear within 2–3 days due to some form of CNS plasticity. Advantages of the vestibular compensation model in relation to other models of CNS lesion-induced plasticity are discussed.

711. Davey, M. J.; Rose, F. D.; Dell, P. A. & Love, S. (1988). **Simultaneous extinction following hemidecortication and unilateral lesions of parietal cortex in the rat.** ***Medical Science Research,*** 16(19), 1043–1044.
30 male rats underwent a bracelet test (removed paper bracelets from left and right forepaws) before and after hemidecorticate, parietal, or sham lesions were surgically inflicted. Results show a simultaneous extinction effect on both the hemidecorticate and parietal lesioned Ss, in that they exhibited a severe delay on bracelet removal from the contralateral, but not the ipsilateral paw. Sham Ss divided attempts equally between paws. Findings contribute to the face validity of T. Shallert and I. Q. Whishaws' (1984) animal model of the human neurological condition of simultaneous extinction following brain injury.

712. Davis, Hasker P.; Baranowski, Judith R.; Pulsinelli, William A. & Volpe, Bruce T. (1987). **Retention of reference memory following ischemic hippocampal damage.** ***Physiology & Behavior,*** 39(6), 783–786.
Evaluated retention of reference memory in an animal model of cerebral ischemia. Male Wistar rats were trained for 70 daily trials on an 8-arm radial maze with 5 arms baited, subjected to 30 min of forebrain ischemia, allowed to recover for 30 days, and then tested for an additional 50 trials. Post ischemic (PI) Ss demonstrated normal retention of reference memory. Working memory was significantly impaired postoperatively. Morphologic analysis showed that PI Ss had primary loss of pyramidal neurons in the CA1 region of hippocampus. It is concluded that PI rats retain information learned prior to ischemic insult that most severely damages CA1 neurons of hippocampus.

713. Dean, Paul; Horlock, Phillip & Strachan, Ian M. (1981). **Meridional variation in visual acuity of hooded rats reared in a carpentered environment.** ***Perception,*** 10(4), 423–430.
Resolution acuity in humans is frequently better for horizontal and vertical gratings than for obliques. An animal model of this oblique effect might be of help in elucidating its underlying neural mechanisms. Rats were chosen because laboratory rats are reared in a "carpentered environment" similar to those proposed to cause the oblique effect in humans, and because electrophysiological experiments suggest that orientation selective units in rats' visual cortex may prefer horizontal and vertical stimuli. The acuity of 8 laboratory-reared hooded rats was measured with high-contrast horizontal, vertical, and oblique gratings. Ss learned to detect low-frequency square-wave gratings with slightly fewer errors if they were horizontal or vertical than if they were oblique, but the effects of grating orientation on acuity were not significant. Refraction of Ss' eyes gave no evidence of astigmatism. Results suggest that the rat may not be a good animal model for studying the mechanisms that underlie

meridional variations in acuity in humans, and raise questions concerning both the neural bases of resolution acuity, and the validity of the "carpentered environment" hypothesis.

714. Denenberg, Victor H.; Sherman, Gordon F.; Schrott, Lisa M.; Rosen, Glenn D. et al. (1991). **Spatial learning, discrimination learning, paw preference and neocortical ectopias in two autoimmune strains of mice. *Brain Research,*** 562(1), 98–104.
Administered a behavioral test battery, including paw preference, water escape, Lashley III maze, and discrimination learning to 79 NZB and 99 BXSB mice and evaluated their brains for cortical ectopias. The incidence of ectopias was 40.5% in NZBs and 48.5% in BXSBs. Left-pawed ectopic NZB Ss and left-pawed ectopic BXSB male Ss had the fastest swimming time in the water escape test, while right-pawed ectopic NZB Ss and male BXSB Ss were the slowest. On discrimination learning the BXSB males had the exact opposite pattern: Right-pawed ectopics were the best learners while left-pawed ectopics were the worst. Male BXSBs and all NZBs were manifesting autoimmune disease at the time of testing, while female BXSBs were not, suggesting that autoimmunity is a necessary background condition for the differential expression of ectopias and paw preference on learning processes.

715. Deuel, R. K. (1977). **Loss of motor habits after cortical lesions. *Neuropsychologia,*** 15(2), 205–215.
After damage to discrete regions of association cortex in man, the condition called "apraxia" may ensue. Apraxia is commonly defined as the inability to perform complex, purposeful movements in the absence of primary motor and sensory deficits. In an attempt to develop an animal model of this condition, 8 monkeys (*Macaca mulatta*) were trained to open a complex latch-box and then had bilateral periarcuate, precentral, or parietal cortical removals. Despite rapid return of other motor functions to normal, latch-box opening remained profoundly impaired after periarcuate or precentral lesions, but not after parietal lesions. Disconnection of the precentral motor system from certain other cortical regions may underlie this apraxia-like syndrome of the monkey. (French & German summaries).

716. DiFiglia, Marian. (1990). **Excitotoxic injury of the neostriatum: A model for Huntington's disease.** Special Issue: Basal ganglia research. ***Trends in Neurosciences,*** 13(7), 286–289.
Examines the role of excitotoxic injury (EI) in Huntington's disease (HD). Intrastriatal lesions with excitatory amino acids mimic some neurochemical and neuropathological characteristics of HD, leading to the hypothesis that an endogenous excitotoxin may be involved in HD. As an experimental model for producing neuronal depletion in the neostriatum, EI has allowed the study of other neuronal characteristics of HD such as progressive atrophy and regeneration, and has permitted exploration of the anatomical and functional recovery induced by intrastriatal grafts. Also, adaptation of the rodent model to the nonhuman primate has enabled investigators to examine lesion-induced motor dysfunctions more comparable to those in HD.

717. Dunnett, Stephen B.; Everitt, Barry J. & Robbins, Trevor W. (1991). **The basal forebrain-cortical cholinergic system: Interpreting the functional consequences of excitotoxic lesions. *Trends in Neurosciences,*** 14(11), 494–501.
Proposes that many functional consequences of ibotenic acid lesions (IALs) on mnemonic tasks observed in rats cannot be attributed to disruption of basal forebrain cholinergic (CH) systems, but may instead result from damage in the globus pallidus to corticostriatal output pathways. Observations of profound performance deficits on discrimination learning (DL) and memory tasks induced by IALs and kainic acid lesions of the basal forebrain have been taken to reflect damage of CH projections from the nucleus basalis magnocellularis to the neocortex, and to provide an animal model of dementia. However, injections of the toxins quisqualic acid and, more recently, α-amino-3-hydroxy-5-methyl-4-isoxazole propionic acid into the same site, which produce at least as extensive CH cell loss, induce only marginal impairments on the same range of cognitive tasks. CH regulation of the neocortex may influence specific aspects of DL and visual attention.

718. Edwards, Emmeline; Harkins, Kelly; Wright, George & Henn, Fritz. (1990). **Effects of bilateral adrenalectomy on the induction of learned helplessness behavior. *Neuropsychopharmacology,*** 3(2), 109–114.
Examined the effects of bilateral adrenalectomy (ADL) on induction of learned helplessness, using 247 male rats. Ss were exposed to 40 min of uncontrollable shock training and subsequently tested in a shock escape paradigm. It was found that 70% of the ADL Ss became helpless, whereas 20–30% of the sham controls and naive, nonoperated Ss became helpless. The increase in behavioral deficits after ADL was reversed by administration of corticosterone, the naturally occurring glucocorticoid in rats. Secretion from the adrenal cortex may be necessary for incorporation of a learned response after stress, and a dysregulation of the hypothalamic-pituitary-adrenal axis could be involved in helpless behavior.

719. England, A. S.; Marks, P. C.; Paxinos, G. & Atrens, D. M. (1979). **Brain hemisections induce asymmetric gastric ulceration. *Physiology & Behavior,*** 23(3), 513–517.
Investigated the possibility that unilateral brain interventions would result in a number of ulcers appearing on the stomach wall innervated by the vagus of the operated-on hemisphere different from that appearing on the wall innervated by the vagus of the intact hemisphere. 54 male Wistar rats were subjected to right or left hemisections or to control operations. Half of the Ss from each group were subjected to the psychological stress of restraint. All Ss were injected with tetrabenazine 5 mg/kg to facilitate ulceration. Hemisected Ss showed on the stomach wall innervated by the vagus originating in the transected hemisphere approximately half the number of ulcers to those they showed on the other stomach wall. This asymmetry in ulceration appeared regardless of whether the right or left hemisphere was transected. It is suggested that this unilateral model of ulceration, since it permits comparison between the 2 walls of the same stomach, could help clarify the relative importance of influences of the brain and of the stomach milieu in the development of ulceration.

720. Engleman, Eric A.; Hingtgen, J. N.; Zhou, F. C.; Murphy, J. Michael et al. (1991). **Potentiated 5-hydroxytryptophan response suppression following 5,7-dihydroxytryptamine raphe lesions in an animal model of depression. *Biological Psychiatry,*** 30(3), 317–320.

Tested the effects of 5,7-dihydroxytryptamine (5,7-DHT)-induced raphe lesions on a 5-hydroxytryptophan (5-HTP) animal model of depression. 5-HTP response suppression was measured following 5,7-DHT raphe lesions in adult male rats trained to press a lever for milk reinforcement. The 5,7-DHT induced lesions of the median and dorsal raphe reduced 5-hydroxytryptamine (5-HT) and 5-hydroxyindoleacetic acid (5-HIAA) levels in specific brain regions and potentiated 5-HTP induced response suppression. Findings suggest that a decrease of 5-HT innervation and an upregulation of 5-HT receptors may be responsible for the potentiation of the 5-HTP induced response suppression.

721. Fibiger, Hans C. (1991). **Cholinergic mechanisms in learning, memory and dementia: A review of recent evidence.** ***Trends in Neurosciences,*** 14(6), 220–223.
Although cholinergic cells in the basal forebrain are among the groups of neurons that degenerate in Alzheimer's disease (AD), there is no direct evidence that damage to these neurons is responsible for cognitive decline in AD. While antimuscarinic drugs can impair learning in humans, these drugs do not model the deficits seen in AD. Thus, programs aimed at developing cholinergic pharmacotherapies for cognitive deficits in AD may be based more on faith than on facts. Traditional pharmacological replacement strategies are unlikely to succeed in AD because many neurochemically distinct systems degenerate and because structures such as the hippocampus and cortex, which are among the presumed postsynaptic targets for cholinergic drugs, are damaged in AD.

722. Fields, Jeremy Z.; Drucker, George E.; Wichlinski, Lawrence & Gordon, John H. (1991). **Neurochemical basis for the absence of overt "stereotyped" behaviors in rats with up-regulated striatal D_2 dopamine receptors.** ***Clinical Neuropharmacology,*** 14(3), 199–208.
Investigated a mechanism by which female rats suppress dyskinetic movements normally associated with elevated D_2 dopamine (DA) receptor density. Neurochemical changes were correlated with behavioral changes using several animal models, including nonneuroleptic ones, which elicit varied levels of DA receptor upregulation. There was a significant positive correlation between striatal DA receptor density and apomorphine-induced stereotypic behaviors. In contrast, there was a significant negative correlation between increased DA receptor density and synthesis capacity for striatal DA (V_{max} for tyrosine hydroxylase). This decrease in V_{max} appears to be a compensatory adjustment of the nigrostriatal DA tract for the increased DA receptor density induced in the animal models. An observed increase in receptor density may not predict a functional change because compensatory neural mechanisms exist.

723. Flicker, Charles et al. (1985). **Animal and human memory dysfunctions associated with aging, cholinergic lesions, and senile dementia.** ***Annals of the New York Academy of Sciences,*** 444, 515–517.
Discusses the results of research on the cognitive abilities of the aged in several species. Although not all behavioral paradigms reveal age-specific deficits, significant impairments have been identified on a number of behavioral tasks, the most consistent of which involve tests of recent spatial memory. Aged nonhuman primates have been previously found to exhibit deficient recall of spatial location in a delayed response task. Other studies suggest that certain neurobehavioral disturbances of aged human and nonhuman primates may be shared by aged rodents as well. The presence of comparable biochemical, neurodegenerative, and behavioral changes in humans with senile dementia of the Alzheimer type and rats with lesions of the nucleus basalis of Meynert is consistent with the hypothesis that the degeneration of cortically projecting cholinergic neurons in the nucleus basalis of Meynert is partly responsible for the cognitive loss associated with senile dementia of the Alzheimer type. These results support the usefulness of aged animals and the applicability of brain lesion techniques to the development of animal models of age-related cognitive dysfunction.

724. Friedman, Eitan; Lerer, Barbara & Kuster, Joan. (1983). **Loss of cholinergic neurons in the rat neocortex produces deficits in passive avoidance learning.** ***Pharmacology, Biochemistry & Behavior,*** 19(2), 309–312.
Bilateral kainic acid lesions of the ventral globus pallidus produced a significant and selective cortical decrease in choline acetyltransferase activity in the brains of male Sprague-Dawley rats. When lesioned and control Ss were compared on performance in a step-through passive avoidance task, lesioned Ss showed a marked retention deficit 24 hrs after the initial training trial. This experimentally induced memory deficit associated with a cortical cholinergic neuronal loss resembles the deficits in senile dementia of the Alzheimer type and may provide a useful animal model for studying the disease.

725. Friehs, G. M.; Parker, R. G.; He, L. S.; Haines, S. J. et al. (1991). **Lesioning of the striatum reverses motor asymmetry in the 6-hydroxydopamine rodent model of Parkinsonism.** ***Journal of Neural Transplantation & Plasticity,*** 2(2), 141–156.
Assessed the efficacy of creating a stereotaxic lesion (STL) in the rat striatum (parkinsonian model) on enhancement of adrenal medullary graft (AMG) survival and behavioral function. Four groups of 6-hydroxydopamine (6-OHDA) lesioned rats (4 rats in each group) receiving various treatments were compared in terms of the behavioral outcome. Treatments included AMG, radiofrequency lesion (RFS), or both. The STL-induced, counterclockwise rotational behavior for each group prior to any treatment was similar. In Ss with AMG without the RFL, there was a decrease in the number of rotations and also a change in the time course of rotations. The RFLs in the original series of Ss destroyed an estimated 40% of the caudate-putamen complex. Selective destruction of caudate-putamen complex without tissue transplantation produced a 92% reduction in Ss' motor asymmetry.

726. Gallagher, Michela & Pelleymounter, Mary A. (1988). **Spatial learning deficits in old rats: A model for memory decline in the aged.** Special Issue: Experimental models of age-related memory dysfunction and neurodegeneration. ***Neurobiology of Aging,*** 9(5–6), 549–556.
Reviews the research that has examined both the nature of age-related impairments in rats on spatial tasks and the relation of such deficits to underlying neurobiological mechanisms. The review supports the notion that hippocampal dysfunction underlies the mild/moderate cognitive decline that often accompanies normal aging. It is concluded that the spatial learning deficit in aged rodents is a promising model for understanding the effect of age on brain systems that serve a memory function in humans.

727. González-Darder, José M.; Barberá, José & Abellán, Maria J. (1986). **Effects of prior anaesthesia on autotomy following sciatic transection in rats.** ***Pain,*** 24(1), 87–91.
Studied the animal model for chronic pain following sciatic nerve section in 30 male Sprague-Dawley rats by varying the sensory afferents prior to nerve section and using the anesthetic blocking of the sciatic nerve. The experimental parameters used were the day of onset of autotomy and the time course of autotomy. Results show that the anesthetic blocking prior to nerve section significantly reduced the degree of autotomy.

728. Gordon, John H. & Fields, Jeremy Z. (1989). **A permanent dopamine receptor up-regulation in the ovariectomized rat.** ***Pharmacology, Biochemistry & Behavior,*** 33(1), 123–125.
Investigated the effects of ovariectomy (OVX) of female rats on their sensitivity to apomorphine (APO). By 3 mo post-OVX, Ss were hypersensitive to the behavioral effects of APO. This hypersensitivity was permanent and was accompanied by an increase in D2 dopamine receptor density in the striatum. The usefulness of this model for understanding postmenopausal tardive dyskinesia and schizophrenia is noted, as both involve dopamine receptor regulation.

729. Górka, Z.; Earley, Bernadette & Leonard, B. E. (1985). **Effect of bilateral olfactory bulbectomy in the rat, alone or in combination with antidepressants, on the learned immobility model of depression.** ***Neuropsychobiology,*** 13(1–2), 26–30.
Compared the responsiveness of 8 bilaterally olfactory bulbectomized, 8 sham-operated, and 8 intact male Sprague-Dawley rats to the acute and chronic administration of nomifensine and trazodone in a learned immobility test. No difference in the duration of immobility was found between the bulbectomized and the sham-operated or intact Ss. The duration of immobility was significantly attenuated in both the bulbectomized and intact Ss after nomifensine treatment; the effect on the bulbectomized Ss was greater than on the intact Ss. Trazodone significantly attenuated the duration of immobility of the intact Ss only after chronic (14 days) administration. Both nomifensine and trazodone significantly affected the turnover of serotonin in the amygdaloid cortex following acute administration, but no changes could be detected in the parameters after chronic drug treatment.

730. Gray, J. A. (1979). **Anxiety and the brain: Not by neurochemistry alone.** ***Psychological Medicine,*** 9(4), 605–609.
The discovery of receptors in the CNS that specifically bind benzodiazepines (BDs) may provide a clue to what brain systems are involved in anxiety, but it might also be misleading if other sources of evidence are not considered. The ubiquity of BD receptors may be linked to their association with gamma-amino butyric acid (GABA) receptors and with the muscle-relaxant and anticonvulsant properties of the drugs. Studying the effects of BDs along with other anti-anxiety drugs (alcohol, barbiturates) could help identify the specific brain areas involved in anxiety. The existence of an animal model of anxiety will be helpful in this research. The septo-hippocampal system seems to be the most likely candidate for the anxiety center, and it seems likely that GABA is intimately involved in its action.

731. Guan, Linchu; Robinson, Terry E. & Becker, Jill B. (1984). **Animal model of rotational behavior with 6-OHDA lesions.** ***Acta Psychologica Sinica,*** 16(4), 416–421.
Observed 43 female albino Holtzman rats, in which the nigrostriatum was damaged on one side of the brain with the neurotoxin hydroxydopamine. Results show a 85–100% dopamine depletion in 38 of the Ss. Three Ss had less than 60% dopamine depletion, and 2 Ss had only a 60–80% depletion. Cocaine induced rotational behavior in the lesioned Ss. High-performance liquid chromatography with electrochemical detection was utilized in this research. (English abstract).

732. Guile, Michael N. (1985). **Hypophysectomy promotes gastric lesions in rats.** ***Physiological Psychology,*** 13(2), 111–113.
Examined the effects of hypophysectomy on ulcer formation in 16 male Long-Evans rats while behavioral indices (tail-flick latencies and flinch and vocalization levels) of nociception were concomitantly taken. Hypophysectomized Ss were found to be more susceptible to restraint-produced gastric lesions than were sham-operated controls (4.5 vs 0 mean lesions). Behavioral measures of nociception did not distinguish between the 2 groups. Results suggest that the differences in ulcer sores are not due to differential pain thresholds between the groups.

733. Gurguis, George N.; Klein, Ehud; Mefford, Ivan N. & Uhde, Thomas W. (1990). **Biogenic amines distribution in the brain of nervous and normal pointer dogs: A genetic animal model of anxiety.** ***Neuropsychopharmacology,*** 3(4), 297–303.
Nervous pointer dogs have been suggested as an animal model for pathological anxiety. To study possible disturbances in neurotransmitter functions in this model, this study measured brain biogenic amines and their metabolites in 8 nervous and 6 normal dogs, who were behaviorally tested and later anesthetized and killed. Compared to normal Ss, nervous Ss had higher norepinephrine in the reticular formation and lower serotonin (5-hydroxytryptamine [5-HT]) and its metabolite 5-hydroxyindoleacetic acid (5-HIAA) in the septal nuclei, indicating possible important differences in noradrenergic and serotonergic functions in the nervous Ss. There was a trend for lower homovanillic acid and dihydroxyphenylacetic acid (DOPAC) levels and a significantly lower DOPAC/dopamine ratio in the nervous Ss, suggesting decreased dopaminergic function.

734. Hellhammer, Dirk H. et al. (1983). **Serotonergic changes in specific areas of rat brain associated with activity-stress gastric lesions.** ***Psychosomatic Medicine,*** 45(2), 115–122.
To study serotonergic involvement in the development of gastric lesions following activity wheel stress, 3 groups of Wistar albino rats (gastric lesions, no gastric lesions, and home-cage controls) were killed following exposure to the experimental procedures. The brains were dissected into 8 specific areas and subjected to analyses for serotonin (5-HT) and 5-hydroxyindoleacetic acid (5-HIAA), using high performance liquid chromatography with electrochemical detection. Lower levels of 5-HT were found in the midbrain, cortex, and hippocampus of Ss with gastric lesions compared to either the no-lesion group, those subjected to shorter periods of activity-stress, or the home-cage control group. Levels of 5-HT and 5-HIAA were elevated in the pons/medulla oblongata of both the lesion and the no-lesion groups

compared to the home-cage controls. Corticosterone levels in blood were also significantly elevated in the lesion group. Data on serotonin changes in the CNS suggest a possible role for this neurotransmitter in stress-induced gastric pathology.

735. Henke, Peter G. (1980). **The amydala and restraint ulcers in rats.** ***Journal of Comparative & Physiological Psychology,*** 94(2), 313–323.
Five experiments with a total of 174 Wistar rats investigated the role of the amygdala and its connections with hypothalamic areas in gastric pathology induced by immobilization. Results show that lesions in the medial amygdala and the ventral amygdalofugal pathway reduced the stomach pathology induced by restraint. Lesions in the stria terminalis (ST), on the other hand, increased the severity of stomach pathology. It is concluded that the lesions in the medial nuclei and the ventral pathway attenuated the effectiveness of the noxious stimulus to produce gastrointestinal abnormalities, whereas the lesions in the ST interfered with inhibitory effects.

736. Henke, Peter G. (1981). **Attenuation of shock-induced ulcers after lesions in the medial amygdala.** ***Physiology & Behavior,*** 27(1), 143–146.
40 male Wistar rats either received bilateral radio-frequency lesions in the medial amygdala, served as operated controls, or were unoperated. Results show that bilateral lesions in the medial amygdala attenuated the effects of electric shock stimulation on gastric pathology. There was no significant difference between double-avoidance conflict and yoked control Ss. It is concluded that the lesions attenuated the aversiveness of noxious inputs, similar to the effects seen in aversively motivated behaviors.

737. Henke, Peter G. (1982). **Septal lesions, emotionality, and restraint-induced stomach pathology in rats.** ***Physiology & Behavior,*** 28(4), 739–741.
Bilateral septal lesions in male Wistar rats did not differentially affect the stomach pathology induced by physical restraint. This was found when septal Ss were hyperemotional or when the so-called septal syndrome had abated. Results of 2 studies suggest that the emotionality changes produced by septal lesions are apparently unrelated to the gastric pathology induced by immobilization.

738. Henke, Peter G. (1990). **Hippocampal pathway to the amygdala and stress ulcer development.** ***Brain Research Bulletin,*** 25(5), 691–695.
Exp 1 localized the reported aggravation of stress ulcers found after large bilateral hippocampal lesions in male rats. Lesions in the ventral hippocampus produced a similiar increase in the severity of gastric erosions after cold-restraint, as was seen after large bilateral lesions. Dorsal hippocampal damage produced no differential effects. In Exp 2, high-frequency electrical stimulation of the ventral CA1 region of the hippocampus, a procedure known to induce long-term potentiation, increased evoked potentials in the lateral central nucleus of the amygdala and in adjacent parts of the lateral and basolateral nuclei. The increase in the efficacy of synaptic transmission in this pathway attenuated stress ulcer development. Data suggest that a physiologically relevant connection, linking ventral hippocampus to central amygdala, exists during stressful experiences.

739. Henke, Peter G. & Savoie, Roger J. (1982). **The cingulate cortex and gastric pathology.** ***Brain Research Bulletin,*** 8(5), 489–492.
Performed bilateral lesions in the anterior or posterior cingulate cortex of male Wistar rats. After recovery, Ss were immobilized for 24 hrs. Results show that posterior lesions increased the severity of the stomach pathology under restraint and nonrestraint conditions. Anterior lesions attenuated the effects of immobilization on gastric pathology. Findings are discussed with respect to telencephalic limbic mechanisms and stress.

740. Hurwitz, Barry E.; Dietrich, W. Dalton; McCabe, Philip M.; Watson, Brant D. et al. (1990). **Sensory-motor deficit and recovery from thrombotic infarction of the vibrissal barrel-field cortex.** ***Brain Research,*** 512(2), 210–220.
Two experiments with rats examined the behavioral consequences of photochemical infarction of the posteromedial barrel subfield of the primary somatosensory cortex in 2 learning tasks requiring sensory–motor integration. A reliable decrement in task performance was observed irrespective of whether the task required active (Exp 1) or passive (Exp 2) vibrissal sensory discrimination. Findings suggest a noninvasive model of stroke in the rat that is amenable to the study of behavioral recovery.

741. Ingram, Donald K. (1988). **Complex maze learning in rodents as a model of age-related memory impairment.** Special Issue: Experimental models of age-related memory dysfunction and neurodegeneration. ***Neurobiology of Aging,*** 9(5–6), 475–485.
Reviews the research on the age-related learning deficit observed in a 14-unit T-maze (Stone maze). Rats and mice of several strains representing different adult age groups were first trained to criterion in 1-way active avoidance in a straight runway; then training in the Stone maze was conducted. Results indicate a robust age-related impairment in acquisition observed in males and females and in outbred, inbred, and hybrid strains. Pharmacological studies using scopolamine in young and aged rats indicated cholinergic involvement for accurate encoding during acquisition of this task. Retention aspects of storage and retrieval were not affected by scopolamine. Future research is outlined to provide more thorough psychological characterization of maze performance, to analyze the specificity of cholinergic involvement in the task, and to test possible therapeutic interventions for alleviating the age-related impairments observed.

742. Itakura, Toru; Yokote, Hideyoshi; Yukawa, Shuya; Nakai, Mitsukazu et al. (1990). **Transplantation of peripheral cholinergic neurons into Alzheimer model rat brain.** Xth Meeting of the World Society for Stereotactic and Functional Neurosurgery (1989, Maebashi, Japan). ***Stereotactic & Functional Neurosurgery,*** 54–55, 368–372.
Results from 18 rats reveal that Ss with a lesion of the nucleus basalis Meynert (NBM) showed abnormal increase of spontaneous activity, disturbance of memory retention, and disturbance of learning acquisition. Ss that had received transplantation of cholinergic neurons into the cerebral cortex displayed amelioration of abnormal behavior produced by the destruction of the NBM.

743. Jaffard, Robert; Durkin, Thomas; Toumane, Abdoulaye; Marighetto, Aline et al. (1989). **Experimental dissociation of memory systems in mice: Behavioral and neurochemical aspects.** Symposium: Memory and aging (1988, Lausanne, Switzerland). ***Archives of Gerontology & Geriatrics,*** 1, 55–70.
In experiments with mice, changes in choice accuracy have been observed to occur both as a function of the conditions of testing and of specifically altered brain function; many changes resemble those observed both in normal people and amnesics. Two major central cholinergic pathways have been shown to exhibit specific changes in their activity as a function of the type of memory process involved in a given task. Since age-related memory dysfunction is also associated with a parallel attenuation of central cholinergic activation and since both of these impairments can be mimicked by local intraseptal injection of an α-noradrenergic antagonist, it is suggested that the memory deficits of old animals might result primarily from a severe attenuation of phasically active noradrenergic excitatory input to cholinergic neurons.

744. Kadar, Tamar; Silbermann, Michael; Brandeis, Rachel & Levy, Aharon. (1990). **Age-related structural changes in the rat hippocampus: Correlation with working memory deficiency.** ***Brain Research,*** 512(1), 113–120.
Monitored hippocampal morphological changes in young, mature, middle-aged, and aged male Wistar rats (aged 3, 12, 17, and 24 mo, respectively) whose memory performance had been previously measured. Ss' cognitive performance was evaluated using an 8-arm radial maze. Significant memory impairments were observed at age 12 mo in all measured parameters (correct choices, errors, total time). Morphometric analysis revealed a decrease in the area of cells within the hippocampus and the number of cells in the CA3 subfield. Findings suggest that 12-mo-old Wistar rats may serve as the animal model of choice for the study of specific age-related behavioral deficits and that the hippocampal CA3 region might play a major role in age-dependent cognitive decline.

745. Kandel, Eric R. (1979). **Small systems of neurons.** ***Scientific American,*** 241(3), 66–76.
Because humans share many behavioral patterns with simpler animals, it has been useful to develop animal models to study how interacting patterns of neurons give rise to behavior. Simple invertebrates are attractive for such investigation because their nervous systems consist of between 10,000 and 100,000 cells instead of the billions in more complex animals. Studies of the neurons of the abdominal ganglion of the marine snail *Aplysia* are examined in terms of cell invariance, functions, excitatory or inhibitory features, strength of synaptic connections, and sensitization and habituation responses.

746. Keefe, Kristen A.; Salamone, John D.; Zigmond, Michael J. & Stricker, Edward M. (1989). **Paradoxical kinesia in Parkinsonism is not caused by dopamine release: Studies in an animal model.** ***Archives of Neurology,*** 46(10), 1070–1075.
Male albino rats made akinetic and cataleptic by dopamine-depleting brain lesions and control Ss were exposed to activating situations (ASTs) to determine the effect of dopamine-receptor antagonists on paradoxical kinesia (PDK). PDK was not dependent on enhanced functioning of residual nigrostriatal dopaminergic neurons under stressful conditions. Appearance of appropriate motor responses under ASTs may reflect continued striatal integration of more intense sensory input without dopaminergic modulation or sensorimotor integration elsewhere in the brain. Delineation of the mechanism involved in restoration of motor function under ASTs may provide insight into stress and performance interactions in Parkinson's disease patients.

747. Kelsey, John E. (1983). **Ventromedial septal lesions in rats reduce stomach erosions produced by inescapable shock.** ***Physiological Psychology,*** 11(4), 283–286.
Exposure to inescapable shock produced fewer and less extensive stomach erosions in male Sprague-Dawley rats with ventromedial septal lesions than in the sham-operated controls ($N = 33$). This finding, in conjunction with previous findings that ventromedial septal lesions also reduce other physiological and behavioral responses to inescapable shock, indicates that this area of the limbic system is importantly involved in mediating several responses to inescapable shock.

748. Kelsey, John E. & Baker, Margaret D. (1983). **Ventromedial septal lesions in rats reduce the effects of inescapable shock on escape performance and analgesia.** ***Behavioral Neuroscience,*** 97(6), 945–961.
In 2 experiments with 104 male Sprague-Dawley rats, lesions of the ventromedial septum (VMS) reduced or eliminated several effects of exposure to inescapable shock, but lesions of the dorsolateral septum did not. Exp I demonstrated that VMS lesions reduced the loss in body weight produced by inescapable shock and eliminated the subsequent (24 hrs later) interference with escape performance (learned helplessness). Exp II demonstrated that VMS lesions reduced the analgesia that occurs immediately following inescapable shock and the analgesia reinstated by exposure to escapable shock 24 hrs later. Findings indicate that VMS lesions reduce several responses to inescapable shock and suggest the possibility that all of these effects may reflect a unitary deficit. It is hypothesized that VMS lesions reduce these effects of exposure to inescapable shock either by reducing the ability of the rats to learn that their responses and shocks were uncorrelated or by reducing the emotional impact of this lack of correlation.

749. Kentridge, Robert W. & Aggleton, John P. (1990). **Emotion: Sensory representation, reinforcement, and the temporal lobe.** Special Issue: Development of relationships between emotion and cognition. ***Cognition & Emotion,*** 4(3), 191–208.
Explores the reciprocal relationships between reinforcers and the central representations for extrinsic stimuli. A synthetic model demonstrates how adaptive stimulus representations may be formed and how such representations may come to elicit emotions. It is proposed that the evolutionary origin of emotions, which are primarily seen as energizing signals of survival value, arose from the need to sensitize an animal in the absence of appropriate signals of reinforcement. Evidence from research on nonhuman primates suggests that many of these functions take place in the temporal lobe.

750. Kimble, Daniel P. (1990). **Functional effects of neural grafting in the mammalian central nervous system.** ***Psychological Bulletin,*** 108(3), 462–479.
Grafts of fetal and nonfetal brain tissues have been successfully implanted into the mammalian central nervous system (CNS). The functional effects of neural grafting in the CNS of rodents and nonhuman primates in a variety of situations

are reviewed. Research areas discussed include the effects of dopamine-rich grafts in animal models of Parkinson's disease and acetylcholine-rich grafts in animals with lesions of the cholinergic pathways to the neocortex and hippocampus. Graft effects also are examined in aged animals and genetic mutants. In addition, the effects of neural grafts on circadian rhythmicity, reproductive functions, and conditioned taste aversion are discussed. The beneficial functional effects of neural grafts and the possible mechanisms and implications for these effects are discussed, including the possibility that the CNS exhibits a regional biochemical specificity that influences the outcome of neural graft procedures.

751. Kimura, Naoto; Yoshimura, Hiroyuki & Ogawa, Nobuya. (1987). **Sex differences in stress-induced gastric ulceration: Effects of castration and ovariectomy.** ***Psychobiology,*** 15(2), 175–178.
Gastric mucosal lesions were produced in male and female mice and rats through the use of a paradigm in which some Ss were shocked and others exposed to stimuli arising from the shocked Ss. Under these conditions, shocked and nonshocked Ss showed an approximately equal incidence of gastric lesions. There were significant differences between the sexes in the incidence of lesions following stress, with males of both species showing a higher incidence of ulceration. Castration reduced the severity of ulceration in males, but ovariectomy had no effect on ulceration in females. Data suggest that male sex hormones play an important role in the development of gastric lesions in both mice and rats.

752. Kiyota, Yoshihiro; Miyamoto, Masaomi & Nagaoka, Akinobu. (1987). **Characteristics of memory impairment in cerebral embolized rats at the chronic stage.** ***Pharmacology, Biochemistry & Behavior,*** 28(2), 243–249.
Assessed the effect of cerebral embolization on learning and memory in male JCL/Wistar rats at the chronic stage (8 wks or more). Ss exhibited an increase in ambulation in an open-field test, and marked impairment of the passive avoidance response at the early stage of cerebral infarction, which gradually diminished by the chronic stage. In a 2-way active avoidance task, embolized Ss showed accelerated acquisition of the avoidance response compared with controls. However, at the chronic stage, Ss exhibited marked impairment of light–dark discrimination learning. Spatial memory impairment was also observed in Ss; there was a significant decrease in initial correct responses and an increase in total errors in a radial maze task. Multifocal necroses were detected in several brain regions, particularly the hippocampus and internal capsule of the embolized hemisphere. The use of embolized rats in modeling vascular-type dementia is suggested.

753. Kiyota, Yoshihiro; Miyamoto, Masaomi; Nagaoka, Akinobu & Nagawa, Yuji. (1986). **Cerebral embolization leads to memory impairment of several learning tasks in rats.** ***Pharmacology, Biochemistry & Behavior,*** 24(3), 687–692.
Studied the effects of cerebral embolization, produced by injecting microspheres into the left internal carotid artery, on passive and active avoidance tasks and a water-filled multiple **T**-maze task in male JCL/Wistar rats. Ss with cerebral embolization were markedly impaired in the acquisition and retention of the 1-trial passive avoidance response. The impairment depended on the number of microspheres injected and continued for 2 wks. The cerebral-embolized Ss were also impaired in the acquisition of 2-way active avoidance response in a shuttlebox. These impairments were not due to decrease in shock sensitivity because there was no significant change in the flinch-jump threshold. The embolized Ss also exhibited a significant disturbance in performance of water-filled multiple **T**-maze learning. Results suggest that Ss with cerebral embolization are impaired in 3 different types of learning tasks and may be useful as an animal model for the vascular type of dementia.

754. Koliatsos, Vassilis E.; Applegate, Michael D.; Kitt, Cheryl A.; Walker, Lary C. et al. (1989). **Aberrant phosphorylation of neurofilaments accompanies transmitter-related changes in rat septal neurons following transection of the fimbria-fornix.** ***Brain Research,*** 482(2), 205–218.
Investigated changes in cytoskeletal elements and transmitter-related enzymes in medial septal nucleus neurons following fibria-fornix (FF) transection in male rats. Results indicate that FF transection provides a useful animal model for further investigations of complex disorders of the central nervous system (CNS) that involve degeneration of transmitter-specific pathways (e.g., Alzheimer's disease).

755. Kordower, Jeffrey H. & Gash, Don M. (1986). **Animals and experimentation: An evaluation of animal models of Alzheimer's and Parkinson's disease.** ***Integrative Psychiatry,*** 4(2), 64–70.
Evaluates animal models of Alzheimer's and Parkinson's diseases, 2 classical neurologic syndromes. Animal studies have focused mainly on specific neurochemical systems involved in these disease states. The advantages and limitations of the various rodent models are examined. The need to establish valid nonhuman primate models is discussed.

756. Kosten, Therese & Contreras, Robert J. (1985). **Adrenalectomy reduces peripheral neural responses to gustatory stimuli in the rat.** ***Behavioral Neuroscience,*** 99(4), 734–741.
The adrenalectomized rat, because of excessive body sodium loss, has been an important animal model for studying the physiological mechanisms underlying salt ingestion. To investigate the mediation by peripheral taste responsivity of changes in salt intake, multiunit responses of the chorda tympani nerve to various concentrations of NaCl, KCl, and LiCl, hydrochloric acid, and quinine hydrochloride were recorded from 18 adrenalectomized or intact male Sprague-Dawley rats. To control for a generalized decrease in sensory sensitivity, recordings from this auriculotemporal nerve to tactile stimulation of the pinna were also performed. There were no group differences in amplitude of the integrated neural responses to tactile stimulation. The largest decrease in gustatory responsivity occurred for suprathreshold concentrations of NaCl and LiCl. Data are discussed with reference to possible mechanisms underlying this neural alteration and the role that reductions in salt taste responsivity play in mediating increases in salt intake.

757. Lal, Harbans & Forster, Michael J. (1988). **Autoimmunity and age-associated cognitive decline.** Special Issue: Experimental models of age-related memory dysfunction and neurodegeneration. ***Neurobiology of Aging,*** 9(5–6), 733–742.
Studied the relationship between brain-reactive antibodies (BRA) and age-associated cognitive dysfunction. A parallel relationship between BRA increases with age and decline of avoidance learning capacity is also described in mouse models. Transfer of immunity from old (aged 22–24 mo) to young (aged 3–4 mo) mice was found to accelerate both age-related formation of brain-reactive antibodies and age-related decline of avoidance learning capacity. Short-lived mouse

genotypes with accelerated autoimmunity showed accelerated age-related declines in their ability to acquire an avoidance response when compared with nonautoimmune mice. Overall, findings indicate that the immune system could be an important target for development of intervention strategies aimed at extending the intellectually competent period of life. It is suggested that mice in which autoimmunity is accelerated may be useful as models for the development of such interventions.

758. Langston, J. William. (1986). **"Animals and experimentation: An evaluation of animal models of Alzheimer's and Parkinson's disease": Commentary.** ***Integrative Psychiatry,*** 4(2), 78–80.
Discusses J. H. Kordower and D. M. Gash's (1986) paper on animal models of Alzheimer's and Parkinson's disease. The use of primates as opposed to rodents in developing animal models, the creation of models of these diseases at a time in the animal life span that corresponds to the typical age of onset in humans, and future models are considered.

759. Leshner, Alan I. & Segal, Menahem. (1979). **Fornix transection blocks "learned helplessness" in rats.** ***Behavioral & Neural Biology,*** 26(4), 497–501.
Male Wistar rats were subjected to fornix transection or control procedures, exposed to unavoidable preshock or not, and then tested for escape/avoidance performance in a shuttlebox. Controls receiving unavoidable preshock exhibited longer escape latencies in the helplessness test than nonpreshocked Ss. However, fornix-transected Ss did not show this effect of preshock.

760. Levine, Joel D.; Strauss, Lisa R.; Muenz, Larry R.; Dratman, Mary B. et al. (1990). **Thyroparathyroidectomy produces a progressive escape deficit in rats.** ***Physiology & Behavior,*** 48(1), 165–167.
Presents data showing that hypothyroid status can produce an escape deficit in rats similar to that produced by inescapable shock, supporting a link between thyroid status and depression suggested by an animal model. While sham-operated Ss improved their performance on a simple escape task over 3 days of testing, thyroparathyroidectomized Ss showed a pronounced decrease in their responses. Markov transition analysis was used to obtain conditional probabilities of escaping given a prior escape or failure to escape for the 2 groups. This analysis showed similar data set structures for the 2 groups. Results suggest that if intact rats learn to escape, then hypothyroid rats may learn not to escape.

761. Levisohn, L. F. & Isacson, O. (1991). **Excitotoxic lesions of the rat entorhinal cortex: Effects of selective neuronal damage on acquisition and retention of a non-spatial reference memory task.** ***Brain Research,*** 564(2), 230–244.
The neurotoxin *N*-methyl-D-aspartate was used to induce selective bilateral neuronal loss in the entorhinal cortex (EC) to model one aspect of the neurodegeneration observed in Alzheimer's disease, Down syndrome, and aging. 21 lesioned, sham-lesioned, and intact control male rats learned a reference memory task involving a brightness discrimination for water reward. The 2 major effects of EC lesions were (1) an impaired retention of reference memory over a 10-day interval and (2) the reduced preference for spatially guided responses. The behavioral results of the excitotoxic model EC lesions and their implications for models of memory and Alzheimer's disease are discussed.

762. Lindvall, Olle. (1991). **Prospects of transplantation in human neurodegenerative diseases.** Special Issue: Transplantation in the nervous system. ***Trends in Neurosciences,*** 14(8), 376–384.
Discusses experimental data obtained from animals suggesting that restoration or preservation of function through cell transplantation into the central nervous system (CNS) might be developed into a useful therapeutic approach in human neurodegenerative disorders. Discussion focuses on clinical trials with transplantation in Parkinson's disease (PD), problems for development of a transplantation therapy in PD, prospects for clinical trials with transplantation in Huntington's disease, and transplantation in other neurodegenerative disorders. Further progress can be made only by systematic studies in animals, but will also eventually require clinical trials in a few well-monitored patients.

763. MacRae, Priscilla G.; Spirduso, Waneen W. & Wilcox, Richard E. (1988). **Reaction time and nigrostriatal dopamine function: The effects of age and practice.** ***Brain Research,*** 451(1–2), 139–146.
Developed a rodent model of human reaction time (RT) in which RT performance correlates highly with neurochemical measures of nigrostriatal dopamine (DA) integrity, using 15 young and 10 old male rats. Data suggest that the plasticity of the motor system allows age-related RT impairments to be overcome with extensive practice. A regulatory mechanism for RT involving D_2DA receptors is suggested.

764. Mayo, Willy; le Moal, Michel & Simon, Hervé. (1988). **A model of central cholinergic dysfunction in the rat: Lesion of the nucleus basalis magnocellularis.** 2nd Conference: Information about Alzheimer's disease (1988, Marseilles, France). ***Psychologie Medicale,*** 20(13), 1909–1914.
Studied the behavioral consequences of bilateral ibotenic acid lesions of the nucleus basalis magnocellularis (NBM) in the rat. Six rats treated with lesions of the NBM and 7 nonlesioned controls were in Study I. 10 rats treated with lesions of the NBM and 9 nonlesioned controls were in Study II. In Study I, spontaneous hoarding, exploratory, and locomotor behaviors were observed. In Study II, the Ss' performances in learning situations involving spatial memory were observed. Differences between lesioned and nonlesioned Ss were analyzed. Implications of results for cognitive deficits associated with Alzheimer's disease are discussed. (English abstract).

765. McGaugh, James L. (1985). **Peripheral and central adrenergic influences on brain systems involved in the modulation of memory storage.** ***Annals of the New York Academy of Sciences,*** 444, 150–161.
Discusses the modulation of memory storage in animals in terms of hormonal modulation of memory, amygdala involvement in memory modulation, and adrenergic involvement in memory dysfunction. Evidence is presented that suggests that, at least in rats and mice, memory storage processes are subject to experimental and endogenous modulating influences. Some of the findings appear to fit reasonably well with the findings of studies of memory modulation and memory dysfunction in humans. Whether these animal models are adequate models of human memory remains to be seen.

766. Messer, William S.; Stibbe, Jennifer R. & Bohnett, Mark. (1991). **Involvement of the septohippocampal cholinergic system in representational memory.** ***Brain Research,*** 564(1), 66–72.
To develop an animal model for testing muscarinic agonists, the effects of cholinergic lesions with the ethylcholine aziridinium ion (AF64A) on 2 types of memory tasks were examined. The tasks provided a distinction between representational and dispositional memory as measured in a single paradigm. Young male rats were trained in a T-maze to learn both a discrimination task and a paired-run alternation (ALT) task. Once Ss learned the tasks, they were administered either saline or AF64A. One week following surgery, saline-treated Ss exhibited comparable performances on both tasks. In contrast, AF64A-treated Ss showed a marked impairment on the ALT task. AF64A-treated Ss also exhibited significant impairments on the ALT task compared to control Ss. Data suggest that AF64A can be used to produce selective lesions of the septohippochemical cholinergic system.

767. M'Harzi, M.; Jarrard, L. E.; Willig, F.; Palacios, A. et al. (1991). **Selective fimbria and thalamic lesions differentially impair forms of working memory in rats.** ***Behavioral & Neural Biology,*** 56(3), 221–239.
Two experiments with 120 male rats compared the effects of selective lesions of the fimbria or of thalamic nuclei on object recognition, place recognition, and the radial arm maze test. Fimbria lesions produced deficits in the radial maze. Object recognition was spared or even facilitated, whereas place recognition was impaired. Electrolytic lesions of either centro-median-parafascicularis (CM-PF) or dorsomedialis nuclei produced highly significant deficits in the radial maze test but spared object and place recognition. Ibotenate lesions of the CM-PF had no effect on any test, which means that the critical structure was the fasciculus retroflexus (FSR). Data may contribute 2 main points to animal models of hippocampal and thalamic amnesia: (1) Different forms of working memory in rats might have different neural bases, and (2) the FSR may be involved in learning and memory processes.

768. Mitchell, I. J. (1990). **Striatal outputs and dyskinesia.** ***Journal of Psychopharmacology,*** 4(4), 188–197.
Reviews recent animal model studies, primarily using the 2-deoxyglucose uptake procedure, to describe the pharmacological, physiological, and anatomical features of striatal outputs in mediating several forms of dyskinesia (i.e., choreiform, ballism, tardive dyskinesia) and Parkinsonism. Results from these studies are synthesized, and the potential for designing novel therapeutic strategies for movement disorders based on the selective manipulation of striatal outputs is stressed.

769. Miyamoto, Masaomi; Kato, Junko; Narumi, Shigehiko & Nagaoka, Akinobu. (1987). **Characteristics of memory impairment following lesioning of the basal forebrain and medial septal nucleus in rats.** ***Brain Research,*** 419(1–2), 19–31.
Biochemical assay revealed that lesioning of the basal forebrain (BF) and medial septal nucleus (MS) of male rats resulted in marked and selective decreases in both choline acetyltransferase and acetylcholinesterase activities in the cerebral cortex and hippocampus, respectively. Ss with BF lesions exhibited a severe deficit in a passive avoidance task; however, only slight impairment of passive avoidance was observed in Ss with MS lesions. Memory impairment in Ss with BF or MS lesions was also investigated using 2 spatial localization tasks, the Morris water task and the 8-arm radial maze task. Results suggest that BF lesions may lead to substantial long-term memory impairment while MS lesions may primarily produce short-term or working memory impairment, indicating a qualitatively different contribution of the 2 cholinergic systems to memory. It is also suggested that these 2 experimental animal models may be useful for evaluation of therapeutic drugs for senile dementia of the Alzheimer type.

770. Morrison, S. D. (1976). **Control of food intake in cancer cachexia: A challenge and a tool.** ***Physiology & Behavior,*** 17(4), 705–714.
Reviews information on the breakdown in the control of feeding and the extent to which studies of the cachectic syndrome might throw light on normal control of feeding. The components of control that break down appear to be those not mediated by the hypothalamus. The cancer cachectic organism thus offers a potential animal model for extra-hypothalamic controls of feeding and for the interaction between hypothalamic and extra-hypothalamic controls.

771. Moss, Mark B.; Rosene, Douglas L. & Peters, Alan. (1988). **Effects of aging on visual recognition memory in the rhesus monkey.** Special Issue: Experimental models of age-related memory dysfunction and neurodegeneration. ***Neurobiology of Aging,*** 9(5–6), 495–502.
Performance by 6 rhesus monkeys 26–27 yrs of age was compared with that of 6 young adult monkeys 4–5 yrs of age on a trial unique delayed nonmatching to sample task. This task assessed a S's ability to identify a novel from a familiar stimulus over a delay and resembles clinical tests used to assess memory function in geriatric patients. The task was presented in 3 stages: acquisition, delays, and lists. As a group, aged Ss were impaired relative to the young adult Ss on all 3 conditions. However, within the aged group, individual cases of efficient performance were observed. Error analyses of item positions of the lists condition revealed the absence of enhanced performance for items presented at the end of a list by aged animals, suggesting an abnormal sensitivity to proactive interference. The finding of a recognition impairment with age suggests that the rhesus monkey is a suitable animal model of human aging.

772. Mukhin, E. I. (1988). **Structural and functional organization of thalamic insufficiency.** ***Zhurnal Nevropatologii i Psikhiatrii imeni S.S. Korsakova,*** 88(7), 3–7.
Studied the role of the thalamic parafascicular complex (TPC) in the intellectual processes of abstraction and generalization in 10 cats. The formation of preverbal concepts during solution of alternative-choice tasks based on the principle of generalization of similar signals was analyzed in freely moving animals. Performance was analyzed before and after destruction of the lateral and medial parts of the TPC. (English abstract).

773. Nagayama, Haruo; Tsuchiyama, Kounosuke; Yamada, Kenji & Akiyoshi, Jotaro. (1991). **Animal study on the role of serotonin in depression.** ***Progress in Neuro-Psychopharmacology & Biological Psychiatry,*** 15(6), 735–744.
Conducted a series of experiments with animals to clarify the role of serotonin (5-hydroxytryptamine [5-HT]) in depression (DP). Rats injected with tetrabenazine showed a rapid decrease in serotonin reflected in levels of 5-hydroxy-

indoleacetic acid (5-HIAA). Results of experiments with rats that received 5-hydroxytryptophan (5-HTP) and antidepressants suggest that behavioral DP may be induced by excessive transmission (TRM) of serotonin at the synapse. Other experiments with rats examined serotonin metabolism and receptor binding and inhibition of forskolin-stimulated adenylate cyclase activity by 8-OH-DPAT. Overall, DP may be caused by excessive TRM of serotonin at the synapse, although it is not clear which subtype system of serotonin receptors is concerned with this excessive TRM.

774. Nariai, Tadashi; DeGeorge, Joseph J.; Lamour, Yvon; Rapoport, Stanley I. et al. (1991). **In vivo brain incorporation of [1-^{14}C]arachidonate in awake rats, with or without cholinergic stimulation, following unilateral lesioning of nucleus basalis magnocellularis.** ***Brain Research,*** 559(1), 1–9.
Measured regional brain incorporation of a radiolabled unsaturated fatty acid, [1-^{14}C]arachidonic acid (^{14}C-AA), in awake male rats following unilateral lesioning of the nucleus basalis magnocellularis. This animal model of Alzheimer's disease shows that the iv ^{14}C-AA technique combined with cholinergic stimulation can be used to detect compensatory regulation of phospholipid-coupled signal transduction caused by a deficit in cholinergic input into the cerebral cortex.

775. Nath, Chandishwar; Gulati, Anil; Dhawan, Keshav N. & Gupta, Gyan P. (1988). **Role of central histaminergic mechanism in behavioural depression (swimming despair) in mice.** ***Life Sciences,*** 42(24), 2413–2417.
A study of mice in a behavioral model of depression indicates a possible link between the central histaminergic system and depression. It is concluded that central H_2 receptors facilitate depression, and antidepressant drugs block central H_2 receptors. The present results and past studies highlight the importance of central H_2 receptors in behavioral control.

776. Navarro Guzmán, José I. (1987). **Behavioral effects of sciatic nerve injury in rats: A contribution to the study of the phantom limb syndrome.** ***Revista de Análisis del Comportamiento,*** 3(2), 165–169.
Studied the effects of sciatic nerve injury on the acquisition of an operant response to assess the possible involvement of higher cognitive processes in phantom limb syndrome. Ss were 73 male Wistar rats (175–275 g body weight). In a factorial design involving 3 levels of intervention (surgical injury, surgical stimulation, and nonintervention) and 2 levels of postsurgical habitat (isolated vs group), Ss were randomly assigned to treatment conditions, and postsurgical acquisition of an FR, water-reinforced operant response was assessed. (English abstract).

777. Neilson, Peter D. (1986). **Do the α and λ models adequately describe reflex behavior in man?** ***Behavioral & Brain Sciences,*** 9(4), 616–617.
Questions the adequacy of the λ and the α animal models discussed by M. B. Berkinblit et al (1986) for describing reflex behavior in humans. An alternative servo-assist model is proposed, with supporting experimental data.

778. Newman, Joseph P.; Gorenstein, Ethan E. & Kelsey, John E. (1983). **Failure to delay gratification following septal lesions in rats: Implications for an animal model of disinhibitory psychopathology.** ***Personality & Individual Differences,*** 4(2), 147–156.
It has been proposed that dysfunction within the medial septum, the posterior hippocampus, and the orbito-frontal cortex (SHF system) may constitute the physiological basis of several disinhibitory syndromes in humans—psychopathy, hyperkinesis, alcoholism, and extraversion. Consequently, the syndrome produced by lesions of the SHF system in animals is offered as a tentative behavioral model of human disinhibitory psychopathology. As predicted from this model, Sprague-Dawley rats with septal lesions, like disinhibited humans, were less likely to delay gratification than controls when given a choice between waiting 10 sec for an assured reinforcement and an immediately available, though infrequently delivered, reinforcement. Inquiry into the nature of this deficit suggested that the Ss were subject to an interference effect, such that the influence of future rewards on behavior could be disrupted or eclipsed by the presence of more immediate, prominent, motivationally significant cues. The possibility that various disinhibitory syndromes in humans may also be due to a similar rigid focus of attention upon the most immediate or prominent motivationally significant event is briefly discussed.

779. Nielsen, Donald W.; Franseen, Laura & Fowler, Deliah. (1984). **The effects of interruption on squirrel monkey temporary threshold shift to a 96-hour noise exposure.** ***Audiology,*** 23(3), 297–308.
Exposed 6 squirrel monkeys (*Saimiri scuireus*) to an octave band of noise with a center frequency of 500 Hz under both continuous and interrupted conditions. Continuous exposures lasted for 4, 8, 16, 24, 48, 72, and 96 hrs, with complete recovery of hearing before the next exposure. The interrupted exposure was 96 hrs long with 5-min interruptions so that temporary threshold shift (TTS) could be measured after 2, 4, 8, 12, 16, 24, 36, 48, 60, 72, and 84 hrs of exposure. There were no differences in TTS from 8 to 96 hrs of exposure between the 2 conditions. However, at 4 hrs of exposure, the interrupted exposure showed significantly less TTS than the continuous exposure. The only difference between the exposures was a 5-min interruption at 2 hrs of exposure during the interrupted exposure. It is therefore suggested that the interrupted-exposure method is valid for making generalizations about continuous exposures of 8 hrs or longer. Results also confirm the 1st author and colleagues' (1978) finding that there was no asymptotic threshold shift for the squirrel monkey, although the exposure time was extended to 96 hrs in the present experiment. The similarity of human and squirrel monkeys TTS growth functions supports the suitability of the squirrel monkey as an animal model for noise-induced hearing loss in humans. (French abstract).

780. Oades, Robert D. (1982). **Search strategies on a hole-board are impaired in rats with ventral tegmental damage: Animal model for tests of thought disorder.** ***Biological Psychiatry,*** 17(2), 243–258.
For 5 days, 8 male hooded food-deprived rats with damage to the ventral tegmentum area (VTA) were given 9 sessions of 10 trials each on a 16-hole board. They searched for food pellets placed consistently in 4 holes. During testing, the control (C) group of 7 Ss reduced the number of empty holes visited more than the group with VTA damage. The proportion of repeated visits to relevant (food-containing) holes vs irrelevant (empty) holes increased for the C, but not for the VTA, group. The frequency with which a preferred sequence of food-hole visits was repeated during a session

increased over sessions for the C, but not the VTA, group. VTA Ss changed their preference between sessions more often. Ss with VTA damage were capable of simple learning, but were impaired when complexity increased. This may be due in part to a deficit in attention-related mechanisms. Results should lead to further study of the contribution of the VTA to attentional dysfunction and the use of the hole-board search task as a model for the study of cognitive function and dysfunction.

781. Olton, David S. (1983). **The use of animal models to evaluate the effects of neurotoxins on cognitive processes.** ***Neurobehavioral Toxicology & Teratology,*** 5(6), 635–640.
Describes testing procedures that are sensitive to hippocampal cholinergic function and to the type of memory that is often impaired in amnesic syndromes. Animal models are useful to determine the pathological changes underlying these behavioral symptoms and the psychological mechanisms responsible for them. Criteria for the development of productive animal models are presented. These are illustrated in the context of an animal model to evaluate changes in short-term memory that occur following damage to the CNS, particularly the hippocampus.

782. Owens, Michael J.; Overstreet, David H.; Knight, David L.; Rezvani, Amir H. et al. (1991). **Alterations in the hypothalamic-pituitary-adrenal axis in a proposed animal model of depression with genetic muscarinic supersensitivity.** ***Neuropsychopharmacology,*** 4(2), 87–93.
Determined concentrations of corticotropin-releasing factor (CRF) in the anterior pituitary and in various brain regions and plasma adrenocorticotropic hormone (ACTH) and corticosterone concentrations in rats from the Flanders Sensitive Line (FSL) and the Flanders Resistant Line (FRL). These lines have been bred for differences in sensitivity to cholinergic agonists. Cholinergically hypersensitive FSL Ss had lower concentrations of CRF in the median eminence, locus ceruleus, and prefrontal cortex, and FSL Ss had significantly lower plasma ACTH concentrations. In a 2nd study, the density of anterior pituitary CRF receptor binding sites was elevated in FSL Ss. FSL rats may have diminished hypothalamic-pituitary-adrenal activity.

783. Peinado-Manzano, Maria A. (1988). **Animal models of medial temporal lobe amnesia.** ***Medical Science Research,*** 16(18), 963–965.
Reviews the literature on the nature of medial temporal lobe (MTL) amnesia and consequently on the nature of memory processes, their structure and dynamics, and the role of the MTL in them. Temporal characteristics of MTL amnesic disorder, mechanisms involved, and sensory modality of memory impairment are outlined.

784. Pepeu, Giancarlo; Casamenti, Fiorella; Pedata, Felicita; Cosi, Cristina et al. (1986). **Are the neurochemical and behavioral changes induced by lesions of the nucleus basalis in the rat a model of Alzheimer's disease?** ***Progress in Neuro-Psychopharmacology & Biological Psychiatry,*** 10(3–5), 541–551.
Reviews work on the neurochemical, EEG, and behavioral changes induced in the rat by lesions of the nucleus basalis. Similarities and differences between the lesions' effects and the neurochemical and clinical alterations characterizing senile dementia of Alzheimer type (SDAT) are pointed out. It is concluded that lesions of the nucleus basalis only partly mimic the complex clinical picture of SDAT. They offer, nevertheless, a useful tool for understanding the critical role of central cholinergic pathways in cognitive processess and identifying potentially useful pharmacological treatments.

785. Pisa, Michele; Sanberg, Paul R. & Fibiger, Hans C. (1980). **Locomotor activity, exploration and spatial alternation learning in rats with striatal injections of kainic acid.** ***Physiology & Behavior,*** 24(1), 11–19.
Bilateral injections of kainic acid (3 nmoles) into the striatum induced temporary aphagia and adipsia, abnormal gait, acute and chronically recurrent seizures, interference with onset of ambulatory activity, and impairments in performance of instrumental spatial alternation in 20 male Wistar rats. Kainate injections often resulted in extrastriatal neuronal loss, most consistently involving the pyramidal neurons of the hippocampus. Similar regulatory, locomotor, and learning alterations were found both in kainate-treated Ss with combined striatal and hippocampal lesions and in those with no detectable hippocampal damage, suggesting that the striatal degeneration accounted for the behavioral impairments. Although striatal injections of kainic acid may fail to produce selective neuronal degeneration in the striatum, they result in behavioral disorders that appear to be similar, at least superficially, to those of patients with Huntington's disease, thus encouraging further use of this neurotoxin as a tool for reproducing some aspects of this disease.

786. Pitman, Roger K. (1989). **Animal models of compulsive behavior.** ***Biological Psychiatry,*** 26(2), 189–198.
Draws on animal research to explain how disturbances in the basal ganglia or limbic systems may produce compulsive behavior. Possible models include stimulation of the reinforcement mechanism, manipulation of the striatal "comparator" function, production and blockade of displacement behavior, and interference with the hippocampus' modulation of the stereotypy-inducing effect of reward. The common denominator of these models is a relative excess of dopaminergic activity in the basal ganglia.

787. Plioplys, Audrius V. & Bedford, H. Melanie. (1989). **Murine trisomy 16 model of Down's syndrome: Central nervous system electron microscopic observations.** ***Brain Research Bulletin,*** 22(2), 233–243.
Electron microscopic (EM) observations were made of the cortical plate in the developing telencephalic vesicle of mice at the gestational age of E17. Microtubular observations lend credence to the hypothesis that abnormal cytoskeletal interactions may underline the mental deficiency seen in Down's syndrome (DS) and may predispose to the eventual development of Alzheimer's disease (AD) in DS individuals. The cellular membrane findings may be related to reported central nervous system (CNS) membrane lipid abnormalities in DS. Nuclear morphologic data may relate to the reported differences in chromatin and nuclear histone expression in AD. Results strengthen the role of the trisomy 16 mouse as a model for DS and potentially for AD.

788. Pohl, Peter. (1984). **Ear advantages for temporal resolution in baboons.** ***Brain & Cognition,*** 3(4), 438–444.
Employed a gap detection task that required that resolution of brief silent intervals in bursts of noise to test the hypothesis that an ear advantage for a temporal resolution task should correlate precisely with an ear advantage for the discrimination of consonant-vowel syllables that differ in their temporal features. Four baboons, a 6-mo-old female, a

17-mo-old female, a 5-yr-old male, and a 7-yr-old male, served as Ss. The Ss were trained to perform a same–different discrimination task using a go/no-go procedure. Two measures of ear advantage, accuracy of discrimination and latency, were obtained for each ear separately. Findings offer support for the hypothesis and thus increase the feasibility of an animal model of functional asymmetry in the auditory system.

789. Port, Richard L.; Sample, John A. & Seybold, Kevin S. (1991). **Partial hippocampal pyramidal cell loss alters behavior in rats: Implications for an animal model of schizophrenia.** ***Brain Research Bulletin,*** 26(6), 993–996.
Evaluated acquisition of shuttlebox avoidance responses in male rats with partial damage to the hippocampus. Intraventricular microinjections of kainic acid were used to partially destroy the pyramidal cell population. Damaged Ss acquired the conditioned response (CR) at rates superior to controls. This facilitation of simple learning is consistent with the finding of faster acquisition in schizophrenic patients. Selective pyramidal cell loss due to intraventricular microjection of kainic acid appears to be a viable model of schizophrenia.

790. Poshivalov, V. P.; Dorokhova, L. N. & Sorokoumov, V. A. (1988). **Intraspecies behavior of animals based on a model of experimental brain ischemia.** ***Neuroscience & Behavioral Physiology,*** 18(5), 375–383.
Investigated the features of the intraspecies behavior of 84 male rats after ligation of the right, left, or both common carotid arteries. Findings show that certain characteristics of behavior change differently following ligation of the right or left common carotid artery. It is suggested that the approach used may be used for the assessment of the severity of ischemic brain damage.

791. Price, Donald L. et al. (1985). **The functional organization of the basal forebrain cholinergic system in primates and the role of this system in Alzheimer's disease.** ***Annals of the New York Academy of Sciences,*** 444, 287–295.
The authors discuss the anatomy and physiology of the basal forebrain cholinergic system in primates in terms of (1) the effects of aging on this system; (2) the consequences of experimental lesions of the primate basal forebrain; and (3) the changes occurring in this neuronal population in Alzheimer's disease, in aged individuals with Down's syndrome, and in certain demented patients with Parkinson's disease.

792. Richards, Sarah-Jane. (1991). **The neuropathology of Alzheimer's disease investigated by transplantation of mouse Trisomy 16 hippocampal tissues.** Special Issue: Transplantation in the nervous system. ***Trends in Neurosciences,*** 14(8), 334–338.
Evaluates the novel application of neural transplantation as a model for studying the neuropathological events associated with Alzheimer's disease and those that have subsequently also been observed in Trisomy 21 (Down's syndrome). The mouse Trisomy 16 model can be used in transplantation studies in an attempt to examine the neuropathological consequences of over-expression of the amyloid precursor protein, whose gene is located on murine chromosome 16 and human chromosome 21.

793. Rieke, Garl K. (1980). **Kainic acid lesions of pigeon paleostriatum: A model for study of movement disorders.** ***Physiology & Behavior,*** 24(4), 683–687.
Stereotaxically injected kainic acid (2.5 or .426 μg/μl) unilaterally into the paleostriatal complex of 22 *Columba livia* pigeons. Ss demonstrated behavioral disturbances similar to those induced in mammals. More specifically these included episodes of rapid rotation toward the side of the injection, involuntary movement, and postural disturbances. The pigeon may therefore serve as another useful experimental model system to assess behavioral and histochemical changes associated with movement disorders.

794. Riekkinen, Paavo; Miettinen, Riitta; Sirviö, Jouni; Aaltonen, Minna et al. (1990). **The correlation of passive avoidance deficit in aged rat with the loss of nucleus basalis choline acetyltransferase-positive neurons.** ***Brain Research Bulletin,*** 25(3), 415–417.
Investigated the effects of aging on the number of choline acetyltransferase (ChAT)-positive neurons in the nucleus basalis (NB) and the correlation between the number of ChAT-positive neurons and passive avoidance (PAV) retention in 5 young (3-mo-old) and 8 aged (26-mo-old) rats. The number of ChAT-positive neurons decreased in aged Ss, and the degree of loss of NB neurons was related to the degree of PAV retention deficit in aged Ss. Aged rats may provide a useful model for studying drugs aimed at reversing cognitive deficits related to age-associated cholinergic deficits in humans.

795. Robinson, Robert G. & Justice, Alan. (1986). **Mechanisms of lateralized hyperactivity following focal brain injury in the rat.** 15th Annual Meeting of the Society for Neuroscience: Locomotor behavior: Neuropharmacological substrates of motor activation (1985, Dallas, Texas). ***Pharmacology, Biochemistry & Behavior,*** 25(1), 263–267.
Describes recent work with rats that explored the differential behavioral and neurochemical change following left vs right cerebral damage and proposes neural mechanisms that may account for this asymmetry. The lateralized responses observed suggest that the neural mechanisms that mediate this phenomenon include both cortical and subcortical components. It is proposed that the neuroanatomical asymmetry is in either accumbal efferents or their postsynaptic connections.

796. Rodin, Barbara E. & Kruger, Lawrence. (1984). **Deafferentation in animals as a model for the study of pain: An alternative hypothesis.** ***Brain Research Reviews,*** 7(3), 213–228.
A literature review on neurotoxins does not support the notion that postdeafferentation autotomy is a pain response. The selective massive destruction of a fiber system considered essential to normal nociception, unmyelinated primary afferent axons, prior to deafferenting nerve lesions did not significantly impede postdenervation damage infliction (DI) despite human and animal evidence that pain after nerve lesions originates in the periphery and is generated by abnormal discharges in the injured nerve. When a reduction in abnormal impulse discharges of injured sensory axons could be inferred following neonatal sympathectomy, DI was not reduced. This observation supports a dissociation between DI and pain and suggests that DI also may be unrelated to nonpainful sensory pathology attributable to abnormal activity in the thick-diameter fiber population. It is concluded that DI may not be a manifestation of deafferentation pain and that perhaps this animal model for the experimental

study of pain should be discarded. An alternative view of DI is that it reflects a proclivity in some species and circumstances to shed a functionally impaired insensate appendage.

797. Roeltgen, David P. & Schneider, Jay S. (1991). **Chronic low-dose MPTP in nonhuman primates: A possible model for attention deficit disorder.** ***Journal of Child Neurology,*** 6(Suppl), S82–S89.
Examined similarities in cognitive performance and behavior between children with attention deficit disorder (ADD) and monkeys exposed to chronic low-dose *N*-methyl-4-phenyl-1,2,3,6-tetrahyrdropyridine (MPTP) via an operant behavior task (OBT) and several frontal lobe tasks (FLTs). After several days of MPTP administration, the monkeys showed typical parkinsonian motor signs and would no longer perform the OBT. On the FLTs, while cognitive deficits were observed Ss became more irritable and restless and appeared less attentive to the tasks. Ss appeared frustrated and became agitated at their inability to correctly perform delayed response and delayed alternation trials. Such behaviors are characteristic of children with ADD.

798. Rose, F. D.; Davey, M. J.; Love, S. & Attree, E. A. (1990). **Water maze performance in rats with bilateral occipital lesions: A model for use in recovery of function research.** ***Medical Science Research,*** 18(5), 167–169.
Describes a pattern discrimination training procedure used with lesioned and sham-operated rats that was fast and yielded a clear bilateral occipital lesion deficit. The training procedure involved placing an S in a water maze and measuring the time taken to reach a platform. Ss underwent 40 trials preoperatively and 15 trials postoperatively (starting 32–34 days after surgery) at the rate of 5 trials per day. Bilateral occipital lesions in this test situation caused a quantitative reduction in performance efficiency rather than a qualitative change in performance strategy.

799. Rose, F. D.; Dell, P. A.; Davey, M. J. & Love, S. (1987). **Hemidecortication in the rat as a model for investigating recovery from brain injury.** ***Medical Science Research: Psychology & Psychiatry,*** 15(1–4), 157–158.
Used preoperative and postoperative training to examine recovery of function in go–no go discrimination and reversal performance of 20 hemidecorticated Lister rats. Results demonstrate that performance is impaired in preoperatively, as well as postoperatively, trained Ss, and show a clear hemidecorticate impairment in discrimination and discrimination reversal.

800. Ryan, S. M.; Watkins, L. R.; Mayer, D. J. & Maier, S. F. (1985). **Spinal pain suppression mechanisms may differ for phasic and tonic pain.** ***Brain Research,*** 334(1), 172–175.
The dorsolateral funiculus (DLF) spinal pathway has previously been identified as a major pathway involved in descending modulation of pain. In the present study, while bilateral lesions of the DLF attenuated systemic morphine analgesia as measured by the tailflick test in 40 male Holtzman-derived rats, they failed to attenuate analgesia as measured by the formalin test. Results suggest that phasic and tonic pain, modeled by the tailflick and formalin tests, respectively, may use different pain suppression mechanisms.

801. Sagen, Jacqueline; Sortwell, Caryl E. & Pappas, George D. (1990). **Monoaminergic neural transplants prevent learned helplessness in a rat depression model.** ***Biological Psychiatry,*** 28(12), 1037–1048.
Serotonin-containing pineal gland tissue, catecholamine-containing adrenal medullary tissue, a combination of both, and a control of striated muscle tissue were implanted into the frontal neocortex of male rats. The monoamine-containing transplants, but not the control transplants, were able to prevent the development of learned helplessness, a model for depression. Immunocytochemical and ultrastructural studies revealed that the grafted monoaminergic tissues survived and continued to produce high levels of monoamines. Neural transplants may provide a long-term local source of monoamines as a potentially new approach for alleviating some forms of depression.

802. Salamone, John D. (1986). **Behavioural functions of nucleus basalis magnocellularis and its relationship to dementia.** ***Trends in Neurosciences,*** 9(6), 256–258.
Discusses the basal nucleus of Meynert's provision of a major cholinergic innervation to neocortex (consistently reduced in Alzheimer's disease) and its rat homolog—the nucleus basalis magnocellularis (NBM). Behavioral studies of NBM in rats show that the nucleus basalis is involved in the performance of tasks related to learning and memory. Lesions of this structure may provide an animal model for dementia and data on how the brain processes information and performs higher mental functions.

803. Sanberg, Paul R.; Giòrdano, Magda; Henault, Mark A.; Nash, David R. et al. (1989). **Intraparenchymal striatal transplants required for maintenance of behavioral recovery in an animal model of Huntington's disease.** ***Journal of Neural Transplantation,*** 1(1), 23–31.
Examined the issue of transplant integration in producing behavioral recovery. In Exp 1, kainic acid-lesioned rats with transplants located within the lateral ventricle were compared against parenchymally transplanted rats. Unless the ventricular transplant grew into the lesioned striatum, no recovery occurred. Exp 2 demonstrated that electrolytic destruction of a successful fetal striatal transplant (FST) could reverse the transplant-induced behavioral recovery. Results suggest that the integrity of the transplant is important in maintaining behavioral recovery. A continuing functional interaction between the host brain and transplanted tissue may be a vital element in the success of the FST.

804. Sanberg, Paul R.; Henault, Mark A.; Hagenmeyer-Houser, Starr H.; Giòrdano, Magda et al. (1987). **Multiple transplants of fetal striatal tissue in the kainic acid model of Huntington's disease: Behavioral recovery may not be related to acetylcholinesterase.** ***Annals of the New York Academy of Sciences,*** 495, 781–785.
Four wks after kainic acid lesions, male rats received injections of fetal Day 17 striatal ridge tissue along the ventral-dorsal plane into each striatum and were tested for locomotor behavior along with sham Ss. Hyperactivity exhibited on many aspects of ambulatory, stereotypical, and rotational behaviors by the striatal transplant group before transplantation and decreased gradually after transplantation in 7 Ss until they reached control levels. Histological results indicated that although cell bodies were distributed abundantly throughout the nontrypsinized transplanted material, few stained for acetylcholinesterase.

805. Sanberg, Paul R.; Henault, Mark A. & Deckel, A. Wallace. (1986). **Locomotor hyperactivity: Effects of multiple striatal transplants in an animal model of Huntington's disease.** 15th Annual Meeting of the Society for Neuroscience: Locomotor behavior: Neuropharmacological substrates of motor activation (1985, Dallas, Texas). ***Pharmacology, Biochemistry & Behavior,*** 25(1), 297–300.
Investigated the effects of multiple homotopic transplantations of normal fetal Day 17 striatal ridge tissue into the lesioned striatum of male kainic acid-treated rats. Nine weeks after transplantation, the spontaneous nocturnal hyperkinetic locomotor abnormalities were attenuated in the striatal transplanted animals. Striatal transplants reconstructed much of the gross morphology of the lesioned striatum. Findings are discussed in relation to the possible treatment of Huntington's disease.

806. Sanberg, Paul R.; Lehmann, John & Fibiger, Hans C. (1978). **Impaired learning and memory after kainic acid lesions of the striatum: A behavioral model of Huntington's disease.** ***Brain Research,*** 149(2), 546–551.
Studied the effects of kainic acid-induced lesions of the caudate-putamen (CP) on locomotor activity and on passive avoidance learning and memory (step-down apparatus) in male Wistar rats to evaluate the usefulness of this procedure as a behavioral and a biochemical model of Huntington's disease (HD). 11 Ss received bilateral stereotaxic injections of kainic acid into the CP. 12 controls received saline injections. Compared to the control group, the lesioned Ss took significantly longer to reach the learning criterion and also stepped off the platform significantly more times before reaching criterion. Biochemical changes that were observed in the CP after intrastriatal kainic acid injections were qualitatively similar to those found in HD. Results suggest that the behavioral changes in HD are related to the pathology identified in the CP of HD patients.

807. Sara, Susan J. (1989). **Noradrenergic–cholinergic interaction: Its possible role in memory dysfunction associated with senile dementia.** Symposium: Memory and aging (1988, Lausanne, Switzerland). ***Archives of Gerontology & Geriatrics,*** 1, 99–108.
Contends that, over the past 10 yrs, the cholinergic hypothesis of memory dysfunction associated with degenerative diseases of aging has been the dominant one. Lack of efficacy of direct or indirect cholinergic drugs has led to research in other neurotransmitter systems. Miscellaneous results in noradrenergic (NE) and metabolite levels and in monoamine enzyme activities have led to discussions about the homogeneity of senile dementia. A series of experiments was performed with rats in which hippocampal acetylcholine (ACh) activity was reduced (fornix section or destruction of cells in the medial septum) and the NE system modified by clonidine or neurotoxic lesions. Results support the hypothesis that a lesion of the cholinergic pathway leads to an enhancement of NE activity that inhibits spared ACh neurons.

808. Schmajuk, Nestor A. (1987). **Animal models for schizophrenia: The hippocampally lesioned animal.** ***Schizophrenia Bulletin,*** 13(2), 317–327.
Evaluates how animals with hippocampal lesions meet the requirements of a model of schizophrenia. Such models seem to comply with the following criteria: (1) similarity of inducing conditions, (2) similarity of behavioral states, (3) similarity of underlying neurobiological mechanisms, and (4) reversibility by usual pharmacological treatment. It is shown that animals with bilateral lesions exhibit many of the behavioral and biological characteristics of schizophrenic patients.

809. Seltzer, Ze'ev; Dubner, Ronald & Shir, Yoram. (1990). **A novel behavioral model of neuropathic pain disorders produced in rats by partial sciatic nerve injury.** ***Pain,*** 43(2), 205–218.
Presents an animal model of partial nerve injury and causalgiform pain disorders in humans. After unilateral ligation of the sciatic nerve in the thigh, male rats developed guarding behavior of the ipsilateral hind paw and licked it often, suggesting spontaneous pain. The plantar surface of the foot was evenly hyperesthetic to non-noxious and noxious stimuli. None of the Ss autotomized. Withdrawal thresholds decreased bilaterally in response to repetitive Von Frey hair stimulation at the plantar side and to CO_2 laser heat pulses. Pin-prick evoked mechanical hyperalgesia bilaterally. The immediate onset and perpetuation of touch-evoked allodynia and hyperalgesia and the resemblance of the contralateral phenomena to "mirror image" pains in some humans with causalgia suggest that this may serve as a model.

810. Shimizu, K.; Yamada, M.; Matsui, Y.; Tamura, K. et al. (1990). **Neural transplantation in mouse Parkinson's disease.** Xth Meeting of the World Society for Stereotactic and Functional Neurosurgery (1989, Maebashi, Japan). ***Stereotactic & Functional Neurosurgery,*** 54–55, 353–357.
A complete recovery from the methamphetamine-induced rotational response was shown in C57BL/6 (H-2^b) mice that had had unilateral lesions in the nigrostriatal pathway about 60 days after transplantation of dopamine-rich cells from syngeneic or allogeneic mouse embryos without immunosuppressive agents. Morphological examination showed tyrosine-hydroxylase-immunoreactive cell clusters around the needle tract in Ss that were transplanted with syngeneic and allogeneic cells.

811. Shiosaka, S.; Yamano, M.; Tsuchiyama, M.; Kudo, Y. et al. (1991). **Catecholamine and acetylcholine in the rat cerebral cortex with special reference to pathogenetic mechanisms of Alzheimer's disease.** Annual Meeting of the Japan Gerontological Society (1990, Kochi, Japan). ***Gerontology,*** 37(Suppl 1), 17–23.
A direct synapse between catecholamine fibers and neuropeptide Y-containing neurons was demonstrated in rat cerebral cortex using an immunohistochemical double-staining method under the electron microscope. A new method to produce a selective reduction in cholingergic neurons in the basal forebrain without damage to the noncholinergic neurons, passing fibers or other cholinergic systems, is described. This animal model seems to be useful to analyze the pathogenesis of Alzheimer's disease and to examine the function of cholinergic neurons of the basal forebrain. In Ss that received an injection of the toxin into the cerebral cortex, acquisition of learning and retention were significantly decreased in sham-operated Ss.

812. Shor-Posner, Gail; Azar, Anthony P.; Insinga, Salvatore & Leibowitz, Sarah F. (1985). **Deficits in the control of food intake after hypothalamic paraventricular nucleus lesions.** ***Physiology & Behavior,*** 35(6), 883–890.

To determine the influence of the hypothalamic paraventricular nucleus (PVN) in monitoring and controlling responses to physiological and pharmacological challenges, PVN electrolytic lesioned male Sprague-Dawley rats were tested for their behavioral responsiveness to agents known to affect the alpha-2 noradrenergic system as well as release of corticosterone, and to short- and long-term periods of food deprivation. Results show that discrete lesions of the PVN produced enhanced feeding, particularly of carbohydrate, in freely feeding Ss maintained on a macronutrient self-selection paradigm. Lesioned Ss demonstrated a behavioral deficit in food intake regulation (a decrease in carbohydrate ingestion) in response to 5- and 24-hr fasts, showed a disturbance in circadian feeding, and exhibited a dramatic decrease in circulating corticosterone. However, feeding in response to 2-deoxy-dextro-glucose and insulin remained intact, suggesting that noradrenergic receptors within the PVN are not involved in the mediation of glucoprivic-induced feeding.

813. Simonov, P. V. (1989). **Experimental neuropsychology and its importance for human brain research.** ***Human Physiology,*** 15(3), 161–167.
Examined the extraversion and introversion of 7 rats by comparing the effectiveness of 2 aversive stimuli: increased intensity of illumination and tonal acoustic stimulation or signals of the partners' defensive excitation. Five rats were retested after bilateral coagulation of the frontal zone of the neocortex and hippocampus. Data suggest that individual features of relations between informational systems (frontal cortex and hippocampus) and the motivational system (amygdala and hypothalamus) were parameters of extraversion and introversion. The relation between frontal cortical hypothalamic and amygdalo-hippocampal systems in the manifestation of emotional stability is demonstrated.

814. Sonsalla, Patricia K. & Heikkila, Richard E. (1986). **"Animals and experimentation: An evaluation of animal models of Alzheimer's and Parkinson's disease": Commentary.** ***Integrative Psychiatry,*** 4(2), 70–72.
In response to J. H. Kordower and D. M. Gash's (1986) article, the present authors discuss the use of 1-methyl-4-phenyl-1,2,3,6-tetrahydropyridine (MPTP) in creating animal models for Parkinson's disease (PD). The available information on the effects of MPTP in various experimental animals is summarized, and mechanisms involved in its neurotoxic actions are described. The impact of MPTP on basic and clinical research associated with PD is considered.

815. Squire, Larry R. (1986). **Mechanisms of memory.** ***Science,*** 232(4758), 1612–1619.
Discusses the organization of memory in the brain and the recent development of an animal model of human amnesia in the monkey. Topics addressed include distributed and localized memory storage, the neuropsychological-neural systems approach, short- and long-term memory, declarative and procedural knowledge, and memory consolidation and retrograde amnesia. It is suggested that the animal model, together with newly available neurological information from a well-studied human amnesiac patient, has permitted the identification of brain structures and connections involved in memory functions. It is emphasized that although animal studies are essential, they cannot illuminate the clinical significance of the observed memory impairments unless the severity of the impairments can be understood in terms of human memory dysfunction.

816. Squire, Larry R. & Zola-Morgan, Stuart. (1985). **The neuropsychology of memory: New links between humans and experimental animals.** ***Annals of the New York Academy of Sciences,*** 444, 137–149.
The authors discuss the neuropsychology of memory in terms of (1) the concept of memory consolidation; (2) the idea that there are 2 types of memory, only one of which is affected in amnesia; and (3) the question of which structures in the medial temporal region must be damaged to produce the strikingly selective deficit that is termed amnesia. Amnesia is characterized by an impaired ability to acquire new information and by difficulty remembering at least some information that was acquired prior to the onset of amnesia. It is concluded that the availability of animal models for the clinical syndrome of amnesia is a cause for both optimism and excitement. The way is now clear for identifying the structures, both in the medial temporal region and in the diencephalic midline, that cause amnesia.

817. Squire, Larry R. & Zola-Morgan, Stuart. (1991). **The medial temporal lobe memory system.** ***Science,*** 253(5026), 1380–1386.
Studies of human amnesia and studies of an animal model of human amnesia in the monkey have identified the anatomical components of the brain system for memory in the medial temporal lobe (MTL) and have illuminated its function. This neural system consists of the hippocampus and adjacent anatomically related cortex, including entorhinal, perirhinal, and parahippocampal cortices. These structures are essential for establishing long-term memory for facts and events (declarative memory). The MTL memory system is needed to bind together the distributed storage sites in neocortex that represent a whole memory. However, the role of this system is only temporary. As time passes after learning, memory stored in neocortex gradually becomes independent of MTL structures.

818. Strupp, Barbara J. & Levitsky, David A. (1990). **An animal model of retarded cognitive development.** ***Advances in Infancy Research,*** 6, 149–185.
Discusses the need for an animal model (AM) of retarded cognitive (COG) development and suggests that the apparent disparity in the degree of COG impairment between AMs and the analogous human condition should not be interpreted as an indictment of the use of AMs. The methodological approach used to assess COG functioning in animals may have led to an underestimation of their COG pathology. An alternative methodological and conceptual framework addresses (1) the failure to design COG tests in accord with the COG profile of mentally retarded humans, (2) the induction of heightened motivational states to assess learning ability, and (3) the assessment of COG functioning on the basis of the animal's ability to solve a single, isolated learning task rather than to accumulate learning.

819. Sutton, Richard L.; Fox, Robert A. & Daunton, Nancy G. (1988). **Role of the area postrema in three putative measures of motion sickness in the rat.** ***Behavioral & Neural Biology,*** 50(2), 133–152.
After thermal cauterization of the area postrema (AP) in rats, the absence of conditioned taste aversion (CTA) to sucrose paired with lithium chloride was used as an index of AP damage. The effects of AP lesions on 3 measures proposed as species-relevant measures of motion sickness were also studied, using off-vertical rotation of 150°/sec for either 30 or 90 min. The initial acquisition of CTA to a novel

solution paired with motion was not affected by AP lesioning, but these CTAs extinguished more slowly in lesioned Ss than in sham-operates or controls. Results are discussed in terms of proposed humoral factors that may induce motion sickness.

820. Taghzouti, K.; Simon, H.; Hervé, D.; Blanc, G. et al. (1988). **Behavioural deficits induced by an electrolytic lesion of the rat ventral mesencephalic tegmentum are corrected by a superimposed lesion of the dorsal noradrenergic system.** ***Brain Research,*** 440(1), 172–176.
Bilateral electrolytic lesion of the ventral mesencephalic tegmentum (VMT) induced, in the rat, behavioral deficits such as locomotor hyperactivity and disappearance of spontaneous alternation ("VMT syndrome"). When a specific 6-hydroxydopamine (6-OHDA) destruction of the dorsal noradrenergic ascending pathway was superimposed to an electrolytic lesion of the VMT, Ss recovered a normal locomotor activity and the possibility to alternate. Relevance to the etiology of Gilles de la Tourette's syndrome and Parkinson's disease is noted.

821. Thompson, Robert; Bjelajac, Victor M.; Huestis, Peter W.; Crinella, Francis M. et al. (1989). **Inhibitory deficits in rats rendered "mentally retarded" by early brain damage.** ***Psychobiology,*** 17(1), 61–76.
Determined whether weanling, male rats with lesions to the parietal cortex, globus pallidus, median raphe, hippocampus, or amygdala would subsequently manifest a global deficit in behavioral inhibition and whether this inhibitory deficit would correlate with learning ability. The parietal and median raphe groups showed losses on at least 6 of 7 tests of inhibition, whereas the remaining groups showed losses on no more than 3 tests. However, the correlations between inhibition scores and learning scores did not exceed .43. Two factors were extracted from the data on the inhibition tests: The first (associated with learning ability) appeared to underlie the suppression of an initially preferred response in the face of nonreinforcement or punishment, and the second appeared to underlie the suppression of an approach response to the negative cue in approach–avoidance and go-no-go discrimination situations.

822. Thompson, Robert; Huestis, P. W.; Crinella, F. M. & Yu, J. (1987). **Further lesion studies on the neuroanatomy of mental retardation in the white rat.** ***Neuroscience & Biobehavioral Reviews,*** 11(4), 415–440.
Assessed the learning ability of groups of young rats with bilateral lesions to various brain regions (e.g., nucleus accumbens, ventral pallidum, inferior colliculus, or red nucleus). Tests included appetitively and aversively motivated learning tasks. Ss with lesions to the posterodorsal caudatoputamen, ventral tegmental area of Tsai, or superior colliculus were deficient in acquiring the problems, suggesting a generalized learning impairment. Findings are interpreted within a framework based on C. Spearman's (1927) 2-factor theory of intelligence. A brain-injured animal model of mental retardation is also discussed.

823. Thompson, Robert & Yu, Jen. (1983). **Specific brain lesions producing nonspecific (generalized) learning impairments in weanling rats.** ***Physiological Psychology,*** 11(4), 225–234.
Recent studies have shown that lesions of the globus pallidus, lateral thalamus, ventrolateral thalamus, parafascicular thalamus, substantia nigra, or midbrain central gray area in adult rats are associated with a nonspecific generalized learning impairment. The present study with 106 weanling male albino Sprague-Dawley rats showed that lesions to any of these structures, except the ventrolateral and parafascicular thalami, also produced a generalized learning impairment as shown by significant deficits on a white–black discrimination, a nonvisual 11-degree-incline plane discrimination, and a 3-cul maze. A brain-injured animal model of mental retardation is outlined.

824. Torigoe, Ryuichiro; Yoshida, Masafumi; Ohtsuru, Katsuyasu; Kuga, Shigehi et al. (1990). **Adrenal medullary graft to brain in 6-hydroxydopamine-lesioned rotation rats: Comparison of two different graft sites (striatum versus lateral ventricle) by behavioral observation and fluorescence histochemistry.** Xth Meeting of the World Society for Stereotactic and Functional Neurosurgery (1989, Maebashi, Japan). ***Stereotactic & Functional Neurosurgery,*** 54–55, 347–352.
Using 6-hydroxydopamine-lesioned rotation rats as a model of Parkinson's disease, an investigation was made of the optimal site for transplantation of adrenal medulla. Six weeks after grafting, the rotational behavior in the striatum and ventricle group was reduced by 43% and 35%, respectively, as compared to the control group. The striatum was slightly superior to the lateral ventricle as the site for transplantation.

825. Treit, Dallas & Pesold, Christine. (1990). **Septal lesions inhibit fear reactions in two animal models of anxiolytic drug action.** ***Physiology & Behavior,*** 47(2), 365–371.
Studied the role of the septum in anxiety using 2 different animal models of antianxiety drug action (the shock probe-burying test and the elevated plus maze test). Antianxiety effects were observed in both paradigms (a decrease in probe burying and an increase in open arm activity) after lesions of the entire septum in rats compared with sham-lesioned controls (Exp I). No differences between lesioned and sham-lesioned Ss were found in general activity, shock reactivity, or handling reactivity at the time of the antianxiety tests. Exp II showed that the antianxiety effects observed in the 2 paradigms were anatomically specific. Results provide convergent evidence that posterior regions of the septum play an important role in the control of anxiety in the rat.

826. Tsainer, B.; Orlova, E. I. & Belichenko, P. V. (1991). **On the biochemical aspects of the pathogenesis of vegetative and emotional disorders in dysfunction of the dopamine system.** ***Zhurnal Nevropatologii i Psikhiatrii imeni S.S. Korsakova,*** 91(12), 39–43.
Studied the synthesis and decomposition of choline acetyltransferase and acetylcholinesterase in the hypothalamus and nucleus of the solitary tract under normal conditions and conditions of experimental activation of the dopamine system in male Wistar rats. L-dopa was administered 3 times. Emotional-behavioral and vegetative responses were compared in Ss with high and low levels of motor activity. (English abstract).

827. Van Petten, Cyma; Roberts, William J. & Rhodes, Dell L. (1983). **Behavioral test of tolerance for aversive mechanical stimuli in sympathectomized cats.** ***Pain,*** 15(2), 177–189.

Developed a device and methodology that allowed the humane testing of tolerance for intense mechanical stimulation of the hindlegs in 6 adult cats. Behavioral tolerance was measured quantitatively before and after unilateral sympathectomy. Results are remarkably similar to those reported for humans. It appears that the new methodology provides a relatively stable, quantitative measure of tolerance for aversive stimulation, and the cat shows promise as an animal model for postsympathectomy hyperalgesia.

828. Weiss, Jay M. et al. (1981). **Behavioral depression produced by an uncontrollable stressor: Relationship to norepinephrine, dopamine, and serotonin levels in various regions of rat brain.** ***Brain Research Reviews,*** 3(2), 167–205.
Conducted 2 experiments to determine the neurochemical changes responsible for stress-induced behavioral depression in 162 male Holtzman albino rats. The experiments measured active motor behavior in a swim tank as well as levels of norepinephrine (NE), dopamine (DA), and serotonin (5-HT) in various brain regions of Ss after they had (a) been exposed to electric shocks they could control (avoidance-escape condition), (b) received the same shocks with no control over them (yoked condition), or (c) received no shock (no-shock condition). Results suggest that large stress-induced depletion of NE in the locus coeruleus (LC) is involved in mediating behavioral depression brought on by severe stress. It is further suggested that the time course for behavioral recovery and for the disappearance of NE depletion in the LC that was seen in yoked Ss after stress parallels the time course previously reported by other investigators for induction of catecholamine-synthesizing enzymes—tyrosine hydroxylase (TH) and dopamine-beta-hydroxylase (DBH)—in the LC, so that induction of TH and DBH activity may be a neurochemical mechanism to bring about recovery from poststress behavioral depression.

829. Wenk, Gary L. (1990). **Animal models of Alzheimer's disease: Are they valid and useful?** XXXI International Congress of Physiological Sciences: Recovery from brain damage: Behavioral and neurochemical approaches (1989, Warsaw, Poland). ***Acta Neurobiologiae Experimentalis,*** 50(4–5), 219–223.
Reviews the usefulness and validity of animal models of Alzheimer's disease (AD) designed to reproduce various components of the pathological, biochemical, and behavioral characteristics of AD. Animal models of AD involve the production of lesions in the nucleus basalis magnocellularis to reproduce the loss of basal forebrain cells found in AD. Animal models of AD have provided much information on the function of the basal forebrain system and have been used to investigate the potential effectiveness of various pharmacotherapies designed to reverse specific symptoms. The loss of cholinergic cells may not be sufficient to produce AD.

830. Woodruff-Pak, Diana S. (1988). **Aging and classical conditioning: Parallel studies in rabbits and humans.** Special Issue: Experimental models of age-related memory dysfunction and neurodegeneration. ***Neurobiology of Aging,*** 9(5–6), 511–522.
Explores how aging in the neural circuitry affects associative learning and memory (i.e., classical conditioning [CC]) over the life span. Parallels between humans and animals in aging and CC of the eyelid response are identified. Research indicates that the loss of Purkinje cells with age is one of the cerebellar age changes likely to affect CC in aged mammals. It is suggested that the model system of CC is of demonstrated utility in extending understanding of the neurobiology of learning, memory, and aging in humans as well as animals.

831. Zbinden, G. (1989). **Neurobehavioral toxicology.** ***Farmakologiya i Toksikologiya,*** 52(4), 5–9.
Discusses the use of 2 experimental methods in neurobehavioral toxicology: models with animals simulating human disorders, and qualitative analysis of behavioral elements and changes that chemical substances cause in these elements. Focus is on J. Bureš' model of an organism's dynamic adaptation to the environment. Experiments concerning the development of behavioral habits in rats and the effects of alcohol, caffeine, and methylmercury on the habits show that neurobehavioral changes in models of normal behavior can be observed even in small laboratory animals. (English abstract).

832. Ziegler, Dewey K. (1986). **"Animals and experimentation: An evaluation of animal models of Alzheimer's and Parkinson's disease": Commentary.** ***Integrative Psychiatry,*** 4(2), 73.
Discusses J. H. Kordower and D. M. Gash's (1986) review of the contributions of animal research to the study of Alzheimer's and Parkinson's disease and the problems associated with conclusions drawn from these studies. Emphasis is on the difficulty of studying these disorders in primates.

833. Zigmond, Michael J.; Abercrombie, Elizabeth D.; Berger, Theodore W.; Grace, Anthony A. et al. (1990). **Compensations after lesions of central dopaminergic neurons: Some clinical and basic implications.** Special Issue: Basal ganglia research. ***Trends in Neurosciences,*** 13(7), 290–296.
Examines why the neurological (NEU) symptoms of Parkinson's disease (PD) do not emerge until the degeneration of the dopaminergic (DAergic) component of the nigrostriatal pathway is nearly complete. A comparable phenomenon observed in animal models is discussed, in which PD is produced by the administration of 6-hydroxydopamine (6-OHDA). Studies using such models suggest that the loss of DAergic neurons is compensated by increased synthesis and release of DA from remaining DA neurons and by a reduced rate of DA inactivation. Implications include (1) synaptic homeostasis possibly prolonging the preclinical phase of NEU disorders and (2) some disorders representing regulatory deficits rather than classical degenerative processes.

834. Zola-Morgan, Stuart & Squire, Larry R. (1985). **Medial temporal lesions in monkeys impair memory on a variety of tasks sensitive to human amnesia.** ***Behavioral Neuroscience,*** 99(1), 22–34.
Relative to 3 unoperated controls, 4 cynomolgus monkeys with conjoint bilateral lesions of the hippocampus and amygdala were impaired on 4 tests of memory—delayed retention of object discriminations, concurrent discrimination, delayed response, and delayed nonmatching to sample. In 3 of the tasks, relatively long-delay intervals between training and test trials were used, and in 2 tasks, distraction was introduced during the delay intervals. The severity of the impairment increased with the length of the delay, and distraction markedly increased the memory impairment. For 1 task given on 2 occasions (delayed nonmatching to sample), the severity of the impairment was unchanged over 1.5 yrs. It is concluded that monkeys with medial temporal lesions constitute an animal model of human amnesia and

that the 4 tasks used in the present study appear to constitute a sensitive and appropriate battery that could be used in other studies of the neuroanatomy of memory functions in the monkey.

Electrophysiology

835. Adamec, R. E. & Stark-Adamec, Cannie. (1985). **Kindling and interictal behavior: An animal model of personality change.** Kindling and Clinical Psychiatry Conference (1983, Toronto, Canada). ***Psychiatric Journal of the University of Ottawa,*** 10(4), 220–230.
Reviews research concerning the impact of repeated experimental limbic seizures on interictal behavior. Two types of functional traces left by seizures (limbic permeability and dynamic process traces) are described. It is argued that repeated limbic seizures in animals enhance the normal functioning of some limbic circuits interictally while simultaneously attenuating the function of others. Alterations in normal limbic function may be mediated interictally by interictally maintained functional traces. The behavioral effects of seizure-induced functional traces, in some cases, appear as an alteration of personality traits. The clinical implications of the animal findings are discussed within the context of research investigating the interictal behavioral and electrophysiological effects of seizure disorders in humans and the effects of subconvulsant electrical stimulation of the human limbic system.

836. Adrien, Joëlle; Dugovic, Christine & Martin, Patrick. (1991). **Sleep–wakefulness patterns in the helpless rat.** ***Physiology & Behavior,*** 49(2), 257–262.
Studied sleep–wakefulness patterns during a 15-day period in relation to induction of helplessness in male rats. After a session of inescapable electric footshocks, Ss exhibited escape deficits in avoidance conditioning as classically described, and their spontaneous sleep–wakefulness patterns were not different from those of controls. However, reduced paradoxical sleep (PS) latency and increased PS amounts were observed in the helpless group after shuttle-box sessions, especially during the initial period after the induction of helplessness. Such modifications of PS latency and PS amounts are evocative of the sleep impairments classically observed in endogenous depression.

837. Ault, Brian; Caviedes, Pablo & Rapoport, Stanley I. (1989). **Neurophysiological abnormalities in cultured dorsal root ganglion neurons from the trisomy-16 mouse fetus, a model for Down syndrome.** ***Brain Research,*** 485(1), 165–170.
Examined the electrical membrane properties of cultured dorsal root ganglion (DRG) neurons from trisomy-16 (Ts16) mice and matched controls, using the patch clamp technique. Ts16 neurons had significantly accelerated rates of action potential depolarization and repolarization compared with normal, diploid neurons, resulting in decreased spike duration. These changes resemble those reported in human trisomy-21 DRG neurons, suggesting that the Ts16 mouse is an appropriate model of Down syndrome (DS). It is concluded that impairment of cellular functions associated with decreased spike duration may contribute to the mental retardation associated with DS.

838. Balleine, Bernard & Job, R. Soames. (1991). **Reconsideration of the role of competing responses in demonstrations of the interference effect (learned helplessness).** ***Journal of Experimental Psychology: Animal Behavior Processes,*** 17(3), 270–280.
In Exp 1a, rats trained to escape shock by performing a 2-sec inactive response were less impaired on a subsequent 2-way shuttle response than their yoked counterparts that received inescapable shock. In contrast, in Exp 1b, rats trained to escape shock by performing a longer duration inactive response were more impaired on the subsequent escape task than their inescapably shocked counterparts. In Exp 2, the results of Exps 1a and 1b were replicated, and the inactive responses performed during pretreatment by both the escapable and inescapable shock groups were assessed and correlated with test stage 2-way shuttle escape performance. These activity data indicate that inactivity during pretreatment shock in both escapable and inescapable shock groups was a highly reliable predictor of subsequent 2-way shuttle performance, irrespective of the pretreatment shock contingency to which these Ss were exposed.

839. Berridge, Kent C. & Valenstein, Elliot S. (1991). **What psychological process mediates feeding evoked by electrical stimulation of the lateral hypothalamus?** ***Behavioral Neuroscience,*** 105(1), 3–14.
Because electrical stimulation of the lateral hypothalamus (ESLH) can elicit both feeding and reward, most investigators have concluded that stimulation does not evoke the aversive cues associated with hunger. It has been hypothesized, instead, that ESLH primes ingestion by evoking pleasurable taste sensations. A direct test of this hedonic hypothesis was undertaken in rats that showed stimulus-bound feeding. Contrary to the prediction, it was found that the taste reactions (gapes, tongue protrusions) during ESLH were more aversive than hedonic. It is suggested that the stimulation influences behavior by potentiating the salience, but not the hedonic value, of external stimuli. The advantages of this incentive salience hypothesis are that it circumvents the need to postulate a hedonic sensory experience during stimulation and that it can explain how evoked feeding may switch to other behaviors when conditions are altered.

840. Berridge, Kent C. & Zajonc, Robert B. (1991). **Hypothalamic cooling elicits eating: Differential effects on motivation and pleasure.** ***Psychological Science,*** 2(3), 184–189.
Examined the attractiveness and pleasure of food to 17 male rats during hypothalamic cooling or hypothalamic heating by measuring feeding and hedonic and aversive facial reactions to taste in 2 experiments. Hypothalamic cooling administered through a permanently implanted thermode elicited feeding; hypothalamic heating did not. Hedonic and aversive reactions to taste were unaltered by hypothalamic cooling or heating, even for Ss that ate during cooling.

841. Brett, C. W.; Burling, Thomas A. & Pavlik, W. B. (1981). **Electroconvulsive shock and learned helplessness in rats.** ***Animal Learning & Behavior,*** 9(1), 38–44.
Three experiments evaluated the effects of a single ECS in alleviating the learned helplessness effect in 152 Long-Evans male hooded rats. In Exp I, ECS was given following helplessness training and testing and was evaluated during a retesting phase; in Exp II, ECS was given either immediately after helplessness training or immediately before helplessness testing; and, in Exp III, ECS was given prior to helplessness training. In all 3 experiments, significant helplessness effects

occurred for Ss not receiving ECS but were absent in Ss receiving ECS. Data are compared with expectations arising from both amnesia-inducing and biochemical-change interpretations of the effects of ECS. Results lend support to the learned helplessness (LH) model of human depression by showing that an effective therapeutic intervention with depression (ECS) also functions under some conditions to alleviate the LH effect in laboratory Ss.

842. Brown, Gary E.; Hughes, Gary D. & Jones, Andrew A. (1988). **Effects of shock controllability on subsequent aggressive and defensive behaviors in the cockroach (*Periplaneta americana*).** ***Psychological Reports,*** 63(2), 563–569.
Male cockroaches were exposed to either escapable, inescapable, or no shock in an escape task for 3 consecutive days. 24 hrs later, these Ss were placed individually in an aquarium with a naive S and the frequency of aggressive behavior and defensive behavior was recorded by an observer blind to assignment. The inescapable shock group of Ss displayed less aggressive behavior and a greater tendency to retreat from social encounter than did the escapable shock or no shock groups.

843. Brown, Gary E.; Howe, Angela R. & Jones, Thomas E. (1990). **Immunization against learned helplessness in the cockroach (*Periplaneta americana*).** ***Psychological Reports,*** 67(2), 635–640.
24 American cockroaches were assigned to an immunization group, an inescapable shock-group, or a no-shock control group to assess their response to inescapable shock with or without prior exposure. Ss exposed to 1 day of escapable shock prior to 3 days of inescapable shock did not become helpless on a shuttlebox-escape task. Like dogs and rats, cockroaches are immunized against learned helplessness by prior experience with escapable shock.

844. Carlson, Jeffrey N. & Glick, Stanley D. (1991). **Brain laterality as a determinant of susceptibility to depression in an animal model.** ***Brain Research,*** 550(2), 324–328.
Assessed differences in the ability to respond to stress and in susceptibility to learned helplessness between left- and right-rotating male rats. Ss having different directional biases of brain laterality, as indicated in tests of rotational behavior, differed greatly in their response to stressors and to the lack of stressor control. Differences in brain laterality appear to be an important source of variability within the animal model of depression. As with humans, only some rats are vulnerable to depression-like symptoms. Findings are relevant to biological theories of depression that are based on lateralized specialization of the human brain for affect.

845. Chen, B. M. & Buchwald, J. S. (1986). **Midlatency auditory evoked responses: Differential effects of sleep in the cat.** ***Electroencephalography & Clinical Neurophysiology: Evoked Potentials,*** 65(5), 373–382.
Middle latency responses (MLRs) in the 10–100 msec latency range, evoked by click stimuli, were studied in 8 adult cats during sleep/wakefulness to determine whether such changes in state were reflected by any MLR component. Data lend support to a functional relation between "wave A" and the ascending reticular activating system and suggest that this potential may provide a unique and dynamic probe of tonic brain activity. Moreover, this animal model provides a hypothetical basis for expecting a similar surface recorded potential in the human, a potential which has subsequently been discovered. (French abstract).

846. Chen, Jaw-sy & Amsel, Abram. (1977). **Prolonged, unsignaled, inescapable shocks increase persistence in subsequent appetitive instrumental learning.** ***Animal Learning & Behavior,*** 5(4), 377–385.
Studied a total of 100 albino Holtzman rats in 3 experiments. In Exp I, a prolonged period of intermittent, unsignaled shocks preceded appetitive runway acquisition, under either continuous (CRF) or partial reinforcement (PRF) and extinction. In Exp II, the shock treatment came between CRF or PRF acquisition and extinction; and in the 3rd experiment, the shocks intervened between appetitive CRF acquisition and shock-punishment extinction. The main finding was that compared with an unshocked control, shock facilitated acquisition in Exp I and led to increased resistance to extinction and/or punishment in all experiments. In Exp I, the shock effect in appetitive extinction was seen mainly in the CRF group; in Exp II, the effect was to increase persistence in both the CRF and PRF groups; and in Exp III, shock treatment produced stronger resistance to punished extinction. The discussion is in terms of habituation and a general theory of persistence, and the concept of helplessness.

847. Cotton, M. M. & Smith, G. M. (1990). **Prenatal stress and learned helplessness.** ***Australian Journal of Psychology,*** 42(1), 47–55.
18 prenatally stressed rats and 18 nonstressed controls were given prior avoidance experience in a Skinner box before being trained in a 2-way shuttlebox avoidance task within a standard triadic learned helplessness (LH) paradigm. Results indicate no effect of prenatal stress on the acquisition of a lever press in a Skinner box, but prenatal stress had significant adverse effects on the acquisition of a more difficult 2-way shuttlebox avoidance response. The phenomenon of LH was demonstrated for nonstressed but not for prenatally stressed Ss.

848. Cubero Talavera, Inmaculada & Maldonado López, Antonio. (1988). **The leap-into-the-air response as a technique for measuring learned helplessness following incontrollable and predicatable aversive stimuli.** ***Psicológica,*** 9(1), 11–25.
Studied learned helplessness effects on a sample of laboratory rats divided into 3 groups of 8 Ss each. In pretraining, one group received escapable and predictable shocks ("escapable group"); the second received the same number of predictable, but inescapable, shocks (yoked inescapable group); the 3rd (control) group received no training at that stage. The actual test (a discriminating avoidance task) involved the leap-into-the-air response technique. Study results included proof of the unmistakable effects of interference with the subsequent avoidance training of the inescapable group, compared with the 2 other groups. (English abstract).

849. Dess, Nancy K. & Chapman, Clinton D. (1990). **Individual differences in taste, body weight, and depression in the "helplessness" rat model and in humans.** Summer Symposium: Brain, behavior and stress (1989, Los Angeles, California). ***Brain Research Bulletin,*** 24(5), 669–676.
In Exp 1, exposure of rats to unsignaled, inescapable shock resulted in finickiness about drinking a weak quinine solution. In contrast, exposure to escapable shock resulted in marked individual differences in finickiness that were predicted by prestress body weight. A more sensitive index of finickiness was used in Exp 2, and a correlation between body weight and finickiness was observed in nonshocked

rats. In Exp 3, measures of quinine reactivity and body weight predicted depressive symptomatology in a nonclinical human sample of 37 undergraduates. Although research in the helplessness paradigm usually focuses on environmental determinants of distress, the paradigm may help identify and explain individual differences in, or intrinsic modulation of, stress and clinical depression.

850. Dorworth, Thomas R. & Overmier, J. Bruce. (1977). **On learned helplessness: The therapeutic effects of electroconvulsive shocks.** ***Physiological Psychology,*** 5(3), 355–358.
Administered a series of unavoidable, inescapable electric shocks to 19 healthy dogs while they were confined in a harness. Later, these Ss were tested for escape/avoidance learning in a shuttlebox. Compared to 2 control groups of 8 Ss each, the preshocked Ss showed marked impairment of escape/avoidance learning. Indeed, 10 of the preshocked Ss never escaped, showing maximum helplessness. These 10 Ss were subsequently divided into 2 groups for the main experimental manipulation. One group of 6 received a series of 6 electroconvulsive shock (ECS) treatments over a 3-day period. The 2nd group of 4 Ss served as controls; they received exactly the same manipulations with the exception of the ECS. Four days later, these groups were tested again for escape/avoidance responding. The group that received ECS showed marked improvement, while the controls did not.

851. Drugan, Robert C.; Ader, Deborah N. & Maier, Steven F. (1985). **Shock controllability and the nature of stress-induced analgesia.** ***Behavioral Neuroscience,*** 99(5), 791–801.
Three experiments, with 160 male albino rats, examined the impact of the escapability of shock on the nature of the analgesia produced by shock. In Exp I, Ss were exposed to 0, 20, 40, or 80 escapable or yoked inescapable shocks. Tail-flick testing revealed a double-peak pattern in which analgesia was present after 20 and 80 shocks for both escapable and inescapable shock. Analgesia after 20 escapable or inescapable shocks was insensitive to subcutaneous naltrexone (14 mg/kg), as was the analgesia after 80 escapable shocks. However, the analgesia after 80 inescapable shocks was completely blocked by naltrexone. In Exp II, the analgesia following 80 inescapable shocks persisted for at least 2 hrs, whereas it dissipated rapidly following 80 escapable shocks. The analgesia produced by escapable shock even dissipated with the continued occurrence of escapable shock. Shock controllability altered the analgesia produced by subsequent exposure to shock. Prior experience with controllable shock completely blocked the late-appearing naltrexone reversible analgesia; prior experience with uncontrollable shock led it to appear sooner.

852. Drugan, Robert C. & Maier, Steven F. (1986). **Control vs. lack of control over aversive stimuli: Nonopioid–opioid analgesic consequences.** ***National Institute on Drug Abuse: Research Monograph Series,*** 74, 71–89.
Discusses the effect of controllability of shock on nonopioid- and opioid-induced analgesia. Exposure to uncontrollable shock leads to several pathologies, including stress-induced analgesia (SIA), while exposure to controllable shock results in few stress effects and transient SIA. It is suggested that fear and anxiety play a role in precipitating stress pathologies and opioid SIA. A model of the effects of controllable/uncontrollable shock on SIA involving gamma-aminobutyric acid (GABA) is presented. The model is intended to provide a perspective on the protective effects of coping with stress. It is suggested that the experience of uncontrollability may precipitate the drug addiction process.

853. Eggers, Howard M. & Blakemore, Colin. (1978). **Physiological basis of anisometropic amblyopia.** ***Science,*** 201(4352), 264–266.
In the visual cortex of 5 kittens that had received their only visual experience while wearing a high-power lens before one eye, most neurons were dominated by input from the normal eye. Moreover, contrast sensitivity and resolving power were lower for stimulation through the originally defocused eye, mimicking psychophysical results from human anisometropic amblyopes.

854. Garrick, Thomas; Minor, Thomas R.; Bauck, Sally; Weiner, Herbert et al. (1989). **Predictable and unpredictable shock stimulates gastric contractility and causes mucosal injury in rats.** ***Behavioral Neuroscience,*** 103(1), 124–130.
The effects of tailshock on gastric contractility and lesions were investigated in rats exposed to 100 1-mA tailshocks while confined inside plastic tubes. A light preceded each shock in one group and was randomly presented with respect to shock in the other. Contractility of the corpus of the stomach was measured by means of chronically implanted extraluminal force transducers. Contractility was measured in 10-min blocks and analyzed by computer. Signaled ($n = 13$) and unsignaled ($n = 17$) shock stimulated high-amplitude gastric contractions in fasted rats, which continued for 2 hr after the shock session. Cumulative contractile activity (1.5-hr shock plus 2-hr rest) in shocked animals was twice that in restrained and unrestrained control animals ($n = 19$, $p < .05$), and contractile activity had a 30%–40% greater average amplitude than after a meal. Compared with unrestrained controls, shocked rats had visibly more mucosal injury. Larger cumulative contractile activity was associated with a larger area of erosions. Frequency and duration of contractions did not distinguish between shocked and unshocked groups.

855. Glenthøj, B.; Hemmingsen, R. & Bolwig, T. G. (1988). **Kindling: A model for the development of tardive dyskinesia?** ***Behavioural Neurology,*** 1(1), 29–40.
Describes the major characteristics of kindling, theories of tardive dyskinesia (TD), and the role of multiplicity in the development of TD. Interruption of neuroleptic therapy may be a risk factor for development of irreversible TD. Induction of dyskinesia in nonhuman primates has been demonstrated after repeated administration of haloperidol. Several experimental results link TD with kindling: both conditions involve repeated stimulations, both seem to involve increased receptor responsiveness, and in both conditions depression in gamma-aminobutyric acid (GABA) transmission in the substantia nigra pars reticulata plays an important role.

856. Golda, V. & Petr, R. (1987). **Animal model of depression: Retention of motor depression not predictable from the threshold of reaction to the inescapable shock.** ***Activitas Nervosa Superior,*** 29(2), 113–114.
Demonstrated in normo- and hypertensive rats that behavioral response to electric shocks was influenced more by psychological factors than by physical properties of the stressors.

857. Guile, Michael N. (1987). **Differential gastric ulceration in rats receiving shocks on either fixed-time or variable-time schedules. *Behavioral Neuroscience,*** 101(1), 139–140.
Research suggests that predictable electric shocks produce less stress than unpredictable shocks. In this experiment, predictability was manipulated by using fixed-time (FT, predictable) and variable-time (VT, unpredictable) schedules of shock delivery. Rats receiving 3-mA, 1-s electric shocks on a FT 45-s schedule developed less gastric pathology than another group that was administered identical shocks on a VT 45-s schedule. It is argued that this finding represents a more subtle effect of predictability than has been obtained heretofore.

858. Hamamura, Yoshihisa. (1982). **Does punishment for biting attenuate gastric lesions in the rat? *Japanese Psychological Research,*** 24(4), 195–199.
Examined the effect of punishment on gastric lesions induced by cold water stress in 40 Wistar albino rats. Ss were restricted in tubes with metal biting targets and exposed to stress for 48 hrs under the following conditions: (1) Ss in punishment group were punished by electric shock for their biting responses; (2) Ss in yoked-control group received shock independently of their own biting responses, but whenever Ss in the punishment group were punished; (3) Ss in biting group were allowed to bite the targets, but received no shock; (4) no target was provided for the control group, which received no shock. The punishment group showed significantly fewer stomach lesions than the other 3 groups. It is concluded that the stress-induced gastric lesions were attenuated by the punishment for biting responses, probably by means of exercising some control over punitive shock through passive avoidance.

859. Henke, Peter G. (1990). **Granule cell potentials in the dentate gyrus of the hippocampus: Coping behavior and stress ulcers in rats. *Behavioural Brain Research,*** 36(1–2), 97–103.
Evoked population potentials of the granule cells in the dentate gyrus of the hippocampus were increased in stress-resistant rats and decreased in stress-susceptible rats, as indexed by restraint-induced gastric ulcers. Inescapable, uncontrollable shock stimulation also suppressed granule cell population spikes and interfered with subsequent coping responses when escape was possible (i.e., the so-called helplessness effect). Data were interpreted to indicate that the hippocampus is part of a coping system in stressful situations.

860. Henke, Peter G. & Sullivan, Ronald M. (1985). **Kindling in the amygdala and susceptibility to stress ulcers. *Brain Research Bulletin,*** 14(1), 5–8.
32 male Wistar rats were either kindled in the centro-medial amygdala to full Stage 5 seizures or partially kindled until the 1st electrographic after-discharges appeared. Following kindling treatment, Ss were stressed for 24 hrs in a restraining apparatus, and their stomachs were inspected to determine the degree of pathology. Results show that kindling facilitated the subsequent development of restraint-induced stomach ulcers, suggesting that the neuronal hyperexcitability produced by kindling led to an increased susceptibility to gastric pathology in response to stress.

861. Hyeon, Seong-yong & Kim, Ki-suk. (1983). **The effect of electroconvulsive shock treatment on learned helplessness in rats. *Korean Journal of Psychology,*** 4(1), 1–10.
Studied the therapeutic effect of electroconvulsive shock therapy (ECT) on learned helplessness in 32 Sprague-Dawley albino rats. 16 Ss received 80 trials of 5 sec inescapable shock, while the other 16 Ss received no shock. All Ss received 30 trials of Y-maze escape training 24 hrs later. Immediately after completion of escape training, 8 randomly selected Ss in each group received a single electroconvulsive shock, while the remaining Ss received no shock. All Ss were tested for Y-maze escape responding 24 hrs later. Among Ss receiving 80 trials of inescapable shock, those that received ECT showed marked improvement, while those that received no ECT did not. The improvement is attributed to the retrograde amnesic effect of ECT. (English abstract).

862. Inoue, M.; Peeters, B. W.; Van Luijtelaar, E. L.; Vossen, J. M. et al. (1990). **Spontaneous occurrence of spike-wave discharges in five inbred strains of rats. *Physiology & Behavior,*** 48(1), 199–201.
Studied the number and duration of cortical spike-wave discharges (SWDs) in 4 inbred strains of rats (G/Cpb, B/Cpb, BN/BiRij, ACI) compared with those of the WAG/Rij strain, a model of absence epilepsy. Eight Ss of each strain were recorded during a continuous 48-hr period. An analysis of variance (ANOVA) showed significant strain differences in both the number and the mean duration of SWDs. The values of both parameters increased the most in the WAG/Rij strain, followed by the G/Cpb, B/Cpb, BN/BiRij and ACI strains. The WAG/Rij strain showed a considerable number of SWDs per hour. The study of biological and genetic factors underlying these strain differences may provide useful information on absence epilepsy in humans.

863. Kesner, Raymond P.; Dixon, David A.; Pickett, Diane & Berman, Robert F. (1975). **Experimental animal model of transient global amnesia: Role of the hippocampus. *Neuropsychologia,*** 13(4), 465–480.
In 3 experiments, different groups of a total of 136 Long-Evans rats were given a hippocampal seizure afterdischarge 1 or 7 days after acquiring a passive avoidance, active avoidance, or barpressing habit. At various delays after the cessation of the seizure afterdischarge, Ss were tested for retention of the previously acquired habit. Results indicate that 1 day, but not 7 days, after learning hippocampal seizures were capable of producing a temporary retrograde amnesia for well-learned responses and an anterograde amnesia for experiences that occurred during the retrograde amnesia period. Data suggest that hippocampal seizures can serve as an experimental prototype of "transient global amnesia" and that the hippocampus is critically involved in retrieval of information from long-term memory. (French & German summaries).

864. Kirk, Raymond C. & Blampied, Neville M. (1985). **Activity during inescapable shock and subsequent escape avoidance learning: Female and male rats compared. *New Zealand Journal of Psychology,*** 14(1), 9–14.
Compared the activity of 12 male and 12 female Wistar rats during inescapable shock and on subsequent escape-avoidance learning. Ss were given 60 1-mA inescapable tailshocks. At first, activity during shock was high immediately after shock onset and decreased during the shock. Over trials, this initial burst of activity decreased for both males and females. After 50 shocks, males were essentially inactive during shock. Females were generally more active than males, particularly during the first and last block of trials. Equal numbers of controls remained in their home cages. 24 hrs

later, all Ss were trained on an escape-avoidance task with a tone as the warning stimulus and 1-mA scrambled footshock as the aversive stimulus. Inescapably shocked males exhibited learned helplessness in that their escape-avoidance responding (2 crossings of the midline of the shuttlebox per trial) was significantly impaired. Inescapably shocked females were not impaired in their performance.

865. Landfield, Philip W. (1988). **Hippocampal neurobiological mechanisms of age-related memory dysfunction.** Special Issue: Experimental models of age-related memory dysfunction and neurodegeneration. ***Neurobiology of Aging,*** 9(5–6), 571–579.
Reviews studies that indicate that hippocampal frequency potentiation (the growth of neural responses during repetitive synaptic stimulation) is impaired in aged rats and that this impairment may be important in learning and memory deficits found in these Ss. Intracellular recording and ultrastructural studies suggest that both hippocampal frequency potentiation and the age deficit in such potentiation are synaptic processes (probably presynaptic) and that the deficit may be due to an age-related increase in calcium influx during depolarization. The latter may result from alterations in the function of a Ca-mediated inactivation of Ca current mechanism recently found in hippocampal neurons. Since major hippocampal changes occur with aging in both rodents and humans, these data may be relevant to human brain aging. It is suggested that Alzheimer's disease results from an acceleration of normal age-related neuronal calcium conductance changes by some unknown process (e.g., viruses, aluminum, genetic factors) leading to a rapid deterioration of brain structure.

866. Landfield, Philip W.; McGaugh, James L. & Lynch, Gary. (1978). **Impaired synaptic potentiation processes in the hippocampus of aged, memory-deficient rats. *Brain Research,*** 150(1), 85–101.
A series of neurophysiological experiments was performed on the Schaffer-commissural system of the hippocampus of aged and young anesthetized male Fischer rats. The aged Ss were previously found to exhibit retention performance deficits. No obvious differences were found between aged and young Ss in amplitude, latency, stimulation threshold, or wave forms of typical synaptic responses when these were elicited by control (0.3 Hz) stimulation pulses. However, aged and young synapses showed different responses during repetitive stimulation. Aged synapses "exhausted" more rapidly during continuous 4-Hz stimulation. Throughout the studies a biphasic pattern of potentiation was observed during repetitive stimulation (brief potentiation, depression, renewed potentiation). Aged Ss were deficient primarily in development of the 2nd phase of potentiation. The possibility that the impaired hippocampal synaptic plasticity may be related to reported deficient behavioral plasticity in the aged Ss is considered.

867. Laudenslager, Mark L.; Fleshner, M.; Hofstadter, P.; Held, P. E. et al. (1988). **Suppression of specific antibody production by inescapable shock: Stability under varying conditions. *Brain, Behavior & Immunity,*** 2(2), 92–101.
Studied the effect of uncontrollable shock on the production of antibodies (ABs) to a novel antigen, keyhole limpet hemocyanin (KLH), in 84 adult male rats (aged 60–90 days). Groups of Ss were tested under 1 of 4 experimental conditions involving testing during either the light or dark portions of their light cycles and following either 1 or 3 daily exposures to tailshock. All shock-exposed Ss showed reduced levels of immunoglobulin G ABs in response to KLH. AB levels were highest among Ss immunized during the dark phase of their cycle for both control and shocked Ss. AB production in response to a novel antigen may serve as a measure of behavioral modulation of immunity.

868. Lerer, B.; Stanley, M.; Keegan, M. & Altman, H. (1986). **Proactive and retroactive effects of repeated electroconvulsive shock on passive avoidance retention in rats. *Physiology & Behavior,*** 36(3), 471–475.
Results of 4 experiments show that electroconvulsive shock (ECS) administered daily for 1–7 days to male albino Sprague-Dawley rats cumulatively impaired retention of passive avoidance when Ss were trained 24 hrs after the last ECS and were intact by 21 days. It is suggested that these findings parallel the effects of ECS and tested 24 hrs after training. Retention was directly proportional to the interval between training and testing; Ss trained 24 hrs after ECS and tested 1 hr later showed no deficit while Ss tested after 24 hrs showed maximal impairment. Retention was significantly improved by 10 days following the last ECS on memory function in humans. The retrograde effects of ECS also paralleled those demonstrated in humans; while retention of a passive-avoidance task learned 24 hrs before ECS was grossly impaired, retention was intact if learning took place 7 days before the ECS course. The application of these findings as an animal model of ECS-induced memory impairment in humans is discussed.

869. Leung, Lai-wo S. (1988). **Hippocampal interictal spikes induced by kindling: Relations to behavior and EEG. *Behavioural Brain Research,*** 31(1), 75–84.
Recorded hippocampal spontaneous interictal spikes (SISs) in 40 rats during daily tetanization of afferent fibers to the hippocampal CA1 region (detected after 3–10 tetanizations). Clear variation of SIS rate with behavior was observed. SIS rate was high during slow-wave sleep (SWS), waking immobility, face-washing, and chewing and low during REM sleep (REMS), walking, and rearing. Scopolamine hydrochloride increased the SIS rate during walking. Despite negative correlation of SIS occurrence with theta rhythm in normal Ss, abolishing theta rhythm by medial septal lesions did not affect suppression of SISs during REMS compared with SWS. When spikes were seen with the theta rhythm, they tended to occur at about 240° after the positive peak of the alvear surface rhythm.

870. Mactutus, Charles F. & Wise, Nancy M. (1985). **The inaccessible, but intact engram: A challenge for animal models of memory dysfunction. *Annals of the New York Academy of Sciences,*** 444, 465–468.
Examined cue-induced memory reactivation in Fischer-344 rats using a multiple-measure passive avoidance task. Following a noncontingent footshock reactivation treatment, inhibitory behavior was restored to a level approximating that observed at a comparable interval after original learning. While vacillatory behavior after reactivation treatment was also restored toward that noted 24 hrs after original training, vacillation remained significantly greater than that observed for a newly acquired memory. The inclusion of sham-trained/noncontingent footshock controls, who also received the noncontingent footshock reactivation treatment, indicated that these observed changes were not an artifact of alterations in behavioral activation per se. Overall findings are consistent with many reports on cue-induced facilitation

of retention performance and imply that memory dysfunction in infrahumans may often be more appropriately viewed as a loss of access to, rather than a loss of, the "engram."

871. Maier, Steven F. & Keith, Julian R. (1987). **Shock signals and the development of stress-induced analgesia.** ***Journal of Experimental Psychology: Animal Behavior Processes,*** 13(3), 226–238.
We report five experiments in which we investigated the effects of "feedback signals" on the pattern of hypoalgesia produced by inescapable shocks. A 5-s lights-out stimulus coincident with shock termination had no effect on the naltrexone-insensitive (nonopioid) hypoalgesia, which occurred after 10 inescapable shocks, but completely blocked the naltrexone-sensitive (opioid) hypoalgesia, which followed 100 inescapable shocks. The stimulus prevented the development of the opioid hypoalgesia rather than merely masking its measurement. This effect did not depend on the use of lights-out as the stimulus but did depend on the temporal relation between the stimulus and shock. Stimuli immediately preceding or simultaneous with shock had no effect. Surprisingly, stimuli randomly related to shock also blocked the opioid hypoalgesia. Simultaneous measurement of both hypoalgesia and fear conditioned to contextual cues revealed that the level of fear did not predict the blockade of hypoalgesia. Different backward groups received different temporal gaps between shock termination and the signal. An interval between 2.5 s and 7.5 s eliminated the effect of the signal on fear, but 12.5–17.5 s were required to eliminate the effect of the signal on hypoalgesia. The opioid hypoalgesia blocking power of the random stimulus was entirely attributable to those stimuli occurring within 15 s of the termination of the preceding shock. The implications of these results for the explanation of stimulus feedback effects and for stress-induced analgesia are discussed.

872. Maier, Steven F. & Laudenslager, Mark L. (1988). **Inescapable shock, shock controllability, and mitogen stimulated lymphocyte proliferation.** ***Brain, Behavior & Immunity,*** 2(2), 87–91.
Comments on the robustness and repeatability of the initial finding by M. L. Laudenslager et al (1983) that inescapable shock treatment in rats was associated with reduced proliferative responses to both phytohemagglutinin and concanavalin, even though the shock parameters were relatively mild. They also found that escapable shock treatment did not reduce the proliferative response relative to restrained or untreated controls. Results of 6 replications were generally unreliable. Further study led to the conclusion that altered proliferation could not be reliably reproduced by administering inescapable shocks.

873. Maier, Steven F.; Ryan, Susan M.; Barksdale, Charles M. & Kalin, Ned H. (1986). **Stressor controllability and the pituitary-adrenal system.** ***Behavioral Neuroscience,*** 100(5), 669–674.
Stressor controllability can alter both behavior and pituitary-adrenal (PAD) activity. Potential mediation of these behavioral effects by differential PAD output requires that the precise conditions that lead to differential behavioral consequences also produce differential PAD activity. In 2 experiments, plasma adrenocorticotropic hormone (ACTH) and corticosterone levels were measured in 152 male Holtzman rats at various times following escapable and yoked inescapable electric shock conditions known to produce differential behavioral outcomes. The escapable and inescapable shock procedures did not produce a detectable differential effect. Both shock conditions produced equivalent elevation of ACTH and corticosterone. Neither decay rates nor the ACTH and corticosterone response to shock reexposure differed among shocked groups. Implications for alterations in immune function and for conceptions of the relation between stress and PAD activity are discussed.

874. Maier, Steven F. & Warren, Donald A. (1988). **Controllability and safety signals exert dissimilar proactive effects on nociception and escape performance.** ***Journal of Experimental Psychology: Animal Behavior Processes,*** 14(1), 18–25.
In the present experiments we assess the ability of exteroceptive safety signals to proactively mimic shock controllability. In Experiment 1, animals were given escapable shock, yoked inescapable shock, restraint, or yoked shock with a stimulus coincident with shock offset on Day 1 and then given inescapable shock on Day 2. Safety signals attenuated the hypoalgesia following the first shock session. On Day 2, however, animals that had been preshocked with safety signals were not less hypoalgesic than those previously restrained. Conversely, shock controllability did not attenuate the hypoalgesia following the first session but proactively attenuated hypoalgesia during the second. Unlike animals preexposed to controllable shock, those given safety signals evidenced shuttle escape deficits 24 hr following the Day 2 inescapable shock treatment. Safety signals therefore failed to exert "immunization" effects. By employing identical preshock and shuttle test procedures, in Experiment 2 we demonstrated that safety signals completely block development of the escape deficit which otherwise results from the initial inescapable shock exposure. Thus, safety signals effectively reduce the impact of shock delivered in the same session but are ineffective in reducing the effect of subsequent shock. These results suggest that distinct processes underlie the effects of controllability and safety signals.

875. Mareš, P.; Chocholová, L. & Schickerová, R. (1984). **Models of minor nonconvulsive seizures.** European C.I.A.N.S. Conference (1983, Olomouc, Czechoslovakia). ***Activitas Nervosa Superior,*** 26(2), 154–155.
Attempted to verify the similarity of afterdischarges to human temporal seizures in 8 adult rats with chronically implanted electrodes. Findings indicate that the behavioral correlates of hippocampal-cortical afterdischarges are compatible with the clinical pattern of human partial seizures with complex symptomatology. An investigation of the behavioral correlates of rhythmic metrazol activity in 12 rats with implanted electrodes indicated that rhythmic metrazol activity can be used as a model of human primary generalized seizures of the absence type.

876. Marshall, John & McCutcheon, Bruce. (1976). **Reduction of stomach ulceration by hypothalamic stimulation in the unrestrained rat.** ***Physiology & Behavior,*** 16(4), 391–393.
Gastric lesions were produced in 16 unrestrained male Long-Evans rats subjected to a 6-hr shock stress session followed by a 2-hr rest period. Electrical stimulation of the brain (posterior lateral hypothalamus) administered during the 2-hr poststress rest period significantly decreased ulcer development (i.e., percentage of Ss showing lesions, number of lesions per S, and severity of lesions) in the test group of 8 Ss.

877. McCutcheon, N. Bruce; Guile, Michael N. & McCormick, Robert. (1986). **Electrical stimulation of the medial forebrain bundle-posterior lateral hypothalamus attenuates gastric lesions. *Physiology & Behavior,*** 37(3), 435–440.
In Exp I, using 45 male Long-Evans rats, it was found that Ss receiving electrical stimulation of the medial forebrain bundle-posterior lateral hypothalamus had significantly less gastric pathology than control Ss not receiving stimulation. Further experiments sought to examine characteristics of the stimulation that would account for this finding. Exps II and III, with a total of 27 male Long-Evans rats, studied signal and analgesic properties, respectively, of the brain stimulation, but no evidence was found for their involvement in the effect.

878. Moshé, Solomon L. & Albala, Bruce J. (1982). **Kindling in developing rats: Persistence of seizures into adulthood. *Developmental Brain Research,*** 4(1), 67–71.
Kindling induced in immature Sprague-Dawley rats produced persistent alterations in brain function despite the continuing brain growth, the progression of myelination, and increasing levels of neurotransmitters. Findings suggest that kindling immature animals may be used as a developmental model of epilepsy and further substantiate its use as a model for the study of neural plasticity, learning, and memory.

879. Musty, Richard E.; Jordan, Mark P. & Lenox, Robert H. (1990). **Criterion for learned helplessness in the rat: A redefinition. *Pharmacology, Biochemistry & Behavior,*** 36(4), 739–744.
Examined potential criteria for learned helplessness (LH) in rats to improve the reliability, reproducibility, and validity of the procedure. 260 male rats were tested in shuttleboxes with or without a copper barrier with an open doorway during inescapable preshock or no preshock. 24 hrs after the pretest, Ss were tested for acquisition of an escape response; Ss were retested 7–14 days later. Ss that showed a latency of 45 sec at least once in the 1st 5 trials and 45-sec latencies on 12 of 15 remaining trials were labeled as exhibiting a failure pattern. A barrier in the shuttlebox combined with the failure pattern criterion increased the reliability and validity of obtaining LH.

880. Ottenweller, John E.; Natelson, Benjamin H.; Pitman, David L. & Drastal, Susan D. (1989). **Adrenocortical and behavioral responses to repeated stressors: Toward an animal model of chronic stress and stress-related mental illness. *Biological Psychiatry,*** 26(8), 829–841.
Explored criteria for chronic stress in 24 adult male rats (1) exposed to 2 hrs of tailshock/day, (2) kept in the same room as shocked Ss without being shocked (bucket rats), or (3) kept in home cages (controls). Shocked Ss exhibited elevated prestress corticosterone levels and abnormal behavior. Bucket rats did not develop elevated corticosterone levels until several days after the shocked Ss, and their behavioral changes were less striking and consistent. Shocked Ss showed partial habituation to the stress procedure but still displayed behavioral aberrations for several days. This animal model may be useful for studying factors that contribute to chronic stress and posttraumatic stress disorder (PTSD).

881. Overmier, J. Bruce. (1986). **Reassessing learned helplessness. *Social Science,*** 71(1), 27–31.
Studied learned helplessness in dogs, focusing on the hypothesis that 2 separate factors contribute to the helplessness syndrome. The ability of Ss previously exposed to uncontrollable or controllable shocks to escape stresses in a new situation was studied. Ss previously exposed to predictable or unpredictable shocks were similarly studied. Unpredictable, uncontrollable shocks resulted in motivational interference but unpredictable, controllable shocks did not. Predictable, controllable shocks resulted in a somewhat improved ability to learn new associations. If Ss were paired so that only 1 controlled the shocks, control was seen to reduce emotional reactivity measured through serum cortisol. Uncontrollability was found to cause a motivational-performance deficit when new stresses were introduced. Unpredictability was found to cause the associative-learning deficit.

882. Overmier, J. Bruce & Murison, Robert. (1989). **Poststress effects of danger and safety signals on gastric ulceration in rats. *Behavioral Neuroscience,*** 103(6), 1296–1301.
Gastric ulceration of rats stressed by restraint in 19°C water for 75 min was markedly increased by allowing a 75-min postrestraint room-temperature rest period during which the rat was exposed to cues that had previously been associated with the delivery of 80 5-s uncontrollable electric shocks distributed over four sessions. This effect obtained equally without regard to whether "danger cues" were punctate signals or constant contextual cues or whether contextual ones were interrupted by punctate safety signals. The experimental treatments used were unusual in that they equated the groups on their total conditioning history and thus allowed a more pure look at the poststress effect than heretofore. Other groups provided controls for prior shocks, rest, and their interaction as well as handling. Analyses of corticosterone after the stress or stress-rest cycle revealed only a general decline in corticosterone levels with rest undifferentiated across groups.

883. Overmier, J. Bruce & Murison, Robert. (1991). **Juvenile and adult footshock stress modulate later adult gastric pathophysiological reactions to restraint stresses in rats. *Behavioral Neuroscience,*** 105(2), 246–252.
Rats were stressed with (1) signaled footshocks, (2) unsignaled footshocks, or (3) handled without footshocks as a prepubertal juvenile (28–36 days old), as an adult (96 days old), or both. This yielded 9 treatment groups (3 × 3). Two days after the adult treatment, all animals were challenged by restraint and partial immersion in water (19°C) to assess their relative susceptibility to gastric erosions ("ulcers"). It was found that any prior exposure to footshock stress increased the amount of ulcers; juvenile and adult experiences each produced equal increases but the combination of the two was less ulcerogenic than either alone. The predictability of the footshocks did not modulate ulcerogenicity. Adult corticosterone responses to (1) adult stress and (2) ulcer induction were not related to the observed ulcer severity; however, juvenile footshock stress appeared to reduce the corticoid response to the ulcerogenic challenge but not to the adult footshock stress.

884. Parra, A.; Padilla, M.; Segovia, S. & Guillamón, A. (1990). **Sexual differences in learned helplessness in the rat. *Revista de Psicología General y Aplicada,*** 43(1), 17–22.

Reports on 2 experiments on learned helplessness, using male and female rats exposed to inescapable electric shock (IES), escapable electric shock (EES), or nonshock (NS). Exp 1 showed that female rats were more sensitive to IES than male rats. In Exp 2, male and female rats, from both control and gonadectomized conditions, were run in a triadic design. They received IES, EES, or NS; all Ss were registered in latencies of nosing response. No effects of gonadectomy were found. The females that received IES showed higher latencies than the other shock-treated groups. (English abstract).

885. Previc, Fred H. & Allen, Ralph G. (1987). **A comparison of visual evoked potential and behavioral measures of flashblindness in humans.** ***USAF School of Aerospace Medicine Technical Report,*** 87-21, 14 p.
Compared visual evoked potential (VEP) and behavioral measures of flashblindness following exposure to intense but eyesafe xenon flashes to further validate the animal model of laser flashblindness based on VEP recordings in anesthetized rhesus monkeys. Monopolar VEPs were recorded from the posterior scalp of 6 adult human Ss in response to square-wave gratings of 3 differential spatial frequencies (1, 4, and 12 c/deg). The VEPs were recorded prior and subsequent to the presentation of a xenon flash. Results show that the moment of initial postflash visibility of the grating, as assessed by the VEP's recovery above its baseline, was highly comparable to that measured behaviorally, and that the predictive ability of the VEP depended in part on its signal-to-noise ratio.

886. Prince, Christopher R. & Anisman, Hymie. (1984). **Acute and chronic stress effects on performance in a forced-swim task.** ***Behavioral & Neural Biology,*** 42(2), 99–119.
Four studies with 400 male CD-1 mice assessed the effects of uncontrollable stressors on performance in a subsequent forced-swim paradigm. Results show that uncontrollable shock initially induced behavioral invigoration; however, within 24 hrs of stressor application, swimming behavior was depressed relative to nonstressed Ss. The controllability of the stressor did not influence the initial invigoration, being present among escapably shocked Ss and Ss that received (yoked) inescapable shock. In contrast, the depression of responding evident 24 hrs after stressor application was related to the availability of behavioral coping methods. Following repeated exposure to footshock, there was no indication of adaptation to the behavioral changes ordinarily induced by acute shock stress. Findings are discussed in terms of the effects of uncontrollable stressors on escape performance and the use of this preparation as an animal model of human depression.

887. Robinson, David L. & Rugg, Michael D. (1988). **Latencies of visually responsive neurons in various regions of the rhesus monkey brain and their relation to human visual responses.** Special Issue: Event related potential investigations of cognition. ***Biological Psychology,*** 26(1–3), 111–116.
Discusses the temporal characteristics of visually responsive neurons in various regions of the monkey brain to provide a means for researchers to compare human data with data from animal studies. Brain regions discussed include areas that are thought to play a role in (1) visual attention (pulvinar, posterior parietal cortex, area V4, and perifrontal cortex), (2) form vision (areas 17 and 18, area V4, and inferotemporal cortex), and (3) visually guided eye movements (superior colliculus and frontal eye fields).

888. Salvi, Richard J. & Ahroon, William A. (1983). **Tinnitus and neural activity.** ***Journal of Speech & Hearing Research,*** 26(4), 629–632.
Measured the spontaneous discharge rates of auditory nerve fibers in a group of normal chinchillas and in 4 chinchillas with high-frequency, noise-induced hearing loss. In contrast to normal units, the high-frequency units in the noise-exposed Ss tended to have elevated spontaneous discharge rates, high thresholds, and a lack of 2-tone inhibition. The change in spontaneous discharge rate across the distribution of nerve fibers is related to models of tinnitus and to human psychophysical data.

889. Sershen, Henry; Wolinsky, Toni; Douyon, Richard; Hashim, Audrey et al. (1991). **The effects of electroconvulsive shock on dopamine-1 and dopamine-2 receptor ligand binding activity in MPTP-treated mice.** ***Journal of Neuropsychiatry & Clinical Neurosciences,*** 3(1), 58–63.
Examined possible therapeutic use of electroconvulsive shock therapy (ECT) in Parkinson's disease by studying the biochemical effects of electroconvulsive shock (ECS) on dopaminergic (DA) systems in mice injected with the neurotoxin 1-methyl-4-phenyl-1,2,3,6-tetrahydropyridine (MPTP). Selective changes occurred after ECS or DA neurotoxins that altered the balance of D_1 and D_2 receptor populations. These changes should be considered in the treatment of certain behavioral states and mood disorders. There seems to be a dissociation of antiparkinsonian and antidepressant effects early in the course of ECT.

890. Shanks, Nola; Zalcman, Steve; Zacharko, Robert M. & Anisman, Hymie. (1991). **Alterations of central norepinephrine, dopamine and serotonin in several strains of mice following acute stressor exposure.** ***Pharmacology, Biochemistry & Behavior,*** 38(1), 69–75.
Exposure to inescapable footshock provoked region-specific alterations of norepinephrine (NE), dopamine (DA), and serotonin (5-HT) activity across 6 strains of mice (A/J, BALB/cByJ, C3H/HeJ, C57BL/6J, DBA/2J, and CD-1). 45 males of each strain served as Ss. The stressor provoked reductions of hypothalamic NE and increased 3-methoxy-4-hydroxyphenylglycol (MHPG) accumulation in all strains. In contrast, the effects of the stressor on NE activity in the hippocampus and locus ceruleus varied appreciably across strains. In the mesocortex and nucleus accumbens, shock induced an increase of 3,4-dihydroxyphenylacetic acid (DOPAC) accumulation and pronounced reductions of DA in some strains, while in others these variations were less pronounced or entirely absent. Stressor-provoked alterations of 5-HT and 5-hydroxyindoleacetic acid (5-HIAA) were most evident in the mesocortex. Strain-specific neurochemical alterations following footshock are discussed relative to stressor-induced behavioral disturbances and animal models of depression.

891. Sines, J. O.; Patterson, K. & Rusch, L. (1977). **The experimental production of resistance to stress-induced stomach lesions in the rat.** ***Journal of Psychosomatic Research,*** 21(6), 457–461.
Results of 3 studies with 250 male hooded rats show that a significant degree of stress-specific protection against the ulcerogenic effects of either restraint or enforced running could be produced by exposing Ss to several brief, daily episodes of either restraint or enforced running. The specificity of the protection suggests that somewhat different

physiological factors are involved in the production of stress ulcers by these 2 procedures. Further research is planned to determine whether generalized immunity can be produced by nonpharmacological nonsurgical techniques.

892. Sonoda, Akihito. (1990). **Effects of uncontrollable shock on subsequent appetitive discrimination learning in rats. *Japanese Journal of Psychonomic Science,*** 8(2), 95–100.
Rats, previously trained to press 2 bars in a Skinner box for food, were divided into 3 groups of 10 each and given either escapable or yoked inescapable foot-tail shock, or no shock, for 80 trials. In the subsequent training to press 2 bars in single alternation, the inescapable group made smaller number of responses in the first 20 min of training, suggesting the presence of a motivational deficit. In a later stage of training, the rats in the inescapable group made smaller number of correct responses, though they were not different from other groups of rats in the total number of responses. There was an associational deficit apart from the motivational deficit. Results support the cognitive interpretation of learned helplessness, which generalized to an appetitive learning situation. (English abstract).

893. Stevens, Janice R. & Livermore, Arthur. (1978). **Kindling of the mesolimbic dopamine system: Animal model of psychosis. *Neurology,*** 28(1), 36–46.
To examine the behavioral and physiologic consequences of chronic activation of the mesolimbic dopamine system, the nucleus of origin in the ventral tegmental area was stimulated electrically for 2 sec daily through chronically implanted intracranial electrodes in 9 adult cats at the same point where instillation of the gamma-aminobutyric acid blocking agent bicuculline induced a characteristic fear, staring, searching, and withdrawal response. None of the Ss developed sustained afterdischarge or seizures following daily stimulation for 2 mo. Progressive fearfulness, hiding, loss of social behavior, and EEG spike or slow activity in the ipsilateral nucleus accumbens developed in 3 of 6 intact Ss. In these 3 Ss with chronic behavioral disturbances, only haloperidol or pimozide (of 6 substances tested) decreased fearfulness. Two Ss with prior 6-hydroxydopamine lesions of catecholamine pathways did not develop behavioral change in response to local bicuculline or daily electrical stimulation of the ventral tegmental area but demonstrated pronounced afterdischarge or EEG spike propagation during the kindling procedure.

894. Stevens, Karen E.; Fuller, Laura L. & Rose, Greg M. (1991). **Dopaminergic and noradrenergic modulation of amphetamine-induced changes in auditory gating. *Brain Research,*** 555(1), 91–98.
Examined the mechanism underlying amphetamine-induced alterations in auditory gating in male rats. The drug-induced reduction in gating paralleled a similar observation in schizophrenic humans. The psychotomimetic amphetamine was used to reproduce the lack of auditory-evoked-potential gating observed in schizophrenic humans. This animal model of amphetamine-induced deficits in sensory gating offers a viable technique for assessing possible mechanisms associated with sensory gating deficits observed in schizophrenia. Data from these animal experiments support the idea that both adrenergic and dopaminergic systems are altered in human schizophrenia.

895. Stuckey, Jeffrey; Marra, Susan; Minor, Thomas & Insel, Thomas R. (1989). **Changes in mu opiate receptors following inescapable shock. *Brain Research,*** 476(1), 167–169.
Examined opiate receptor binding to evaluate the role of endogenous opiates in learned helplessness and long-term hypoalgesia. The binding of a selective μ-opiate receptor agonist in the brains of male rats exposed to inescapable shock was reduced relative to Ss exposed to escapable or no shock. The decrease in binding appeared to result from a decrease in the number of μ-receptors and not a change in their affinity. Results support the hypothesis that inescapable shock produces long-term changes in endogenous opiate systems.

896. Troisi, Joseph R.; Bersh, Philip J.; Stromberg, Michael F.; Mauro, Benjamin C. et al. (1991). **Stimulus control of immunization against chronic learned helplessness. *Animal Learning & Behavior,*** 19(1), 88–94.
In Exp 1, male rats were exposed to 50 pairings per session of a white-noise stimulus with escapable shock during the immunization phase. Subsequently, they were exposed to 50 pairings per session of a different (houselight) stimulus with inescapable shock. Shock-escape performance in a shuttlebox test with constant illumination revealed no evidence of immunization relative to the performance of Ss given prior sessions of light-signaled inescapable shock only. Exp 2 was identical to Exp 1, except that both the escapable- and the inescapable-shock phases for Ss in the immunization treatment group involved the same stimulus (houselight) as a shock signal. Under these circumstances, the prior escapable-shock training significantly reduced the shuttle box escape deficit engendered by chronic exposure to signaled inescapable shock; performance in the shuttlebox was not reliably different from that of Ss exposed to signaled escapable shock alone.

897. Tsuda, Akira & Hirai, Hisashi. (1976). **Effects of signal-shock contingency probability on gastric lesions in rats as a function of shock region. *Japanese Journal of Psychology,*** 47(5), 258–267.
Effects of signal-shock contingency probability on stress pathology in 77 rats were studied with 2 regions of electric shock and with only sound signal presentations (NS), unrelated presentations of the signal and tail shock (TS) or foot shock (FS), signal presentations, all of which were paired with the TS or FS (100 PS), or signal presentations, 50% of which were paired with a TS or FS (50 PS). Ss in NPS and 50 PS groups with TS showed more stress-induced pathology than Ss in the 100 PS group with TS; however, the 100 PS with FS resulted in greater gastric lesions than NPS, 50 PS, or NS groups. It is concluded that the interaction between shock region and signal-shock contingency probability plays an important role.

898. Tsuda, Akira; Tanaka, Masatoshi; Hirai, Hisashi & Pare, William P. (1983). **Effects of coping behavior on gastric lesions in rats as a function of predictability of shock. *Japanese Psychological Research,*** 25(1), 9–15.
54 male Wistar rats received shock that was predicted by either a light signal alone, light-tone complex signal, or no signal. In the light signal alone and light-tone complex signal conditions, experimental Ss that could avoid and/or escape shock developed less gastric ulceration than did yoked "helpless" Ss that had the same shock but had no control over it. In the nonsignal condition, however, experimental

Ss did not differ from matched yoked Ss. Ulceration of 27 nonshock control Ss was negligible as compared to experimental and yoked Ss in each of the 3 warning conditions. Presence or absence of a safety signal had an effect on ulcerogenesis. The light-signal-alone condition was effective in reducing ulceration for "helpless" Ss as well as Ss that could control shock. Thus, the effectiveness of a coping behavior depends on the reliable prediction of shock in a stressful situation.

899. Van Luijtelaar, E. L.; Van der Werf, S. J.; Vossen, J. M. & Coenen, A. M. (1991). **Arousal, performance and absence seizures in rats.** ***Electroencephalography & Clinical Neurophysiology,*** 79(5), 430–434.
Investigated whether the execution of a task influenced the number of spike-wave discharges (SWDs) in rats of the WAG/Rij strain. 11 rats (aged 4–12 mo) were trained to press a lever for food in a fixed (60 sec) interval task until a stable response pattern emerged. EEG electrodes were implanted, and baseline EEGs were made, before and after the 1st and 5th test sessions. During the task execution, a significantly smaller number of SWDs was found compared with the preceding and succeeding baseline hours. The postreinforcement pause was significantly enhanced in trials with SWDs compared with trials without discharges, indicating a clear change in performance. Both results agree with what could be expected in patients with absence epilepsy and provide further evidence for the validation of the SWDs as genuine epileptic phenomena.

900. Vaughan, Herbert G. (1982). **The neural origins of human event-related potentials.** ***Annals of the New York Academy of Sciences,*** 388, 125–138.
Reviews current knowledge on the neural origins of human event-related potentials (ERPs) and the applications of the 3 main approaches to ERP investigations—analysis of surface topography of electrical and magnetic fields, observations of the effects of selective brain lesions on surface distribution, and intracranial mapping of the potential field. Issues related to the selection of appropriate animal models for the comparative analysis of ERP generators; distinctions between internal fields, recorded within an active structure, and external fields, which are volume-conducted beyond the region that generates them; and the need for computerized methods of chronotopographic analysis of scalp potentials are discussed.

901. Visintainer, Madelon A.; Volpicelli, Joseph R. & Seligman, Martin E. (1982). **Tumor rejection in rats after inescapable or escapable shock.** ***Science,*** 216(4544), 437–439.
Male Sprague-Dawley rats experienced inescapable, escapable, or no electric shock 1 day after being implanted with a Walker 256 tumor preparation. Only 27% receiving inescapable shock rejected the tumor, whereas 63% receiving escapable shock and 54% receiving no shock rejected the tumor. These results imply that lack of control over stressors reduces tumor rejection and decreases survival.

902. Vorob'eva, T. M. & Shevereva, V. M. (1990). **Neurophysiological mechanisms of the effects of transcranial micropolarization in normal rats and rats with emotional disturbances.** ***Human Physiology,*** 16(3), 185–192.
Studied the neurophysiological mechanisms of the effects of transcranial anodal micropolarization (TCAMP) in 50 normal noninbred mature male rats and in similar rats with experimental emotional disturbances. In the model of emotional disorders TCAMP had an antistressor effect, exerting its influence through the system of positive emotions and the mechanisms of general physiological activity.

903. Wahlestedt, Claes; Blendy, Julie A.; Kellar, Kenneth J.; Heilig, Markus et al. (1990). **Electroconvulsive shocks increase the concentration of neocortical and hippocampal neuropeptide Y (NPY)-like immunoreactivity in the rat.** ***Brain Research,*** 507(1), 65–68.
Examined whether treatment of 32 male rats with electroconvulsive shock (ECS) would produce effects similar to antidepressant drugs. Ss were divided into 4 groups. Group 1 received ECS once a day for 13–14 days. Group 2 was handled, but received no current. Group 3 was handled, but was given low current shocks, and Group 4 was not handled. Group 3 developed behavioral signs reminiscent of shock-induced learned helplessness. Evidence suggests that NPY may be a marker of major depression in humans. Since the reduced NPY concentrations in the cerebrospinal fluid (CSF) of Ss were correlated with anxiety, NPY may be linked to the pathogenesis of depression.

904. Warren, Donald A.; Castro, Carl A.; Rudy, Jerry W. & Maier, Steven F. (1991). **No spatial learning impairment following exposure to inescapable shock.** ***Psychobiology,*** 19(2), 127–134.
Examined whether inescapable shock (IS) produces a learning deficit in a task known to be sensitive to disruption in hippocampal function (i.e., the Morris water-maze spatial-learning task). Four experiments with rats showed that a single session of IS of the type used in learned helplessness experiments had no apparent effect on performance, regardless of whether shock was given 30 min or 6 hrs prior to maze training, or whether Ss were tested at 1 of 4 retention intervals. Similarly, no effect of IS was observed when shock and maze training were distributed over 3 sessions. Results are inconsistent with the hypothesis that the effects of IS on cognitive performance result from disturbances in hippocampal function.

905. Wilson, Josephine F. & Cantor, Michael B. (1987). **An animal model of excessive eating: Schedule-induced hyperphagia in food-satiated rats.** ***Journal of the Experimental Analysis of Behavior,*** 47(3), 335–346.
19 male rats were maintained on ad libitum wet mash and water and were trained to press a lever on FI or FR schedules of reinforcement with electrical brain stimulation. 14 Ss ate at least 150% more during intermittent reinforcement sessions than during baseline, massed reinforcement control, and/or extinction sessions. In a 3-hr session, 11 of those 14 Ss consumed the equivalent of nearly half their daily food intake. In subsequent control sessions, the electrodes did not support stimulus-bound eating despite attempts to make stimulation parameters optimal. Results indicate that the eating was schedule-induced or adjunctive and suggest that the procedure may provide an animal model of excessive nonregulatory eating that contributes to obesity in humans.

906. Zacharko, Robert M.; Bowers, Walter J.; Kokkinidis, Larry & Anisman, Hymie. (1983). **Region-specific reductions of intracranial self-stimulation after uncontrollable stress: Possible effects on reward processes.** ***Behavioural Brain Research,*** 9(2), 129–141.
Rates of responding for intracranial self-stimulation from the medial forebrain bundle, nucleus accumbens, and substantia nigra were evaluated in 54 male CD-1 mice that had been exposed to either escapable shock, yoked inescapable shock, or no-shock treatment. Whereas performance was unaffected by escapable shock, marked reductions of responding from the medial forebrain bundle and nucleus accumbens were evident following the uncontrollable shock treatment. It is suggested that uncontrollable shock reduces the rewarding value of responding for electrical brain stimulation from those brain regions in which stressors are known to influence dopamine activity. Extrapolating to human depressive disorders, it is possible that stressors may produce reward change by reductions in catecholamine activity or an inappropriate response to environmental stimuli related to excessive catecholamine activity.

907. Zacharko, Robert M.; Lalonde, Gerald T.; Kasian, Marilyn & Anisman, Hymie. (1987). **Strain-specific effects of inescapable shock on intracranial self-stimulation from the nucleus accumbens.** ***Brain Research,*** 426(1), 164–168.
Assessed responding for electrical stimulation from the nucleus accumbens in 3 inbred strains of mice (DBA/2J, C57BL/6J, and BALB/cByJ) following exposure to uncontrollable footshock. While the operant response was most readily acquired in the DBA/2J strain, exposure to inescapable shock in this strain induced a marked deterioration of self-stimulation responding, which tended to dissipate over a 168-hr period. The stressor did not affect self-stimulation responding in the C57BL/6J strain and produced a transient enhancement of responding in BALB/cByJ Ss. It is concluded that although uncontrollable aversive events may engender an anhedonic effect, such an outcome is strain-dependent, suggesting the importance of considering individual and genetic differences in the development of animal models of depression.

Physiological Processes

908. Bowen, Deborah J. (1989). **Possible explanations for excess weight gains in pregnancy: An animal model.** ***Physiology & Behavior,*** 46(6), 935–939.
48 adult female rats were divided into 2 diet groups, (1) bland chow and (2) chow plus glucose solution. Half of each group was impregnated. Body weight, consumption of both foods, and water consumption were measured daily before, during, and after pregnancy. During pregnancy and after delivery adipose tissue analyses were performed on Ss from each group. During pregnancy and during lactation, pregnant Ss with 2 foods consumed more glucose solution and gained more adipose tissue with larger cells than did any other group. The opportunity to consume sweet-tasting food while pregnant was accompanied by excess consumption of this food and led to excess body fat gains.

909. Chafetz, Michael D. (1985). **Biological factors in anorexia.** ***Southern Psychologist,*** 2(3), 12–18.
Discusses 3 types of animal models of anorexia (comparative, hypothalamic dysfunction, micronutrient) in terms of their contributions to anorexia nervosa. The comparative model seeks to show the adaptive strategies of naturalistic animal anorexia. The hypothalamic model is cited most frequently because of the history of studies showing hypothalamic control of feeding. The micronutrient model is discussed in relation to actual deprivation and metabolic deprivation. Evidence suggests a causal relationship between zinc deprivation and anorexia in humans.

910. Coe, Christopher L.; Mendoza, Sally P.; Smotherman, William P. & Levine, Seymour. (1978). **Mother–infant attachment in the squirrel monkey: Adrenal response to separation.** ***Behavioral & Neural Biology,*** 22(2), 256–263.
Evaluated the pituitary-adrenal response following separation in mother and infant squirrel monkeys. Four mother–infant pairs and a pregnant female, living in a social group, were the Ss. The plasma cortisol levels of the mothers and infants were determined after the following conditions: (a) basal levels at 1100 hrs, (b) 30 min after momentary separation and reunion, (c) 30 min after infant removal from the group, and (d) 30 min after mother removal from the group. Levels of plasma cortisol were significantly elevated in both mothers and infants following separation, and the response was not reduced by the presence of familiar Ss (e.g., separated infants which were "aunted" by the pregnant female). Separation followed by immediate reunion did not result in elevated values. These data indicate that a specific attachment relationship develops between mother and infant and that the agitation following separation is reduced only by reunion with the object of attachment.

911. Coe, Christopher L.; Mendoza, Sally P. & Levine, Seymour. (1979). **Social status constrains the stress response in the squirrel monkey.** ***Physiology & Behavior,*** 23(4), 633–638.
Evaluated the influence of dominance on the pituitary-adrenal and gonadal systems in male squirrel monkeys. Basal and stress levels of plasma cortisol and testosterone were determined in 8 male pairs across a 5-wk period. Findings indicate that squirrel monkeys have unusually high levels of steroid hormones in comparison with other species. Dominant Ss had higher levels of cortisol and testosterone and showed a smaller stress response than did subordinate Ss.

912. Collier, Timothy J. & Coleman, Paul D. (1991). **Divergence of biological and chronological aging: Evidence from rodent studies.** Special Issue: Animal models for aging research. ***Neurobiology of Aging,*** 12(6), 685–693.
A review of the literature on aging populations of rodents supports the view that significant functional variation exists among like-aged, elderly individuals, and that chronological age as a solitary measure is a poor indicator of biological age. Various studies are highlighted which classify aged rodents based on genetic or behavioral similarities. Beyond their descriptive value for gerontological research, these studies suggest ways in which biological aging can be manipulated to promote good function in aged individuals.

913. Corson, S. A. & Corson, Elizabeth O. (1986). **Visceral orienting reflexes as significant components of psychophysiologic personality assessment.** International Conference on Preventive Cardiology (1985, Moscow, USSR). ***Activitas Nervosa Superior,*** 28(2), 117–122.

Reviewed animal models aiding the study of psychophysiological substrates of Type A (coronary prone) and Type B (noncoronary prone) behavior. Idiographic and nomothetic research with repeated measures found stable constitutional differences in reactions to repeated exposures to a psychologically stressful environment. Some dogs (e.g., beagles, hounds, some mongrels) showed rapid physiological adaptation, while others (e.g., fox terriers, cocker spaniels, German shepherds) showed low adaptation (LA) characterized by almost indistinguishable psychophysiological reactions, including tachycardia; polypnea; profuse salivation; increased energy metabolism, rectal temperature, and electromyograph (EMG) reactions; vasopressin release; and increased urinary catecholamine. LA dogs showed higher frequency, more intense, persistent, and poorly modulating cardiac and respiratory orienting responses.

914. Fleming, Alison S. & Corter, Carl. (1988). **Factors influencing maternal responsiveness in humans: Usefulness of an animal model.** Special Issue: Psychoneuroendocrine aspects of maternal behavior. ***Psychoneuroendocrinology,*** 13(1–2), 189–212.
Reviews the hormonal, sensory, and experiential factors that regulate the onset and early maintenance of maternal responsiveness in rat and human mothers. Studies indicate that changes in feelings and attitudes associated with pregnancy in humans are probably not hormonally mediated, but associated with a variety of psychosocial factors. It is suggested that when women give birth, they undergo a period of elevated responsiveness that may be influenced by puerperal hormones. Following the early postpartum period, various factors influence human maternal responsiveness, including the Ss' affective state, social relationships, and experiences caring for young. The relative contributions of psychological and physiological influences to maternal responsiveness at different stages of the maternity cycle are discussed.

915. Fleshner, M.; Laudenslager, Mark L.; Simons, L. & Maier, S. F. (1989). **Reduced serum antibodies associated with social defeat in rats.** ***Physiology & Behavior,*** 45(6), 1183–1187.
Examined the effect of a social stressor, defeat associated with territorial defense, on serum antibodies to a specific protein, keyhole limpet hemocyanin (KLH). Pairs of male rats formed colonies, and experimental rats were intruders. Experimental Ss were immunized with KLH prior to exposures to territorially defensive colonies. Controls were placed into colonies but separated from residents by a plexiglass barrier. Behavioral measures, including number of bites and total time spent in submissive postures, were taken for colony–intruder interactions. Serum antibody levels were determined from blood samples taken 1, 2, and 3 wks following immunization. Experimental Ss had significantly less serum antibodies to KLH than controls. A stressful social encounter may thus affect immune function in a manner independent of the influence of physical (nociceptive) stressors.

916. Fokkema, Dirk S. & Koolhaas, Jaap M. (1985). **Acute and conditioned blood pressure changes in relation to social and psychosocial stimuli in rats.** ***Physiology & Behavior,*** 34(1), 33–38.
The naturally occurring tendency to compete with other rats for territorial space was used to study individual behavior characteristics and blood pressure reactivity to social stimuli in 11 Tyron Maze Dull-S3 rats. The competitive characteristics of the individual Ss were consistent in 2 social situations (victory and defeat). Blood pressure responses during the victory of home territory Ss over intruders was more pronounced in the more competitive Ss. In addition to defeat by a trained fighter rat, the experimentals were also psychosocially stimulated by the fighter while it was confined in a small wire mesh cage. The blood pressure response to this event was enhanced by the prior defeat of the test S by the one now confined to the small cage. This response was more pronounced in competitive Ss. It is suggested that this approach has potential as an animal model of etiological processes in socially induced hypertension.

917. Friedman, Richard. (1988). **Environmental–genetic interactions in experimental hypertension: The Dahl rat model.** ***Health Psychology,*** 7(2), 149–158.
Discusses the relationship between psychological factors and hypertension raised by L. K. Dahl's (1961) animal model of hypertension. In this model, Dahl-selected lines of salt-sensitive rats respond to hypertensinogenic stimuli such as sodium or behavioral stressors with marked elevations in blood pressure while Dahl salt-resistant rats do not. The 2 lines differ in terms of heart-rate reactivity and central nervous system (CNS) organization.

918. Gomez, Roberto E.; Pirra, G. & Cannata, M. A. (1989). **Open field behavior and cardiovascular responses to stress in normal rats.** ***Physiology & Behavior,*** 45(4), 767–769.
Studied the relationship between open field behavior and cardiovascular responses to stress in normal rats with patterns of hypo- or hyperactivity in open field testing. Findings indicate that elevations in blood pressure in response to stress produced by placement in a new environment differed in Ss with different open field behavior patterns. Hypo- and hyperactive rats may represent a useful animal model for studying behavior-related cardiovascular hyperreactivity to stress.

919. Greenberg, Lawrence; Edwards, Emmeline & Henn, Fritz A. (1989). **Dexamethasone Suppression Test in helpless rats.** ***Biological Psychiatry,*** 26(5), 530–532.
Measured hypothalamic-pituitary-adrenal (HPA) axis function in 73 male rats, using the dexamethasone suppression test (DST). Dexamethasone (DEX) injection produced significant effects on corticosterone levels in all S groups. Helpless Ss were distinguished from nonhelpless and naive Ss by their impaired ability to suppress corticosterone following DEX. Psychological stress can produce altered HPA function possibly via the hippocampal formation, and this coupling may depend on a combination of factors, including the affective state.

920. Häfner, Heinz; Behrens, Stephan; de Vry, Jean & Gattaz, Wagner F. (1991). **An animal model for the effects of estradiol on dopamine-mediated behavior: Implications for sex differences in schizophrenia.** ***Psychiatry Research,*** 38(2), 125–134.
Used animal models in which the effects of the hormones on behavioral changes induced by the dopamine (DA) antagonist haloperidol (catalepsy) and by the DA agonist apomorphine (oral stereotypies, grooming, and sitting behavior) were investigated in neonatal and in adult treated rats. No consistent effects of testosterone were observed. Estradiol significantly reduced the behavioral changes induced by both haloperidol and apomorphine; this effect was more pronounced in neonatally treated animals. Results suggest a

downward regulation of DA neurotransmission by estradiol. Estradiol might act as a protective modulator in schizophrenia by enhancing the vulnerability threshold for psychosis through the downward regulation of DA neurotransmission. Such mechanism could explain, at least in part, the later onset and the more favorable course of schizophrenia in female patients.

921. Haracz, John L.; Minor, Thomas R.; Wilkins, Jeffery N. & Zimmermann, Emery G. (1988). **Learned helplessness: An experimental model of the DST in rats.** ***Biological Psychiatry,*** 23(4), 388–396.
Employed the learned helplessness model of depression to test, using 72 male albino rats, the effectiveness of psychological stress in inducing a resistance of plasma corticosterone levels to dexamethasone suppression. Inescapably shocked Ss exhibited corticosterone levels that were significantly more resistant to dexamethasone suppression than were the levels of Ss receiving an equivalent amount of escapable shock or no shock. Results confirm the hypothesis that hypothalamic-pituitary-adrenal cortical resistance to dexamethasone suppression was enhanced by the distress associated with the inefficacy of behavioral coping responses. Present findings represent an analog of the dexamethasone suppression test (DST) in the learned helplessness model of depression.

922. Hazzard, Dewitt G. (1991). **Relevance of the rodent model to human aging studies.** Special Issue: Animal models for aging research. ***Neurobiology of Aging,*** 12(6), 645–649.
Rodents have proven to be a useful general model for aging research. Although they are not necessarily appropriate for the study of such specific human age-associated diseases as atherosclerosis, rodents have provided the basis for important age-related findings in many diverse areas, including nutrition, behavior, immunology, physiology, oncology, biochemistry, and neurobiology. This article reviews the literature supporting the relevance of the rodent model to human aging in each of these areas.

923. Hennessy, John W.; King, Maurice G.; McClure, Thomas A. & Levine, Seymour. (1977). **Uncertainty, as defined by the contingency between environmental events, and the adrenocortical response of the rat to electric shock.** ***Journal of Comparative & Physiological Psychology,*** 91(6), 1447–1460.
The pituitary-adrenal system is thought to be sensitive to the degree of uncertainty in a situation. In addition, there is some question whether the pituitary-adrenal system can be conditioned in a Pavlovian sense. Three experiments were conducted with 64 male Long-Evans rats; the 1st and 3rd sought to define uncertainty in terms of conditioned CS-UCS and UCS-UCS contingencies, which varied the amount of information that could be used to predict the occurrence of discrete shocks. The 2nd experiment examined the possibility that the adrenocortical system was subject to the laws of Pavlovian conditioning, by using a conditioned emotional response paradigm. Results show that the magnitude of the pituitary-adrenal response varied in a curvilinear manner along the dimension of uncertainty. Very low and very high degrees of uncertainty resulted in greater corticosterone elevations than did moderate levels. No evidence for Pavlovian conditioning of the adrenocortical system was found, although behavioral measures showed fear conditioning. The data are supportive of the hypothesis that the pituitary-adrenal response reflects the operation of an arousal system.

924. Hilakivi, Leena A. & Lister, Richard G. (1990). **Correlations between behavior of mice in Porsolt's swim test and in tests of anxiety, locomotion, and exploration.** ***Behavioral & Neural Biology,*** 53(2), 153–159.
Examined whether the behavior of 78 male mice in a putative animal model of depression, a swim test (R. D. Porsolt et al, 1977), is related to that in other behavioral tests. The other tests were the plus-maze test of anxiety, the holeboard test of exploration and locomotor activity, and a test of seizure threshold to bicuculline. Immobility of Ss in the swim test did not correlate with their behavior in any of the other tests used. The only significant correlations occurred between individual measures in the holeboard and plus-maze tests. Immobility in the swim test is not related to behavior in the tests of anxiety, directed exploration, locomotor activity, or seizure threshold.

925. Ingram, Donald K.; London, Edythe D. & Reynolds, Mark A. (1982). **Circadian rhythmicity and sleep: Effects of aging in laboratory animals.** ***Neurobiology of Aging,*** 3(4), 287–297.
Reviews the literature on age-related differences in sleep and rhythmic phenomena in laboratory animals covering 3 general areas: (1) age-related differences in biorhythms in general; (2) age-related differences in sleep patterns as assessed by psychophysiological measures; and (3) neurobiological correlates of biorhythms and sleep, including considerations of possible morphological, chemical, and endocrine bases of age-related defects in animal models. It is concluded that while several promising areas have been explored, systematic research bridging these areas is lacking.

926. Klein, Ehud H.; Tomai, Thomas & Uhde, Thomas W. (1990). **Hypothalamo-pituitary-adrenal axis activity in nervous and normal pointer dogs.** ***Biological Psychiatry,*** 27(7), 791–794.
Evaluated hypothalamo-pituitary-adrenal axis activity in genetically nervous pointer dogs, which have been suggested as an animal model for maladaptive anxiety, and in normal dogs. Findings indicate that nervous and normal dogs did not differ in plasma cortisol and adrenocorticotropic hormone (ACTH) levels or in cerebrospinal fluid (CSF) ACTH and corticotropin-releasing hormone levels.

927. Laudenslager, Mark L. (1988). **The psychobiology of loss: Lessons from humans and nonhuman primates.** ***Journal of Social Issues,*** 44(3), 19–36.
Suggests that an animal model with similarities to the grief process is available in young nonhuman primates, which have demonstrated a number of immunological, endocrine, and physiological responses to brief mother–infant separation experiences. Many of these changes occur in the absence of some of the confounding factors noted in the human research. Results of relevant studies of biological correlates in both humans and nonhuman primates are reviewed, and social intervention strategies that may have a significant impact on the biological system and modulate the risk for illness are discussed.

928. Laudenslager, Mark L. et al. (1983). **Coping and immunosuppression: Inescapable but not escapable shock suppresses lymphocyte proliferation.** ***Science,*** 221(4610), 568–570.
40 rats were given series of escapable shocks, identical inescapable shocks, or no shock. Ss were reexposed to a small amount of shock 24 hrs later, after which an in vitro measure of the cellular immune response was examined. Lymphocyte proliferation in response to the mitogens phytohemagglutinin and concanavalin A was suppressed in the inescapable shock group but not in the escapable shock group. This suggests that the controllability of stressors is critical in modulating immune functioning. Results also suggest that the immune system might be altered by the sorts of variables known to modulate consequences of uncontrollability such as learned helplessness. Possible mechanisms for the change in immune response include elevation of circulating corticosteroids and release of endogenous opioids.

929. Lawler, James E.; Barker, Gregory F.; Hubbard, John W. & Allen, Michael T. (1980). **The effects of conflict on tonic levels of blood pressure in the genetically borderline hypertensive rat.** ***Psychophysiology,*** 17(4), 363–370.
The failure to elicit large tonic elevations in blood pressure (BP) in animals may be due either to stressors that are insufficiently potent and/or to an inadequate physiological model. The present study sought to maximize the probability of producing large tonic changes in BPs by using a conflict paradigm in a genetic strain of rats that develops systolic BPs in the borderline hypertensive area. 48 male F_1 generation offspring of spontaneously hypertensive rats mated with normotensive controls (Kyoto-Wistar) were randomly split into 3 groups: experimental (subjected to 3 wks of avoidance training and 12 wks of conflict in conditioning cages), mild restraint control (placed in conditioning cages daily but not shocked), and maturation control (neither shocked nor restrained) groups. Results show that Ss subjected to conflict gradually developed tonic levels of systolic BP well into the hypertensive range. Restraint controls also showed some elevation, but maturational controls showed no change. The saliency of this animal model for the study of the etiology of stress-induced hypertension is discussed.

930. Leibowitz, Sarah F. & Shor-Posner, Gail. (1986). **Brain serotonin and eating behavior.** ***Appetite,*** 7(Suppl), 1–14.
Proposes that serotonin acts, in part, through a satiety mechanism of the medial hypothalamus to reduce ingestion of carbohydrates (CARs) while sparing protein intake in rats. In controlling the ratio of CAR to protein intake, this serotonergic system, which is responsive to the anorexic agent fenfluramine, is believed to function in direct opposition to the alpha$_2$-noradrenergic system of the paraventricular nucleus, which inhibits satiety for CAR and thereby potentiates the size of CAR meals. This serotonergic system may also indirectly oppose the catecholaminergic systems of the lateral hypothalamus, which mediate amphetamine anorexia and inhibit a hunger-stimulating system for protein intake, thereby delaying the initiation of protein meals. Examination of rats' normal eating patterns has indicated specific points in the circadian eating cycle where these hypothalamic monoamine systems, along with changes in circulating hormones and nutrients, may be physiologically activated.

931. Leigh, Hoyle & Hofer, Myron A. (1975). **Long-term effects of preweaning isolation from littermates in rats.** ***Behavioral Biology,*** 15(2), 173–181.
12 male Wistar rats who were isolated from littermates on the 12th postnatal day were weaned on the 21st day and housed individually till adulthood. Comparison groups consisted of 9 Ss isolated only after weaning and housed with littermates before and after weaning. In adulthood, reaction to handling, social preference, resting heart rate (HR), HR response to an air blast, HR response to a strange rat introduced to home cage, and gastric ulceration after immobilization were tested. The prewean isolates were significantly different from the comparison groups in the reaction to handling, social preference, and in the immediate HR change following the introduction of an intruder. The prewean isolates were more reactive to handling, preferred to be near a single adult rat, and showed less cardioacceleratory responses to intrusion. Results indicate that the altered early social relationships produced by reduction of litter size to a single pup during the socialization period before weaning can have distinct long-term effects on behavior, social preference, and physiologic response in adult rats.

932. Levine, Seymour & Wiener, S. G. (1988). **Psychoendocrine aspects of mother–infant relationships in nonhuman primates.** Special Issue: Psychoneuroendocrine aspects of maternal behavior. ***Psychoneuroendocrinology,*** 13(1–2), 143–154.
A review of the literature on the physiological and behavioral responses of squirrel monkeys and rhesus macaques following disruptions of mother–infant relationships shows that circulating levels of plasma cortisol increase following mother–infant separation. The presence of familiar Ss during the time of separation reduces the pituitary-adrenal response, compared to that elicited by total isolation. Visual access to the mother during separation ameliorates the plasma cortisol response and increases vocalization. Data support the hypothesis that vocalizations may serve as a coping response that reduces the physiological indices of arousal. Social interaction with familiar conspecifics may serve as a nonvocal coping response (e.g., proximity contact to other Ss) that also reduces the behavioral and physiological responses to maternal separation.

933. Manuck, Stephen B.; Kaplan, Jay R.; Adams, Michael R. & Clarkson, Thomas B. (1988). **Studies of psychosocial influences on coronary artery atherogenesis in cynomolgus monkeys.** ***Health Psychology,*** 7(2), 113–124.
Discusses studies (e.g., J. R. Kaplan et al, 1982) indicating that cynomolgus monkeys maintained on a cholesterol-containing diet provide a suitable primate model for studying how psychosocial factors contribute to coronary artery atherogenesis. Interactions with diet, sex, and behaviorally elicited cardiovascular reactivity are described.

934. Martin, Robert E.; Sackett, Gene P.; Gunderson, Virginia M. & Goodlin-Jones, Beth L. (1988). **Auditory evoked heart rate responses in pigtailed macaques (*Macaca nemestrina*) raised in isolation.** ***Developmental Psychobiology,*** 21(3), 251–260.
Studied heart rate (HR) responses evoked by 1 sec of 85-db white noise in 12 1-yr-old pigtailed macaques, 6 of which were raised in social isolation and 6 with mothers and peers. Tests were given for 5 days. Although baseline HR did not differ between groups, the pattern of change from baseline was not the same. Isolates showed only HR acceleration.

Socially reared Ss had a 10–11 sec biphasic response of acceleration followed by deceleration. It is concluded that early rearing experiences may affect later physiological processes involving autonomic nervous system balance. This conclusion is related to observations of persistent individual differences in HR by human children classified as inhibited.

935. McClintock, Martha K. (1978). **Estrous synchrony and its mediation by airborne chemical communication (*Rattus norvegicus*).** ***Hormones & Behavior,*** 10(3), 264–276.
Results of 2 experiments with 50 Ss show that the estrous cycles of female laboratory rats that lived together became significantly more synchronized than either the cycles of solitary Ss or the cycles of Ss randomly selected from different living groups. Airborne chemical communication between otherwise isolated groups of Ss was sufficient to produce the same level of estrous synchrony found among Ss that were actually living together.

936. McClintock, Martha K. (1984). **Estrous synchrony: Modulation of ovarian cycle length by female pheromones.** ***Physiology & Behavior,*** 32(5), 701–705.
Identified 3 hormonally distinct phases in the estrous cycles of virgin female Sprague-Dawley rats, aged 125–140 days, as points in the cycle that could produce different chemosignals: diestrus (preovulatory or follicular phase), proestrus (ovulatory phase), and metestrus (postovulatory phase). Each of these phases was used to generate a constant odor that was presented to female rats living in either an upwind or a downwind tunnel. Airborne chemosignals from the different phases had opposing effects on the timing of the estrous cycle that were consistent with a coupled oscillator model of ovarian synchrony. Preovulatory odors shortened or phase-advanced the ovarian cycle, whereas ovulatory odors lengthened or phase-delayed the cycle.

937. McGaugh, James L. (1983). **Hormonal influences on memory.** ***Annual Review of Psychology,*** 34, 297–323.
Examines recent studies of the effects on memory of hormones, as well as treatments affecting hormonal functioning, focusing on adrenergic catecholamines (norepinephrine and epinephrine), pituitary peptides (ACTH, vasopressin, and oxytocin), and opiate peptides (endorphin and enkephalin). Experiments in which the treatments are administered shortly after training are considered. Also considered is whether treatments known to modulate memory storage act through influences on hormones. There is extensive evidence that hormones can modulate memory and moderately convincing evidence that endogenous hormones modulate memory storage. Findings suggest that peripheral processes may be involved in the physiology of memory, but the studies have not revealed the details of the mechanisms by which hormones modulate memory. Such understanding, when obtained, should contribute significantly to a more general understanding of the physiology and cellular neurobiology of memory.

938. Meisel, Robert L.; Hays, Timothy C.; del Paine, Stephanie N. & Luttrell, Vickie R. (1990). **Induction of obesity by group housing in female Syrian hamsters.** ***Physiology & Behavior,*** 47(5), 815–817.
Examined the effects of group housing on body weight in adult female Syrian hamsters housed in groups of 5 per cage over 10 wks. Ss increased their body weight by 61% compared with an 18% increase in body weight for females housed individually. Group-housed females were significantly longer, had a higher percentage of body fat, and had larger adrenal glands than individually housed Ss. Results are discussed in the context of social stress mediating obesity in Syrian hamsters and support a socially based animal model of obesity.

939. Meyer-Bahlburg, Heino F. (1977). **Sex hormones and male homosexuality in comparative perspective.** ***Archives of Sexual Behavior,*** 6(4), 297–325.
Reviews psychoendocrine studies of human homosexuality—the effects of hormone treatments on sexual orientation, the association of clinical endocrine syndromes with homosexuality, sex hormone measurements in homosexual Ss, and the issue of prenatal endocrine influences on human sexual orientation. The available studies, often deficient in methodology, have produced conflicting and largely negative results as to a hormonal theory of human homosexuality. It is noted that there is hardly any overlap between psychoendocrine studies on sexual orientation in man and on sex-dimorphic mating behavior in subhuman mammals that would allow systematic comparisons. A major alternate theory of human sexuality has been derived from learning theory, and this position has been strengthened by the recent development of partially successful behavior therapy approaches to the change of sexual orientation. At present the author sees a lack of animal experimentation that could assist in formulating a learning-based etiological theory of human homosexuality. Since the influence of hormones relative to learning on sexual behavior appears to diminish along the evolutionary scale, it is suggested that animal models need to include manipulation of both hormonal and learning conditions.

940. Mrosovsky, N. (1988). **Seasonal affective disorder, hibernation, and annual cycles in animals: Chipmunks in the sky.** Special Issue: Seasonal affective disorder: Mechanisms, treatments, and models. ***Journal of Biological Rhythms,*** 3(2), 189–207.
Discusses 2 aspects of hibernation that suggest it may be useful as a model of depression (particularly seasonal affective disorder [SAD]): the apparent withdrawn state that accompanies hibernation and the periodicity of its manifestations. Topics considered include animal models of SAD, dissimilarities between SAD and hibernation, and animal models of the temporal aspects of SAD.

941. Murison, Robert & Overmier, J. Bruce. (1989). **Animal models: Promise and pitfalls.** ***Biological Psychiatry,*** 26(4), 431–433.
Asserts that the facts reported by J. L. Haracz et al (1988) do not demonstrate a parallel between depressive syndrome and the learned helplessness animal model. The present authors claim that the dexamethasone-plus-challenge test should have been administered after the animals were in a helpless state. A response by Haracz et al follows and addresses the differences between functional and biological parallelism.

942. Paré, William P. (1989). **"Behavioral despair" test predicts stress ulcer in WKY rats.** ***Physiology & Behavior,*** 46(3), 483–487.
Spontaneously hypertensive rats (SHR), Wistar Kyoto (WKY) rats, and Wistar rats were exposed to Porsolt's forced-swimming test of "behavioral despair." In addition to floating time, which was the measure of despair, headshakes, bobbing, diving, and struggling time were also recorded. Ss

were subsequently exposed to the activity stress (A-S) ulcer procedure. Wistar Ss had the highest struggling time scores and the fewest A-S ulcers. WKY Ss were judged as more depressed and their ulcer severity scores were significantly greater as compared to SHR and Wistar Ss. In addition, a within-strains analysis revealed that WKY Ss with high despair scores also had the most severe stress-ulcer scores.

943. Paré, William P. (1989). **Stress ulcer susceptibility and depression in Wistar Kyoto (WKY) rats.** ***Physiology & Behavior,*** 46(6), 993–998.
Compared 16 Wistar Kyoto (WKY) rats with 16 Fisher 344 rats, 16 spontaneously hypertensive rats, and 16 Wistar rats to determine whether any behavioral differences were related to WKY rats' greater vulnerability to stress-ulcer disease. WKY Ss were deficient in several behavioral tasks. Prevalence of freezing behavior in a shuttlebox task and low ambulation scores in an open-field test suggested that depressive behavior was a WKY behavior characteristic. WKY Ss were judged more depressed in the Porsolt forced-swim test as compared with other strains. A depression–ulcer relationship may exist in WKY Ss making them a good model for studying relationships between depression and stress-induced disease.

944. Paré, William P. & Vincent, George P. (1989). **Environmental enrichment, running behavior and activity-stress ulcer in the rat.** ***Medical Science Research,*** 17(1), 35–36.
Two experiments tested the hypothesis that Sprague-Dawley rats exposed to an enriched developmental environment would develop more activity-stress (A-S) ulcers compared with rats reared in an impoverished environment. Results underline the notion that the etiology of stress ulcers is a function of the interaction of predisposing factors (e.g., genetic factors, environmental complexity) and precipitating factors (e.g., the A-S procedure).

945. Paré, William P. & Vincent, George P. (1986). **Stomach acid secretion in submissive and aggressive rats.** ***Behavioral Neuroscience,*** 100(3), 381–389.
Conducted 4 experiments with 60 female and 162 male Long-Evans rats to determine whether aggressive and submissive behavior are related to either an increase or a decrease in gastric secretion. In Exp I, intruder rats placed in an established male–female colony and attacked by a dominant alpha male secreted less acid than intruders exposed to nonaggressive males and females. In Exp II, intruders exposed to attack and subsequently returned to the encounter site, but protected from physical attack, still demonstrated a gastric hyposecretion. Ss with chronic gastric cannulas in Exp III also revealed an acid inhibition when attacked and later when exposed to, but protected from, attack. Both intruders and attacking males were prepared with gastric cannulas in Exp IV. Both demonstrated secretory inhibition following attack and attack-protected sessions. The inhibitory effect was greater and more persistent for intruders than for aggressive Ss. It is suggested that the inhibition occurring during the attack-protected sessions may have been mediated by some conditioning processes, and other possible associative mechanisms, including a learning model or a direct sensory model, are discussed.

946. Paré, William P.; Vincent, George P.; Isom, Kile E. & Reeves, Jesse M. (1978). **Restricted feeding and incidence of activity-stress ulcers in the rat.** ***Bulletin of the Psychonomic Society,*** 12(2), 143–146.
In the 1st experiment with 24 male Sprague-Dawley rats, results show that Ss housed in running-wheel activity cages and fed 1 hr daily exhibited excessive running and subsequently died, revealing large stomach ulcers. However, prior experience with a 1-hr feeding schedule for 15 days before exposure to the activity-stress procedure did significantly extend survival time. A 2nd study with 60 male Sprague-Dawley rats illustrated that feeding schedules of .5, 1.0, 1.5, or 2.0 hrs daily did not yield differences in terms of stomach ulcers or survival time, but a 3-hr daily feeding did significantly reduce ulcer incidence and increase survival time.

947. Paré, William P. (1975). **Coping behavior, punishment and gastric secretion in the rat.** ***Physiology & Behavior,*** 15(5), 627–629.
One group of 4 pairs of male Sprague-Dawley rats could avoid continuous grid shock by depressing a treadle manipulandum. In another group of 4 pairs, the escape response was punished by shock administered via the treadle. Chronic gastric cannulas allowed the collection of gastric secretion. Total acid output was not significantly different between shock-yoke control Ss and their respective experimental mates. Gastric secretion and total acid output were significantly depressed during the shock period as compared to the preshock period, and Ss whose coping response was punished secreted less acid as compared to Ss which were not punished for eliciting the coping response. Gastric secretion, as a response to environmental stress, is discussed.

948. Pellis, Vivien C.; Pellis, Sergio M. & Teitelbaum, Philip. (1991). **A descriptive analysis of the postnatal development of contact-righting in rats (*Rattus norvegicus*).** ***Developmental Psychobiology,*** 24(4), 237–263.
Studied the development of contact-righting by rats from birth to weaning. Such righting involved both vestibular and tactile forms, which matured at different rates. Vestibularly triggered righting from the supine position and asymmetrical contact of the body surface with the ground are explained in terms of the maturation process. Findings are discussed with reference to the reverse sequence that occurs in some parkinsonian patients, damage to the nervous system, and fetal alcohol syndrome. This study shows that the postnatal development of contact-righting involves a complex interaction of righting subsystems.

949. Petty, F. & Sherman, A. D. (1983). **Learned helplessness induction decreases *in vivo* cortical serotonin release.** ***Pharmacology, Biochemistry & Behavior,*** 18(4), 649–650.
Frontal neocortices of freely moving, unanesthetized male Sprague-Dawley rats were perfused before, during, and after exposure to 40 min of uncued pulsed random footshock. Ss developing nontransient learned helplessness had lower levels of serotonin in cortical perfusate than those failing to develop helplessness.

950. Sherman, A. D. & Petty, F. (1984). **Learned helplessness decreases [^{3}H]imipramine binding in rat cortex.** ***Journal of Affective Disorders,*** 6(1), 25–32.
235 rats were divided into control and footshock exposure groups. Footshock Ss received training to induce learned helplessness, an animal model of depression. Ss exhibited a range of helplessness behavior after learned helplessness induction, with some performing in the control range and others demonstrating more profound deficits in escape behavior. Specific binding of [^{3}H]imipramine decreased in the

frontal neocortex from Ss demonstrating learned helplessness. The decrease was in maximal binding but not in affinity for the receptor site. No change in [^{3}H]imipramine binding was found in septum or hippocampus. The receptor changes found in frontal neocortex parallel behavioral and neurochemical changes produced by learned helplessness in this region. These changes are also similar to those found in the frontal neocortex from suicides and in platelets of patients with depression. Results demonstrate the utility of the learned helplessness model in preclinical studies of the mechanism of action of antidepressant drugs.

951. Shiromani, Priyattam J.; Klemfuss, Harry; Lucero, Sam & Overstreet, David H. (1991). **Diurnal rhythm of core body temperature is phase advanced in a rodent model of depression.** ***Biological Psychiatry,*** 29(9), 923–930.
Examined the diurnal rhythm of core body temperature in a strain of rats with an upregulated central muscarinic receptor system. The Flinders-Sensitive Line (FSL) was derived by selectively breeding rats for sensitivity to cholinergic agonists. When compared with 6 male control rats, the 6 male FSL rats showed a strong phase advance of the acrophase in body temperature during a standard light–dark schedule. Some patients with some types of depression also show phase advances in a number of circadian rhythms, including temperature. The finding of a phase advance in a rodent model with a known upregulated muscarinic receptor system is compatible with both the phase advance and the muscarinic overdrive theories of depression. Findings also further validate the usefulness of the FSL rats in the study of depression.

952. Somsen, Riek J.; Molenaar, Peter C.; Van der Molen, Maurits W. & Jennings, J. Richard. (1991). **Behavioral modulation patterns fit an animal model of vagus–cardiac pacemaker interactions.** ***Psychophysiology,*** 28(4), 383–399.
Used a computer model of the dynamic interaction between the vagus nerve and the sinoatrial pacemaker membrane potential in the heart of the rabbit to reconstruct heart rate (HR) changes under vagal excitation conditions. The aim was to determine whether a hypothetical pattern of vagal acetylcholine (ACh) release, which was based on human HR results in a reaction time (RT) task, could be fit to this model. The reconstructed HR results showed changes that were highly consistent with experimental human HR changes. The model reliably reproduced effects of parameters such as intrinsic HR level, ACh stimulus intensity, and ACh stimulus duration. The effects of anticipatory vagal ACh release, stimulus-induced ACh, and subsequent blocking of ACh could be untangled in the reconstructed HR results.

953. Vincent, George P. & Paré, William P. (1976). **Activity-stress ulcer in the rat, hamster, gerbil and guinea pig.** ***Physiology & Behavior,*** 16(5), 557–560.
Conducted an experiment with 30 male Sprague-Dawley rats, 20 male LHC/LoK hamsters, 20 male DUB/Hart guinea pigs, and 20 male gerbils. Ss were housed in activity cages and fed 1 hr/day. By the end of the 21-day period, 86, 100, 70 and 70% of rats, hamsters, gerbils, and guinea pigs had developed lesions in the glandular stomach. This procedure was thus capable of producing lesions in species other than the rat, thereby increasing the value of the procedure as an ulcerogenic technique.

954. Weiner, Herbert. (1991). **From simplicity to complexity (1950–1990): The case of peptic ulceration: II. Animal studies.** ***Psychosomatic Medicine,*** 53(5), 491–516.
Discusses progress in the understanding of the formation of gastric erosions in rats and considers the role of gastric acid secretion in their pathogenesis. With several experimental procedures, the body temperature falls; preventing this decrease averts erosions. A fall in body temperature or exposure to cold is associated with the secretion of thyrotropin-releasing hormone (TRH) and with both an increase and a decrease in corticotropin-releasing factor (CRF) in discrete regions of rat brains. TRH increases in acid secretion, producing gastric erosions and slow contractions, while CRF has the opposite effects. A major site of interaction of the 2 peptides is in the dorsal motor complex of the vagus nerve. TRH increases serotonin (5-hydroxytryptamine [5-HT]) secretion into the stomach, and 5-HT counterregulates acid secretion and slow contractions. Many other peptides stimulate or inhibit gastric acid secretion.

955. Weiss, Jay M.; Pohorecky, Larissa A.; Salmon, Sherry & Gruenthal, Michael. (1976). **Attenuation of gastric lesions by psychological aspects of aggression in rats.** ***Journal of Comparative & Physiological Psychology,*** 90(3), 252–259.
In a study with 80 male albino rats, Ss that fought with each other in response to electric shock showed reduced gastric lesions in comparison with Ss that received the same shocks alone so that fighting behavior did not occur. Also, gastric lesions were similarly reduced in Ss that fought even though they could not physically contact one another because of a barrier between them. In this case, the "protective" effect of fighting derived from the release or display of fighting behavior and did not require physical combat. A 2nd experiment with 48 rats showed that Ss that received shock together but did not engage in fighting behavior showed no reduction of gastric lesions, so that the protective effect of fighting was not an artifact of Ss receiving shock together.

Psychophysiology

956. Aluja, A. (1990). **Psychopathology of disinhibition: A model for the study of psychopathy.** ***Revista de Psiquiatría de la Facultad de Medicina de Barcelona,*** 17(3), 130–141.
Describes an animal model of human disinhibitory behavior. Based upon the behavior of animals with septal lesions, it is proposed that limbic system irregularities are the biological foundation of disinhibitory psychopathological syndromes, of which psychopathy is the prototype. Other topics discussed include: disinhibition as a personality trait, platelet monoamine oxidase (MAO) activity, androgens, estrogens, and monoaminergic brain functioning. (English abstract).

957. Anisman, Hymie & Zacharko, Robert M. (1982). **Depression: The predisposing influence of stress.** ***Behavioral & Brain Sciences,*** 5(1), 89–137.
A research review suggests that the effects of stressful experiences on affective states may be related to the depletion of several neurotransmitters, including norepinephrine, dopamine, and serotonin. Aversive experiences give rise to behavioral attempts to cope with the stressor, coupled with increased use and synthesis of brain amines to contend with environmental demands. However, when there is no perceived control over the aversive stimuli, amine utilization may exceed synthesis, depleting amine stores and promoting affective disorder. The processes governing the depletions

may be subject to sensitization or conditioning, affecting later reactions to related stimuli. It is suggested that either the initial amine depletion provoked by aversive stimuli or a dysfunction of the adaptive process, resulting in persistent amine depletion, contributes to depression. Other organismic, experiential, and environmental variables also influence the effects of aversive experiences on neurochemical activity and may thus influence vulnerability to depression. (28 peer commentaries and the authors' response are appended.).

958. Cox, Ronald H. (1991). **Exercise training and response to stress: Insights from an animal model.** ***Medicine & Science in Sports & Exercise,*** 23(7), 853–859.
Examined the effect of exercise training (ET) on the sympathoadrenal and cardiovascular responses to stress in borderline hypertensive rats (BHRs). Ss trained with swimming showed reduced heart rates and plasma norepinephrine and epinephrine levels in response to novel footshock stress. In contrast, systolic blood pressure was significantly higher in trained BHRs during the stress. Cholinergic blockade with atropine (ATR) sulfate had little effect on heart rate during stress in untrained Ss. In contrast, parasympathetic blockade with ATR sulfate significantly elevated the heart rate of trained Ss during stress. Findings illustrate the dilemma of trying to draw conclusions about the effect of ET on stress reactivity.

959. Dantzer, Robert. (1986). **Psychobiology of adaptive behavior: A French contribution to the issue of stress, coping, and health.** ***Advances,*** 3(4), 36–44.
Discusses the French contribution to the problem of behavior and health, concentrating on holistic studies that specifically address the interaction between at least 2 different levels of response of the organism (e.g., physiology and behavior). Two main perspectives from which the problem of behavior and health may be viewed are discussed (i.e., the influences of the central nervous system [CNS] on the responses of the body and the body's influences on brain and behavior). Animal models are discussed to illustrate the mechanisms of relationships among stress, coping, and health.

960. Doerries, Lee E.; Stanley, Eric Z. & Aravich, Paul F. (1991). **Activity-based anorexia: Relationship to gender and activity-stress ulcers.** ***Physiology & Behavior,*** 50(5), 945–949.
Determined the susceptibility of 24 male and 24 female rats to activity-based anorexia (ABA) and activity-stress ulcer (ASU) following 25% and 30% losses of their original body weights. Males reached both weight loss criteria in fewer days than did females. None of the Ss sacrificed at the 25% weight loss criterion evidenced gastric lesions; 52% of the Ss sacrificed at the 30% weight loss criterion had 1 or more lesions. No gender differences with respect to gastric lesions were observed at the 30% weight loss criterion; however, at both weight loss criteria, females ate and ran more than males. It is concluded that ASUs are a consequence rather than a cause of ABA and that there is a sexually dimorphic susceptibility to ABA but not ASUs.

961. Glick, Stanley D. & Greengard, Olga. (1980). **Exaggerated cerebral lateralization in rats after early postnatal hyperphenylalaninemia.** ***Brain Research,*** 202(1), 243–248.
Consistent with observations on phenylketonuric children (J. DelValle, 1978), hyperphenylalaninemia in rats, which was induced by phenylalanine plus alphamethylphenylalanine, led to hyperactivity (HY) and diminished learning ability (DLA) in later life. Results suggest that exaggerated lateralization may in part be responsible both for the HY and the DLA in maze paradigms of animal models of phenylketonuria.

962. Hara, Chiaki; Ogawa, Nobuya & Imada, Yoshiro. (1981). **The activity-stress ulcer and antibody production in rats.** ***Physiology & Behavior,*** 27(4), 609–613.
Divided 40 male Wistar rats into 4 groups: a control group fed freely for 7 days and 3 stressed groups fed 1 hr each day for 3, 5, and 7 days. All Ss were housed in activity-wheel cages under a reversed light–dark cycle. They were immunized twice with sheep erythrocytes before beginning the stress schedule. After Ss were sacrificed, their stomach, lungs, thymus, spleen, and adrenals were removed and examined histologically. The incidence of Ss with ulcer reached 60% in the 3-day stress group, although antibody production was unaffected and no Ss died in this group. In the 5- and 7-day stress groups, the incidence of ulcer reached 90%, antibody production was significantly inhibited, and the degree of mortality increased to more than 50%. The severity of ulcer was in proportion to the length of stress periods. Data suggest that activity-stress Ss have 2 pathological phases: stomach ulcer and immunodeficiency. The present study emphasizes that immunological response can serve as an indicator of the biological response for stress.

963. McGaugh, James L. (1989). **Involvement of hormonal and neuromodulatory systems in the regulation of memory storage.** ***Annual Review of Neuroscience,*** 12, 255–287.
A review of the literature indicates that retention of recently acquired information can be altered by different posttraining treatments that affect hormonal and neuromodulatory systems (H&NSs). Support is given to the general hypothesis that endogenous H&NSs activated by learning may play an important role in regulating the storage of information. Recent research has investigated the role of the amygdala and the interaction of the adrenergic, cholinergic, and opioid-peptide systems.

964. McGaugh, James L. (1990). **Significance and remembrance: The role of neuromodulatory systems.** ***Psychological Science,*** 1(1), 15–25.
A review of experiments (e.g., J. L. McGaugh and P. E. Gold; 1989) with memory provide strong evidence in support of the view that the neuromodulatory systems activated by experience affect memory storage through effects involving noradrenergic receptors within the amygdaloid complex and consequent influences on other brain regions mediated by the stria terminalis. These endogenous physiological processes appear to play a central role in enabling the significance of events to influence their remembrance.

965. Mempel, Eugeniusz & Wieczorek, Marek. (1990). **Parkinson's syndrome induced in cats by the use of 6-hydroxydopamine: Observations of behavior and motor disorders.** XXXI International Congress of Physiological Sciences: Recovery from brain damage: Behavioral and neurochemical approaches (1989, Warsaw, Poland). ***Acta Neurobiologiae Experimentalis,*** 50(4–5), 269–279.
Investigated the possibility of inducing Parkinson's syndrome in 6 cats by microinjection of 6-hydroxydopamine (6-OHDA) into the (1) pars compacta of substantia nigra (SNC), (2) SNC and globus pallidus, and (3) SNC and caput nuclei caudati. In all 3 kinds of lesions of the dopaminergic

system, disturbances of behavior involving especially the motor system were obtained. These disturbances corresponded to the parkinsonism syndrome of bradykinesia-akinesia, increased muscle tonus of plastic type, vegetative disorders (sialorrhea, pupils), and psychic disorders such as the lack of interest in the surroundings and food. The character of the enhanced muscle tonus typical for extrapyramidal disturbances was confirmed by electromyogram (EMG). The parkinsonismlike syndrome was transient and receded after several weeks.

966. Mine, Kazunori et al. (1981). **A new experimental model of stress ulcers employing aggressive behavior in 6-OHDA-treated rats.** ***Physiology & Behavior,*** 27(4), 715–721.
30 male Wistar rats receiving intraventricular injection of 6-hydroxydopamine (6-OHDA, 250 µg) were housed in isolation for 1 mo and placed in the same cage as an untreated rat while being subjected to continuous tail pinching. 6-OHDA-treated Ss violently attacked the untreated S and stabilized fighting between the 2 persisted for over 1 hr. The 6-OHDA-treated Ss consistently played the dominant role in the dominant–subordinate relationship established between Ss. By allowing this fighting to continue for 1 hr, gastric erosion associated with severe hemorrhage occurred with a high incidence in untreated Ss. These erosions were comparable in Ss examined immediately after fighting and in those examined after 1 and 3 hrs. Poststress rest was not required for erosion to occur. The presence of bite wounds incurred during fighting was unrelated to the incidence of gastric erosion. In addition, it was unnecessary to fast the untreated Ss beforehand. Based on these findings, the present model apparently does not rely on physical stimulation. These factors and the ease of its execution make the present new experimental model of stress ulcers very useful.

967. Morini, Giuseppina & Impicciatore, Mariannina. (1985). **A new experimental model for reproducing chronic ulcer in rats.** ***IRCS Medical Science: Psychology & Psychiatry,*** 13(9–10), 806.
Generated chronic mucosal lesions in 40 individually housed and 20 group-housed female Wistar rats to determine whether isolation-induced aggression is correlated with an altered gastric function. Results indicate that barrier mucus values in the isolated groups were significantly higher than those of group-living Ss; these differences are explained as a defense reaction of the stomach.

968. Murison, Robert & Olafsen, Kåre. (1991). **Stress ulceration in rats: Impact of prior stress experience.** ***Neuroscience & Biobehavioral Reviews,*** 15(3), 319–326.
Studies are lacking on the significance of shock characteristics (i.e., controllability, predictability) for later activity stress ulcers. Further studies are required on the significance of psychological characteristics of the prestress and on the effects of these prestressors at different stages of the life cycle. The identification of these factors, and a clearer picture of the protective and exacerbating effect of prior stress, will allow exploration of the physiological (central and peripheral) mechanisms underlying ulcer development and ulcer susceptibility.

969. Natelson, Benjamin H. & Cagin, Norman A. (1981). **The role of shock predictability during aversive conditioning in producing psychosomatic digitalis toxicity.** ***Psychosomatic Medicine,*** 43(3), 191–197.
Subjected 30 male Hartley albino guinea pigs to Pavlovian fear conditioning and then infused them with ouabain (6 µg/kg/min, iv) on a day when all experimental stimuli except shock were delivered. A significant shortening in latency to the onset of life-threatening digitalis toxicity was found when comparisons were made to controls that had never been shocked. This effect was not found in other Ss infused with ouabain after exposure to sessions of unsignaled shock. Data indicate that psychological factors, when divorced from physical factors, may produce lethal digitalis toxicity in an organism that would otherwise be asymptomatic. It is suggested that the intense arousal associated with a signal that had previously been paired with shock is sufficient to precipitate cardiac arrhythmias in an animal with a predisposition to cardiac automaticity such as is produced by digitalis. Conversely, the less intense arousal associated with unsignaled shock is inadequate to produce this effect. This interpretation indicates that the state of an animal's health will affect its psychosomatic response to signaled or unsignaled shock.

970. Paré, William P.; Vincent, George P. & Isom, Kile E. (1979). **Age differences and stress ulcer in the rat.** ***Experimental Aging Research,*** 5(1), 31–42.
In 2 experiments, male Sprague-Dawley rats, aged 2, 7, and 12 mo, were exposed to supine body restraint plus cold for 3 hrs. The mean cumulative length of lesions for the 2-, 7-, and 12-mo-old Ss was 22, 43, and 16 mm, respectively. The same experimental design was used in Exp II, but the pylorus was ligated prior to restraint. Total acid output was 134.6, 178.2, and 64.7 µ Eq/60 min, respectively, for the 3 groups. Older Ss were not more susceptible to stress ulcer, and gastric acid was not significantly related to degree of ulceration.

971. Unis, Alan S.; Petracca, Frances & Diaz, Jaime. (1991). **Somatic and behavioral ontogeny in three rat strains: Preliminary observations of dopamine-mediated behaviors and brain D-1 receptors.** ***Progress in Neuro-Psychopharmacology & Biological Psychiatry,*** 15(1), 129–138.
Compared the acquisition of developmental milestones and maturational motor reflexes in Fischer (F344), Buffalo, and Sprague-Dawley rat strains. Open-field behavior on Postnatal Day 21 was scored for locomotor activity, rears, and center entries. F344 Ss, which may model attention deficit-hyperactivity disorder (ADHD), were the slowest in acquiring developmental milestones and in gaining weight, but they were intermediate in scores for locomotor activity and rearing. Preliminary autoradiographic data using the D-1 specific ligand [^{3}H]-SCH 23390 suggested that D-1 receptors, which display age-dependent changes in concentration and distribution, were relevant to Day 21 open-field behavior. F344 Ss demonstrated developmental dysmaturation consistent with that observed in children with ADHD in that somatic growth was disproportionately delayed in comparison with neurological and motor maturation.

972. Zalcman, Steve; Irwin, Jill & Anisman, Hymie. (1991). **Stressor-induced alterations of natural killer cell activity and central catecholamines in mice.** ***Pharmacology, Biochemistry & Behavior,*** 39(2), 361–366.
Determined natural killer (NK) cell cytotoxicity at various intervals (0.5, 24, or 48 hrs) following exposure to uncontrollable footshock in 3 strains of mice. Stressor application provoked reductions of NK activity, but the time course of the NK changes varied across strains. Whereas NK

cytotoxicity was reduced in C57BL/6J mice 0.5–48 hrs following stressor exposure, this effect was delayed in C3H/HeJ mice, being evident 24–48 hrs following stressor application. In BALB/cByJ mice, NK activity was significantly reduced 24 hrs after footshock but returned to control levels within 48 hrs of stressor exposure. Central norepinephrine and dopamine concentrations and activity were influenced by the stressor treatment in a strain-dependent fashion.

Psychopharmacology

973. Adamec, Robert E. (1991). **Acute and lasting effects of FG-7142 on defensive and approach-attack behavior in cats: Implications for models of anxiety which use response suppression.** ***Journal of Psychopharmacology,*** 5(1), 29–55.
Investigated the effect of the anxiogenic β-carboline, FG-7142, on defense and approach–attack behavior in adult male cats. FG-7142 enhanced defensive responses to rats and mice when tested 10 and 20 min after injection. FG-7142 suppressed approach and attack responses to prey, and the behavioral effects of FG-7142 were blocked by the specific benzodiazepine receptor blocker RO-15-1788. One or two administrations of FG-7142 produced a lasting (up to 105 days) enhancement of defensive response to rats. Defensive response to conspecific threat vocalizations was also lastingly increased. Findings have implications for rodent models of a form of benzodiazepine receptor-dependent anxiety that use response suppression to assess fearful "anxiogenic" effects of drugs.

974. Adamec, Robert E. (1991). **Corticotropin releasing factor: A peptide link between stress and psychopathology associated with epilepsy?** ***Journal of Psychopharmacology,*** 5(2), 96–104.
Examines the effects of corticotropin releasing factor (CRF) on the link between stress and psychopathology associated with limbic epilepsy. Data indicate that CRF does not play a simple modulatory role in behavior. CRF reproduces many of the central effects initiated by environmental stressors. One primary effect of limbic and locus coeruleus CRF systems seems to be to modulate neural circuitry involved in anxiety. The functioning of CRF as a modulator of animal anxiety can vary from anxiogenic, to neutral to anxiolytic, depending on the prior experience of the animal. Handling stress eliminates the anxiogenic action of CRF, and repeated limbic seizures change CRF into an anxiolytic. CRF is an important link between stress and affective disturbance, but the nature of that link is complex and poorly understood.

975. Adams, Jane; Vorhees, Charles V. & Middaugh, Lawrence D. (1990). **Developmental neurotoxicity of anticonvulsants: Human and animal evidence on phenytoin.** Special Issue: Qualitative and quantitative comparability of human and animal developmental neurotoxicity. ***Neurotoxicology & Teratology,*** 12(3), 203–214.
Reviews literature on the effects of gestational phenytoin exposure. Focus is on the human and animal evidence for the teratogenicity of phenytoin and on animal models of fetal anticonvulsant syndromes. The Fetal Hydantoin Syndrome (FHS) consists of craniofacial defects and any 2 of the following: pre/postnatal growth deficiency, limb defects, major malformations, and mental deficiency. Data suggest a prevalence of FHS of 10–30% in infants of women ingesting phenytoin during the 1st trimester or beyond. Gestational exposure to phenytoin produces serum-related morphological and behavioral abnormalities in both humans and animals. Effects include vestibular dysfunction, hyperactivity, and learning and memory deficits. Difficulties with existing studies include the absence of thorough behavioral assessments of prenatally exposed children and of comparable end points for human and animal studies.

976. Adams, Lynne M. & Geyer, Mark A. (1985). **Patterns of exploration in rats distinguish lisuride from lysergic acid diethylamide.** ***Pharmacology, Biochemistry & Behavior,*** 23(3), 461–468.
Evaluated an animal model of the effects of LSD in humans, proposed by the present authors (in press), by studying the effects of doses of 5, 15, 30, and 60 μg/kg lisuride (a nonhallucinogenic congener of LSD) in male Sprague-Dawley rats, using a behavioral pattern monitor (BPM). The BPM provided both quantitative measures of crossovers, rearings, and holepokes and qualitative measures of spatial patterns of locomotion. A holeboard chamber connected to a home cage provided 2 test situations. Ss were tested either with or without access to the home cage. In both situations, lisuride exhibited a biphasic dose–response curve for horizontal locomotion (low dose suppression and high dose enhancement), while rearing was significantly reduced at all doses. Lisuride also produced a dose-dependent increase in the perseverative quality of locomotor patterns. A comparison of these results with previous studies with LSD indicates that, with the exception of rearings, lisuride failed to mimic LSD's characteristic effects on exploratory activity. Rather, lisuride exhibited many similarities to the dopamine agonist apomorphine.

977. Adams, Lynne M. & Geyer, Mark A. (1985). **Effects of DOM and DMT in a proposed animal model of hallucinogenic activity.** ***Progress in Neuro-Psychopharmacology & Biological Psychiatry,*** 9(2), 121–132.
Tested an animal model of LSD's effects in humans for its applicability to other hallucinogens (2,5-dimethoxy-4-methylamphetamine [DOM] and N,N-dimethyltryptamine [DMT]). Findings from studies with male Sprague-Dawley rats suggest that the enhanced avoidance of novel and central areas in an exploration test is a valid indicator of hallucinogenic activity, since all 3 hallucinogens produced this effect. The behavioral effects in rats may be analogous to the enhanced responsiveness to environmental settings induced by hallucinogens in humans.

978. Adams, Lynne M. & Geyer, Mark A. (1985). **A proposed animal model for hallucinogens based on LSD's effects on patterns of exploration in rats.** ***Behavioral Neuroscience,*** 99(5), 881–900.
Examined the utility of various measures of exploratory activity in rats in an animal model of hallucinogens. A hole board chamber, connected by a door to a home cage, provided 2 test situations. Ss either were placed directly into the hole board with the door closed (forced exploration) or were placed in the home cage and, following adaptation, the door was opened (free exploration). The monitoring system provided both quantitative measures (crossovers, rearings, and hole pokes) and qualitative measures of locomotor patterns. Four experiments, in which 196 male Sprague-Dawley rats received subcutaneous injections of saline or 2–160 μg/kg LSD, revealed 3 major categories of effects, distinguishable on the basis of dose dependency, time course, or response to environmental manipulation: (a) increased avoidance of novel and central areas, (b) disruption of the spatial

patterning of locomotion, and (c) suppression of rearing. All 3 effects exhibited partial tolerance 24 hrs after 1 injection of 30 μg/kg LSD and complete tolerance after 5 daily injections. The possibility that LSD's enhancement of neophobia in rats may be a valid analog model of its intensification of affective reactions in humans is discussed.

979. Adler, Lawrence E.; Rose, Greg & Freedman, Robert. (1986). **Neurophysiological studies of sensory gating in rats: Effects of amphetamine, phencyclidine, and haloperidol.** ***Biological Psychiatry,*** 21(8–9), 787–798.
Recorded auditory evoked potentials (AEPs) in a conditioning–testing paradigm from unanesthetized freely moving male Sprague-Dawley albino rats to initiate development of an animal model of abnormal sensory gating. Middle latency (15–50 msec) AEPs were recorded, and gating mechanisms were assessed by measuring the suppression of response to a 74 db click test stimulus following an earlier identical conditioning stimulus at 0.5-sec intervals. Ss demonstrated significant suppression of the N50 response to the auditory stimulus. Amphetamine (AMP) significantly interfered with the suppression of the response to the stimulus; haloperidol, injected after AMP, returned the conditioning–testing ratio toward normal values. Phencyclidine caused a similar decrease in suppression. Results with psychotomimetic drugs in an animal model parallel abnormalities in sensory gating previously observed in psychotic human Ss.

980. Adrien, Joëlle & Martin, Patrick. (1990). **Animal models sensitive to antidepressants: What sort of sleep?** ***Psychiatrie Française,*** 21(1), 46–48.
Discusses the utility of the learned-helplessness paradigm for animal models of depression, for studies on the correlation between sleep and depression, and for studies on the behavioral effects of long-term psychopharmacotherapy. Results are reported of a study on the correlation between sleep parameters and experimentally induced depression (i.e., learned helplessness) in rats. (English abstract).

981. Altenor, Aidan & DeYoe, Edgar A. (1977). **The effects of DL-*p*-chlorophenylalanine on learned helplessness in the rat.** ***Behavioral & Neural Biology,*** 20(1), 111–115.
Exp I studied the replicability of the learned helplessness (LH) effect, using 16 male Long-Evans hooded rats as Ss. For 1 hr, Ss were either exposed to a series of inescapable electric shocks in a restraining tube or simply restrained (controls). All Ss received 20 trials of shock–escape in a 2-way shuttlebox 24 hrs later. Ss that received prior inescapable shock were significantly slower in escaping than were controls, thus replicating the LH effect. Exp II, which used 32 rats, studied the effect of para-chlorophenylalanine (PCPA)-induced depletion of brain serotonin (5-hydroxytryptamine) levels on the LH effect. 48 hrs after pretreatment with drug plus vehicle or vehicle alone, Ss received inescapable shock or no-shock pretreatment and were tested 24 hrs later for shock–escape in the shuttlebox. PCPA pretreatment increased Ss' reactivity to shock but had no effect on the LH phenomenon.

982. Amit, Zalman; Sutherland, Ann & White, Norman. (1976). **The role of physical dependence in animal models of human alcoholism.** ***Drug & Alcohol Dependence,*** 1(6), 435–440.
Critically examines the concept that physical dependence is a necessary attribute for animal models of human alcoholism. On the basis of a review of the literature, it is argued that since the production of physical dependence requires the presence of continuous high blood-alcohol levels, and since the production of preference for alcohol requires intermittent presentation of alcohol, the 2 cannot, in principle, be established in the same organism at the same time. It is further argued that physical dependence does not play a role in the development of high alcohol intake in animals. The implications of these observations for human alcoholism are discussed.

983. Ando, Kiyoshi; Johanson, Chris E. & Schuster, Charles R. (1987). **The effects of ethanol on eye tracking in rhesus monkeys and humans.** ***Pharmacology, Biochemistry & Behavior,*** 26(1), 103–109.
Conducted 2 experiments to study the effects of ethanol administration on eye tracking in 3 adult male Rhesus monkeys (Exp I) and 3 male and 3 female human Ss (aged 23–52 yrs [Exp II]). Doses of ethanol and intervals between ethanol administration during posttraining trials were identical for both groups. Nevertheless, blood ethanol levels were higher in humans than in monkeys, possibly due to slight procedural differences in the administration of ethanol and training techniques. Results are reported on levels of ethanol related to the disruption of pursuit eye performance and switch-pressing behavior. Although it appears that monkeys may be good analogs for humans in performance-related ethanol studies, the use of human Ss provides for reports of mood changes and other subjective states.

984. Andreev, B. V.; Ignatov, Yu. D.; Nikitina, Z. S. & Sytinsky, I. A. (1982). **Anti-stress role of the GABAergic system of the brain.** ***Zhurnal Vysshei Nervnoi Deyatel'nosti,*** 32(3), 511–519.
It has been shown in experiments with rats that painful stress disturbs the emotional and somatic state of the Ss and is accompanied by a rise of GABA level and by inhibition of GABA-transaminase in the forebrain, but not in the brain stem. In the present experiments, both gamma-vinyl GABA and gamma-acetylenic GABA-inducing changes in metabolism, as well as muscimol and diazepam-activating postsynaptic GABA receptors, reduced the number of gastric mucosal erosions. Gamma-vinyl GABA and diazepam also reduced behavioral depression following stress. GABA-negative compounds (picrotoxin and thiosemicarbazide) impaired Ss' resistance to stress. It is suggested that directed activation of GABAergic transmission is one method of stress regulation.

985. Andronova, L. M.; Barkov, N. K. & Vikhlyaev, Yu. I. (1978). **Effect of some psychotropic agents on changes in the behavior of mice caused by acetaldehyde.** ***Farmakologiya i Toksikologiya,*** 41(6), 660–665.
Employed the method of A. Oritz et al (1973) to examine the effects of 26 psychotropic drugs on a set of behavioral changes induced in mice by the introduction of acetaldehyde. D. B. Goldstein's (1972) 4-point scale system was used to measure the degree of reduction of convulsions in Ss. Results show that tranquilizers, hypnotics, sodium oxybutyrate, and ethanol were most effective, and neuroleptics and antidepressants were least effective. It is concluded that this model may be used in the screening of drugs for treating abstinence syndromes in chronic alcoholism.

986. Archer, Trevor. (1988). **Ethopharmacological approaches to aggressive behavior.** ***Nordisk Psykiatrisk Tidsskrift,*** 42(6), 471–477.
Reviews the aggression-enhancing effects of low to moderate doses of alcohol, as studied over different animal models and species within the context of an ethopharmacological approach to aggressive behavior. The interaction between alcohol dosages, testosterone levels, and male status is discussed. The involvement of 5-hydroxytryptamine (5-HT) and 5-hydroxyindoleacetic acid (5-HIAA), as monitored by turnover and levels in the central nervous system (CNS), in the aggressive responses of animals and humans implicate these pathways for the eventual alleviation of clinical problems relating to this behavior. Data on substances that increase 5-HT neurotransmission are presented to demonstrate the serenic effects of these agents.

987. Arendt, Thomas; Hennig, Dirk; Gray, Jeffrey A. & Marchbanks, Roger. (1988). **Loss of neurons in the rat basal forebrain cholinergic projection system after prolonged intake of ethanol.** ***Brain Research Bulletin,*** 21(4), 563–569.
A reduction in the number of acetylcholinesterase (AChE)-positive neurons in the basal nucleus of Meynert complex to 83% of control values was observed in rats after ethanol intake for 12 wks. Activity of choline acetyltransferase and AChE in the basal forebrain was simultaneously reduced to 74% and 81% and content of acetylcholine (ACh) to 56% of control values, respectively. ACh content and activity of AChE were significantly reduced in the cortex, hippocampus, and amygdala. Chronic intake of ethanol in rats is suggested to represent an animal model suitable to test the cholinergic hypothesis of postalcoholic dementia and Alzheimer's disease.

988. Arletti, Rossana & Bertolini, Alfio. (1987). **Oxytocin acts as an antidepressant in two animal models of depression.** ***Life Sciences,*** 41(14), 1725–1730.
Investigated the action of oxytocin in 2 animal models sensitive to antidepressant treatments. In the behavioral despair test in mice, oxytocin injected 60 min before testing significantly reduced the duration of immobility, an effect similar to that of the antidepressant imipramine. A more powerful effect was obtained with a 10-day treatment schedule. In the learned helplessness test, oxytocin reduced escape failures and latency significantly and more powerfully than imipramine. Results support the role of regulatory peptides in the central nervous system (CNS).

989. Arushanyan, E. B. & Baturin, V. A. (1988). **Changes in the circadian rhythm of the rest-activity cycles in "depressive" rats induced by tricyclic antidepressants.** ***Farmakologiya i Toksikologiya,*** 51(3), 5–8.
Studied the effect of the tricyclic antidepressants imipramine and amitriptyline on the structure of circadian rhythms in rats subjected to the depressogenic effect of reserpine and in those with signs of depressiveness in the open field. Ss were 45 white rats. Imipramine (10 and 25 mg/kg) and amitriptyline (10 mg/kg) were administered daily for 2–3 wks. One week before receiving imipramine and amitriptyline, some rats received 4 injections of reserpine (1 mg/kg every other day). (English abstract).

990. Attal, N.; Chen, Y. L.; Kayser, V. & Guilbaud, G. (1991). **Behavioural evidence that systemic morphine may modulate a phasic pain-related behaviour in a rat model of peripheral mononeuropathy.** ***Pain,*** 47(1), 65–70.
Investigated the effects of various iv doses of morphine on the vocalization thresholds elicited by paw pressure and compared the effects obtained with the same doses in normal rats. In neuropathic rats, morphine (0.1 and 0.3 mg/kg) produced a significant analgesic effect on the lesioned hind paw, maximum at 15 min post injection with a recovery at 20–25 min. For doses of 0.6 and 1 mg/kg, a modification of the kinetics was observed, with maximum effect at 20–30 min post injection and a recovery at 50–80 min. An analgesic effect was also observed on the unlesioned side, significantly less potent than that observed on the lesioned paw. The effects induced by morphine appeared higher than in normal rats following 1 mg/kg morphine, whereas they were comparable to those obtained from the sham-operated paw. Data show that morphine induces potent antinociceptive effects in a rat model of neuropathy.

991. Avdulov, N. A. et al. (1982). **Behavior and neurochemical study of a number of atypical antidepressants.** ***Trudy Leningradskogo Nauchno-Issledovatel'skogo Psikhonevrologicheskogo Instituta im V M Bekhtereva,*** 101, 15–23.
Investigated the effects of trazodone, zimelidine, viloxazil, and pyrazidol, as well as other substances with presumed antidepressant action (e.g., the derivatives of 1,3-benzthiazine). These drugs were compared with classic tricyclic compounds. Mice and rats were first exposed to unavoidable stress in a cylinder filled with water in order to examine the effects of antidepressants on "behavioral despair." Ss were then subjected to unavoidable stress (electrical shock) to examine the effects of antidepressants on learned helplessness. During the single introduction of antidepressants, there were no changes in the average time of the realization of the avoidance reaction. Results show that simple behavioral tests are adequate in experimental simulations to determine antidepressant effects even with so-called atypical antidepressants.

992. Aylmer, C. G.; Steinberg, Hannah & Webster, R. A. (1987). **Hyperactivity induced by dexamphetamine/chlordiazepoxide mixtures in rats and its attenuation by lithium pretreatment: A role for dopamine?** ***Psychopharmacology,*** 91(2), 198–206.
Results of 4 experiments with female hooded rats suggest that (1) **Y**-maze hyperactivity following chlordiazepoxide (CDZP) and CDZP/dexamphetamine (DEX) mixtures (compared with DEX alone) and (2) the attenuating effect of lithium on mixture-induced increases, were not due to an action on striatal dopamine (DA) neurons. Changes in rearing observed with DEX alone may, however, have reflected an action on DA neurons. Animal models of "mania" are discussed.

993. Bailey, William H. & Weiss, Jay M. (1978). **Effect of $ACTH_{4-10}$ passive avoidance of rats lacking vasopressin (Brattleboro strain).** ***Hormones & Behavior,*** 10(1), 22–29.
The hypothesis that the effects of $ACTH_{4-10}$ on avoidance are mediated via the release of endogenous vasopressin was investigated. To test this hypothesis, the effect of $ACTH_{4-10}$ on the passive avoidance of 20 Brattleboro rats with diabetes insipidus resulting from a total genetic deficiency of vasopressin (DI) and 23 Brattleboro rats without diabetes insipidus (HE) was studied. 20 normal Long-Evans rats (LE) were also included for comparison purposes. Results do not

support the hypothesis. $ACTH_{4-10}$ did influence the passive avoidance of DI rats; this should not have occurred if the release of endogenous vasopressin is necessary for $ACTH_{4-10}$ to influence avoidance.

994. Baizman, E. R.; Ezrin, A. M.; Ferrari, R. A. & Luttinger, D. (1987). **Pharmacologic profile of fezolamine fumarate: A nontricyclic antidepressant in animal models.** ***Journal of Pharmacology & Experimental Therapeutics,*** 243(1), 40–54.
Investigated neurochemical, behavioral, and pharmacologic properties of fezolamine fumarate (FF) in a series of studies with male albino Swiss-Webster mice (behavioral), Charles River mice (behavioral despair), male Sprague-Dawley rats (behavioral and biochemical), and dogs and cats. Results show that in vitro, it was 3 to 4 times more selective in blocking synaptosomal uptake of [^{3}H]norepinephrine than uptake of [^{3}H]serotonin or [^{3}H]dopamine. In classical behavioral tests using monoamine-depleted animals, it prevented the depressant effects of reserpine and tetrabenazine. In addition, it was active in the behavioral despair procedure. Its potency in 3 of these models was similar to that of standard tricyclics (e.g., imipramine, amitriptyline) or newer nontricyclic antidepressants (e.g., bupropion). It is concluded that FF may show antidepressant efficacy in man with minimal anticholinergic or cardiovascular side effects common to tricyclic antidepressants.

995. Bakay, Roy A.; Barrow, Daniel L.; Fiandaca, Massimo S.; Iuvone, P. Michael et al. (1987). **Biochemical and behavioral correction of MPTP Parkinson-like syndrome by fetal cell transplantation.** ***Annals of the New York Academy of Sciences,*** 495, 623–638.
9 rhesus monkeys (aged 2–16 yrs) received injections of 1-methyl-4-phenyl-1,2,3,6-tetrahydropyridine (MPTP) at a level that would ensure behavior resembling Parkinson's disease. Ss' behavior was followed for at least 2 mo after the final MPTP dose and biochemical studies were conducted on cerebrospinal fluid (CSF). Nontransplanted monkeys with the Parkinson-like syndrome served as controls. Ss treated with MPTP developed a Parkinson-like syndrome. The MPTP-treated primate seems to be a good animal model to study transplant plasticity in the central nervous system (CNS). Also demonstrated was the anatomical integration of fetal mesencephalic cells into the caudate nucleus of the host.

996. Baker, Timothy B. & Cannon, Dale S. (1982). **Alcohol and taste-mediated learning.** ***Addictive Behaviors,*** 7(3), 211–230.
Taste-mediated learning is relevant to the alcohol consumption patterns of animals. A literature review suggests that taste aversion learning has thus far prevented development of an animal model of alcoholism. The presence of a taste cue, lack of control over alcohol administration, and high alcohol concentrations or dosages all facilitate the development of alcohol aversions. There is little evidence that taste-preference learning is involved in the development of alcohol dependence. Data from taste-mediated learning research with animals are consistent with drinking patterns of human alcoholics.

997. Baldwin, Helen A. & File, Sandra E. (1989). **Flumazenil prevents the development of chlordiazepoxide withdrawal in rats tested in the social interaction test of anxiety.** ***Psychopharmacology,*** 97(3), 424–426.
Male rats were chronically treated with chlordiazepoxide (CDP) or vehicle for 27 days. 24 hrs after their last dose, Ss received flumazenil or vehicle and were tested for social interaction in a low-light, familiar arena. CDP withdrawal significantly reduced the time spent in social interaction compared with controls, indicating an anxiogenic withdrawal response. This was completely reversed by flumazenil. A 2nd group of Ss received CDP for 27 days and, in addition, received a single dose of flumazenil 6 days before testing. Flumazenil prevented the development of the anxiogenic withdrawal response in these Ss.

998. Balster, Robert L. (1989). **Behavioral pharmacology of PCP, NMDA and sigma receptors.** 51st Annual Meeting of the Committee on Problems of Drug Dependence (1989, Keystone, Colorado). ***National Institute on Drug Abuse: Research Monograph Series,*** 95, 270–274.
Discusses the neuropharmacological rationale for using the effects of phencyclidine (PCP) in animals as a model of schizophrenia. Studies are reviewed of the neural basis for PCP intoxication in animals that may provide clues to the underlying pathology present in psychosis. The possible role of the sigma receptor in the effects of PCP is addressed, along with the possible role of *N*-methyl-*D*-aspartate (NMDA) antagonism by PCP in PCP intoxication.

999. Bankiewicz, K. S.; Oldfield, E. H.; Chiueh, C. C.; Doppman, J. L. et al. (1986). **Hemiparkinsonism in monkeys after unilateral internal carotid artery infusion of 1-methyl-4-phenyl-1,2,3,6-tetrahydropyridine (MPTP).** ***Life Sciences,*** 39(1), 7–16.
Presents a hemiparkinsonian animal model. Infusion of MPTP (0.2–0.8 mg/kg) into the right internal carotid artery of 14 monkeys produced toxin-induced injury to the right nigro-striatal pathway with sparing of other dopaminergic neurons on the infused side and with negligible or little injury to the opposite, untreated side. There were contralateral limb dystonic postures, rigidity, and bradykinesia, but the Ss were able to eat and maintain health without drug treatment.

1000. Bannet, J.; Belmaker, R. H. & Ebstein, R. P. (1980). **The effect of drug holidays in an animal model of tardive dyskinesia.** ***Psychopharmacology,*** 69(2), 223–224.
Intermittent haloperidol treatment in Sabra wild-type mice increased ^{3}H-spiroperidol binding to the same degree as continual haloperidol feeding. Results do not support the concept that drug holidays can reduce the incidence of tardive dyskinesia.

1001. Bareggi, S. R.; Becker, R. E.; Ginsburg, Benson E. & Genovese, E. (1979). **Paradoxical effect of amphetamine in an endogenous model of the hyperkinetic syndrome in a hybrid dog: Correlation with amphetamine and *p*-hydroxyamphetamine blood levels.** ***Psychopharmacology,*** 62(3), 217–224.
A telomian-beagle hybrid has been studied as a model for the hyperkinetic syndrome in children. Using 13 hybrids and 6 beagles, the present study showed that hybrids, like children, exhibit hyperactivity, impulsiveness, and impaired learning. Two groups of hybrid could be differentiated; the behavior of one improved (responders) after dextroamphetamine (DAM) (1.2–2.0 mg/kg, orally), while that of the other did not (nonresponders). Hybrids were less responsive than beagles to other DAM effects such as stereotyped behavior and hyperthermia. Measurement of blood levels of DAM and its active metabolite parahydroxyamphetamine (POA)

showed that hybrids form less POA. The lesser response of hybrids to toxic effects of DAM may be due to this difference in DAM metabolism. Responders showed higher peak blood levels of DAM than nonresponders, and their improvement on DAM correlated with blood levels of DAM. High DAM levels appear to be necessary for its "paradoxical" effect in this model. This suggests that DAM acts by activating both noradrenergic and dopaminergic neuronal systems in the CNS.

1002. Barron, Susan; Zimmerberg, Betty; Rockwood, Gary A. & Riley, Edward P. (1987). **Prenatal alcohol exposure: Recent work on behavioral dysfunctions in pre-weanling rats.** ***Advances in Alcohol & Substance Abuse,*** 6(4), 59–72.
Reviews data examining the behavioral effects of prenatal alcohol exposure in preweanling rats. The behavioral alterations that are discussed include suckling dysfunctions, overactivity, and early learning deficits. Similar behavioral dysfunctions have also been reported in children exposed to alcohol during prenatal development. These similarities are discussed as well as possible insights that might be gained from animal models of fetal alcohol effects.

1003. Barros, H. M. & Leite, J. R. (1987). **The effects of carbamazepine on two animal models of depression.** ***Psychopharmacology,*** 92(3), 340–342.
Investigated the effects of carbamazepine (CBZ) on the potentiation of amphetamine-induced anorexia and the behavioral despair models of depression, using 179 male Wistar rats. CBZ neither modified the methamphetamine anorectic effect nor induced anorexia when administered alone. Subacute and chronic administration of imipramine decreased immobility in the behavioral despair model. Subacute and chronic administration of CBZ (40 mg/kg) also decreased immobility, while 10 mg/kg CBZ was effective only after chronic treatment. It is concluded that CBZ is similar to atypical antidepressants, since it did not potentiate amphetamine-induced anorexia but did affect behavioral despair.

1004. Bartus, Raymond T. & Dean, Reginald L. (1988). **Tetrahydroaminoacridine, 3,4 diaminopyridine and physostigmine: Direct comparison of effects on memory in aged primates.** ***Neurobiology of Aging,*** 9(4), 351–356.
Evaluated the effects of tetrahydroaminoacridine (THA); 3,4 diaminopyridine (3,4 DAP); and physostigmine for their ability to reduce memory impairments in 10 aged, test-sophisticated cebus monkeys (18–26 yrs old). Several po doses of each drug were tested po in each S, allowing for direct and extensive comparison of each drug's efficacy in this model. Results show that (1) all drugs produced improvement in a portion of the Ss tested; (2) wide variations in most effective dose/S, were observed; (3) different Ss responded more effectively to one drug than another; and (4) under these tightly controlled conditions, physostigmine produced the most reliable and robust effects, in more Ss, than did either THA or 3,4 DAP.

1005. Beardslee, S. L.; Papadakis, E.; Fontana, D. J. & Commissaris, Randall L. (1990). **Antipanic drug treatments: Failure to exhibit anxiolytic-like effects on defensive burying behavior.** ***Pharmacology, Biochemistry & Behavior,*** 35(2), 451–455.
Prior to testing, female rats were placed in a chamber containing clay bedding material (5 cm deep) for 30-min periods on each of 4 consecutive days. On the 5th day, a wire-wrapped prod was placed at one end of the chamber. Ss were placed in the chamber individually and a shock was delivered upon contact with the prod. Defensive burying behavior (DBB) was recorded for 15 min postshock. In a dose-dependent manner, acute treatment with chlordiazepoxide reduced the frequency of occurrence of DBB, increased the latency to initiation of DBB, and decreased the duration of DBB. Chronic treatment with imipramine, desipramine, or pargyline failed to exhibit anxiolytic-like effects on any measure of DBB. The DBB paradigm may not be an "animal model" for the study of panic disorder and potential antipanic agents.

1006. Becker, Howard C.; Randall, Carrie L. & Middaugh, Lawrence D. (1989). **Behavioral teratogenic effects of ethanol in mice.** Conference of the Behavioral Teratology Society, the National Institute on Drug Abuse, and the New York Academy of Sciences: Prenatal abuse of licit and illicit drugs (1988, Bethesda, Maryland). ***Annals of the New York Academy of Sciences,*** 562, 340–341.
C57BL mice were sensitive to the behavioral teratogenic effects of ethanol (EtOH). EtOH-exposed Ss exhibited hyperactivity and impaired active and passive avoidance behavior. This strain may provide an animal model of EtOH's effects.

1007. Bell, Joanne et al. (1982). **Behavioral effects of early deprivation of nerve growth factor: Some similarities with familial dysautonomia.** ***Brain Research,*** 234(2), 409–421.
Hypothesized that rats exposed to nerve growth factor (NGF) might represent a model of many symptoms found in familial dysautonomia. Female rats immunized with mouse NGF develop an antibody (anti-NGF) which reaches offspring through the placenta and via the milk. Pups exposed to maternal anti-NGF have fewer dorsal root and sympathetic neurons. Two such female rats and their offspring were examined on behavioral tests (including thermoregulation, the flinch-jump test, passive avoidance, and swimming ability) and were found to exhibit severe deficits in response to stress (ulceration, corticosterone levels) and mild deficits on some sensory and cognitive tasks. The pathologic and behavioral profiles of the Ss closely mimicked the sensory and sympathetic aspects of familial dysautonomia.

1008. Bennett, Debra A. et al. (1982). **Comparison of the actions of trimethadione and chlordiazepoxide in animal models of anxiety and benzodiazepine receptor binding.** ***Neuropharmacology,*** 21(11), 1175–1179.
Trimethadione (TMD) was compared with chlordiazepoxide (CDP) for antianxiety activity in 2 behavioral tests known to predict the anxiolytic action of drugs. In the drug-discrimination test, male hooded Long-Evans rats were trained to discriminate the anxiogenic action of pentylenetetrazol (PTZ) from saline by responding for food reinforcement on 1 of 2 levers after treatment with PTZ (1,450 μmol/kg) and on the other lever after injection of saline. Pretreatment with either CDP (2.8–33 μmol/kg) or TMD (559–2,236 μmol/kg) prior to the injection of PTZ produced a dose-dependent antagonism of the anxiogenic stimulus. In the other test, male Wistar rats were trained to respond for milk reinforcement in a conflict procedure in which some of the reinforced responses resulted in the delivery of footshock. Treatment of these Ss with CDP (17–67 μmol/kg) or TMD (1,118–2,236 μmol/kg) antagonized the footshock-induced

suppression of responding. In a receptor binding study, TMD failed to inhibit flunitrazepam binding. Data suggest that TMD is an effective anxiolytic agent whose action does not directly involve benzodiazepine receptors.

1009. Bhattacharya, Salil K.; Jaiswal, Arun K.; Mukhopadhyay, Meeta & Datla, Krishna P. (1988). **Clonidine-induced automutilation in mice as a laboratory model for clinical self-injurious behaviour.** ***Journal of Psychiatric Research,*** 22(1), 43–50.
Clonidine (20, 50, and 100 mg/kg, ip) produced dose-related self-biting and automutilation in mice that had been isolated and food-deprived for 24 hrs. This clonidine-induced behavior was significantly attenuated by pharmacological treatments that selectively augment central serotonergic or reduce central dopaminergic activity. The clonidine effect was potentiated by pharmacological agents that selectively reduce or enhance central serotonergic and dopaminergic activity, respectively. Drug-induced alterations in central noradrenergic or cholinergic activity had no significant effect on the self-mutilatory behavior induced by clonidine. Clonidine-induced automutilation appears to be an acceptable model for clinical self-injurious behavior that may be linked to a serotonin deficiency and/or excess dopamine.

1010. Bhattacharya, Salil K. & Sen, Ananda P. (1991). **Effects of muscarinic receptor agonists and antagonists on swim-stress-induced "behavioural despair" in rats.** ***Journal of Psychopharmacology,*** 5(1), 77–81.
Investigated the effects of some selective agonists and antagonists of cholinergic muscarinic (M) receptor subtypes on swim-stress-induced immobility (SI) in rats. This paradigm has been proposed to assess "behavioral despair" in rodents as a laboratory model for clinical depression. All the Ss were pretreated with atropine ethoiodide ip to obviate any peripheral cholinergic influence. M_1 receptor agonists induced dose-related increases in SI, where M_1 receptor antagonists decreased SI. One M_2 receptor agonist had a dose-dependent dual effect, whereas M_2 receptor antagonists increased SI. The data suggest that muscarinic M_1 receptors may function to accentuate depression, whereas muscarinic M_2 receptors may exert an inhibitory modulatory effect.

1011. Bignami, Giorgio. (1991). **Possibilities and limitations of models for studying animal behavior.** ***Psychologie Française,*** 36(3), 233–240.
Discusses the history, advantages, and limitations of using animal behavior models for evaluating the effects of pharmacological treatment on depression and anxiety. The etiopathogenesis of psychiatric disorders, sources of interpretation errors in animal pharmacotherapeutic studies, and differences among depression and anxiety models are considered. (English abstract).

1012. Blackburn, James R. & Phillips, Anthony G. (1989). **Blockade of acquisition of one-way conditioned avoidance responding by haloperidol and metoclopramide but not by thioridazine or clozapine: Implications for screening new antipsychotic drugs.** ***Psychopharmacology,*** 98(4), 453–459.
Compared the effects of haloperidol, 2 atypical neuroleptics, thioridazine and clozapine, and a substituted benzamide, metoclopramide, on one-way avoidance by rats. Thioridazine and clozapine disrupted both acquisition and performance of conditioned avoidance responding (CAR). Haloperidol and metoclopramide completely blocked the acquisition of CAR, yet initially produced only a slight disruption in the performance of a previously acquired response. The ineffectiveness of the atypical neuroleptics in producing a complete disruption of acquisition of CAR may be due to the anticholinergic properties of these drugs. Results suggest caution in using CAR as an animal model for assessing the antipsychotic potential of new pharmacological agents.

1013. Blanchard, D. Caroline; Shepherd, Jon K.; de Padua Carobrez, Antonio & Blanchard, Robert J. (1991). **Sex effects in defensive behavior: Baseline differences and drug interactions.** Ethopharmacology Conference: Advances in ethopharmacology (1991, Lisek, Czechoslovakia). ***Neuroscience & Biobehavioral Reviews,*** 15(4), 461–468.
Female rats consistently show a pattern of differences in defensive behaviors compared to males that parallels the effects of exposure to a nonpainful threat stimulus (cat or cat odor) in the same tests and measures. These indications of greater defensiveness for females are particularly common in situations involving potential, as opposed to actual and present, threat, a factor that probably also reflects ceiling or floor effects in situations involving very intense defensiveness. In addition, pharmacological studies (e.g., D. C. Blanchard; in press) indicate sex differences in the effects of selective serotonin (5-hydroxytryptamine [5-HT]) receptor agonists and antagonists on defensive responding. These findings indicate that sex effects must be considered in studies of the pharmacological control of defensive behaviors, and suggest that responsivity to sex effects may be an additional criterion for the suitability of animal models of anxiety.

1014. Bodnoff, Shari R.; Suranyi-Cadotte, Barbara; Quirion, Remi & Meaney, Michael J. (1989). **A comparison of the effects of diazepam versus several typical and atypical anti-depressant drugs in an animal model of anxiety.** ***Psychopharmacology,*** 97(2), 277–279.
Examined the anxiolytic effects of a variety of anti-depressant drugs, administered either acutely or chronically, in an animal model of anxiety involving novelty-suppressed feeding in food-deprived rats. Following a single injection of desipramine, amitriptyline, mianserin, fluoxetine, buspirone, gepirone, or nomifensine, there was no decrease in the latency to begin eating in the novel environment such as occurred with diazepam. An increased latency was observed for desipramine, amitriptyline, fluoxetine, and nomifensine. In contrast, chronic (21 days) treatment with each of the above-mentioned drugs, except nomifensine, significantly reduced the latency to begin eating relative to vehicle controls. Findings suggest that a variety of tricyclic and novel anti-depressant drugs acquire anxiolytic properties following chronic administration.

1015. Bodnoff, Shari R.; Suranyi-Cadotte, Barbara; Aitken, David H.; Quirion, Remi et al. (1988). **The effects of chronic antidepressant treatment in an animal model of anxiety.** ***Psychopharmacology,*** 95(3), 298–302.
Examined the anxiolytic activity of acute and chronic antidepressant treatment involving novelty-suppressed feeding. Rats were food deprived for 48 hrs, placed into a novel environment containing food, and the latency to begin eating was recorded. Chronic but not acute injections of desipramine (DMI) and amitriptyline significantly reduced the latency to begin eating compared to controls, but the decrease was not as great as that seen with treatment with diazepam or adinazolam. A time course study indicated that at least 2 wks of treatment was necessary to observe a

significant anxiolytic effect of antidepressants. A single dose of the central benzodiazepine receptor antagonist, Ro15-1788 did not block the anxiolytic effects of chronic DMI, while it completely eliminated the effect of chronic diazepam treatment. Data suggest that antidepressants acquire anxiolytic properties following chronic administration.

1016. Bonthius, Daniel J. & West, James R. (1990). **Alcohol-induced neuronal loss in developing rats: Increased brain damage with binge exposure.** ***Alcoholism: Clinical & Experimental Research,*** 14(1), 107–118.
Using a rat model of 3rd-trimester fetal alcohol (AL) exposure, it was determined that a smaller dose of AL can be more damaging to the developing central nervous system (CNS) than a larger dose if the smaller dose is consumed in such a way that it produces a higher blood AL concentration than the larger dose. It was also found that there are regional differences within the CNS in the teratogenic effects of AL. Some neuronal populations are more vulnerable than others to AL-induced cell death during the brain growth spurt. Within the cerebellum, there appears to be a developmental variable that influences AL-induced cell death. In the neonatal rat, more mature Purkinje cells are more vulnerable to AL-induced cell death than are less mature Purkinje cells. Data caution against binge AL consumption during pregnancy.

1017. Borison, Richard L. & Diamond, Bruce I. (1978). **A new animal model for schizophrenia: Interactions with adrenergic mechanisms.** ***Biological Psychiatry,*** 13(2), 217–225.
Proposes amphetamine-induced stereotyped behavior in animals as a model for schizophrenia. Chronic amphetamine administration produces stereotyped behavior and a paranoid schizophreniform syndrome in humans, whereas in animals a behavioral sensitization to stereotypy is evoked. It is shown that phenylethylamine (PEA), an amphetamine-like stimulant concentrated in the limbic system of human brain, produces stereotypy in rats with a behavioral sensitization when chronically administered. In comparing amphetamine-induced stereotypy with PEA-induced stereotypy, it was found that the alpha-adrenergic blocking agents phentolamine and phenoxybenzamine selectively antagonize PEA stereotypy, whereas the beta-adrenergic blocking agent propranolol fails to alter significantly stereotypies evoked by PEA or amphetamine administration. Catecholamine depletion by alpha-methylparatyrosine administration blocks stereotypies induced by both PEA and amphetamine, whereas selective norepinephrine (NE) depletion antagonizes only PEA stereotypy: the amino acid precursors of both NE and dopamine potentiate stereotypies. Therefore, PEA-elicited, but not amphetamine-elicited, stereotypy is dependent on NE; the significance of this for the PEA animal model of schizophrenia is discussed.

1018. Borison, Richard L.; Havdala, Henri S. & Diamond, Bruce I. (1977). **Chronic phenylethylamine stereotypy in rats: A new animal model for schizophrenia?** ***Life Sciences,*** 21(1), 117–122.
In a study with male Sprague-Dawley rats, the chronic administration of phenylethylamine (50 mg/kg) or *d*-amphetamine (3.75 mg/kg) produced stereotyped behavior. The chronic administration induced a behavioral sensitization consisting of an increase in intensity of stereotypies and a decrease in their latency to onset. The substitution of phenylethylamine for amphetamine, or vice versa, maintained not only the stereotypies but also the sensitization effect. Acute pretreatments with the antipsychotic dopamine blockers, haloperidol and pimozide, blocked the phenylethylamine- and amphetamine-induced stereotypy, while thioridazine and clozapine, antipsychotics with fewer extrapyramidal effects, preferentially blocked phenylethylamine-induced behavior. Results suggest that phenylethylamine effects may represent a better animal model for schizophrenia in that the former does not depend so greatly on neostriatal mechanisms.

1019. Borison, Richard L.; Havdala, Henri S. & Diamond, Bruce I. (1978). **A new animal model for schizophrenia: Cholinergic and serotonergic modulation.** ***Communications in Psychopharmacology,*** 2(3), 209–214.
In a study with male Sprague-Dawley rats, chronic dextroamphetamine, but not phenylethylamine-induced stereotyped behavior, was enhanced by the anticholinergic agent trihexyphenidyl and antagonized by the cholinesterase inhibitor physostigmine. The serotonin amino acid precursor, 5-hydroxytryptophan, failed to alter either amphetamine or phenylethylamine stereotypy, whereas the serotonin receptor blocker methysergide selectively antagonized phenylethylamine-evoked stereotyped behavior. These interactions are consistent with previous suggestions that phenylethylamine-induced stereotypy represents a valid animal model for schizophrenia.

1020. Borsini, F.; Nowakowska, E. & Samanin, R. (1984). **Effect of repeated treatment with desipramine in the behavioral "despair" test in rats: Antagonism by "atypical" but not "classical" neuroleptics or antiadrenergic drugs.** ***Life Sciences,*** 34(12), 1171–1176.
Administered a 7-day treatment of desipramine (DM [20 mg/kg/day]) to approximately 77 CD-COBS rats to examine its effect on immobility time (IT) in the behavioral "despair" test. Results indicate that DM produced a reduction in IT. The effect of DM was antagonized by sulpiride, metoclopramide, and clopazine, but not by haloperidol or chlorpromazine. Alpha-adrenoreceptor blockers (prazosin, aceperone, azapetine, and phentolamine), dl-propranolol and clonidine failed to modify the anti-immobility effect of DM. Data suggest that a particular subtype of dopamine receptors is involved in the anti-immobility effect of a 7-day treatment with DM in the behavioral despair test in rats.

1021. Borsini, Franco; Volterra, Giovanna & Meli, Alberto. (1986). **Does the behavioral "despair" test measure "despair"?** ***Physiology & Behavior,*** 38(3), 385–386.
Compared behavior of male CD-COBS rats subjected to various experimental conditions (4, 15, 30 cm of water) to assess whether or not "despair" was the cause of immobility. Results indicate that the S's behavior in response to exposure to a dangerous situation, such as that represented by 15 or 30 cm water, depended on previous knowledge of the environment rather than despair. It is concluded that this test is far from reproducing behavior changes that characterize depressive illness in humans.

1022. Bourin, Michel; Colombel, Marie-Claude; Malinge, Myriam & Bradwejn, Jacques. (1991). **Clonidine as a sensitizing agent in the forced swimming test for revealing antidepressant activity.** ***Journal of Psychiatry & Neuroscience,*** 16(4), 199–203.

Investigated the potential for clonidine to render the forced swimming test (FST) sensitive to antidepressants in mice by using a behaviorally inactive dose of this agent (0.1 mg/kg). All antidepressants studied (tricyclics, 5-hydroxytryptamine [5-HT] uptake inhibitors, iprindole, mianserin, viloxazine, and trazodone) showed either activity at lower doses or activity at previously inactive doses. The effect appeared specific because it did not appear with drugs other than antidepressants (diazepam, chlorpromazine, sulpiride, and atropine), except for amphetamine and apomorphine, which have a strong effect on the dopaminergic system. The use of behaviorally subactive doses of clonidine may thus provide an important means of increasing the sensitivity of the forced swimming test.

1023. Bowden, Douglas M. et al. (1983). **A periodic dosing model of fetal alcohol syndrome in the pig-tailed macaque (*Macaca nemestrina*).** ***American Journal of Primatology,*** 4(2), 143–157.
A nonhuman primate on a periodic ethanol dosing schedule should provide a model of fetal alcohol syndrome (FAS) most relevant to the majority of pregnant women who are "social drinkers" and can exercise reasonable control over their ethanol intake. In the present study, 4 pregnant pig-tailed macaques received ethanol 1 time/week from 40 days' gestation in moderate doses (MD, 2.5 mg/kg) or high doses (HD, 4.1 mg/kg). One of the 3 MD Ss aborted after the 1st dose. The other 3 pregnancies were compared with 8–10 control pregnancies, and the infants' development over the 1st 6 mo was compared with that of the control offspring. The fetal heart rate response to maternal restraint was absent in the HD S. Gestational duration and simian Apgar scores were normal. All 3 infants were abnormally large, and 2 were also abnormally heavy. Skeletal maturation was not accelerated. The HD infant was scaphocephalic, with an underdeveloped cranial base and midface, and its brain was small and dysplastic; its reflex, motor, and cognitive development was retarded. One MD infant had some brain abnormalities, was hyperkinetic, and showed developmental retardation on several behavioral measures. It is concluded that the periodic model offers an effective means of investigating FAS and that when nutrition is maintained, intermittent intake of ethanol by the pregnant primate does not necessarily retard fetal growth.

1024. Bowman, Robert E.; Heironimus, Mark P. & Harlow, Harry F. (1979). **Pentylenetetrazol: Posttraining injection facilitates discrimination learning in rhesus monkeys.** ***Physiological Psychology,*** 7(3), 265–268.
In the 1st 2 of 4 experiments, monkeys given immediate posttraining injections of pentylenetetrazol (PTZ) exhibited improved acquisition of difficult discrimination-learning problems (Exp II) but not of easy problems (Exp I). Optimal facilitation occurred at a dosage of 10 mg/kg. In Exps III and IV, the interval between training and injection of 10 mg/kg of PTZ was varied, and optimal facilitation occurred at the 15-min interval. At the 30-min interval, facilitation was absent and error rates were as high as in saline controls, but at the 60-min interval a lower error rate was again seen. All findings are similar to those previously reported in mice, suggesting a common mechanism for the effect of PTZ on simple discrimination learning in both species.

1025. Bozarth, Michael A.; Murray, Aileen & Wise, Roy A. (1989). **Influence of housing conditions on the acquisition of intravenous heroin and cocaine self-administration in rats.** ***Pharmacology, Biochemistry & Behavior,*** 33(4), 903–907.
In 2 experiments, group-housed and individually housed male rats were tested for the acquisition of a leverpressing response reinforced by iv heroin or cocaine. Ss in each housing condition quickly learned to self-administer the drugs. In Exp 1, isolated Ss learned to self-administer heroin earlier than group-housed Ss, but the 2 groups self-administered similar levels of heroin by the 5th wk of testing. In Exp 2, cocaine self-administration was learned with equal speed in the 2 groups, and similar levels of cocaine were self-administered by both groups throughout the experiment. While social isolation can influence levels of heroin self-administration, isolation is not a necessary condition for heroin or cocaine injections to be reinforcing.

1026. Brady, Joseph V. (1991). **Animal models for assessing drugs of abuse.** Symposium on Animal-To-Human Extrapolation (1990, San Antonio, Texas). ***Neuroscience & Biobehavioral Reviews,*** 15(1), 35–43.
Several converging lines of evidence testify to the reliability and broad generality of observations concerning drug abuse liability in humans based on animal laboratory models. The most important point of contact that characterizes the interaction between such animal assessment models and the human drug abuse arena is the demonstrated relationship between the biochemical/pharmacological/toxic properties of drugs on the one hand and their environmental/behavioral stimulus functions on the other. As a result of these developments and the consequent advances in knowledge of drug action, an operational basis has been provided for redefining the bewildering range of phenomena and experiential pseudo-phenomena loosely associated with such terms as "addiction," "dependence," and "abuse." Two reasonably exclusive categories based on operational criteria distinguish between events that occur before and after actual substance intake.

1027. Braff, David L. & Geyer, Mark A. (1980). **Acute and chronic LSD effects on rat startle: Data supporting an LSD–rat model of schizophrenia.** ***Biological Psychiatry,*** 15(6), 909–916.
Animal models of human schizophrenia using LSD and related hallucinogens have been challenged on several grounds. One compelling argument against the LSD model is that, while schizophrenia can be chronically debilitating, animal and human effects of LSD exhibit behavioral tolerance following chronic administration. The present study tested the effects of acute and chronic LSD on measures of rat startle, a widely used behavioral measure of reactivity and habituation. Results with 20 male Sprague-Dawley rats suggest that behavioral tolerance after chronic LSD administration is incomplete, with tolerance exhibited to the acute impairment of habituation but potentiation of startle magnitude on both the 1st response and the 1st block of 30 trials. These results are interpreted as supporting the viability of LSD as a model for one or more of the group of schizophrenias.

1028. Breese, George R. et al. (1984). **Neonatal-6-hydroxydopamine treatment: Model of susceptibility for self-mutilation in the Lesch-Nyhan syndrome.** ***Pharmacology, Biochemistry & Behavior,*** 21(3), 459–461.

In an attempt to disrupt catecholamine-containing fibers, 5-day-old rats were given 100 μg of 6-hydroxydopamine (6-OHDA) intracisternally to reduce brain catecholamines, and adult male rats were treated with 200 μg 6-OHDA after 50 mg/kg pargyline on 1 occasion, with an additional 200 μg 6-OHDA administered 1 wk later. Controls received appropriate vehicle and drug pretreatment. Neonatal Ss were then administered levodihydroxyphenylalanine (L-DOPA [100 mg/kg, ip]) after 50 mg/kg of RO-4-4602 (a decarboxylase inhibitor), and they showed a high incidence of self-mutilation behavior (SMB) and self-biting. These behaviors were not observed in adult 6-OHDA-treated Ss or in controls. Since inhibition of dopamine-beta-hydroxylase did not prevent or inhibit the SMB exhibited in neonatal 6-OHDA-treated Ss after L-DOPA, norepinephrine is not likely to be contributing to this response. The age-dependent effects are consistent with the hypothesis that neonatal reduction of dopamine-contraining fibers is responsible for the SMB susceptibility observed in Lesch-Nyhan disease (described by M. Lesch and W. L. Nyhan [1964]), making the neonatal Ss a model of this neurological syndrome.

1029. Breese, George R. et al. (1978). **An alternative to animal models of central nervous system disorders: Study of drug mechanisms and disease symptoms in animals.** ***Progress in Neuro-Psychopharmacology,*** 2(3), 313–325.
Attempts to produce complete animal models of human disorders of the CNS have had limited success. By having alternative approaches to the production of "true" models of CNS disease, animal research has been able to make significant contributions to the understanding of central disease mechanisms. One alternative includes an examination of the mechanism(s) of action of drugs that alter symptoms of disorders of the CNS. Another approach has been the study of underlying neurobiological mechanisms of individual functions that are abnormal in central diseases. This overview provides examples of research that have extended current knowledge about CNS disorders and outlines some of the difficulties in interpretation encountered when these approaches are used.

1030. Breese, George R.; Criswell, Hugh E. & Mueller, Robert A. (1990). **Evidence that lack of brain dopamine during development can increase the susceptibility for aggression and self-injurious behavior by influencing D_1-dopamine receptor function.** ***Progress in Neuro-Psychopharmacology & Biological Psychiatry,*** 14(Suppl), S65–S80.
Lesch-Nyhan syndrome (LNS) has a defined neurological lesion that is accompanied by symptoms that include mental retardation, abnormal motor function, aggression, and self-injurious behavior (SIB). The dopamine deficiency in LNS has been modeled by destroying dopamine-containing neurons in neonatal rats with 6-hydroxydopamine (6-OHDA). Because D_1-dopamine antagonists will block SIB induced by L-DOPA in neonatal 6-OHDA-lesioned rats, these antagonists are proposed as a potential therapy for patients with symptoms of aggression and SIB. The determination that the drug SCH-12679, which exhibited effectiveness against aggressiveness in mentally retarded patients, is a D_1-dopamine antagonist supports the view that new D_1-dopamine antagonists being developed will be an effective therapy for some types of aberrant behavior in this population.

1031. Breese, George R.; Criswell, Hugh E.; McQuade, Robert D.; Iorio, Louis C. et al. (1990). **Pharmacological evaluation of SCH-12679: Evidence for an *in vivo* antagonism of D_1-dopamine receptors.** ***Journal of Pharmacology & Experimental Therapeutics,*** 252(2), 558–567.
Evaluated the potential mechanism of action of SCH-12679, a benzazepine that has been shown to have clinical efficacy against aggressive and self-injurious behavior in mentally deficient patients, using 3-day-old rat pups treated with 6-hydroxydopamine (6-OHDA). The effects of SCH-12679 on increased activity and behaviors produced by D_1- and D_2-dopamine antagonists, and on self-mutilation and other behaviors induced by l-dopamine, were examined. Findings demonstrate the D_1-dopamine antagonist properties of SCH-12679. It is concluded that at least some forms of aggressive and self-injurious behavior may be alleviated by compounds with D_1-dopamine antagonist activity.

1032. Britton, Donald R. & Indyk, Elzbieta. (1990). **Central effects of corticotropin releasing factor (CRF): Evidence for similar interactions with environmental novelty and with caffeine.** ***Psychopharmacology,*** 101(3), 366–370.
Three experiments investigated how centrally administered rat/human corticotropin-releasing factor (rCRF) increases low levels of locomotor activity by rats tested in a familiar environment but suppresses the higher levels of activity associated with exposure of the animals to a novel environment. These opposing responses do not appear to be manifestations of a simple rate-dependent effect, since icv-administered rCRF did not lower the higher levels of locomotor activity associated with the dark (active) phase of the animal's activity cycle. Caffeine, which has anxiogenic effects in man, produces effects in rats which are similar to those of rCRF. That is, both compounds elevate activity in a familiar environment but lower activity in a novel environment.

1033. Brocco, M. J.; Koek, Wouter; Degryse, A.-D. & Colpaert, Francis C. (1990). **Comparative studies on the anti-punishment effects of chlordiazepoxide, buspirone and ritanserin in the pigeon, Geller-Seifter and Vogel conflict procedures.** ***Behavioural Pharmacology,*** 1(5), 403–418.
Compared the responsiveness to effective anxiolytics of 2 major conflict procedures in rats (I. Geller and J. Seifter's procedure [1960] and the procedure by J. R. Vogel, B. Beer, et al [1971]) and of a newer pigeon conflict procedure (PCPR). Data obtained in the PCPR confirm and extend findings by J. E. Barrett et al (1986) showing chlordiazepoxide to produce significant and dose-dependent increases of punished responding. Data also replicate earlier findings (e.g., Barrett et al) that buspirone and ritanserin can increase punished responding in the pigeon. All the drugs depressed the rate of unpunished responding. Results show that data obtained with the PCPR are reproducible across different laboratories. The PCPR (1) allows the behavioral response to be highly defined and permits the experimenter to rigidly control the stimulus events and (2) is responsive to benzodiazepines, as well as to anxiolytic agents with effects that are not mediated by benzodiazepine receptors.

1034. Brochet, Denis M.; Martin, Patrick; Soubrié, Philippe & Simon, Pierre. (1987). **Triiodothyronine potentiation of antidepressant-induced reversal of learned helplessness in rats.** ***Psychiatry Research,*** 21(3), 267–275.

Examined the relationship between thyroid function (represented through L-Triiodothyronine [T_3] given at a .03 mg/kg dose) and the response to 3 antidepressants—imipramine (16 and 32 mg/kg), desipramine (16 and 32 mg/kg), and nomifensine (.5 and 2 mg/kg)—administered intraperitoneally to male rats using the learned helplessness model. Data show that the reversal by antidepressants of depressive-like behavior was significantly hastened in Ss given daily T_3. Enhancement of antidepressant action by T_3 did not appear to involve an elevation of antidepressant drug brain levels. It is suggested that the learned helplessness paradigm might be a useful model for approaching in animals the neurohormonal correlates of affective disorders and the neurobiochemical bases of the reported T_3 enhancement of antidepressants.

1035. Broekkamp, Chris L.; Berendsen, H. H.; Jenck, François & Van Delft, Anton M. (1989). **Animal models for anxiety and response to serotonergic drugs.** 16th Collegium Internationale Neuro-Psychopharmacologicum Congress Satellite Conference: New Findings with Anxiolytic Drugs (1988, Munich, Federal Republic of Germany). ***Psychopathology,*** 22(Suppl 1), 2–12.
Reviews information on serotonin and anxiety in animal experiments and applications in humans. It is contended that in examining anxiety and the response of animal models to serotonergic drugs, 4 aspects should be taken into account: (1) The serotonin receptor is subdivided into at least 6 receptor subtypes; (2) benzodiazepines have acute anxiety-relieving effects, whereas antidepressants, serotonin-uptake inhibitors, buspirone, and serotonin antagonists have antianxiety effects only after prolonged administration; (3) diagnostic criteria differentiate several distinguishable anxiety disorders that have different responsiveness to serotonin-related drugs; and (4) various types of animal models exist, each responding differently to serotonin-related drugs. The role of 5-hydroxytryptamine (5-HT) receptor subtypes in anxiety disorders and anxiety models is discussed.

1036. Brooks-Eidelberg, Barbara A.; Fuchs, Albert F. & Finocchio, Dom. (1986). **Saccadic eye movement deficits in the MPTP monkey model of Parkinson's disease.** ***Brain Research,*** 383(1–2), 402–407.
Saccadic eye tracking was studied in an adolescent rhesus monkey given intravenous injections of N-methyl-4-phenyl 1,2,3,6-tetrahydropyridine (MPTP). The Parkinson-like symptoms that appeared in S's general motor behavior (akinesia, bradykinesia, hypokinesia) were also observed in its eye tracking. Similar oculomotor deficits are seen in patients with idiopathic Parkinsonism. The MPTP model is seen as offering possibilities for studying the mechanisms underlying the motor disabilities of Parkinson's disease.

1037. Brown, Loren; Rosellini, Robert A.; Samuels, Owen B. & Riley, Edward P. (1982). **Evidence for a serotonergic mechanism of the learned helplessness phenomenon.** ***Pharmacology, Biochemistry & Behavior,*** 17(5), 877–883.
Examined the role of the serotonergic system in the learned helplessness (LH) phenomenon in 5 experiments with 202 male Holtzman rats. In Exp I, 200 mg/kg of levotryptophan injected 30 min prior to testing disrupted acquisition of FR2 shuttle escape behavior. In Exp II, 100 mg/kg of 5-HTP produced interference with the acquisition of the escape response. This interference was prevented by treatment with the serotonergic antagonist methysergide (MS). In Exp III, Ss were pretreated with a subeffective dose of levotryptophan in combination with subeffective exposure to inescapable shock. These Ss showed a deficit in the acquisition of FR2 shuttle escape. In Exp IV, combined exposure to a subeffective dose of 5-HTP and inescapable shock (40 trials) resulted in an acquisition deficit. This deficit was reversed by MS. Exp V showed that the detrimental effects of exposure to prolonged (80 trials) inescapable shock can be prevented by treatment with MS. Results implicate the serotonergic system as a possible mediator of the LH phenomenon. Results also attest to the utility of the LH paradigm in the examination of the psychophysiological bases of human depression.

1038. Brown, Patricia A.; Brown, Thomas H. & Vernikos-Danellis, Joan. (1976). **Histamine H-sub-2 receptor: Involvement in gastric ulceration.** ***Life Sciences,*** 18(3), 339–344.
Results from female Sprague-Dawley rats given metiamide, an H-sub-2 receptor antagonist, support the hypothesis that histamine mediates both stress and stress plus aspirin-induced ulceration by a mechanism involving the histamine H-sub-2 receptor.

1039. Brugge, Karen L.; Hingtgen, Joseph N. & Aprison, M. H. (1987). **Potentiated 5-hydroxytryptophan induced response suppression in rats following chronic reserpine.** ***Pharmacology, Biochemistry & Behavior,*** 26(2), 287–291.
Tested an animal model of depression in an experiment designed to measure the effects of chronic reserpine treatment on 5-hydroxytryptophan (5-HTP)-induced behavioral depression in male Wistar rats trained on a food reinforcement operant schedule. Rats were trained on a VI 1 reinforcement schedule and then divided into 3 chronic treatment groups. One received daily injections of a placebo, another 0.025 mg/kg reserpine, and the 3rd 0.05 mg/kg reserpine. It was found that 5-HTP-induced behavioral depression was potentiated in rats chronically treated with reserpine. Results support the hypothesis that in some types of human depression a decreased release of 5-hydroxytryptamine (5-HT) occurs of sufficient duration to permit the subsequent development of supersensitive 5-HT receptors.

1040. Burbacher, Thomas M.; Sackett, Gene P. & Mottet, N. Karle. (1990). **Methylmercury effects on the social behavior of *Macaca fascicularis* infants.** ***Neurotoxicology & Teratology,*** 12(1), 65–71.
Observations of the social behavior of 12 infant monkeys exposed in utero to methylmercury (MeHg) and 13 nonexposed control infant monkeys were performed as part of a study of the toxic, reproductive and developmental effects of maternal MeHg intake. MeHg-exposed offspring exhibited a decrease in social play behavior and a concomitant increase in nonsocial passive behavior. The MeHg effect on social play behavior tended to decrease with age, while the group differences in nonsocial passive behavior tended to increase.

1041. Burunat, Enrique; Castro, Rafael; Diaz-Palarea, Maria D. & Rodriguez, Manuel. (1987). **Conditioned response to apomorphine in nigro-striatal system-lesioned rats: The origin of undrugged rotational response.** ***Life Sciences,*** 41(15), 1861–1866.
Explored the development and time-course characteristics of early rotational response (ER) to apomorphine in 6-hydroxydopamine-lesioned rats. It is concluded that this ER can be considered a nonpharmacological conditioned response (CR) from repeated drug administration, whereas ER in response

to saline injections 2 wks after drug treatment can be considered CR to the environment associated with drug treatment. The undrugged CR is seen as an antiparkinsonian placebo effect.

1042. Cadet, Jean L.; della Puppa, Andrea & London, Edythe. (1989). **Involvement of nigrotecto-reticulospinal pathways in the iminodipropionitrile (IDPN) model of spasmodic dyskinesias: A 2-deoxy-D-[1-^{14}C]glucose study in the rat.** ***Brain Research,*** 484(1–2), 57–64.
Identified brain areas with abnormal functioning in IDPN-treated adult male rats with persistent spasmodic dyskinetic syndrome (PSDS [characterized by lateral and vertical head twitches, random circling, and increased tactile and acoustic startle responses]). An autoradiographic 2-deoxy-D-[1-^{14}C]glucose method was used to map cerebral glucose utilization in these Ss. Findings suggest that deleterious effects of IDPN on the nigrotectal pathways affecting head and neck movements and circling behaviors via the brainstem reticulospinal tracts may play an important role in IDPN-induced PSDS in rats.

1043. Caine, S. B.; Geyer, Mark A. & Swerdlow, N. R. (1991). **Carbachol infusion into the dentate gyrus disrupts sensorimotor gating of startle in the rat.** ***Psychopharmacology,*** 105(3), 347–354.
Prepulse inhibition (PPI) is the decrease in a startle response that occurs when the startling stimulus is preceded by a weaker stimulus or "prepulse." Carbachol infusion into the dentate gyrus of the hippocampal formation of male rats disrupted PPI. This disruption of sensorimotor gating occurred when the startling stimulus was acoustic or tactile. While pretreatment with the D2 dopamine receptor antagonist spiperone reversed disruption of PPI caused by systemic administration of apomorphine, this pretreatment failed to reverse disruption of PPI induced by carbachol infusion into the hippocampus. Prepulse inhibition of the startle reflex is an animal model in which pharmacologic stimulation of the hippocamopus mimics the deficits in sensorimotor gating observed in schizophrenic patients.

1044. Campbell, A.; Baldessarini, Ross J. & Cremens, M. C. (1988). **Dose-catalepsy response to haloperidol in rat: Effects of strain and sex.** ***Neuropharmacology,*** 27(11), 1197–1199.
Haloperidol induced catalepsy in 4 species of rats, with females showing greater and more varied drug sensitivity than males. Inconsistent findings regarding a biphasic dose-effect relationship in Sprague-Dawley males and Long-Evans females led to additional testing of both species. Findings indicate that this paradigm is not an adequate model for a proposed relationship between high potency doses of neuroleptics and acute dystonia in humans.

1045. Canon, Jeffrey G. & Houser, Vincent P. (1978). **Squirrel monkey active conflict test.** ***Physiological Psychology,*** 6(2), 215–222.
Subjected 10 female squirrel monkeys to a discrete trial conflict procedure which contained an active avoidance contingency. Failure to respond during conflict trials was therefore punished. Several classes of clinically active anxiolytics, including the benzodiazepines (chlordiazepoxide and diazepam), carbamates (meprobamate), and barbiturates (sodium phenobarbital and sodium pentobarbital), were tested to determine if these drugs could alter behavior generated by this schedule. The potency and efficacy of the above anxiolytic drugs compared favorably with the clinical activity of these agents in humans. Several other nonanxiolytic compounds (chlorpromazine, amitriptyline, and dextroamphetamine) were tested to ascertain the specificity of the above procedure in detecting the anxiolytic properties of drugs. None of these latter compounds produced an anxiolytic effect in this conflict procedure. It thus appears that this particular animal model may be a useful tool in evaluating the potency and efficacy of drugs commonly used for the treatment of anxiety in man.

1046. Cappeliez, Philippe & Moore, Elizabeth. (1990). **Effects of lithium on an amphetamine animal model of bipolar disorder.** ***Progress in Neuro-Psychopharmacology & Biological Psychiatry,*** 14(3), 347–358.
Examined the effects of chronic Li administration on changes induced by amphetamine administration and withdrawal on open-field locomotor activity of rats, considered to be an animal model of behaviors displayed in bipolar disorders. For 21 days, 36 male rats were administered either single daily injections of 0.9% saline, 0.15 mEq/kg LiCl, or 1.5 mEq/kg LiCl. From Day 7 to Day 16, half of the animals in each group were administered twice daily injections of either 1.5 mg/kg d-amphetamine (DAMP) or 0.9% saline. From Days 17 to 21, DAMP was withdrawn. Neither dose of LiCl significantly altered DAMP-induced increases in activity levels, which suggests that the influence of Li and DAMP on activity are mediated by different neurotransmitter systems.

1047. Cardeal, J. O. & Cavalheiro, E. A. (1987). **Occurrence of the alcoholic tolerance phenomenon and its relation to daily frequency of ethanol doses and to the presence of a withdrawal interval in an experimental model of alcoholism in rats.** ***Arquivos de Neuro-Psiquiatria,*** 45(1), 1–6.
Studied the effect of (1) the number of daily toxic doses of ethanol and (2) the duration of the withdrawal period on alcohol tolerance. Animal subjects: 40 male white Wistar rats. Ss received 1–3 daily doses of ethanol (20 ml/kg). Each S received a total of 21 doses. Ss receiving doses at 8-hr intervals underwent a 2nd 21-dose intoxication cycle beginning 15 days after the end of the 1st cycle. (English abstract).

1048. Carey, Robert J. (1991). **Pavlovian conditioning between co-administered drugs: Elicitation of an apomorphine-induced antiparkinsonian response by scopolamine.** ***Psychopharmacology,*** 104(4), 463–469.
Male rats with unilateral 6-hydroxydopamine (6-OHDA) substantia nigra lesions were given combined scopolamine (SCP) and apomorphine treatments. When the drugs were coadministered, Ss rotated in the contralateral direction, creating the opportunity for the stimulus effect of SCP to become associated with the response effect of apomorphine. In tests with SCP, Ss that previously had SCP and apomorphine coadministered rotated contralaterally in the test chamber, behaving as if they had received apomorphine; thus, SCP exhibited a functionally acquired conditioned stimulus/stimuli (CS) property. Data suggest that, through conditioning, SCP can activate antiparkinsonian effects through nondopaminergic mechanisms and that interoceptive cues may provide an important source of conditioning.

1049. Carli, M. & Samanin, R. (1988). **Potential anxiolytic properties of 8-hydroxy-2-(di-*n*-propylamino)tetralin, a selective serotonin$_{1A}$ receptor agonist. *Psychopharmacology,*** 94(1), 84–91.
Investigated the effects of peripherally administered 8-hydroxy-2-(Di-*n*-propylamino)tetralin (8-OH-DPAT) and of 8-OH-DPAT microinjections into the midbrain raphe nuclei of 45 male rats in 2 animal models of anxiety. Two experiments examined the effects of peripherally administered 8-OH-DPAT on the drinking time of water-deprived rats, naive or habituated to the test apparatus, to distinguish between effects on punished responses and those on primary drive states. The effects of systemically administered 8-OH-DPAT were studied in rats in which exploratory behavior was suppressed by immobilization stress. Data show that brain serotonin is involved in the mechanisms mediating behavioral suppression in the presence of aversive stimuli.

1050. Carlson, Jeffrey N. & Rosellini, Robert A. (1987). **Exposure to low doses of the environmental chemical dieldrin causes behavioral deficits in animals prevented from coping with stress. *Psychopharmacology,*** 91(1), 122–126.
Conducted 2 experiments to assess the effects of low doses of an environmental contaminant in conjunction with various forms of stress. Male Sprague-Dawley albino rats were given acute doses (0, 0.5, 1.5, or 4.5 mg/kg) of the chemical dieldrin and subsequently exposed to a series of 40 escapable shocks, identical inescapable shocks, or no shock in an operant chamber. Eight hours later, the Ss were re-exposed to escapable footshock. Escape deficits related to the size of the dieldrin dose were found in the inescapable shock group only. Data suggest that experience with the lack of control over stress is critical in determining the behavioral effects of the agent and that the behavioral effects caused by uncontrollable stress may be exacerbated by concurrent exposure to such compounds. The response to uncontrollable stress and the common neuronal systems that may be involved are discussed.

1051. Carnoy, P.; Ravard, S.; Hervé, D.; Tassin, J.-P. et al. (1987). **Apomorphine-induced operant deficits: A neuroleptic-sensitive but drug- and dose-dependent animal model of behavior. *Psychiatrie & Psychobiologie,*** 2(4), 266–273.
Investigated the effects of apomorphine (APO; i.e., at doses thought to stimulate dopaminergic autoreceptors) on operant behavior in male rats to assess alterations that might underly behavioral deficits associated with a reduced dopaminergic transmission. Results show that low doses of APO caused a reward deficit when Ss were shifted from continuous reinforcement to FR schedules of food delivery. This effect could be accounted for by a decreased ability of secondary reinforcers to sustain responding and/or by a disruption of cognitive processes. The APO-induced reward deficit in the FR-4 schedule was reversed by disinhibitory neuroleptics, including amisulpride, pimozide, pipotiazine, and sulpiride, at low to moderate doses. Results obtained following 6-hydroxydopamine (6-OHDA) lesions suggest that the APO-induced behavioral deficit was related to a functional imbalance between mesolimbic and mesocortical dopaminergic systems. (French abstract).

1052. Carnoy, P.; Soubrie, P.; Puech, A. J. & Simon, P. (1986). **Performance deficit induced by low doses of dopamine agonists in rats: Toward a model for approaching the neurobiology of negative schizophrenic symptomatology? *Biological Psychiatry,*** 21(1), 11–22.
In searching for reliable animal models of negative schizophrenic symptomatology, the possibility that a deficient response to rewarding stimuli might be the basis for some features of the disease was considered. Apomorphine (0.015 and 0.03 mg/kg) and *N-n*-propyl-3(3-hydroxyphenyl)-piperidine (3-PPP [1 mg/kg]) caused such a reward deficit when male Wistar AF rats were shifted from continuous reinforcement to an FR schedule of food delivery. Further experiments indicated that this effect could be accounted for by a decreased ability of secondary reinforcers to sustain responses, rather than by motor impairment, appetite loss, or reduced reward value of the food. It is suggested that if this deficit is due to decreased dopaminergic transmission produced by low doses of dopamine agonists, some symptoms of schizophrenia (e.g., anhedonia) are not incompatible with deficient dopaminergic transmission. Low to moderate doses of sulpiride, amisulpride, pimozide, and pipotiazine, but not fluphenazine, metoclopramide, haloperidol, thioridazine, and chlorpromazine, reversed the apomorphine-induced reward deficit. It is tentatively proposed that only some neuroleptics, at dosages insufficient to block dopamine transmission postsynaptically, are effective in reducing negative schizophrenic symptoms.

1053. Carolei, A.; Margotta, V. & Palladini, G. (1975). **Proposal of a new model with dopaminergic-cholinergic interactions for neuropharmacological investigations. *Neuropsychobiology,*** 1(6), 355–364.
Observations of the *Dugesia gonocephala s.l.* planaria indicate that their motor system shows a striking similarity to the extrapyramidal system of high vertebrates and of humans, with evidence of correlations between dopaminergic and cholinergic neurons. The utilization of this model seems to be useful in testing drugs which presumably act on dopaminergic or cholinergic transmission. In this model, the quantification of animal behavior seems considerably easier when compared with the difficulties met in other animal models commonly employed.

1054. Case, Todd C.; Snider, Stuart R.; Hruby, Victor J. & Rockway, Todd. (1985). **Active and inactive L-prolyl-L-leucyl glycinamide synthetic analogs in rat models of levodopa-treated Parkinson's disease. *Life Sciences,*** 36(26), 2531–2537.
Evaluated the interaction of levo-prolyl-levo-leucyl-glycinamide (PLG) and its synthetic analogs with levodopa in 2 animal models of Parkinson's disease. In Exp I, which used rats with chronic unilateral lesions of the nigrostriatal dopamine pathway, PLG and Cbz-pro-leu-gly-NH_2 (Z-PLG) potentiated the contraversive rotation elicited by levodopa with carbidopa (L/C). In a 2nd experiment with reserpinized rats, PLG, Z-PLG, and cyclo-leu-gly potentiated L/C reversal of hypokinesia.

1055. Casey, Daniel E.; Gerlach, Jes & Christensson, Erik. (1980). **Dopamine, acetylcholine, and GABA effects in acute dystonia in primates. *Psychopharmacology,*** 70(1), 83–87.
Treated 8 monkeys (*Cercopithecus aethiops*) with haloperidol at doses sufficient to evoke dystonia (5–10 mg/kg/day, orally). The effects of agents that influenced dopaminergic, cholinergic, or GABA-ergic neurotransmitters were evaluated. Apomorphine, a dopamine (DA) agonist, and biperiden, an acetylcholine (ACh) antagonist, decreased acute dystonia, whereas alpha-methylparatyrosine, an inhibitor of DA synthesis, and physostigmine, an ACh agonist, increased symp-

toms. Muscimol, a GABA agonist, increased dystonia in a dose-dependent way. GABA inhibition with picrotoxin also aggravated dystonia, complicated by systemic intoxication and seizures. The reciprocal interaction between DA and ACh influences is consistent with clinical findings and animal models of dyskinesias. Data suggest complex interactions among DA, ACh, and GABA neurotransmission.

1056. Cavoy, A.; Ennaceur, A. & Delacour, J. (1988). **Effects of piracetam on learned helplessness in rats.** ***Physiology & Behavior,*** 42(6), 545–549.
The effects of Piracetam (P), the prototype of nootropic drugs, were studied in male rats, using the learned helplessness (LH) phenomenon. Exposure to uncontrollable and unsignalled shocks impaired subsequent escape-avoidance learning. In Exp I, this deficit was abolished by 200 mg/kg of P, and to a lesser extent by a 100 mg/kg dose administered before the training session. In nonstressed Ss, no dose of P had a facilitatory effect on escape-avoidance. In Exp II, the administration of P before the stress had no effect on the LH phenomenon, regardless of the dose.

1057. Chan, Arthur W. (1984). **Effects of combined alcohol and benzodiazepine: A review.** ***Drug & Alcohol Dependence,*** 13(4), 315–341.
Presents a review of both human and animal studies relating to the combined effects of alcohol and benzodiazepine (BZD). Although the combination of alcohol and BZD is sometimes associated with drug-induced deaths, drug overdoses, and traffic accidents or fatalities, epidemiological information is lacking on the true extent of the combined abuse and on the patterns and prevalence of use of these 2 drugs. Since BZD is widely used for the short- and long-term treatment of alcoholics, these patients are deemed more at risk of developing BZD or alcohol/BZD dependence than the general population. It is suggested that there is a need for large-scale controlled studies concerning the efficacy of BZD in the long-term treatment of alcoholics. Compared to males, females are at a higher risk as far as the potential for BZD addiction is concerned, since they tend to use BZD more often. Epidemiologic studies on the patterns of use of BZD, alcohol, or alcohol/BZD in pregnant women are called for. Animal models are also needed to ascertain whether prenatal exposure to both alcohol and BZD can impart long-lasting behavioral changes in the progeny, because it is possible that BZD can exacerbate the damaging prenatal effects of alcohol.

1058. Chan, Arthur W. & Siemens, Albert J. (1979). **Development of acute tolerance to pentobarbital: Differential effects in mice.** ***Biochemical Pharmacology,*** 28(4), 549–552.
Investigated whether the selective development of acute tolerance in C57BL mice was specific to ethanol or whether it would occur with other hypnotics such as pentobarbital. Results support the use of C57BL and DBA mice as animal models to determine the relationships between voluntary ethanol intake and neural sensitivity to the acute effects of ethanol, and the rates of development of tolerance to and physical dependence on ethanol.

1059. Chapman, Joab; Feldon, Joram; Alroy, Gil & Michaelson, Daniel M. (1989). **Immunization of rats with cholinergic neurons induces behavioral deficits.** ***Journal of Neural Transplantation,*** 1(2), 63–76.
Examined the effect of repeated immunization of rats with perikarya (PK) and nerve terminals of the electric fish *Torpedo* cholinergic neurons as antigens for 1 yr. Immunoblot studies revealed that sera of cholinergic PK immunized (CPKI) Ss contained a high level of antibodies to cholinergic PK proteins. Sera from Ss immunized with cholinergic synaptosomes and from controls contained very low levels of these antibodies. Behavioral studies revealed that the CPKI Ss were impaired in spatial learning and memory tasks when compared to controls and that the synaptosome-immunized Ss showed no such deficit. The association of antibodies to cholinergic neurons with cognitive deficits in this rat model suggests that such antibodies may be involved in the pathogenesis of Alzheimer's disease.

1060. Chiu, Simon; Paulose, C. S. & Mishra, Ram K. (1981). **Neuroleptic drug-induced dopamine receptor supersensitivity: Antagonism by L-prolyl-L-leucyl-glycinamide.** ***Science,*** 214(4526), 1261–1262.
Used an animal model of tardive dyskinesia to evaluate the potential antidyskinetic properties of the neuropeptide levo-prolyl-levo-leucyl-glycinamide (PLG). In male Sprague-Dawley rats, PLG administered concurrently with the neuroleptic drug haloperidol or chlorpromazine antagonized the enhancement of specific [^{3}H]spiroperidol binding in the striatum that is associated with long-term neuroleptic treatment. Results are discussed in relation to a possible functional coupling of the putative PLG receptor with neuroleptic–dopamine receptor complex and clinical implications for tardive dyskinesia.

1061. Chiueh, Chuang C. et al. (1985). **Primate model of parkinsonism: Selective lesion of nigrostriatal neurons by 1-methyl-4-phenyl-1,2,3,6-tetrahydropyridine produces an extrapyramidal syndrome in rhesus monkeys.** Meeting of the American Society for Pharmacology & Experimental Therapeutics: Pharmacological features of the dopaminergic neurotoxin MPTP (1984, Indianapolis, Indiana). ***Life Sciences,*** 36(3), 213–218.
Administered 1-methyl-4-phenyl-1,2,3,6-tetrahydropyridine (MPTP) systemically to adult rhesus monkeys. Results show that doses of 1.0–2.5 mg/kg produced irreversible damage to nigrostriatal neurons. Dopaminergic neurons in the dorsolateral part of the striatum were the most vulnerable. The major clinical signs of an extrapyramidal syndrome, but not resting tremor, appeared only in MPTP-treated Ss suffering from more than 80% reduction in striatal dopamine. No chronic changes in the mesolimbic dopaminergic system were observed. Immunocytochemical staining of the midbrain with a tyrosine hydroxylase antiserum indicated that MPTP produced a significant decrease of dopaminergic cell bodies in the A9, but not in the A10, ventrotegmental area. Despite greater than 80% decrease in A9 nigral cell bodies, the dopamine content decreased only by 50%. Behavioral, neurochemical, and histological results indicate that MPTP produces an ideal primate model for studying parkinsonism.

1062. Claridge, Gordon. (1978). **Animal models of schizophrenia: The case for LSD-25.** ***Schizophrenia Bulletin,*** 4(2), 186–209.
Some difficulties in establishing an animal model of schizophrenia are considered, and a review is made of the evidence on the experimental psychopathology of schizophrenia, particularly that concerned with attention and arousal. It is concluded that the core feature that needs to be modeled in animals is some aspect of "input dysfunction." Of

the pharmacological strategies, LSD-25 comes nearest to meeting that requirement, for 2 reasons: (a) The phenomenology of an LSD "model psychosis" closely parallels that of the natural disease. (b) The experimental effects of the drug, both in animals and humans, are very similar to or can be closely aligned theoretically with those of schizophrenia. LSD has been found to produce psychophysiological effects virtually identical to those observed occurring naturally in acute psychotic patients and in normal Ss high in psychotic personality traits. The rejection of LSD as a drug model was premature, especially as the currently popular preference for amphetamine has not been vindicated, either by the latter's ability to mimic an important central feature of the psychotic state or by work on dopamine as a specific common mediator of amphetamine psychosis and schizophrenia.

1063. Collins, Allan C. & Marks, Michael J. (1991). **Progress towards the development of animal models of smoking-related behaviors.** ***Journal of Addictive Diseases,*** 10(1–2), 109–126.
Human twin studies (e.g., E. Raaschou-Nielsen, 1960) have indicated that genetic factors influence whether people smoke and may influence amount of tobacco used. Research also shows that inbred mouse strains (MSs) differ in sensitivity to actions of a 1st-challenge dose of nicotine. These strain differences are due, in part, to differences in the number of brain nicotinic receptors. MSs also differ in the development of tolerance to nicotine and subtle differences in chronic nicotine-induced increases in the number of brain nicotinic receptors have been detected. Data suggest that MSs differ in oral self-selection of nicotine containing solutions, a finding that suggests genetic influences on rewarding effects on nicotine. Humans may also differ, for genetic reasons, in sensitivity to nicotine, in the development of tolerance to nicotine, and in rewarding effects of nicotine.

1064. Colton, Carol A.; Yao, Jibin; Gilbert, Daniel & Oster-Granite, Mary-Lou. (1990). **Enhanced production of superoxide anion by microglia from trisomy 16 mice.** ***Brain Research,*** 519(1–2), 236–242.
Investigated whether disruption of normal oxygen radical metabolism in the central nervous system (CNS) contributes to neuropathological changes associated with Down syndrome (trisomy 21) and its mouse counterpart, the trisomy 16 (Ts16) mouse. Primary glial cultures were prepared from cerebral cortices of Ts16 and normal littermate (NL) mice. Stimulation by either opsonized zymosan or phorbol myristate acetate produced significantly higher levels of superoxide (SPX) per mg protein in Ts16 cultures. Astrocyte enriched cultures exhibited low levels of SPX production that was higher in Ts16 Ss. SPX production in the Ts16 media treated rat microglia was significantly higher than in those treated with NL conditioned media.

1065. Commissaris, Randall L.; Ellis, Donna M.; Hill, Timothy J.; Schefke, Diane M. et al. (1990). **Chronic antidepressant and clonidine treatment effects on conflict behavior in the rat.** ***Pharmacology, Biochemistry & Behavior,*** 37(1), 167–176.
Examined the effects of chronic treatment with several antidepressants and clonidine on conflict behavior, using the conditioned suppression of drinking paradigm with female rats. In daily 10-min sessions, water-deprived Ss were trained to drink from a tube that was occasionally electrified. Electrification was signaled by a tone. Chronic desipramine or clonidine treatment resulted in time-dependent anticonflict effects, with a latency to onset of approximately 3–4 wks. In contrast, chronic buproprion, mianserin, or trazodone resulted in, at best, only a weak anticonflict effect. The efficacy of these antidepressants and clonidine to increase punished responding when administered chronically correlates well with their efficacy as antipanic agents in humans.

1066. Commissaris, Randall L.; Harrington, Gordon M. & Altman, Harvey J. (1990). **Benzodiazepine anti-conflict effects in Maudsley Reactive (MR/Har) and Non-Reactive (MNRA/Har) rats.** ***Psychopharmacology,*** 100(3), 287–292.
Compared the effects of diazepam and alprazolam on conditioned suppression of drinking, a model behavior for the study of anxiety and/or emotionality in MR/Har and MNRA/Har rat strains. Ss were trained to drink from a tube that was occasionally electrified. Both diazepam and alprazolam increased punished responding in a dose-related manner. MNRA/Har strain Ss exhibited a significantly greater anti-conflict effect following diazepam or alprazolam treatment than did MR/Har strain Ss. The behavioral differences exhibited by MR/Har and MNRA/Har rat strains may constitute a genetically based animal model for the study of emotionality and/or anxiety.

1067. Commissaris, Randall L.; McCloskey, Timothy C.; Harrington, Gordon M. & Altman, Harvey J. (1989). **MR/Har and MNRA/Har Maudsley rat strains: Differential response to chlordiazepoxide in a conflict task.** ***Pharmacology, Biochemistry & Behavior,*** 32(3), 801–805.
In daily 10-min sessions, water-deprived rats were trained to drink from a tube that was occasionally electrified. After several weeks of conditioned suppression of drinking (CSD) testing, MNRA/Har Ss accepted significantly more shocks than did MR/Har Ss during control (nondrug) sessions. In both strains, the number of shocks accepted was inversely related to the intensity of the shock used, with MNRA/Har Ss accepting significantly more shocks than MR/Har Ss at all intensities. The effects of various doses (1.25–28.4 mg, ip) of chlordiazepoxide were determined in Ss of the MNRA/Har strain at the original training intensity (0.5 mA), while a lower intensity (0.25 mA) was utilized in MR/Har Ss. MNRA/Har Ss were more responsive to the anticonflict effects of chlordiazepoxide than Ss of the MR/Har strain.

1068. Commissaris, Randall L.; Vasas, Rita J. & McCloskey, Timothy C. (1988). **Convulsant versus typical barbiturates: Effects on conflict behavior in the rat.** ***Pharmacology, Biochemistry & Behavior,*** 29(3), 631–634.
Compared the effects on rats of the convulsant barbiturate cyclohexylideneethyl-5-barbituric acid (CHEB) and of the barbiturates in the conditioned suppression of drinking paradigm, an animal model for the study of anxiety and anti-anxiety agents. Water-deprived females were trained to drink from a tube that was electrified after a tone and were then administered drug tests. Consistent with previous reports, typical barbiturates (pentobarbital, secobarbital, phenobarbital) produced dose-dependent increases in the number of shocks received at doses that did not depress water intake. Subconvulsant doses of CHEB (0.3–2.5 mg/kg) produced a dose-dependent depression of both punished responding and water intake. Results suggest that (1) convulsant and typical barbiturates have markedly different effects on conflict behavior in the rat and (2) CHEB has no barbiturate antagonist qualities.

1069. Concannon, James T. & Schechter, Martin D. (1982). **Failure of amphetamine isomers to decrease hyperactivity in developing rats.** ***Pharmacology, Biochemistry & Behavior,*** 17(1), 5–9.
Investigated possible amphetamine-induced changes in locomotor activity in developing Sprague-Dawley rats administered intracisternal injections of 6-hydroxydopamine (6-OHDA) or its vehicle at 5 days of age. Administration of the dopamine neurotoxin resulted in a significant depletion of whole-brain dopamine to 44% of control levels, whereas norepinephrine levels were not significantly reduced. In normal and 6-OHDA-treated Ss, activity increased from moderately low levels at 15 days of age to moderately high levels at 25 days. However, 6-OHDA-treated Ss were hyperactive at 20 days. At 25 days, activity in both groups was equal and declined to levels typical for adults. Administration of graded doses of dextro- (DAM) and levoamphetamine (LAM) generally increased activity in both groups, with DAM being more potent than LAM. No dose of either isomer decreased activity in 6-OHDA-treated, hyperactive Ss. Hence, no convincing evidence was found for a "paradoxical calming" effect of amphetamine in hyperactive rats, supporting other recent reports. Results suggest that the neonatal dopamine-depleted rat does not provide an accurate model system for preclinical investigation of the human hyperkinetic syndrome.

1070. Cooper, Barrett R.; Howard, James L. & Soroko, Francis E. (1983). **Animal models used in prediction of antidepressant effects in man.** ***Journal of Clinical Psychiatry,*** 44(5, Sect 2), 63–66.
The discovery of bupropion's potential antidepressant activity resulted from studies of its behavioral effects in a number of animal models of depression. The present authors review these animal models (e.g., the tetrabenazine model, behavioral despair model) and data pertaining to their selectivity for other standard antidepressant drugs.

1071. Coper, H.; Rommelspacher, H. & Wolffgramm, J. (1990). **The "point of no return" as a target of experimental research on drug dependence.** Special Issue: Research and policy. ***Drug & Alcohol Dependence,*** 25(2), 129–134.
Argues for an approach to the study of drug-taking behavior using animal models (ANMs) that emphasizes the efforts an individual is willing to make to get access to the drug. This focus on stages of drug taking raises the issue of "a point of no return" (PONR) in drug dependence. The concept of PONR suggests the succession of 2 states of drug consumption with a transition zone (TZ) between them. The criteria for discrimination between these states are control and reversibility of drug taking. Long-lasting changes in neuronal systems as a result of drug dependence are also discussed. The PONR as a research area has a basis in functional and neuronal levels; ANMs of dependence should permit study of the PONR.

1072. Corson, Samuel A. & Corson, Elizabeth O. (1988). **Psychopharmacologic facilitation of psychosocial therapy of violence and hyperkinesis.** ***Activitas Nervosa Superior,*** 30(1), 22–39.
Low dosages of amphetamine (AM) administered to 11 hyperkinetic and violent dogs and cats to inhibit nonadaptive forms of behavior permitted the development of discriminated Pavlovian and operant conditioned responses (CRs). The beneficial effects of this drug persisted in the no-drug state. AM also ameliorated conditional emotional visceral responses in Ss with low adaptation to psychologically stressful situations. The same AM dosage that improved behavior and learning in experimental Ss produced disorientation in normal dogs with previously stable CRs.

1073. Corwin, Rebecca L.; Woolverton, William L.; Schuster, Charles R. & Johanson, Chris E. (1987). **Anorectics: Effects on food intake and self-administration in rhesus monkeys.** ***Alcohol & Drug Research,*** 7(5–6), 351–361.
In Exp I, the effects of benzphetamine (BENZ [20 mg]), chlorphentermine (CHLOR [40 mg]), clortermine (CLOR [40 mg]), mazindol (2 mg), phendimetrazine (PHD [10 mg]), and phenmetrazine (10 mg) on food intake were compared with the effects of dextroamphetamine (5 mg) in 6 rhesus monkeys given daily access to food pellets. In Exp II, the ability of these drugs to maintain intravenous self-administration under a FR-10 schedule was determined. All were effective anorectics; CHLOR, CLOR, and PHD were not positive self-administered reinforcers and might be less subject to abuse; BENZ was more potent as a positive reinforcer than as an anorectic and may be therapeutically undesirable.

1074. Costall, B.; Jones, B. J.; Kelly, M. E.; Naylor, R. J. et al. (1990). **Ondansetron inhibits a behavioural consequence of withdrawing from drugs of abuse.** ***Pharmacology, Biochemistry & Behavior,*** 36(2), 339–344.
Examined the ability of the selective 5-hydroxytryptamine (5-HT_3) receptor antagonist ondansetron (OND) to influence the behavioral consequences of withdrawal from chronic treatment with ethanol, nicotine, or cocaine in the light/dark exploration test in the mouse and social interaction test in the rat. In both tests acute and chronic (7 days) treatments with OND disinhibited suppressed behavior; withdrawal from chronic treatment did not exacerbate the behavioral suppression. OND administration during the period of ethanol, nicotine, and cocaine withdrawal prevented the exacerbation in suppressed behavior. OND potently reduces behavioral suppression during acute and chronic treatments in the rodent models, does not cause a rebound exacerbation of behavioral suppression following withdrawal, and is an effective inhibitor of the increased behavioral suppression following withdrawal from the drugs of abuse: ethanol, nicotine, and cocaine.

1075. Costall, B.; Jones, B. J.; Kelly, M. E.; Naylor, R. J. et al. (1989). **Exploration of mice in a black and white test box: Validation as a model of anxiety.** ***Pharmacology, Biochemistry & Behavior,*** 32(3), 777–785.
The validity of a black and white test box to measure changes in mouse exploratory behavior relevant to assessment of anxiety was investigated by variation of the illumination within the test box, the use of different strains of mice, holding conditions and drug treatments. The suppression of exploratory activity in the white section caused by bright illumination was antagonized by anxiolytic agents from the benzodiazepine series, buspirone, 5-HT_3 receptor antagonists, alcohol, nicotine, morphine and SCH23390. The anxiogentic agent FG7142 exacerbated the behavioral suppression. Black C57/BL/6, brown DBA_2 and albino BKW mice were sensitive to the effects of drug treatments, whereas albino Tuck mice were less responsive. It is concluded that the characteristic change caused by anxiolytic agents is to preferentially increase exploratory behavior of mice.

1076. Costall, Brenda; Kelly, M. Elizabeth; Naylor, Robert J. & Onaivi, Emmanuel S. (1989). **The actions of nicotine and cocaine in a mouse model of anxiety.** ***Pharmacology, Biochemistry & Behavior,*** 33(1), 197–203.
Determined whether nicotine and cocaine as drugs of abuse could induce a spectrum of behaviors indicative of anxiolytic and anxiogenic action during and following withdrawal. Similar effects have been reported for ethanol and diazepam in the mouse black and white test box model (J. M. Barry et al; 1987; B. Costall et al, 1988). Withdrawing nicotine produced a behavioral profile similar to that of ethanol and diazepam (anxiolytic action during acute/chronic administration and anxiogenesis during withdrawal). Cocaine produced a profile similar to that from alcohol or nicotine, except that an acute 1- or 2-dose treatment failed to produce an anxiolytic response. Anxiogenesis may be associated with an increased 5-hydroxytryptamine (5-HT) function.

1077. Cowan, Alan & Gmerek, Debra E. (1982). **Defining the antinocisponsive actions of neurotensin.** ***Annals of the New York Academy of Sciences,*** 400, 438–439.
Although neurotensin (NT) is active in behavioral analgesic tests with mice, it is inactive in several of these tests with rats. To develop additional animal models for the antinocisponsive effects of NT, the present experiment studied the attenuating effects of NT and morphine on thyrotropin-releasing hormone (TRH)-induced wet-dog shakes in male Sprague-Dawley rats. Results show that small icv doses of NT antagonized TRH-induced shakes. Naloxone did not block nor was tolerance developed to the suppressant action of NT. NT was 16 times more potent than morphine in attenuating the TRH-induced shakes.

1078. Crabbe, John C.; Merrill, Catherine M.; Kim, Daniel & Belknap, John K. (1990). **Alcohol dependence and withdrawal: A genetic animal model.** ***Annals of Medicine,*** 22(4), 259–263.
Describes development of lines of mice genetically susceptible to severe (withdrawal seizure prone [WSP]) or mild (withdrawal seizure resistant [WSR]) withdrawal after chronic alcohol inhalation. Studies with the WSP/WSR genetic animal model suggest that genes influencing alcohol withdrawal severity are distinct from those mediating sensitivity and tolerance. On the other hand, some traits differ markedly between WSP and WSR mice, and those traits may be related to underlying genetic susceptibility to dependence. In particular, differences in seizure susceptibility between WSP and WSR mice are beginning to be defined more clearly. Indications of differences in hippocampal systems are accumulating and include zinc content in mossy fibers, as well as numbers of dihydropyridine-sensitive calcium channel binding sites and N-methyl-D-aspartate receptor sites.

1079. Crawley, Jacqueline N.; Glowa, John R.; Majewska, Maria D. & Paul, Steven M. (1986). **Anxiolytic activity of an endogenous adrenal steroid.** ***Brain Research,*** 398(2), 382–385.
Investigated the behavioral profile of 3 alpha,5 alpha-tetrahydrodeoxycorticosterone (THDOC) with male Swiss-Webster mice and Sprague-Dawley rats, using 2 animal models of anxiety—the 2-chambered mouse exploration test and the lick suppression conflict test. THDOC showed anxiolytic activity in both models, with an anxiolytic dose range of 5–15 mg/kg, intraperitoneally (ip), separable from the sedative dose range of above 20–30 mg/kg, ip. These dose ranges are compared with those of diazepam, chlordiazepoxide, and sodium pentobarbital, as determined by the 1st author (1981) in a previous study.

1080. Critchley, M. A. & Handley, S. L. (1987). **Effects in the X-maze anxiety model of agents acting at 5-HT$_1$ and 5-HT$_2$ receptors.** ***Psychopharmacology,*** 93(4), 502–506.
Three 5-hydroxytryptamine (5-HT) agonists (8-hydroxy-2-[di-*n*-propylamino]tetralin, 5-methoxy-N,N-dimethyltryptamine, and RU 24969) produced in male rats a dose-related fall in open/total arm entry ratio in the elevated **X**-maze model of anxiety at doses that did not affect total entries. The relative potency of the agonists suggests that 5-HT$_1$ rather than 5-HT$_2$ receptors may be involved in anxiety. Results indicate involvement of 5-HT receptors in anxiety separately from any change in the ability to withhold responding.

1081. Czyrak, A.; Mogilnicka, E. & Maj, J. (1989). **Dihydropyridine calcium channel antagonists as antidepressant drugs in mice and rats.** ***Neuropharmacology,*** 28(3), 229–233.
Tested a pharmacological profile of the effects of nimodipine, nifedipine, and nitrendipine in several models indicative of possible antidepressant activity in mice and rats. These compounds, as well as verapamil (short-lasting effect), but not diltiazem, reduced the hypothermia induced by apomorphine in mice. Various antidepressants (imipramine, amitriptyline, citalopram, mianserin) used in the behavioral despair test in mice, in doses not effective by themselves, increased the immobility-reducing effect when given jointly with 1,4-dihydropyridine calcium channel antagonists. Results indicate that the psychopharmacological profile of nimodipine, nifedipine, and nitrendipine resembles that of antidepressants in only some tests. Results also support the assumption that concomitant administration of antidepressants and 1,4-dihydropyridine calcium channel antagonists may result in a greater antidepressant efficacy.

1082. Dahl, Carl B. & Götestam, K. Gunnar. (1989). **Lack of self-administration of different fenfluramine isomers in rats.** ***Addictive Behaviors,*** 14(3), 239–247.
12 rats were tested in an animal model for self-administration of dl-, d-, and l-fenfluramine. Amphetamine and saline were used as reference substances. Each S was only tested on 1 drug and dose. Analyses of variance (ANOVAs) were performed to assert that high rates of self-administration were maintained on amphetamine whereas saline gave low rates of responding. Results show that all 3 forms of fenfluramine differed significantly from amphetamine, but not from saline. As d-fenfluramine is both more effective in reducing food intake and has less sedative action than dl-fenfluramine, it may be an improvement in the pharmacotherapy of obesity.

1083. Danysz, W.; Fowler, C. J. & Archer, T. (1987). **Experimental methods employed in antidepressant research: A critical review.** ***New Trends in Experimental & Clinical Psychiatry,*** 3(1), 59–88.
Reviews the basic experimental methods employed in the search for potential antidepressant (AD) compounds. The aspects discussed include considerations and treatment design for animal studies of ADs, investigations into different individual aspects of AD action, screening tests, and models of depression. It is concluded that the potential utility of

animal models in AD research remains valid, since the application of integrated behavioral models of proven predictive value may permit the postulation of a common mechanism of AD action.

1084. Davis, J. Michael; Otto, David A.; Weil, David E. & Grant, Lester D. (1990). **The comparative developmental neurotoxicity of lead in humans and animals.** Special Issue: Qualitative and quantitative comparability of human and animal developmental neurotoxicity. ***Neurotoxicology & Teratology,*** 12(3), 215–229.
Reviews human and animal studies on the effects of lead on neurobehavioral development. This research provides a basis for comparing the developmental neurobehavioral toxicity of lead across species and for assessing the validity of animal models of developmental neurotoxicity. Comparisons of human and animal findings suggest that the greatest qualitative similarities involve cognitive and relatively complex behavioral processes such as learning. Quantitative comparisons based on dose-response relationships for these end points are difficult to make because the relationships are sometimes nonmonotonic (**U**-shaped) and because blood lead levels may not be directly comparable between species. However, the lowest levels of exposure at which developmental neurobehavioral effects have been observed are similar in children, primates, and rodents.

1085. Davis, Kenneth L. et al. (1979). **Dimethylaminoethanol (deanol): Effect on apomorphine-induced stereotypy and an animal model of tardive dyskinesia.** ***Psychopharmacology,*** 63(2), 143–146.
Deanol was administered acutely to male Sprague-Dawley rats subsequently injected with apomorphine. While 80 mg of deanol had no effect on the severity of apomorphine-induced stereotypy, 160 mg significantly diminished the severity of the stereotypy. This dose of deanol did not significantly alter spontaneous locomotor activity. Deanol did not reduce apomorphine-induced stereotypy in Ss previously exposed to haloperidol and presumed to have post-synaptic dopamine receptor supersensitivity. These results with deanol are contrasted with the effects of choline chloride and suggest that choline chloride may be more effective than deanol at augmenting striatal cholinergic activity.

1086. Daws, Lynette C.; Schiller, Grant D.; Overstreet, David H. & Orbach, Joe. (1991). **Early development of muscarinic supersensitivity in a genetic animal model of depression.** ***Neuropsychopharmacology,*** 4(3), 207–217.
Examined the sensitivity of Flinders-Sensitive Line (FSL) rats and Flinders-Resistant Line (FRL) rats to the hypothermic and locomotor inhibitory effects of the muscarinic agonist, oxotremorine. These findings were compared to the regional development of muscarinic receptor binding in similarly aged rats. FSL Ss were significantly more sensitive to oxotremorine-induced hypothermia than FRL Ss. The magnitude of this difference reached a maximum at the latest age tested (60 days old). A significant difference in muscarinic acetylcholine receptor (mAChR) concentrations was found only in the older Ss (61 days), where the FSL Ss had significantly higher numbers. Findings suggest that behavioral cholinergic supersensitivity need not be directly associated with increases in mAChR number. Implications for human depressive disorders are elaborated.

1087. de la Gándara Martín, J. J.; Redondo Martinez, A. L.; Hernandez Herrero, H.; Navarrette Lopez, M. et al. (1987). **Kindling: From experimental model to psychiatric clinic.** ***Archivos de Neurobiología,*** 50(6), 358–378.
Discusses the relevance for clinical psychiatry of animal experimentation concerning the kindling phenomenon in epilepsy. Bioelectrical, neurochemical, neurocytologic, and behavioral aspects of affective disorders, psychoses, and other psychopathologies are discussed. A model for etiopathogenic research for psychiatry is outlined. (English abstract).

1088. Dellinger, John A. (1991). **Pharmacologic challenges for establishing interspecies extrapolation models in neurotoxicology.** Symposium on Animal-To-Human Extrapolation (1990, San Antonio, Texas). ***Neuroscience & Biobehavioral Reviews,*** 15(1), 21–23.
Reports data from various studies that estimated vagal tone in atropine-treated commercial student pilots, cardiovascular-tethered baboons, telemetry-equipped dogs, and atropine-treated rhesus monkeys. Respiratory sinus arrhythmias were quantified following atropine sulfate administration, and the resulting vagolytic blockade was used as a pharmacologic challenge technique. Dose/response studies in humans and rhesus monkeys served as a basis for interspecies comparisons using ED_{50} calculations for a 30% decrease in the respiratory sinus arrhythmias following atropine.

1089. DeNoble, Victor J. (1987). **Vinpocetine enhances retrieval of a step-through passive avoidance response in rats.** ***Pharmacology, Biochemistry & Behavior,*** 26(1), 183–186.
Studied effects of vinpocetine, vincamine, apovincaminic acid, vinconate, aniracetam, Hydergine, and pemoline on retrieval of a step-through passive avoidance response in male Sprague-Dawley rats (aged 90–120 days). Results indicate that vinpocetine has cognition-activating abilities as defined in an animal model of memory retrieval; vinpocetine, which increases memory recall in human patients and volunteers, promoted memory retrieval in a passive avoidance paradigm in the Ss.

1090. de Pablo, Juan M.; Ortiz-Caro, Javier; Sanchez-Santed, Fernando & Guillamón, Antonio. (1991). **Effects of diazepam, pentobarbital, scopolamine and the timing of saline injection on learned immobility in rats.** ***Physiology & Behavior,*** 50(5), 895–899.
Used the rat forced-swimming test (FST) for screening substances with a potential antidepressant effect. Rat immobility shown in the FST has been interpreted as behavioral despair and has been suggested as an animal model of human depression. In 2 experiments, pentobarbital and scopolamine administered immediately after the 1st phase and diazepam administered 15 min before the 1st phase behave as false positives in the FST. Also, sensitivity of the FST was affected by the fact that the last saline injection, 1 hr before the 2nd phase, increased Ss' mobility. It is concluded that a learning-memory hypothesis copes better with the behavior of rats during the FST than does a behavioral despair hypothesis.

1091. de Ryck, Marc. (1990). **Animal models of cerebral stroke: Pharmacological protection of function.** 4th International Symposium on Calcium Antagonists: Basis for the application of calcium antagonists in neurology (1989, Florence, Italy). ***European Neurology,*** 30(Suppl 2), 21–27.

Investigated whether neocortical thrombotic infarcts reliably produce long-lasting sensorimotor deficits in rats and, if so, whether such deficits can be counteracted by flunarizine (FLZ). A photochemical technique was used which induces thrombotic infarction by iv injection of the fluorescein derivative Rose Bengal and focal illumination of the intact skull surface. Posttreatment with FLZ, a class IV calcium antagonist, within a critical period of the 1st 6 hrs after infarction resulted in marked sparing of sensorimotor function (tactile/proprioceptive limb placing reactions), while Ss remained normoglycemic and were free of drug-induced behavioral toxicity. This could reflect FLZ-induced coping of neuronal tissue with ischemia-related ionic shifts.

1092. de Souza, Helenice; Trajano, Eleonora; de Carvalho, Fernando V. & Palermo Neto, João. (1978). **Effects of acute and long-term cannabis treatment on restraint-induced gastric ulceration in rats.** ***Japanese Journal of Pharmacology,*** 28(3), 507–510.
Wistar rats of both sexes were either (a) given 2 doses of *Cannabis sativa* extract (CSE), with doses ranging from 5 to 60 mg/kg ip, or saline, and were then immobilized; (b) given CSE doses of 40 or 60 mg/kg/day or saline for 20 days and immobilized; or (c) given 60 mg/kg/day of CSE for 25 days and left unrestrained. Ss were then killed and their stomachs examined for ulcers. Acute administration of 40 or 60 mg/kg CSE significantly reduced ulceration, but long-term administration did not for the immobilized Ss. Long-term administration of CSE significantly increased ulceration in unrestrained Ss.

1093. de Vito, Michael J.; Brooks, William J. & Wagner, George C. (1987). **Behavioral deficits following intracranial administration of MMP$^+$ to the rat.** ***Research Communications in Psychology, Psychiatry & Behavior,*** 12(2), 65–74.
Evaluated the long-term behavioral and neurochemical consequences of the administration of 40µg of 1-methyl-4-phenylpyridinium ion (MPP$^+$), the metabolite of the dopaminergic neurotoxin 1-methyl-4-phenyl-tetrahydropyridine (MPTP), using 25 female rats. In the rotorod paradigm, the MPP$^+$-treated Ss accumulated less time on the rod than the sham-treated Ss. In the passive avoidance task, the MPP$^+$-treated Ss required significantly more time to make an escape response than the sham-treated Ss. These behavioral deficits are discussed in reference to the rodent model of Parkinson's disease using MPTP and MPP$^+$.

1094. Díaz Palarea, María D.; Castro Fuentes, R.; Aréualo Garcia, R. M.; Pérez Hernández, J. et al. (1986). **Evaluation of the therapeutic effectiveness of dopaminergic agonists in Parkinsonism: Utility of animal models.** ***Psiquis: Revista de Psiquiatría, Psicología y Psicosomática,*** 7(3–86), 48–57.
Describes an animal model for studying the effects of dopaminergic agents, including L-Dopa (dopamine precursor), apomorphine (dopamine receptor agonist), and cocaine and nomifensine (dopamine uptake inhibitors). The advantage of the model is that it permits discrimination of the locus of action for each agonist and its relative potency for intact or lesioned sides of the brain. (English abstract).

1095. di Lorenzo, Rosaria; Bernardi, Mara; Genedani, Susanna; Zirilli, Enrico et al. (1987). **Acute alkalosis, but not acute hypocalcemia, increases panic behavior in an animal model.** ***Physiology & Behavior,*** 41(4), 357–360.
Tested nonpretrained, randomized adult rats in a panic-inducing model of passive avoidance. IV treatment with alkalinizing agents but not with a hypocalcemic dose of ethylendiamine tetra-acetic acid (EDTA) 3 min before testing, significantly increased panic behavior. Data may support the hypothesis that panic attacks are due to alkalosis.

1096. D'Mello, G. D. (1986). **Effects of sodium cyanide upon swimming performance in guinea-pigs and the conferment of protection by pretreatment with p-aminopropiophenone.** ***Neurobehavioral Toxicology & Teratology,*** 8(2), 171–178.
Reports that the swimming performance of guinea pigs was degraded following administration of sodium cyanide (NaCN) at doses that were not lethal for individual Ss. Data suggest that NaCN may affect both motor and cognitive function in guinea pigs. The relevance of this animal model for predicting the behavioral effects of cyanide poisoning and for assessing the protective efficacy of pretreatment with para-aminopropiophenone in humans is discussed.

1097. Dole, Vincent P. (1986). **On the relevance of animal models of alcoholism in humans.** ***Alcoholism: Clinical & Experimental Research,*** 10(4), 361–363.
Suggests that it is not yet possible to induce pharmacologically motivated drinking in C57BL mice. Three objective tests of pharmacologically motivated drinking are described—analysis of intake patterns, determination of elasticity of appetite, and monitoring the concentration of alcohol in the blood during spontaneous drinking. Conditions that might make alcohol rewarding to laboratory animals are discussed.

1098. Dolphin, A. C.; Jenner, P. & Marsden, C. D. (1976). **The relative importance of dopamine and noradrenaline receptor stimulation for the restoration of motor activity in reserpine or a-methyl-p-tyrosine pre-treated mice.** ***Pharmacology, Biochemistry & Behavior,*** 4(6), 661–670.
Two animal models of parkinsonism have been employed to investigate the role of noradrenaline (NA) in the motor effects of levodopa. Pretreatment with reserpine or alphamethylparatyrosine (AMPT) causes cerebral amine depletion and reduction of motor activity, which can be reversed by levodopa. The present experiments studied the effect of inhibitors of NA synthesis and antagonists of NA and dopamine (DA) receptors on the action of levodopa in male Swiss S or P strain mice. For comparison, the effects of such treatments on apomorphine action were investigated. Reversal of reserpine (10 mg/kg) induced akinesia by levodopa plus the peripheral decarboxylase inhibitor MK-486 was inhibited by prior administration of phenoxybenzamine, haloperidol, pimozide, or the dopamine-beta-hydroxylase inhibitor FLA-63. Apomorphine (2 mg/kg) reversal of reserpine akinesia was similarly inhibited by haloperidol and pimozide but not by phenoxybenzamine or FLA-63. Apomorphine (5 mg/kg) reversal of reserpine akinesia was enhanced by simultaneous administration of the noradrenergic agonist clonidine and this effect was not significantly altered by prior administration of FLA-63. Clonidine, however, reversed the FLA-63 induced inhibition of the levodopa effect on reserpine akinesia. Results suggest that full restoration of motor activity in reserpine or AMPT pretreated animals requires stimulation of both DA and NA receptors.

1099. Domer, Floyd R. & Wolf, Carlos L. (1979). **Drugs, lead and the blood-brain barrier.** ***Research Communications in Psychology, Psychiatry & Behavior,*** 4(2), 135–148.
CD-1 mice were exposed from birth to either .5% sodium acetate or lead acetate in drinking water. Locomotor activity (A) evaluation after weaning showed that lead-exposed Ss (LS) were hyperactive. Amphetamine and caffeine decreased A whereas methylphenidate and pemoline increased A. It is concluded that lead-induced hyperactivity (H) is not a valid model for evaluating drugs for use in children with H or minimal brain dysfunction.

1100. Donát, P. & Kršiak, M. (1985). **Effects of a combination of diazepam and scopolamine in animal model of anxiety and aggression.** 27th Annual Psychopharmacology Meeting (1985, Jeseník, Czechoslovakia). ***Activitas Nervosa Superior,*** 27(4), 307–308.
Compared the effects of administration of diazepam and scopolamine separately and in combination on the behavior of aggressive and timid male mice. Diazepam significantly reduced timid defensive escape behavior and increased sociable behavior, and scopolamine markedly reduced aggressive behavior and stimulated sociability. Combined treatment led to a broader spectrum of behavior effects, but the reduction of timid and aggressive activities was not more pronounced than when the drugs were administered alone.

1101. Dourish, C. T.; Hutson, P. H.; Kennett, G. A. & Curzon, G. (1986). **8-OH-DPAT-induced hyperphagia: Its neural basis and possible therapeutic relevance.** ***Appetite,*** 7(Suppl), 127–140.
Examined the pharmacological and neurochemical bases of hyperphagia induced by the serotonin agonist and novel ergot drug 8-hydroxy-2-(di-n-propylamino) tetralin (8-OH-DPAT) and assessed the possible therapeutic potential of 8-OH-DPAT and related drugs in the treatment of anorexic pathology in an animal model of anorexia (as induced by acute immobilization stress). In normal rats, 8-OH-DPAT elicited feeding after peripheral injection and after intracerebral application to the brain-stem raphe nuclei. Feeding elicited by peripheral injection of the drug was attenuated by pretreatment with the serotonin synthesis inhibitor parachlorophenylalanine. Following a hyperphagic dose of 8-OH-DPAT, brain serotonin metabolism was reduced. 8-OH-DPAT elicited feeding via an agonist action on serotonin autoreceptors in the raphe nuclei. 8-OH-DPAT and other 5-hydroxytryptamine$_{1A}$ (5-HT$_{1A}$) agonists attenuated the anorexia and body weight loss caused by immobilization stress.

1102. Downs, Nancy S.; Britton, Karen T.; Gibbs, Daniel M.; Koob, George F. et al. (1986). **Supersensitive endocrine response to physostigmine in dopamine-depleted rats: A model of depression?** ***Biological Psychiatry,*** 21(8–9), 775–786.
Examined whether specific abnormal neuroendocrine responses demonstrated by depressed patients could be incorporated into an animal model in 2 experiments with albino male Wistar rats. The effects of acetyl cholinesterase physostigmine (PHY) treatments on the plasma concentrations of prolactin (PRL) and adrenocorticotrophic hormone (ACTH) were examined. Physostigmine (0–0.6 mg/kg) produced a dose-dependent increase in PRL and ACTH immunoreactivity in unoperated Ss. Following depletion of brain dopamine, but not norepinephrine, Ss exhibited a "supersensitive" increase in plasma ACTH values. Results suggest parallels between the abnormal endocrine response to PHY demonstrated by depressed patients and that demonstrated by rats following depletion of central nervous system (CNS) DA levels.

1103. Drago, Filippo; Pulvirenti, Luigi; Spadaro, Francesco & Pennisi, Giovanni. (1990). **Effects of TRH and prolactin in the behavioral despair (swim) model of depression in rats.** ***Psychoneuroendocrinology,*** 15(5–6), 349–356.
The neuropeptides thyrotropin releasing hormone (TRH) and prolactin (PRL), which affect various behaviors in animals, showed antidepressant properties in an experimental model of depression. In male rats, sc administration of TRH reduced total immobility in the despair (constrained swim) test and potentiated the anti-immobility effect of ip desimipramine (DMI). Hyperprolactinemia induced by pituitary homografts under the kidney capsule and icv injection of PRL also potentiated the DMI-induced reduction of total immobility time of Ss in the despair test and exerted antidepressant effects in aged Ss.

1104. Drago, Filippo; Spadaro, F.; D'Agata, Velia; Valerio, Carmela et al. (1991). **Protective action of phosphatidylserine on stress-induced behavioral and autonomic changes in aged rats.** ***Neurobiology of Aging,*** 12(5), 437–440.
Administered phosphatidylserine (PS) ip in aged (24-mo-old) male rats subjected to stressor stimuli to evaluate its effect on grooming behavior, core temperature, and gastric ulcers. Novelty-induced grooming appeared to be higher in aged Ss, compared with young (3-mo-old) male control rats. The subchronic treatment with PS decreased grooming activity in aged Ss, but it did not affect that of young controls. Restraint stress induced hyperthermia in both aged and young Ss. However, 90 min after the beginning of restraint, PS-treated old Ss showed a normalization of core temperature. Furthermore, restraint-plus-cold stress induced gastric ulcers in both aged and young Ss. The treatment with PS was followed by a decreased incidence of gastric lesions in aged, but not in young Ss.

1105. Driscoll, Cynthia D.; Streissguth, Ann P. & Riley, Edward P. (1990). **Prenatal alcohol exposure: Comparability of effects in humans and animal models.** Special Issue: Qualitative and quantitative comparability of human and animal developmental neurotoxicity. ***Neurotoxicology & Teratology,*** 12(3), 231–237.
Compares the human and animal literature about the potential consequences of prenatal alcohol exposure with respect to qualitative and quantitative similarities and differences. Focus is on the effects reported in humans following moderate levels of alcohol exposure and the neurobehavioral effects detected using animal models. Reference is also made to clinical reports of children with fetal alcohol syndrome. General functional categories, such as deficits in learning, inhibition, attention, regulatory behaviors, and motor performance were reported to be affected in both animals and children. Quantitatively, although the dose required to produce an effect differs across species, the resultant circulating blood alcohol levels are quite similar. In addition, the magnitude of the observed effects appear to be dose-related for both humans and animals.

1106. Drugan, Robert C. & Maier, Steven F. (1983). **Analgesic and opioid involvement in the shock-elicited activity and escape deficits produced by inescapable shock.** ***Learning & Motivation,*** 14(1), 30–47.
Three experiments with 104 male albino rats examined the involvement of analgesic processes and endogenous opioids in the production of the shuttlebox escape acquisition and unconditioned activity deficits that follow exposure to inescapable shock. Exp I showed that the opiate antagonist naltrexone administered before the inescapable shock session interfered with the shuttlebox escape acquisition deficit that would normally follow. Exp II showed that naltrexone completely prevented the unconditioned activity deficit. Exp III revealed that dexamethasone, a synthetic glucocorticoid that abolishes the analgesia produced by inescapable shock, reversed the activity deficit. These results indicate that endogenous opioids may be involved in the production of both the escape acquisition and activity deficits. They also suggest that the analgesia produced by these opioids may participate in the mediation of the activity deficit, even though analgesia is not involved in producing the shuttlebox acquisition deficit.

1107. Drugan, Robert C.; Ryan, Susan M.; Minor, Thomas R. & Maier, Steven F. (1984). **Librium prevents the analgesia and shuttlebox escape deficit typically observed following inescapable shock.** ***Pharmacology, Biochemistry & Behavior,*** 21(5), 749–754.
In 2 experiments with 100 Holtzman male albino rats, 4 daily injections of librium (chlordiazepoxide HCl [CDP]; 10 mg/kg, ip) prior to exposure to inescapable shock prevented both the long-term analgesia and the shuttle-escape deficit typically observed following inescapable shock. If given only prior to testing, CDP had little effect. The protective effects of CDP were not a result of state dependency or a general facilitatory effect on the drug on escape performance. It is suggested that the induction of anxiety or fear by inescapable shock is critical in mobilizing endogenous changes such as transmitter depletion that are thought to be responsible for the deficits observed. The possible role of GABA in these deficits is also discussed.

1108. Duffy, Orla & Leonard, B. E. (1991). **Changes in behaviour and brain neurotransmitters following pre- and post-natal exposure of rats to ethanol.** ***Medical Science Research,*** 19(9), 279–280.
Examined central nervous system (CNS) changes in rats that have been chronically exposed to ethanol in utero to establish a rodent model of the fetal alcohol syndrome (FAS). The offspring of 10 female rats were divided into controls (not exposed to alcohol), exposed to alcohol prenatally only, exposed to alcohol postnatally only, and exposed to alcohol pre- and postnatally. Prenatal exposure was associated with changes in the behavior and in neurotransmitter concentrations in the brain of the offspring. The changes were more pronounced in the female than the male offspring. Pre- and postnatal exposure were associated with a fall in the gamma-aminobutyric acid (GABA) concentration in the female pups and prenatal exposure was associated with a rise in the GABA concentration of male pups. The rat may be of value in developing a model of the FAS.

1109. Dworkin, Barry R. & Miller, Neal E. (1986). **Failure to replicate visceral learning in the acute curarized rat preparation.** ***Behavioral Neuroscience,*** 100(3), 299–314.
Attempted to replicate a series of experiments reported to demonstrate robust visceral learning (autonomic instrumental learning) in rats during acute (2–4 hr) pharmacological paralysis. The results of exploratory procedures involving more than 2,000 animals are described, and 6 complete experiments are presented. In the 1st 3 experiments (with 72 male albino Sprague-Dawley rats), which closely followed the original procedures, the characteristics of the preparation were reproduced with the exception of initial heart rhythm and visceral learning. In the 2nd 3 experiments (with 74 male Sprague-Dawley rats), the respiration procedure was modified to satisfactorily reproduce the heart rhythm, and the Pa_{O2}, Pa_{CO2}, and pH, were verified to be within the range of freely moving, normally behaving Ss; nevertheless, visceral learning was not observed in these experiments either. It is concluded that the original visceral learning experiments are not replicable and that the existence of visceral learning remains unproven. However, neither the original experiments nor the replication attempt included the necessary controls to support a general negative conclusion about visceral learning. Possible explanations for a failure to replicate the original findings are discussed in terms of inadequate statistical power, unsatisfactorily reproduced experimental conditions, and spurious or artifactual results of the original experiments.

1110. Egli, Mark & Thompson, Travis. (1989). **Effects of methadone on alternative fixed-ratio fixed-interval performance: Latent influences on schedule-controlled responding.** ***Journal of the Experimental Analysis of Behavior,*** 52(2), 141–153.
Examined the effects of methadone (MET) on 15 adult female pigeons' key pecking under alternative FR FI schedules. Control by the interval contingency was greatest following extensive exposure to the interval component embedded within the alternative schedule (Condition 1), but was apparent to a lesser degree with even very limited exposure to the alternative FR FI schedule (Condition 2). Inter-reinforcement intervals comparable to those under an FI schedule were not observed under the FR schedules presented alone (Condition 3). Repeated exposure to the FI contingency outside the context of the alternative FR FI schedule did not engender performance changes under an FR schedule which would mimic those of increased FI contingency control (Condition 4). Drug administration can be used to reveal influences of both past and present environmental variables in maintaining and modifying behavior.

1111. Eison, Michael S. (1984). **Use of animal models: Toward anxioselective drugs.** Symposium of the World Psychiatric Association, Section Clinical Psychopathology: Physiological basis of anxiety (1983, Vienna, Austria). ***Psychopathology,*** 17(Suppl 1), 37–44.
Discusses developments in the research of anxioselective drugs for anxiety treatment. While currently available drugs successfully alleviate the distressing symptoms suffered by anxious patients, they also carry potential liabilities that must be considered in their use. Through the use of predictive animal models, researchers hope to elucidate a set of structure-activity relationships through which anxiolytic and, most particularly, anxioselective compounds can be rationally designed. Such models should predict side-effect potential as well as efficacy, so that unwanted ancillary effects of potential anxiolytics can be eliminated by appropriate structural modifications. The state of the art in preclinical testing

for anxiolytic potential is discussed with particular emphasis on the need to design test systems capable of detecting anxiolytic activity in diverse, nontraditional chemical series. The contribution of such methodology to the discovery and development of one anxioselective, nonbenzodiazepine anxiolytic, buspirone, is detailed. Findings from research on the treatment effects of buspirone are presented.

1112. Eison, Michael S.; Wilson, W. Jeffrey & Ellison, Gaylord. (1978). **A refillable system for continuous amphetamine administration: Effects upon social behavior in rat colonies.** ***Communications in Psychopharmacology,*** 2(2), 151–157.
36 male Long-Evans rats implanted with a refillable sc Silastic loop delivery system that was filled twice daily with 11 mg/kg dextroamphetamine base exhibited a 3-phase behavioral syndrome during 10 days of continuous drug administration. Initially exploratory and hyperactive, Ss then entered a prolonged stereotypy phase lasting 4 days. A distinct behavioral transition out of stereotypy, followed by exaggerated social behaviors and increased frequency of spontaneous startle responses, occurred 6 days after the 1st loop fill. These data corroborate a previously described behavioral syndrome observed after implantation of a nonrefillable amphetamine pellet in rats; they imply that continuous administration of amphetamine to rats can provide a useful animal model of amphetamine psychosis.

1113. Ellenbroek, B. et al. (1985). **Muscular rigidity and delineation of a dopamine-specific neostriatal subregion: Tonic EMG activity in rats.** ***Brain Research,*** 345(1), 132–140.
Examined catalepsy and spontaneous muscle tone in male Wistar rats receiving haloperidol (250–750 ng/0.5 µl and 0.5–2.5 µg/0.5 µl) injections into various parts of the neostriatum. Results show that (1) even relatively low doses of haloperidol led to the occurrence of catalepsy, (2) both catalepsy and rigidity might be closely related phenomena in at least certain regions of the brain, and (3) only attenuation of the dopaminergic activity within the most rostral part of the neostriatum resulted in a tonic electromyogram (EMG) activity in the gastrocnemius-soleus muscle, considered to be a measure for muscular rigidity. It is suggested that haloperidol-induced tonic EMG activity might prove to be a useful animal model for studying the mechanisms underlying muscular rigidity in parkinsonian patients.

1114. Ellenbroek, B. & Cools, A. R. (1988). **The PAW TEST: An animal model for neuroleptic drugs which fulfils the criteria for pharmacological isomorphism.** ***Life Sciences,*** 42(12), 1205–1213.
Describes the paw test, a recently developed animal model that can discriminate between classical and atypical neuroleptic drugs by measuring the ability of rats to spontaneously withdraw fore- and hindlimbs. The ability of a drug to affect rats' hindlimb retraction time was associated with the antipsychotic efficacy of the drug.

1115. Ellenbroek, Bart A.; Willemen, Annette P. & Cools, Alexander R. (1989). **Are antagonists of dopamine D_1 receptors drugs that attenuate both positive and negative symptoms of schizophrenia? A pilot study of Java monkeys.** ***Neuropsychopharmacology,*** 2(3), 191–199.
Amphetamine-induced social isolation (ASI) has been suggested as an animal model for the negative symptoms of schizophrenia (i.e., features present in normal Ss that are markedly reduced due to the disease). The effects of the selective D_1 dopamine receptor antagonist SCH 23390 was effective in antagonizing both stereotyped behavior and social isolation. SCH 23390 was able to reinstate normal behavior in these Ss. Results have important consequences for understanding the functional significance of the D_1 receptor as well as for the clinical treatment of positive and negative symptoms of schizophrenia.

1116. Ellis, Donna M.; Fontana, David J.; McCloskey, Timothy C. & Commissaris, Randall L. (1990). **Chronic anxiolytic treatment effects on conflict behavior in the rat.** ***Pharmacology, Biochemistry & Behavior,*** 37(1), 177–186.
Examined the effects of chronic posttest treatment with the antipanic agent alprazolam (ALP) or the traditional anxiolytic agents chlordiazepoxide (CDP) and phenobarbital (PHB) on conflict behavior, using the conditioned suppression of drinking paradigm with female rats. In daily 10-min sessions, water-deprived Ss were trained to drink from a tube that was occasionally electrified. Electrification was signaled by a tone. Chronic ALP, CDP, PHB, or vehicle were injected after conflict testing (in some experiments again 12–16 hrs later) for a minimum of 6 wks. Only chronic ALP resulted in a time-dependent increase in punished responding, with a latency to onset of 3–4 wks; this effect was not antagonized by the benzodiazepine antagonist Ro15-1788. Data support the hypothesis that conflict paradigms may serve as animal models for the study of antipanic agents. Moreover, not all anxiolytics will exhibit antipanic efficacy.

1117. Ellison, Gaylord D. (1977). **Animal models of psychopathology: The low-norepinephrine and low-serotonin rat.** ***American Psychologist,*** 32(12), 1036–1045.
A variety of evidence implies that alterations in the midbrain neuronal systems that use the monamines norepinephrine and serotonin as transmitters are involved in emotional imbalances and psychopathology in humans. The effects of alterations in these circuits can be studied in animals administered selective monoamine neurotoxins. From 7 yrs of research on the effects of alterations in norepinephrine and serotonin on rat behavior, the author notes that rats with norepinephrine depletions are inactive, have hunger deficits, and model aspects of depression, while rats with serotonin depletions are hyperactive, are frightened in novel environments, and model aspects of anxiety. When kept in isolation cages following lesioning, these animals recover but develop paradoxical, overcompensatory behavioral syndromes. When allowed to recover in social rat-colony environments, these animals become progressively more similar to controls, but their presence in the social colony leads to gradually increasing social disruptions.

1118. Ellison, Gaylord. (1987). **Stress and alcohol intake: The socio-pharmacological approach.** Workshop on the Interaction of Ethanol and Stress (1986, Helsinki, Finland). ***Physiology & Behavior,*** 40(3), 387–392.
Male rats housed in seminaturalistic colony environments and given access to ad lib water and 10% ethanol showed rhythms of alcohol consumption that do not develop in caged isolates and that are similar to those that develop in human populations. A subpopulation of Ss developed extreme preferences for alcohol. Compared to nonconsumers, these Ss were relatively inactive and low in dominance. Implications for a new animal model of alcoholism are discussed.

1119. Ellison, Gaylord D. & Eison, Michael S. (1983). **Continuous amphetamine intoxication: An animal model of the acute psychotic episode. *Psychological Medicine,*** 13(4), 751–761.
Reports that when amphetamines are administered to humans every few hours for several days, either during the "speed runs" of addicts or in controlled laboratory settings, the psychosis that reliably results is similar to paranoid schizophrenia in a number of important aspects. This continuous presence of stimulants over a prolonged period of time can be simulated in animals using sc implanted slow-release silicone pellets containing dextroamphetamine base. Monkeys and rats implanted with these pellets develop stages of behavioral alterations somewhat similar in sequence to those observed in humans who have received frequent doses of amphetamine. An initial period of hyperactivity and exploratory behavior is followed by the gradual development of motor stereotypies that become virtually incessant. A period of relative inactivity then appears that is followed, at 4–5 days after pellet implantation, by a late stage. This final stage is characterized by wet-dog shakes, parasitoticlike grooming episodes, and a variety of other forms of hallucinatorylike behavior. At about the same time, there are distinctive and partially irreversible alterations in dopaminergic innervations of the caudate nucleus, but not in mesolimbic dopamine innervation of the nucleus accumbens or in several other neurotransmitter systems. Continuous amphetamine administration may reproduce some aspects of the prolonged excitation that accompanies an acute psychotic episode and may be a fruitful model for the clarification of the dopamine theory of schizophrenia.

1120. Ellison, Gaylord; Eison, Michael S. & Huberman, Harris S. (1978). **Stages of constant amphetamine intoxication: Delayed appearance of abnormal social behaviors in rat colonies. *Psychopharmacology,*** 56(3), 293–299.
36 male Long-Evans hooded rats in colonies were observed for 7 days after half of them were implanted with slow-release silicone pellets containing dextroamphetamine base. The drug-implanted Ss were initially hyperactive and exploratory, but this gradually evolved over the next 24 hrs into motor stereotypies of an increasingly more circumscribed nature. On the 4th day after amphetamine implantation, Ss transiently withdrew to the burrows area; thereafter they were characterized by heightened startle responses and increased social behaviors such as fighting and fleeing. During the last phase some of the drug-implanted Ss tended to focus their fighting behaviors on one other drug-implanted S. This late phase of constant amphetamine intoxication in Ss has a number of similarities to amphetamine psychosis in humans and can serve as a useful animal model for the study of its biochemical correlates.

1121. Ellison, Gaylord D.; Levy, Andrew & Lorant, Nir. (1983). **Alcohol-preferring rats in colonies show withdrawal, inactivity, and lowered dominance. *Pharmacology, Biochemistry & Behavior,*** 18(Suppl 1), 565–570.
Male hooded rats with free access to water and 10% alcohol were raised in enriched, social colonies for prolonged periods of time. Those Ss that had developed extreme alcohol or water preferences were identified and returned to the colony. Both high- and low-alcohol consumers showed increased alcohol consumption just prior to feeding, but only the high consumers had a peak of alcohol consumption during the early morning hours. Compared to low consumers, high-alcohol consumers ate less food, ran less in the activity wheel, spent more time in the burrows, and ranked low on several dominance measures. When access to alcohol was removed in the colony, these high-alcohol consumers became more active but remained low in dominance. When tested in photocell cages, they showed a pattern of hyperactivity suggesting withdrawal effects. This subpopulation of animals from rat colonies that voluntarily prefer alcohol to water represent a novel and social animal model of chronic alcohol consumption.

1122. Ellison, Gaylord & See, Ronald E. (1989). **Rats administered chronic neuroleptics develop oral movements which are similar in form to those in humans with tardive dyskinesia. *Psychopharmacology,*** 98(4), 564–566.
Recorded oral movements in 24 rats chronically administered haloperidol, fluphenazine (FLU), or no drug using a computerized video analysis system that measured the distance between 2 fluorescent dots painted above and below the rat's mouth. Data were analyzed using fast-fourier analysis. Following an initial period of sedation (decreased energy at all frequencies), the drugged Ss (and especially the FLU Ss) began to show increased oral movements of 1–2 Hz, an effect that increased substantially on drug withdrawal. This is the same altered energy spectrum observed in humans with tardive dyskinesia.

1123. Erickson, Carlton K. & Kochhar, A. (1985). **An animal model for low dose ethanol-induced locomotor stimulation: Behavioral characteristics. *Alcoholism: Clinical & Experimental Research,*** 9(4), 310–314.
Describes a rat strain, Maudsley reactive (MR), that reliably showed enhanced locomotor stimulation in an apparatus after low doses of ethanol (0.75 g/kg, intraperitoneally). Sprague-Dawley and Wistar inbred strains did not show stimulation, and Maudsley nonreactive rats showed a less dramatic and more variable response to ethanol, compared to the MR strain. MR females showed greater stimulation than MR males, and the response was dose-, age-, and apparatus-related. It is suggested that low-dose, ethanol-induced locomotor stimulation in the MR rat strain could be a valuable rodent model for studying central neurochemical correlates of alcohol intoxication.

1124. Eroglu, Lütfiye & Esin, Yesin. (1990). **Effects of long-term nifedipine treatment in rats. *Psychiatry Research,*** 32(2), 203–205.
Studied the behavioral effects of long-term nifedipine treatment using the forced swimming test as a model of depression in male rats. Nifedipine had antidepressant properties in that it, like imipramine, significantly decreased immobility time.

1125. Ervin, Frank R.; Palmour, Roberta M.; Young, Simon N.; Guzman-Flores, Carlos et al. (1990). **Voluntary consumption of beverage alcohol by vervet monkeys: Population screening, descriptive behavior and biochemical measures. *Pharmacology, Biochemistry & Behavior,*** 36(2), 367–373.
17% of 196 feral vervet monkeys spontaneously drank appreciable quantities of beverage alcohol in 3% sucrose in preference to 3% sucrose alone. Ethanol consumption increased over time, as did the concentration of ethanol tolerated. Individual patterns of drinking and behavioral responses to ethanol were variable. Some Ss drank to ataxia and unconsciousness; signs of withdrawal, including tremulousness, pacing, irritability, and increased aggression, fol-

lowed the abrupt discontinuation of ethanol availability. A variety of changes in social interaction, including increased orientation to external stimulus, increased incidence of stereotyped aggression and of other stereotyped behaviors, and decreased frequency of affiliative behaviors, were observed during ethanol periods. The alcohol-preferring vervet may provide a useful means of studying alcohol abuse in humans.

1126. Estrada Robles, Uriel. (1976). **Intravenous self administration of amphetamine, phencamphamine and phenproporex in rhesus monkeys.** ***Cuadernos Cientificos CEMEF,*** 6, 114–130.
Discusses amphetamines in general, especially their toxicity and their capacity to cause dependence. Studies are reported of the patterns of iv self-administration of dextroamphetamine, phencamphamine, and phenproporex by *Macaca mulatta* and of behavioral alterations in such animals. The usefulness of this animal model in predicting the risk of dependence on these substances is indicated by the similar findings of laboratory research and clinical experiences. (English summary).

1127. Ettenberg, Aaron & Geist, Timothy D. (1991). **Animal model for investigating the anxiogenic effects of self-administered cocaine.** ***Psychopharmacology,*** 103(4), 455–461.
Seven male rats were trained to traverse a straight alley for a reward of 5 iv injections of cocaine. Animals were tested for 1 trial per day. While start latencies remained short and stable, running times tended to increase over days. This effect was apparently related to a concomitant increase in the number of retreats occurring in the alley. Retreats tended to occur in close proximity to the goal box, suggesting that Ss working for cocaine came to exhibit a form of conflict behavior putatively stemming from the drug's rewarding and anxiogenic properties. Diazepam pretreatment dose dependently reduced the incidence of retreat behaviors in the alley.

1128. Evans, Eric B. & Wenger, Galen R. (1990). **The effects of cocaine in combination with other drugs of abuse on schedule-controlled behavior in the pigeon.** ***Pharmacology, Biochemistry & Behavior,*** 37(2), 349–357.
Determined the effects of cocaine (COC) alone and in combination with *d*-amphetamine, caffeine, morphine (MOR), or delta-9-tetrahydrocannabinol (THC) in 5 male pigeons. Under a multiple FR FI schedule, when COC was combined with inactive doses of the drugs, the FR and FI response rate dose–response curves were not shifted relative to the COC-alone curves. When COC was combined with an active dose of a drug that decreased response rate when given alone (.3 mg/kg THC and 3 mg/kg MOR), the position of the response rate dose–response curves shifted compared with the COC-alone curves. Data indicate that the potential consequences of coabusing COC with the drugs tested can most often be predicted from the effects of each drug when taken alone.

1129. Evans, H. L.; Bushnell, P. J.; Pontecorvo, Michael J. & Taylor, J. D. (1985). **Animal models of environmentally induced memory impairment.** ***Annals of the New York Academy of Sciences,*** 444, 513–514.
Examined the effects of toxicants and drugs on memory in macaques and pigeons using a 3-choice, variable-delay matching-to-sample (DMS) procedure. Delayed matching of previously trained pigeons was evaluated following daily inhalation of toluene or n-hexane. Inhalation of 3,000 ppm toluene reduced matching accuracy after 1–2 wks of daily exposure; recovery occurred within 2 wks after the exposure stopped. Acute inhalation of toluene by monkeys during performance of the DMS task produced immediate decrements in accuracy and reaction time (RT). A decrement in DMS also occurred in monkeys and pigeons after acute exposure to trimethyltin.

1130. Feldon, Joram & Weiner, Ina. (1991). **The latent inhibition model of schizophrenic attention disorder: Haloperidol and sulpiride enhance rats' ability to ignore irrelevant stimuli.** ***Biological Psychiatry,*** 29(7), 635–646.
Two experiments investigated the effects of haloperidol (0.02, 0.1, and 0.5 mg/kg) and sulpiride (100 mg/kg) administration on latent inhibition (LI) in 175 male rats. The investigation was carried out using a conditioned emotional response (CER) procedure consisting of 3 stages (preexposure, conditioning, testing) conducted 24 hrs apart. In the preexposure stage, only 10 nonreinforced stimulus preexposures were given, a procedure known to be insufficient to yield LI in normal animals. In both experiments, LI was absent in the placebo Ss. In marked contrast, Ss treated with haloperidol (Exp 1) as well as with sulpiride (Exp 2) exhibited LI. Results demonstrate that both typical and atypical neuroleptics enhance animals' capacity to ignore irrelevant stimuli. Implications of this finding for an animal model of schizophrenia and for a novel screening test for antipsychotic drugs are discussed.

1131. Fernández Teruel, A.; Jiménez López, P.; Segarra Tomás, J. & Tobeña Pallarés, A. (1985). **Effects of diazepam, Ro 15-1788, and muscimol in an animal model of frustration.** ***Revista de Psiquiatría y Psicología Médica,*** 17(2), 109–117.
Investigated the effects of diazepam (DZ), muscimol (MU), and Ro 15-1788 (RO) in 30 90-day-old Sprague-Dawley rats using an animal model of anxiety (extinction after continuous reinforcement) to obtain behavioral data about the GABAergic mediation for the anxiolytic effects of benzodiazepines. Ss were trained for 10 days (15 min daily) on a continuous reinforcement schedule using a standard Skinner box. Extinction tests were administered to previously treated Ss on Days 11 and 12. Ss were divided into the following treatment groups: DZ—2 mg/kg intraperitoneally (ip); MU—0.00125 mg/kg ip; DZ—2 mg/kg ip plus RO—2 mg/kg ip; MU—0.00125 mg/kg ip plus RO—2 mg/kg ip; and saline group—control. Data for the different groups were submitted to an analysis of variance (ANOVA). Results indicate that low doses of MU produced an anxiolytic effect by increasing responding during extinction; such action was not blocked by RO. DZ treatment resulted in a decrease in responding during extinction and in ambulatory responses masking the expected anxiolytic disinhibition effect. RO effectively blocked the sedative DZ actions in both extinction and activity tests. Data partially support the validity of the frustration/extinction model in studying anxious behavior in animals. Results also support data confirming an anxiolytic action for GABAagonists such as MU at low doses. (English abstract).

1132. Ferris, Robert M.; Cooper, Barrett R. & Maxwell, Robert A. (1983). **Studies of bupropion's mechanism of antidepressant activity.** ***Journal of Clinical Psychiatry,*** 44(5, Sect 2), 74–78.

A review of pertinent studies indicates that the antidepressant activity of bupropion (BP) cannot be explained by its ability to inhibit MAO present in brain or to increase the release of biogenic amines from nerve endings, since the drug possesses neither of these properties. It is also unlikely that the weak properties of the drug as an inhibitor of dopamine (DA) uptake in brain can explain its antidepressant activity. It is clear, however, that DA neurons must be present for the CNS properties of BP to be manifested in animal models; at antidepressant doses of the drug, DA turnover is reduced in brain. The antidepressant properties of BP have been dissociated from down-regulation of postsynaptic beta-receptors. BP is the first clinically effective antidepressant whose mechanism of action cannot be explained on the basis of alterations in either presynaptic events or postsynaptic receptor-mediated events in catecholamine or serotonin pathways. Thus, BP is a novel antidepressant whose mechanism of action must still be elucidated.

1133. Fielding, Stuart & Szewczak, Mark R. (1984). **Pharmacology of nomifensine: A review of animal studies.** Hoechst-Roussel Pharmaceuticals Nomifensine (Merital) Symposium: Nomifensine: The preclinical and clinical profile of a second generation antidepressant (1983, San Diego, California). ***Journal of Clinical Psychiatry,*** 45(4, Sect 2), 12–20.
A review of animal research shows that nomifensine has demonstrated efficacy in several animal models that have been found to be predictive of clinical antidepressant activity and has also been found to have a low potential for both cardiovascular and anticholinergic side effects. A comparison of nomifensine's profile with those of standard antidepressant agents shows this drug to possess clear advantages that may make it an attractive choice for the treatment of endogenous depression.

1134. File, Sandra E. (1987). **The contribution of behavioural studies to the neuropharmacology of anxiety.** 25th Anniversary Symposium: Frontiers in neuropharmacology (1986, Tarpon Springs, Florida). ***Neuropharmacology,*** 26(7-B), 877–886.
Discusses animal tests of anxiety and examines the contribution of behavioral studies to the neuropharmacology of anxiety. Sites of action for anxiogenic drugs are discussed. It is suggested that while progress has been made in determining the functioning of the gamma-aminobutyric acid (GABA)-benzodiazepine receptor complex, there has been little advance in understanding other neurochemical pathways concerned with mediating changes in anxiety (e.g., noradrenergic sites, peptidergic modulation). The claim to have developed a nonsedative anxiolytic should be based on firm behavioral evidence. Putative nonsedative anxiolytics discussed include a triazolopyridazine and 2 phenylquinolines. Drugs that exert their anxiolytic effects away from the benzodiazepine receptors (e.g., buspirone) are discussed.

1135. File, Sandra E. (1989). **Chronic diazepam treatment: Effect of dose on development of tolerance and incidence of withdrawal in an animal test of anxiety.** ***Human Psychopharmacology Clinical & Experimental,*** 4(1), 59–63.
The effect of the treatment dose of diazepam was assessed on the rate of development of tolerance to diazepam's effects on rat behavior in the elevated plus-maze test of anxiety. Tolerance developed after 14 days when Ss were given 2.5 mg/kg/day, but not until 21 days when they were given 1 mg/kg/day. When Ss were tested undrugged 36 hrs after the last of 14 daily doses of diazepam (2.5 mg/kg), they showed behavioral changes indicating increased anxiety. Ss tested at this time with a lower dose of diazepam (1 mg/kg) also showed changes indicating increased anxiety compared with the control scores. This indicates that a more gradual tapering of doses would be necessary to avoid withdrawal responses.

1136. File, Sandra E. & Baldwin, Helen A. (1987). **Effects of β-carbolines in animal models of anxiety.** Sixth European Winter Conference on Brain Research: Bidirectional effects of β-carbolines in behavioral pharmacology (1986, Avoriaz, France). ***Brain Research Bulletin,*** 19(3), 293–299.
Reviews findings on the effects of beta-carbolines (BCs) on animal models of anxiety. Types of models include those based on conflict or conditioned fear; those exploiting the anxiety produced by novelty; those in which anxiety or aversion is chemically induced. Many of the BCs are anxiogenic in the tests, however ZK-91296 and ZK-93423 appear to have anxiolytic properties, and ZK-93426 has a similar profile to that of the benzodiazepine (BDZ) receptor antagonist Ro-15-1788. The reliability and sensitivity of the tests are assessed. The evidence that the anxiogenic and anxiolytic actions of the BCs are mediated by the BDZ binding sites is also discussed.

1137. File, Sandra E. & Hyde, J. R. (1978). **Can social interaction be used to measure anxiety? *British Journal of Pharmacology,*** 62(1), 19–24.
In a study with 272 male hooded rats, pairs of Ss were placed in a test box for 10 min, and the time they spent in active social interaction was scored. Maximum active interaction was found when Ss were tested under low light in a familiar box. When the light level was increased or when the box was unfamiliar active social interaction decreased. Exploration decreased in the same way in relation to test conditions as did social interaction. As these decreased, defecation and freezing increased. Anosmic controls showed that the decrease in social interaction across test conditions could not be attributed to olfactory changes in the partner. Chlordiazepoxide (CDP, 5mg/kg, ip) given chronically prevented or significantly reduced the decrease in social interaction that occurred in undrugged Ss as light level or unfamiliarity of the test box was increased. Controls showed that this effect could not be entirely attributed to CDP acting selectively to increase low levels of responding. The effect of chronic CDP contrasted with its action when given acutely; in the latter case it had only sedative effects. The use of this test as an animal model of anxiety is discussed and compared with existing tests of anxiety.

1138. Fischer, W.; Wictorin, K.; Björklund, A.; Williams, L. R. et al. (1987). **Amelioration of cholinergic neuron atrophy and spatial memory impairment in aged rats by nerve growth factor. *Nature,*** 329(6134), 65–68.
As in Alzheimer-type dementia in humans, degenerative changes in the forebrain cholinergic system may contribute to age-related cognitive impairments in rodents. Recent studies have shown that neurotrophic protein nerve growth factor (NGF) can prevent retrograde neuronal cell death and promote behavioral recovery after damage to the septo-hippocampal connections in rats. A study is reported in which continuous intracerebral infusion of NGF over 4 wks partly reversed the cholinergic cell body atrophy and improved retention of a spatial memory task (the Morris water maze) in 24 behaviorally impaired aged rats.

1139. Fischman, Marian W. (1988). **Behavioral pharmacology of cocaine.** APT Foundation North American Conference: Cocaine abuse and its treatment (1987, Washington, DC). ***Journal of Clinical Psychiatry,*** 49(Suppl), 7–10.
Contends that animals self-administer cocaine in patterns similar to those seen in humans and that conditions of availability and species are unimportant determinants of cocaine self-administration (SA). Studies are reviewed showing that animals will continue SA of the drug, even in the face of severe toxic side effects. Data indicate that cocaine is a potent reinforcer with significant abuse liability. It is concluded that the animal SA model and laboratory data concerning SA in humans show cross-species generality and provide useful information about cocaine's behavioral mechanisms of action.

1140. Fisher, Abraham; Brandeis, Rachel; Karton, Ishai; Pittel, Zipora et al. (1991). **(±)-*cis*-2-methyl-spiro(1,3-oxathiolane-5,3')quinuclidine, an M1 selective cholinergic agonist, attenuates cognitive dysfunctions in an animal model of Alzheimer's disease. *Journal of Pharmacology & Experimental Therapeutics,*** 257(1), 392–403.
Examined AF102B [(±)-*cis*-2-methyl-spiro(1,3-oxathiolane-5,3')quinuclidine], a structurally rigid analog of acetylcholine, in neurochemical, pharmacological, and behavioral tests related to cholinergic functions in mice and guinea pigs. AF102B induced atropine-sensitive contractions of isolated guinea pig ilea and trachea preparations. Binding studies using radioligands in rat cerebral cortex and quinuclidinyl benzilate in cerebellar homogenates indicate that AF102B is a potent and highly selective M1-type muscarinic probe. Repetitive administrations of AF102B improved AF64A-induced working memory deficits in the Morris water maze test. Data show that the selective M1 agonist AF102B can restore AF64A-induced cognitive impairments, without producing adverse central and peripheral side effects and indicate potential use for treating Alzheimer's disease.

1141. Fisher, Abraham & Hanin, Israel. (1986). **Potential animal models for Senile Dementia of Alzheimer's Type, with emphasis on AF64A-induced cholinotoxicity. *Annual Review of Pharmacology & Toxicology,*** 26, 161–181.
Reviews the literature on ethylcholine aziridinium ion (AF64A), which has recently been proposed as a potential tool for developing an animal model for Senile Dementia of the Alzheimer's type (SDAT), a disorder in which a central cholinergic hypofunction has been implicated. The clinical, behavioral, and neuropathological features of SDAT are discussed to ascertain the relevance of AF64A-induced cholinotoxicity to the animal model of SDAT. The AF64A model is compared with other experimental models of SDAT (e.g., scopolamine- or hemicholinium-treated Ss, anoxic or hypoxic rodents, aluminum-treated Ss, aged rodents or monkeys, and excitotoxin-lesioned rats or monkeys). Overall data show that AF64A is a valuable clinical tool with which a persistent cholinergic deficiency of presynaptic origin can be induced. Implications of its apparent selectivity of action are noted.

1142. Fontana, D. J.; Schefke, D. M. & Commissaris, R. L. (1990). **Acute versus chronic clonidine treatment effects on conflict behavior in the rat. *Behavioural Pharmacology,*** 1(3), 201–208.
Examined the effects of acute and chronic treatment with the alpha-2-adrenoceptor agonist clonidine on behavior in the conditioned suppression drinking conflict paradigm, an animal model for the study of anti-anxiety treatments. In daily 10-min sessions, water-deprived female rats were trained to drink from a tube that was occasionally electrified. Acute treatment with clonidine (1.25–40 μg/kg; 10-min pretreatment) did not exert an anticonflict effect, with doses greater than 5 μg/kg significantly depressing unpunished responding (i.e., water intake). Chronic posttest treatment with clonidine resulted in a dramatic and time-dependent increase in punished responding. The response to an acute challenge with chlordiazepoxide or the benzodiazepine antagonist Ro 15-1788 did not differ in chronic clonidine-treated vs saline-treated Ss.

1143. Fontana, David J.; Carbary, Timothy J. & Commissaris, Randall L. (1989). **Effects of acute and chronic anti-panic drug administration on conflict behavior in the rat. *Psychopharmacology,*** 98(2), 157–162.
Evaluated the conditioned suppression of drinking (CSD) conflict paradigm as an animal model for the study of panic disorder and anti-panic agents. In daily sessions, water-deprived female rats were trained to drink from an occasionally electrified tube. Desipramine, amitriptyline, or phenelzine was administered in acute (10-min pretreatment) and chronic (twice daily for up to 9 wks) regimens. Acute administration resulted in no change or a decrease in the number of shocks accepted and a decrease in water intake at higher doses. Chronic administration of each agent resulted in a gradual (2–4 wk latency) increase in the number of shocks received in CSD sessions over several weeks. The time-dependent anti-conflict effect observed in the CSD following drug treatment relates to the anti-panic effects of the drugs rather than their anti-depressant effects.

1144. Fontana, David J. & Commissaris, Randall L. (1989). **Effects of cocaine on conflict behavior in the rat. *Life Sciences,*** 45(9), 819–827.
Examined the effects of cocaine (CC) treatment (acute, chronic, and withdrawal) on female rat behavior in the Conditioned Suppression of Drinking (CSD) conflict paradigm, an animal model for the study of anxiety and anti-anxiety treatments. Chronic CC treatment, CC withdrawal, and, to a lesser extent, acute CC treatment resulted in anxiogenic effects in the CSD paradigm. Conflict paradigms may be sensitive behavioral tests to study the neuropharmacological basis for the anxiogenic effects of CC and potential treatments for CC-induced anxiety states.

1145. Fontana, David J. & Commissaris, Randall L. (1988). **Effects of acute and chronic imipramine administration on conflict behavior in the rat: A potential "animal model" for the study of panic disorder? *Psychopharmacology,*** 95(2), 147–150.
Investigated the effects of imipramine in a potential animal model for panic disorder, the conditioned suppression of drinking (CSD) paradigm. In daily 10-min sessions, water-deprived rats were trained to drink from a tube that was occasionally electrified (0.5 mA). Imipramine was administered both in an acute (3.5–20 mg/kg) and a chronic (2.5 mg/kg, twice daily for 5 wks) regimen. Results show that acute administration resulted in a decrease in the number of

shocks accepted and a decrease in water intake. In contrast, chronic administration resulted in a gradual increase in the number of shocks received in CSD sessions over the course of several weeks of testing.

1146. Fontana, David J.; McCloskey, Timothy C.; Jolly, Surindar K. & Commissaris, Randall L. (1989). **The effects of beta-antagonists and anxiolytics on conflict behavior in the rat.** ***Pharmacology, Biochemistry & Behavior,*** 32(3), 807–813.
In daily 10-min sessions, water-deprived rats were trained to drink from a tube that was occasionally electrified. Within 2–3 wks, control responding had stabilized; drug tests were then conducted at weekly intervals. Diazepam and phenobarbital administration resulted in a marked and dose-dependent increase in punished responding at doses that did not markedly alter background responding (water intake). Neither propranolol nor the beta-1-selective antagonist atenolol significantly affected punished responding in the conditioned suppression of drinking. Both propranolol and atenolol produced significant beta-1-adrenoceptor blockade, as evidenced by the production of significant bradycardic effects in conscious Ss at the doses employed.

1147. Fowler, Stephen C.; Liao, Ruey-ming & Skjoldager, Paul. (1990). **A new rodent model for neuroleptic-induced pseudo-Parkinsonism: Low doses of haloperidol increase forelimb tremor in the rat.** ***Behavioral Neuroscience,*** 104(3), 449–456.
Rats were trained to use the right forelimb to exert continuous downward pressure on a force transducer and simultaneously to drink sweetened milk from a dipper controlled by the emitted force. Oscillations in forelimb force during this performance were spectrally analyzed to describe the tremorogenic effects of haloperidol (0.04, 0.08, or 0.16 mg/kg). Haloperidol reduced time-on-task and increased the variance of force oscillations in the 10.0–25.0 Hz frequency band. When atropine sulfate (5 mg/kg) was given along with haloperidol, time-on-task was partially restored, and the effects of haloperidol on the 10.0–25.0 Hz band were diminished. These data suggest that the behavioral deficits produced by relatively low doses of haloperidol in rats are analogous (and possibly homologous) to neuroleptic-induced Parkinsonian symptoms in humans.

1148. Francès, Henriette. (1988). **New animal model of social behavioral deficit: Reversal by drugs.** ***Pharmacology, Biochemistry & Behavior,*** 29(3), 467–470.
Studied escape attempts of socially isolated (SI) and group-reared (GR) mice before and after treatment with 11 drugs, including clomipramine, imipramine, and diazepam. SI Ss were isolated for 1 wk or longer and placed in a transparent beaker alone or with GR Ss. SI Ss made significantly more escape attempts when they were tested alone than when observed with GR Ss. SI Ss were again isolated for a period of 7–9 days before ip or sc drug administration. Drugs acting on the serotonergic system, namely the 5-hydroxytryptamine (5-HT) 1B receptors, were found to reverse the behavioral deficit.

1149. Freemark, Michael et al. (1978). **Testosterone-attenuated stereotypy and hyperactivity induced by β-phenylethylamine in pargyline-pretreated rats.** ***Biological Psychiatry,*** 13(4), 455–463.
Testosterone pretreatment (1.0–4.0 mg/kg) attenuated, in a dose–response fashion, the induction of stereotyped behavior and hyperactivity by pargyline (0.25, 4.0 mg/kg) and beta-phenylethylamine (8.0, 16.0 mg/kg) in prepubertal, male Sprague-Dawley rats. The dyskinetic movements induced by pargyline and beta-phenylethylamine are proposed as a possible animal model for tardive dyskinesias. Attenuation by testosterone of these effects suggested a hormonal involvement consistent with the reported predominant occurrence of tardive dyskinesias in women and in the elderly.

1150. Friedman, H.; Redmond, D. Eugene & Greenblatt, David J. (1991). **Comparative pharmacokinetics of alprazolam and lorazepam in humans and in African green monkeys.** ***Psychopharmacology,*** 104(1), 103–105.
Six African green monkeys received a single 250 μg/kg oral dose of alprazolam and of lorazepam on 2 separate occasions. 22 healthy male volunteers received a single 1 mg oral dose of alprazolam; another 24 Ss received a single 2 mg oral dose of lorazepam. All human Ss were aged 18–40 yrs. Mean values of elimination half-life in humans were substantially longer than corresponding values in the primate animal model, and human values of clearance likewise were much lower. However, in humans, kinetic differences between the 2 drugs were much smaller than in the monkeys. Thus, comparative studies of the behavioral effects of these 2 drugs in African green monkeys should utilize relative dosages that reflect the pharmacokinetic properties of the drugs in that species. Use of dosage ratios analogous to those used in humans may lead to results that cannot be extrapolated to humans.

1151. Friedman, Steven; Sunderland, Gayle S. & Rosenblum, Leonard A. (1988). **A nonhuman primate model of panic disorder.** ***Psychiatry Research,*** 23(1), 65–75.
Reproduced several features of panic attacks following sc administration of sodium lactate to 12 unrestrained macaques. Ss were evaluated by an observer without knowledge of the Ss' treatment with either sodium lactate or a dextrose control solution. The lactate produced temporally circumscribed episodes of agitation, wariness, and motor responses, normally elicited under stressful or threatening conditions. In an initial pharmacological intervention, pretreatment with imipramine blocked the response to lactate. It is suggested that the model offers promise for the systematic examination of etiological factors in susceptibility to lactate induction of panic attacks, the physiological basis of the response, and new modes of treatment of panic disorder.

1152. Froehlich, J. C. & Li, T.-K. (1991). **Animal models for the study of alcoholism: Utility of selected lines.** ***Journal of Addictive Diseases,*** 10(1–2), 61–71.
An approach to identifying factors, which when inherited increase the risk for alcoholism, is to analyze how organisms with genetic predisposition toward ethanol drinking differ from organisms without genetic risk. Among alcohol-preferring (AP) and alcohol-nonpreferring (ANP) rat lines selected for high and low voluntary ethanol intake, line differences in ethanol-related traits (e.g., ethanol sensitivity) have appeared in original and replicate lines. Behavioral characterizations of the AP and ANP lines are discussed. Rapid induction of tolerance to the aversive effects of ethanol with repeated bouts of voluntary ethanol drinking and persistence of ethanol tolerance in rats of the AP line may increase and maintain alcohol-seeking behaviors. These mechanisms may serve to promote and maintain ethanol self-administration.

1153. Fuentes, Jose A.; Oleshansky, Marvin A. & Neff, Norton H. (1976). **Comparison of the apparent antidepressant activity of (-) and (+) tranylcypromine in an animal model.** ***Biochemical Pharmacology,*** 25(7), 801–804.
The isomers of tranylcypromine (TCP) readily entered the brains of male Sprague-Dawley rats after ip administration and reached peak concentrations within 15 min. Apparently (-)TCP entered the brain more rapidly and reached somewhat higher concentrations than (+)TCP. After a dose of 2.5 mg/kg of (-) or (+)TCP, there was significantly more drug in brain than has been reported necessary to block the reuptake of amines by synaptosomes. Both isomers blocked MAO in vivo and in vitro. (+)TCP was between 10 and 60 times more active than (-)TCP, depending on the amine substrate evaluated, and both isomers were better inhibitors of type-B MAO activity than type-A activity. The (+) isomer was more active in preventing reserpine-induced sedation. The ability to prevent the reserpine syndrome was apparently related to the ability of the drugs to block MAO activity rather than the blockade of amine reuptake.

1154. Fung, Yiu K. & Lau, Yuen-sum. (1989). **Effects of prenatal nicotine exposure on rat striatal dopaminergic and nicotinic systems.** ***Pharmacology, Biochemistry & Behavior,*** 33(1), 1–6.
Demonstrated that prenatal exposure to nicotine induced growth deficit in rat offspring. Although nicotine did not modify the characteristics of nicotinic receptor binding sites (RBSs) in the striatum, male offspring did show a change in the number and affinity of dopaminergic RBSs. Results provide a useful animal model of the effects of maternal smoking in humans.

1155. Gallaher, Edward J.; Gionet, Susanne E. & Feller, Daniel J. (1991). **Behavioral and neurochemical studies in diazepam-sensitive and -resistant mice.** ***Journal of Addictive Diseases,*** 10(1–2), 45–60.
Selective breeding is becoming an increasingly important technique for producing animal models for behavioral and pharmacological research. Underlying principles of selecting breeding in the development of diazepam-sensitive (DS) and diazepam-resistant (DR) mice are discussed. Behavioral and neurochemical studies conducted with DS and DR mice are reviewed. A conceptual framework is presented that illustrates the potential role of selective breeding in the eludication of drug mechanisms.

1156. Gambill, John D. & Kornetsky, Conan. (1976). **Effects of chronic d-amphetamine on social behavior of the rat: Implications for an animal model of paranoid schizophrenia.** ***Psychopharmacology,*** 50(3), 215–223.
Eight male hooded rats in a social colony were given increasing daily doses of d-amphetamine up to 8 mg/kg. Time-lapse, photographically recorded behavior was analyzed for grooming, feeding, sex, sleeping, resting, stereotypy, agonistic behavior, muricidal activity, and the location and movement of each S. Subordinant Ss receiving amphetamine actively withdrew from social interactions by retreating to strategically defensible locations in the environment. They remained hypervigilant of other Ss and overreacted to their approaches by either fleeing or defensively rearing and "boxing." On the other hand, when the dominant S received the maximum dose, it seemed totally oblivious to the other Ss. The responses to drug treatment in subordinant rats may provide a model for the social behavior of frightened paranoid schizophrenics.

1157. Garcin, Francoise; Radouco-Thomas, Simone; Tremblay, R. & Radouco-Thomas, C. (1977). **Experimental narcotic dependence models of primary dependence, abstinence and relapse obtained by intravenous self-administration in the rat.** ***Progress in Neuro-Psychopharmacology,*** 1(1–2), 61–81.
Evaluated the usefulness of 2 animal models of morphine dependence, one based on iv voluntary self-administration and one based on an automated or forced infusion model. The pattern of self-administration observed during primary dependence development in drug-naive female Sprague-Dawley rats stabilized after about 14 days with a daily drug intake of 240 mg/kg. On the basis of the self-administration data, a 6-day accelerated model of primary dependence was produced by passive, automatic administration. In these conditions, Ss displayed a similar degree of physical dependence to that observed with self-administration. A triphasic abstinence period was then implemented. Data on the duration, symptoms, and propensity to relapse were analyzed for 3 phases (early, intermediate, and protracted). The secondary dependence or relapse data obtained in ex-addicted rats allowed to readminister morphine showed that in comparison with the primary dependence, a higher drug intake was sustained when tested at regular intervals between 1 and 6 mo following morphine removal. Naloxone administered before and in association with morphine during a 7-day period blocked morphine self-administration. This effect was more potent in preventing secondary dependence in ex-addicted Ss than the primary dependence in drug-naive Ss.

1158. Gardner, C. R. (1986). **Recent developments in 5 HT-related pharmacology of animal models of anxiety.** ***Pharmacology, Biochemistry & Behavior,*** 24(5), 1479–1485.
Reviews recently developed anxiolytic drug candidates and compounds specific for subtypes of 5-hydroxytryptamine (5-HT) binding sites in investigations of the role of 5-HT in anxiety. The proposed anxioselective drug buspirone interacts with 5-HT_1 receptors. An analog, MJ 13805, produces a 5-HT behavioral syndrome blocked by central 5-HT pathway lesions. Ritanserin, a selective 5-HT_2 antagonist, shows activity in an emergence test but not conflict models. Preliminary clinical reports indicate qualitatively different anxiolytic activity from that of benzodiazepines.

1159. Gauvin, David V. & Holloway, Frank A. (1991). **Cross-generalization between an ecologically relevant stimulus and a pentylenetetrazole-discriminative cue.** ***Pharmacology, Biochemistry & Behavior,*** 39(2), 521–523.
Trained 12 male rats in a 2-choice pentylenetetrazole (PTZ) vs saline drug discrimination (DD) task under an FR 10 schedule of food reinforcement. Ss were injected with saline and then exposed for 20 min to the presence of a domestic cat pretreated with catnip. Following the predator exposure, Ss were tested for stimulus generalization in the DD task. The predator/prey interaction engendered 92% PTZ-appropriate responding. Data suggest that the interoceptive state associated with species-specific defense reactions in rats is similar to the cues produced by a pharmacological agent within a behavioral assay that has been suggested as an animal model of human anxiety.

1160. Geoffroy, Marianne; Scheel-Krüger, Jørgen & Christensen, Anne V. (1990). **Effect of imipramine in the "learned helplessness" model of depression in rats is not mimicked by combinations of specific reuptake inhibitors and scopolamine.** ***Psychopharmacology,*** 101(3), 371–375.

Examines how imipramine (IMI), which blocks noradrenergic, serotonergic, and cholinergic reuptake, administered to rats for 4 days counteracts the shuttlebox escape failures otherwise seen in rats that have been exposed to inescapable shock (the learned helplessness model of depression). The effects of the more selective reuptake inhibitors talsupram (noradrenergic), citalopram (serotonergic), and the anticholinergic compound scopolamine (SCO) were assessed alone and in combination after acute or 4 days' administration on escape behavior. Talsupram and citalopram were ineffective, whereas SCO counteracted the escape failures. IMI retains a pharmacological effect that is not mimicked by SCO alone or by combining the specific reuptake inhibitors with SCO.

1161. Geoffroy, Marianne; Tvede, Kirsten; Christensen, Anne V. & Schou, Jens S. (1991). **The effect of imipramine and lithium on "learned helplessness" and acetylcholinesterase in rat brain.** ***Pharmacology, Biochemistry & Behavior,*** 38(1), 93–97.
Investigated the effect of short- and long-term treatment with imipramine and lithium on shock stress-induced escape failures in shuttlebox, using 72 male rats. Acetylcholinesterase (AChE) activity was measured in the frontal cortex, hippocampus, and striatum after the shuttlebox test. Imipramine was found to normalize escape behavior, whereas lithium further aggravated escape behavior. Lithium did not have an antidepressant effect on learned helplessness, and AChE activity was not correlated to escape behavior. However, both imipramine and lithium normalized the decreased level of AChE activity in striatum in rats exposed to shock stress.

1162. George, Frank R. (1991). **Is there a common biological basis for reinforcement from alcohol and other drugs?** ***Journal of Addictive Diseases,*** 10(1–2), 127–139.
Demonstrated differences between genetically distinct (GD) rodent stocks in drug-seeking behavior (DSB), as well as within-strain commonalities in DSB across ethanol, cocaine, and opioids. Four GD rodent strains were used: the LEWIS/CRLBR and Fischer 344 inbred rat strains and the C57BL/6J and DBA/2J inbred mouse strains. All Ss were male. Results suggest a high degree of qualitative commonality across these genotypes and drugs. Genotypes that showed high intakes of ethanol also showed the greatest amounts of intake of and reinforcement from cocaine. Conversely, genotypes that showed weak or no reinforcement from ethanol showed similarly low levels of intake of cocaine. Demonstration of genetic differences in animal models of DSB suggests that human populations have differing degrees of biological risk for drug abuse.

1163. Geyer, Mark A. et al. (1979). **A characteristic effect of hallucinogens on investigatory responding in rats.** ***Psychopharmacology,*** 65(1), 35–40.
The disruption of the temporal distribution of investigatory responses by adult male Sprague-Dawley rats in a novel holeboard following LSD-25 as described in a companion paper by M. A. Geyer and R. K. Light (1979) was found to be a characteristic effect of a variety of hallucinogens. Similar effects were produced by indoleamine hallucinogens, such as LSD, N,N-dimethyltryptamine, and psilocin, and by phenylethylamine hallucinogens, such as mescaline or 2,5-dimethoxy-4-methylamphetamine (DOM). Congeners of DOM that are inactive in humans had no significant effects. Furthermore, of a variety of other psychoactive drugs tested, only apomorphine produced an effect similar to that of the hallucinogens. Results suggest that a simple behavioral measure of exploration in a holeboard may provide a useful animal model with which to examine the common effects of hallucinogens.

1164. Giardina, William J. & Ebert, Donn M. (1989). **Positive effects of captopril in the behavioral despair swim test.** ***Biological Psychiatry,*** 25(6), 697–702.
Captopril, an angiotensin II converting enzyme (ACE) inhibitor, was evaluated for potential antidepressive activity on the forced swim-induced behavioral despair (immobility) test in mice. Captopril (10.0 and 30.0, mg/kg, ip) significantly reduced immobility and mimicked the effects of the antidepressants imipramine (30.0 mg/kg, ip) and mianserin (3.0, 10.0, and 30.0 mg/kg, ip). Captopril increased the motor activity of mice at these same dosages. Naloxone (20.0 mg/kg, ip) blocked the effects of captopril (30.0 mg/kg, ip) in the swim test. Data suggest that captopril has potentiated antidepressive activity; however, the positive effects may be related to motor stimulation. The blockade of the captopril effects by naloxone suggests that brain opioid peptides play a role in this behavioral effect of captopril.

1165. Gibson, E. L. & Booth, David A. (1988). **Fenfluramine and amphetamine suppress dietary intake without affecting learned preferences for protein or carbohydrate cues.** ***Behavioural Brain Research,*** 30(1), 25–29.
Assessed the effects of pharmacological modulation of monoamine transmitter activity on selection by measuring drug-induced changes in nutrient-specific dietary choice behavior using rats that had learned to select an odor cueing protein or carbohydrate content. Anorexigenic doses of fenfluramine and amphetamine did not affect selection.

1166. Gilliam, David M. & Collins, Allan C. (1986). **Quantification of physiological and behavioral measures of alcohol withdrawal in long-sleep and short-sleep mice.** ***Alcoholism: Clinical & Experimental Research,*** 10(6), 672–678.
Tested the utility of a quantitative, multidimensional animal model of alcohol withdrawal syndrome (AWS) for studying individual differences in susceptibility to alcohol dependence. Following exposure to control or ethanol diets for 7 or 14 days, various parameters of AWS were measured in Long-Sleep (LS) and Short-Sleep (SS) mice to determine how initial alcohol sensitivity influences dependence liability. SS mice consumed more of the ethanol diet and exhibited more severe AWS than did LS mice. AWS severity resulted from an interaction of genotype with duration of ethanol exposure. Initial alcohol sensitivity influenced the rate of serum alcohol increase that, in turn, influenced AWS severity. This model incorporates several discriminative measures that independently assess AWS reactions and provides a useful animal model of the AWS.

1167. Gilliam, David M. & Irtenkauf, Keith T. (1990). **Maternal genetic effects on ethanol teratogenesis and dominance of relative embryonic resistance to malformations.** ***Alcoholism: Clinical & Experimental Research,*** 14(4), 539–545.
A reciprocal cross study was conducted in an animal model using C57BL/6J (B6) and long-sleep (LS) mice. B6 and LS dams were reciprocally mated to B6 or LS males producing 4 embryonic genotype groups: the true-bred B6B6 and LSLS genotypes, and the genetically similar B6LS and LSB6 genotypes (the F_1 genotype). Dams were intubated with either 5.8 g/kg ethanol or an isocaloric amount of sucrose on Day 9 of pregnancy. Fetuses were removed on Gestation

Day 18, weighed, and assessed for soft tissue or skeletal malformations. Results showed a greater litter weight deficit and increased total malformation rate in ethanol-exposed F_1 litters carried by B6 mothers compared to ethanol-exposed F_1 litters carried by LS mothers. Maternal genetic factors were important determinants of susceptibility to prenatal alcohol effects.

1168. Ginsburg, Benson E.; Becker, Robert E.; Trattner, Alice & Bareggi, Silvio R. (1984). **A genetic taxonomy of hyperkinesis in the dog.** ***International Journal of Developmental Neuroscience,*** 2(4), 313–322.
To develop an animal model of the hyperkinetic syndrome seen in children, the ability of 6 purebred beagles and 14 hybrids (all over 6 mo of age) to learn an inhibitory task (the standard "sit–stay" command) and to perform this task once it was learned using positive and negative reinforcement was tested. Hybrids would not stay in the sitting position for the test period and had to be frequently repositioned by voice command. Dextroamphetamine (DAM; 1.2–2 mg/kg) was administered to the hybrids. Findings show 2 categories of reaction to the drug: Five Ss showed a dramatic improvement, and 9 Ss were not helped. Neurochemical studies showed that hybrids had lower medullary and pons dopamine (DA) neuron levels than the purebred Ss. DAM-response hybrids differed from nonresponders by lower CSF and caudate nucleus homovanillic acid levels, lower midbrain DA neuron levels, and lower brain-stem norepinephrine neuron levels. Both DA neuron and homovanillic acid levels were lower in the hippocampal region of the responders. Improvement after DAM in the responders paralleled the time-course of changes in the DAM levels in the plasma. It is suggested that the effect of DAM was to facilitate performance of previously learned behavior, rather than allowing the acquisition of behavior.

1169. Giralt, M.; Garcia-Marquez, C. & Armario, A. (1987). **Previous chronic ACTH administration does not protect against the effects of acute or chronic stress in male rats.** ***Physiology & Behavior,*** 40(2), 165–170.
Studied the effect of previous chronic adrenocorticotropin (ACTH) administration on the physiological response to acute and chronic immobilization stress in male Sprague-Dawley rats. Chronic ACTH administration slightly reduced food intake and drastically inhibited body weight gain. Serum corticosterone levels were similar in saline- and ACTH-treated Ss 20 hrs after the last administration. However, the corticosterone response to 1 hr immobiization was greatly reduced by previous ACTH administration. When the exposure to the stressor was prolonged up to 18 hrs, the corticosterone response was similar in saline, and ACTH-treated Ss. Body weight loss caused by starvation and acute stress was lower in ACTH-treated Ss, but stomach ulceration was greater in the latter Ss. Pituitary-adrenal adaptation to the repeated stressor was the same in saline- and ACTH-treated Ss.

1170. Glavin, Gary B. (1985). **Methylphenidate effects on activity-stress gastric lesions and regional brain noradrenaline metabolism in rats.** ***Pharmacology, Biochemistry & Behavior,*** 23(3), 379–383.
Administered methylphenidate (5, 10, or 20 mg/kg) or saline to 120 male Wistar rats in the activity-stress ulcer paradigm to investigate methylphenidate's effects on activity-stress gastric lesions and on regional brain noradrenaline (NA) metabolism. Running-wheel activity and food consumption did not differ among groups. Methylphenidate produced dose-related increases in gastric ulcer severity, decreases in hypothalamic NA, and increases in 3-methoxy-4-hydroxyphenylethyleneglycol sulfate in the hypothalamus, amygdala, hippocampus, and thalamus. These results differ markedly from the effects seen with amphetamine and suggest different mechanisms of action for these drugs.

1171. Glavin, Gary B. & Krueger, Hans. (1985). **Effects of prenatal caffeine administration on offspring mortality, open-field behavior and adult gastric ulcer susceptibility.** ***Neurobehavioral Toxicology & Teratology,*** 7(1), 29–32.
Administered caffeine (0, 0.017, 0.034, or 0.05%) to 40 pregnant Wistar rats in drinking water throughout gestation. Offspring were cross-fostered to non-caffeine-treated mothers at birth. A dose-related increase in offspring mortality was observed at 24 hrs and at 10 days postpartum. Prenatal caffeine exposure did not significantly influence open-field ambulation or defecation when Ss were tested at 48, 68, or 196 days of age. A significant dose-related increase in restraint-stress gastric ulcer susceptibility was detected at 200 days of age. Offspring from Ss treated with 0.05% caffeine during prgnancy developed significantly more frequent and significantly more severe gastric lesions than did offspring from control Ss or from Ss prenatally exposed to 0.017 and 0.034% caffeine. Results suggest that prenatal caffeine exposure may predispose organisms to increased gastric disease susceptibility as adults and interfere with neonatal feeding ability, thereby producing infant mortality.

1172. Glavin, Gary B. & Mikhail, Anis A. (1976). **Role of gastric acid in restraint-induced ulceration in the rat.** ***Physiology & Behavior,*** 17(5), 777–780.
In 50 male Sprague-Dawley rats, gastric hyperacidity in response to starvation and to restraint stress was selectively blocked during either or both of these periods to assess the relative contributions of these treatments to ulcer formation. Results show that when the antacid drug (aluminum hydroxide) was administered during food deprivation, regardless of subsequent treatment, both rumenal and glandular ulcer incidence and severity were markedly reduced. It is suggested that food deprivation produces increased stomach acidity which, if prevented, reduces ulceration ultimately resulting from a subsequent stressor. If such acidity increases are unaltered, stress-induced gastric pathology is accentuated.

1173. Glavin, Gary B.; Pinsky, Carl & Bose, Ranjan. (1990). **Domoic acid-induced neurovisceral toxic syndrome: Characterization of an animal model and putative antidotes.** Summer Symposium: Brain, behavior and stress (1989, Los Angeles, California). ***Brain Research Bulletin,*** 24(5), 701–703.
A rodent model of neurovisceral toxic syndrome induced by the neuroexcitant amino acid, domoic acid, is described, along with the activity of a putative antidote, the nonselective excitotoxin antagonist, kynurenic acid. Both an extract of contaminated mussels and pure domoic acid induced a characteristic syndrome in mice, including sluggishness, scratching stereotypy, convulsions, and death. Autopsy revealed gastric and duodenal lesions and peritoneal ascites. Kynurenic acid significantly obtunded these behavioral and physiological effects, particularly when given 60–75 min after the toxic insult. Probenecid, a blocker of organic acid transport, and tryptophan, a precursor of endogenous brain kynurenic acid, increased the time frame in which kynurenic acid exerted its protective effects.

1174. Glavin, Gary B. & Rockman, Gary E. (1985). **Acute ethanol administration: Effects on stress-induced gastric and duodenal ulcer in rats.** ***Alcohol,*** 2(5), 651–653.
Investigated the effect of a short-term forced choice presentation of ethanol (1.5–2.5 g/kg/day [low]; 2.5–4.5 g/kg/day [medium]; 4.5–6.0 g/kg/day [high]) on restraint-stress-induced gastric injury, using 128 male Wistar rats. Ss were given 6% ethanol as their only source of liquid for 4 days. On the basis of ethanol consumption (g/kg/day), Ss were divided into high, medium, and low ethanol consuming groups. A nonethanol exposed control group was included. Following 24-hr food deprivation, Ss were restrained for 3 hrs. No differences in gastric ulcer frequency or severity were noted with the exception of a slight tendency toward a lower incidence among ethanol consuming Ss relative to controls. An unusual observation was the high incidence of duodenal ulcer observed only among ethanol consuming Ss. This ethanol/stress interaction is discussed in terms of an animal's history of ethanol exposure. It is suggested that for low-dose ethanol exposure, the method of administration rather than the dose may account for a marked increase in stress ulcer severity.

1175. Gold, Paul E. & Stone, William S. (1988). **Neuroendocrine effects on memory in aged rodents and humans.** Special Issue: Experimental models of age-related memory dysfunction and neurodegeneration. ***Neurobiology of Aging,*** 9(5–6), 709–717.
Reviews findings demonstrating that epinephrine and glucose treatments attenuate age-related memory impairments in rodents and humans. Additional results suggest that, in aged human and animal Ss, poor glucose regulation predicts memory performance of individual Ss.

1176. Golda, V. & Petr, R. (1988). **Genetically based animal model of depression: Cholinergic supersensitivity.** 23rd Conference of the Higher Nervous Functions (1987, Mariánské Lázně, Czechoslovakia). ***Activitas Nervosa Superior,*** 30(4), 292–294.
Traced cholinergic supersensitivity under oxotremorine treatment in 3 groups of 10-wk-old rats. Data suggest that genetically hypertensive corpulent Koletsky (KOL) rats and noncorpulent siblings of KOL Ss may serve as a genetically based animal model of depression states in humans.

1177. Golda, V. & Petr, R. (1989). **Animal model of anxiety: Interaction of nicergoline and scopolamine in the genetically hypertensive rats.** ***Activitas Nervosa Superior,*** 31(1), 76–77.
Traced the effect of nicergoline and scopolamine on genetically hypertensive rats of the S. Koletsky (1975) type in the elevated plus maze. A high degree of aversion to open space and height was accompanied by low turnover-rate of diencephalo-mesencephalic norepinephrine (NE), by elevated cholinergic supersensitivity, and vice versa. The alleviation of aversion to open space and height by nicergoline and/or scopolamine was proportional to NE and/or cholinergic characteristics of individual groups of Ss. Application of both drugs together showed mutual potentiation in male Ss.

1178. Golda, V. & Petr, R. (1989). **Animal model of anxiety: Effect of diazepam in the rats with cholinergic supersensitivity and reduced turnover-rate of brain norepinephrine.** 31st Annual Psychopharmacology Meeting (1989, Jeseník, Czechoslovakia). ***Activitas Nervosa Superior,*** 31(4), 297–299.
Reports data on the effect of diazepam in rats with cholinergic supersensitivity and reduced turnover rate of brain norepinephrine. Testing drugs that are or can be effective in affective disorders by using animal models is useful.

1179. Golda, V. & Petr, R. (1987). **Animal model of depression: Effect of nicotergoline and metergoline.** 28th Psychopharmacological Conference (1986, Jeseník, Czechoslovakia). ***Activitas Nervosa Superior,*** 29(2), 115–117.
Demonstrated the effectiveness of ergoline derivatives in alleviating acquisition and reacquisition of experimentally induced motor depression in rats, noting effects dependent on sex, drug, dosage, and strain.

1180. Golda, V.; Petr, R.; Rozsíval, V. & Šuba, P. (1986). **Animal model of depression: Imipramine, bromocriptine and lisuride alleviate motor depression.** 26th Annual Psychopharmacology Meeting (1984, Jeseník, Czechoslovakia). ***Activitas Nervosa Superior,*** 28(1), 26–27.
Verified the validity of an animal model of human reactive depression using imipramine and tested the antidepressant potency of the ergoline derivatives bromocriptine and lisuride. Experiments were conducted with adult normotensive Wistar rats and adult genetically hypertensive rats. Imipramine and acute lisuride significantly alleviated the acquisition and retention of motor depression, and chronic bromocriptine alleviated retention of motor depression, in all rats.

1181. Golda, V.; Petr, R. & Šuba, P. (1986). **Animal model of depression: Tranylcypromine alleviates motor depression whereas caffeine does not.** 27th Annual Psychopharmacology Meeting (1985, Jeseník, Czechoslovakia). ***Activitas Nervosa Superior,*** 28(1), 27–29.
Compared the effects of the typical monoamine oxidase (MAO) inhibitor tranylcypromine and the typical stimulant caffeine on motor depression in normotensive Wistar rats and genetically hypertensive rats to check the drug specificity of animal models of depression. Tranylcypromine alleviated generalized and proper motor depression in all Ss and acquisition of motor depression only in normotensive Ss. Caffeine had no alleviating effect. The findings support M. E. Seligman's (1975) learned helplessness model and the authors' model.

1182. Golda, V. & Petr, R. (1987). **Animal model of depression: Drug induced changes independent of changes in exploratory activity.** ***Activitas Nervosa Superior,*** 29(2), 114–115.
Documented the independence of drug-induced changes in the acquisition and retention of motor depression from changes in Ss' exploratory activity in normo- and hypertensive rats.

1183. Golub, Mari S. (1990). **Use of monkey neonatal neurobehavioral test batteries in safety testing protocols.** Special Issue: Methods in behavioral toxicology and teratology. ***Neurotoxicology & Teratology,*** 12(5), 537–541.
Describes the Neurobehavioral Test Battery, a 2-wk test for monkey neonates, which examines simple elicited behaviors and assesses muscle tone, environmental responsiveness, presence of age-appropriate reflexes and behavior patterns, and rigor of responding. Such test batteries cover a number of aspects of neurobehavioral function and employ an animal model similar to the human in complexity and maturity of brain function at birth. However, monkey neonate test

batteries produce a data base with a small number of degrees of freedom (due to small sample sizes) and a larger number of endpoints to be evaluated for group differences. With attention to experimental design and statistical analysis, data from such protocols can produce valuable information concerning safety.

1184. Gómez, Ariel & Troncoso, Edgardo. (1984). **Animal model of tardive dyskinesia: Hypersensitivity to apomorphine and binding of H^3-spiroperidol to membranes of haloperidol-treated rat striatum.** ***Revista Chilena de Neuro-Psiquiatría,*** 22(1), 55–59.
Describes an original animal model of tardive dyskinesia in which behavioral hypersensitivity (stereotypy) evoked by subthreshold doses of apomorphine and kinetic parameters (KD and Bmax) of striatal D2 receptors are measured in the same animals. 10 female Sprague-Dawley rats received daily intravenous injections of 0.6 mg/kg haloperidol diluted in isotonic saline for 15 days, and control rats received saline solution. All Ss then received 0.15 mg/kg apomorphine subcutaneously. Experimental Ss demonstrated involuntary behavior ranging from intermittent sniffing to continuous sniffing and nibling. Rat pairs were killed 7, 8, and 14 days after treatment and striatal dopaminergic D2 receptors were examined. Results show that treatment with haloperidol produced a considerable increase in receptor density (Bmax). KD affinity showed no clear increase. Both stereotypy and increases in Bmax (D2 site) values run a parallel time course. (English abstract).

1185. Gonzalez, Fernando A. & Byrd, Larry D. (1977). **Physiological effects of cocaine in the squirrel monkey.** ***Life Sciences,*** 21(10), 1417–1423.
Doses of cocaine (0.03, 0.3, and 3.0 mg/kg) that have previously been shown to have behavioral effects produced dose-related increases in arterial blood pressure, heart rate, and core temperature in 8 catheterized, unanesthetized male squirrel monkeys. The pressor effect was immediate, but heart rate increased more gradually after cocaine injection. The onset of hyperthermia was substantially delayed. It is suggested that the squirrel monkey may be a good animal model of the physiological concomitants of cocaine abuse in humans.

1186. Goodwin, G. M. & Wood, A. J. (1987). **Captopril as an antidepressant: Lack of effect in animal models of serotonergic function.** ***Biological Psychiatry,*** 22(10), 1274–1276.
In a study of 16 male C57/black/6/Ola mice, administration of captopril for 24 hrs or 14 days had no effect on the head-twitch response evoked by precursor loading; the hypothermic response to 8-hydroxy-2-(di-*n*-propylamino)tetralin was unchanged following chronic captopril treatment; and plasma potassium was elevated in the treated Ss.

1187. Gordon, John H. & Diamond, Bruce I. (1981). **Antagonism of dopamine supersensitivity by estrogen: Neurochemical studies in an animal model of tardive dyskinesia.** ***Biological Psychiatry,*** 16(4), 365–371.
Demonstrated that the withdrawal from chronic haloperidol or estradiol benzoate (EB) treatment results in a behavioral supersensitivity to dopamine agonists in ovariectomized Sprague-Dawley rats. The administration of EB during withdrawal from haloperidol or continuous treatment with EB attenuated or prevented development of a supersensitivity to dopamine agonists. Results indicate that exogenous estrogens may modulate the number of dopamine receptors in the CNS and, as such, may decrease the incidence and/or relieve the symptoms of tardive dyskinesia.

1188. Graeff, Frederico G.; Audi, Elisabeth A.; Almeida, Sebastiao S.; Graeff, Eneida O. et al. (1990). **Behavioral effects of 5-HT receptor ligands in the aversive brain stimulation, elevated plus-maze and learned helplessness tests.** Second Brazilian Symposium: Neurosciences and behavior (1989, Florianópolis, Brazil). ***Neuroscience & Biobehavioral Reviews,*** 14(4), 501–506.
Describes experiments to illustrate the use of animal models in the study of the anxiolytic and antidepressant properties of drugs acting on 5-hydroxytryptamine (5-HT) receptors. With electrical stimulation of the midbrain central gray (CG), an aversive area of the brain, the 5-HT-1 receptor antagonist propranolol raised the aversive threshold in a dose-dependent way in rats, following its microinjection into the CG. This antiaversive effect of propranolol, which is similar to that of benzodiazepine anxiolytics, was prevented by microinjection into the same brain site of the 5-HT-2 receptor blocker ritanserin. In another animal model of anxiety, the elevated plus-maze, intra-CG propranolol also caused an anxiolytic-like effect, antagonized by ritanserin, indicating a 5-HT mediation.

1189. Grant, Kathleen A. & Samson, Herman H. (1985). **Oral self administration of ethanol in free feeding rats.** International Society for Biomedical Research on Alcoholism (ISBRA) (1984, Sante Fe, New Mexico). ***Alcohol,*** 2(2), 317–321.
Examined whether ethanol ingestion could be induced and maintained in 14 male Long-Evans rats without the use of food deprivation. A conditioning paradigm was used to establish oral ethanol self-administration in free-feeding Ss. Through initial reinforcement of 5% ethanol consumption with 20% sucrose solution, Ss were trained to work for and consume concentrations of ethanol up to and including 40%. Blood ethanol levels above 100 mg ethanol/dl blood were frequently found. A control group, induced to drink quinine with the same procedures, indicated the relative importance of ethanol's pharmacological effect in maintaining high levels of self-administration. Results show that free-feeding rats can maintain oral self-administration of intoxicating quantities of high ethanol concentrations when its initial consumption is paired with an additional reinforcer. The procedure of initially reinforcing oral ethanol consumption with access to sucrose appears to be a simple, efficient method capable of fulfilling many of the criteria outlined for an animal model of alcohol abuse.

1190. Grasing, Kenneth W. & Miller, Neal E. (1989). **Self-administration of morphine contingent on heart rate in the rat.** ***Life Sciences,*** 45(21), 1967–1976.
Trained male rats to self-administer iv morphine infusions through changes in heart rate (HR), controlling for unconditioned drug effects by reversing the direction of change in HR required for infusions and addition of a yoked control S. Ss were given continuously available morphine based on tachycardia, and 1 S was also tested on a schedule of limited drug availability and with continuously available morphine based on bradycardia. In opioid naive Ss, a single noncontingent iv infusion of morphine depressed HR. Ss exposed to .1 mg/kg infusions of morphine contingent on tachycardia showed trends for elevated HR with increased locomotor

and grooming activity preceding infusions. Increases in HR were most pronounced during daytime, normally inactive periods. Findings suggest that behavioral changes with tachycardia-contingent morphine were instrumentally conditioned.

1191. Green, A. R.; Davies, E. M.; Little, H. J.; Whittington, M. A. et al. (1990). **Action of chlormethiazole in a model of ethanol withdrawal.** ***Psychopharmacology,*** 102(2), 239–242.
Demonstrated that male mice withdrawn from exposure for 14 days to ethanol inhalation showed signs of ethanol withdrawal, including convulsive behavior. Injection of chlormethiazole (CMZ) 5 hrs after the start of withdrawal (the time that the convulsive behavior was near maximal) resulted in the virtual disappearance of the withdrawal-induced behavior) in 30 min, with its reappearance by 60 min. The time course of the effect of CMZ in the withdrawal test was similiar to its effect in raising seizure threshold and decreasing locomotor activity. CMZ, used clinically to treat ethanol withdrawal is effective in this animal model of withdrawal. CMZ is likely to work by increasing GABAergic function.

1192. Griffith, Neil; Engel, Jerome & Bandler, Richard. (1987). **Ictal and enduring interictal disturbances in emotional behaviour in an animal model of temporal lobe epilepsy.** ***Brain Research,*** 400(2), 360–364.
Investigated long-term changes in emotionality in 17 adult cats given unilateral microinjections of kainic acid in the dorsal hippocampus to induce seizures. During a 4-mo period, 8 Ss demonstrated ictal emotional behavior similar to a defensive rage reaction. Interictally, these Ss demonstrated emotional lability. Thresholds for electrical brain stimulation-induced defensive rage were lowered. During periods when no spontaneous seizures were observed for several days, Ss were less emotionally reactive, and thresholds for stimulation-induced defense reactions returned to baseline. Results suggest that emotional disturbances in epileptic patients result from pathophysiological mechanisms related to the epileptogenic process and that such emotional disturbances might be alleviated if the epileptic seizures could be controlled.

1193. Grindlinger, Howard M. & Ramsay, Ed. (1991). **Compulsive feather picking in birds.** ***Archives of General Psychiatry,*** 48(9), 857.
Discusses a preliminary study of the effectiveness of clomipramine in feather-picking disorder, a behavioral abnormality in birds that may represent an avian analog to obsessive-compulsive disorder in humans. Five of 10 treated birds (parrots, cockatiels, and cockatoos) picked their feathers significantly less with treatment.

1194. Gruol, D. L. (1991). **Chronic exposure to alcohol during development alters the membrane properties of cerebellar Purkinje neurons in culture.** ***Brain Research,*** 558(1), 1–12.
Used intracellular recordings to investigate the effects of chronic alcohol exposure (CAE) on the development of physiological function in cultured Purkinje neurons obtained from the cortical region of cerebella from 20-day-old rat embryos. CAE caused multiple changes in the physiological properties of the developing Purkinje neurons. These neuronal actions may contribute to behavioral deficits observed in animal models of fetal alcohol syndrome (FAS). Similar target sites of ethanol action may be present in the human central nervous system (CNS) neurons and may be involved in human FAS.

1195. Guile, Michael N. (1989). **Arousal as an explanation for differences in rats selectively bred for differential alcohol sensitivity.** ***Journal of Psychology,*** 123(3), 279–284.
Interprets published findings about 2 lines of rats selectively bred by E. P. Riley et al for locomotor impairment in response to a subhypnotic dose of ethanol in light of a hypothesis suggesting adventitiously selected differences in central arousal between the 2 lines. These lines, designated "most affected" and "least affected," were compared in a variety of tests and showed differences in a number of phenotypic traits and locomotor impairment to ethanol. Interpretation of these differences shows that their usefulness as animal models of alcoholism is seriously compromised.

1196. Guillet, Ronnie & Kellogg, Carol. (1991). **Neonatal exposure to therapeutic caffeine alters the ontogeny of adenosine A1 receptors in brain of rats.** ***Neuropharmacology,*** 30(5), 489–496.
Investigated the effects of early developmental exposure to caffeine on the ontogeny of the adenosine A1 receptor, using an animal model (R. Guillet, 1990) designed to approximate human neonatal exposure to therapeutic caffeine. Rats received caffeine over Days 2–6 of life and were observed during Days 14–90. There was a significant effect of exposure on specific binding of the A1 receptor in the cortex, cerebellum, and hippocampus, compared with control Ss. Kinetic analysis of binding to the A1 site in cortical tissue suggests that this increase was due to an increased maximum binding density; binding affinity did not vary. Early developmental exposure to caffeine may accelerate the development of adenosine A1 receptors with attainment of adult densities of receptors at earlier ages, thus suggesting an ontogenic change.

1197. Gunne, Lars M.; Andersson, Ulf; Bondesson, Ulf & Johansson, Per. (1986). **Spontaneous chewing movements in rats during acute and chronic antipsychotic drug administration.** ***Pharmacology, Biochemistry & Behavior,*** 25(4), 897–901.
Single intraperitoneal doses of the antipsychotic drugs clozapine, sulpiride, haloperidol, and fluphenazine induced a depression of the spontaneous chewing movement (SCM) rate in Sprague-Dawley rats during the 1st 6–8 hrs. Haloperidol and fluphenazine elicited a rebound increase in SCM on Days 2–5. Atropine reduced the SCM rate. Chronic administration for 10 mo of thioridazine, chlorpromazine, fluphenazine, and haloperidol produced highly significant increases in SCM rates. The present animal model may prove useful for monitoring the risk of tardive dyskinesia with individual drugs.

1198. Gunne, Lars-M. & Barany, Sven. (1976). **Haloperidol-induced tardive dyskinesia in monkeys.** ***Psychopharmacology,*** 50(3), 237–240.
In 3 cebus monkeys, the chronic daily administration of haloperidol (0.5 mg/kg/day orally) created sedation and parkinsonism during the 1st 5–7 wks. Later the Ss developed signs reminiscent of acute dystonia, as seen in the clinic during treatment with neuroleptics. These signs were dose-dependent and, in extreme cases, included widespread tonic and clonic seizures. After 3 and 12 mo, respectively, 2 of the Ss developed buccolingual signs (grimacing and tongue pro-

trusion), similar to tardive dyskinesia. The tardive dyskinesia symptoms were reduced in a dose-dependent manner after each haloperidol administration, being most pronounced in the morning before haloperidol was given. Biperiden reduced acute dystonia but reinstated signs of tardive dyskinesia, which had been abolished by haloperidol. It is suggested that cebus monkeys may provide a useful animal model for the study of long-term neurologic complications from neuroleptic drugs.

1199. Gunne, Lars M. & Johansson, P. (1989). **Chronic melperone administration does not enhance oral movements in rats.** Workshop: Melperone: An atypical neuroleptic (1986, Malmö, Sweden). ***Acta Psychiatrica Scandinavica,*** 80(352, Suppl), 48–50.
Compared melperone and haloperidol in a rat model for tardive dyskinesia in which 48 rats were given the drugs chronically for 1 yr and frequency of vacuous chewing movement (VCM) rates were measured at monthly intervals. Drugs that induce tardive dyskinesia have previously been observed to increase VCM. Melperone did not induce an increase in VCM rate, while haloperidol did, supporting the notion that haloperidol and melperone have different long-term effects on VCM. Melperone may represent a potent antipsychotic with a low risk of tardive dyskinesia.

1200. Haefely, W.; Martin, J. R. & Schoch, P. (1991). **Novel anxiolytics that act as partial agonists at benzodiazepine receptors.** ***Giornale di Neuropsicofarmacologia,*** 13(3), 83–88.
Reviews recent research on the development and use of benzodiazepine agonists as anxiolytic medications, including bretazenile, clonazepam, FG8205, imidazo(1,2-A) pyrimidine, abecarnile, and alpidem. The relationship between gamma-aminobutyric acid (GABA) and benzodiazepine receptors and the role of partial and selective agonism for a single receptor type are also discussed. An animal model used to evaluate the effects of partial benzodiazepine receptor agonists is described.

1201. Häfner, Heinz; Behrens, Stephan; de Vry, Jean & Gattaz, Wagner F. (1991). **Oestradiol enhances the vulnerability threshold for schizophrenia in women by an early effect on dopaminergic neurotransmission: Evidence from an epidemiological study and from animal experiments.** ***European Archives of Psychiatry & Clinical Neuroscience,*** 241(1), 65–68.
Examined whether (1) the effect of estradiol on the dopaminergic system enhances the vulnerability threshold for schizophrenia, which is lowered again during menopause, and (2) testosterone reduces the vulnerability threshold and thus furthers the earlier onset of schizophrenia in males. Three animal models were used to examine the effect of gonadal hormones on haloperidol-induced catalepsy and on apomorphine-induced stereotypies in neonatal and adult rats. No clear influence by testosterone was shown. Estradiol caused a significant reduction of both dopamine-agonist and dopamine-antagonist induced behavior; the effects were stronger in neonatal Ss. Estradiol caused the dopamine receptor affinity for sulpiride to be reduced; this suggests that the behavioral changes due to estradiol were accounted for by a down-regulation of dopamine receptor sensitivity.

1202. Hall, Frank S.; Stellar, James R. & Kelley, Ann E. (1990). **Acute and chronic desipramine treatment effects on rewarding electrical stimulation of the lateral hypothalamus.** ***Pharmacology, Biochemistry & Behavior,*** 37(2), 277–281.
Two wks of chronic desipramine HCl (DMI) treatment did not alter reward or motor/performance components of intracranial self-stimulation (ICSS) as assessed with the rate-frequency method in male rats. Acute DMI treatment produced an ICSS reward decrement relative to saline control treatment, which was similar in size on Day 1 and Day 15. Failure to find a chronic DMI effect on ICSS reward suggests that ICSS in normal rats may not be a valid animal model of depression. A better paradigm may be to test the ability of antidepressants to reverse a chronic reduction in ICSS reward function that is first produced by some other method.

1203. Hampson, Robert E.; Foster, Thomas C. & Deadwyler, Sam A. (1989). **Effects of Δ-9-tetrahydrocannabinol on sensory evoked hippocampal activity in the rat: Principal components analysis and sequential dependency.** ***Journal of Pharmacology & Experimental Therapeutics,*** 251(3), 870–877.
Assessed trial specific influences (sequential dependency and serial position effects) on the outer molecular layer of the dentate gyrus (OMDG) auditory evoked potential (AEP) to determine consequences for cognitive functioning resulting from acute exposure to Δ-9-tetrahydrocannabinol (Δ-9-THC) in 5 male rats during performance of a 2-tone discrimination task. Δ-9-THC disrupted the trial-to-trial sequential dependency of the OMDG AEPs. Δ-9-THC selectively influenced the serial dependence of the OMDG AEP. Results imply that Δ-9-THC is a potent disruptor of temporally specific information as it is processed by the hippocampus. Such disruption may be the basis of Δ-9-THC effects on memory processes in humans.

1204. Hannigan, John H. & Riley, Edward P. (1988). **Prenatal ethanol alters gait in rats.** ***Alcohol,*** 5(6), 451–454.
Assessed changes in gait for rats (aged 55 days) exposed to ethanol in utero. Ethanol-exposed Ss had significantly shorter stride lengths, more open step angles, and less gait symmetry than controls. This pattern of changes in motor function indicates that prenatal exposure to ethanol produced long-lasting ataxia in Ss. Results resemble previous findings of altered gait following neonatal ethanol exposure in rats, as well as clinical findings in some fetal alcohol syndrome children. Results are consistent with a hypothesis of prenatal ethanol-induced disruption of functional hippocampal and/or cerebellar development.

1205. Hansen, Thor W.; Sagvolden, Terje & Bratlid, Dag. (1987). **Open-field behavior of rats previously subjected to short-term hyperbilirubinemia with or without blood–brain barrier manipulations.** ***Brain Research,*** 424(1), 26–36.
Investigated (1) whether bilirubin encephalopathy with lasting sequelae could be created in a rat model and (2) putative differences in brain toxicity between bound and unbound bilirubin. Hyperbilirubinemia was produced by infusing bilirubin into 6-wk-old male rats. Different groups were exposed to hyperosmolality, hypercarbia, and sulfisoxazole. Ss were later studied in an open-field apparatus for 7 measures of activity (e.g., crossings in cage). Level of activity was higher in the bilirubin-treated Ss. Data show that both unbound and albumin-bound bilirubin were neurotoxic, with unbound bilirubin having a more pronounced effect. The

sequelae of bilirubin brain toxicity appear to include changes in stimulus processing; this is compatible with findings from neuropsychological tests of children who have had significant neonatal hyperbilirubinemia.

1206. Hanulak, Ann T. & Hull, Elaine M. (1987). **Behavioral deficits in a rat model of maternal PKU.** ***Psychobiology,*** 15(1), 75–78.
Examined behavioral development in 49 Long-Evans rats following exposure of their dams to alpha-methyl-phenylalanine and phenylalanine on Days 10–21 of gestation to induce phenylketonuria (PKU). The pups' development was compared with that of 67 offspring of pair fed or ad-lib fed controls. Prenatally exposed Ss were deficient on measures of swimming development, active avoidance acquisition, and initiation of masculine sexual behavior compared with controls. It is noted that this experimental model of maternal PKU exhibits fewer toxic effects than earlier models. Results suggest that this model is useful for studying the biochemical mechanisms of and treatment for maternal PKU.

1207. Harlow, Henry J.; Phillips, John A. & Ralph, Charles L. (1982). **Circadian rhythms and the effects of exogenous melatonin in the nine-banded armadillo, *Dasypus novemcinctus*: A mammal lacking a distinct pineal gland.** ***Physiology & Behavior,*** 29(2), 307–313.
Circadian rhythms of body temperature, activity, and oxygen consumption were recorded in 23 armadillos. The rhythms were entrained by light–dark cycles and appeared to be phase-coupled. Under constant illumination, Ss displayed free-running circadian rhythms with period lengths of slightly less than 24 hrs. Melatonin implants caused a lengthening of the free-running period of activity and body temperature. The occurrence of a "normal" circadian rhythmicity in the armadillo indicates that the pineal organ is not necessary for the circadian organization of this animal, whereas melatonin of nonpineal origin may have a role.

1208. Harrell, Ernest H.; Haynes, Jack R.; Lambert, Paul L. & Sininger, Rollin A. (1978). **Reversal of learned helplessness by peripheral arousal.** ***Psychological Reports,*** 43(3, Pt 2), 1211–1217.
Many studies have demonstrated similarities between human depression and learned helplessness. However, pilot studies have suggested that sympathomimetic drugs that do not alleviate depression may alter learned helplessness. The present study was designed to replicate these preliminary data. 32 male Sprague-Dawley rats were exposed to either no-shock or a sequence of inescapable shock previously shown to produce learned helplessness. 24 hrs later metaraminol bitartrate (Aramine) or physiological saline was injected prior to the testing in a barpress, shock-escape task. Inescapable shock 1 day prior to testing produced a severe response deficit in saline controls but not in the drug group. Reversal of learned helplessness by peripheral autonomic arousal indicates that a reevaluation of the learned helplessness model of human depression may be necessary.

1209. Hartley, P.; Neill, D.; Hagler, M.; Kors, D. et al. (1990). **Procedure- and age-dependent hyperactivity in a new animal model of endogenous depression.** ***Neuroscience & Biobehavioral Reviews,*** 14(1), 69–72.
Replicated the findings of M. Mirmiran et al (1983) that neonatal administration of the antidepressant clomipramine (CLI) to male rats results in hyperactivity in open-field tests in adulthood. This effect did not reliably occur in a "Digiscan" activity device. The difference in effect between the 2 activity measuring devices may be due to more stress being present in the open-field test, and CLI-treated rats may be more reactive to stress. This hypothesized enhanced reactivity to stress may be similar to the proposed vulnerability of depressed humans to stress. In addition, the open-field effect did not occur until Ss were at least 4 mo old; this delayed effect may be analogous to the progressive onset of endogenous depression in humans.

1210. Hasey, Gary & Hanin, Israel. (1991). **The cholinergic–adrenergic hypothesis of depression reexamined using clonidine, metoprolol, and physostigmine in an animal model.** ***Biological Psychiatry,*** 29(2), 127–138.
Three experiments examined the role of central nervous system (CNS) cholinergic and noradrenergic mechanisms in the pathogenesis of depression and hypothalamic-pituitary-adrenal axis hyperactivity, using the behavioral despair rat model of depression. Immobility (IM), the analog of depression in this model, and plasma corticosterone (CCS) were increased by physostigmine (PHYSO) in male rats. Neostigmine produced the same peripheral cholinomimetic effects and motor inhibition as PHYSO, but did not change IM. PHYSO's effects on CCS and IM were blocked by metoprolol pretreatment and partially blocked by clonidine pretreatment. PHYSO increased acetylcholine in the striatum. In this rat model of depression, PHYSO produced a behavioral and endocrine syndrome analogous to the syndrome of depression with glucocorticoid hypersecretion in humans. Findings are consistent with the hypothesis that depression is due to an imbalance of cholinergic and adrenergic functioning in the CNS.

1211. Heilig, Markus; Söderpalm, Bo; Engel, Jörgen A. & Widerlöv, Erik. (1989). **Centrally administered neuropeptide Y (NPY) produces anxiolytic-like effects in animal anxiety models.** ***Psychopharmacology,*** 98(4), 524–529.
Examined the effects of intracerebroventricular neuropeptide Y (NPY) and its C-terminal 13–36 amino acid fragment with respect to anxiolytic properties in 2 rat anxiety models, Montgomery's conflict test (MT), and Vogel's drinking conflict test (VT). In the MT, 1.0 and 5.0 nmol NPY abolished the normal preference for the closed arms of the maze. In the VT, both 0.2 and 1.0 nmol NPY increased the number of shocks accepted. Findings are discussed in relation to the noradrenaline hypothesis of anxiety and in relation to observations indicating involvement of NPY in the pathophysiology of major depression.

1212. Held, Irene R.; Sayers, Scott T. & McLane, Jerry A. (1991). **Acetylcholine receptor gene expression in skeletal muscle of chronic ethanol-fed rats.** ***Alcohol,*** 8(3), 173–177.
Evaluated the expression of the neuromuscular acetylcholine receptor (AChR) alpha-subunit gene in soleus muscles from an animal model of chronic alcoholism. At 8 wks of age, male rats were placed on a nutritionally complete liquid diet containing 6.7% ethanol. Matched controls were pair-fed an isocaloric liquid diet. After a 16-wk diet period, soleus muscles were obtained and total RNA and poly(A)$^+$ RNA were isolated. Muscle RNA levels from ethanol-fed and control Ss were comparable. AChR alpha-subunit mRNA was detected

by hybridization of muscle poly(A)$^+$ RNA with a ^{32}P-labeled, complementary riboprobe. The steady-state level of AChR alpha-subunit mRNA was reduced by 39% in soleus muscles from the ethanol-fed Ss compared with pair-fed controls.

1213. Hemingway, R. B. & Reigle, T. G. (1987). **The involvement of endogenous opiate systems in learned helplessness and stress-induced analgesia.** ***Psychopharmacology,*** 93(3), 353–357.
Investigated the role of endogenous opiate systems in learned helplessness (LH) and stress-induced analgesia (SIA) in rats exposed to escapable and inescapable footshock. Following an initial footshock, analgesia was observed in Ss that could not control their stress exposure; this SIA was prevented by the opiate antagonist naloxone. Exposure to a shuttlebox escape task at 48 hrs reinstated SIA. Escape deficits indicative of LH were exhibited by Ss receiving inescapable footshock prior to testing. SIA and LH observed in the inescapable group were prevented by prior treatment with naloxone. It is concluded that LH and SIA can occur in the presence of opiate antagonism and may require the participation of additional transmitter systems.

1214. Henke, Peter G. (1987). **Chlordiazepoxide and stress tolerance in rats.** ***Pharmacology, Biochemistry & Behavior,*** 26(3), 561–563.
Evaluated the effectiveness of chlordiazepoxide (CDP; 5, 10, or 20 mg/kg, intraperitoneally) in attenuating stomach ulcers induced by restraint in 40 male Wistar rats. CDP attenuated gastric stress pathology when given prior to restraint, but it interfered with adaptation to chronic stress. Results are discussed in terms of stress adaptation and drug withdrawal effects.

1215. Henke, Peter G. (1990). **Potentiation of inputs from the posterolateral amygdala to the dentate gyrus and resistance to stress ulcers formation in rats.** ***Physiology & Behavior,*** 48(5), 659–664.
In 3 experiments, physical restraint increased the activity of a number of multiple units in the lateral amygdala of male rats. High-frequency electrical stimulation of units in the posterolateral amygdala increased the amplitudes of granule cell potentials in the dentate gyrus. This bilateral long-term potentiation (LTP) of inputs from posterior areas of the lateral amygdala also attenuated the severity of stress ulcers produced by physical restraint. This effect was reversed by iv injections of the selective N-methyl-D-aspartate receptor blocker, aminophosphonovaleric acid. LTP in this pathway also reduced "struggling" behavior during restraint. Results indicate that LTP in this temporal lobe pathway increased the coping ability because of faster habituation to stressors.

1216. Heym, James & Koe, B. Kenneth. (1988). **Pharmacology of sertraline: A review.** Symposium: Serotonin in behavioral disorders (1987, Zurich, Switzerland). ***Journal of Clinical Psychiatry,*** 49 Suppl, 40–45.
Reviews the pharmacological profile of the serotonin (5-hydroxytryptamine [5-HT]) reuptake inhibitor sertraline (STRL) and discusses the mechanism of action and potential indications for clinical use. Reduction of immobility or inactivity in animal models of behavioral despair has become a standard screening test for antidepressant activity. STRL produces a marked reduction in the immobility score of mice placed in an inescapable swim tank; STRL does not reverse reserpine-induced hypothermia in mice (another antidepressant screening method). Animal studies have demonstrated that serotonin acts to suppress feeding behavior. STRL reduces food intake by rats in a dose-dependent fashion. The influence of STRL on energy balance is translated into a weight loss after chronic administration.

1217. Heyser, Charles J.; Chen, Wei-jung; Miller, James; Spear, Norman E. et al. (1990). **Prenatal cocaine exposure induces deficits in Pavlovian conditioning and sensory preconditioning among infant rat pups.** ***Behavioral Neuroscience,*** 104(6), 955–963.
Offspring derived from Sprague-Dawley dams that received daily sc injection of 40 mg/kg 3 cc^{-1} cocaine hydrochloride (C40) or saline (LC) from Gestational Days 8–20 were tested for first-order Pavlovian conditioning and sensory preconditioning at Postnatal Days 8 (P8), P12, and P21. Although C40 dams gained significantly less weight than LC dams, pup body weights did not differ between the 2 groups. Significant sensory preconditioning was obtained at P8 and P12 (but not at P21) in LC offspring, confirming previous reports (e.g., L. P. Spear et al; 1989) of decline in performance in this task during ontogeny. In contrast, C40 offspring failed to exhibit sensory preconditioning at any test age. In addition, C40 pups tested at P8 did not display significant first-order conditioning. Taken together these results suggest a more general deficit in cognitive functioning rather than a delay in cognitive development in prenatally cocaine-exposed offspring.

1218. Hilakivi, Leena A.; Durcan, Michael J. & Lister, Richard G. (1989). **Effects of ethanol on fight- or swim-stressed mice in Porsolt's swim test.** ***Neuropsychopharmacology,*** 2(4), 293–298.
Examined the effects of ethanol in Porsolt's swim test on mice preexposed to fight- or swim-stressors. Low doses of ethanol reversed lengthened immobility of Ss preexposed to a stressor. This suggests that ethanol either has antidepressant-like properties, or it improves animals' ability to cope with a stressful situation, or both. Because antidepressant drugs decrease and stressors increase immobility in the swim test, the test may serve as a putative animal model of depression.

1219. Hilakivi, Leena A. & Hilakivi, Ilkka. (1987). **Increased adult behavioral "despair" in rats neonatally exposed to desipramine or zimeldine: An animal model of depression?** ***Pharmacology, Biochemistry & Behavior,*** 28(3), 367–369.
18 male Wistar rats were daily given either desipramine or zimeldine from the 7th to the 18th postnatal days. At 2 and 5 mo of age, desipramine-treated and zimeldine-treated Ss expressed lengthened immobility times in a swim test.

1220. Hilakivi-Clarke, L. A.; Wozniak, K. M.; Durcan, Michael J. & Linnoila, M. (1990). **Behavior of streptozotocin-diabetic mice in tests of exploration, locomotion, anxiety, depression and aggression.** ***Physiology & Behavior,*** 48(3), 429–433.
Examined behavior of streptozotocin-diabetic male mice in Porsolt's swim test, the holeboard test of exploration and locomotor activity, in the plus maze test of anxiety, and in the resident–intruder paradigm of aggression. Two weeks after an ip injection of streptozotocin, which caused a 20% weight loss and increased fluid consumption and urination, Ss showed lengthened duration of immobility in the swim test. One week of insulin treatment partially antagonized this change. The locomotor activity scores in the streptozotocin-treated Ss were lower in the holeboard but

higher in the plus maze than in the controls; therefore, the lengthened immobility was not likely due to a general motor impairment. No significant changes in the time spent in social interaction or aggressive behavior were found in the streptozotocin-treated mice.

1221. Himnan, Donald J. (1984). **Tolerance and reverse tolerance to toluene inhalation: Effects on open-field behavior.** ***Pharmacology, Biochemistry & Behavior,*** 21(4), 625–631.
Exposed 20 male Long-Evans rats by inhalation to extremely high concentrations of toluene vapors twice daily for 6 wks, as an animal model of organic solvent abuse. Six Ss were in a sham-exposure group. At preset intervals during repeated exposure, Ss were exposed to test concentrations of toluene and effects on behavior in an open field were measured. Concentration–effect curves were determined during Weeks 4–6 of repeated exposure. Tolerance to toluene was measured as a decreased response to the test exposure and a shift of the concentration–effect curve to the right. Reverse tolerance was measured as an increased response to the test exposure and a shift of the concentration–effect curve to the left. Results demonstrate that the effects of repeated exposure to toluene showed behavioral selectivity: Tolerance developed to ataxia, hindlimb myoclonus, and inhibition of rearing; reverse tolerance developed to headshakes and increased locomotor activity.

1222. Hingtgen, J. N.; Hendrie, H. C. & Aprison, M. H. (1984). **Postsynaptic serotonergic blockade following chronic antidepressive treatment with trazodone in an animal model of depression.** ***Pharmacology, Biochemistry & Behavior,*** 20(3), 425–428.
Studied both the chronic and acute effects of a recently introduced antidepressant, trazodone, a triazolopyridine compound. Male Wistar rats working for milk reinforcement and exhibiting behavioral depression following administration of 50 mg/kg 5-HTP were pretreated (1 hr before 5-HTP) with 1, 2, or 4 mg/kg trazodone with resulting blockade of 5-HTP induced depression of 35, 62, and 70% respectively. Chronic administration of trazodone (2 mg/kg/day) also resulted in a significant blockade of the 5-HTP effect (75%). Neither 2 mg/kg or 4 mg/kg trazodone potentiated the shorter period of depression following 25 mg/kg 5-HTP. Chronic treatment with the antidepressant drugs, amitriptyline or mianserin, also blocked 5-HTP depression. Data suggest an important postsynaptic mechanism associated with chronic administration of trazodone, amitriptyline, and mianserin, which could be implicated in the therapeutic effectiveness of these drugs. The potency of trazodone in relation to other antidepressant drugs in the behavior model of depression paralleled their potency in displacing radioligand binding to 5-HT receptors, and additional support is given for the hypersensitive postsynaptic serotonin receptor theory of depression.

1223. Hingtgen, Joseph N.; Fuller, Ray W.; Mason, Norman R. & Aprison, M. H. (1985). **Blockade of a 5-hydroxytryptophan-induced animal model of depression with a potent and selective 5-HT$_2$ receptor antagonist (LY53857).** ***Biological Psychiatry,*** 20(6), 592–597.
Tested the hypothesis that a new potent and selective 5-hydroxytryptamine$_2$ (5-HT$_2$) receptor antagonist (LY53857) would be an excellent blocker of 5-hydroxytryptophan (5-HTP)-induced response suppression in an animal model of depression. LY53857 was administered 60 min prior to 5-HTP injections into male Wistar rats working on an operant schedule for mild reinforcement. As predicted, LY53857 pretreatment significantly blocked 5-HTP depression (90%) in doses as low as 0.1 mg, intraperitoneally. When the dose was further reduced to 0.025 mg, blockade of 5-HTP-induced depression was still greater than 30%. In doses as high as 5.0 mg, LY53857 alone had no effect on the baseline performance of Ss working a VI 1 schedule. Pretreatment with desipramine (2.5 mg), an antidepressant characterized as having major noradrenergic effects, did not significantly block the 5-HTP-induced depression. Data suggest that the 5-HTP-induced depression is mediated by serotonergic mechanisms involving 5-HT$_2$ receptors, as LY53857 is a selective antagonist of these receptors. Data also support the suggestion, based on other published data from this laboratory, that some antidepressants are antagonizing 5-HT$_2$ receptors in the animal model of depression and may also act in a similar manner in depressed patients. Results suggest the potential value of LY53857 as an antidepressant agent.

1224. Hingtgen, Joseph N.; Shekhar, Anantha; DiMicco, Joseph A. & Aprison, M. H. (1988). **Response suppression in rats after bilateral microinjection of 5-hydroxytryptophan in lateral hypothalamus.** ***Biological Psychiatry,*** 23(7), 711–718.
Rats were implanted with bilateral cannulae in the lateral hypothalamus and received microinjections of 100–500 ng of d,l-5-hydroxytryptophan (5-HTP) 15 min after the start of a VI operant session (milk reinforcement). Significant decreases in responding were observed that were comparable to those obtained after a systemic injection of 50 mg/kg d,l-5-HTP. Ss receiving a microinjection of 5-HTP in the posterior hypothalamus did not exhibit a behavioral effect. Ss working on shock-avoidance schedules did not demonstrate response suppression following microinjection of 5-HTP into the lateral hypothalamus. Data support the important role previously assigned to central 5-hydroxytryptamine (5-HT) mechanisms in the 5-HTP animal model of depression.

1225. Hinson, Riley E. & Siegel, Shepard. (1982). **Nonpharmacological bases of drug tolerance and dependence.** ***Journal of Psychosomatic Research,*** 26(5), 495–503.
Reviews research demonstrating a role of nonpharmacological factors in drug tolerance as well as research documenting a role of similar nonpharmacological factors in posttreatment relapse. It is suggested that there may be a common mechanism involved in nonpharmacological influences on tolerance and relapse. Such a common mechanism, involving Pavlovian-conditioned drug-compensatory responses, is described. Evidence is summarized implicating such CRs in tolerance and posttreatment relapse. The treatment implications of the Pavlovian conditioning analysis are discussed.

1226. Hjorth, S.; Carlsson, A. & Engel, J. A. (1987). **Anxiolytic-like action of the 3-PPP enantiomers in the Vogel conflict paradigm.** ***Psychopharmacology,*** 92(3), 371–375.
Investigated the effect of the (+)- and (–)-enantiomers of 3-PPP (conventional and atypical dopamine [DA]-receptor active agents, respectively) on a licking-conflict animal model of anxiety by J. R. Vogel et al (1971), using male Sprague-Dawley rats. Findings indicate that low doses of the 3-PPP enantiomers, in particular (–)-3-PPP, attenuated anxiety-elicited increases in neurotransmission in certain mesocortical/limbic DA pathways that are consistent with the preferentially "limbic" net antidopaminergic profile of action of (–)-

3-PPP. Results support an active role for DA in conditions associated with anxiety and reinforce the view that novel putative anxiolytics might be found among selective DA-modulating agents such as (-)-3-PPP.

1227. Hliňák, Z. & Krejčí, I. (1990). **Long-term behavioural consequences of sodium nitrite hypoxia: An animal model.** 31st Annual Psychopharmacology Meeting (1989, Jeseník, Czechoslovakia). ***Activitas Nervosa Superior,*** 32(1), 48–49.
Two experiments examined spontaneous behavior and radial maze performance in male rats to determine alterations developing after sodium nitrite-induced hypoxia. Long-term behavioral consequences were found in addition to a short-term inhibition effect. The model used is proposed for evaluation of therapeutic agents in the treatment of consequences of cerebral impairment induced by hypoxia.

1228. Hock, Franz J. (1987). **Drug influences on learning and memory in aged animals and humans.** ***Neuropsychobiology,*** 17(3), 145–160.
Reviews research investigating the effects of neurotransmitter systems and specific drugs on cognitive functions of aged animals and humans. Animal models of age-related changes in learning and memory are discussed. When studying age-related changes in human memory, S characteristics, such as health, must be considered in relation to performance of memory tasks. Pertinent pharmacological topics include neuropeptides, acetylcholine, catecholamines, compounds influencing the central nervous system (CNS), stimulants, and nootropics. It is suggested that while there have been many different pharmacologic and behavioral approaches used to treat and test cognitive deficits in aged animals and geriatric or demented patients, there is still the question of the validity of animal models of human disorders. Attempts are made to show parallels and similarities between animal and human disorders.

1229. Holtzman, Stephen G. (1985). **Drug discrimination studies.** Committee on Problems of Drug Dependence Symposium: Mixed agonist–antagonist analgesics (1983, Innisbrook, Florida). ***Drug & Alcohol Dependence,*** 14(3–4), 263–282.
Reviews studies in which animal models were developed to study components of action of opioids that bear on their subjective effects in humans. Findings show that opioid agonists and agonist–antagonists comprise a heterogeneous body of compounds that can be partitioned into at least 3 groups on the basis of their discriminative stimulus properties in several animal species; (1) stimulus effects similar to those of morphine or fentanyl and blocked completely by low doses of antagonists, such as naloxone and naltrexone; (2) stimulus effects similar to those of ethylketocyclazocine or nalorphine and blocked by higher doses of antagonists; and (3) stimulus effects similar to those of *N*-allylnormetazocine or phencyclidine and not blocked by antagonists. This diversity of stimulus properties is consistent with other evidence that multiple populations of receptors mediate the actions of opioids. In humans, drugs in Group 1 produce subjective effects that are entirely morphinelike and highly reinforcing whereas drugs in Groups 2 and 3 produce dysphoric and psychotomimetic subjective effects.

1230. Holtzman, Stephen G. (1985). **Discriminative stimulus effects of morphine withdrawal in the dependent rat: Suppression by opiate and nonopiate drugs.** ***Journal of Pharmacology & Experimental Therapeutics,*** 233(1), 80–86.
Tested morphine and 7 other opiate agonists varying in potency and chemical structure for their ability to block the stimulus control of behavior and loss of body weight induced by naltrexone in 20 male morphine-dependent Sprague-Dawley rats trained to discriminate between saline and 0.1 mg/kg of the antagonist in a 20-trial avoidance paradigm. Subcutaneous administration of the 7 opiates—which included morphine sulfate, levomethorphan HBr, and oxymorphone HCl—blocked dose dependently the discriminative effects of naltrexone and loss of body weight. Effects were stereoselective for levorotatory isomers. Loperamide blocked loss of body weight but not discriminative effects, suggesting that discriminative effects are mediated centrally. Clonidine (0.01–1.0 mg/kg) blocked discriminative effects of naltrexone partially and weight loss completely. The blockade by morphine (30 mg/kg) of naltrexone-induced discriminative effects and weight loss was surmounted by increasing the dose of naltrexone, whereas the blockade by clonidine (0.1 mg/kg) was not. Thus, blockade by opiates of effects of naltrexone appears to be due to a competitive interaction at the mu opioid receptor; clonidine has a different mechanism of action. This discrimination paradigm may afford a specific animal model for studying fundamental processes underlying physical dependence on opiates and for evaluating novel pharmacologic approaches for treating opiate withdrawal in humans.

1231. Holtzman, Stephen G.; Shannon, Harlan E. & Shaefer, Gerald J. (1976). **Discriminative properties of narcotic antagonists.** ***Psychopharmacology Communications,*** 2(4), 315–318.
Hypothesized that the component of drug action responsible for the subjective effects of narcotic antagonists in humans is related to the component of action responsible for discriminative effects in animals. This hypothesis was tested by evaluating the effects of 10 narcotic antagonists in discrimination paradigms with rats and squirrel monkeys. Of the 10 antagonists tested in rats, only the 4 with prominent morphine-like activity (butorphanol, nalmexone, pentazocine, and profadol) substituted for the training dose of morphine (3.0 mg/kg). Cyclazocine and agonist-antagonists with prominent cyclazocine-like activity failed to substitute for morphine. The extent of stimulus generalization to the dysphoric narcotic antagonists was a function of the morphine training dose. Assessment of the cyclazocine-like discriminative properties of narcotic antagonists in monkeys yielded data complementary with those obtained in rats. On the other hand, many narcotic antagonists that completely or partially substituted for 3.0 mg/kg of morphine in rats showed less activity in monkeys also trained to discriminate between morphine and saline. Results show that, in general, there is a good correspondence between patterns of discriminative effects of narcotic antagonists relative to the training drugs in infrahuman species and the subjective effects that these drugs produce in humans. Drug discrimination paradigms may be useful as animal models for preclinical evaluations in humans.

1232. Hrdina, Pavel D.; von Kulmiz, Paul & Stretch, Roger. (1979). **Pharmacological modification of experimental depression in infant macaques.** ***Psychopharmacology,*** 64(1), 89–93.
Provides evidence that mother–infant separation in macaques is a useful experimental model of depression. At the age of 6–8 mo, 7 macaque infants underwent 2 consecutive separations from their mothers lasting 21 and 15 days. The frequency and duration of a set of individual and social behaviors were recorded throughout baseline, separation, and reunion conditions. In response to maternal separation, the infants showed marked increases in frequency of behaviors reflecting distress, self-directed activity, or anxiety (e.g., vocalization, locomotion, body play). Both individual and social play behaviors were markedly suppressed in separated infants. During the 2nd separation, one group of Ss was given, in a double-blind fashion, daily doses of 5 mg/kg of desipramine (DMI) im. DMI markedly diminished most of the behavioral alterations induced by separation. In particular, increases in distress and self-directed behaviors as well as the suppression of play activities were prevented or antagonized. Plasma levels of DMI after 5 days of administration were in the range of 50–150 ng/ml.

1233. Huberman, Harris S.; Eison, Michael S.; Bryan, Karen S. & Ellison, Gaylord. (1977). **A slow-release silicone pellet for chronic amphetamine administration.** ***European Journal of Pharmacology,*** 45(3), 237–242.
Describes (a) a slow-release amphetamine pellet consisting of a silicone capsule containing base in polyethylene glycol and (b) its effects in male Long-Evans rats. Because sc amphetamine pellets closely approximate the regimen of drug administration used in human studies of amphetamine psychosis, they promise to be useful tools in the development of animal models of schizophrenia. Studying the long-term mechanisms by which animals respond to the challenge of sustained amphetamine treatment may clarify the physiological mechanisms underlying the amphetamine model psychosis.

1234. Hubner, Carol B.; Bain, George T. & Kornetsky, Conan. (1987). **The combined effects of morphine and d-amphetamine on the threshold for brain stimulation reward.** ***Pharmacology, Biochemistry & Behavior,*** 28(2), 311–315.
Analyzed the effect of morphine (M) and dextroamphetamine (A) co-administration on the rate-independent threshold for rewarding intracranial electrical stimulation of the medial forebrain bundle–lateral hypothalamic or ventral tegmental area in male albino rats. In Exp I ($n = 5$), M and A caused a dose-related lowering of reward threshold. In Exp II, with 3 Ss from Exp I, combined M and A lowered the reward threshold more than separate doses. Results are consistent with animal models of drug-induced euphoria and with users' reports of increase euphoria when amphetamine is combined with opiate drugs.

1235. Hull, Elaine M.; Franz, Jonathan R.; Snyder, Abigail M. & Nishita, J. Ken. (1980). **Perinatal progesterone and learning, social and reproductive behavior in rats.** ***Physiology & Behavior,*** 24(2), 251–256.
Administered progesterone (0–12 mg/kg) to 298 Long-Evans rats perinatally via either maternal Silastic implants (Exp I) or daily maternal injections (Exp II). Ss were tested at 14 days of age on an active avoidance task, and in adulthood on a Lashley III maze task, active and passive avoidance tasks, and open field activity (Exp I) and on social and reproductive behavior measures (Exp II). Adult males' performance on the Lashley III task was significantly impaired by progesterone treatment in Exp I, as were male copulatory and aggressive behaviors in Exp II. Perinatal progesterone as administered in these experiments does not result in an animal model for the reported enhancement of human performance consequent to prenatal progesterone treatment. It is, however, consistent with an interpretation of demasculinization of behavior patterns.

1236. Hurlbut, B. J.; Lubar, Joel F.; Switzer, R.; Dougherty, J. et al. (1987). **Basal forebrain infusion of HC-3 in rats: Maze learning deficits and neuropathology.** ***Physiology & Behavior,*** 39(3), 381–393.
10 adult male Sprague-Dawley rats were infused with hemicholinium (HC-3) over a 14-day period through bilateral, chronically implanted cannulas in the nucleus basalis magnocellularis (nbm), and 10 matched controls were infused with saline. HC-3 Ss receiving implants demonstrated a significant deficit in maze-learning ability compared with individual and group performances before receiving the implants and compared with control Ss. In HC-3 Ss, cholinergic cell bodies were destroyed with concurrent degeneration of terminal fields in the cortex, whereas the cholinergic neurotransmitter system appeared unharmed in controls. Stains for neuritic plaques and neurofibrillary damage were negative in both groups. The memory deficit in experimental Ss supported by the demonstrated destruction of nbm cholinergic neurons suggests that HC-3 may be useful in the development of an animal model for Alzheimer's disease.

1237. Inomata, Kenichirou; Nasu, Fumio & Tanaka, Harumi. (1987). **Decreased density of synaptic formation in the frontal cortex of neonatal rats exposed to ethanol *in utero.*** ***International Journal of Developmental Neuroscience,*** 5(5–6), 455–460.
Examination of the developing synaptic junctions in the rat frontal cortex in cases of fetal alcohol syndrome found that on Day 21 of gestation, the ultrastructural synaptic junction revealed no obvious differences between the ethanol-exposed and control rats. However, the number of synapses in ethanol-exposed rats was one-third that of controls. The possible relationship between synaptic density in the frontal cortex and mental development is considered.

1238. Introini-Collison, Ines B. & McGaugh, James L. (1988). **Modulation of memory by post-training epinephrine: Involvement of cholinergic mechanisms.** ***Psychopharmacology,*** 94(3), 379–385.
Examined the interaction of peripheral epinephrine (EPI) and cholinergic drugs in memory modulation in mice that were trained on an inhibitory avoidance response or a **Y**-maze discrimination response. Findings indicate that, when administered in low doses, EPI interacted with oxotremorine and physostigmine to enhance memory. Cholinergic drugs blocked both the memory-enhancing and the memory-impairing effects of EPI.

1239. Iorio, Louis C.; Eisenstein, Norman; Brody, Philip E. & Barnett, Allen. (1983). **Effects of selected drugs on spontaneously occurring abnormal behavior in beagles.** ***Pharmacology, Biochemistry & Behavior,*** 18(3), 379–382.
A subpopulation of "depressed" beagles with abnormal behavioral patterns (e.g., reduced approach behavior, lack of barking) was identified, isolated, and tested for responsivity to selected classes of psychoactive drugs. The abnormal be-

havior was ameliorated by anxiolytics and antidepressants but not by antipsychotics, and antihistaminic (chlorpheniramine), an alpha-adrenergic blocker (phentolamine), a beta-adrenergic blocker (propranolol), or an anticholinergic (scopolamine). Improvement occurred after a single dose of the anxiolytic drugs but did not occur until 10–18 days after daily dosing with standard tricylic antidepressants and the MAO inhibitor isocarboxazid. This delayed onset in beagles resembles that seen in humans given these drugs. Results suggest that the abnormal behaviors of the beagles are related to anxiety and are in part depressive in nature. This colony provides an animal model of abnormal behavior that allows evaluation of the drugs' effects and estimates of the onset of action.

1240. Irle, Eva & Markowitsch, Hans J. (1983). **Widespread neuroanatomical damage and learning deficits following chronic alcohol consumption or vitamin-B_1 (thiamine) deficiency in rats.** ***Behavioural Brain Research,*** 9(3), 277–294.
Investigated consequences of long-term consumption of alcohol (20 mo) and of pyrithiamine-induced blockade of vitamin-B_1-uptake on the shape of individual brain structures and on the acquisition of 2 learning tasks in 3 groups of 26 male Sprague-Dawley rats (alcohol group [AL], thiamine-deficient group [TH], and control group). Groups AL and TH were allowed an 8-wk or 3-wk recovery period, respectively, with normal food and water available before behavioral testing started. This consisted of training for an active 2-way avoidance task and a spatial reversal task. Ss of both experimental groups were, compared to controls, significantly impaired in acquiring the avoidance task and in acquiring the original discrimination of the spatial reversal task. No differences were found among the 2 experimental groups. Histological and microscopical examinations of the brains of Ss with a history of thiamine deficiency or of chronic alcohol consumption revealed a variety of severely affected brain areas. In both groups, hippocampal and cerebellar damage was prominent. Furthermore, the mamillary nuclei, certain brain-stem regions situated around the ventricles, and a few cortical areas contained loss or damage of neurons. It is concluded that the anatomical changes can be related to those seen in chronic alcoholics and that, consequently, animal models can be established to investigate in detail the multiple interactions of alcohol consumption, thiamine deficiency, brain damage, and behavioral deterioration.

1241. Jacobs, Barry L.; Trulson, Michael E. & Stern, Warren C. (1976). **An animal behavior model for studying the actions of LSD and related hallucinogens.** ***Science,*** 1944(266), 741–743.
Adult female cats given ip injections of LSD (10, 25, 50, or 100 μg/kg) exhibited a group of behaviors that appear to be specific to hallucinogenic drugs. Two of these behaviors, limb flick and abortive grooming, have an extremely low frequency of occurrence in normal cats, but often dominated the behavior of LSD-treated cats. The frequency of occurrence of this group of behaviors was related to the dose of LSD. The behavioral changes were long-lasting following a single injection of LSD and showed tolerance following the repeated LSD administration. They were not elicited by a variety of control drugs, but were elicited by other indole nucleus hallucinogens (e.g., psilocybin). Because the behavioral effects are specific, reliable, easy to score, and quantifiable, they represent an animal model that can be used in studies of the effects of LSD and related hallucinogens.

1242. Jarrard, Leonard E.; Levy, Aharon; Meyerhoff, James L. & Kant, G. Jean. (1985). **Intracerebral injections of AF64A: An animal model of Alzheimer's disease?** ***Annals of the New York Academy of Sciences,*** 444, 520–522.
Two experiments investigated the effects of ethylcholine aziridinium ion (AF64A) on cholinergic systems of the brain in rats. Exp I tested the effects of intracerebroventricular injections of AF64A on motivation, memory, and neurotransmitters. Exp II tested the effects of injecting AF64A into a predominantly noncholinergic brain area. Results of the 2 experiments were similar, suggesting that the influence of AF64A on memory is due to a nonspecific lesion effect rather than to a specific effect on cholinergic systems.

1243. Jastreboff, Pawel J.; Brennan, James F.; Coleman, John K. & Sasaki, Clarence T. (1988). **Phantom auditory sensation in rats: An animal model for tinnitus.** ***Behavioral Neuroscience,*** 102(6), 811–822.
To measure tinnitus induced by sodium salicylate injections, 84 rats were used in a conditioned suppression paradigm. In Exp 1, Ss were trained with a conditioned stimulus/stimuli (CS) consisting of the offset of a continuous background noise. One group began salicylate injections before Pavlovian training, a 2nd group started injections after training, and a control group received daily saline injections. Resistance to extinction was profound when injections started before training but minimal when initiated after training, suggesting that salicylate-induced effects acquired differential conditioned value. In Exp 2, salicylate treatments were mimicked by substituting a 7 kHz tone in place of respective injections, resulting in effects equivalent to salicylate-induced behavior. A 3rd experiment included a 3 kHz CS, and again replicated the salicylate findings. In Exp 4, we decreased the motivational level, and the sequential relation between salicylate-induced effects and suppression training was retained. Findings support the demonstration of phantom auditory sensations in animals.

1244. Javitt, Daniel C. (1987). **Negative schizophrenic symptomatology and the PCP (phencyclidine) model of schizophrenia.** ***Hillside Journal of Clinical Psychiatry,*** 9(1), 12–35.
Maintains that PCP, which has been shown to induce a schizophreniform psychosis consisting of both productive and deficit symptomatology, may provide a useful model of schizophrenia. The literature concerning the PCP model of schizophrenia is reviewed, and a study of male Long-Evans hooded rats is reported that confirms the ability of PCP to modulate mesocortical dopaminergic activity. Since PCP appears to mediate its central nervous system (CNS) effects via a subclass of glutamate receptors, a possible glutamate theory of schizophrenia is proposed.

1245. Johanson, C. E. & Schuster, C. R. (1981). **Animal models of drug self-administration.** ***Advances in Substance Abuse,*** 2, 219–297.
Discusses the usefulness of the animal model that demonstrates that psychotropic drugs can serve as reinforcers maintaining operant behavior in laboratory animals. One variable affecting self-administration in these experiments has been the type of drug. Psychomotor stimulants, narcotic agonists and antagonists, depressants, hallucinogens, and anesthetics have been used in animal studies. Administration schedules, magnitude of reinforcement, and food deprivation have also been found to affect self-administration. Behavioral and pharmacological variables that decrease drug-maintained responding in animals have included punishment,

reinforcement, and pharmacological blockade. Results of these drug experiments with animals indicate that the principal drugs of abuse are those that can interact with the normal brain mechanisms developed through evolution to mediate biologically essential behaviors directed toward food, water, and sex. The psychopathology of many drug abusers may thus be a form of toxicity produced either by the drug itself or by the life-style of the user. Also, the psychopathology may reinforce properties of certain drugs.

1246. Johanson, Chris-Ellyn. (1990). **Behavioral pharmacology, drug abuse, and the future.** ***Behavioural Pharmacology,*** 1(4), 385–393.
Emphasizes the importance of behavioral pharmacology to understanding the behavioral and pharmacological mechanisms contributing to the maintenance of drug-seeking behavior in humans and discusses the contributions of laboratory-based drug self-administration studies in nonhumans that measure the reinforcing properties of drugs. Complex behavioral studies using drug self-administration methods need to incorporate mechanisms for improving the quantitative precision of measures of reinforcing efficacy and developing a strategy for efficiently investigating the variables that could influence behavior.

1247. Johnson, Kenneth M.; Gordon, Marc B. & Ziegler, Michael G. (1978). **Phencyclidine: Effects on motor activity and brain biogenic amines in the guinea pig.** ***Pharmacology, Biochemistry & Behavior,*** 9(4), 563–565.
Phencyclidine (PCP) possesses anesthetic, psychomotor stimulant, and sedative-hypnotic properties in animals; in humans, its psychotomimetic properties cause bizarre emergence reactions. The present study, using male albino guinea pigs, attempted to determine whether future behavioral or neuropharmacological studies of PCP utilizing the guinea pig could be better correlated with studies in higher species than studies using the rat. Findings show that the guinea pig would afford no such advantage, because PCP produced behavioral effects in the guinea pig similar to those observed in rats and mice.

1248. Johnston, Amanda L. & File, Sandra E. (1988). **Profiles of the antipanic compounds, triazolobenzodiazepines and phenelzine, in two animal tests of anxiety.** ***Psychiatry Research,*** 25(1), 81–90.
Employed 2 animal tests of anxiety (the elevated plus-maze and the social interaction test) to investigate the effects of several antipanic agents. In the elevated plus-maze, the triazolobenzodiazepines (TBDs) adinazolam and alprazolam, tested after 5 days of pretreatment, demonstrated significant anxiolytic effects in male rats, while phenelzine, after 21 days of pretreatment, demonstrated nonsignificant anxiolytic effects. In the social interaction test, the TBDs generally did not produce an anxiolytic profile, and phenelzine even revealed significant anxiogenic activity. The antipanic agents distinguished between the 2 tests of anxiety. Results suggest a distinction between generalized anxiety and panic disorder.

1249. Johnston, Amanda L. & File, Sandra E. (1988). **Can animal tests of anxiety detect panic-promoting agents?** ***Human Psychopharmacology Clinical & Experimental,*** 3(2), 149–152.
Evaluated sodium lactate, isoproterenol, and yohimbine in the elevated "plus" maze and the social interaction test in male rats. None of the drugs changed the time spent in social interaction or evoked paniclike behavior. The highest dose of isoproterenol (0.6 mg/kg) reduced the time spent on the open arms; yohimbine reduced both the percentage of entries and the time spent on the open arms, indicating anxiogenic activity. The lack of a strong anxiogenic profile with the propanic compounds suggested that anxiety and panic may be biologically distinct.

1250. Kan, Jean-Paul; Steinberg, Regis; Mouget-Goniot, Claire; Worms, Paul et al. (1987). **SR 95191, a selective inhibitor of type A monoamine oxidase with dopaminergic properties: II. Biochemical characterization of monoamine oxidase inhibition.** ***Journal of Pharmacology & Experimental Therapeutics,*** 240(1), 251–258.
Describes the interaction of SR 95191 with monoamine oxidase (MAO)-A and MAO-B in the rat brain, liver, and duodenum. The effects of SR 95191 on the endogenous levels of 5-hydroxytryptamine (5-HT), norepinephrine, dopamine, and their metabolites were also examined. The activity of SR 95191 was compared to that of various MAO inhibitors, including clorgyline, harmaline, levodeprenyl, moclobemide, and cimoxatone. Results show that SR 95191 is a selective and short-acting type A MAO. Findings are in agreement with those in behavioral models of SR 95191—that it is active in most animal models of depression, with a profile of activity comparable to that of moclobemide.

1251. Kaplan, Joel N.; Hennessy, Michael B. & Howd, Robert A. (1982). **Oral ethanol intake and levels of blood alcohol in the squirrel monkey.** ***Pharmacology, Biochemistry & Behavior,*** 17(1), 111–117.
Oral alcohol ingestion and blood alcohol levels were examined in 3 studies with 81 adult female squirrel monkeys to assess the feasibility of using this primate as a model for fetal alcohol effects. Results show that nonpregnant Ss drank ethanol at concentrations of 5–10% and that the amount of ethanol consumed was related to the concentration and length of time ethanol was available. When given access to a 5% ethanol solution, pregnant animals drank quantities that varied between individuals and subtypes, with maximum blood levels ranging from 1 to 196 mg%, measured up to 6 hrs after presentation.

1252. Karevina, T. G. & Shevchuk, I. M. (1988). **Effect of biogenic amines receptor blockers on experimental ulcerogenesis.** ***Farmakologiya i Toksikologiya,*** 51(4), 56–60.
Studied the effects of the serotonin antagonist cyproheptadine, cholinolytic atropine, the H2-receptor blocker cimetidine, and dopaminergic H2-receptor blocker metoclopramide on social stress-induced and exogenous serotonin-induced gastric ulcerogenesis. Animal subjects: 151 mongrel white rats. The number and type of lesions in Ss receiving the different substances were recorded. Drugs used: Cyproheptadine, cholinolytic atropine, cimetidine, and metoclopramide were administered ip in doses of 5, 2.5, 15, and 0.3 mg/kg, respectively. (English abstract).

1253. Kasarskis, Edward J. et al. (1981). **Abnormal maturation of cerebral cortex and behavioral deficit in adult rats after neonatal administration of antibodies to ganglioside.** ***Developmental Brain Research,*** 1(1), 25–35.

48 5-day-old male Sprague-Dawley rats received a single injection (50 μl) of antiserum to ganglioside into the cisterna magna and were compared to control Ss injected with the antiserum which had been absorbed with pure G_{M1} ganglioside to remove the specific antibodies. Both groups showed normal rates of body growth. However, 80-day-old Ss who received antiganglioside serum had impaired performance when tested on a complex learning task (DRL). The results provide an animal model in which an immunologically mediated disturbance of cortical development is associated with chronic behavioral impairment.

1254. Kastin, A. J.; Zadina, J. E.; Ehrensing, R. H. & Schwartzenburg, D. (1987). **MIF-1 and Tyr-MIF-1 can act differently from amitriptyline in an animal model of depression.** ***New Trends in Experimental & Clinical Psychiatry,*** 3(1), 49–58.
The potentiation of thyrotropin-releasing hormone (TRH)-induced hyperthermia has been used as a test for screening antidepressants that activate alpha-adrenergic systems. MIF-1 (Pro-Leu-Gly-NH_2) and Tyr-MIF-1 (Tyr-Pro-Leu-Gly-NH_2), peptides with antidepressant activity, were tested in this model using male mice. The tricyclic antidepressant amitriptyline was significantly more active than either peptide in potentiating the TRH-induced hyperthermia. Results demonstrate that the actions of amitriptyline can differ from those of MIF-1 and Tyr-MIF-1.

1255. Kastin, Abba J. et al. (1985). **Tyr-MIF-1, identified in brain tissue, and its analogs are active in two models of antinociception.** ***Pharmacology, Biochemistry & Behavior,*** 23(6), 1045–1049.
Five studies with male albino Swiss-Webster mice tested the anti-opiate activities of tyr-pro-leu-gly-NH/2 (Tyr-MIF-1) and some of its representative analogs in 2 animal models of antinociception. Doses of the tetrapeptides as low as 0.001 mg/kg injected peripherally blocked the analgesic effects of morphine in both the tailflick test of mild thermal pain induced by heat and the scratching test of mild chemical pain induced by hypertonic saline. These tetrapeptides showed cross-reactivity in the radio-immunoassay used to identify the presence of Tyr-MIF-1 in brain extracts and in the brain membrane binding assay. Only Tyr-MIF-1, however, eluted at the position of the immunoreactive peak after gel filtration chromatography and high performance liquid chromatography. Results support the concept that peptides with anti-opiate activity can exist in the brain.

1256. Kastin, Abba J. et al. (1978). **Enkephalin and other peptides reduce passiveness.** ***Pharmacology, Biochemistry & Behavior,*** 9(4), 515–519.
Enkephalin and other brain peptides have been shown previously to be active in the dopa potentiation test that may be considered an animal model of mental depression. A recently described model of passive immobility during swimming, also sensitive to tricyclic antidepressants, was therefore used to study a large number of naturally occurring peptides and some of their analogs. Results of an experiment with approximately 1,000 male albino rats indicate that several enkephalins with no opiate activity after peripheral injection reduced the immobility and thus increased the activity of swimming rats. α-MSH, but not its 4–10 core or a 4–9 analog, also caused significantly more swimming than did the diluent control. As previous studies have found, a smaller dose of MIF-I was more effective than larger doses. Results confirm our concept of the CNS actions of brain peptides and support the suggestion that some of them, like the enkephalins, might be useful after peripheral administration in mental depression or other CNS disorders.

1257. Kastin, Abba J.; Honour, Lynda C. & Coy, David H. (1981). **Effects of MIF-1 and three related peptides on reserpine-induced hypothermia in mice.** ***Pharmacology, Biochemistry & Behavior,*** 15(6), 983–985.
Pro-Leu-Gly-NH_2 (MIF-1), Try-Pro-Leu-Gly-NH_2 (Tyr-MIF-1), pGlu-Leu-Gly-NH_2, and cyclo-Leu-Gly were tested at 60, 120, and 180 min at 3 doses (0.1, 1.0, 10.0 mg/kg, ip). 435 male albino Sprague-Dawley mice pretreated 18 hrs earlier with reserpine (2.5 mg/kg, ip). Reversal of hypothermia was significant only with MIF-1 at 120 and 180 min but not with the 2 analogs (pGlu-Leu-Gly-NH_2 and cyclo-Leu-Gly). Results demonstrate that the relative potency of related peptides can differ depending on the experimental situation. Implications of the reversal effects of reserpine for animal models of depression and the clinical treatment of symptoms of mental depression are discussed.

1258. Kastin, Abba J.; Schwartzenburg, Debra; Tsui, Lori; Miller, Lawrence G. et al. (1989). **Differential effects of Tyr-MIF-1 and naloxone in two animal models involving benzodiazepine.** ***Brain Research Bulletin,*** 23(6), 443–446.
Examined the effects of the endogenous brain peptide Tyr-MIF-1 (Tyr-Pro-Leu-Gly-NH_2) in 2 models in which the antiopiate naloxone has been reported to decrease the activity of benzodiazepines (M. Sansone and J. Vetulani and P. Soubrie et al; 1988 and 1980, respectively). Tyr-MIF-1 can act as an antiopiate and can also increase binding and function at the $GABA_A$/benzodiazepine receptor complex. Unlike naloxone, Tyr-MIF-1 and MIF-1 neither prevented chlordiazepoxide-induced locomotor hyperactivity in male albino mice on a tilting floor nor suppressed chlordiazepoxide-induced eating in male albino rats. Tyr-MIF-1 did not act as an antiopiate or alter the effects of a benzodiazepine, indicating a selectivity in the actions of Tyr-MIF-1.

1259. Katz, J. L. (1990). **Models of relative reinforcing efficacy of drugs and their predictive utility.** ***Behavioural Pharmacology,*** 1(4), 283–301.
Studies of drugs as reinforcers in animals have indicated a close agreement between the drugs that function as reinforcers in animals and those that are abused by humans. This agreement has prompted the use of results of studies in animals as a model to predict the abuse of newly synthesized drugs. While this model has widely accepted nominal predictive utility, attempts have been made to extend its predictive capability to assess the relative reinforcing efficacy of drugs, implying that the reinforcing effects of drugs can be compared on ordinal, interval, or ratio scales. Nominal scaling may represent the best information that can be provided for predicting abuse of drugs.

1260. Katz, R. J. (1984). **Effects of zometapine, a structurally novel antidepressant, in an animal model of depression.** ***Pharmacology, Biochemistry & Behavior,*** 21(4), 487–490.
Zometapine (ZM), a pyrazolodiazepine, bears a close structural resemblance to benzodiazepines. It possesses an unusual pharmacological profile and is active in some, but not all, tests of antidepressant activity. In clinical tests it appears to be an extremely effective pharmacotherapeutic agent and may represent a new class of antidepressant. Because the

preclinical profile of ZM is unusual, its effects were examined in a behavioral test of antidepressant potential. The 1st factor studied was the presence or absence of chronic stress, and the 2nd was presence vs absence of acute stress. In addition, vehicle was compared to 2 doses of ZM. Ss were 72 male Sprague-Dawley rats that were subjected to 3 wks of a chronic stress procedure that consisted of unpredictable stress administration. Following 3 wks of treatment, the drug selectively reversed a behavioral depression following chronic stress. Drug-induced reversal was seen only in Ss activated by acute noise exposure and was dose related. Reversal was confirmed by a 2nd measure, defecation, and partially confirmed by the normalization of an elevated basal corticosterone response.

1261. Katz, Richard J. (1982). **Animal model of depression: Pharmacological sensitivity of a hedonic deficit.** ***Pharmacology, Biochemistry & Behavior,*** 16(6), 965–968.
A reduction in sucrose and saccharin consumption following chronic stress is reported for the rat. This deficit may be related to consummatory deficits seen in endogenous depression. To further examine this state pharmacologically, stressed male Sprague-Dawley rats were treated with the antidepressant imipramine. Despite a general absence of appetitive effects (or in some cases mild anorexia), imipramine significantly restored saccharine consumption in a variety of tests. The pharmacological similarity of the deficit to the changes accompanying affective disorders further supports the potential applicability of the chronic stress model.

1262. Katz, Richard J. & Baldrighi, G. (1982). **A further parametric study of imipramine in an animal model of depression.** ***Pharmacology, Biochemistry & Behavior,*** 16(6), 969–972.
Proposes that chronic stress may produce motivational, behavioral, and neuroendocrine symptoms in rats resembling endogenous depression in humans. The chronic stress model has proved responsive to chronic treatment by antidepressant drugs. Two issues concerning this effect remain unresolved: the requirement of drug chronicity, and treatment outcome to different drug doses. The present experiment with 144 male Sprague-Dawley rats examined both issues in a factorial design in which vehicle and 2 doses of the tricyclic antidepressant imipramine were varied across 2 treatment periods: acute (1 hr) and chronic (3 wks). Both factors were found to interact significantly with treatment outcome, suggesting that chronic treatment is necessary for recovery and that this outcome depends on drug level.

1263. Katz, Richard J. & Sibel, Michael. (1982). **Further analysis of the specificity of a novel animal model of depression: Effects of an antihistaminic, antipsychotic and anxiolytic compound.** ***Pharmacology, Biochemistry & Behavior,*** 16(6), 979–982.
Previously reported (R. J. Katz, 1981; Katz and S. Hersch, 1981; K. A. Roth and Katz, 1981) that chronic stress-elicited reductions in selected forms of open field activity resembled endogenomorphic depression on behavioral, motivational, neuroendocrine, and neuropharmacological grounds. In particular, the loss of acute stress-elicited activity proved to be exclusively reversible by antidepressant treatments. Insofar as clinically ineffective compounds were tested, the deficit proved refractory to treatment, suggesting that the model reflected just those processes that were disrupted in depression. A number of ineffective compounds are known to yield false positives on other related tests, but have yet to be examined in the present model. Three such compounds, an antihistamine (tripelennamine), a neuroleptic (haloperidol), and an anxiolytic (oxazepam), were examined for their behavioral and neuroendocrine effects in 144 male Sprague-Dawley rats. Although other stress-related phenomena were replicated, none of the above compounds was effective in restoring the activation deficit or in eliminating the endocrine abnormality. This suggests that the depression model is relatively selective pharmacologically and not critically dependent on receptor blocking properties of the above drugs.

1264. Katz, Richard J. & Sibel, Michael. (1982). **Animal model of depression: Tests of three structurally and pharmacologically novel antidepressant compounds.** ***Pharmacology, Biochemistry & Behavior,*** 16(6), 973–977.
Previous studies have identified behavioral and neuroendocrine abnormalities in chronically stressed rats that resemble some of the more prominent features of clinical depression. These abnormalities have proved responsive to pharmacotherapy by standard antidepressant drugs and related somatic treatments. Several structurally and pharmacologically atypical compounds, resembling neither standard agents nor each other, have recently been identified as clinically effective antidepressants. These drugs do not show typical preclinical response profiles in other drug screening tests and, therefore, represent critical instances for evaluating the selectivity of the chronic stress model. Three drugs were tested: iprindole, bupropion, and mianserin; a tricyclic indole, propriophenone, and tetracyclic compound, respectively. Four behavioral measures and a measure of circulating corticosterone were obtained for 144 male Sprague-Dawley rats. All compounds proved capable of reversing chronic stress-induced behavioral deficits, and all but one reversed the attendant basal hypersecretion of corticosterone. Findings argue that the chronic stress model provides an accurate and selective assessment of the therapeutic potential of both standard and structurally novel compounds.

1265. Keane, Bernadette & Leonard, B. E. (1989). **Rodent models of alcoholism: A review.** ***Alcohol & Alcoholism,*** 24(4), 299–309.
Reviews the literature on rodent (mainly rat) models of alcoholism, focusing on the establishment of a cheap, reliable model in easily available strains of rodents to assess the chronic effects of alcohol (AL) on behavioral and neurochemical processes. A description is included of the models in which rodents are forced to consume increasing quantities of AL as part of a nutritionally enriched milk diet that is easy to administer, reliable, and inexpensive.

1266. Keck, Paul E.; Seeler, David C.; Pope, Harrison G. & McElroy, Susan L. (1990). **Porcine stress syndrome: An animal model for the neuroleptic malignant syndrome?** ***Biological Psychiatry,*** 28(1), 58–62.
Investigated porcine stress syndrome as a possible animal model for neuroleptic malignant syndrome in 2 ways. Haloperidol and lithium carbonate were administered, alone and in combination, to susceptible and resistant swine. An attempt to prevent the syndrome by pretreating animals with bromocriptine was also made. Porcine stress syndrome was induced in 2 of 3 susceptible and 1 of 3 resistant swine by combined treatment with lithium and haloperidol but was not triggered by treatment with lithium or haloperidol alone. Pretreatment with bromocriptine conferred no protection against the syndrome.

1267. Kelley, Ann E.; Lang, Christopher G. & Gauthier, Andrea M. (1988). **Induction or oral stereotypy following amphetamine microinjection into a discrete subregion of the striatum. *Psychopharmacology,*** 95(4), 556–559.
Contends that amphetamine and other psychostimulant drugs induce perseverative motor behavior in rodents, such as compulsive sniffing, licking, and biting. It is known that this behavior, termed stereotypy, is a consequence of dopaminergic stimulation of the striatum, but that the precise localization of the site of activation is unclear. It is reported that microinjection of amphetamine into a circumscribed subregion of the striatum specifically produces intense oral stereotypy. It is proposed that this region, which corresponds to a small area within the ventrolateral striatum, contains motor circuitry critical to oral behavior, including feeding. It is suggested that the behavior elicited by amphetamine-induced stimulation of this area may represent a simple animal model in which to study certain orofacial dyskinesias.

1268. Khonicheva, N. M. & Danchev, N. (1985). **Manifestation of passivity in rats: Effect of tranquilizing and antidepressant drugs. *Zhurnal Vysshei Nervnoi Deyatel'nosti,*** 35(2), 339–347.
Investigated the notion that passivity as a feature of depressionlike behavior in rats develops as a result of unavoidable painful stimulation by using antidepressant drugs to eliminate the passivity manifested in an almost complete absence of motor searching reactions in an open field and a maze. However, tranquilizing drugs rather than antidepressants induced to a greater extent the presupposed effect. Findings suggest that this type of passivity corresponds more to neurotic behavior than to a special depressionlike behavioral manifestation. In a 2nd series of experiments, the action of multiple injections of antidepressants on similar manifestations of passivity as well as on alimentary instrumental conditioned responses in rats with initially expressed passive character of behavior was examined. Neither the presupposed increase of motor searching reaction, significant changes in the rate of instrumental conditioning, nor elimination of its failures of a "refuse" type were observed.

1269. Kilfoil, T.; Michel, A.; Montgomery, D. & Whiting, R. L. (1989). **Effects of anxiolytic and anxiogenic drugs on exploratory activity in a simple model of anxiety in mice. *Neuropharmacology,*** 28(9), 901–905.
Examined the effects of anxiolytic and anxiogenic drugs (chlordiazepoxide, pentylenetetrazole, phenobarbital, N-methyl-β-carboline-3-carboxamide [FG-7142], buspirone and the novel 5-hydroxytryptamine^3 [5-HT^3] antagonist, 3-tropanyl-indole-3-carboxylate [ICS 205-930]) on mice in J. N. Crawley and F. K. Goodwin's (1980) 2-compartment exploratory model. Results indicate that utilizing the time Ss spent in the dark side of the apparatus as an index of anxiety increased the sensitivity of the model and enabled both anxiolytic and anxiogenic agents to be detected.

1270. Kilts, C. D.; Commissaris, R. L. & Rech, Richard H. (1981). **Comparison of anti-conflict drug effects in three experimental animal models of anxiety. *Psychopharmacology,*** 74(3), 290–296.
Describes a new procedure for experimentally inducing conflict behavior—the conditioned suppression of drinking (CSD)—and compares it with 2 conventional animal models of human anxiety—a modified Geller-Seifter procedure and an Estes-Skinner procedure (the conditioned emotional response). Ss were male Sprague-Dawley rats. CSD offers significant advantages over the 2 operant procedures in that the session is short (10 min), and the establishment of stable behavioral baselines is rapid (about 2 wks). Like the more conventional procedures, CSD permits the simultaneous determination of drug effects on shock-suppressed and nonsuppressed responding as indicators of anti-anxiety and sedative properties, respectively. With the CSD, the anticonflict profiles of the benzodiazepines were highly correlated with their clinical anti-anxiety potency. The CSD appears to be a valuable tool in screening for possible anti-anxiety agents as well as in testing hypotheses regarding the mechanisms of action of such agents.

1271. King, G. A. & Burnham, W. M. (1980). **Effects of *d*-amphetamine and apomorphine in a new animal model of petit mal epilepsy. *Psychopharmacology,*** 69(3), 281–285.
Results with Long-Evans hooded rats support the hypothesis that the flash-evoked afterdischarge is a valid model of the petit mal seizure. Furthermore, evidence is provided indicating that norepinephrine is necessary for the seizure-suppressant action of dextroamphetamine.

1272. Kinon, Bruce J. & Kane, John M. (1989). **Difference in catalepsy response in inbred rats during chronic haloperidol treatment is not predictive of the intensity of behavioral hypersensitivity which subsequently develops. *Psychopharmacology,*** 98(4), 465–471.
Examined the association between a history of significant neuroleptic-induced parkinsonism and an increased incidence for the development of tardive dyskinesia. Catalepsy-sensitive Fisher rats and catalepsy-resistant Brown Norway rats were treated with haloperidol (1 mg/kg or 5 mg/kg daily). Despite significant interstrain difference in catalepsy response to either neuroleptic dose, Brown Norway Ss treated with 5 mg/kg developed behavioral hypersensitivity and D-2 receptor supersensitivity equivalent to that of the Fisher Ss. Catalepsy did not predict the intensity of behavioral and receptor changes considered to result from chronic antagonism of striatal dopamine receptors and to possibly underlie tardive dyskinesia.

1273. Klawans, Harold L.; Weiner, W. J. & Nausieda, P. A. (1977). **The effect of lithium on an animal model of tardive dyskinesia. *Progress in Neuro-Psychopharmacology,*** 1(1–2), 53–60.
Suggests that chronic administration of neuroleptics results in a decreased threshold for both amphetamine- and apomorphine-induced stereotyped behavior; this prolonged drug-induced dopamine receptor site hypersensitivity may be a workable model of tardive dyskinesia. Male guinea pigs treated with haloperidol alone developed a significantly prolonged decrease in the threshold for both amphetamine- and apomorphine-induced stereotyped behavior. Those treated with haloperidol and concurrent lithium developed no alteration in threshold. Lithium given after haloperidol-induced hypersensitivity of dopamine receptors had developed had no effect on threshold for either amphetamine or apomorphine. Results suggest that while prophylactic treatment with lithium may decrease the incidence of tardive dyskinesia, it would probably be of no value in the treatment of tardive dyskinesia.

1274. Kleven, Mark S.; Schuster, Charles R. & Seiden, Lewis S. (1988). **Effect of depletion of brain serotonin by repeated fenfluramine on neurochemical and anorectic effects of acute fenfluramine.** ***Journal of Pharmacology & Experimental Therapeutics,*** 246(3), 822–828.
By examining neurochemical and anorexic effects of acute fenfluramine (FNL) in male rats 2 and 8 wks after a 4-day regimen of FNL, it was demonstrated that tolerance to the anorexic and neurochemical effects of acute administration of FNL is likely the result of long-term depletions of central 5-hydroxytryptamine (5-HT) caused by repeated administration of high doses of FNL.

1275. Kleven, Mark S. & Sparber, Sheldon B. (1989). **Modification of quasi-morphine withdrawal with serotonin agonists and antagonists: Evidence for a role of serotonin in the expression of opiate withdrawal.** ***Psychopharmacology,*** 98(2), 231–235.
Examined the role of serotonin (5-hydroxytryptamine [5-HT]) in the rate-decreasing effects of 3-isobutyl-1-methylxanthine (IBMX) on operant behavior in male rats. 5-Hydroxytryptophan (5-HTP) and the 5-HT reuptake blocker fluoxetine were administered in combination with IBMX to Ss performing an FR-30 operant for food reinforcement. Both drugs failed to reverse the behavioral suppression caused by low doses of IBMX, suggesting that elevated 5-HT neurotransmission contributes to the quasi-morphine withdrawal syndrome (QMWS). The selective 5-HT_2 antagonists mianserin and pirenperone blocked the IBMX-induced suppression, whereas the classic 5-HT antagonist methysergide had no effect. Operant behavioral effects of IBMX and possibly the QMWS may be mediated by serotonergic mechanisms.

1276. Kleven, Mark S.; Woolverton, William L.; Schuster, Charles R. & Seiden, Lewis S. (1988). **Behavioral and neurochemical effects of repeated or continuous exposure to cocaine.** 49th Annual Scientific Meeting of the Committee on Problems of Drug Dependence, Incorporated: Problems of drug dependence, 1987 (1987, Philadelphia, Pennsylvania). ***National Institute on Drug Abuse: Research Monograph Series,*** 81, 86–93.
Investigated whether prolonged exposure to cocaine (COC) produced physical dependence by measuring behavioral effects of continuous infusion in monkeys and neurochemical effects in rats. For monkeys, the rate of responding for food under an FR schedule was reduced by COC in dose-related manner. Locomotor stimulation and stereotyped behaviors were also observed. As dosage increased, tolerance developed repeatedly until the monkeys were responding at baseline rates with a continuous infusion of 32 mg/kg/day. When the drug was terminated at this dosage, signs and symptoms of withdrawal syndrome appeared (hyporesponsive, reduced motor activity, sitting in hunched posture). In rats, high doses of COC did not produce the damage to monoamine containing neurons in brains as has been seen with amphetamine-like compounds.

1277. Kline, Joseph & Reid, Kenneth H. (1985). **The acute periventricular injury syndrome: A possible animal model for psychotic disease.** ***Psychopharmacology,*** 87(3), 292–297.
Developed a new experimental brain syndrome involving localized periventricular damage induced by intracerebroventricular injections of lysophosaphtidyl choline in male Sprague-Dawley Cox rats. The acute periventricular injury syndrome is characterized by transient weight loss, decreased emotionality, extreme postural indifference (catalepsy), inappropriate aggressive responses, impaired grooming, cerebral ventricular enlargement, and periventricular damage to both cells and fiber sheaths. This syndrome appears to simulate several features of schizophrenia, and it may prove useful in the study of psychotic disorders in humans.

1278. Knoll, J. (1975). **Predictive values of pharmacological models to study opiate dependence.** ***Neuropharmacology,*** 14(12), 921–926.
Used a battery of tests (hot-plate, writhing, tail-flick, and algolytic) in rats and mice and a test on the longitudinal muscle strip of the guinea pig ileum (a) to measure the analgesic activity of morphine, hydromorphone, and oxymorphone and (b) to compare it with that of 2 newly synthesized opiate analgesics, azidomorphine (AM) and 14-hydroxyazidomorphine. The latter were more potent than hydromorphone and oxymorphone and more effective in man in clinical trials for chronic intractable pain. The comparable physical dependence capacity of the 5 analgesics was evaluated in mice, rats, and rhesus monkeys, and results were compared with observations in 10 59–75 yr old patients treated for 14–77 days with subcutaneous AM ad lib. The low dependence capacity of AM was demonstrated, and no abstinence signs were observed after nalorphine or naloxone precipitation. Thus, data obtained with the animal model were in good agreement with the observations in man. It is concluded that the animal test systems seem to be adequate for structure-activity relationship studies aiming at the development of therapeutically useful structures with high analgesic potency and low dependence capacity.

1279. Knusel, Beat; Jenden, Donald J.; Lauretz, Sharlene D.; Booth, Ruth A. et al. (1990). **Global in vivo replacement of choline by N-aminodeanol: Testing a hypothesis about progressive degenerative dementia: I. Dynamics of choline replacement.** ***Pharmacology, Biochemistry & Behavior,*** 37(4), 799–809.
Rat pups weaned at 29 days were placed on experimental and control diets ad lib. Data are presented that show a substantial and progressive replacement of free and phospholipid-bound choline by the novel choline isostere N-amino-N,N-dimethylaminoethanol during its dietary administration in place of choline. Free choline in blood fell to – 20% of controls after 10–30 days on diet. Phospholipid-bound choline in plasma was reduced to less than 15%, and in erythrocytes to about 22%. After 120 days of diet, free and bound choline were reduced in most tissues to approximately 30% of controls. Acetylcholine was decreased to 33–50% of control. Findings are discussed with regard to Alzheimer's disease in that a moderate reduction of choline acetyltransferase activity was seen in striatum and myenteric plexus.

1280. Koller, W. C.; Fields, J. Z.; Gordon, J. H.; Perlow, M. J. et al. (1986). **Evaluation of ciladopa hydrochloride as a potential anti-Parkinson drug.** ***Neuropharmacology,*** 25(9), 973–979.
Investigated the effects of ciladopa hydrochloride in animal models of dopaminergic activity, using white male guinea pigs and Sprague-Dawley rats. Results show that ciladopa caused stimulation of central dopaminergic receptors and that the drug acted as a partial dopamine agonist with di-

rect-acting properties. It is suggested that ciladopa differs from other available dopaminergic drugs and may possess therapeutic advantages for the treatment of Parkinson's disease.

1281. Koller, William C.; Curtin, J. & Fields, J. (1984). **Molindone compared to haloperidol in a guinea-pig model of tardive dyskinesia.** ***Neuropharmacology,*** 23(10), 1191–1194.
Treated white male guinea pigs with molindone for 14 days. Doses of 3, 6, 20, and 40 mg/kg enhanced the stereotyped behavioral response induced by apomorphine and increased the number of D-2 dopamine receptors in the striatum labeled by high affinity binding or [^{3}H]spiroperidol, but doses of 1 mg/kg had no effect. Chronic administration of haloperidol (0.1, 0.5, and 5.0 mg/kg) also increased both the behavioral response to apomorphine and the number of dopamine receptors but had no effect at 0.02 and 0.004 mg/kg. It is suggested that the use of molindone, like other neuroleptics, may result in tardive dyskinesia.

1282. Kolpakov, V.; Barykina, N. & Chepkasov, I. (1981). **Genetic predisposition to catatonic behaviour and methylphenidate sensitivity in rats.** ***Behavioural Processes,*** 6(3), 269–281.
To study the relationship between 3 animal models of schizophrenia (genetically determined akinetic catatonia, stereotypies induced by amphetamine-like psychostimulators, and behavioral changes in chronic intoxication with such stimulators), the frequency of different types of reactions to a functional amphetamine analog, methylphenidate (M; 24 mg/kg, ip), was studied in wild Norway rats, nonselected Wistar rats, and Wistar rats bred for predisposition to akinetic catatonia. A positive relationship between the predisposition to catatonia and the level of stereotypies in a single M administration was found in wild Ss, but not in Wistar bred for catatonia. A closer study of catatonia in laboratory Ss permitted subdivision into several types—occurring in selected and nonselected Ss both naturally and as a result of chronic intoxication with amphetamines. It was found in nonselected Wistar Ss that there was a positive relationship between some of these types and an increased stereotypy level in repeated M administration. It is concluded that the natural akinetic catatonia and the chronic intoxication with amphetamines are 2 homologous varieties of the same model of schizophrenia, while the stereotypies are characteristics of this model. Studies of MAO activity imply a cortical component in the predisposition to akinetic catatonia.

1283. Kolpakov, V. G.; Gilinsky, M. A.; Alekhina, T. A.; Barykina, N. N. et al. (1987). **Experimental studies on genetically determined predisposition to catatonia in rats as a model of schizophrenia.** ***Behavioural Processes,*** 14(3), 319–341.
Compared neurophysiologic and neurochemical changes in rats with a genetic predisposition to catalepsy (RCs) to analogous changes found in schizophrenia or chronic amphetamine intoxication. In RCs, the threshold of audiogenic seizures was elevated and the activity of tryptophan hydroxylase in the striatum was higher. Noradrenaline content and noradrenaline/dopamine ratio were lower in the diencephalon. RCs had a higher frequency of inversion of hemispheric asymmetry. The effects of haloperidol and apomorphine on motor activity of cataleptic and normal animals point to a higher sensitivity of postsynaptic dopamine receptors in the former. The changes are analogous to those known to be present in schizophrenia and/or chronic intoxication with amphetamine or its pharmacological analogs.

1284. Kornetsky, Conan; Esposito, Ralph U.; McLean, Stafford & Jacobson, Joseph O. (1979). **Intracranial self-stimulation thresholds: A model for the hedonic effects of drugs of abuse.** ***Archives of General Psychiatry,*** 36(3), 289–292.
Presents the thesis that many drugs of abuse are used for their hedonic effects and that a relevant animal model for the study of these effects is the action of these drugs on the pathways that support rewarding intracranial self-stimulation. In a study with male Charles River CDF rats, a relationship between abuse potential of a drug and its ability to lower the threshold for rewarding brain stimulation was found. Of all compounds studied, morphine and cocaine were the drugs that caused the maximum lowering of the rewarding threshold. Phencyclidine hydrochloride and the mixed agonist–antagonist pentazocine lowered the threshold to a lesser degree, while the mixed agonist–antagonists cyclazocine and nalorphine hydrochloride had inconsistent effects. Naloxone hydrochloride had no effect on the threshold. Further, there was no evidence that tolerance develops to the threshold-lowering effect of morphine, suggesting that continued use of narcotics by the physically dependent individual is not simply due to an effort to avoid the pain of withdrawal.

1285. Kossenko, A. F. & Melnik, L. A. (1982). **Functional activity of the hypothalamo–neurohypophyseal secretory system under the effect of adrenoblocking agents in development of stress-induced gastric ulcers.** ***Fiziologicheskii Zhurnal SSSR im I.M. Sechenova,*** 68(5), 667–672.
Alpha- and beta-adrenoreceptors and the hypothalamo-neurohypophyseal secretory system (HNSS) were shown to contribute to the formation of stomach ulcers in guinea pigs. Adrenoblocking agents suppressed activity in the HNSS and caused a decrease in sympathetic impulses that aided in preventing stress.

1286. Kostowski, Wojciech; Danysz, Wojciech; Dyr, Wanda; Jankowska, Ewa et al. (1991). **MIF-1 potentiates the action of tricyclic antidepressants in an animal model of depression.** ***Peptides,*** 12(5), 915–918.
Investigated the effect of simultaneous treatment of male rats with small doses of MIF-1 (a chain of 3 amino acids linked by peptide bonds [Pro-Leu-Gly-NH_2]) and tricyclic antidepressants on rat behavior in a forced swim test. MIF-1 stimulated, in a dose-dependent manner, active escape-directed behavior of Ss in this paradigm. The effect of MIF-1 appeared to be independent of changes in Ss' locomotion in an open field test. The combined treatment of Ss with MIF-1 and amitriptyline and desipramine significantly stimulated active behavior in the forced swim test above the level obtained with each of the drugs given separately. Data indicate the potential clinical efficacy of small combined doses of MIF-1 and tricyclic antidepressants for treating depressed patients.

1287. Kovalenko, V. S.; Zvartau, E. E. & Bershadsky, B. G. (1984). **Behavioural manifestations of abstinence syndrome in rats.** ***Zhurnal Vysshei Nervnoi Deyatel'nosti,*** 34(3), 581–583.

Elaborated a specialized animal model to predict the nature of development of the withdrawal syndrome in male rats. Animals were made dependent on an opiate analgesic. The formation of the state of dependence was evaluated based on changes in pain and emotional reactions. A mathematical model of the developmental dynamics of the withdrawal syndrome was formulated. Repeated injections of the analgesic altered indicators of the pain reaction and subsequent emotional reactions to painless stimuli. There was a general tendency toward progressive sensitization to nociceptive and emotogenic stimuli. The threshold of aggressive reactions to a metal rod did not diminish in animals with initially low emotional reactivity, while initially high reactivity led to more expressed manifestations of aggression in the withdrawal period. The mathematical model helped to identify early symptoms of withdrawal, which may characterize the stage of emotional rather than physical dependence, and to link these with specific initial animal traits.

1288. Kozlowski, Michael R. & Arbogast, Roni E. (1986). **Specific toxic effects of ethylcholine nitrogen mustard on cholinergic neurons of the nucleus basalis of Meynert.** ***Brain Research,*** 372(1), 45–54.
Injected the putative cholinergic neurotoxin ethylcholine aziridinium ion (AF64A [0.01–0.05 nmol]) unilaterally into the nucleus basalis of Meynert (nbM) in male Sprague-Dawley rats to determine whether it would produce specific damage to the cholinergic cell bodies. Results suggest that AF64A can produce specific lesions of cholinergic neurons and therefore may be useful in developing animal models of human disorders involving cholinergic hypofunction, such as senile dementia of the Alzheimer type. However, there was a narrow dose range for producing these specific effects.

1289. Kraemer, Gary W. & McKinney, William T. (1979). **Interactions of pharmacological agents which alter biogenic amine metabolism and depression: An analysis of contributing factors within a primate model of depression.** ***Journal of Affective Disorders,*** 1(1), 33–54.
Examined the degree to which factors such as prior rearing condition, repeated peer separation, and housing environment can interact with behavioral effects produced by biogenic amine depleting agents. Emphasis placed on studies utilizing alphamethylparatyrosine, an inhibitor of tyrosine hydroxylase, to reduce levels of the catecholamine neurotransmitters norepinephrine and dopamine. Results provide quantitative estimates, in terms of dose–effect relationships, of the degree to which several factors can combine to produce despair-like behavior in rhesus monkeys. Findings may be of value in studying the contribution of similar factors to the precipitation of human depression. Analysis of literature relating alterations in behavior to changes in biogenic amine metabolism in animals suggests that there are important differences between rodent and primate species. These differences, when fully established, may indicate that research is needed to examine mechanisms whereby modest alterations in biogenic amine metabolism can interact with environmental and social stress.

1290. Krejčí, I. (1987). **Effect of nootropic drugs on the disruption of conditioned taste aversion by ECS.** 29th Annual Psychopharmacology Meeting (1987, Jeseník, Czechoslovakia). ***Activitas Nervosa Superior,*** 29(3), 217–218.
Examined the possible disruption of conditioned taste aversion induced by electroconvulsive shock (ECS) to evaluate the effects of piracetam and cyclo (1-amino-1-cyclopentane-carbonyl-L-alanyl) on amnesia in male rats. Both types of shock resulted in partial disruption of conditioned taste aversion. Results confirm the proactive effects of ECS on conditioned stimulus (CS)–unconditioned stimulus (UCS) association.

1291. Krijzer, F.; Snelder, M. & Bradford, D. (1984). **Comparison of the (pro)convulsive properties of fluvoxamine and clovoxamine with eight other antidepressants in an animal model.** ***Neuropsychobiology,*** 12(4), 249–254.
Compared the (pro)convulsive potencies of 2 newly developed nontricyclic antidepressants, fluvoxamine and clovoxamine, with other clinically effective antidepressants (imipramine HCl, amitriptyline HCl, desmethylimipramine HCl, mianserin HCl, viloxazine HCl, maprotiline HCl, zimelidine HCl, and nomifensine maleate) in freely moving male Wistar rats who were implanted with cortical, caudal, thalamic, and reticular electrodes. Drugs were infused intravenously at a constant rate up to a final cumulative dose of 40, 50, or 60 mg/kg. Ranking the tested antidepressants in decreasing order in accordance with their relative (pro)convulsive properties yields the following equation: amitriptyline > mianserin >> imipramine > desmethylimipramine > viloxazine >> maprotiline >> zimelidine > clovoxamine > nomifensine = fluvoxamine.

1292. Kudryavtseva, N. N.; Bakshtanovskaya, I. V. & Popova, N. K. (1989). **Development of pathological forms of behaviour in submissive male mice of C57BL/6J line in the process of agonistic zoosocial interactions: Possible model of depression?** ***Zhurnal Vysshei Nervnoi Deyatel'nosti,*** 39(6), 1134–1141.
Studied the development of depressive pathological states in male mice as a result of defects in zoosocial interaction. Animal subjects: Male C57BL/6J mice (aged 2.5–3.5 mo) (weighing 22–25 g). Drugs used: Chronic administration of imipramime (10 mg/kg ip, twice/day for 2 wks). A test developed by R. D. Porsalt was used. (English abstract).

1293. Kuhar, M. J.; Ritz, M. C. & Boja, J. W. (1991). **The dopamine hypothesis of the reinforcing properties of cocaine.** ***Trends in Neurosciences,*** 14(7), 299–302.
A review of the evidence suggests a dopamine (DA) hypothesis for the reinforcing properties of cocaine. This hypothesis proposes that cocaine binds at the DA transporter and mainly inhibits neurotransmitter reuptake; the resulting potentiation of dopaminergic neurotransmission in the limbic pathways ultimately causes reinforcement. Evidence for and against the DA hypothesis is discussed, noting that the hypothesis suggests medications for treating cocaine abuse and dependence. Animal models might be more effective in studying biological variables related to cocaine self-administration, since they are not confounded by differential or unknown drug histories or psychological and social factors that have strong influences on human drug-seeking behavior and drug dependence.

1294. Lacey, Daniel J. (1986). **Cortical dendritic spine loss in rat pups whose mothers were prenatally injected with phenylacetate ("maternal PKU" model).** ***Developmental Brain Research,*** 27(1–2), 283–285.

Applied an animal model that parallels the biochemical, behavioral, and pathologic features of human phenylketonuria (PKU) to the issue of maternal PKU by injecting pregnant albino rats subcutaneously with 3.5 µmol/gm phenylacetate or saline from Day 7 of gestation until delivery. Pups exposed to phenylacetate had structurally abnormal cortical pyramidal cell dendrites. Findings are related to the high incidence of microcephaly, congenital anomalies, mental retardation, and seizures among children born to PKU mothers.

1295. LaHoste, Gerald J.; Mormède, Pierre; Rivet, Jean-Michel & le Moal, Michel. (1988). **Differential sensitization to amphetamine and stress responsivity as a function of inherent laterality.** ***Brain Research,*** 453(1–2), 381–384.
Rats differentiated on the basis of their preferred direction of rotation following peripheral administration of amphetamine (AM) were found to differ in their sensitization to AM in 2 different behavioral paradigms. Ss that displayed leftward rotational biases developed greater sensitization and greater hormonal response to stress following sensitization.

1296. Lancaster, F.; Spiegel, K. & Zaman, M. (1987). **Voluntary beer drinking in rats.** ***Alcohol & Drug Research,*** 7(5–6), 393–403.
30 female Long-Evans rats were tested for individual preference for beer (BR) and were assigned to either BR or control (CT) groups according to preference. BR Ss were allowed ad libitum access to BR, food, and water; CT Ss were allowed ad libitum access to food and water. Over a 3-wk period, BR Ss initially ate more and drank more total water than CTs; body weights were not affected. Changes in body temperatures, tailflick latencies, hyperactivity, shivering, and tremoring indicated that BR Ss had become physically dependent on alcohol. The model is suggested for the study of mechanisms for initation of drinking behavior, reinforcement of taste, and effects of BR drinking on other factors.

1297. Landauer, Michael R. & Balster, Robert L. (1982). **Opiate effects on social investigatory behavior of male mice.** ***Pharmacology, Biochemistry & Behavior,*** 17(6), 1181–1186.
Examined the effects of morphine (MP) and naloxone (NAL), alone and in combination, on social investigatory behavior and motor activity in CD-1 male mice. Tests were conducted in a plexiglas apparatus in which a center area was separated from 2 stimulus compartments by wire mesh screens. One compartment housed a female conspecific; the other remained empty and served as a control for nonspecific investigatory responses. Ss were placed individually into the center area, and the time spent investigating each screen was recorded during the 15-min test. In Exp I, 11 Ss received saline or 0.1, 1.0, or 10.0 mg/kg MP ip 20 min prior to testing. The high dose significantly decreased investigation of the female compartment, while investigation of the uninhabited chamber and motor activity were not significantly affected. In Exp II (16 Ss), 3, 10, and 30 mg/kg NAL administered 30 min prior to testing had no significant effect on any measure. In Exp III (16 Ss), 3 mg/kg NAL reversed the decrease in female investigation time observed with 10 mg/kg MP, indicating an opiate mechanism and providing evidence that an animal model can be used to study the disruption of sociosexual behavior produced by opiates.

1298. Lankford, M. F.; Roscoe, A. K.; Pennington, S. N. & Myers, R. D. (1991). **Drinking of high concentrations of ethanol versus palatable fluids in alcohol-preferring (P) rats: Valid animal model of alcoholism.** ***Alcohol,*** 8(4), 293–299.
A genetically based animal model of alcoholism has been characterized in Wistar-derived rats in terms of their preference (P rats) or lack of preference (NP rats) for 10% ethanol over water. When concentrations of 3–30% were presented, the mean absolute intake of ethanol of P rats was 6.7 g/kg per day, with a maximum intake of 10.9 g/kg per day at the 25% concentration. These levels were significantly higher than those found with the commonly used constant concentration of 10%. The mean absolute intake of ethanol by NP rats was also elevated significantly at concentrations of 15–30% above that consumed at the 10% concentration. The mean absolute intake by each P rat of the maximally preferred solution of ethanol tested in the presence of an artificially sweetened fluid or a nutritionally fortified, highly palatable chocolate drink rose significantly higher than the mean intakes during the 10% and 3–30% preference tests.

1299. Lathers, Claire M.; Schraeder, Paul L. & Carnel, Shirley B. (1984). **Neural mechanisms in cardiac arrhythmias associated with epileptogenic activity: The effect of phenobarbital in the cat.** ***Life Sciences,*** 34(20), 1919–1936.
Sudden unexplained death accounts for 5–17% of mortality in epileptic persons; autonomic dysfunction is thought to be a contributing factor. The present study examined the effect of intravenous phenobarbital (PB [20 mg/kg]) 1 hr prior to administration of 6 increasing doses of pentylenetetrazol (10–2,000 mg/kg), given at 10-min intervals, on autonomic parameters in 18 cats. Results indicate that PB prevented only some forms of autonomic dysfunction associated with epileptogenic activity in this model.

1300. Lecci, Alessandro; Borsini, Franco; Volterra, Giovanna & Meli, Alberto. (1990). **Pharmacological validation of a novel animal model of anticipatory anxiety in mice.** ***Psychopharmacology,*** 101(2), 255–261.
Examined the action of anxiolytics, antidepressants, neuroleptics, antipyretics, muscle relaxants, antihypertensives, and naloxone (NAL) in an animal model of anxiety, using male mice, based on the notion that mice removed last from their cage develop hyperthermia (stress-induced hyperthermia [SIH]) when compared with those removed first. Alprazolam, chlordiazepoxide, estazolam, phenobarbital, ethanol, buspirone, and prazosin, as well as repeatedly administered diazepam, inhibited SIH. Findings indicate that SIH was prevented by anxiolytics (with the exception of tofisopam) and not by antidepressants, neuroleptics, muscle relaxants, antihypertensives (with the exception of prazosin), antipyretics, or NAL.

1301. Lehr, E. (1989). **Distress call reactivation in isolated chicks: A behavioral indicator with high selectivity for antidepressants.** Second International Meeting of the European Behavioural Pharmacology Society (1988, Athens, Greece). ***Psychopharmacology,*** 97(2), 145–146.
Tested the sensitivity and specificity of distress call behavior in 4-day-old male chicks to 33 neuroleptics. Chicks were placed in isolation in individual sound-shielded chambers, allowing parallel measurement of controls and Ss treated with different drugs at all doses. Controls responded to isolation with a high frequency of distress calls, which decreased during 2 hrs of isolation. 11 compounds counteracted the decrease in distress calling rate; 15 compounds

caused an inhibition of distress calling frequency down to less than 50% of the control rate. A decrease in distress calling during isolation seemed to reflect some general aspect of depressive syndrome, and this appeared sensitive to pharmacologic manipulation. Diverse clinically active antidepressants were able to counteract this decrease with high selectivity; nonantidepressants had either no influence or further inhibited the distress calling.

1302. Leith, Nancy J. & Barrett, Robert J. (1980). **Effects of chronic amphetamine or reserpine on self-stimulation responding: Animal model of depression?** ***Psychopharmacology,*** 72(1), 9–15.
Tested the effects of chronic reserpine (RE), a treatment that produces depression in humans, on self-stimulation responding in 18 male Sprague-Dawley rats. Separate groups of Ss implanted with stimulating electrodes in the medial forebrain bundle were administered daily injections of saline, dextroamphetamine (5 mg/kg for 7 days and then 10 mg/kg for another 7 days) or RE (.05 mg/kg for 18 days). At treatment termination, both drug groups showed a significant elevation of the reinforcement threshold, with no recovery occurring during 18 subsequent days. Results suggest that drug-induced depression of self-stimulation responding may serve as an animal model of the physiological basis for clinical or drug-induced depression.

1303. Lemberger, Louis; Kellams, Jeffrey J.; Small, Joyce G. & Rowe, Howard. (1977). **The effect of L-dopa and lergotrile mesylate on the interaction of fluphenazine decanoate and amphetamine-induced stereotypy and mortality.** ***Communications in Psychopharmacology,*** 1(5), 501–507.
Clinical studies conducted by the authors and their associates (1977) in schizophrenic patients revealed an interaction between fluphenazine decanoate and lergotrile, a dopamine agonist. An animal model utilizing amphetamine stereotypy was designed to simulate the clinical situation. Male Sprague-Dawley rats were treated with fluphenazine decanoate (10 mg/kg) 24 hrs prior to the administration of amphetamine (20 mg/kg). The phenothiazine completely protected Ss against the lethal effects of amphetamine. In addition, these Ss showed no stereotypic behavior. Certain groups treated with fluphenazine decanoate and amphetamine also received either levodopa, a dopamine precursor, or lergotrile, a dopamine agonist. In Ss given levodopa, the protective effect of fluphenazine on lethality was reversed in a dose-related manner. In contrast, lergotrile-treated Ss did not demonstrate any increase in mortality. Moreover, levodopa produced a dose-dependent increase in amphetamine-induced stereotypy, whereas lergotrile produced only a minimal increase in amphetamine effects.

1304. Leonard, Brian E. (1988). **Pharmacological effects of serotonin reuptake inhibitors.** Symposium: Serotonin in behavioral disorders (1987, Zurich, Switzerland). ***Journal of Clinical Psychiatry,*** 49 Suppl, 12–17.
Suggests that changes in neurotransmitter receptor numbers and function of blood cells in depressed patients may be state markers and that antidepressant efficacy may not be directly associated with specificity of amine reuptake inhibition. Bilaterally bulbectomized rats showed deficits in platelet and synaptosomal serotonin (5-hydroxytryptamine [5-HT]) transport resembling those in depressed patients.

1305. Leshner, Alan I.; Hofstein, Raphael & Samuel, David. (1978). **Intraventricular injection of antivasopressin serum blocks learned helplessness in rats.** ***Pharmacology, Biochemistry & Behavior,*** 9(6), 889–892.
Using the learned helplessness paradigm as the memory testing situation, male Wistar rats treated intraventricularly with control rabbit serum immediately after an initial period of inescapable preshocks showed subsequent deficits in learning an escape task, a replication of the learned helplessness phenomenon. However, Ss treated intraventricularly with antiserum to vasopressin after the initial inescapable shocks did not show later escape deficits. Findings support the suggestion that endogenous vasopressin is involved in long-term memory.

1306. Leshner, Alan I.; Remler, Helga; Biegon, Anat & Samuel, David. (1979). **Desmethylimipramine (DMI) counteracts learned helplessness in rats.** ***Psychopharmacology,*** 66(2), 207–208.
In rats, desipramine (desmethylimipramine) attenuated the deficits in escape responding that ordinarily follow prior exposure to inescapable preshocks, and it did so in a dose-dependent fashion. These findings support the position that the learned helplessness phenomenon is mediated by catecholamine changes.

1307. Levin, Edward D.; McGurk, Susan R.; South, David & Butcher, Larry L. (1989). **Effects of combined muscarinic and nicotinic blockade on choice accuracy in the radial-arm maze.** ***Behavioral & Neural Biology,*** 51(2), 270–277.
Investigated the effects of separate vs combined blockade of muscarinic and nicotinic acetylcholine (ACh) receptors on the spatial memory performance of 19 female rats in a radial-arm maze. The muscarinic receptor blocker scopolamine and the nicotinic receptor blocker mecamylamine each moderately impaired choice accuracy. Combined treatment with scopolamine and mecamylamine significantly decreased choice accuracy relative to either drug alone. This combination treatment lowered choice accuracy to chance levels. Data indicate that nicotinic and muscarinic blockade have at least additive effects in producing an anterograde memory deficit. Concurrent blockade of these 2 components of ACh systems may provide a better animal model of cognitive impairments due to the loss of cholinergic neurons, such as Alzheimer's disease.

1308. Levis, Donald G. & Ford, J. Joe. (1989). **The influence of androgenic and estrogenic hormones on sexual behavior in castrated adult male pigs.** ***Hormones & Behavior,*** 23(3), 393–411.
Three experiments evaluated the influence of testosterone propionate (TP), estradiol cypionate (EC), dihydrotestosterone propionate (DHTP), EC plus TP, EC plus DHTP, and TP plus DHTP on traits of masculine sexual behavior in castrated adult male pigs. Masculine sexual behavior was restored and maintained by TP, whereas EC initially activated sexual behavior, including copulation and ejaculation, but was unable to sustain copulatory behavior for 8–18 wk periods. Treatment with DHTP was ineffective for stimulation of sexual behavior. Testosterone promoted some aspects of masculine sexual behavior via aromatization to estrogen, but both androgen and estrogen were required for maintenance of the full complement of masculine sexual behavior traits.

1309. Ley, Michael F. & Crow, Lowell T. (1979). **The effects of alcohol on learned helplessness.** ***Physiological Psychology,*** 7(4), 387–390.
56 Sprague-Dawley rats were subjected to inescapable shock or no-shock pretreatments with or without 1.5 g/kg ip ethanol injections. 24 hrs later, all Ss were trained in an FR-2 shuttlebox avoidance task with or without alcohol in a 2^3 design. Alcohol accentuated learned helplessness (LH), but LH was observed in the shuttlebox after the shock pretreatment with or without alcohol if similar drug states accompanied each experience. LH was not seen when avoidance conditioning took place in an alcohol state different from that of the inescapable shock experience. Results are discussed in terms of the role of the retrieval of emotionally conditioned cues in LH.

1310. Li, T.-K. et al. (1979). **Progress toward a voluntary oral consumption model of alcoholism.** ***Drug & Alcohol Dependence,*** 4(1–2), 45–60.
With the goal of obtaining a suitable animal model for voluntary oral consumption of ethanol, alcohol-preferring and alcohol-nonpreferring rats from the Wistar strain were selectively used, with preference considered as a function of the concentration of ethanol ingested. Studies with these Ss showed that (a) drinking is voluntary and not contingent on caloric restriction; (b) they will work to obtain ethanol even when food and water are freely available, and in so doing, show psychological or behavioral tolerance; and (c) the amount of ethanol voluntarily consumed approaches their apparent maximum capacity for ethanol elimination. This amount of ethanol was capable of altering brain neurotransmitter content, thus exerting a CNS pharmocologic effect. In addition, Ss will barpress for iv administration of ethanol, and with prolonged, free-choice consumption, ethanol intake increases to as much as 12 g/kg daily without producing behavioral deficits, suggesting the development of tolerance.

1311. Li, Ting-kai; Lumeng, Lawrence; McBride, William J.; Murphy, James M. et al. (1988). **Pharmacology of alcohol preference in rodents.** ***Advances in Alcohol & Substance Abuse,*** 7(3–4), 73–86.
Suggests that since alcohol-seeking behavior is the final common pathway in alcoholism, regardless of etiology, exploration of its neuroanatomical, neurophysiological, and neurochemical substrates is key to understanding the biology of alcoholism. The present authors describe a genetic approach to developing animal models for alcoholism research, including the development of a pair of rat lines, high preference (P) and low preference (LP), for use in laboratory studies. Other topics of discussion include the differences in response to low and high dose ethanol between the P and LP rats and neurochemical and neuropharmacological differences between ethanol-naive P and LP rats.

1312. Li, Ting-kai; Lumeng, Lawrence; McBride, William J. & Murphy, James M. (1987). **Alcoholism: Is it a model for the study of disorders of mood and consummatory behavior?** ***Annals of the New York Academy of Sciences,*** 499, 239–249.
Presents an animal model of alcoholism developed by selectively breeding free-fed rats for the traits of alcohol-preference (P) and nonpreference (NP). Findings concerning differences between these lines are summarized and may shed some light on the possibility of shared mediating pathways for depression, eating disorders, and alcoholism. With sedative-hypnotic doses of ethanol, P rats develop acute tolerance more quickly than NP rats. One major difference between the lines is the lowered content of serotonin in certain brain regions of the P rats. Fluoxetine curbs the alcohol-seeking behavior of the P rats; however, variation in dietary carbohydrate content does not modify voluntary alcohol intake. P rats are similar in body weight to NP rats but are more active in a novel environment than the NP rats.

1313. Liebman, Jeffrey & Neale, Robert. (1980). **Neuroleptic-induced acute dyskinesias in squirrel monkeys: Correlation with propensity to cause extrapyramidal side effects.** ***Psychopharmacology,*** 68(1), 25–29.
In adult male squirrel monkeys given repeated treatment with haloperidol (1.25 mg/kg) at intervals of 7–14 days, subsequent acute administration of haloperidol induced dystonia and dyskinesias. This effect was dose-related and occurred at the same doses that impaired Sidman avoidance performance (0.3–1.25 mg/kg). Chlorpromazine, fluphenazine, metoclopramide, tetrabenazine, and Su-23397, all of which have extrapyramidal side effects, reliably produced dyskinesia. Dyskinesia was less marked after thioridazine and absent after clozapine. Baclofen and diazepam failed to elicit dyskinesias. In contrast to dyskinesia, catalepsy or tremor did not accurately predict extrapyramidal symptomatology. Acute dyskinesia in squirrel monkeys may serve as an animal model of the ability of antipsychotics to cause extrapyramidal dysfunction and clarify the mechanisms of drug-induced motor disorders.

1314. Lillrank, S. M.; Oja, Simo S. & Saransaari, Pirjo. (1991). **Animal models of amphetamine psychosis: Neurotransmitter release from rat brain slices.** ***International Journal of Neuroscience,*** 60(1–2), 1–15.
Examined the effects in male rats of 4 types of amphetamine (AMP) treatment on neurotransmitter release from brain slices using a superfusion system. Sensitization developed after long-term intermittent injection ip of AMP, and tolerance developed after chronic continuous AMP administered sc with osmotic minipumps. The acute injection of AMP caused a long-lasting decrease in the stimulated release of dopamine (DA) from both cortical and striatal slices. The exposure of brain slices to AMP stimulated the release of DA from both frontal cortical and striatal slices in vitro. The potassium-stimulated release of DA also was enhanced in both brain regions in vitro. Changes in brain DA systems alone cannot explain the behavioral manifestations in AMP-induced psychosis.

1315. Lindner, M. D. & Schallert, T. (1988). **Aging and atropine effects on spatial navigation in the Morris water task.** ***Behavioral Neuroscience,*** 102(5), 621–634.
A recent animal model that has been particularly useful in the neurobiology of aging has been the age-related decline of spatial information processing capacity in Sprague-Dawley rats measured in the place-learning water task developed by Morris (1981). In the first experiment of the present study, place behavior was examined in young (6 months), old (23–24 months), and very old (28 months) rats of another strain, Long-Evans. As an analogue of aging-related cholinergic dysfunction the effects of atropine sulfate (5–50 mg/kg), an anticholinergic drug that is known to disrupt behavior in this task, also was determined. Place navigation was not impaired in undrugged rats, even those in the oldest age group. Rats treated with atropine showed dose-dependent deficits. In a second experiment, young (4–5 months), old (18–20 months), and very old (28 months) Fischer-344 rats were examined. Place navigation was impaired in the old

rats. The very old (28 months) rats could not swim well enough to be tested adequately. Although nonspatial deficits associated with aging may be found across most strains tested, there appear to be very large strain-related differences in spatial processing ability as a function of age.

1316. Lipman, Jonathan J. & Tolchard, Stephen. (1989). **Comparison of the effects of central and peripheral aluminum administration on regional 2-Deoxy-D-glucose incorporation in the rat brain.** ***Life Sciences,*** 45(21), 1977–1987.
Surveyed and compared the *in vivo* regional cerebral glucose uptake capacity of 56 male rats injected with aluminum tartrate 7 or 14 days previously. The widespread involvement of aluminum in dementing, encephalopathic states, including Alzheimer's disease, has prompted a search for a suitable laboratory model in which to study primary aluminum cerebral neurointoxication. The rat, which is amenable to learning and memory studies, has been found to be the most suitable subject.

1317. Lister, Richard G. (1987). **The use of a plus-maze to measure anxiety in the mouse.** ***Psychopharmacology,*** 92(2), 180–185.
Investigated whether an elevated plus-maze (PM) with 2 open and 2 closed arms could be used as a model of anxiety in the mouse. NIH Swiss mice were tested in the PM following a holeboard test. Factor analysis yielded 3 factors assessing anxiety, directed exploration and locomotion. The anxiolytics chlordiazepoxide, sodium pentobarbital and ethanol increased time spent in open arms. The anxiogenics FG 7142, caffeine, and picrotoxin reduced this measure. Results suggest that the PM is useful as a test of anxiolytic and anxiogenic agents.

1318. Llorens, Jordi; Tusell, Josep M.; Suñol, Cristina & Rodríguez-Farré, Eduard. (1990). **On the effects of lindane on the plus-maze model of anxiety.** Second Meeting of the International Neurotoxicology Association (1989, Sitges, Spain). ***Neurotoxicology & Teratology,*** 12(6), 643–647.
The behavioral effects of an acute subconvulsant dose of the insecticide γ-hexachlorocyclohexane (lindane) were compared in the plus-maze (PLM) animal model of anxiety to those elicited by the anxiogenic gamma-aminobutyric $acid_A$ ($GABA_A$) gamma-aminobutyric acid (GABA) antagonist pentylenetetrazole (PTZ) and the anxiolytic benzodiazepine diazepam. Effects of the coadministration of diazepam with lindane or PTZ were also studied. 100 male rats served as Ss. The effects of subconvulsant doses of lindane and PTZ on Ss' behavior in the PLM were similar and opposite to those elicited by diazepam, and the administration of the convulsants antagonized the effects of diazepam.

1319. Lloyd, K. G.; Zivkovic, B.; Sanger, D.; Depoortere, H. et al. (1987). **Fengabine, a novel antidepressant GABAergic agent: I. Activity in models for antidepressant drugs and psychopharmacological profile.** ***Journal of Pharmacology & Experimental Therapeutics,*** 241(1), 245–250.
Fengabine (at a minimal dose of 25 mg/kg, ip) reversed the passive avoidance deficit in olfactory bulbectomized male albino rats, antagonizing the escape deficit in the learned helplessness model and decreasing paradoxical sleep in the Ss. In contrast to tricyclic antidepressants, fengabine antagonized 5-hydroxytryptophan (5-HTP) induced head twitches and only weakly reversed reserpine-induced ptosis. Fengabine inhibits neither monoamine uptake nor monoamine oxidase (MAO). A GABAergic mechanism of fengabine is indicated as bicuculline reversed its action in the olfactory bulbectomy and learned helplessness models. The wide-spectrum anticonvulsant action of fengabine was consistent with a gamma-aminobutyric acid (GABA)-mimetic action and was in contrast to the proconvulsant effect of most classical antidepressants.

1320. Loew, Gilda et al. (1984). **Pyrazolo[1,5-a]pyrimidines: Receptor binding and anxiolytic behavioral studies.** ***Pharmacology, Biochemistry & Behavior,*** 20(3), 343–348.
Pyrazolo[1,5-a]pyrimidines (PZP) have been reported to be specific anxiolytic agents that do not potentiate ethanol or barbiturates. To further investigate these compounds, 3 of the most promising analogs were synthesized and a tritium-labeled analog of one of them prepared by a new synthetic procedure. These analogs did not compete with [^{3}H]flunitrazepam or [^{3}H]Beta-carboline ethyl ester binding nor did they potentiate the [^{3}H]flunitrazepam binding. Receptor binding studies with the [^{3}H]PZP revealed a low affinity receptor site, distinct from that of the benzodiazepines, but with only a small fraction (20%) of specific binding. Behavioral tests using 3 animal models for anxiety—muricide (with 24 male Long-Evans rats), approach/avoidance conflict (male Sprague-Dawley rats), and 2-chamber exploration (male NIH rats) tests—gave conflicting results, positive in the 1st and negative in the latter 2. These compounds were not antagonists of diazepam's anticonvulsant activity. Results do not support evidence that these analogs are promising specific anxiolytic agents.

1321. Lohr, James B.; Cadet, Jean L.; Wyatt, Richard J. & Freed, William J. (1988). **Partial reversal of the iminodipropionitrile-induced hyperkinetic syndrome in rats by α-tocopherol (vitamin E).** ***Neuropsychopharmacology,*** 1(4), 305–309.
In 2 experiments, male rats were treated with the neurotoxin iminodipropionitrile (IDPN), which causes an irreversible movement disorder accompanied by arousal damage similar to that seen in Vitamin E deficiency. Results show that Vitamin E administered either concurrently or following IDPN significantly reduced the severity of IDPN-induced dyskinesia compared to those receiving IDPN alone. Findings suggest a possible involvement of free radical formation in the neurotoxicity of IDPN. IDPN has been proposed as a model for hyperkinetic movement disorders in humans, including Tourette's syndrome, Huntington's disease, spasmodic dystonias, and tardive dyskinesia.

1322. Lopez, Maria C.; Huang, Dennis S.; Watzl, Bernhard; Chen, Guan-jie et al. (1991). **Splenocyte subsets in normal and protein malnourished mice after long-term exposure to cocaine or morphine.** ***Life Sciences,*** 49(17), 1253–1262.
Developed an experimental model that resembles human drug addiction to study the effect of chronic drug (cocaine or morphine) administration on the immune system. A low-protein diet was evaluated for its contribution to the impairment of the immune system during addiction. Female mice received a 20% or 4% casein diet and were administered cocaine or morphine daily for 11 wks in increasing daily doses. Cocaine administration reduced body weight and spleen weight in both groups of mice but particularly in protein-malnourished mice. Results suggest that cocaine, morphine, and saline injection altered the immune system in a stress-dependent way.

1323. Löscher, Wolfgang. (1985). **Influence of pharmacological manipulation of inhibitory and excitatory neurotransmitter systems on seizure behavior in the Mongolian gerbil.** ***Journal of Pharmacology & Experimental Therapeutics,*** 233(1), 204–213.
Studied the relationship between different neurotransmitter systems and seizure susceptibility in Mongolian gerbils with genetically determined epilesy and examined the effects of drugs that manipulate inhibitory or excitatory neurotransmitter systems. Comparison of anticonvulsant potencies of various drugs, such as apomorphine HCl and muscimol HBr, with potencies reported in other genetic animal models of epilepsy, such as audiogenic seizure-susceptible mice, indicated that drugs that increase GABA and dopamine levels in the brain are strikingly more effective in gerbils than in other species in blocking generalized seizures. Seizures in gerbils appear not to be sensitive to alterations in acetylcholine, noradrenaline, 5-HT, glycine, and excitatory amino acid-mediated neurotransmission.

1324. Löscher, Wolfgang & Hönack, Dagmar. (1991). **Anticonvulsant and behavioral effects of two novel competitive N-methyl-D-aspartic acid receptor antagonists, CGP 37849 and CGP 39551, in the kindling model of epilepsy: Comparison with MK-801 and carbamazepine.** ***Journal of Pharmacology & Experimental Therapeutics,*** 256(2), 432–440.
Evaluated the N-methyl-D-aspartate (NMDA) receptor antagonists CGP 37849 and its ethyl ester CGP 39551 in amygdala-kindled female rats. Anticonvulsant and behavioral effects of these novel compounds were compared with those of the noncompetitive NMDA receptor antagonist MK-801 and carbamazepine. In contrast to carbamazepine, CGP 37849, CGP 39551, and MK-801 exerted only weak anticonvulsant effects in fully kindled Ss and did not increase the focal seizure threshold. The weak anticonvulsant effects of the NMDA receptor antagonists in kindled Ss were associated with profound untoward behavioral effects. Findings suggest that these compounds would not be clinically useful against partial and secondary generalized seizures. These compounds also produce a behavioral syndrome with phencyclidine-like effects.

1325. Lowy, Martin T.; Nash, J. Frank & Meltzer, Herbert Y. (1990). **Reserpine-induced DST nonsuppression in rats.** ***Biological Psychiatry,*** 27(5), 546–548.
Two experiments with male rats examined the effect of reserpine (RSP) on hypothalamic-pituitary-adrenal (HPA) function in a rat model of the dexamethasone suppression test (DST) and investigated its effect on the concentration of biogenic amines in the hypothalamus, RSP administration increased serum corticosterone levels, decreased neuronal monoamine levels, and reproduced DST levels frequently observed in depression. Depletion of biogenic amines can reproduce the HPA abnormalities typically seen in depressed patients. Biogenic amines play a role in the regulation of the HPA axis, and DST abnormalities are present in an animal model of depression.

1326. Lumeng, Lawrence & Li, Ting-kai. (1986). **The development of metabolic tolerance in the alcohol-preferring P rats: Comparison of forced and free-choice drinking of ethanol.** ***Pharmacology, Biochemistry & Behavior,*** 25(5), 1013–1020.
Conducted 3 experiments with a total of 68 female Wistar and P rats. Results indicate that the P Ss on chronic free-choice drinking of alcohol developed metabolic tolerance to much the same degree as Ss forced fed ethanol contained in liquid diets. It is suggested that the acquisition of ethanol tolerance with free-choice drinking is a necessary criterion for an animal model of alcoholism.

1327. Lunn, Robert J. et al. (1981). **Anesthetics and electroconvulsive therapy seizure duration: Implications for therapy from a rat model.** ***Biological Psychiatry,*** 16(12), 1163–1175.
Investigated the effects of methohexital, Innovar (fentanyl and droperidol), and ketamine on seizure duration following ECS in a rat model of ECT. 12 male albino Holtzman rats were ip injected with each anesthetic before undergoing ECS. Compared to unanesthetized controls, methohexital shortened seizure duration by 42%, ketamine tended to increase seizure duration, and Innovar had no effect on duration of seizures.

1328. Lurie, Scott; Kuhn, Cynthia M.; Bartolome, Jorge & Schanberg, Saul. (1989). **Differential sensitivity to dexamethasone suppression in an animal model of the DST.** ***Biological Psychiatry,*** 26(1), 26–34.
Attempted to use a model of dexamethasone (DEX) feedback suppression of corticosterone (CST) secretion in rats that provides 24-hr suppression of serum CST to block CST secretion resulting from various stimuli. The authors examined whether defined neural controls of corticotropin-releasing factor would show differential sensitivity to DEX blockade. Findings indicate that CST responses to morphine were well-suppressed by a dose regimen of DEX that did not influence CST responses to physostigmine. The suppressibility of responses to 2 stressors, ether and immobilization, resembled that of morphine. Findings suggest that the dexamethasone suppression test (DST) can detect changes in neurochemical processes associated with affective illness and may be useful in the study of animal models of depression.

1329. Lynch, Minda R. & Carey, Robert J. (1988). **Sensitization of chronic neuroleptic behavioral effects.** ***Biological Psychiatry,*** 24(8), 950–951.
Reports that chronic dosing with low-dose haloperidol in rats was associated with a progressive enhancement of drug effects on exploratory behavior in the open field, mimicking the delayed onset seen with the clinical use of neuroleptics.

1330. Lynch, Minda R.; Kuhn, Hans-Georg & Carey, Robert J. (1988). **Chronic haloperidol-amphetamine interactions and mesolimbic dopamine.** ***Neuropsychobiology,*** 19(2), 97–103.
Male rats received 21 days of chronic treatment with amphetamine, haloperidol, a combination of these 2 drugs, or saline. On day 21, mesolimbic (but not striatal) dopamine (DA) concentrations were positively related to locomotor activity in an open field. DA metabolites in this region were inversely correlated with the behavior. The combined drug group showed saline-like levels of both behavioral activity and mesolimbic DA. Metabolic indices in this group suggested that increased DA availability partially competed with the neuroleptic receptor blockade in mesolimbic regions. 21 days of haloperidol did not induce behavioral or biochemical tolerance. Findings are consistent with the lack

of tolerance development to antipsychotic effects and suggest that animal models incorporating chronic low-dose neuroleptic regimens may be useful for the study of chronic treatment issues.

1331. Lyubimov, B. I. et al. (1983). **Chronic alcoholic intoxication in animals as a model for studying the safety of new antialcoholic agents.** ***Farmakologiya i Toksikologiya,*** 46(2), 98–102.
Experiments on 22 male rats demonstrated that daily administration of ethanol (8 mg/kg, po) for 1 mo produced pathological changes in organs and systems of the body similar to manifestations of chronic alcoholic intoxication in humans. The model under consideration may be used during preclinical study of the safety of new anti-alcoholic agents. (English abstract).

1332. Macenski, Mitchell J.; Cleary, James & Thompson, Travis. (1990). **Effects on opioid-induced rate reductions by doxepine and bupropion.** ***Pharmacology, Biochemistry & Behavior,*** 37(2), 247–252.
12 adult female pigeons key-pecked under a multiple VI 15-sec, VI 150-sec schedule of food reinforcement. Effects of 2 opioid drugs, buprenorphine (BUP) and methadone, were determined alone and in combination with daily administration of the antidepressants doxepin or bupropion. Methadone initially produced dose-dependent key-pecking rate reductions when administered acutely, prior to the session; BUP produced key-pecking rates that reached a plateau at 50–80% of baseline rate and were not reduced further by higher doses. Unlike bupropion, doxepin interfered with the development of opioid tolerance. Neither antidepressant systematically altered effects of BUP on key pecking.

1333. Maggi, Adriana & Pérez, Jorge. (1985). **Role of female gonadal hormones in the CNS: Clinical and experimental aspects.** ***Life Sciences,*** 37(10), 893–906.
Reviews the data supporting a widespread effect of estrogens and progesterone in the central nervous system (CNS) of mammals. The primary target areas for sex hormones in the CNS in mammals include the preoptic, hypothalamic, and amygdaloid areas. Clinical evidence of estrogen and progestin involvement in the regulation of the extrapyramidal functions is discussed in terms of steroid hormones and chorea and steroid hormones and tardive dyskinesia. Behavioral, biochemical, and electrophysiological models supporting the hypothesis of estrogen modulation of extrapyramidal functions are presented. Clinical evidence of estrogen and progesterone involvement in the manifestation of epilepsy is summarized, and animal models for the study of the influence of sex hormones on epilepsy are described. Evidence of estrogens and progesterone involvement in the manifestation of affective disorders is considered. Findings suggest that estrogens and progesterone act in numerous regions of the CNS to regulate motor and limbic functions and to modulate neuronal activity through a wide variety of functions.

1334. Maier, Donna M. & Pohorecky, Larissa A. (1987). **The effect of repeated withdrawal episodes on acquisition and loss of tolerance to ethanol in ethanol-treated rats.** ***Physiology & Behavior,*** 40(4), 411–424.
72 2–5 mo old male rats were administered ethanol via an intragastric catheter (8.0–12.0 g/kg/day) either continuously for 8 wks or on a binge schedule with 4 2-wk cycles of drug administration separated from each successive cycle by a 2-wk period of no drug treatment. Older Ss were administered ethanol for 2 wks to provide an age control for the binge-treated Ss. Acquisition and loss of tolerance to ethanol-induced motor impairment were measured, and loss of tolerance to ethanol-induced hypothermia was assessed. Acceleration of tolerance development to both ethanol-induced motor impairment and hypothermia was observed in Ss subjected to repeated withdrawal episodes (binge-Study 1) but not in the controls, who experienced withdrawal only once (continuous-Study 2). In Ss exposed to prolonged ethanol treatment, persistent changes in responding to the drug were found.

1335. Maier, Donna M. & Pohorecky, Larissa A. (1986). **The effect of stress on tolerance to ethanol in rats.** ***Alcohol & Drug Research,*** 6(6), 387–401.
Studied the effects of stress on tolerance to ethanol in 44 male Holtzman Sprague Dawley rats. Ss were given 6 g per kg ethanol each day for 28 days, with 8 g per kg subsequently to 41 days. Footshock stress was administered intermittently for 15 min every other day to some Ss. Other Ss observed different Ss being shocked while a control group had no stress experience. Tolerance was tested every 4 days through performance on a dowel task and body temperature after ethanol challenge. Footshock stress accelerated the development of tolerance with tolerance occurring more rapidly and with a smaller daily dose of ethanol. Nonstressed Ss required the 8-g dose to show tolerance. Observer-stressed Ss showed no functional tolerance at all. It is concluded that by enhancing tolerance, stress may contribute to alcohol abuse.

1336. Maier, Donna M. & Pohorecky, Larissa A. (1987). **Repeated withdrawal from ethanol on radial arm acquisition in rats.** ***Alcohol,*** 4(6), 433–436.
Male rats were treated with ethanol for 4 2-wk periods interrupted every 2 wks by a 2-wk period of no drug treatment. Other rats were treated with ethanol for 8 wks with no interruptions. Acquisition of an 8-arm radial maze response when daily ethanol treatment was ended was not affected by either the experience of 4 withdrawals from ethanol or by 8 wks of ethanol treatment, contrary to the expectation that repeated withdrawal would impair learning.

1337. Maier, Donna M. & Pohorecky, Larissa A. (1987). **The effect of ethanol treatment on social behavior in male rats.** ***Aggressive Behavior,*** 13(5), 259–268.
Adult male rats were treated with ethanol (ETH) or equicaloric dextrin maltose for 2 or 8 wks. Social interaction was assessed before and after chronic drug treatment. It was found that 0.50 g/kg of ETH increased aggressive behavior and time spent interacting with stimulus juvenile male from its first presentation to the second presentation (20 min apart). Saline injection decreased aggressive behavior. After chronic drug treatment ended, Ss treated chronically for 2 wks with ETH were more aggressive when they were not intoxicated than when they had been treated with ETH. Aggressive behavior and time spent interacting with the juvenile were greater in Ss treated chronically with ETH, regardless of whether they were injected with saline or ETH.

1338. Maier, Steven F. (1990). **Diazepam modulation of stress-induced analgesia depends on the type of analgesia.** ***Behavioral Neuroscience,*** 104(2), 339–347.
Factors that determine the impact of diazepam on the hypoalgesia produced by electric shocks were investigated. Tail-shocks (1, 5, & 20) were followed by an initial hypoalgesia, lasting 2–4 min, that was unaffected by prior administration

of diazepam. This hypoalgesic reaction was followed by a second hypoalgesia if subjects were allowed to remain in the shock environment during testing, and this reaction was reduced or eliminated by prior diazepam. If subjects were removed from the shock situation, this second reaction did not occur. In contrast, 80 shocks were followed by a single hypoalgesia that was sensitive to blockade by diazepam throughout its entire course and was not affected by removing subjects from the shock environment. These results have implications for the perceptual–defensive–recuperative, working memory, and unconditioned response (UCR)-learned helplessness interpretations of shock-produced analgesia.

1339. Maier, Steven F. (1990). **Role of fear in mediating shuttle escape learning deficit produced by inescapable shock.** ***Journal of Experimental Psychology: Animal Behavior Processes,*** 16(2), 137–149.
The relation between the shuttlebox escape deficit produced by prior inescapable shock (IS) and fear during shuttlebox testing as assessed by freezing was investigated in rats. IS rats learned to escape poorly and were more fearful than either escapably shocked subjects or controls, both before and after receiving shock in the shuttlebox. However, fear and poor escape performance did not covary with the manipulation of variables designed to modulate the amount of fear and the occurrence of the escape deficit. A 72-hr interval between IS and testing eliminated the escape deficit but did not reduce preshock freezing. Diazepam before testing reduced both preshock and postshock fear in the shuttlebox but had no effect on the escape deficit. Naltrexone had no effect on fear but eliminated the escape deficit. This independence of outcome suggests that the shuttlebox escape deficit is not caused by high levels of fear in IS subjects.

1340. Mair, Robert G.; Otto, Timothy A.; Knoth, Russell L.; Rabchenuk, Sharon A. et al. (1991). **Analysis of aversively conditioned learning and memory in rats recovered from pyrithiamine-induced thiamine deficiency.** ***Behavioral Neuroscience,*** 105(3), 351–359.
Rats that had recovered from pyrithiamine-induced deficiency (PTD) were trained on tasks motivated by escape from mild footshock. On postmortem examination, the PTD model showed 2 consistent lesions: a bilaterally symmetrical lesion of the medial thalamus, which was centered on the internal medullary lamina (IML), and a lesion centered on the medial mammillary nuclei. PTD rats with IML lesions were impaired in learning a spatial nonmatching-to-sample task that was mastered without error by controls and PTD animals without IML lesions. These animals were able to perform as well as controls on discrimination tasks based on either place or visual (light–dark) cues. They made more errors than controls in reaching criterion in the initial place discrimination problem. These findings are consistent with findings from appetitively motivated tasks that PTD rats with IML lesions have an impaired capacity for working memory but not for reference memory.

1341. Mandell, Arnold J. & Knapp, Suzanne. (1975). **A model for the neurobiological mechanisms of action involved in lithium prophylaxis of bipolar affective disorder.** ***National Institute on Drug Abuse: Research Monograph Series,*** 3, 97–107.
The effects of chronic administration of lithium chloride on the serotonin synthesizing apparatus in male Sprague-Dawley rat brain suggest a theoretical model that could explain how chronic sc treatment with lithium is prophylactic against both poles of affect in manic-depressive disorder. After 3–5 days of lithium chloride the uptake of (^{14}C)tryptophan into striate synaptosomes increased to 140% of control values, and tryptophan-to-serotonin conversion activity increased to about the same degree. After 21 days of drug administration, (^{14}C)tryptophan uptake remained above control levels and soluble midbrain and solubilized striate synaptosomal enzyme activity remained below, but synaptosomal conversion activity had returned to control levels. In vitro, drug concentrations from 10 to 53 mM did not affect the enzyme activity but did enhance uptake and conversion measures. Increasing tryptophan levels either by pre-incubation with *l*-tryptophan in vitro or by the administration of *l*-tryptophan (20 to 60 mg/kg) in vivo enhanced uptake and conversion measures. Data suggest the possibility that lithium pushes 2 complementary adaptive mechanisms to their capacities, and the net result is the restricted but balanced function of serotonergic transmission in the brain.

1342. Mantione, Charles R.; Fisher, Abraham & Hanin, Israel. (1981). **The AF64A-treated mouse: Possible model for central cholinergic hypofunction.** ***Science,*** 213(4507), 579–580.
Observed a loss in the number of functional, sodium ion-dependent, high-affinity choline transport sites in the cortex and hippocampus of mice given an icv injection of 65 nanomoles of ethylcholine mustard aziridinium ion (AF64A) 3 days earlier. The effect was not observed in the striatum. This effect of AF64A represents a long-term neurochemical deficit at cholinergic nerve terminals in some brain regions that can lead to a persistent deficiency in central cholinergic transmission. The AF64A-treated animal may thus be a model for certain psychiatric or neurological disorders that appear to involve central cholinergic hypofunction.

1343. Mantione, Charles R.; Fisher, Abraham & Hanin, Israel. (1984). **Possible mechanisms involved in the presynaptic cholinotoxicity due to ethylcholine aziridinium (AF64A) *in vivo.*** ***Life Sciences,*** 35(1), 33–41.
Investigated the potential of an in vivo active cholinotoxin by administering the toxin directly into the brains of rats and mice. The neurochemical and behavioral consequences of AF64A administration were reminiscent of similar measures in patients with Alzheimer's disease. It is tentatively suggested that the AF64A-treated animal may be explored as a potential animal model of this debilitating disease state.

1344. Marco, Luis A.; Reed, Timothy F.; Joshi, Rajani S.; Aldes, Leonard D. et al. (1989). **Metoclopramide fails to suppress linguopharyngeal events in a rat dyskinesia model.** Annual Meeting of the American Psychiatric Association (1988, Montreal, Canada). ***Journal of Neuropsychiatry & Clinical Neurosciences,*** 1(1), 53–56.
Examined the effects of metoclopramide (MET) on ketamine-induced linguopharyngeal events exemplified by tongue retrusions, protrusions, and swallowing acts in ketamine-anesthetized female rats. Ss were mounted on a stereotaxic frame specially designed to monitor retrusion, protrusion, and swallowing. MET at doses of 0.5–50 mg/kg im failed to decrease protrusions, retrusions, or swallowing; all 3 events

increased for up to 2.5 hrs. Results differ from those of J. M. Karp et al (1981) indicating that MET suppressed dyskinetic activity in humans treated for a minimum of 3 mo with the drug.

1345. Marcy, René; Quermonne, Marie-Anne; Raoul, Josette & Nammathao, Bounsay. (1987). **Recovery of normobaric hypoxia-lowered skin conductance response (SCR) in mice: SCR-hypoxia test, an animal model for testing drugs against brain hypoxia.** ***Progress in Neuro-Psychopharmacology & Biological Psychiatry,*** 11(1), 35–43.
Developed an animal model of an SCR-hypoxia test and assessed its validity through the study of 15 drugs, including antihypoxic drugs, cerebral vasodilators (including dopamine agonists), metabolic modifiers (expense slackeners and energy metabolizers), using male and female Swiss Orl mice. It is suggested that the SCR-hypoxia test might help develop new potentially antihypoxic drugs, since SCR was easily recorded in Ss.

1346. Marczynski, T. J. & Urbancic, M. (1988). **Animal models of chronic anxiety and "fearlessness."** United States Air Force School of Aerospace Medicine Symposium: Basic questions in neuroscience (1987, San Antonio, Texas). ***Brain Research Bulletin,*** 21(3), 483–490.
Describes feline and rodent models of chronic anxiety and a rodent model of "fearless" behavior. The models were obtained by pre- or perinatal exposure of animals to diazepam or RO 15-1788, which produced enduring postnatal deficits or enrichment, respectively, of brain benzodiazepine receptors in their progenies. Receptor-deficient 1-yr-old cat progenies showed hyperarousal, unabated restless behavior, delayed acquisition of instrumentally conditioned behavior, bizarre escape responses, and absence or reduced alpha-like EEG activity. Receptor-deficient rat progenies (aged 5–6 mo) showed a reduction of time spent in deep slow wave sleep and inability to habituate to novel environments such as radial arm maze. Receptor-enriched (i.e., "fearless") progenies were superior to controls and to the receptor-deficient groups in exploratory behavior, particularly when Ss were challenged by novel and intimidating visual and/or auditory stimuli.

1347. Markou, Athina & Koob, George F. (1991). **Postcocaine anhedonia: An animal model of cocaine withdrawal.** ***Neuropsychopharmacology,*** 4(1), 17–26.
Developed an animal model of postcocaine depression or anhedonia and studied the time course of this cocaine withdrawal symptom. 21 rats were allowed to self-administer cocaine iv for prolonged periods of time, and their brain reward thresholds were assessed using intracranial self-stimulation (ICSS) thresholds. ICSS thresholds were used operationally as a measure of the "hedonic–anhedonic" state. During cocaine withdrawal, ICSS thresholds were elevated compared to predrug baseline levels and control animal thresholds, reflecting an "anhedonic" state. The magnitude and duration of the anhedonic state was proportional to the amount of cocaine consumed during the binge. A measure of response latency provided evidence that this postcocaine elevation of thresholds is due to a desensitization of the reward pathways mediating ICSS reward and not to any nonspecific effects of cocaine exposure.

1348. Marquis, K. L. & Moreton, J. E. (1987). **Animal models of intravenous phencyclinoid self-administration.** ***Pharmacology, Biochemistry & Behavior,*** 27(2), 385–389.
Investigated the self-administration of phencyclidine (PCP), ketamine, and other phencyclinoid drugs in female rats, and characterized the self-administration of higher unit doses of PCP than previously reported. Preliminary results of the assessment of the reinforcing efficacy of some PCP analogs measured by the progressive ratio procedure are also presented.

1349. Martin, Joan C. (1984). **Perinatal psychoactive drug use: Effects on gender, development, and function in offspring.** ***Nebraska Symposium on Motivation,*** 32, 227–266.
Discusses the effects of selective sedative and stimulant psychoactive drugs of abuse in animal models (in studies involving rats), specifically nicotine, alcohol, barbiturates, and the amphetamines. A rationale is presented for using animal models in behavioral teratology (the study of modified function in offspring following maternal, and sometimes paternal exposure to noxious extrinsic agents before and during the perinatal period). Principles of teratology and experimental design considerations (e.g., subject selection, drug amount) are explained. Studies are cited that report sex-ratio change following drug exposure, multiple drug use, and paternal drug use before pregnancy. It is argued that there is sufficient clinical and correlational evidence to implicate tobacco, alcohol, and amphetamines as human teratogens.

1350. Martin, P.; Brochet, D.; Soubrie, P. & Simon, P. (1985). **Triiodothyronine-induced reversal of learned helplessness in rats.** ***Biological Psychiatry,*** 20(9), 1023–1025.
Investigated the effect of triiodothyronine (TI) on the learned-helplessness paradigm with male Wistar AF rats. Findings indicate that Ss preexposed to inescapable electric footshocks and treated with TI (0, 0.015, 0.03, or 0.06 mg/kg, intraperitoneally) for 4 consecutive days did not exhibit escape and avoidance deficits when tested in a shuttlebox paradigm. This protective antidepressantlike effect seemed to affect deficits specifically, because TI neither caused intertrial shuttling nor did it facilitate shuttlebox responses in Ss not trained for learned helplessness.

1351. Martin, P.; Massol, J.; Belon, J. P.; Gaudel, G. et al. (1987). **Thyroid function and reversal by antidepressant drugs of depressive-like behavior (escape deficits) in rats.** ***Neuropsychobiology,*** 18(1), 21–26.
Investigated in male rats the effect of hypothyroidism (induced by propylthiouracil and of L-triiodothyromine [T_3] administration) on the ability of antidepressant drugs to reverse helpless behavior (escape deficits following exposure to inescapable shock). Findings show that the reversal by clomipramine, desipramine, imipramine, and nialamide of depressive-like behavior in rats was markedly attenuated in hypothyroid rats. Conversely, the effect of these same antidepressants was significantly hastened in euthyroid rats given daily T_3. These findings support the notion of intricate thyroid/central nervous system (CNS) interactions in the mechanisms of action of antidepressant drugs.

1352. Martin, P.; Thiébot, M. H. & Puech, A. J. (1990). **Animal models sensitive to antidepressant drugs: Involvement of central serotoninergic and noradrenergic systems.** ***Psychiatrie & Psychobiologie,*** 5(3), 209–217.
Discusses the results of recent studies on the behavioral effects of antidepressants with excitatory or inhibitory serotoninergic or noradrenergic properties in 2 rat behavior models: (1) the learned helplessness paradigm and (2) the

test for waiting capacity in a T-maze. The effects of imipramine, clomipramine, desipramine, citalopram, indalpine, fluvoxamine, zimelidine, buspirone, gepirone, ritanserin, and ipsapirone on reversal of escape deficit and tolerance to reward delay are described. (English abstract).

1353. Martin, Patrick. (1991). **1-(2-Pyrimidinyl)-piperazine may alter the effects of the 5-HT1A agonists in the learned helplessness paradigm in rats.** ***Psychopharmacology,*** 104(2), 275–278.
Investigated the role of 1-(2-pyrimidinyl)-piperazine (1-PP) in learned helplessness to determine whether it affects the reversal of helpless behavior induced by 5-hydroxytryptamine 1A (5-HT1A) agonists at high doses in male rats. 1-PP was evaluated alone and in combination with 8-hydroxy-(di-n-propylamino)tetralin (8-OH-DPAT) and buspirone. In addition, buspirone was examined at a higher dose in the presence of proadifen, which inhibits oxidative metabolism. Results show that (1) daily injections of 1-PP did not reverse helpless behavior; (2) the reversal of helpless behavior by 8-OH-DPAT or an active dose of buspirone was antagonized by daily coadministration of 1-PP; and (3) in Ss pretreated with proadifen, the highest "inactive" dose of buspirone induced a reversal of helpless behavior.

1354. Martin, Patrick; Beninger, Richard J.; Hamon, Michel & Puech, Alain J. (1990). **Antidepressant-like action of 8-OH-DPAT, a 5-HT$_{1A}$ agonist, in the learned helplessness paradigm: Evidence for a postsynaptic mechanism.** ***Behavioural Brain Research,*** 38(2), 135–144.
Investigated the ability of 8-hydroxy-2-(di-*n*-propylamino)tetralin (8-OH-DPAT), a serotonin (5-hydroxytryptamine [5-HT]) agonist, to reduce helpless behavior (HB) in male Wistar rats following (1) ip administration in Ss whose ascending 5-HT neurons were partially destroyed by previous 5,7-dihydroxytryptamine (5,7-DHT) injection into the raphe nuclei or (2) after local microinjection into the raphe nuclei or into the septum. The reversal of HB by 8-OH-DPAT (ip) was still observed in 5,7-DHT-treated Ss with telencephalic 5-HT uptake reduced by 50–75% depending on the region. 8-OH-DPAT microinjected into the raphe nuclei did not reverse HB; in contrast, 8-OH-DPAT microinjected into the septum reversed HB. Results suggest that the ability of 8-OH-DPAT to reverse HB probably involved the stimulation of postsynaptic rather than presynaptic 5-HT$_{1A}$ receptors.

1355. Martin, Patrick; Massol, Jacques & Puech, Alain J. (1990). **Captopril as an antidepressant? Effects on the learned helplessness paradigm in rats.** ***Biological Psychiatry,*** 27(9), 968–974.
Captopril, an angiotensin-coverting enzyme inhibitor (ACEI) currently used as an antihypertensive agent, may exhibit antidepressant properties in humans. The present experiment evaluated potential antidepressive activity of captopril on the learned helplessness paradigm in rats. Captopril (8, 16, 32 mg/kg/day, ip) induced a reversal of escape deficits but did not affect significantly the motor activity, suggesting that this effect was not due to motor stimulation. This antidepressant-like activity was comparable to that of imipramine (16, 32 mg/kg/day, ip). Naloxone (0.5; 1 mg/kg, ip) blocked the effect of captopril (16 mg/kg, ip) in this test. An opioid mediation could thus be responsible at least in part for its behavioral effect.

1356. Martin, Patrick; Pichat, Philippe; Massol, Jacques; Soubrié, Philippe et al. (1989). **Decreased GABA B receptors in helpless rats: Reversal in tricyclic antidepressants.** ***Neuropsychobiology,*** 22(4), 220–224.
Investigated the role of chronic administration of imipramine (IMI) and desipramine (DES) on gamma-aminobutyric acid (GABA) B binding in the frontal cortex and hippocampus of male rats in a learned helplessness (LH) paradigm. The procedure included 2 phases: helpless induction and inescapable shock preconditioning and conditioned avoidance training. Ss were randomly treated according to the following protocols: controls with no shock were given vehicle; experimental Ss with inescapable shocks were injected with DES or IMI. Data demonstrate the ability of DES or IMI to reduce the escape failures in Ss trained for LH. Helpless behavior in Ss subjected to inescapable shocks was associated with a decrease in the number of GABA B receptors in the frontal cortex. This reduction in GABA B receptors was counteracted by DES or IMI only in Ss considered responders to the treatment.

1357. Martin, Patrick; Soubrié, Philippe & Puech, Alain J. (1990). **Reversal of helpless behavior by serotonin uptake blockers in rats.** ***Psychopharmacology,*** 101(3), 403–407.
Tested the effects of citalopram, fluvoxamine, indalpine, and zimelidine in rats subjected to helplessness training. Reversal of escape deficit by serotonin uptake blockers was observed only when the drugs were administered after shuttlebox sessions. At higher doses, the 4 serotonin uptake blockers were without effect. These data suggest that serotonin uptake blockers exert antidepressant-like effects in animals but only when they produce a moderate stimulation of serotonin neurotransmission.

1358. Martin, Patrick; Soubrié, Philippe & Simon, Pierre. (1987). **The effect of monoamine oxidase inhibitors compared with classical tricyclic antidepressants on learned helplessness paradigm.** ***Progress in Neuro-Psychopharmacology & Biological Psychiatry,*** 11(1), 1–7.
Male Wistar rats were exposed to inescapable shock pretreatment, and 48 hrs later, shuttlebox training was initiated to evaluate interference effect. Ss with inescapable shocks exhibited escape and avoidance deficits when tested for subsequent responding in a shuttlebox. Daily intraperitoneal injections of the monoamine oxidase inhibitors (MAOIs) nialamide (8 and 16 mg/kg), toloxatone (16 and 32 mg/kg), levodeprenyl (32 and 64 mg/kg), and tricyclic antidepressants (clomipramine, 16 and 32 mg/kg; desipramine, 16 and 24 mg/kg; imipramine, 16 and 32 mg/kg) eliminated escape deficits. In Ss exposed to inescapable shocks and treated with levodeprenyl (16 mg/kg/day), nialamide (32 mg/kg/day), or toloxatone (64 mg/kg/day), avoidance responses were significantly increased as compared with nondrugged Ss preexposed to inescapable shocks. Data extend previous results by A. D. Sherman et al (1982) concerning the similarity of action of MAOIs of the A type and antidepressants in learned helplessness paradigms.

1359. Martin, Patrick; Tissier, Marie-Hélène; Adrien, Joëlle & Puech, Alain J. (1991). **Antidepressant-like effects of buspirone mediated by the 5-HT1A post-synaptic receptors in the learned helplessness paradigm.** ***Life Sciences,*** 48(26), 2505–2511.

Investigated whether pre- or postsynaptic 5-hydroxytryptophan$_1$≅ (5-HT$_1$≅) receptors were involved in the antidepressant action of buspirone in the learned helplessness paradigm in rats. The ability of buspirone compared with 8-hydroxy-2(di-n-propylamino)tetralin (8-OH-DPAT) to reduce helpless behavior was investigated after local microinjections into the raphe nuclei or the septum. Microinjections of buspirone or 8-OH-DPAT into the raphe nuclei did not reverse helpless behavior. In contrast, microinjections of both 5-HT$_1$≅ agonists into the septum reverse helpless behavior. Results suggest that antidepressant-like properties of buspirone and 8-OH-DPAT may be mediated by the postsynaptic 5-HT$_1$≅ receptors through functional enhancement of the 5-HT transmission.

1360. Massol, Jacques; Martin, Patrick; Chatelain, F. & Puech, Alain J. (1990). **Tricyclic antidepressants, thyroid function, and their relationship with the behavioral responses in rats.** ***Biological Psychiatry,*** 28(11), 967–978.
Studied the effects of tricyclic antidepressants (TCAs) on thyroid function in rats in the learned helplessness paradigm. TCAs (clomipramine 32 mg/kg; desipramine 16, 24 mg/kg; or imipramine 8, 16, 32 mg/kg per day) were injected ip for 5 consecutive days. Blood samples were collected 1 hr after the last administration of the antidepressant for radioimmunoassay determination of triiodothyronine (T_3) and thyrotropin. TCA therapy dose dependently decreased the T_3 levels without changing thyroid-stimulating hormone (TSH) levels in helpless Ss and in naive controls. Using 2 models of experimentation, one involving diabetes induction, the other using food deprivation, TCAs further decreased the T_3 levels in diabetic and food-restricted Ss. This study confirms that TCAs decrease thyroid function and suggests that the antidepressant effect of TCAs is not related to their T_3 decreasing effects.

1361. Massol, Jacques; Martin, Patrick; Belon, Jean-Paul; Puech, Alain J. et al. (1989). **Helpless behavior (escape deficits) in streptozotocin-diabetic rats: Resistance to antidepressant drugs.** ***Psychoneuroendocrinology,*** 14(1–2), 145–153.
Compared the ability of antidepressants (clomipramine, desipramine, imipramine) and a central beta-receptor agonist (clenbuterol) to reverse helpless behavior (i.e., restore operant escape responding) in diabetic and nondiabetic rats trained for learned helplessness. Diabetes did not alter learned helplessness induction in that diabetic and nondiabetic Ss subjected to inescapable shocks exhibited identical escape deficits when tested for subsequent shuttle-box responding. Data indicate that experimental diabetes is associated with a reduced response to antidepressant drugs, possibly linked to altered central beta-adrenergic function. It is suggested that there may be a similar resistance to conventional antidepressants in depressed diabetic patients.

1362. Masur, Jandira. (1981). **Animal models in psychobiologic research: Perspectives and limits.** ***Jornal Brasileiro de Psiquiatria,*** 30(5), 411–412.
Discusses basic mechanisms of alcohol tolerance in terms of animal models. The fetal alcoholism syndrome comprises problems that are species-specific and that hinder the extrapolation of results to humans. (English abstract).

1363. Matthysse, Steven. (1983). **Making animal models relevant to psychiatry.** ***Annals of the New York Academy of Sciences,*** 406, 133–139.
Argues that a major cause of the lack of progress in the development of better drugs for controlling psychoses is the unimaginativeness of the screening tests in animals used to develop new antischizophrenic drugs. The greatest need in developing animal models for screening drugs used to treat schizophrenia is to go beyond the limits of the "dopamine hypothesis" (screening tests based on dopamine blockade in animals) and to mimic the loss of motivation that is the plague of chronic schizophrenic patients whose "positive" symptoms have been successfully treated by neuroleptic drugs. For example, radiation damage (in animals) and chronic marihuana use (in humans) are known to cause an "amotivational syndrome." Studies of the deficits associated with lesions in the frontal lobes would lead to other models of amotivational syndrome. It is concluded that when drug screening tests are unimaginative and only lead to the production of minor variants with essentially the same spectrum of action and side effects as existing drugs, there is neither an ethical use of animals nor an ethical exercise of responsibility for patients.

1364. McCloskey, Timothy C.; Beshears, James F.; Halas, Nancy A. & Commissaris, Randall L. (1988). **Potentiation of the anticonflict effects of diazepam, but not pentobarbital and phenobarbital, by aminooxyacetic acid (AOAA).** ***Pharmacology, Biochemistry & Behavior,*** 31(3), 693–698.
In daily 10-min sessions water-deprived female rats were trained to drink from a tube that was occasionally electrified, electrification being signaled by a tone. Within 2–3 wks control conditioned suppression of drinking (CSD) responding had stabilized; drug tests were conducted at weekly intervals. Diazepam, pentobarbital, and phenobarbital alone markedly increased the number of shocks received at doses that did not depress background responding (i.e., water intake). Treatment with the gamma-amino-butyric acid (GABA)-transaminase inhibitor AOAA alone had no anticonflict effect on CSD behavior. However, pretreatment with AOAA significantly potentiated the effects of diazepam. By contrast, the anticonflict effects of pentobarbital and phenobarbital were unaffected by this AOAA pretreatment. Thus, while increases in GABA transmission alone do not appear to affect CSD behavior, the anticonflict effects of benzodiazepines, but not barbiturates, appear to be potentiated by increases in GABA transmission.

1365. McGaugh, James L. (1989). **Dissociating learning and performance: Drug and hormone enhancement of memory storage.** 18th Annual Meeting of the Society for Neuroscience: Neural mechanisms of behavior: Performance versus learning (1988, Toronto, Canada). ***Brain Research Bulletin,*** 23(4–5), 339–345.
Reviews studies examining the enhancing effects of drugs and hormones on learning and memory. Many strategies have been used in an effort to dissociate drug effects on learning from drug effects on other processes affecting the performance of responses. These strategies include the use of tasks with various motivational and response requirements, the use of studies explicitly examining drug influences on performance, the use of posttraining drug administration, and the use of various forms of latent learning tasks. The dissociation of learning and performance effects of drugs cannot rest on one task or one experiment. Data suggest that drugs can and do enhance retention and that the effects are due to influences on memory storage rather than to other factors that influence performance.

1366. McKinney, William T. & Kraemer, Gary W. (1989). **Effects of oxaprotiline on the response to peer separation in rhesus monkeys.** ***Biological Psychiatry,*** 25(6), 818–821.
Hypothesized that administration of oxaprotiline, a relatively specific norepinephrine uptake inhibitor, would result in a less severe behavioral reaction to social separation in rhesus monkeys, whereas CGP 12103A, the (-) form of oxaprotiline, which has no effects on norepinephrine, would have no effect. 10 monkeys (approximately 5 yrs of age) were treated with oxaprotiline, CGP 12103A, or placebo in a repeated measures crossover design. All Ss exhibited age-appropriate social behaviors and had been housed with 4-monkey peer groups. Data on (1) stereotypy, (2) huddling, (3) locomotion, (4) self-directed behavior, and (5) inactivity were collected and analyzed when Ss were housed in separate cages 4 days/week. The hypothesis was not supported.

1367. McMillan, D. E. & Snodgrass, S. H. (1991). **Effects of acute and chronic administration of Δ^9-tetrahy-drocannabinol or cocaine on ethanol intake in a rat model.** ***Drug & Alcohol Dependence,*** 27(3), 263–274.
Examined the effects of acute and chronic administration of Δ^9- tetrahydrocannabinol (Δ^9-THC) or cocaine in 14 adult rats. Ss were trained to obtain all of their daily food by leverpressing during 4 equally spaced 30-min periods with water and 5% or 7.5% ethanol solutions freely available. Ethanol intake increased in rats during chronic Δ^9-THC administration and during withdrawal from chronic Δ^9-THC administration. Ethanol intake also increased during chronic cocaine administration (but not withdrawal). However, increases in ethanol intake during chronic cocaine administration may be slow to disappear when cocaine administration is discontinued.

1368. Means, Larry W.; Burnette, Mary A. & Pennington, Sam N. (1988). **The effect of embryonic ethanol exposure on detour learning in the chick.** ***Alcohol,*** 5(4), 305–308.
Assessed the effects of embryonic ethanol exposure on survivability, posthatching growth, and detour learning in 4 groups of 30 fertile sex-linked eggs injected with a solution of water and 0–50% ethanol immediately preceding incubation. A 5th group of 33 eggs served as noninjected controls. Results indicate that compared with controls, a smaller percentage of Ss receiving 37.5 or 50% ethanol hatched and survived for behavioral testing. Ss receiving 50% ethanol required more trials to reach criterion on the detour learning problem than did Ss receiving no ethanol. It is concluded that chicks are a good model for studying developmental and behavioral effects resulting from embryonic ethanol exposure.

1369. Mellanby, Jane H. (1986). **A comparison of the effects of epilepsy and ageing on learning and hippocampal physiology.** Fourth Workshop on Memory Functions (1985, Marstrand, Sweden). ***Acta Neurologica Scandinavica,*** 74(Suppl 109), 123–128.
Describes a model of hippocampal epilepsy that has the unusual feature among animal models of chronic epilepsy involving spontaneous fits that is eventually reversible. Research by C. A. Barnes (1979), who investigated changes in learning ability and in hippocampal physiology in senescent rats, is reviewed. It is concluded that both senescent rats and postepileptic rats are impaired in learning tasks that require hippocampal function.

1370. Mello, Nancy K. (1976). **Animal models for the study of alcohol addiction.** ***Psychoneuroendocrinology,*** 1(4), 347-357.
Animal models of alcoholism which fulfill the pharmacological criteria of addiction, tolerance, and physical dependence have been developed only recently. Both pharmacological (forced alcohol administration) and behavioral (self-administration) models are now possible in rodents and primates. Studies using these models are reviewed, noting that the relative advantages and disadvantages of pharmacological and behavioral models depend on the types of questions to be investigated. It is now possible to examine the neurophysiological, endocrinological, biochemical, and behavioral correlates of the development of alcohol addiction and to study the alcohol withdrawal syndrome in experiments which are neither feasible nor ethical in man.

1371. Mello, Nancy K.; Mendelson, Jack H.; King, Norval W.; Bree, Mark P. et al. (1988). **Alcohol self-administration by female macaque monkeys: A model for study of alcohol dependence, hyperprolactinemia and amenorrhea.** ***Journal of Studies on Alcohol,*** 49(6), 551–560.
In an amenorrheic alcohol-dependent monkey, prolactin levels increased during chronic, high-dose alcohol self-administration. Four amenorrheic cycles (85–194 days) from 2 other alcoholic female monkeys were also studied. Data suggest that both alcohol intoxication and relative alcohol withdrawal may alter basal prolactin levels and that hypothalamic amenorrhea is associated with suppression of gonadotropin secretory activity.

1372. Mello, Nancy K.; Mendelson, Jack H.; Bree, Mark P. & Lukas, Scott E. (1989). **Buprenorphine suppresses cocaine self-administration by rhesus monkeys.** ***Science,*** 245(4920), 859–862.
Daily administration of buprenorphine (BUP), an opioid mixed agonist-antagonist, significantly suppressed cocaine self-administration (CSA) by rhesus monkeys for 30 consecutive days. The effects of BUP were dose-dependent. The suppression of CSA by BUP did not reflect a generalized suppression of behavior. These data suggest that BUP would be a useful pharmacotherapy for treatment of cocaine abuse.

1373. Menéndez Abraham, Emilia; Menéndez Patterson, Angeles & Marín, Bernardo. (1986). **Rat model of sexual behavior of the male alcoholic: Sexual behavior and alcoholism.** ***Revista de Psicología General y Aplicada,*** 41(5), 961–973.
Studied parameters involved in sexual behavior of male rats after prolonged brandy consumption. 252 male and female Wistar rats (mean age 6 mo) (males' mean bodyweight 480 g) drank water or 10% ethanol solution ad lib for 6 mo. Copulatory behavior and other sexual behaviors of males were repeatedly tested. (English abstract).

1374. Menon, M. K.; Kodama, C. K.; Kling, A. S. & Fitten, J. (1986). **An *in vivo* pharmacological method for the quantitative evaluation of the central effects of alpha$_1$ adrenoceptor agonists and antagonists.** ***Neuropharmacology,*** 25(5), 503–508.
Describes a method for the quantitative evaluation of alpha$_1$-adrenoceptor agonists and antagonists that consists of recording the myoclonic twitch activity of the suprahyoideal muscle of anesthetized Sprague-Dawley rats. Results indicate that the method is an easily quantifiable procedure for the evaluation of the central effects of alpha$_1$-adrenoceptor ago-

nists and antagonists. It is anticipated that this new animal model would also prove valuable in assessing the changes in the functional state of the alpha$_1$-adrenergic system resulting from acute and chronic treatment with neuroleptic and antidepressant drugs.

1375. Merali, Zulfiquar; Ahmad, Qadeer & Veitch, Jennifer. (1988). **Behavioral and neurochemical profile of the spontaneously diabetic Wistar BB rat.** ***Behavioural Brain Research,*** 29(1–2), 51–60.
Investigated the effects of dopamine agonists and circadian cycle on locomotion, floor activity, rearing frequency, and rearing duration in male spontaneously diabetic rats. Results demonstrated a blunted response to D-amphetamine and lower levels of spontaneous locomotor and rearing activity in the latter part of the dark cycle.

1376. Mine, Kazunori et al. (1983). **Autonomic drug effects and gastric secretion in a new experimental model of stress ulcers in rats.** ***Pharmacology, Biochemistry & Behavior,*** 19(2), 359–364.
A psychological procedure that does not involve physical stimulation was used to produce gastric ulcers in male Wistar rats. Ulceration was induced by exposing the Ss to the aggressive attacks of other rats treated with 6-hydroxydopamine (6-OHDA). Gastric secretion and the effects of autonomic drugs on ulcer formation were investigated. Atropine methylbromide did not significantly inhibit the occurrence of erosions. Phentolamine or hexamethonium bromide significantly inhibited the production of erosions, and combined administration of an anticholinergic agent and an alpha-blocking agent led to a complete inhibition, with no notable behavioral changes. In case of pylorus ligation, gastric secretion during exposure to attack of 6-OHDA-treated Ss was significantly less than that in the controls. Data suggest that the sympathetic nervous system plays an important role in the production of gastric erosions.

1377. Miyamoto, Masaomi; Yamazaki, Naoki; Nagaoka, Akinobu & Nagawa, Yuji. (1989). **Effects of TRH and its analogue, DN-1417, on memory impairment in animal models.** ***Annals of the New York Academy of Sciences,*** 553, 508–510.
Studied the effects of thyrotropin-releasing hormone (TRH) and γ-butyrolactone-γ-carbonyl-histidyl-prolinamide citrate (DN-1417) on passive avoidance impairment induced by cycloheximide and CO_2 in rats. Results suggest that TRH and DN-1417 may improve memory in some animal models and may be useful for treatment of senile dementia.

1378. Morrissey, Thomas K.; Pellis, Sergio M.; Pellis, Vivien C. & Teitelbaum, Philip. (1989). **Seemingly paradoxical jumping in cataleptic haloperidol-treated rats is triggered by postural instability.** ***Behavioural Brain Research,*** 35(3), 195–207.
Examined the hypothesis that jumping in haloperidol (HAL)-treated male rats is triggered by instability using 64 male rats. Ss were administered HAL in a concentration of 5 mg/ml ip and control injections of 0.5 ml physiological saline. Results show that jumping by HAL-treated Ss is another allied reflex consistent with the defense of postural stability. Jumping by HAL-treated Ss only occurred when stability was challenged. Besides being important as an animal model of the behavioral effects of dopamine deficiency, the HAL-treated rat is a useful preparation for the function analysis of behavior. It provides a way of isolating and decomposing intermediate-level functionally organized subsystems of allied reflexes involved in exploration and locomotion, which are common to many forms of complex motivated behavior.

1379. Mueller, Kathyrne & Nyhan, W. L. (1982). **Pharmacologic control of pemoline induced self-injurious behavior in rats.** ***Pharmacology, Biochemistry & Behavior,*** 16(6), 957–963.
Three experiments with male Long-Evans hooded rats studied the effects of haloperidol, pimozide, diazepam, and serotonin depletion by pretreatment with parachlorophenylalanine (PCPA) or chronic pretreatment with parachloroamphetamine (PCA) on abnormal behavior produced by pemoline. Diazepam increased the duration of stereotyped behavior and reduced licking/biting and self-biting. Pretreatment with PCA had negligible effects on stereotyped behavior. Pretreatment with PCPA increased locomotion and rearing without affecting other components: stereotyped head movements, licking/biting, and self-biting. Haloperidol (0.2 and 0.3 mg/kg) produced a dose-related normalization of pemoline-induced behaviors, including elimination of self-biting. Pimozide (0.5, 0.8, and 1.3 mg/kg) had little or no effect on behaviors such as locomotions, rears, licking/biting, or stereotyped head movements, but eliminated self-biting at 1.3 mg/kg. Data suggest that pemoline, like amphetamine, produces stereotyped behavior through central dopaminergic mechanisms. Dopaminergic mechanisms also appear to be involved in pemoline-induced self-biting. Pemoline is apparently pharmacologically and behaviorally similar to amphetamine and may provide an animal model for syndromes characterized by self-injurious behavior and other repetitive behaviors.

1380. Mueller, Kathyrne; Saboda, Stephanie; Palmour, Roberta & Nyhan, W. L. (1982). **Self-injurious behavior produced in rats by daily caffeine and continuous amphetamine.** ***Pharmacology, Biochemistry & Behavior,*** 17(4), 613–617.
Two experiments with 71 male Long-Evans hooded rats compared self-biting (SB) produced by continuous amphetamine (AM) to SB produced by daily caffeine, which has been proposed as an animal model for self-injurious behavior (SIB) in the Lesch-Nyhan syndrome. Silicone pellets containing AM base were implanted for 4.5 days; caffeine (140 mg/kg, sc) was administered daily for 10 days. AM produced a higher rate of SB (75 vs 40%) than caffeine, with fewer toxic effects (no deaths vs 3 deaths). Neither drug produced stereotypy. Haloperidol (.2 mg/kg, ip) was only marginally effective in controlling SB produced by daily caffeine, but pimozide (1.5 mg/kg, ip), which has a longer duration of action, prevented SB by AM. Continuous release AM pellets may provide an alternative to the caffeine model of SIB in humans.

1381. Murphy, Michael R.; Blick, Dennis W. & Brown, G. Carroll. (1989). **Effects of hazardous environments on animal performance.** ***USAF School of Aerospace Medicine Technical Report,*** 88-40, 26 p.
Conducted animal experiments to estimate the threat of the nerve agent soman to US Air Force aircrews and ground personnel, along with the effects of pharmacological countermeasures to soman. An animal model system consisting of the primate equilibrium platform test and a rodent test battery was developed to study the effects of chemical warfare and defense drugs. The performance safety of the prophylactic use of pyridostigmine bromide was tested and

supported by the model. The nerve agent soman had an extremely steep dose–effect function on performance; the effects of soman were reliable and could be replicated in the same Ss at exposures separated by 6 wks. The levels of prophylactic and therapeutic drugs studied conferred only minor protection from soman-induced performance deficits.

1382. Muscat, Richard; Sampson, David & Willner, Paul. (1990). **Dopaminergic mechanism of imipramine action in an animal model of depression.** ***Biological Psychiatry,*** 28(3), 223–230.
In 2 experiments, a total of 124 male rats were subjected chronically (10–12 wks) to a variety of mild, unpredictable stressors or to no stressors. Stressed Ss showed a decrease in their consumption of weak sucrose solutions; normal behavior was restored by chronic (5–9 wks) treatment with tricyclic antidepressant imipramine. Acute administration of the dopamine receptor antagonist pimozide or the specific dopamine D2 receptor antagonist raclopride had no effect in nonstressed Ss or in vehicle-treated stressed Ss, but both drugs selectively reversed the improvement of performance in imipramine-treated stressed Ss. The 5-hydroxytryptamine (5-HT) antagonist metergoline increased sucrose consumption in all groups. Data suggest that the mechanism of action of imipramine in this model is an increase in functional activity at dopamine synapses.

1383. Myers, Michael M.; Musty, Richard E. & Hendley, Edith D. (1982). **Attenuation of hyperactivity in the spontaneously hypertensive rat by amphetamine.** ***Behavioral & Neural Biology,*** 34(1), 42–54.
Demonstrated that spontaneously hypertensive rats (SHR) are behaviorally hyperactive when compared with their normotensive Wistar Kyoto (WKY) progenitor strain. Behavioral hyperactivity was present in 3-min open-field tests and 1-hr tests using an automated activity recording chamber. Under certain conditions, dextroamphetamine (1.25–3.5 mg/kg) decreased activity in the SHR while inducing the expected increase in activity in the WKY. Analysis showed that the attentuation of SHR behavioral hyperactivity by amphetamine can be predicted based upon rate dependency of the actions of amphetamine. The SHR may provide a valuable animal model for studying spontaneous hyperactivity and investigating the neurochemical basis of the "paradoxical response" to amphetamine seen in children.

1384. Nabeshima, Toshitaka; Tohyama, Keiko; Murase, Kenshi; Ishihara, Sei-ichi et al. (1991). **Effects of DM-9384, a cyclic derivative of GABA, on amnesia and decreases in $GABA_A$ and muscarinic receptors induced by cycloheximide.** ***Journal of Pharmacology & Experimental Therapeutics,*** 257(1), 271–275.
Examined the effects of *N*-(2,6-dimethyl-phenyl)-2-(2-oxo-1-pyrrolidinyl)-acetamide (DM-9384), a cyclic derivative of gamma-aminobutyric acid (GABA), in the cycloheximide (CXM)-induced amnesia animal model, using the passive avoidance task with male mice. Pre- and posttraining and preretention test administration of DM-9384 attenuated the CXM-induced amnesia as indicated by prolongation of step-down latency. Aniracetam, another cyclic derivative of GABA, also showed antiamnesic effects. CXM decreased the number of $GABA_A$ and muscarinic acetylcholine (ACh) receptor binding sites. DM-9384 not only inhibited this effect but increased the latter. Results suggest that DM-9384 attenuates CXM-induced amnesia by interacting with GABAergic and AChergic neuronal systems and enhancing protein synthesis in the brain.

1385. Nadal, Roser; Pallarés i Año, Marc & Ferré, Nuria. (1991). **The reinforcing value of alcohol.** Special Issue: Alcohol and alcoholism. ***Avances en Psicología Clínica Latinoamericana,*** 9, 107–149.
Reviews the experimental animal literature on the reinforcing value of alcohol. Topics discussed include: free access paradigm, operant self-administration, conditioned taste preference, intracranial self-stimulation, conditioned place preference, tolerance, individual differences, route of alcohol administration, neurochemical basis of reward, monoaminergic system, opioid receptors, limbic system, and inadequacies of animal models of alcoholism. (English abstract).

1386. Neale, Robert; Fallon, Scott; Gerhardt, Susan & Leibman, Jeffrey M. (1981). **Acute dyskinesias in monkeys elicited by halopemide, mezilamine and the "antidyskinetic" drugs, oxiperomide and tiapride.** ***Psychopharmacology,*** 75(3), 254–257.
Oxiperomide and tiapride are dopamine receptor antagonists claimed to have "antidyskinetic" properties in animal models and in the clinic. Halopemide and mezilamine are other dopamine antagonists predicted to lack extrapyramidal side effects in humans on the basis of animal studies. Acute dyskinesias, a neuroleptic-induced acute extrapyramidal syndrome, were elicited in male squirrel monkeys by oxiperomide (1 mg/kg), tiapride (30 mg/kg), and halopemide (10 mg/kg). The dyskinesias were virtually indistinguishable from those caused by a standard behaviorally equivalent dose of haloperidol (1.25 mg/kg, po) in the same Ss. Mezilamine (0.3 mg/kg) also induced dyskinesias, which appeared to be less pronounced than those following haloperidol. The antidyskinetic properties of oxiperomide and tiapride evidently do not confer protection against dyskinetic movements induced by dopamine antagonism.

1387. Neill, D.; Vogel, Gerald W.; Hagler, M.; Kors, D. et al. (1990). **Diminished sexual activity in a new animal model of endogenous depression.** ***Neuroscience & Biobehavioral Reviews,*** 14(1), 73–76.
Examined whether 84 adult male rats show decreased sexual activity, a behavioral abnormality found in endogenous depression (EDP), after neonatal treatment with the antidepressant clomipramine (CLI) or with saline. After neonatal CLI, Long-Evans rats had a pervasive diminution of sexual activities, including decreased mounts, intromissions, ejaculations, and increased mount latencies and postejaculatory pause. Sprague-Dawley and Wistar strains also tended to show decreased intromissions and ejaculations, but their baseline sexual activity was too low to give interpretable data. Results with the sexually active Long-Evans strain are consistent with the hypothesis that neonatal CLI produces adult rats that model human EDP.

1388. Nelson, William T. et al. (1981). **Brain site variations in effects of morphine in electrical self-stimulation.** ***Psychopharmacology,*** 74(1), 58–65.
Pairs of bipolar electrodes were stereotaxically aimed at 2 of 3 sites: the locus coeruleus (LC), the substantia nigra, pars compacta (SNC), and the medial forebrain bundle (MFB). 41 male Holtzman Sprague-Dawley rats were shaped to bar-press for trains of intracranial electrical stimulation pre-

sented as pairs of monophasic pulses. The 1st pulse of a pair (conditioning pulse) was followed by a 2nd pulse (test pulse) after a parametrically varied interval. The effects of chronic morphine administration were tested in a paradigm of 7 days saline, 7 days morphine, 1 day morphine + naloxone, and 6 days postdrug saline. High doses of morphine depressed response rates for intracranial self-stimulation (ICSS). LC placements and those just lateral or ventral to the LC showed large increases in ICSS rates under morphine. This area was delimited on either side by tips that showed response rate depressions under morphine. MFB placements yielded response rate facilitations under morphine. Sites medial to the MFB and ventral within the MFB showed rate depressions under morphine. Dorsal substantia nigra placements showed facilitated rates, whereas placements ventral and within the SNC and substantia nigra, pars reticulata produced more variable results, with rates tending to be depressed by morphine. It is concluded that the ICSS procedure may be a useful animal model for detecting the abuse potential of drugs.

1389. Nemeroff, Charles B. et al. (1978). **Cholecystokinin inhibits tail pinch-induced eating in rats.** ***Science,*** 200(4343), 793–794.
Found that peripheral administration of the COOH-terminal octapeptide of cholecystokinin (1–100 μ/kg) significantly antagonized tail pinch-induced eating in male Sprague-Dawley rats, an animal model for stress-induced human hyperphagia. Centrally administered cholecystokinin was effective only in high doses. The finding that the minimal effective dose of cholecystokinin in suppressing stress-induced appetitive behavior is smaller after peripheral than central administration suggests that the peptide is acting on peripheral, as opposed to CNS substrates.

1390. Nemova, E. P.; Lyubimov, B. I.; Smolnikova, N. M.; Krylova, A. M. et al. (1991). **Correction with lithium hydroxybutyrate of disturbances of CNS function in offspring of male rat alcoholics.** ***Farmakologiya i Toksikologiya,*** 54(1), 65–67.
Administered alcohol intragastrically in a dose of 8 g/kg/day for 4 wks, and studied effects in offspring with and without corrective doses of lithium hydroxybutyrate in male rats (immature) (fathers) (experimental and control Ss); 60 rats (offspring). From Days 8–14 of life, 20 offspring were administered lithium hydroxybutyrate in a dose of 44 mg/kg/day sc. 20 other offspring were administered a sodium chloride solution. 20 offspring of control rats also received sodium chloride. Central nervous system (CNS), motor activity, emotionality, exploratory activity, extrapolation behavior, learning, and memory, were studied in offspring. (English abstract).

1391. Newton, Michael W.; Crosland, Richard D. & Jenden, Donald J. (1986). **Effects of chronic dietary administration of the cholinergic false precursor N-amino-N,N-dimethylaminoethanol on behavior and cholinergic parameters in rats.** ***Brain Research,*** 373(1–2), 197–204.
The choline analog, N-amino-N,N-dimethylaminoethanol (NADe [0.5%]) was fed ad libitum to weanling male Sprague-Dawley rats in a low choline, low methionine synthetic diet. Control Ss were fed choline chloride (0.5%). Experimental and control Ss gained weight more slowly than Ss fed standard lab chow. After 25 days on the diet, the performance of Ss fed NADe in a 1-trial passive avoidance test was significantly impaired compared to control Ss. These behavioral results were similar in 2 separate feeding experiments using deuterium-labeled and unlabeled NADe. The twitch response of isolated phrenic nerve-diaphragms during stimulation did not show any impairment of neuromuscular function in Ss fed NADe. Acetylcholinesterase and choline acetyltransferase activities in cortex were similar in all Ss. Results suggest that dietary NADe is not cytotoxic but may be useful in producing an animal model of central cholinergic hypofunction.

1392. Nielsen, Erik B.; Lee, Tong H. & Ellison, Gaylord. (1980). **Following several days of continuous administration *d*-amphetamine acquires hallucinogenlike properties.** ***Psychopharmacology,*** 68(2), 197–200.
Female albino rats injected ip with LSD (8 or 25 μg/kg) or mescaline (5 or 10 mg/kg) showed the behavioral syndrome—limb flicks and whole body shakes—previously reported after injections of hallucinogens in higher mammals. Although the behaviors were not elicited by acute dextroamphetamine (0.3, 1, or 2 mg/kg), they were present in Ss pretreated for 108 hrs with a slow-release amphetamine pellet, given a 12-hr rest period, and then injected with dextroamphetamine. Pellet-pretreated Ss also groomed excessively. It is proposed that this novel syndrome can serve as an animal model of the amphetamine psychosis produced in humans by a similar drug regimen.

1393. Niemegeers, C. J. (1974). **The predictive value of the anti-apomorphine test in dogs for neuroleptic therapy in man.** ***Activitas Nervosa Superior,*** 16(1), 56–58.
Presents a study which utilized the fact that most neuroleptics are potent inhibitors of apomorphine-induced emesis in dogs to demonstrate that the neuroleptics haloperidol and pimozide prevent the emesis due to apomorphine. It is suggested that, using this model in dogs, it is possible to predict in man the maintenance dose, optimal dosage schedule, and regularity of action of more than 20 clinically available neuroleptics belonging to the butyrophenones, phenothiazines, diphenylbutylpiperidines, thioxanthenes, and dibenzazepines.

1394. Nijssen, A. & Schelvis, P. R. (1987). **Effect of an anti-anxiety drug in a learned helplessness experiment.** ***Neuropsychobiology,*** 18(4), 195–198.
Investigated whether the inactivity of helpless rats caused by electric shock of moderate duration can be considered a symptom of anxiety. Results are in agreement with the inactivity hypothesis of J. Weiss et al (1976) and with the results of the study by R. C. Drugan et al (1984), who also emphasized the role of anxiety or fear in learned helplessness experiments.

1395. Nikulina, Ella M.; Skrinskaya, Julia A. & Popova, Nina K. (1991). **Role of genotype and dopamine receptors in behaviour of inbred mice in a forced swimming test.** ***Psychopharmacology,*** 105(4), 525–529.
The role of genotype in the effects of selective D1 and D2 dopamine agonists and antagonists on behavioral despair (Porsolt's test) was studied. Mice of 9 inbred strains showed significant interstrain differences in duration of immobility. The influence of dopaminergic drugs was assessed in 6 strains characterized by different levels of swimming activity. SKF 38393 (10 mg/kg), an agonist at D1 dopamine receptors, increased swimming activity, while the D1 antagonist SCH 23390 (0.2 and 0.5 mg/kg) reduced it, the effects being genotype dependent. The D2 agonist quinpirole (2.5

mg/kg) increased immobility in the majority of the mouse strains studied, while in CBA mice it resulted in a marked reduction of immobility. The D2 antagonist sulpiride (20 mg/kg) decreased immobility and increased active swimming in 2 strains. Results suggest a different role for D1 and D2 dopamine receptors in the regulation of swimming in the mouse.

1396. Ninan, Philip T. et al. (1982). **Benzodiazepine receptor-mediated experimental "anxiety" in primates.** ***Science,*** 218(4579), 1332–1334.
Previous findings indicate that the ethyl ester of beta-carboline-3-carboxylic acid has a high affinity for benzodiazepine (BDP) receptors in the brain. The present study demonstrated that in the rhesus monkey, this substance produced an acute behavioral syndrome characterized by dramatic elevations in heart rate, blood pressure, plasma cortisol, and catecholamines. The effects were blocked by BDPs and the specific BDP receptor antagonist Ro 15-1788. The BDP receptor may consist of several subsites or functional domains that independently recognize agonists, antagonists, or "active" antagonists such as beta-carboline-3-carboxylic acid ethyl ester. Results suggest that the BDP receptor is involved in both the affective and physiological manifestations of anxiety and that the administration of beta-carboline-3-carboxylic acid ethyl ester to monkeys may provide a reliable and reproducible animal model of human anxiety.

1397. Njung'e, Kung'u & Handley, Sheila L. (1991). **Evaluation of marble-burying behavior as a model of anxiety.** ***Pharmacology, Biochemistry & Behavior,*** 38(1), 63–67.
On individual placement in a cage with 20 evenly spaced glass marbles, 155 untreated female mice buried 7.8 ± 0.2 marbles. Olfactory stimuli from experimenters' hands and sex of mice had no influence on number buried, but most marbles were buried when they were evenly spaced. There was no habituation to these novel objects on serial testing or prehousing with marbles. The anxiogenic agents yohimbine and ethyl-beta-carboline-3-carboxylate did not enhance burying, and yohimbine decreased burying at doses also reducing locomotor activity. Diazepam effects depended on dose: 0.1 mg/kg increased burying, 0.25 mg/kg had no effect, and 1.0–5.0 mg/kg reduced it. Zimeldine reduced burying but not locomotor activity. Inhibition of marble burying may be a correlational model for detection of anxiolytics rather than an isomorphic model of anxiety.

1398. Nobrega, José N. & Wiener, Neil I. (1983). **Effects of catecholamine agonist and antagonist drugs on acute stomach ulceration induced by medial hypothalamic lesions in rats.** ***Pharmacology, Biochemistry & Behavior,*** 19(5), 831–838.
To investigate the involvement of catecholamines (CAs) in acute stomach ulceration induced by hypothalamic lesions, male Wistar rats were given bilateral electrolytic anodal lesions in the medial hypothamalmus followed by a single sc injection (3–36 mg/kg) of CA agonist or antagonist drugs. Lesioned Ss that received no postoperative drug treatment showed extensive gastric damage when examined 24 hrs after the brain lesion. Chlorpromazine, amphetamine, desipramine hydrochloride, and isoproterenol caused significant reductions in the extent (total length) and/or number of erosions induced by the brain lesion. Results are consistent with evidence indicating a protective role for catecholamines in acute ulcer formation, and suggest a commonality between acute stress ulcers and those induced by the lesions.

1399. Nurnberger, John I. (1987). **Pharmacogenetics of psychoactive drugs.** Munich Genetic Discussion International Symposium (1986, Berlin, Federal Republic of Germany). ***Journal of Psychiatric Research,*** 21(4), 499–505.
Contends that response to psychoactive drugs is subject to genetic variation. Heritable differences in metabolism of antidepressants have been demonstrated and appear to be clinically relevant. More complex genetic variations in drug response are important in alcoholism and possibly other forms of substance abuse. Pharmacologic challenge studies in volunteer twins have shown familial variation in responses to the cholinergic agonist arecoline and to the monoaminergic agonist amphetamine. Arecoline responses also differentiate well-state bipolar patients from controls; this is consistent with a hypothesis of muscarinic supersensitivity in affective disorder. It is suggested that genetically characterized animal models may be useful in further analyzing such neurochemical differences.

1400. Nuzhny, V. P.; Abdrashitov, A. Kh.; Listvina, V. P. & Uspensky, A. E. (1986). **Pattern of alcohol administration and alcohol loading in rats.** ***Farmakologiya i Toksikologiya,*** 49(1), 96–100.
Studied the effectiveness of voluntary and semivoluntary methods as techniques for modeling alcoholism in normal male white mongrel rats (1.5–8 mo). Water and ethanol consumption and blood and urine ethanol concentrations were measured in the Ss. The Ss had ethanol solutions (5–25%) and water as an alternate fluid (2-bottle choice) or a 10% ethanol solution as a sole water source. (English abstract).

1401. Ogawa, Norio et al. (1984). **Potential anti-depressive effects of thyrotropin releasing hormone (TRH) and its analogues.** ***Peptides,*** 5(4), 743–746.
The antidepressive effects of thyrotropin releasing hormone (TRH) and its analogs (DN-1417: gamma-butyrolactone-gamma-carbonyl-histidyl-prolinamide citrate; MK-771: levo-pyro-2-aminoadipyl-histidyl-thiazolidine-4-carboxamide) were examined in behavioral despair rats, an animal model of depression. TRH, DN-1417, MK-771, amitriptyline, and diazepam were injected 3 times after the 1st forced swimming. One hour after the last injection, a 5-min swimming test was performed. Experimental Ss were placed in a Hall's type open-field apparatus immediately before and after the 5-min test, and their locomotor activities were determined. In the 5-min swimming test, TRH, D-1417, and MK-771 caused a dose-dependent decrease in immobility, showing an antidepressive effect similar to amitriptyline. After the swimming test, locomotor activity remarkably decreased in controls, while decreased locomotor activity was partially prevented in the TRH, DN-1417, MK-771, and amitriptyline treated Ss, which exhibited active movement not only during the swimming period but also after it. In terms of the minimum effective dose, TRH and DN-1417 seemed to be of similar potency, while MK-771 was 40-fold stronger than TRH. An examination of a possible correlation between the cross-reactivity of TRH analogs in a radioreceptor assay and the effects of the analogs on despair rats suggested that the structure-binding relationship was proportional to the structure-activity relationship.

1402. Olivier, B.; Tulp, M. Th. & Mos, J. (1991). **Serotonergic receptors in anxiety and aggression: Evidence from animal pharmacology.** ***Human Psychopharmacology Clinical & Experimental,*** 6(Suppl), 73–78.

Describes the effects of 5-hydroxytryptamine$_{1A}$ (5-HT$_{1A}$), 5-HT$_{1B}$, 5-HT$_{1C}$, 5-HT$_{1D}$, 5-HT$_2$, and 5-HT$_3$ ligands in preclinical models of anxiety and aggression in rodents. 5-HT$_{1A}$ agonists show up as strong anxiolytic drugs in some animal paradigms, and their behavioral profile is clearly different from that of benzodiazepines. The 5-HT$_{1B}$, 5-HT$_{1C}$, and 5-HT$_{1D}$ ligands have mixed effects in anxiety paradigms. 5-HT$_3$ antagonists seem to exert anxiolytic effects, at least in some animal anxiety models. In offensive aggression, 5-HT$_{1A}$ agonists are not specifically antiaggressive, presumably due to sedation and interference with serotonergic behavior. Mixed 5-HT$_{1A}$ and 5-HT$_{1B}$ agonists appear to be specific antiaggressive agents, reducing offense without sedative or other unwanted side effects. Data suggest a modulatory role for the 5-HT$_{1B}$ or 5-HT$_{1A/1B}$ combination in offensive aggression.

1403. Olivier, Berend & Mos, Jan. (1991). **Pharmacologic treatment of aggression.** ***Psycholoog,*** 26(11), 490–494.
Discusses the development of a new class of antiaggression drugs, the serenics. Animal models of aggressive behavior in psychopharmacologic research are considered, with emphasis on the behavioral profile of the serenic eltoprazine in preclinical aggression models as compared with current clinically used drugs (e.g., neuroleptics and benzodiazepines). The role of the 5-hydroxytryptamine (5-HT) neurotransmitter system in the central nervous system (CNS) in modulating aggressive behavior is discussed. (English abstract).

1404. Olton, D. S. & Wenk, L. (1990). **The development of behavioral tests to assess the effects of cognitive enhancers.** Second International I.T.E.M.-Labo Symposium on Strategies in Psychopharmacology: Cognition enhancers—from animals to man (1988, Paris, France). ***Pharmacopsychiatry,*** 23(Suppl 2), 65–69.
Proposes a framework to assist in the development and evaluation of specific behavioral tests to determine the effects of drugs on cognition enhancement in animals. This framework requires explicit identification of the goals of the research, the cognitive process, the neural systems involved in the process, the selectivity and sensitivity of tasks that measure the process, the validity of the behavioral tasks that measure the process, and the validity of the behavioral tasks as a model to predict the effects of the drug in humans. The 5 characteristics of a representation of recent memory are discussed to illustrate the importance of specificity in defining the type of cognitive process to be enhanced. Ethical issues relating to enhancing cognitive functions are addressed. (German abstract).

1405. Opitz, Klaus & Weischer, Marie-Luise. (1988). **Volitional oral intake of nicotine in tupaias: Drug-induced alterations.** ***Drug & Alcohol Dependence,*** 21(2), 99–104.
10 nondeprived, unstressed tree shrews drank moderate quantities of a nicotine solution regularly over a period of 14 mo. Self administration of nicotine was attenuated by several drugs, including the nicotine-related alkaloid anabasine, the opioid agonist methadone, and the benzodiazepine chlordiazepoxide. Three serotonergic drugs—5-hydroxytryptophan (5-HTP), quipazine, and ipsapirone—were most effective in reducing nicotine intake. It is suggested that the serotonergic system is important for the reinforcing properties of both nicotine and alcohol.

1406. Orlando, Roy C.; Hernandez, Daniel E.; Prange, Arthur J. & Nemeroff, Charles B. (1985). **Role of the autonomic nervous system in the cytoprotective effect of neurotensin against gastric stress ulcers in rats.** ***Psychoneuroendocrinology,*** 10(2), 149–157.
Investigated the effect of autonomic nervous system activity on the development of cytoprotection against cold plus restraint stress-induced gastric ulcers in male Sprague-Dawley rats and its potential for mediating gastric mucosal cytoprotection by centrally administered neurotensin (NT). Drugs that stimulated alpha- or beta-adrenergic receptors or blocked muscarinic cholinergic receptors reduced the incidence of ulcers to a similar degree as intracisternal NT; alpha- or beta-adrenergic blockade as well as cholinergic stimulation prevented NT's beneficial effect. However, pretreatment with indomethacin blocked only the cytoprotective effect of NT or beta-adrenergic stimulation but not that of muscarinic cholinergic blockade. Pretreatment with reserpine or guanethidine also was effective in preventing cytoprotection by intracisternal NT. These data indicate that the mechanism for cytoprotection by centrally administered NT is mediated at least in part through activation of the sympathetic nervous system. This activation by NT appears to produce cytoprotection by stimulation of gastric mucosal prostaglandin synthesis.

1407. Osipov, V. V. (1988). **Alcoholism model on minipigs.** ***Farmakologiya i Toksikologiya,*** 51(3), 97–99.
Developed an alcoholism model that used voluntary alcohol consumption with free access to water as the main method of forming alcohol motivation. Ss were 295 male and female minipigs (10–24 mo). Ss received different concentrations of ethanol (1) in water and (2) in water with added mixed feed and (3) sweetened ethanol solutions in water. (English abstract).

1408. Otsuki, Taisuke; Nakahama, Hiroshi; Niizuma, Hiroshi & Suzuki, Jiro. (1986). **Evaluation of the analgesic effects of capsaicin using a new rat model for tonic pain.** ***Brain Research,*** 365(2), 235–240.
Produced an animal model of tonic pain in which the effects of analgesics can be determined objectively and quantitatively. Monosodium urate crystals (1–6 mg) were injected into knee joints of the hind paws of male Sprague-Dawley rats. Significant analgesic effects were obtained with capsaicin (50 mg/kg subdermally) when administered to neonatal Ss or locally to peripheral nerves (10 µl of 2% capsaicin inside the covering sheath of each nerve).

1409. Overstreet, David H. (1986). **Selective breeding for increased cholinergic function: Development of a new animal model of depression.** ***Biological Psychiatry,*** 21(1), 49–58.
Two lines of Sprague-Dawley rats that were selectively bred to vary in their sensitivity to the anticholinesterase DFP exhibited different degrees of behavioral depression after injection of the muscarinic agonist arecoline (2 mg/kg). The line of Ss with increased behavioral depression after arecoline (the Flinders sensitive or S-line) also exhibited a greater reduction of activity in an open-field chamber following exposure to footshock and greater immobility in a forced swim test than the line of Ss with reduced behavioral depression after arecoline (the Flinders resistant or R-line). In addition, the Flinders S-line exhibited a better memory on an inhibitory avoidance task. These differences were not related to differences in shock sensitivity between the lines. It is concluded that the Flinders S-line of rats reacts to both

mild stressors and a cholinergic agonist with greater behavioral depression and may, therefore, be a useful new animal model of human depressive disorders, one that focuses on cholinergic supersensitivity.

1410. Overstreet, David H. et al. (1986). **Enhanced elevation of corticosterone following arecoline administration to rats selectively bred for increased cholinergic function.** ***Psychopharmacology,*** 88(1), 129–130.
Measured serum corticosterone levels following subcutaneous administration of the cholinergic agonist arecoline HCl (4 mg/kg) to rats selectively bred for differences in cholinergic function. Eight Ss from Flinders Sensitive Line (FSL) exhibited both greater suppression of behavioral activity and enhanced elevation of serum corticosterone than 8 Flinders Resistant Ss. These enhanced responses to arecoline in the FSL rats parallel those reported in depressed humans, suggesting that these rats may provide a new animal model of affective disorders.

1411. Overstreet, David H.; Double, Kay & Schiller, Grant D. (1989). **Antidepressant effects of rolipram in a genetic animal model of depression: Cholinergic supersensitivity and weight gain.** ***Pharmacology, Biochemistry & Behavior,*** 34(4), 691–696.
Studied the effects of rolipram (ROL), a new generation antidepressant that is a selective inhibitor of phosphodiesterase, on female Flinders Sensitive Line (FSL) rats, a genetic animal model of depression. Acutely, ROL produced comparable decreases in temperature and activity in FSL and female Flinders Resistant Line rats. Chronic treatment produced a trend for ROL to counteract the shock-induced suppression of activity in FSL Ss, suggesting an antidepressantlike effect. However, both groups gained a significant amount of weight. In addition, both groups were significantly more affected by the muscarinic agonist oxotremorine than their vehicle-treated counterparts. Thus, FSL Ss, which are genetically supersensitive to cholinergic agonists, are even more sensitive following chronic treatment with ROL. Findings suggest that ROL may not be appropriate as an antidepressant for humans because of undesirable side effects.

1412. Oxenkrug, Gregory F.; McIntyre, Iain M.; Stanley, Michael & Gershon, S. (1984). **Dexamethasone suppression test: Experimental model in rats, and effect of age.** ***Biological Psychiatry,*** 19(3), 413–416.
Administered the dexamethasone suppression test (DST) to 2-, 4-, and 6-mo-old male Sprague-Dawley rats to determine the minimal dose of dexamethasone that significantly suppressed serum corticosterone. Based on a .2 μg/g dose of dexamethasone and a 1:00 PM sampling time, postdexamethasone corticosterone levels were significantly suppressed in 2- and 4-mo-old Ss, with higher levels in the latter group. Findings are discussed in the context of the development of an experimental animal model of the DST and its implications for use with humans.

1413. Panickar, Kiran S. & McNaughton, Neil. (1991). **Comparison of the effects of buspirone and chlordiazepoxide on successive discrimination.** ***Pharmacology, Biochemistry & Behavior,*** 39(2), 275–278.
Examined the effects of buspirone on successive discrimination, a conflict task employing omission of reward rather than shock, using 30 rats. Buspirone (3.3, 1.1, and 0.3 mg/kg) and chlordiazepoxide (5 and 20 mg/kg) were administered to separate groups of Ss throughout acquisition of a visual successive discrimination. Chlordiazepoxide released nonrewarded responding in a dose-related fashion. The effects of buspirone were similar in releasing response suppression but were both less in magnitude and less clearly related to dose. The action of buspirone in successive discrimination tasks does not depend on the use of shock but appears to be a genuine failure to fully release behavioral inhibition. Current animal models may need to be used with considerable care to ensure detection of novel anxiolytics.

1414. Paré, William P.; Glavin, G. B. & Vincent, George P. (1978). **Effects of cimetidine on stress ulcer and gastric acid secretion in the rat.** ***Pharmacology, Biochemistry & Behavior,*** 8(6), 711–715.
Cimetidine at 25, 50, and 100 mg/kg significantly inhibited gastric acid secretion in 10 male Sprague-Dawley rats with chronic gastric cannulas. Ss receiving either 50 or 100 mg/kg of cimetidine secreted significantly less gastric acid 3 hrs after injection. Cimetidine failed to reduce the number or size of gastric lesions in Ss exposed to the activity-stress procedure, but cimetidine at 100 mg/kg significantly reduced the number and size of gastric lesions in Ss subjected to a supine restraint procedure.

1415. Paré, William P. & Vincent, George P. (1984). **Cimetidine and stress ulcer in aged rats.** ***Physiology & Behavior,*** 33(2), 305–308.
200 male Sprague-Dawley rats (aged 2, 6, 11, 18, and 26 mo) were subdivided into 4 drug-treatment groups and administered either cimetidine (25, 50, or 100 mg/kg) or a placebo control before being exposed to restraint plus cold stress for 3 hrs to investigate the effects of cimetidine on stress ulcer in old vs young Ss. It is noted that the use of cimetidine to treat gastrointestinal disorders in elderly human patients is questionable but widespread. Results indicate that cimetidine at all dose levels significantly reduced ulcer severity only in 2-, 6-, and 11-mo-old Ss. Aged Ss were not more susceptible to restraint-induced stress ulcer than younger Ss. Cimetidine had the least anti-ulcer effect with aged 26-mo-old Ss.

1416. Pellow, Sharon & File, Sandra E. (1984). **Multiple sites of action for anxiogenic drugs: Behavioural, electrophysiological and biochemical correlations.** ***Psychopharmacology,*** 83(4), 304–315.
On the basis of a review of the literature on animal models of anxiety, a summary is provided of (a) known binding sites for anxiogenic drugs and (b) tests in which an anxiogenic action has been identified for drugs thought to act at benzodiazepine-binding or related sites and of drugs that can reverse these effects. There are limitations with all animal models currently in use, but there is considerable agreement among models in identifying anxiogenic drugs and identifying those that produce anxiety in humans. There appear to be 2 sites on the GABA-benzodiazepine-ionophore complex that can mediate anxiogenic effects: the benzodiazepine receptor and the picrotoxin site. Drugs that have intrinsic actions in behavioral tests also have intrinsic actions in electrophysiological preparations, and the patterns of reversal by other drugs are similar. Classification of these compounds according to data from ligand-binding studies does not correlate highly with their pharmacological interactions

in behavioral and electrophysiological procedures. It is not clear at present to what extent it is meaningful to correlate the functional consequences caused by a drug with its site of action.

1417. Peng, Rick Y.; Mansbach, Robert S.; Braff, David L. & Geyer, Mark A. (1990). **A D_2 dopamine receptor agonist disrupts sensorimotor gating in rats: Implications for dopaminergic abnormalities in schizophrenia.** ***Neuropsychopharmacology,*** 3(3), 211–218.
After administration of the D_1 agonist SK&F 38393, the D_2 agonist quinpirole, or a combination of the 2, male rats were tested for prepulse inhibition (PI) of the startle response by presenting acoustic stimuli alone or preceded by a weak prepulse that inhibits startle. The solitary administration of various doses of SF&F 38393 had no effect on PI, while quinpirole significantly reduced PI, at doses of .3 and .9 mg/kg. Findings confirm that a disruption of sensorimotor gating results from D_2 dopaminergic stimulation in the rat and extend applicability of this animal model to similar deficits in sensory gating exhibited by schizophrenic patients.

1418. Petersen, Dennis R. (1983). **Pharmacogenetic approaches to the neuropharmacology of ethanol.** ***Recent Developments in Alcoholism,*** 1, 49–69.
Describes the various animal models that have been used in studying the pharmacogenetics of alcohol's action. The principles of pharmacogenetic models are discussed with reference to inbred strains, F_1 and F_2 populations, heterogeneous stocks, and selectively bred lines. Models have been used to study the effects of alcohol on the CNS by studying sleep time in mice and sensitivity to subhypnotic acute alcohol intoxication. Alcohol withdrawal reactions have been studied in animals selectively bred for the severity of their reactions. The literature demonstrates that many of the behavioral and pharmacological responses to either acute or chronic actions of alcohol are heritable. It is doubtful if any one specific animal model will be developed that will serve as a prototype for human alcoholism. Genetic influence for alcoholism is indicated by the fact that nearly every attempt to breed selectively for an alcohol-related phenotype has been successful.

1419. Petty, Frederick; McChesney, Cheryl & Kramer, Gerald. (1985). **Intracortical glutamate injection produces helpless-like behavior in the rat.** ***Pharmacology, Biochemistry & Behavior,*** 22(4), 531–533.
Studied the behavioral effects of intracortical glutamate injection in 30 Sprague-Dawley rats in 3 experiments. Acute injection of glutamic acid hydrochloride (1–500 µg) into frontal neocortex of Ss produced a subsequent deficit in escape performance behavior that was similar to that produced by exposure to uncontrollable shock. The behavioral deficit was dose-related. The behavioral deficit was similar in time-course to that produced by 15 min (but not 40 min) of exposure to learned helplessness induction. Unlike learned helplessness produced by exposure to inescapable shock, the behavioral deficit produced by intracortical glutamate injection was not prevented by chronic ip administration of imipramine (10 mg/kg). Data confirm the utility of the learned helplessness animal model of depression as a tool for the study of the complex interaction between stress, anxiety, and depression.

1420. Petty, Frederick & Sherman, Arnold D. (1981). **A pharmacologically pertinent animal model of mania.** ***Journal of Affective Disorders,*** 3(4), 381–387.
6-Hydroxydopamine (iv, 450 µg) produced irritability and hyperresponsiveness to environmental stimuli in 34 male Sprague-Dawley rats. Chronic administration of lithium prevented the development of hyperreactivity to mild footshock, while chronic ECS effected a return toward normal behavior. Hydroxydopamine-induced hyperreactivity is proposed as a pharmacological model of mania.

1421. Petty, Frederick; Sacquitne, J. L. & Sherman, Arnold D. (1982). **Tricyclic antidepressant drug action correlates with its tissue levels in anterior neocortex.** ***Neuropharmacology,*** 21(5), 475–477.
In the behavioral reversal of learned helplessness in 24 male Sprague-Dawley rats, induced by imipramine, a strong correlation was found between "cure" of helplessness and drug level in anterior neocortex, the locus of drug action in this model of depression. This suggests that the delayed onset of therapeutic action of antidepressant drugs is due to the time required to achieve adequate drug levels at the site of their action.

1422. Phillips, Anthony G. & Fibiger, Hans C. (1990). **Role of reward and enhancement of conditioned reward in persistence of responding for cocaine.** ***Behavioural Pharmacology,*** 1(4), 269–282.
Advocates the continued use of animal models to study the control of behavior by cocaine (COC) and examines 3 influential techniques to study the rewarding properties of COC in animals: iv self-administration, pharmacological enhancement of brain-stimulation reward, and conditioned place preference. Experiments that exemplify COC's enhancement of the effects of conditioned rewarding stimuli are described along with those that relate to its neural substrates. The compulsion to abuse COC may be based on (1) primary reward related to euphoria and (2) a more subtle but equally potent influence by which the drug amplifies or enhances the effects of conditioned rewarding stimuli. A combined behavioral pharmacologic intervention, involving extinction procedures in conjunction with neuroleptic treatment, is proposed as a potential therapeutic strategy for breaking the COC habit in humans.

1423. Piercey, M. F. & Ray, C. A. (1988). **Dramatic limbic and cortical effects mediated by high affinity PCP receptors.** ***Life Sciences,*** 43(4), 379–385.
Used standard 2-deoxyglucose (2-DG) autoradiographic technique to obtain level of glucose metabolism in rats. The psychotomimetic drug "angel dust" (phencyclidine [PCP]) dramatically increased metabolism in diencephalic and telencephalic brain regions known to be rich in high affinity PCP receptors. Data support J. W. Papez's (1937) assertion concerning the importance of the central limbic circuit in emotional expression; reaffirm the potential importance of dopaminergic function in psychotic-like behavior; provide an animal model for schizophrenia; and establish that high affinity receptor sites are used by PCP to express its psychobiological effects.

1424. Pignatiello, Michael F.; Olson, Gayle A.; Kastin, Abba J.; Ehrensing, Rudolph H. et al. (1989). **MIF-1 is active in a chronic stress animal model of depression.** ***Pharmacology, Biochemistry & Behavior,*** 32(3), 737–742.

MIF-1, a tripeptide, was tested in an animal model of depression that used unpredictable chronic stress. In the paradigm, rats received either no stresses or a daily protocol of a variety of stressors for 20 days, during which time daily, ip injections of various compounds were given. The tricyclic antidepressant imipramine (5 mg) and low doses (0.1 and 1 mg) of MIF-1 significantly increased activity and decreased defecation in an open field on Day 21. No dose of naloxone (0.01–10.0 mg) acted as an antidepressant. A high dose (10 mg) of MIF-1 significantly increased the effects of chronic stress and produced hyperalgesia. Chronically-stressed Ss were significantly more analgesic than controls. Results indicate that MIF-1 can act as an antidepressant in this model.

1425. Pisa, Michele; Sanberg, Paul R.; Corcoran, Michael E. & Fibiger, Hans C. (1980). **Spontaneously recurrent seizures after intracerebral injections of kainic acid in rat: A possible model of human temporal lobe epilepsy.** ***Brain Research,*** 200(2), 481–487.
Intrastriatal injections of kainic acid in male Wistar rats induced repeated episodes of clonic convulsions. Spontaneously recurrent generalized seizures and a potentiation of the convulsant effects of pentylenetetrazol were then observed in most of the treated Ss several weeks after surgery. Loss of striatal neurons, and limbic pathological alterations similar to those found in human temporal lobe epilepsy were observed in the brains of the kainic-acid treated Ss. This preparation might serve as an animal model of human temporal lobe epilepsy.

1426. Poling, Alan; Cleary, James; Berens, Kurt & Thompson, Travis. (1990). **Neuroleptics and learning: Effects of haloperidol, molindone, mesoridazine and thioridazine on the behavior of pigeons under a repeated acquisition procedure.** ***Journal of Pharmacology & Experimental Therapeutics,*** 255(3), 1240–1245.
Examined the effects of haloperidol (HAL), molindone, mesoridazine, and thioridazine on the the behavior of pigeons exposed to a repeated acquisition procedure. At sufficiently high doses, each of these neuroleptics increased error rates (interfered with learning) and reduced rate of responding. When the drugs were compared on the basis of absolute doses administered, HAL disrupted behavior at doses considerably lower than the other drugs. If, however, chlorpromazine equivalent doses were examined, HAL was the least disruptive of the 4 drugs. Comparing the degree of behavioral disruption produced by the 4 drugs with their relative neuroreceptor affinities for dopamine D-2, cholinergic muscarinic, histamine H_1, *alpha*-1 adrenergic, and *alpha*-2 adrenergic receptors suggests that behavioral disruption cannot be attributed in any simple way to dopamine or acetylcholine receptor blockade.

1427. Pollock, Jondavid & Kornetsky, Conan. (1991). **Naloxone prevents and blocks the emergence of neuroleptic-mediated oral stereotypic behaviors.** ***Neuropsychopharmacology,*** 4(4), 245–249.
A common animal model for tardive dyskinesia (TD) is the oral stereotypy expressed by a challenge dose of a dopamine agonist after daily administration of dopamine antagonists (neuroleptics). In Exp 1, the expression of this dopamine agonist-induced oral stereotypy in male rats was prevented by the concomitant administration of the opiate antagonist naloxone. In Exp 2, if the stereotypy was allowed to be expressed, it could be blocked by the administration of naloxone. To the extent that the effects of chronic neuroleptic treatment in rats represent a model of TD, results suggest that administration of naloxone can both prevent and block the dyskinetic syndrome associated with neuroleptic use.

1428. Poncelet, M.; Dangoumau, L.; Soubrié, Philippe & Simon, P. (1987). **Effects of neuroleptic drugs, clonidine and lithium on the expression of conditioned behavioral excitation in rats.** ***Psychopharmacology,*** 92(3), 393–397.
Investigated the effect of neuroleptics, clonidine, and lithium on male Wistar rats treated with amphetamine (AMP) for 21 days and that showed enhanced activity under placebo in their AMP-associated environment. This conditioned effect was reduced by haloperidol, pimozide, and sulpiride at doses similar to or higher than those required to antagonize the unconditioned stimulus/stimuli (UCS) effects of AMP. Clonidine or lithium abolished AMP-conditioned hyperactivity. The hyperactivity induced by daily anticipation of food delivery was similar to the behavioral excitation produced by conditioning with AMP. Findings suggest that incentive activity elicited by AMP may model noradrenergic-dependent aspects of mania.

1429. Porsolt, R. D. (1986). **Psychopharmacological analysis of some behavioural models of depression.** ***Psychiatrie & Psychobiologie,*** 1(2), 150–155.
Describes a pharmacological analysis of separation and social dominance phenomena in monkeys and acute and chronic stress effects in rats to illustrate that judicious testing of known drugs can be an important tool for interpreting behavioral changes observed in the different behavioral models. (French abstract).

1430. Porsolt, R. D.; le Pichon, M. & Jalfre, M. (1977). **Depression: A new animal model sensitive to antidepressant treatments.** ***Nature,*** 266(5604), 730–732.
Describes a new behavioral model of depression that both resembles depressive illness and is selectively sensitive to antidepressant treatment. The method is based on the observation that a rat, when forced to swim in an inescapable situation, will eventually cease to move altogether, making only movements which are necessary to keep its head above water. It is argued that this characteristic and readily identifiable behavioral immobility indicates a state of despair in which the rat has learned that escape is impossible and resigns itself to the experimental conditions. This hypothesis was supported by data indicating that immobility is reduced by different treatments known to be therapeutic in depression including 3 drugs (iprindole, mianserin, and viloxazine) which, although clinically active, show little or no antidepressant activity in standard animal tests. Further, immobility was unaffected by anxiolytics and increased by drugs which induce depressive states in humans.

1431. Porsolt, R. D.; Lenègre, A.; Avril, I.; Stéru, L. et al. (1987). **The effects of exifone, a new agent for senile memory disorder, on two models of memory in the mouse.** ***Pharmacology, Biochemistry & Behavior,*** 27(2), 253–256.
Used male mice to investigate the effects of exifone, hexahydro-2,3,4,3',4',5'-benzophenone, in 2 models of memory (habituation of exploratory activity and antagonism of amnesia induced by scopolamine in a passive avoidance task). Results suggest that both exifone and piracetam improve memory and attenuate experimental amnesia in a manner that could be expected with drugs proposed for improving cognitive functioning in senile patients.

1432. Porsolt, Roger D. (1989). **Animal models of mental illness: The contribution of psychopharmacology.** ***Confrontations Psychiatriques,*** 30, 151–163.
Discusses 4 major criteria used to develop animal models of psychopathology: (1) similarity of triggering factors, (2) similarity of induced behavior states, (3) similarity of underlying neurobiochemical mechanisms, and (4) similarity in therapeutic effectiveness. These criteria are applied to models of depression and psychosis in monkeys and rodents, including the maternal or social separation model in the young monkey, the total social isolation model and the submissive behavior model of the dominated monkey, and the despair model in the rodent. (English, Spanish & German abstracts).

1433. Porsolt, Roger D.; Martin, Patrick; Lenègre, Antoine; Fromage, Sylvie et al. (1990). **Effects of an extract of Ginkgo Biloba (EGB 761) on "learned helplessness" and other models of stress in rodents.** ***Pharmacology, Biochemistry & Behavior,*** 36(4), 963–971.
Investigated the effects of repeated oral administration of EGB 761 on behavioral models of stress in male mice and rats, including learned helplessness, shock-suppressed licking, and forced swimming-induced immobility (for rats), as well as shock-suppressed exploration, spontaneous exploration, and food consumption in a novel situation (for mice). Further tests in rats examined the effects of EGB 761 on memory and responsiveness to shock. Repeated administration of EGB 761 before exposure to unavoidable shock reduced the subsequent avoidance deficits in the learned helplessness procedure but was less effective when administered after helplessness induction. Anxiolyticlike activity was seen in the emotional hypophagia test in mice. Results suggest that EGB 761 reduces the consequences of stress in some experimental situations, but this effect cannot be assimilated to either classical antidepressant or anxiolytic activity.

1434. Pradhan, N.; Arunasmitha, S. & Krishnan, U. (1989). **Study of MPTP induced parkinsonism in bonnet monkeys (macaca radiata).** ***Pharmacopsychoecologia,*** 2(1–2), 31–36.
Created an animal model of parkinsonism in 12 adult bonnet monkeys by systemic administration of the neurotoxin 1-methyl-4-phenyl-1,2,3,6-tetra hydropyridine (MPTP). The behavioral effects and the biochemical and pathological changes in the brain and cerebrospinal fluid (CSF) were studied. Intravenous administration of MPTP produced a severe parkinsonian syndrome within a few days. The acute effects included signs of sympathetic release, abnormal movements, and changes in posture and motor behavior. Histopathological examination of the brain of these Ss showed selective damage to the cells of the zona compacta of the substantia nigra. There was marked depletion of dopamine, dihydroxyphenyl acetic acid, and homovanillic acid in the substantia nigra, caudate nucleus, putamen, and globus pallidus. The neurochemical changes seen in the brain were not reflected in lumbar CSF.

1435. Preskorn, Sheldon H. et al. (1984). **Cerebromicrocirculatory defects in animal model of depression.** ***Psychopharmacology,*** 84(2), 196–199.
In the tetrabenazine (TBZ) model of depression, the cerebromicrocirculation of male Sprague-Dawley rats was discovered to respond abnormally to metabolic demand as mimicked by the administration of CO_2. Altered responsivity of cerebral blood flow and effective permeability of the blood were found. These physiologic defects coincided temporally with TBZ-induced depletion of central norepinephrine and dopamine and with the development of the behavioral effects of TBZ. Pretreatment with amitriptyline prevented the development of these TBZ-induced abnormalities in the cerebromicrocirculation.

1436. Price, Lawrence H.; Charney, Dennis S.; Delgado, Pedro L. & Heninger, George R. (1990). **Lithium and serotonin function: Implications for the serotonin hypothesis of depression.** ***Psychopharmacology,*** 100(1), 3–12.
Reviews the animal literature on the effects of lithium on various aspects of central serotonin function. Preclinical evidence of lithium's effects on 5-hydroxytryptamine (5-HT) function at the levels of precursor uptake, synthesis, storage, catabolism, release, receptors and receptor–effector interactions suggests that lithium's primary actions on 5-HT may be presynaptic, with many secondary postsynaptic effects. Time dependent and region and species specific variations are noted. The cerebrospinal fluid (CSF), metabolite platelet 5-HT function, and neuroendocrine challenge studies in humans generally suggest that lithium has a net enhancing effect on 5-HT function. These actions of lithium may serve to correct as-yet unspecified abnormalities of 5-HT function involved in the pathogenesis of depression.

1437. Pueschel, Siegfried M.; Schrier, Allan M.; Povar, Morris L. & Boylan, Joan M. (1985). **Biological and behavioural assessments of young rhesus monkeys after intrauterine exposure to high phenylalanine concentrations.** ***Journal of Mental Deficiency Research,*** 29(3), 247–256.
Performed biological and behavioral assessments of 20 young rhesus monkeys (*Macaca mulatta*) after intrauterine exposure to high phenylalanine concentrations. 19 pregnant rhesus monkeys were fed a special diet throughout the gestational period in an attempt to render them hyperphenylalaninemic. Group A (control group) received a regular diet, Group B was given a low phenylalanine diet, Group C a median phenylalanine diet, and Group D a high phenylalanine diet. Nearly all Ss had an uncomplicated pregnancy and an uneventful delivery. Biological measurements were obtained shortly after the birth of the infants and behavioral assessments were done when the offspring were between 6 and 18 mo. The results of the biological and behavioral evaluations reveal that there was no statistically significant difference among the respective study groups. It is concluded that a combination of factors inherent in an imperfect animal model may account for the negative results of this study.

1438. Pulvirenti, Luigi & Kastin, Abba J. (1988). **MIF-1 reduces stress-induced eating in rats.** ***New Trends in Experimental & Clinical Psychiatry,*** 4(4), 229–233.
Stress-induced eating in rats is considered a behavioral model of human stress-related hyperphagia. This model is characterized by a mild tailpinch-induced oral syndrome of eating, gnawing, and licking in the presence of food. Since it has been suggested that it may depend on activation of the opiate system, the present study tested the possibility that MIF-1 (Pro-Leu-Gly-NH_2), a brain peptide with antiopiate properties, may affect this behavior. In ip doses of 0.1 and 1.0 mg/kg, the peptide reduced tailpinch-induced eating. Antiopiate agents may contribute to the understanding and treatment of eating disorders.

1439. Puri, S.; Ray, Arunabha; Chakravarty, A. K. & Sen, P. (1991). **Role of histaminergic mechanisms in the regulation of some stress responses in rats.** ***Pharmacology, Biochemistry & Behavior,*** 39(4), 847–850.
Studied the involvement of histaminergic mechanisms in the regulation of some stress responses in inbred male rats. The brain neuronal histamine (HA) depletor, α-fluoromethyl histidine (α-FMH), at doses that markedly lower brain HA, significantly attenuated gastric ulcer formation and the elevation in plasma corticosterone in response to cold restraint stress (CRS). α-FMH also reduced gastric muscosal HA content. The H_1-antagonist pheniramine attenuated both gastric mucosal and endocrine response to CRS, while the effects of the H_2-antagonist cimetidine were on the plasma corticosterone levels. Results are discussed in light of complex HA-ergic mechanisms in the maintenance of physiological homeostasis during stress.

1440. Pycock, C.; Tarsy, D. & Marsden, C. D. (1975). **Inhibition of circling behavior by neuroleptic drugs in mice with unilateral 6-hydroxydopamine lesions of the striatum.** ***Psychopharmacologia,*** 45(2), 211–219.
Studied the development of circling behavior to apomorphine, amphetamine, and levodopa in male Swiss mice with unilateral 6-hydroxydopamine lesions of the dopaminergic nerve terminals in the striatum. The effect of a range of neuroleptic and sedative drugs on this circling behavior also was investigated. Circling induced by all the stimulant drugs was inhibited in a dose-dependent manner by haloperidol, pimozide, chlorpromazine, metoclopramide, and clozapine (in descending rank order of potency), but not by phenoxybenzamine, diazepam, promethazine, and pentobarbitone sodium. This relatively simple animal model appears useful for screening neuroleptic drugs which may block striatal dopamine receptors, thereby predicting their potency to cause unwanted extrapyramidal effects but not their antipsychotic efficacy.

1441. Randall, Carrie L.; Becker, Howard C. & Middaugh, Lawrence D. (1986). **Effect of prenatal ethanol exposure on activity and shuttle avoidance behavior in adult C57 mice.** ***Alcohol & Drug Research,*** 6(5), 351–360.
Examined the behavioral effects of prenatal alcohol exposure in C57 mice, a strain that is known to be highly sensitive to the teratogenic actions of ethanol. Pregnant mice were administered from Day 5 through Day 18 of gestation a liquid diet that contained 25% ethanol-derived calories (EDC). Control animals were pair-fed an isocaloric 0% EDC diet during the same period, with sucrose substituted for ethanol. At 23 days of age, offspring were tested for spontaneous locomotor activity in an open field. At age 70 days, different offspring were tested in a shuttle-avoidance task. Results indicate that the 25% EDC progeny were more active than were the controls. In addition, prenatal alcohol exposure produced a deficit in acquisition and performance of a shuttle-avoidance task. Alcohol-treated offspring made fewer avoidance responses and required more trials to reach a criterion performance of 10/10 avoidances consecutively followed by at least 9/10 avoidances. These results support an animal model of fetal alcohol syndrome in which both the behavioral and morphological consequences of prenatal alcohol exposure may be assessed in the same species.

1442. Rastogi, Santosh K.; Rastogi, Ram B.; Singhal, Radhey L. & Lapierre, Yvon D. (1983). **Behavioural and biochemical alterations following haloperidol treatment and withdrawal: The animal model of tardive dyskinesia reexamined.** ***Progress in Neuro-Psychopharmacology & Biological Psychiatry,*** 7(2–3), 153–164.
Administered haloperidol to male Sprague-Dawley rats for 30 or 37 days. On Day 38, some Ss from both groups received apomorphine and were tested in the open field and for stereotyped behavior. Long-term administration of haloperidol resulted in supersensitivity of dopamine receptors. This was manifested by enhanced stereotypic biting, rearing, and locomotor and floor activity of Ss when challenged to a low dose of apomorphine. Chronic haloperidol treatment significantly decreased dopamine synthesis and release. Ss withdrawn from chronic haloperidol treatment showed significant increases in GABA level and glutamic acid decarboxylase activity. This probably resulted in further inhibition of dopamine release as evidenced by marked accumulation of dopamine in the corpus striatum and midbrain. Data suggest that chronic haloperidol treatment and subsequent withdrawal results in the development of behavioral dopamine supersensitivity as well as biochemical alterations in dopaminergic and GABAergic system. However, the duration of supersensitive responses in animals is generally brief, while tardive dyskinesia in human patients is generally more enduring.

1443. Ray, A.; Henke, P. G. & Sullivan, R. M. (1990). **Effects of intra-amygdalar thyrotropin releasing hormone (TRH) and its antagonism by atropine and benzodiazepines during stress ulcer formation in rats.** ***Pharmacology, Biochemistry & Behavior,*** 36(3), 597–601.
Bilateral intra-amygdalar microinjections of TRH and physostigmine into the central nucleus (CEA) aggravated cold restraint stress induced gastric ulcer formation in male rats, whereas atropine attenuated this phenomenon. Similar stress ulcer reducing effects were seen with chlordiazepoxide (CDP) and midazolam. Pretreatment with atropine or CDP antagonized the ulcerogenic effects of both TRH and physostigmine. When administered intra-CEA, midazolam neutralized the effects of TRH in a dose-related manner.

1444. Ray, Arunabha & Henke, Peter G. (1990). **Enkephalin–dopamine interactions in the central amygdalar nucleus during gastric stress ulcer formation in rats.** ***Behavioural Brain Research,*** 36(1–2), 179–183.
Intra-amygdalar (i/am) microinjections of the enkephalin analog, (D-Ala[2])-Met-enkephalinamide (DAMEA, 3, 10 and 30 μg) into the central amygdalar nucleus (CEA) produced a dose-related, naltrexone-reversible attenuation of cold restraint (3 hrs at 4°C)-induced gastric mucosal lesions in rats. Similarly, gastric stress ulcer formation was also inhibited by i/am dopamine (DA [10 μg]), an effect that was reversed by the DA-antagonist, clozapine (5 mg/kg) pretreatment. Further, pretreatment of Ss with clozapine or the DA-neurotoxin, 6-hydroxydopamine (6-OHDA) (10 μg, i/am) clearly reversed and/or antagonized the gastric cytoprotective effect of DAMEA (30 μg). Results indicate interactions between enkephalinergic and DAergic systems at the level of the CEA in the maintenance of gastric mucosal integrity during immobilization stress.

1445. Rebec, George V. & Bashore, Theodore R. (1982). **Comments on "Amphetamine models of paranoid schizophrenia": A precautionary note.** ***Psychological Bulletin,*** 92(2), 403–409.
L. Kokkinidis et al (1980) reviewed animal research investigating the mechanisms of action of amphetamine and discussed the implications of this research for the development of animal models of paranoid schizophrenic psychosis. On the basis of their review of the literature, they hypothesized that abnormalities in neurotransmitter systems containing norepinephrine and dopamine were associated with this psychotic process. The present authors believe that the involvement of catecholamines in psychosis is determined by complex pre- and postsynaptic events not adequately discussed by Kokkinidis et al. Furthermore, open-field behavioral changes produced in the rat by amphetamines are not as simple nor are their neural substrates as well defined as Kokkinidis et al suggest.

1446. Reid, Larry D.; Delconte, John D.; Nichols, Michael L.; Bilsky, Edward J. et al. (1991). **Tests of opioid deficiency hypotheses of alcoholism.** ***Alcohol,*** 8(4), 247–257.
In 4 experiments, male rats were maintained on a daily regimen involving 22 hrs of deprivation of fluids followed by 2 hrs of access to water and a sweetened alcoholic beverage. A series of injections of opioids was given subsequent to establishing stable daily intakes of ethanol. When morphine was given 0.5 hrs before opportunity to drink (OTD), intake of ethanol was increased. However, when morphine was given 4 hrs before OTD, intake of ethanol was decreased. Nearly opposite effects were observed when naloxone was given. Morphine given 4 hrs before OTD potentiated the effects of naloxone and attenuated the effects of morphine. The effects of morphine were also similar among Ss taking a solution of ethanol and water rather than a sweetened solution. Data support the idea that surfeits in opioidergic activity increase propensity to take alcoholic beverages.

1447. Reis de Oliveira, Irismar; Diquet, Bertrand; Van der Meersch, Véronique; Dardennes, Roland et al. (1990). **Self-inhibiting action of nortriptyline's anti-immobility effect at high plasma and brain levels in mice.** ***Psychopharmacology,*** 102(4), 553–556.
Treated 72 male mice acutely with 0, 2.5, 5, 10, 20, and 40 mg/kg nortriptyline (NT) or with control vehicle 30 min before the tail suspension test. Ss were sacrificed after test for evaluation of plasma and brain levels of NT. The anti-immobility effect increased with increasing doses and concentrations of the drug, reaching significance at a dose of 20 mg/kg, 865 ng/ml in plasma and 11 μg/g in brain tissue. The anti-immobility effect was blocked with the highest nontoxic concentrations: 1,630 ng/ml plasma or 32 μg/g brain tissue. Results indicate a biphasic curvilinear relationship between plasma and brain levels of NT and behavior in mice. This animal model might be considered for research into the hypothesis of OH-metabolite blockade of antidepressant effect.

1448. Rejeski, W. Jack; Brubaker, Peter H.; Herb, Robert A.; Kaplan, Jay R. et al. (1988). **Anabolic steroids and aggressive behavior in cynomolgus monkeys.** ***Journal of Behavioral Medicine,*** 11(1), 95–105.
Employed an animal model to examine the effect of testosterone on aggressive behavior in 10 monkeys. Although the administration of testosterone for 8 wks resulted in a significant increase in aggression by the Ss (compared with sham-treated controls), changes in behavior appeared to be mediated by social status. The incidence of both contact and noncontact aggression in dominant Ss was far greater than the frequency of these behaviors in subordinate Ss. Results suggest that cardiovascular risk associated with anabolic steroid use may have a complex psychophysiological basis involving learned helplessness.

1449. Ricaurte, George A.; Langston, J. William; Delanney, Louis E.; Irwin, Ian et al. (1986). **Fate of nigrostriatal neurons in young mature mice given 1-methyl-4-phenyl-1,2,3,6-tetrahydropyridine: A neurochemical and morphological reassessment.** ***Brain Research,*** 376(1), 117–124.
Examined the effect of 1-methyl-4-phenyl-1,2,3,6-tetrahydropyridine (MPTP) on male C57BL/6J IMR mice nigrostriatal dopaminergic neurons to determine whether this effect is permanent or transient and to ascertain the validity of the MPTP-treated mouse as an experimental model of Parkinson's disease. It is suggested that while the young mature MPTP mouse may not be a valid animal model of Parkinson's disease, it will be valuable for the study of how MPTP destroys dopaminergic nerve terminals and may prove useful as an experimental system for studying recovery of dopaminergic fibers after injury.

1450. Richelle, Marc. (1989). **Methodological critique of animal models in psychopharmacology.** ***Confrontations Psychiatriques,*** 30, 165–177.
Discusses methodological factors that should be considered in developing psychopharmacological animal models, including explicative and predictive goals, individual animal history and behavior, ontogenic factors, interindividual differences in behavior, and differential reactivity to drugs. Difficulties in developing models based on analogies between observed animal behavior and human psychiatric disorders are illustrated, using an example of acquired helplessness. Current methods of studying animal behavior and behavioral responses to drugs are also reviewed. (English, Spanish & German abstracts).

1451. Rick, Garl K.; Scarfe, A. David & Hunter, Jon F. (1984). **L-pyroglutamate: An alternate neurotoxin for a rodent model of Huntington's disease.** ***Brain Research Bulletin,*** 13(3), 443–456.
Divided 91 adult albino male mice into 66 Ss used for behavioral and neuropathological evaluations and 25 Ss used for neuropathological evaluations only. Intrastriatal injections of levopyroglutamate (LPGA) produced behavioral and neuropathological effects that resembled in part the kainate-injected rat striatal model of Huntington's disease (HD). The behavioral responses induced after unilateral injections of LPGA included circling, postural asymmetry of head and trunk, and possible dyskinesias. The neuropil in the injected striatum contained dilated profiles, degenerating neurons and oligodendroglia, and numerous phagocytic microglial-like cells. A dose–response relation existed. LPGA was a weak neurotoxin when compared to kainic acid. Several factors raise interest in the possible role of LPGA in HD, including previous work by the authors (1983) and by the authors and colleagues (1984), which found elevated plasma levels of LPGA in some HD patients.

1452. Ridley, R. M.; Baker, H. F. & Scraggs, P. R. (1979). **The time course of the behavioral effects of amphetamine and their reversal by haloperidol in a primate species.** ***Biological Psychiatry,*** 14(5), 753–765.
Six marmosets were administered amphetamine (AM, Phase 1), AM plus haloperidol (HA, Phase 2), and then AM alone (Phase 3) over consecutive periods of 27, 51, and 33 days after which treatment was terminated (Phase 4). Ss' behavior was monitored during these periods and during a predrug control period. The time course of the effects of AM and HA on the behavioral categories suggests that different mechanisms may be involved in each case. Viewed as a model of schizophrenia, the time course of HA in reversing AM-induced suppression of locomotion most closely resembles the time course of the antipsychotic effect of neuroleptics in man. Some effects of AM (e.g., suppression of social interaction) are not reversed by HA, and some effects of withdrawal of HA (e.g., precipitation of checking movements not present when HA was commenced) do not have an obvious counterpart in the clinical situation.

1453. Rigdon, Greg C. (1990). **Differential effects of apomorphine on prepulse inhibition of acoustic startle reflex in two rat strains.** ***Psychopharmacology,*** 102(3), 419–421.
Apomorphine (APO) disruption of prepulse inhibition (PPI) has been proposed as an animal model of sensorimotor gating deficits exhibited by schizophrenics. The effects of APO on PPI of the acoustic startle reflex in male rats of Wistar and CD (Sprague-Dawley derived) strains were compared under identical test conditions. In Wistar Ss, sc administration of APO blocked PPI without affecting startle amplitude. In CD Ss, APO had no effect on PPI, but increased startle amplitude. Thus, choice of rat strain is an important factor in the design of experiments studying APO effects on PPI.

1454. Rijk, Huub; Crabbe, John C. & Rigter, Henk. (1982). **A mouse model of alcoholism.** ***Physiology & Behavior,*** 29(5), 833–839.
A model of alcoholism should demonstrate self-administration of alcohol by the animal and a withdrawal syndrome when the animal no longer has access to the drug. The present authors describe a mouse model that meets these criteria and enables the induction of "alcoholism" in a large number of animals within a short period of time. 80 male Swiss albino mice were used to test the model. It is concluded that self-administration of ethanol vapor, with resulting tolerance and withdrawal activity, may be used as a model of alcoholism.

1455. Riley, Edward P. (1990). **The long-term behavioral effects of prenatal alcohol exposure in rats.** ***Alcoholism: Clinical & Experimental Research,*** 14(5), 670–673.
Reviews some of the longer lasting behavioral consequences of gestational alcohol exposure in animals, focusing on activity, reactivity, and motor behavior; avoidance and operant behavior; sexually dimorphic behaviors; and spatial deficits. Prenatal alcohol exposure is shown to have long-lasting effects, although some of these effects might only occur under challenging or stressful circumstances. As the animal matures, compensatory mechanisms or strategies may develop to compensate for these dysfunctions. Thus, behavioral problems may only be detected when these compensatory systems break down, either as a result of stress, because of complex testing procedures, or because of old age.

1456. Riley, Edward P. & Barron, Susan. (1989). **The behavioral and neuroanatomical effects of prenatal alcohol exposure in animals.** Conference of the Behavioral Teratology Society, the National Institute on Drug Abuse, and the New York Academy of Sciences: Prenatal abuse of licit and illicit drugs (1988, Bethesda, Maryland). ***Annals of the New York Academy of Sciences,*** 562, 173–177.
Reviews data collected using rodent models of prenatal and neonatal alcohol exposure and relates them to clinical findings in children exposed to alcohol in utero. Topics discussed include response inhibition deficits, feeding and suckling behavior, balance and gait disorders, and early learning abilities. Investigations using rodent models may be useful in delineating the mechanisms underlying the effects of prenatal alcohol exposure, determining risk factors for fetal alcohol effects, and developing therapies for behavior dysfunctions resulting from prenatal alcohol exposure.

1457. Riley, Edward P.; Freed, Earl X. & Lester, David. (1976). **Selective breeding of rats for differences in reactivity to alcohol: An approach to an animal model of alcoholism: I. General procedures.** ***Journal of Studies on Alcohol,*** 37(11), 1535–1547.
Reports a continuing research effort to breed 1 rat strain in which the injection of a standard dose of alcohol would produce an objectively measurable intoxication effect and a 2nd strain in which an identical dose would produce virtually no effect. In Exp I, the spontaneous locomotor behavior of 16 Sprague-Dawley albino and 16 Long Evans hooded rats was tested in a Latin square design balanced for sex, strain, and alcohol dose. Ss received ip injections (15 mg/kg) of 0, 10, or 15% ethyl alcohol (0, 0.75, 1.50, or 2.25 mg/kg; A, B, C, or D, respectively) in isotonic saline. Ss were tested on a weekly basis and received alcohol doses in 1 of the following orders: ABCD, BADC, CDAB, or DCBA. Increasing the alcohol dose generally lowered activity, with greatest impairment at the highest dose, but greatest performance decrement was seen when 1.5 mg/kg alcohol was given. Based on extremes of ratings of alcohol-induced motor impairment, 2 Ss from each cell (formed by the combination of sex and strain) were bred to produce offspring that were either most affected (MA) or least affected (LA) by alcohol consumption. Similar testing of offspring for the effect of alcohol on locomotor behavior followed, and Ss were selectively bred until 9 generations of MA and LA were produced. Significant differences in reactions to alcohol appeared by the 5th generation and were primarily due to performance decrement in the MA Ss. Implications for an animal model of alcoholism are discussed.

1458. Riley, Edward P.; Plonsky, Mark & Rosellini, Robert A. (1982). **Acquisition of an unsignalled avoidance task in rats exposed to alcohol prenatally.** ***Neurobehavioral Toxicology & Teratology,*** 4(5), 525–530.
Determined the influence of prenatal alcohol exposure on a Sidman avoidance schedule in a shuttlebox. Long-Evans rats consumed liquid diets containing either 35 or 0% ethanol-derived calories (EDCs) during Days 6–20 of pregnancy. These liquid diets were isocaloric, and pair feeding was employed. An ad lib lab chow control group was also included. At approximately 60 or 120 days of age, 66 female offspring were tested for 2 hrs on 3 consecutive days. Shock was programmed to occur every 10 sec unless a response was made, in which case the next shock was postponed for 20 sec. Results show that prenatal alcohol exposure facili-

tated the acquisition of this task at both ages. The 35% EDC offspring made significantly more responses than controls and received fewer shocks. Analysis of interresponse times for the older Ss revealed that the 35% EDC group learned to space their responses better than controls. Possible explanations of the facilitation shown by the 35% EDC progeny are discussed in terms of an inhibition deficit.

1459. Riley, Edward P.; Worsham, Elizabeth D.; Lester, David & Freed, Earl X. (1977). **Selective breeding of rats for differences in reactivity to alcohol: An approach to an animal model of alcoholism: II. Behavioral measures.** ***Journal of Studies on Alcohol,*** 38(9), 1705–1717.
An earlier report by the authors (1976) describes 2 lines of rats, those most affected (MA) and those least affected (LA) by a subhypnotic dose of alcohol. The present 5 studies examined other traits that might distinguish the 2 lines: blood alcohol concentrations immediately after motor activity tests, sleeping time after varying amounts of alcohol, open field and running wheel behavior without alcohol, open field behavior in the dark and in the light without alcohol, and amounts of alcohol of varying solutions drunk in proportion to body weight over periods of 12–14 days. Results suggest that the MA and LA lines differ in sensitivity to alcohol over a range of doses; the differences, however, are not reflected in alcohol intake or selection. Differences in running wheel activity without alcohol may suggest a more sensitive neural apparatus in MA than in LA animals. Other differences by line and sex are discussed.

1460. Robinson, Terry E. et al. (1985). **Enduring enhancement in frontal cortex dopamine utilization in an animal model of amphetamine psychosis.** ***Brain Research,*** 343(2), 374–377.
In 2 experiments, 7 daily intraperitoneal injections of dextroamphetamine sulfate (3 mg/kg) and 9 injections at intervals of 3–4 days produced in ovariectomized rats an enduring (at least 10 days) enhancement in medial frontal cortex dopamine utilization, which was not further increased by footshock stress. This change in mesocortical dopamine activity may be involved in the behavioral sensitization produced by psychomotor stimulant drugs and some of the cognitive abnormalities (e.g., amphetamine psychosis) associated with both stimulant drug abuse and stress-precipitated psychopathologies thought to involve brain catecholamine dysfunction in humans.

1461. Robinson, Terry E. & Becker, Jill B. (1986). **Enduring changes in brain and behavior produced by chronic amphetamine administration: A review and evaluation of animal models of amphetamine psychosis.** ***Brain Research Reviews,*** 11(2), 157–198.
Reviews the literature on the effects of chronic amphetamine (AM) treatment on the brain and behavior of animals. Research indicates that chronic AM administration produces 2 syndromes that have been proposed as animal models of psychosis. The 1st syndrome, AM neurotoxicity, is produced by maintaining elevated brain concentrations of AM for prolonged periods of time; the 2nd syndrome, behavioral sensitization, is produced by repeated intermittent administrations of lower doses of AM. It is argued that changes in the brain and behavior associated with the phenomenon of behavioral sensitization provide a better model of AM psychosis than those associated with AM neurotoxicity. The biological basis and generalizability of behavioral sensitization are discussed.

1462. Robinson, Terry E.; Jurson, Phillip A.; Bennett, Julie A. & Bentgen, Kris M. (1988). **Persistent sensitization of dopamine neurotransmission in ventral striatum (nucleus accumbens) produced by prior experience with (+)-amphetamine: A microdialysis study in freely moving rats.** ***Brain Research,*** 462(2), 211–222.
In humans the repeated use of amphetamine (AMPH) produces a hypersensitivity to the drug's psychotogenic effects that persists for months to years after the cessation of drug use. An escalating dose pretreatment regimen was used with female rats to mimic the development of addiction and AMPH psychosis. Escalating doses of dextro-AMPH were not neurotoxic, and 25–30 days after drug treatment Ss showed relatively normal levels of spontaneous motor activity across the day–night cycle. However, AMPH pretreatment produced robust behavioral sensitization. Ss showed a hypersensitivity to the motor stimulant effects of an AMPH challenge, even after 15–20 days of withdrawal. This hyperdopaminergic behavioral syndrome was accompanied by significantly elevated dopamine (DA) release in the ventral striatum. AMPH pretreatment had no effect on the basal extracellular concentrations of DA. Results suggest that the sensitization produced by AMPH use is due to enduring changes in the releasability of DA.

1463. Rockman, G. E.; Hall, A. & Glavin, G. B. (1986). **Effects of restraint stress on voluntary ethanol intake and ulcer proliferation in rats.** ***Pharmacology, Biochemistry & Behavior,*** 25(5), 1083–1087.
High, medium, and low ethanol-consuming male Wistar rats were exposed to daily 1 hr restraint stress for 10 consecutive days. Voluntary ethanol consumption was monitored during the stress period and for 25 days poststress. High ethanol preferring Ss consumed less ethnol in the 1st 5 days of the poststress period compared to nonstressed controls. The medium ethanol preferring group drank more ethanol than controls during Days 1–5 poststress. At the end of the poststress period, stressed Ss exhibited a significantly greater ulcer severity and ulcer frequency than nonstressed groups.

1464. Rockman, Gary E. & Glavin, Gary B. (1984). **Ethanol-stress interaction: Differences among ethanol-preferring rats' responses to restraint.** ***Alcohol,*** 1(4), 293–295.
118 male Wistar rats were screened in an alcohol/water free-choice paradigm and divided into low-ethanol preferring (1.5–2.5 g/kg, daily), medium-ethanol-preferring (2.5–4.5 g/kg, daily), and high-ethanol-preferring (4.5–6 g/kg, daily) groups. A non-ethanol-exposed group was also included. Half of the Ss in each group were food deprived, and the other half had free access to food before being subjected to an acute stressor (cold environment) for 3 hrs. Significant differences in ulcer incidence and severity occurred, suggesting a dose-dependent potentiation of gastric glandular ulcers.

1465. Rockman, Gary E.; Hall, Arleen M.; Markert, Lynn E. & Glavin, Gary B. (1988). **Influence of rearing conditions on voluntary ethanol intake and response to stress in rats.** ***Behavioral & Neural Biology,*** 49(2), 184–191.
Male weanling rats were reared in an enriched environment, with a female or male partner, or individually for 90 days. At 111 days of age, voluntary consumption of ethanol in increasing concentrations was assessed. Ss were then randomly divided into stressed and nonstressed groups and exposed to 3 hrs of immobilization. Enriched Ss consumed greater amounts of ethanol than all other groups. Among environmentally enriched rats, ethanol attenuated stress-ul-

cer development relative to their non-ethanol-exposed but stressed controls. In nonstressed enriched Ss, ethanol alone exacerbated stomach damage. Data suggest that environmental rearing conditions influence the complex interaction between ethanol intake and the response to stess.

1466. Rockman, Gary E.; Hall, Arleen M.; Markert, Lynn; Glavin, Gary B. et al. (1987). **Ethanol-stress interaction: Immediate versus delayed effects of ethanol and handling on stress responses of ethanol-consuming rats.** ***Alcohol,*** 4(5), 391–394.
Rats screened for voluntary ethanol intake were given acute immobilization stress immediately following the ethanol screening procedure or after a 20-day period without access to ethanol. Among Ss examined for stress responses immediately after screening, water-only Ss developed less frequent and less severe gastric stress ulcers than Ss in all ethanol-exposed groups. Results indicate that among high ethanol-consuming Ss, ethanol enhanced ulcer severity, while prior experience with ethanol consumption did not predispose Ss to exacerbated stress gastric pathology.

1467. Rodriguez, L. A.; Moss, Donald E.; Reyes, E. & Camarena, M. L. (1986). **Perioral behaviors induced by cholinesterase inhibitors: A controversial animal model.** ***Pharmacology, Biochemistry & Behavior,*** 25(6), 1217–1221.
Replicated an earlier experiment in which long-term neuroleptic treatment in rats failed to produce observable perioral behaviors. The Ss were 40 female Sprague-Dawley rats. The effects of cholinesterase inhibitors on perioral behaviors in rodents may not be solely attributed to cholinesterase inhibition. Implications for animal models of tardive dyskinesia are discussed.

1468. Rosenblum, Leonard A.; Coplan, Jeremy D.; Friedman, Steven & Bassoff, Trina. (1991). **Dose–response effects of oral yohimbine in unrestrained primates.** ***Biological Psychiatry,*** 29(7), 647–657.
Six unrestrained bonnet macaques were observed after oral administration of 4 doses of yohimbine hydrochloride (0.10, 0.25, 0.50, and 0.75 mg/kg) and a placebo. Yohimbine significantly increased episodes of motoric activation and affective response interspersed with intervals of behavioral enervation. Yohimbine scores correlated closely with baseline levels; there was no dose–response relationship. Response to oral yohimbine differed in several ways from sc and iv sodium lactate infusions, including prominent enervative symptoms and the appearance of sexual arousal. The appearance of cyclic enervative episodes suggests limitations on primate models of panic disorder that use oral yohimbine.

1469. Rupniak, N. M.; Samson, N. A.; Steventon, M. J. & Iversen, S. D. (1991). **Induction of cognitive impairment by scopolamine and noncholinergic agents in rhesus monkeys.** ***Life Sciences,*** 48(9), 893–899.
Compared the effects of scopolamine with a range of noncholinergic agents on spatial delayed response performance in adolescent male rhesus monkeys. A scopolamine-like impairment of spatial delayed response performance was induced using phencyclidine, lorazepam, or tetrahydrocannabinol, but not amphetamine, yohimbine, or morphine. Findings suggest that disruption of specific neurotransmitter systems other than acetylcholine may contribute to cognitive decline in aging and dementia.

1470. Rupniak, N. M.; Tye, Spencer J. & Iversen, Susan D. (1990). **Drug-induced purposeless chewing: Animal model of dyskinesia or nausea?** ***Psychopharmacology,*** 102(3), 325–328.
Compared the ability of drugs that induce oral movements in rodents to induce chewing and retching or emesis in male squirrel monkeys. Acute administration of the muscarinic agonist oxotremorine, the selective D_1 receptor agonist SKF38393, or the emetic ipecacuanha caused dose-related increases in purposeless chewing that was frequently associated with retching and emesis. Treatment with the neuroleptic haloperidol decreased spontaneous chewing at doses of 0.03 and 0.06 mg/kg. Thus, some drug-induced oral behaviors in rodents may reflect nausea rather than dyskinesia.

1471. Rush, Douglas K. (1988). **Scopolamine amnesia of passive avoidance: A deficit of information acquisition.** ***Behavioral & Neural Biology,*** 50(3), 255–274.
Investigated the influence of scopolamine (SLM) administered prior to or immediately following training on 24-hr retention of step-through passive avoidance in mice. In low doses (0.3–3.0 mg/kg, ip), pretraining administration (–5 min) of SLM induced a very strong amnesia. Posttraining SLM induced a significant effect only at the highest dose tested (30 mg/kg). Results indicate that SLM could induce a small posttrial effect, presumably through an influence on consolidation processes. The much larger effect of pretrial SLM, however, indicated a primary influence on processes related to information acquisition. Together with findings from the literature, results suggest that SLM-induced amnesia partially, but not completely, models the memory deficits of human dementia.

1472. Russell, Kristanne H.; Hagenmeyer-Houser, Starr H. & Sanberg, Paul R. (1987). **Haloperidol-induced emotional defecation: A possible model for neuroleptic anxiety syndrome.** ***Psychopharmacology,*** 91(1), 45–49.
Studied the possible roles played by catalepsy, the gastrointestinal tract, and anxiety in the neuroleptic-induced defecation response in adult male Sprague-Dawley rats. Morphine sulfate, domperidone, haloperidol, and diazepam were injected intraperitoneally in each of 3 studies. Haloperidol increased defecation, whereas no significant increases in defecation were detected with morphine treatment. When the peripheral dopamine receptor antagonist domperidone was tested, no significant differences in fecal elimination were recorded. It appears that the cataleptic state or the peripheral effects of haloperidol were not responsible for the increased defecation. Under certain circumstances, normal rats given haloperidol showed "emotional defecation" that seemed to reflect increased anxiety. It is concluded that findings may serve as a basis for the development of an animal model for some of the atypical side effects of major tranquilizers, such as akathisia, dysphoria, and neuroleptic anxiety syndrome.

1473. Russell, Roger W.; Jenden, Donald J.; Booth, Ruth A.; Lauretz, Sharlene D. et al. (1990). **Global in vivo replacement of choline by N-aminodeanol: Testing a hypothesis about progressive degenerative dementia: II. Physiological and behavioral effects.** ***Pharmacology, Biochemistry & Behavior,*** 37(4), 811–820.
Used rat pups weaned at 29 days to examine the progressive effects of replacement of dietary choline (Ch) with N-amino-N,N-dimethylaminoethanol (NADe) for 120 days on a broad spectrum of behavioral and physiological functions involving

the cholinergic system. The magnitudes of these effects tended to increase with time on the NADe diet, but those related to learning and memory were largely confined to the 60–120-day period. Neurochemical effects were concomitant with the replacement of Ch by NADe, being consistent with a hypocholinergic state as found in such progressive degenerative dementias as Alzheimer's disease. More complex behavioral functions were affected progressively, cognitive processes (e.g., learning and memory) being most sensitive and showing the least adaptability. It is proposed that the syndrome generated by NADe replacement of Ch represents an experimental model of progressive degenerative dementia.

1474. Rylov, A. L.; Pak, E. S.; Karelin, A. A.; Ulyashin, V. V. et al. (1990). **Use of a new model of stable state of animals' aggressiveness to test neurotropic drugs with a depressant action.** ***Zhurnal Vysshei Nervnoi Deyatel'nosti,*** 40(4), 783–785.
Tested an experimental model of a stable psychotic state involving aggressiveness induced by administration of endogenous substances. Animal subjects: 250 male mongrel rats. A substance consisting of a heavy metal with nonorganic residue was injected in a dose of 0.150 mg/kg into the lateral brain ventricles. Ss were subdivided into 4 groups and were administered the following drugs 24 hrs later: (1) a physiologic solution ip; (2) 0.5 mg/kg haloperidol ip; (3) 0.1 mg/kg diazepam ip; and (4) 2.0 mg/kg morphine sc. The effects of these drugs were tested 0.5, 1, 1.5, 2, 4, 8, and 24 hrs after administration of the heavy metal complex.

1475. Sackler, A. M. & Weltman, A. S. (1985). **Effects of methylphenidate on whirler mice: An animal model for hyperkinesis.** ***Life Sciences,*** 37(5), 425–431.
Investigated whether a neurological mutant strain of whirler mice, due to its natural intrinsic circling behavior and hyperactivity, could be used as an animal model for hyperkinesis studies. Young male whirler mice were divided into circling activity- and body weight-matched test and control groups, and test mice received paradoxically oral intubation of daily doses of 5 mg/kg of methylphenidate during a 23-wk period. Findings indicate that this treatment significantly decreased circling activity in test Ss. The effects on circling behavior were reversible following cessation of methylphenidate administration. After 18 wks of cessation of the central nervous system (CNS) stimulant, oral administration of a single dose of 2.5 mg/kg of methylphenidate caused a 37.8% increase in circling activity, but the increase compared to control Ss was not significant. It is concluded that use of this strain as an animal model may be especially beneficial in the screening of new drugs for the treatment of hyperkinesis.

1476. Sagimbaeva, Sh. K. & Voronin, L. G. (1981). **An experimental model for Korsakov's Syndrome.** ***Soviet Neurology & Psychiatry,*** 13(4), 85–91.
Studied the morphological changes in the structure of 20 white rat brains occurring after chronic alcohol intoxication. An attempt was made to develop a physiological model for the study of Korsakoff's syndrome, which is characterized by disorientation in time and space. Findings of overall S behavior—space disorientation, inabilities to sustain alertness, motor automatism, and disruptions of CR performance—indicate that this goal was achieved.

1477. Salimov, R. M. & Viglinskaya, L. V. (1991). **Multivariate analysis of behaviour related to alcohol abuse in rats.** ***Drug & Alcohol Dependence,*** 27(2), 135–137.
Examined among 46 male albino rats (aged 8 wks) the relationship between early alcohol consumption and alcoholism. The ethanol withdrawal procedure was used to measure the alcohol abuse in Ss after 2 mo access to a sweet ethanol solution. Alcohol abuse was found to be predicted by the initial level of Ss' locomotor asymmetry. The initial consumption of ethanol was not associated with the probability of alcohol abuse. This study has replicated an earlier study by R. M. Salimov and I. V. Viglinskaya (1990).

1478. Salt, Jeremy S. & Taberner, Peter V. (1984). **Differential effects of benzodiazepines and amphetamine on exploratory behaviour in weanling rats: An animal model for anxiolytic activity.** ***Progress in Neuro-Psychopharmacology & Biological Psychiatry,*** 8(1), 163–169.
Describes an animal behavior model in which the exploratory activity of weanling Wistar rats in an area inaccessible to the mother rat was measured by ultrasound. The model distinguished between benzodiazepines and amphetamine or desipramine when the drugs were given acutely to Ss. Flurazepam, nitrazepam oxazepam, and chlordiazepoxide produced a dose-dependent increase in exploratory behavior and a reduction of activity at higher doses. Chronic administration of chlordiazepoxide in the diet of the mother also produced increased exploratory activity in the weanlings. It is suggested that the model has value in assessing anxiolytic activity in acutely or chronically administered benzodiazepines.

1479. Sampson, David; Willner, Paul & Muscat, Richard. (1991). **Reversal of antidepressant action by dopamine antagonists in an animal model of depression.** ***Psychopharmacology,*** 104(4), 491–495.
Male rats subjected chronically to a variety of mild, unpredictable stressors showed a reduced consumption of sucrose or a sucrose/saccharin mixture in 2-bottle consumption tests. The deficit was apparent within 2 wks of stress: Normal behavior was restored by chronic treatment with desmethylimipramine (DMI) or amitriptyline (AMI). Acute administration of SCH-23390 1 wk after withdrawal or sulpiride 2 wks after withdrawal were without effect in vehicle-treated stressed Ss and in nonstressed Ss. These dopamine antagonists selectively reversed the improvement of performance in DMI- or AMI-treated stressed Ss.

1480. Samson, Herman H. & Grant, Kathleen A. (1990). **Some implications of animal alcohol self-administration studies for human alcohol problems.** Special Issue: Research and policy. ***Drug & Alcohol Dependence,*** 25(2), 141–144.
Examines how environmental manipulations influence the process of establishing and maintaining excessive alcohol drinking, based on the premise that an understanding of these processes is essential to the design of effective alcoholism prevention and treatment programs. Animal models are described that highlight the role of conditioning factors in the initiation of drinking among humans, and animal studies are discussed that demonstrate the interaction of available reinforcers and drinking.

1481. Sanberg, Paul R. & Norman, Andrew B. (1989). **Underrecognized and underresearched side effects of neuroleptics.** ***American Journal of Psychiatry,*** 146(3), 411–412.

Comments on R. D. Bruun (1988) and suggest that changes in emotional defecation in rats may be appropriate for studies with neuroleptics.

1482. Sanberg, Paul R.; Russell, Kristanne H.; Hagenmeyer-Houser, Starr H.; Giordano, Magda et al. (1989). **Neuroleptic-induced emotional defecation: Effects of scopolamine and haloperidol.** ***Psychopharmacology,*** 99(1), 60–63.
In male rats, defecation was measured for a 1-hr test period in their home cage following doses of the central and peripheral anticholinergics scopolamine and *n*-methylscopolamine (MSP), respectively. A decrease in fecal excretions and an attenuation of haloperidol-induced defecation was found following administration of scopolamine. MSP reduced defecation at all doses. When MSP was combined with haloperidol, both fecal mass and number decreased significantly. Results support the view that anticholinergics may be useful for the emotional and dysphoric reactions associated with neuroleptics.

1483. Sanger, D. J. & Joly, Danielle. (1990). **Psychopharmacological strategies in the search for cognition enhancers.** Second International I.T.E.M.-Labo Symposium on Strategies in Psychopharmacology: Cognition enhancers—from animals to man (1988, Paris, France). ***Pharmacopsychiatry,*** 23(Suppl 2), 70–74.
Discusses difficulties in the development of animal tests that will make the discovery of cognition enhancers possible. One research approach involves attempts to create models homologous with the disorder to be treated. Recent studies of the effects of certain brain lesions on learning in monkeys or rats as models of Alzheimer's disease may go some way toward developing an animal model for this disorder. A more pragmatic strategy involves the development of empirical models, according to which any biological and behavioral test can be used if it provides a reasonable prediction of activity in the clinic. For example, the passive avoidance test in rodents does not need to model human cognition if it accurately predicts clinical activity. (German abstract).

1484. Sanger, David J.; Perrault, Ghislaine; Morel, Elaine; Joly, Daniele et al. (1991). **Animal models of anxiety and the development of novel anxiolytic drugs.** Special Issue: Perspectives in Canadian neuro-psychopharmacology: Proceedings of the 13th Annual Canadian College of Neuro-psychopharmacology. ***Progress in Neuro-Psychopharmacology & Biological Psychiatry,*** 15(2), 205–212.
Discusses research on the development of novel anxiolytics without problems of sedation, muscle relaxation, amnesia, and dependence. Novel ω (benzodiazepine) receptor ligands with anxiolytic properties are described, including alpidem, bretazenil, suriclone, and abecarnil, and are shown to have different pharmacological profiles. The differences may be related to low intrinsic activity or to selectivity for ω receptor subtypes. The association of serotonin with neural mechanisms underlying anxiety and actions of anxiolytic drugs are discussed.

1485. Sathananthan, Gregory L.; Sanghvi, Indravadan; Phillips, Neil & Gershon, Samuel. (1975). **MJ 9022: Correlation between neuroleptic potential and stereotypy.** ***Current Therapeutic Research,*** 18(5), 701–705.
Reports results of 2 studies, 1 with acute schizophrenic patients and 1 with dogs. Results show a close correlation between the blockade of amphetamine-induced stereotypy in dogs and the neuroleptic potential in schizophrenia, strengthening the claim that stereotypy in dogs could be used as an animal model for human psychoses.

1486. Schechter, Martin D. & Concannon, James T. (1982). **Haloperidol-induced hyperactivity in neonatal rats: Effect of lithium and stimulants.** ***Pharmacology, Biochemistry & Behavior,*** 16(1), 1–5.
Investigated the effect of chronic sc administration of haloperidol (HAL) into neonatal rats as a model for hyperkinesis in human children in terms of its onset, duration, and offset of hyperactivity. In addition, the ability of chronically administered lithium in the diet of nursing mothers to attenuate the HAL-induced hyperactivity was investigated for half the Ss. Experiments with acute administration of amphetamine (AM) and methylphenidate (MPD) to Ss were also conducted. Ss were assigned to 1 of 3 conditions—2.5 mg/kg HAL, .25 mg/kg HAL, or saline—and then received ip either dextroAM or MPD daily before observation. Results indicate that, although chronic HAL (2.5 mg/kg) produces hyperactivity relative to controls on the 25th day of life, this hyperactive behavior does not return to control levels at 30 days of age. Moreover, neither the stimulants nor lithium attenuates this hyperactivity and lithium produces increased activity. Thus, chronic HAL administered directly into Ss produces hyperactivity possibly by the production of dopaminergic supersensitivity, yet this effect does not model the temporal course seen in hyperkinetic humans. In addition, the administration of drugs that are clinically useful in treating childhood hyperactivity were unable to decrease the hyperactivity produced by HAL in neonatal rats. Taken together, these observations cast doubt on the usefulness of this animal model to mimic the human condition.

1487. Schetinin, E. V.; Baturin, V. A.; Arushanian, E. B.; Ovanesov, K. B. et al. (1989). **Chrono-biological approach to forced swimming test as a model of behavioural depression.** ***Zhurnal Vysshei Nervnoi Deyatel'nosti,*** 39(5), 958–964.
Developed a biorhythmic model of the depressive state to allow the study of the pathophysiology of higher nerve activity in vivo. Animal subjects: 103 noninbred rats (weight 180–260 g). A depressive state was induced in Ss by forced swimming. Ss were visually monitored by a recording device. The duration and rhythmic structure of active and passive swimming were recorded. Agents were administered to induce behavior depression, which were then counteracted with antidepressants. Drugs used: Reserpine, clonidine, imipramine, amitriptyline, and niamide. (English abstract).

1488. Schickerová, R.; Mareš, P. & Trojan, S. (1989). **Rhythmic metrazol activity in rats as a model of human absences.** ***Activitas Nervosa Superior,*** 31(1), 16–20.
Rhythmic activity of the spike-and-wave type was induced by administering metrazol (25 or 50 mg/kg, sc) in 10 adult male rats with implanted cortical electrodes. The Ss were deprived of water and then allowed to lick water from a tube. Under control conditions, Ss licked for 3–4 min without an interruption. Rhythmic metrazol activity deranged the licking. When this pathologic activity was represented

about one-third of the time, the licking was fully blocked. The impairment of highly motivated behavior confirmed the adequacy of the rhythmic metrazol activity as a model of human primary generalized seizures of the absence type.

1489. Schiller, Grant D.; Daws, Lynette C.; Overstreet, David H. & Orbach, Joe. (1991). **Lack of anxiety in an animal model of depression with cholinergic supersensitivity.** ***Brain Research Bulletin,*** 26(3), 433–435.
Investigated the behavior of the Flinders Sensitive Line (FSL) of rats, which are animal models of depression with cholinergic supersensitivity, in the elevated (+)-maze test of anxiety. 58 FSL and 64 Flinders Resistant (control) rats were used. Results indicate that anxiety responses (percentage of open/total arm entries) did not differ between the 2 lines. Treatment with diazepam significantly increased the percentage of open/total scores to a similar degree in both lines, further suggesting that the lines do not differ in anxiety. It is concluded that the FSL rat is an animal model of depression without evidence for inherent alteration in anxiety-related behavior. Data are consistent with human studies in correlating cholinergic supersensitivity with depression without anxiety.

1490. Schlemmer, R. Francis & Davis, John M. (1986). **A primate model for the study of hallucinogens.** American Society for Pharmacology and Experimental Therapeutics Meeting: New perspectives on the pharmacology of hallucinogenic drugs (1984, Indianapolis, Indiana). ***Pharmacology, Biochemistry & Behavior,*** 24(2), 381–392.
Presents an animal model for studying the actions of hallucinogenic drugs using primate social colonies, which is the result of studies occurring over a 10-yr period. Stumptail monkeys (*Macaca arctoides*) were administered 10 different hallucinogens and were tested either in social colonies or in dyads. Behavior was recorded with a checklist of 40 social and solitary behaviors common to the species and several abnormal or emergent behaviors. Although hallucinogens induced a number of behavioral changes, 1 emergent behavior, limb jerks, appeared to be selectively induced by 3 classes of hallucinogens in doses that correlate with those reported to be hallucinogenic in humans. Data suggest that other behavioral changes induced, such as ptosis and social withdrawl, may be useful in studying aspects of hallucinogen intoxication other than hallucinations, or psychosis in general. Tolerance developed to all hallucinogens tested except 2, and cross-tolerance between hallucinogens could be demonstrated.

1491. Schmidt, Michael J.; Fuller, Ray W. & Wong, David T. (1988). **Fluoxetine, a highly selective serotonin reuptake inhibitor: A review of preclinical studies.** Symposium: Progress in antidepressant therapy: Fluoxetine: A comprehensive overview (1987, Telfs, Austria). ***British Journal of Psychiatry,*** 153(Suppl 3), 40–46.
Reviews preclinical studies that have characterized fluoxetine as a selective inhibitor of serotonin uptake, including specificity and duration of amine uptake inhibition, enhanced serotonin function resulting from uptake inhibition, adaptive changes in receptors after treatment, and side-effects of fluoxetine compared to existing antidepressant drugs.

1492. Schmidt, W.; Popham, Robert E. & Israel, Y. (1987). **Dose-specific effects of alcohol on the lifespan of mice and the possible relevance to man.** ***British Journal of Addiction,*** 82(7), 775–788.
To determine the effects on lifespan of daily consumption of alcohol throughout adulthood, 3 groups of 100 male mice each, housed 1 to a cage, were given 3.5, 7.5, and 12% alcohol in distilled water as the only source of drinking fluid. Two control groups of 100 mice each, 1 group singly housed and the other housed 5 to a cage, received distilled water ad libitum. There was no difference between the survival curves of the low-alcohol and water-drinking singly housed controls. The medium-alcohol Ss had the longest mean lifespan of the 5 groups and the high-alcohol Ss had the shortest.

1493. Schneider, J. S. (1990). **Chronic exposure to low doses of MPTP: II. Neurochemical and pathological consequences in cognitively-impaired, motor asymptomatic monkeys.** ***Brain Research,*** 534(1–2), 25–36.
Conducted neurochemical and neuropathological analyses of the brains of 4 adult monkeys which had served as Ss in a study by J. S. Schneider and C. J. Kovelowski (1990). Ss had extensive caudate and putamen dopamine (DA) depletions, with coincident loss of substantia nigra DA neurons. Explanations are offered for the observation that Ss maintained normal movement despite the loss of DA in some striatal regions. Together with results from Schneider and Kovelowski, findings suggest that the monkey exposed to chronic low-dose 1-methyl-4-phenyl-1,2,3,6-tetrahydropyridine (MPTP) which develops cognitive difficulties but no gross motor disorder may present an animal model for the asymptomatic MPTP-exposed human or for the early, compensate form of idiopathic Parkinson's disease.

1494. Schouten, M. Joris; Bruinvels, Jacques; Pepplinkhuizen, Lolke & Wilson, J. Paul. (1983). **Serine and glycine-induced catalepsy in porphyric rats: An animal model for psychosis?** ***Pharmacology, Biochemistry & Behavior,*** 19(2), 245–250.
Investigated whether an increased demand for glycine, as has been postulated to occur in patients with episodic psychoses and multiple perceptual distortions, could evoke psychotic reactions. Catalepsy was used as a measure for psychosis and was observed after injection of serine or glycine in 54 male Wistar porphyric rats. Catalepsy occurred after serine as well as glycine administration in 2-allyl-2-isopropylacetamide (AIA)-pretreated rats, while in lead (Pb) + phenobarbital-pretreated Ss only glycine was effective. Administration of AIA to Ss resulted in a strongly enhanced excretion of porphobilinogen (PBG) in urine, while Pb + phenobarbital-pretreated Ss showed increased excretion of gamma-aminolevulinic acid (ALA). The Pb + phenobarbital-pretreated Ss showed elevated serine plasma levels and lowered glycine plasma levels 18 hrs after injection, while no significant differences in plasma levels of these amino acids were found 24 hrs after AIA. In AIA, or saline-pretreated Ss, glycine formation from serine was elevated. It is concluded that this animal model can be used to investigate episodic psychoses.

1495. Schuster, Charles R. & Fischman, Marian W. (1975). **Amphetamine toxicity: Behavioral and neuropathological indexes.** ***Federation Proceedings,*** 34(9), 1845–1851.

Outlines an animal model for the assessment of a drug's toxicity. The model includes behavioral assays, the results of which can be correlated with other functional and morphological changes occurring simultaneously in the experimental organism. Findings on the actions of amphetamines, the effects of therapeutic doses, the behavioral effects of chronically administered amphetamines in animals and humans, and the effects of withdrawal of amphetamines after chronic administration are reviewed. Data from monkeys given chronic administrations of methamphetamine demonstrate a residual tolerance to the response suppressant effects of the drug. Morphological and biochemical consequences of methamphetamine are also examined. Implications for future research are discussed in terms of the suggestion that functional changes can occur with shorter exposures or with exposure to lower doses than are necessary to produce morphological changes.

1496. Schwarting, R. K.; Bonatz, A. E.; Carey, R. J. & Huston, J. P. (1991). **Relationships between indices of behavioral asymmetries and neurochemical changes following mesencephalic 6-hydroxydopamine injections.** ***Brain Research,*** 554(1–2), 46–55.
Studied behavioral and neurochemical changes in male rats that had received 1 of 3 doses of 6-hydroxydopamine (6-OHDA) injected into the ventral mesencephalon. Behavioral analysis comprised tight turns (TTs), wide turns (WTs), and locomotor activity. 6-OHDA-injected Ss were assigned to 3 groups according to the degree of asymmetry in TTs, in spontaneous behavior, and after the dopamine (DA) receptor agonist apomorphine. Neurochemically, the 3 groups differed with respect to degree of neostriatal DA depletion and increase in DA metabolism in the damaged hemisphere. Ipsiversive asymmetry in TTs was negatively correlated with DA levels in the damaged neostriatum and positively correlated with the increase in metabolism. Analysis of TTs vs WTs may provide distinctive and sensitive indices related to different functional deficits in animal models of hemiparkinsonism.

1497. Schwartz, Jeffrey M.; Ksir, Charles; Koob, George F. & Bloom, Floyd E. (1982). **Changes in locomotor response to beta-endorphin microinfusion during and after opiate abstinence syndrome: A proposal for a model of the onset of mania.** ***Psychiatry Research,*** 7(2), 153–161.
Beta-endorphin (BE; 0.3 or 0.6 nM) was infused into the A10 ventral tegmental area (VTA) of 16 male Wistar rats previously treated for 6 days with either morphine sulfate or lactose via implanted Silastic pellets. BE microinfusions occurred at 24 and 96 hrs after pellets were removed. Profound changes in locomotor response to BE were found, with morphine-pretreated Ss showing a spontaneous switch from hypo- to hyperresponsiveness over 72 hrs, compared to lactose-pretreated controls. These findings may reflect on current biochemical theories regarding the "switch" process in bipolar affective disease. Data can be viewed within a heuristic model of receptor changes that may underlie the transition from depression to mania.

1498. Schwartzman, Robert J. & Alexander, Guillermo M. (1985). **Changes in the local cerebral metabolic rate for glucose in the 1-methyl-4-phenyl-1,2,3,6-tetrahydropyridine (MPTP) primate model of Parkinson's disease.** ***Brain Research,*** 358(1–2), 137–143.
To examine the change in the local cerebral metabolic rate for glucose in the nigrostriatal dopaminergic system in the MPTP model of Parkinson's disease (PD), 3 adult monkeys (*Macaca fascicularis*) were injected intravenously daily with 0.5 mg/kg of MPTP over a 4-day period. One S was injected weekly over a 4-wk period, and 5 Ss served as controls. Findings show that MPTP-treated Ss exhibited symptoms that closely resembled human PD (e.g., rigidity, akinesia, flexed posture) and that they had depressed glucose metabolism in all cortical areas except the cerebellar cortex compared with controls. It is suggested that the 2-deoxydextroglucose analysis of the MPTP primate model of PD is particularly suited to demonstrate areas in the central nervous system (CNS) that are affected by MPTP.

1499. Seale, Thomas W. & Carney, John M. (1991). **Genetic determinants of susceptibility to the rewarding and other behavioral actions of cocaine.** ***Journal of Addictive Diseases,*** 10(1–2), 141–162.
Describes the use of an animal model, the inbred mouse, to identify and to characterize variants with inherently altered susceptibilities to the rewarding and other behavioral actions of cocaine. Studies are reviewed which show that among a battery of inbred strains chosen solely for their genetic diversity, genetic polymorphisms commonly occurred which altered the potency and/or efficacy of cocaine to induce conditioned place preference, oral self-administration, motor activity activation, seizures and lethality. These changes in cocaine sensitivity were of a behavior-specific and pharmacodynamic nature. The existence of different phenotypic classes of variants displaying altered vulnerabilities to cocaine suggests that different genotypic changes underlie each class.

1500. See, Ronald E.; Levin, Edward D. & Ellison, Gaylord D. (1988). **Characteristics of oral movements in rats during and after chronic haloperidol and fluphenazine administration.** ***Psychopharmacology,*** 94(3), 421–427.
Rats were administered either haloperidol (HAL) or fluphenazine (FLU) for 8 mo, given these same drugs in their drinking water for 2 mo, and then withdrawn from the drugs. Ss in both groups showed initial decreases in the number of computer-scored movelets (CSMs). After 6 mo of chronic neuroleptics, HAL-treated Ss showed increased oral movements (OMs) and CSMs; this effect increased during drug withdrawal. FLU-treated Ss showed a more persistent depression of both OMs and large-amplitude CSMs. Small-amplitude CSMs increased in both groups.

1501. Segall, Mark A. & Crnic, Linda S. (1990). **An animal model for the behavioral effects of interferon.** ***Behavioral Neuroscience,*** 104(4), 612–618.
Interferon, which is produced during viral infections, has cognitive and neurological effects in humans. A dose of 1600 U/g of mouse interferon-alpha significantly depressed horizontal activity, head pokes into a food chamber, and food intake in mice 10 hr and 24 hr after injection. An 800 U/g dose had only slight effects on horizontal activity and food intake, whereas a 400 U/g dose had no effect. There was no evidence of sensitization to interferon when a second 400 U/g dose was given after the 1600 U/g dose. The results imply that mouse interferon-alpha can be used in mice as a model for studying the fatigue and anorexia produced by interferon.

1502. Seggie, Jo; Steiner, Meir; Wright, Noel & Orpen, Gail. (1989). **The effect of lithium on pupillary response to pulses of light in sheep.** ***Psychiatry Research,*** 30(3), 305–311.
Administration of Li carbonate (600–1800 mg/day) to female sheep resulted in a linear rise in Li levels in plasma and red blood cells (RBCs). In contrast to the rodent model but in agreement with the human condition, plasma Li levels exceeded those of RBCs, and polydipsia and body weight changes were not evident. Four Li-treated female sheep and 4 female controls were tested for pupillary response to a light pulse. At plasma levels of 0.70–.08 m*M*/l, Li attenuated the ability of the pupil to constrict in response to 30-sec pulses of light in the 25–150 μW/cm² intensity range but not the ability to dilate in the dark. Similar to observations in humans, Li reduced sensitivity to light. Results point to the advantages of the sheep (over the rat) as a model for studying the actions of Li.

1503. Sershen, Henry; Hashim, Audrey & Lajtha, Abel. (1987). **Behavioral and biochemical effects of nicotine in an MPTP-induced mouse model of Parkinson's disease.** ***Pharmacology, Biochemistry & Behavior,*** 28(2), 299–303.
Examined effects of nicotine (0.4 mg/kg, sc) on locomotor activity and on the level of dopamine (DA) and its metabolites 3,4-dihydroxyphenylacetic acid (DOPAC) and homovanillic acid (HVA) in the striatum and olfactory tubercle of adult female BALB/cBY mice treated with the neurotoxin 1-methyl-4-phenyl-1,2,3,6-tetrahydropyridine ([MPTP] 30 mg/kg, sc). Results suggest that nicotine has an influence on locomotor activity in MPTP-treated mice and that this effect is not due to changes in DA receptor activity in the striatum caused by chronic nicotine.

1504. Shanbhogue, Ravindranath; Hrishikeshavan, Hiremagalur J.; Devi, Kshama & Munonyedi, Sam. (1990). **Behavioral neurobiology of inverse agonist FG 7142 induced anxiety syndrome in rats.** ***Progress in Neuro-Psychopharmacology & Biological Psychiatry,*** 14(2), 249–260.
Neurobehavioral survey of N-methyl beta-carboline-3 carboxamide, the inverse agonist FG 7142, was performed in rats, using a novel anxiety paradigm (field response 2-way crossover in a shuttle box). FG 7142 syndrome was similar to learned helplessness following shock treatment. Significant increase in mean latency to escape was observed from 0 to 25th trial. Effect of FG 7142 on the behavioral and neurological profile did not deviate significantly from controls. However, a general increase in arousal, darting, and sideway movement (weaving) of the head was noted. Drugs with specificity at benzodiazepine (BDZ) receptor site were employed as pretreatments to study their influence on FG 7142 induced anxiety syndrome. Diazepam and 5-benzyloxy-4-methoxymethyl beta-carboline-3-carboxylate methyl ester, ZK 91296, significantly blocked the inverse agonist response. A careful combination of a BDZ agonist and inverse agonist may be beneficial in the treatment of generalized anxiety disorders.

1505. Shannon, Harlan E. & Holtzman, Stephen G. (1976). **Blockade of the discriminative effects of morphine in the rat by naltrexone and naloxone.** ***Psychopharmacology,*** 50(2), 119–124.
Evaluated the capacity of the specific narcotic antagonists naltrexone and naloxone to block the discriminative effects produced by morphine in male CFE rats using a 2-choice, discrete-trial avoidance task. The antagonists produced a dose- and time-dependent blockade of morphine's effects as measured by responding on the morphine-appropriate choice lever. Naltrexone and naloxone were equipotent when given sc concomitantly with sc morphine. However, when the antagonists were administered orally at 0, 2, 4, or 8 hrs prior to sc morphine, naltrexone was more potent than naloxone at every time point and had a duration of action at least twice that of oral naloxone. The discriminative effects of the narcotic analgesics morphine and methadone were also compared after oral and sc administration. Both drugs produced dose-related discriminative effects and were one-tenth as potent by the oral as by the sc route. Results suggest that the discriminative effects produced by morphine in the rat can provide an animal model for the quantitative evaluation of the narcotic antagonist properties of drugs that might be considered for use in programs for the treatment of narcotic addiction.

1506. Shapiro, Neil R.; Dudek, Bruce C. & Rosellini, Robert A. (1983). **The role of associative factors in tolerance to the hypothermic effects of morphine in mice.** ***Pharmacology, Biochemistry & Behavior,*** 19(2), 327–333.
Associative learning theories of drug tolerance emphasize the importance of stimuli that predict drug administration. One model holds that drug tolerance is due to the development of a CR that is directionally opposed to the UCR to the drug. By virtue of their opposing natures, the overlapping occurrence of the CR and UCR is seen as a diminished response (i.e., tolerance). The present 2 experiments tested the predictions of this model using 2 doses of morphine and included truly random controls to examine the role of excitatory and inhibitory conditioning in tolerance; 202 male CD-1 mice were used as Ss. Tolerance was greatest in Ss administered morphine in the context of stimuli previously paired with drug administration, intermediate in random controls, and least or absent in Ss administered the drug in the presence of cues paired with vehicle injections. No direct evidence of a compensatory CR that could offset morphine's hypothermic effect was obtained in placebo test sessions, nor was evidence for such a response obtained in cross-drug tests with amphetamine and apomorphine.

1507. Shaw, Walter N.; Mitch, Charles H.; Leander, J. David & Zimmerman, Dennis M. (1990). **Effect of phenylpiperidine opioid antagonists on food consumption and weight gain of the obese Zucker rat.** ***Journal of Pharmacology & Experimental Therapeutics,*** 253(1), 85–89.
Long-term chronic sc administration to meal-fed obese Zucker rats, an animal model for obesity, showed that LY88329 and LY117413 significantly reduced food consumption as long as the drug was administered and significantly decreased body weight gain when compared to nontreated obese controls. Obese Ss did not develop a tolerance to the appetite suppressant effect of these opioid antagonistic agents.

1508. Sherman, A. D. & Petty, Frederick. (1982). **Additivity of neurochemical changes in learned helplessness and imipramine.** ***Behavioral & Neural Biology,*** 35(4), 344–353.
Conducted 2 experiments with 216 male Sprague-Dawley rats to determine the effects of helplessness training, antidepressants, or both on the neurochemical events within the anterior neocortex, hippocampus, and the septum in rats. Exposure to learned helplessness training produced a decrease in the calcium-specific release of serotonin from slices of neocortex, of GABA from hippocampal slices, and of serotonin from septal slices when measured 1 or 4 days

later. Chronic, but not acute, ip administration of imipramine produced opposing changes in control Ss and reversed the decreased release measured in helpless Ss. These neurochemical changes characteristic of learned helplessness required 30 min of training to develop.

1509. Sherman, A. D. & Petty, F. (1980). **Neurochemical basis of the action of antidepressants on learned helplessness.** ***Behavioral & Neural Biology,*** 30(2), 119–134.
In 8 experiments with male Sprague-Dawley rats, learned helplessness, an animal model of depression, could be prevented by administration of desipramine only into the frontal neocortex (FN), hippocampus (H), or lateral geniculate body (LGB), by administration of GABA only into these same areas, or by norepinephrine injected only into the H. It could be reversed by desipramine in the frontal neocortex, GABA in either the H or LGB, or by serotonin in the FN or septum. Other brain areas were inactive. The reversal of learned helplessness by cortical desipramine (DES) restored a deficit in septal release of serotonin and could be prevented by administration of bicuculline into the H but not the LGB. These data are consistent with two types of DES sensitive cells, one regulating a GABA-modulated release of serotonin in the septum, and the other regulating a GABA-modulated release of norepinephrine in the H. These mechanisms are proposed for the antidepressant actions of DES.

1510. Sherman, Arnold D.; Sacquitne, J. L. & Petty, Frederick. (1982). **Specificity of the learned helplessness model of depression.** ***Pharmacology, Biochemistry & Behavior,*** 16(3), 449–454.
The learned helplessness model of depression was tested with Sprague-Dawley rats for its responsiveness to several types of antidepressant therapies and to a number of psychoactive drugs that are not effective in treating depression in humans. Chronic administration of tricyclic antidepressants (imipramine, desipramine, amitryptyline, nortryptyline, or doxepin), atypical antidepressants (iprindole or mianserin), MAO inhibitors (iproniazid or pargyline), or ECS was effective in reversing learned helplessness. Chronic treatment with anxiolytics (diazepam, lorazepam, or chlordiazepoxide), neuroleptics (chlorpromazine or haloperidol), stimulants (amphetamine or caffeine), or depressants (phenobarbital or ethanol) was not. Thus, this model provides a reasonable degree of specificity toward therapies that are successful in humans.

1511. Shibata, Shigenobu; Nakanishi, Hiroshi; Watanabe, Shigenori & Ueki, Showa. (1984). **Effects of chronic administration of antidepressants on mouse-killing behavior (muricide) in olfactory bulbectomized rats.** ***Pharmacology, Biochemistry & Behavior,*** 21(2), 225–230.
Two forms of drug administration—systemic sc administration and microinjection into the medial amygdala—were employed to examine the effect of chronic administration of psychotropic drugs on muricide in 114 olfactory bulbectomized male Wistar King A rats. Muricide inhibition induced by the systemic doses of chlorpromazine (CPZ [10 mg/kg]) and diazepam (10 mg/kg) was reduced with chronic administration, while that by desipramine (DMI [10 mg/kg]) and amitriptyline (30 mg/kg) was augmented with chronic administration. Muricide inhibition induced by microinjection of CPZ was also reduced, while that by DMI was augmented. Results indicate that muricide by olfactory bulbectomized rats is a useful animal model for evaluating antidepressants and that a potential site of action of antidepressants is located in the medial amygdala.

1512. Shideler, Charlotte E.; DeLuca, Donald C.; Newton, Joseph E. & Angel, Charles. (1983). **Effects of naloxone and neuroleptic drugs on muscle rigidity and heart rate of the nervous pointer dog.** ***Pavlovian Journal of Biological Science,*** 18(4), 211–215.
Eight normal (A-line) and 8 genetically nervous (E-line) pointer dogs were pair matched for age, sex, and contrasting behavioral scores. Experimental Ss were given 3 mg/ml of either pimozide or haloperidol iv or 10 mg/ml of naloxone iv. Results reveal that muscle rigidity, a persistent physiological characteristic of the E-line pointer dog, was attenuated not only by haloperidol and pimozide but also by naloxone. Paloxone administration resulted in a modest but significant increase in heart rate in these Ss. Data support the contention that an abnormality in CNS dopaminergic function is involved in the genesis of abnormal behavior of this animal model.

1513. Shimizu, Jun; Tamaru, Masao; Katsukura, Takaharu; Matsutani, Tenhoshimaru et al. (1991). **Effects of fetal treatment with methylazoxymethanol acetate on radial maze performance in rats.** ***Neuroscience Research,*** 11(3), 209–214.
Treated pregnant rats with 15 mg/kg per day of methylazoxymethanol acetate (MAM) on Days 13–15 of gestation. Experimental Ss were 9 male rats randomly selected from the MAM-treated offspring who were examined for their spatial recognition ability by the radial maze technique and compared with control offspring. Although the performances of MAM Ss were inferior to that of controls, MAM Ss could reach the predetermined criterion within 15 trials. Subsequent retention tests revealed the drastic impairment of performance in MAM Ss when the retention interval was over 15 min. The total activity of choline acetyltransferase showed a significant decrease in the hippocampus and cerebral cortex of MAM Ss. Results suggest that working memory disorders of MAM rats on radial maze tasks may be due to the lowering of cholinergic functions in their hippocampus and cerebral cortex. MAM treatment may serve as an animal model for minimal brain dysfunction.

1514. Shippenberg, Toni S.; Stein, C.; Huber, A.; Millan, M. J. et al. (1988). **Motivational effects of opioids in an animal model of prolonged inflammatory pain: Alteration in the effects of κ- but not of μ-receptor agonists.** ***Pain,*** 35(2), 179–186.
A preference conditioning procedure was used to examine the motivational effects of opioids in naive male rats and male rats suffering from prolonged pain associated with Freund's adjuvant (FA)-induced inflammation of a hindlimb. It was found that the μ-opioid agonist morphine functioned as a reinforcer in naive Ss producing marked preferences for the drug-paired place. Ss injected with FA 7 days prior to conditioning exhibited a preference for the morphine place. Administration of the κ-opioid receptor agonist U-69593 to naive Ss produced dose-related place aversions. The aversive effect of this κ-agonist was, however, abolished in FA-treated Ss. Data suggest that κ-agonists may be effective therapeutic agents in the management of chronic pain states.

1515. Shuster, Louis. (1984). **Genetic determinants of responses to drugs of abuse: An evaluation of research strategies.** ***National Institute on Drug Abuse: Research Monograph Series,*** 54, 50–69.
Presents an overview of goals and strategies that can be used to study genetic determinants of drug abuse, noting that genetically defined animal models can aid such studies. Goals of genetic analysis are to explain individual response variations, separate different components of drug response, develop models, establish neurochemical mechanisms of drug action, and identify genetic determinants of particular responses. Research strategies for these goals include the use of selective breeding, inbred strains, recombinant-inbred and congenic lines, defined mutants, and sublines of inbred strains.

1516. Siegel, R. K.; Johnson, C. A.; Brewster, J. M. & Jarvik, M. E. (1976). **Cocaine self-administration in monkeys by chewing and smoking.** ***Pharmacology, Biochemistry & Behavior,*** 4(4), 461–467.
Two rhesus monkeys self-administered cocaine hydrochloride (20 mg/kg/day) in a gum-base vehicle on an FR-10 schedule with performance characterized by frequent pauses and increased intertrial interval responding. Three other monkeys self-administered cocaine base in lettuce cigarette vehicles showed smoking performances marked by shortened puff durations. Urinary benzoyl ecgonine levels correlated with amount of cocaine chewed or smoked. Ss did not prefer cocaine gum in choice tests with plain or procaine gum, but did significantly prefer cocaine cigarettes to plain cigarettes. Results emphasize the importance of route of administration in determining reinforcement efficacy of human coca use and suggest animal models for their further experimental analysis.

1517. Siegel, Shepard; Hinson, Riley E.; Krank, Marvin D. & McCully, Jane. (1982). **Heroin "overdose" death: Contribution of drug-associated environmental cues.** ***Science,*** 216(4544), 436–437.
A model of "overdose" deaths among heroin addicts is proposed that emphasizes recent findings concerning the contribution of drug-associated environmental cues to drug tolerance. Results of experiments with male Wistar rats, performed to evaluate this model, suggest that conditioned drug-anticipatory responses, in addition to pharmacological factors, affect heroin-induced mortality.

1518. Siegel, Shepard & MacRae, James. (1984). **Environmental specificity of tolerance.** ***Trends in Neurosciences,*** 7(5), 140–143.
Reviews research that demonstrates that the display of tolerance is often more pronounced in the usual drug-administration environment than in an alternative environment. This environmental specificity of tolerance is consistent with a Pavlovian conditioning model that suggests that tolerance is partially mediated by drug-compensatory conditional responses elicited by the usual drug-administration environment. On the basis of this model, the administration of a drug in the absence of cues previously associated with drug administration leads to a less tolerant response through the absence of the compensatory anticipatory responses that contribute to tolerance. It is concluded that with the incorporation of such Pavlovian procedures in chronic pharmacological treatment regimens, it may be possible to attenuate the clinical complication of tolerance.

1519. Siegel, Shepard; Sherman, Jack E. & Mitchell, Doreen. (1980). **Extinction of morphine analgesic tolerance.** ***Learning & Motivation,*** 11(3), 289–301.
It has been suggested that the analgesic effect of morphine becomes attenuated over the course of successive administrations by a conditional, compensatory, hyperalgesic response elicited by the administration procedure, thus accounting in part for analgesic tolerance. However, data have both confirmed and refuted the prediction that established tolerance would be extinguished by placebo sessions. The present authors sought to determine the reasons for their divergent findings in an experiment with 64 Sprague-Dawley and Wistar rats. It was found that placebo sessions did consistently attenuate morphine analgesic tolerance. Such extinction was not limited to the experimenter, drug preparation, rat strain, or apparatus used in the original, successful demonstration of the phenomenon, but rather was also demonstrable under conditions similar to those used in subsequent experiments that failed to demonstrate extinction of tolerance. Results suggest that the failures to demonstrate extinction of tolerance were attributable to insufficient extinction training.

1520. Simon, P. & Boutelier, I. (1982). **Pharmacology of hypnotics.** Symposium of the 13th Collegium Internationale Neuropsychopharmacoligicum Congress: Zopiclone: A third generation of hypnotics (1982, Jerusalem, Israel). ***International Pharmacopsychiatry,*** 17(Suppl 2), 39–45.
Reviews the pharmacology of sedative hypnotics, which belong to several chemical classes and thus have heterogeneous pharmacodynamic properties. Conventional pharmacology using animal models is advocated, as it enables the defining of hypnotics on a 5-facet pharmacological spectrum. This spectrum includes the evaluation of anticonvulsant activity, muscle relaxant effects, sedative-hypnotic effects, antiaggressive activity, and anticonflict activity. The animal model remains the best predictive tool of hypnotic efficacy in humans. The study of discriminative properties is a useful extension of these conventional models. It is concluded that electrophysiological and biochemical studies allow for the investigation of the sites and mechanisms of action, but cannot provide a final answer to define a pharmacological profile of hypnotics.

1521. Simson, Prudence G.; Weiss, Jay M.; Ambrose, Monica J. & Webster, Ann. (1986). **Infusion of a monoamine oxidase inhibitor into the locus coeruleus can prevent stress-induced behavioral depression.** ***Biological Psychiatry,*** 21(8–9), 724–734.
Tested whether stress-induced depression can be eliminated by correcting the depletion of norepinephrine (NE) in the locus coeruleus (LC). Behavioral depression produced by exposing albino male Holtzman rats to a stressor that they could not control (uncontrollable shock) was reversed by infusion of the monoamine oxidase (MAO) inhibitor pargyline into the LC region of the brain stem. Ss exposed to uncontrollable shock and then infused with vehicle exhibited significantly less activity in swim test than Ss not exposed and infused. The concentration of NE, dopamine, 5-hydroxytryptamine (5-HT), and 5-hydroxy-indoleacetic acid in 7 brain regions at the conclusion of the swim test showed that pargyline infusion into the LC eliminated the large depletion of NE in the LC that is normally observed after exposure to uncontrollable shock while having no effect on NE levels in the other brain regions examined.

1522. Simson, Prudence G.; Weiss, Jay M.; Hoffman, Laura J. & Ambrose, Monica J. (1986). **Reversal of behavioral depression by infusion of an alpha-2 adrenergic agonist into the locus coeruleus.** ***Neuropharmacology,*** 25(4), 385–389.
Examined the influence of clonidine, piperoxan, and inactive vehicle on reversing depression in 30 stressed and 12 nonstressed albino male Holtzman rats. Results show that stressed Ss infused with vehicle exhibited significantly less active behavior than did nonstressed Ss infused with vehicle. Stressed Ss infused with clonidine showed no difference in active behavior compared to nonstressed Ss infused with vehicle and showed significantly more activity than did the stressed Ss infused with vehicle. Stressed Ss infused with piperoxan showed no significant difference in activity compared to the stressed Ss infused with vehicle and were significantly less active than the nonstressed Ss infused with vehicle and the stressed Ss infused with clonidine. Results support the hypothesis that a deficiency of norepinephrine at alpha-2 receptors in the locus coeruleus is responsible for mediating stress-induced behavioral depression.

1523. Singhal, R. L. & Telner, J. I. (1983). **A perspective: Psychopharmacological aspects of aggression in animals and man.** ***Psychiatric Journal of the University of Ottawa,*** 8(3), 145–153.
A review of the literature indicates that attempts for classifying aggressive behavior have led to useful laboratory models that have yielded some insight into understanding the psychopharmacological mechanisms involved in aggression. While affective aggression is modeled in the laboratory by the use of paradigms involving shock-elicited fighting, isolation-induced fighting, and rage reaction, predatory aggression encompasses the muricidal strategy. A variety of neurotransmitters (e.g., catecholaminergic and cholinergic mechanisms) appear to be involved in aggressive behavior. Research has demonstrated the facilitative role of dopamine, while norepinephrine and 5-HT appear to have an inhibitory function. Data from investigations on psychoactive drugs and aggression in animals have not led to any definitive conclusions due to the multitude of variables involved. There is no specific "anti-aggressive" drug available and various psychotherapeutic agents are prescribed for an underlying disorder with aggression only as a symptom. It is concluded that preclinical investigations utilizing animal models can further identify the important etiological, neural, and therapeutic variables involved in aggressive behavior.

1524. Skolnick, P. et al. (1984). **A novel chemically induced animal model of human anxiety.** Symposium of the World Psychiatric Association, Section Clinical Psychopathology: Physiological basis of anxiety (1983, Vienna, Austria). ***Psychopathology,*** 17(Suppl 1), 25–36.
Attempted to develop a reproducible, chemically induced model of anxiety using the ethyl ester of beta-carboline-3-carboxylic acid (CCE), which has a high affinity for benzodiazepine receptors and can antagonize some of the pharmacologic actions of benzodiazepines in rodents. Administration of CCE (2.5 mg/kg) to chair-adapted, male rhesus monkeys elicited a behavioral syndrome characterized by extreme agitation, head and body turning, distress vocalization, and other behaviors which might be termed anxious. Concomitant increases in plasma cortisol, epinephrine, norepinephrine, heart rate, and mean arterial blood pressure were observed. Pretreatment of animals with the benzodiazepine receptor antagonist Ro 15-1788 (5 mg/kg) antagonized the behavioral, endocrine, and somatic changes produced by CCE, but it did not elicit any significant changes in these parameters when administered alone. It is suggested that the administration of CCE to primates may be a reliable and reproducible model of human anxiety and, as such, may prove valuable for studying the postulated role of stress or anxiety in a variety of human disorders. Results also suggest that benzodiazepine receptors not only mediate the pharmacologic actions of benzodiazepines but may also subserve both the affective and physiologic expression of anxiety.

1525. Slangen, J. L.; Earley, Bernadette; Jaffard, R.; Richelle, M. et al. (1990). **Behavioral models of memory and amnesia.** Second International I.T.E.M.-Labo Symposium on Strategies in Psychopharmacology: Cognition enhancers—from animals to man (1988, Paris, France). ***Pharmacopsychiatry,*** 23(Suppl 2), 81–83.
Discusses criteria for establishing and evaluating animal models for memory, learning, and amnesia. In modeling memory and amnesia, the different forms of cognition must be distinguished. Animal models should be developed for each type of cognition, be based on information from the clinic, and attempt to be specific. An example of a model more specific than the passive avoidance test is the radial maze. In the field of cognition there are no generally recognized reference compounds and therefore no empirical models. There is a need for simulation models that imitate the various aspects of cognition and its pathology. The major criterion for validating this kind of model is that it should show changes similar to those observed in humans either resulting from a particular pathology or from a particular drug treatment. (German abstract).

1526. Slotkin, Theodore A. (1983). **"Preclinical perinatal and developmental effects of methadone: Behavioral and biochemical aspects": Critique.** ***National Institute on Drug Abuse: Treatment Research Monographs,*** 83-1281, 347–359.
Comments on S. B. Sparber's (1983) review of preclinical research on the pharmacology and behavioral effects of perinatal exposure to methadone. Issues related to the applicability of animal models to clinical experience with humans are discussed.

1527. Söderpalm, B. & Engel, J. A. (1988). **Biphasic effects of clonidine on conflict behavior: Involvement of different alpha-adrenoceptors.** ***Pharmacology, Biochemistry & Behavior,*** 30(2), 471–477.
Investigated the effect of the alpha-2-adrenoceptor agonist clonidine on anxiety-related behavior in rats using 2 anxiety models: a modified drinking conflict model and an elevated plus-maze. Results reveal that biphasic dose-response curves were obtained in both models; in a narrow low-dose range (6.25–10.0 μg/kg) the drug produced anxiolyticlike effects, whereas anxiogeniclike properties were found after higher doses (12.5–80.0 μg/kg). Attempts to block the effects obtained were made in an elevated plus-maze. The specific alpha-2-adrenoceptor antagonist idazoxan blocked the anxiolyticlike effect but did not influence the anxiogeniclike activity. The specific alpha-1-adrenoceptor antagonist prazosin blocked the anxiogeniclike effect but did not alter the anxiolyticlike activity. It is suggested that alpha-1- and alpha-2-adrenergic receptor mechanisms are reciprocally involved in anxiety-related behavior.

1528. Söderpalm, Bo; Hjorth, S. & Engel, Jörgen A. (1989). **Effects of 5-HT_{1A} receptor agonists and L-5-HTP in Montgomery's conflict test.** ***Pharmacology, Biochemistry & Behavior,*** 32(1), 259–265.
Investigated the effects of the pyrimidinyl-piperazines buspirone, gepirone, ipsapirone and their common metabolite 1-(2-pyrimidinyl)-piperazine and of 8-hydroxy-2-(di-*n*-propylamino)tetralin and L-5-hydroxytryptophan (5-HTP) in Montgomery's conflict test, an animal anxiety model based on the animal's inborn urge to explore a new environment and its simultaneous fear of elevated, open spaces. Results indicate that anxiolytic- and anxiogenic-like effects of drugs affecting central serotonergic neurotransmission can be obtained in a sensitive rat anxiety model that neither involves consummatory behavior nor punishment. The anxiolytic-like effects of these compounds may be due to their 5-hydroxytryptamine$_{1A}$ (5-HT_{1A}) agonistic properties. Moreover, the present data may provide support for a possible reciprocal association of presynaptic 5-HT_{1A} receptors vs postsynaptic 5-HT_{1A} as well as 5-HT_2 receptors with regard to anxiety.

1529. Solomon, Paul R. & Staton, Donna M. (1982). **Differential effects of microinjections of d-amphetamine into the nucleus accumbens or the caudate putamen on the rat's ability to ignore an irrelevant stimulus.** ***Biological Psychiatry,*** 17(6), 743–756.
Latent inhibition (LI) is an attentional process by which animals learn to ignore an irrelevant stimulus. 64 Sprague-Dawley rats received either 0 or 30 preexposures to a tone that was later used as a CS in a 2-way avoidance task. Tone preexposure resulted in retarded conditioning (i.e., LI) in Ss that received microinjections of saline or dextroamphetamine (DAM; 10 µg) in the caudate-putamen and for those that received microinjections of saline in the nucleus accumbens. This LI effect, however, was not present in Ss that received DAM microinjections in the nucleus accumbens. The failure of CS preexposure to retard conditioning in these Ss was not due to drug-induced changes in either tone or shock sensitivity. Results are discussed in terms of the role of the mesolimbic dopamine system in learning to ignore an irrelevant stimulus and the use of LI as a possible animal model of the attentional deficit that seems to characterize some subpopulations of schizophrenic humans.

1530. Soubrié, P. & Simon, P. (1989). **Animal models in psychopharmacology.** ***Confrontations Psychiatriques,*** 30, 113–129.
Discusses the use of animal psychopharmacological models in understanding and interpreting human mental disorders. Methods of validating human models, criteria used in extrapolating information relevant to humans from animal data, and the use of models to generate new hypotheses regarding human clinical data are discussed. Current models on the behavioral effects of neuroleptic, antidepressant, and anxiolytic medications are reviewed. Clinical and research implications are also examined. (English, Spanish & German abstracts).

1531. Spear, Linda P. (1990). **Neurobehavioral assessment during the early postnatal period.** Special Issue: Methods in behavioral toxicology and teratology. ***Neurotoxicology & Teratology,*** 12(5), 489–495.
Few laboratories investigating neurobehavioral consequences of developmental toxicants assess offspring early in ontogeny other than examining physical maturation, reflex development, and locomotor activity; these measures tap only a limited portion of the neurobehavioral capacities of young organisms. The importance of including a wider range of neurobehavioral assessments during the early postnatal period in developmental toxicology test batteries is discussed. Special considerations for the design of testing early in life are enumerated, and examples are given of suckling, cognitive, and psychopharmacological tests that are sensitive indicators early in life of the effects of gestational drug exposure.

1532. Spear, Linda P.; Kirstein, Cheryl L. & Frambes, Nancy A. (1989). **Cocaine effects on the developing central nervous system: Behavioral, psychopharmacological, and neurochemical studies.** Conference of the Behavioral Teratology Society, the National Institute on Drug Abuse, and the New York Academy of Sciences: Prenatal abuse of licit and illicit drugs (1988, Bethesda, Maryland). ***Annals of the New York Academy of Sciences,*** 562, 290–307.
Reviews studies establishing a rodent model system for gestational cocaine exposure, neurobehavioral teratogenic experiments with cocaine, and behavioral and psychopharmacological assessments of dopaminergic function in cocaine-exposed Ss. Data suggest that sc cocaine administration results in dose-dependent increases in brain and plasma cocaine in both dams and fetuses and produces maternal plasma levels in the range of or above those observed in human cocaine users. Chronic sc cocaine (10, 20, or 40 mg/kg) from Gestational Day 8 to term does not alter litter size, body weight, reflex development, or physical landmarks in the offspring. Offspring exposed to cocaine during gestation show learning and/or retention deficits in some conditioning situations.

1533. Spear, Linda P.; Kirstein, Cheryl L.; Frambes, Nancy A. & Moody, Carole A. (1989). **Neurobehavioral teratogenicity of gestational cocaine exposure.** 51st Annual Meeting of the Committee on Problems of Drug Dependence (1989, Keystone, Colorado). ***National Institute on Drug Abuse: Research Monograph Series,*** 95, 232–238.
Reviews research on postnatal neurobehavioral consequences of gestational exposure to cocaine (GEC), a dopamine (DA) uptake inhibitor, in rats, focusing in general dosing and treatment procedures and maternal/litter data. Despite normal rates of physicial development, offspring who experience GEC may exhibit pronounced cognitive deficits when tested during the neonatal to weanling age periods. Behavioral, pharmacological, and biochemical studies of DA activity and function suggest that there may be a functional attenuation of DA activity in offspring with GEC and that the tuberoinfundibular DA system may be sensitive to the effects of GEC.

1534. Sperk, G. et al. (1985). **Kainic acid-induced seizures: Dose-relationship of behavioural, neurochemical and histopathological changes.** ***Brain Research,*** 338(2), 289–295.
Investigated changes induced by injection of kainic acid (3, 6, and 10 mg/kg, subcutaneously) in male Sprague-Dawley rats. There was a positive correlation between the dose and the extent of (a) the acute neurochemical changes 3 hrs after the injection (increases of 3,4-dihydroxyphenylacetic acid and 5-hydroxyindoleacetic acid levels and a decrease in noradrenaline levels in all brain regions investigated), (b) the acute histopathological changes (shrinkage and condensation of nerve cells and brain edema in the entire forebrain), and (c) the extent of behavioral alterations (immobility, wet dog shakes, and limbic seizures). It is suggested that irreversible

brain lesions in this animal model of limbic (temporal lobe) epilepsy are not induced solely by direct action of kainic acid but may be caused at least in part by secondary pathogenetic mechanisms.

1535. Stanishevskaya, A. V. & Mezentseva, L. N. (1977). **The effect of some pharmacological agents on adaptation under stress.** ***Farmakologiya i Toksikologiya,*** 40(1), 9–12.
Experiments on rats show that the formation of pathological states caused by stressors lead to the development of ulcerative lesions of the gastric mucosa associated with the degree of the catecholamine level decrease in the mesencephalon and hypothalamus. The application of seduxen and the combination of levodopa with seduxen or with a levo-adreno-blocking agent (i.e., pyroxan) tends to reduce the frequency of development of ulcerative lesions of the stomach. The protective effect produced by the combination of levodopa with a levo-adreno-blocking agent (pyroxan) is barred by an additional administration of a beta-adreno-blocking agent (i.e., inderal). It is concluded that seduxen and its combination with levodopa and pyroxan upgrades adaptation to stressors and thus prevents the development of pathological syndromes. (English summary).

1536. Stanton, Mark E. & Spear, Linda P. (1990). **Workshop on the qualitative and quantitative comparability of human and animal developmental neurotoxicity, Work Group I report: Comparability of measures of developmental neurotoxicity in humans and laboratory animals.** Special Issue: Qualitative and quantitative comparability of human and animal developmental neurotoxicity. ***Neurotoxicology & Teratology,*** 12(3), 261–267.
Examines assessment measures used in developmental neurotoxicology for their comparability in humans and laboratory animals and their ability to detect comparable adverse effects across species. Compounds used for these comparisons include substances for abuse, anticonvulsant drugs, ethanol, and ionizing radiation. At the level of functional category (sensory, motivational, cognitive, motor, and social), close agreement was found across species for all neurotoxic agents reviewed, particularly at high exposure levels. This was true even though end points used to assess these functions often varied across species. Discussion focuses on the ability of the Environmental Protection Agency's developmental neurotoxicity test battery to identify the risks of human exposure to these agents, the effects of maternal toxicity in the pre- or postnatal period, and the need for more emphasis on evaluation during development in animal studies.

1537. Stephens, D. N. & Schneider, H. H. (1985). **Tolerance to the benzodiazepine diazepam in an animal model of anxiolytic activity.** ***Psychopharmacology,*** 87(3), 322–327.
Investigated the antipunishment properties of diazepam (DZ) in NMRI mice treated acutely or following 9 daily treatments with either DZ (5 mg/kg per os) or its vehicle. Acutely, or following chronic vehicle treatment, DZ produced a dose-related increase in activity punished by footshock. Following chronic DZ, test doses of DZ given 24 or 48 hrs following the last chronic treatment were no longer, or less effective, in enhancing punished activity. Effects on unpunished activity were unaffected. Tolerance was not seen after 1 or 3 daily treatments but was present after 6 days. Following establishment of tolerance by 9 days of treatment, the antipunishment activity of DZ reappeared after 8 days of withdrawal and was restored to acute levels after 16 days. Tolerance was not associated with changes in benzodiazepine (BZ) receptor affinity or numbers, but the ability of gamma-aminobutyric acid (GABA) to enhance BZ binding was increased. There was no change in the ability of DZP or the convulsant 6,7-dimethoxy-beta-carboline-3-carboxylic acid methyl ester to modulate 35ß-t-butylbicyclophosphorothionate binding. The mechanism of tolerance to the antipunishment properties of DZP therefore remains unknown.

1538. Stephens, D. N.; Schneider, H. H.; Kehr, W.; Andrews, J. S. et al. (1990). **Abecarnil, a metabolically stable, anxioselective β-carboline acting at benzodiazepine receptors.** ***Journal of Pharmacology & Experimental Therapeutics,*** 253(1), 334–343.
Data from experiments with rats and mice show that abecarnil displaced tritiated benzodiazepines (BZDs) from their binding sites in brain with a high affinity, suggesting that abecarnil can be characterized as a partial agonist at central BZD receptors. Abecarnil produced a potent anxiolytic effect in animal models, but showed no or only weak side effects in tests of motor incoordination and muscle relaxation. It also had a relatively weak ability to potentiate the effects of ethanol and barbiturates.

1539. Stephens, D. N.; Schneider, H. H.; Kehr, W.; Jensen, Leif H. et al. (1987). **Modulation of anxiety by β-carbolines and other benzodiazepine receptor ligands: Relationship of pharmacological to biochemical measures of efficacy.** Sixth European Winter Conference on Brain Research: Bidirectional effects of β-carbolines in behavioral pharmacology (1986, Avoriaz, France). ***Brain Research Bulletin,*** 19(3), 309–318.
Reviews the biochemical and behavioral effects of anxiolytic and anxiogenic beta-carbolines and other benzodiazepine (BDZ) receptor ligands, focusing on 4 animal models of anxiety. Findings indicate that BDZ receptor ligands may exert either anxiolytic or anxiogenic effects in animal models of anxiety; also, the qualitative nature of these effects is predicted by biochemical indices of the nature of the interaction of the ligands with the gamma-aminobutyric acid (GABA) receptor (GABA ratio) and events at the chloride channel associated with GABA receptors (TBPS ratio).

1540. Stewart, Karen T.; McEachron, Donald L.; Rosenwasser, Alan M. & Adler, Norman T. (1991). **Lithium lengthens circadian period but fails to counteract behavioral helplessness in rats.** ***Biological Psychiatry,*** 30(5), 515–518.
Examined whether Li-induced lengthening of circadian period would alter escape performance in male rats previously subjected to inescapable shock (IS). Circadian period lengthened over the course of the experiment; this effect was significantly greater in Li-treated Ss than in controls. IS had no effect on circadian period, either alone or in interaction with Li treatment. The period-lengthening effects of Li were not associated with improved escape performance. Thus, treatments that alter circadian period are not necessarily associated with changes in behavioral state.

1541. Stoessl, A. J.; Dourish, C. T. & Iversen, S. D. (1989). **Chronic neuroleptic-induced mouth movements in the rat: Suppression by CCK and selective dopamine D1 and D2 receptor antagonists.** ***Psychopharmacology,*** 98(3), 372–379.
Demonstrated that fluphenazine decanoate resulted in spontaneous vacuous chewing mouth movements (MMs) and jaw tremor, using 250 male rats. These movements were suppressed by the selective D1 or D2 dopamine antagonists

SCH 23390 and raclopride, respectively, by cholecystokinin octopeptide sulphated (CCK), and by scopolamine administered at a dosage that induced hyperactivity. MMs resulting from chronic administration of neuroleptics to the rat may serve as a useful pharmacological model of tardive dyskinesia in the human. A relative increase of D1 activity as well as impaired CCK function may contribute to the pathogenesis of this disorder.

1542. Stoessl, A. Jon & Szczutkowski, Elizabeth. (1991). **Neurotensin and neurotensin analogues modify the effects of chronic neuroleptic administration in the rat.** ***Brain Research,*** 558(2), 289–295.
Examined the effects of neurotensin (NT) and NT-like analogs (neuromedin N and [D-Trp^{11}]NT) in an animal model of tardive dyskinesia. Chronic administration of low doses of fluphenazine decanoate alone failed to elicit vacuous chewing mouth movements (VCMs) in male rats, but VCMs were seen in neuroleptic-treated Ss following additional administration of NT. A higher dose of fluphenazine greatly increased VCM response; this potentiation was suppressed to control levels by [D-Trp^{11}]NT but was unaffected by neuromedin N. Alterations in NT may contribute to the deleterious extrapyramidal effects (e.g., tardive dyskinesia) of long-term neuroleptic administration. [D-Trp^{11}]NT may attenuate these effects by blockade of NT receptors within the central nervous system (CNS).

1543. Stoff, David M.; Moja, Egidio A.; Gillin, J. Christian & Wyatt, Richard J. (1978). **Disruption of conditioned avoidance behavior by *N,N*-dimethyltryptamine (DMT) and stereotype by β-phenylethylamine (PEA): Animal models of attentional defects in schizophrenia.** ***Journal of Psychiatric Research,*** 14(1–4), 225–240.
Describes 2 separate drug-induced behaviors in the rat that satisfy some criteria that should be met for animal models of schizophrenia. DMT, which is present in man although in extraordinarily small quantities, disrupts a simple learned behavior. The rat cannot adjust to repeated administration of DMT (it does not become tolerant) and, like the schizophrenic, performs better when given neuroleptics. PEA, which is also present in human tissue, forces the rat into a limited repertoire of behaviors (stereotypy). Again, repeated administration of PEA in the presence of decreased MAO does not induce tolerance, and stereotypy is blocked by neuroleptics. The 2 models are particularly germane because they deal with substances that are potential endogenous agents of schizophrenia.

1544. Strek, Kimi F.; Spencer, Karen R. & DeNoble, Victor J. (1989). **Manipulation of serotonin protects against an hypoxia-induced deficit of a passive avoidance response in rats.** ***Pharmacology, Biochemistry & Behavior,*** 33(1), 241–244.
Evaluated the serotonin (5-hydroxytryptamine [5-HT]) antagonists ketanserin, mianserin, methysergide, and cyproheptadine and the 5-HT uptake inhibitors fluoxetine and zimeldine for their ability to protect against an hypoxia-induced performance deficit in a passive avoidance (PAV) task in male rats. The ability to retain a PAV response decreased as the oxygen concentration decreased, with the largest retention deficit occurring at 6.5% O_2. Ketanserin and mianserin administered 1 min after PAV training produced dose-dependent increases in retention latencies following exposure to a 6.5% oxygen environment. Inhibition of 5-HT reuptake by fluoxetine produced dose-dependent increases in retention latencies. Modification of 5-HT after exposure to hypoxia can ameliorate a performance deficit in an anima model of learning and memory.

1545. Strupp, Barbara J.; Himmelstein, Sharon; Bunsey Michael; Levitsky, David A. et al. (1990). **Cognitive profile o rats exposed to lactational hyperphenylalaninemia: Correspon dence with human mental retardation.** ***Developmental Psycho biology,*** 23(3), 195–214.
Hyperphenylalaninemia was induced in 28 rats during post natal Days 3–21. Ss then received a learning set test fol lowed by a social learning paradigm or vice versa at age mo. An attentional test was administered at age 10–14 mo Ss evidenced (1) impaired learning set formation; (2) stimu lus perseveration, particularly after an error; and (3) dif ficulty in utilizing the less salient features of their environ ment in mastering discrimination problems. Long-term memory function and the ability to form simple association did not differ from 26 control rats. This pattern of cognitiv functions bears remarkable similarity to that of mentall retarded humans and neonatally hyperphenylalaninemic rhe sus monkeys, thus affirming the use of rats to study menta retardation.

1546. Sudilovsky, A. (1975). **Effects of disulfiram on th amphetamine-induced behavioral syndrome in the cat as model of psychosis.** ***National Institute on Drug Abuse: Re search Monograph Series,*** 3, 109–135.
Examined the amphetamine behavioral syndrome in 26 fe male cats as it was modified by pretreatment with ora disulfiram. Following pretreatment, a faster development o certain end-stage components of the amphetamine syndrom was obtained. On Day 1, development of a reactive attitud and of more prominent behavioral disjunction occurred wit combined drug administration as compared with ip amphet amine alone. In contrast with the facilitation of these behav iors was the absence of dyskinesias and hyperreflexia on tha day. Stereotyped behavior, loss of motor initiative, and hy perkinetic activity were markedly enhanced and appeare with a shorter latency period on subsequent days of th intoxication cycle. During the later days, a particularly hig level of compulsive activity was evident. In general, modi fication of the amphetamine effects on behavior was in direction consistent with comparable features in experimen tal catatonia and the catatonic form of schizophrenia. Th need to integrate such phenomena in any amphetamin model of psychosis is stressed, and analogies are drawn wit similar features reported in animals treated with bu bocapnine or other psychotogenic compounds and wit symptoms of human amphetamine psychosis and schizo phrenia.

1547. Šulcová, A. (1985). **Tranquillizing effects of alpraz lam in animal model of agonistic behaviour.** 27th Annu Psychopharmacology Meeting (1985, Jeseník, Czechoslova kia). ***Activitas Nervosa Superior,*** 27(4), 310–311.
Studied the effects of alprazolam on aggressive and defe sive-escape behavior in male mice. The nonsedative dos (1.25 mg/kg) significantly reduced aggressiveness in singl housed mice. The drug was more effective in selectivel inhibiting timidity (beginning with 0.05 mg/kg) and in stim ulating sociability (0.25 and 1.25 mg/kg). The findings co firm the drug's strong anxiolytic action.

1548. Sullivan, Ronald M.; Henke, Peter G. & Ray, Arunabha. (1988). **The effects of buspirone, a selective anxiolytic, on stress ulcer formation in rats.** ***Pharmacology, Biochemistry & Behavior,*** 31(2), 317–319.
Studied the effects of buspirone hydrochloride on the formation of cold-immobilization gastric stress ulcers in male rats. Low doses significantly attenuated and higher doses potentiated gastric stress pathology. Haloperidol and apomorphine reversed the buspirone effects. The role of dopamine in the expression of buspirone's effects is discussed.

1549. Sunderland, Gayle; Friedman, Steven & Rosenblum, Leonard A. (1989). **Imipramine and alprazolam treatment of lactate-induced acute endogenous distress in nonhuman primates.** ***American Journal of Psychiatry,*** 146(8), 1044–1047.
Studied the response of 10 macaque monkeys to the panicogenic agent sodium lactate after treatment with the tricyclic antidepressant imipramine, the triazolobenzodiazepine alprazolam, or placebo. Both drugs effectively blocked the lactate-induced, acute endogenous distress responses. Only alprazolam significantly reduced the occurrence of conditioned situational anxiety responses observed in the home cage. The alprazolam monkeys appeared to be sedated compared with the imipramine and placebo groups, and tremor and temporary exacerbation of anxious behaviors were observed during alprazolam withdrawal.

1550. Surkova, L. A. & Tyurina, I. V. (1989). **Method for evaluating the reinforcing properties of morphine on the model of addictive behavior.** ***Farmakologiya i Toksikologiya,*** 52(4), 93–95.
Studied the positive reinforcing effect of morphine in rats via a spatial orientation test in a **U**-shaped maze. Animal subjects: Six mongrel white rats (weight 250–270 g) (experimental Ss). Six mongrel white rats (weight 250–270 g) (controls). After a preparatory stage, food was placed in the left arm and morphine in the right arm of the maze (experimental group). Time spent in each arm, the number of shifts from 1 arm to the other, and other movements by Ss were examined. The quantitative positive reinforcing effect of morphine was measured. (English abstract).

1551. Swartzwelder, H. S.; Holahan, W. & Myers, R. D. (1983). **Antagonism by *d*-Amphetamine of trimethyltin-induced hyperactivity: Evidence toward an animal model of hyperkinetic behavior.** ***Neuropharmacology,*** 22(9), 1049–1054.
In 16 male Long-Evans rats, either 7.0 mg/kg of trimethyltin (TMT) or 0.9% NaCl was administered by intragastric gavage. After a period of recovery from the typical signs of TMT toxicity, each S was tested at 72-hr intervals for its locomotor activity in an open-field apparatus, the floor of which was divided into square grids. The baseline activity of each of the TMT-treated Ss was significantly greater than the saline-treated controls. Dextroamphetamine, injected ip in a dose of 0.5 or 2.0 mg/kg, augmented the hyperactivity of TMT-treated Ss. However, a 4.0 mg/kg dose of dextroamphetamine markedly attenuated the hyperactivity of TMT-treated Ss while elevating that of the controls. Since TMT produced an autismlike behavioral disorder involving hyperactivity, perseveration, aggressiveness, and impairment in problem solving and memory function, the placating effect of amphetamine supports the proposition that the pathology due to TMT may represent an experimental analog to the hyperkinetic syndrome in children.

1552. Tachiki, K. H. et al. (1978). **Animal model of depression: III. Mechanism of action of tetrabenazine.** ***Biological Psychiatry,*** 13(4), 429–443.
Studied the biochemical mechanism whereby tetrabenazine (TBZ) produces a sedative effect on locomotor activity in rats. Sprague-Dawley rats injected with levo-5-hydroxytryptophan (5-HTP, 30 mg/kg), the immediate precursor of 5-hydroxytryptamine (5-HT), showed the characteristic bison appearance, pitosis, and catalepsy normally observed after injecting TBZ (30 mg/kg). The treatment of Ss with low doses of 5-HTP (9 mg/kg) plus TBZ (2 mg/kg) significantly decreased locomotor activity, whereas low doses of either drug given alone had no significant effect on locomotor activity. The level of 5-hydroxyindoleacetic acid (5-HIAA) was elevated in the brain of Ss sacrificed 3 hrs after treatment with low doses of 5-HTP (9 mg/kg) plus TBZ (2 mg/kg). No significant changes in the levels of 5-HIAA were observed in rats treated with low doses of either L-5-HTP or TBZ alone. Treatment of Ss with para-chlorophenylalanine to inhibit the synthesis of 5-HT had an inhibitory effect on the duration of sedation following an injection of TBZ (30 mg/kg). Results indicate that the sedative action of TBZ is due to an excess of functional 5-HT.

1553. Talalayenko, A. N. & Zinkovskaya, L. Ya. (1988). **Features of the anxiolytic effect of some derivatives of benzodiazepine and GABA on experimentally modelled anxiety states.** ***Farmakologiya i Toksikologiya,*** 51(4), 20–22.
Studied the anxiolytic effects of the benzodiazepine derivative diazepam and the gamma-aminobutyric acid (GABA) derivatives phenibut and baclofen on various experimentally modeled anxiety states formed by aversive actions. Ss were 88 male rats. Inherent avoidance responses and anxiety induced by conflict situations were studied. Diazepam, phenibut, and baclofen were administered ip in doses of 1, 5, and 2 mg/kg, respectively. (English abstract).

1554. Tamminga, Carol A.; Dale, J. M.; Goodman, L.; Kaneda, H. et al. (1990). **Neuroleptic-induced vacuous chewing movements as an animal model of tardive dyskinesia: A study in three rat strains.** ***Psychopharmacology,*** 102(4), 474–478.
Vacuous chewing movements (VCMs) in 3 rat strains developed at different rates after 19 wks of treatment with the neuroleptic haloperidol (HAL). Sprague Dawley (SD) Ss displayed relatively high rates with low variability, compared to Wistar and Long Evans (LE) Ss. Atropine decreased but did not abolish VCMs in 2 strains (LE > SD). After HAL withdrawal, VCMs remitted gradually in all Ss but least rapidly in SD Ss. In a separate group of SD Ss, VCMs rated weekly showed considerable interindividual variability. Even after 24 wks of continuous HAL, 12 of 32 Ss showed no VCMs, while 13 had intense movements, analogous to the clinical situation in which only some patients treated with neuroleptics develop tardive dyskinesia. Data suggest genetically determined differences in development of tardive dyskinesia.

1555. Tancer, Manuel E.; Stein, Murray B.; Bessette, Brenda B. & Uhde, Thomas W. (1990). **Behavioral effects of chronic imipramine treatment in genetically nervous pointer dogs.** ***Physiology & Behavior,*** 48(1), 179–181.
Treated 17 genetically nervous pointer dogs for 4 wks with imipramine hydrochloride (10 mg/kg), a potent antipanic agent in humans. Although 3 of the Ss showed marked improvement in response to imipramine but not placebo

treatment after short-term exposure, chronic imipramine failed to modify aberrant behavior in any of the Ss. Findings are discussed in the context of the nervous pointer dog as a model for human anxiety disorders.

1556. Tang, Andrew H.; Franklin, Stanley R.; Himes, C. S. & Ho, P. M. (1991). **Behavioral effects of U-78875, a quinoxalinone anxiolytic with potent benzodiazepine antagonist activity.** ***Journal of Pharmacology & Experimental Therapeutics,*** 259(1), 248–254.
In unanesthetized rats implanted with cortical electrodes for EEG recording, ip injections of U-78875 increased EEG power density in frequencies above 12 Hz and decreased EEG power at lower frequencies. This effect was completely antagonized by pretreatment with flumazenil. In animal models measuring central nervous system (CNS), U-78875 is much weaker than diazepam. It produced minimal impairment of rotarod performance in rats at doses up to 30 mg/kg, but at lower doses completely reversed the impairment from 10 mg/kg of diazepam. In rats trained to avoid shocks in a shuttle box, U-78875 increased avoidance responses and antagonized the suppression of avoidance from diazepam. In the mouse 1-trial passive avoidance task, pretreatment with U-78875 before training produced no anterograde amnesia, but completely blocked the amnesic effect from diazepam.

1557. Tang, Maisy; Brown, Charlesetta; Maier, Donna & Falk, John L. (1983). **Diazepam-induced NaCl solution intake: Independence from renal factors.** ***Pharmacology, Biochemistry & Behavior,*** 18(6), 983–984.
Rehydrating albino Holtzman rats injected with diazepam (8 mg/kg, sc) increased their intake of 2% NaCl solution. Neither bilateral nephrectomy nor bilateral ureter ligation interfered with the increased NaCl solution ingestion produced by diazepam. It is concluded that the increased intake of the NaCl solution is not secondary to renal water-electrolyte losses or dependent on intact renal benzodiazepine receptors.

1558. Tang, Maisy & Falk, John L. (1990). **Schedule-induced oral self-administration of cocaine and ethanol solutions: Lack of effect of chronic desipramine.** ***Drug & Alcohol Dependence,*** 25(1), 21–25.
Investigated oral drug abuse in 18 male albino rats drinking either solutions of cocaine HCl, ethanol, or water. Ss drank excessive, equivalent volumes in daily 3-hr sessions of food-pellet delivery under a fixed-time 1-min (FT 1-min) schedule. During single-session exposures to pellet delivery schedules of FT 3- or 5-min, cocaine and ethanol Ss drank more. Chronic desipramine treatment affected neither FT 1-min schedule-induced polydipsia nor the enhanced ingestional response to the greater FT probes for the cocaine and ethanol Ss. Chronic administration of desipramine may be of use in treating cocaine abuse Ss through withdrawal.

1559. Tang, Siu W.; Helmeste, Daiga M. & Stancer, Harvey C. (1979). **The effect of clonidine withdrawal on total 3-methoxy-4-hydroxyphenylglycol in the rat brain.** ***Psychopharmacology,*** 61(1), 11–12.
Cessation of clonidine treatment in male Wistar rats resulted in an elevation in the level of brain total 3-methoxy-4-hydroxyphenylglycol. The suitability of the clonidine withdrawal syndrome as a model of bipolar depression is discussed.

1560. Testar, X.; López, D.; Llobera, M. & Herrera, E. (1986). **Ethanol administration in the drinking fluid to pregnant rats as a model for the fetal alcohol syndrome.** ***Pharmacology, Biochemistry & Behavior,*** 24(3), 625–630.
Adding ethanol (ETH) to the drinking fluid of pregnant rats has been questioned as an experimental model for the fetal alcohol syndrome (FAS). The present study used a modified version of previous protocols to overcome the major defects. Female Wistar rats were given 10% ETH in drinking fluid for 1 wk, 15% for the 2nd wk, 20% for the 3rd, and 25% for the 4th, at the end of which they were mated with non-treated males and given 25% throughout gestation. Three groups of non-ETH-treated sex- and age-matched rats were studied: normal controls receiving solid diet ad lib, pair-fed Ss, and Ss fed ad lib the solid diet mixed with 50% fiber. Findings demonstrate that this protocol is a suitable animal model for the FAS and indicate that rats on 50% fiber diet are better control Ss than pair-fed rats.

1561. Thiébot, Marie-Hélène; Dangoumau, Laure; Richard, Gwenaëlle & Puech, Alain J. (1991). **Safety signal withdrawal: A behavioural paradigm sensitive to both "anxiolytic" and "anxiogenic" drugs under identical experimental conditions.** ***Psychopharmacology,*** 103(3), 415–424.
Male rats were trained under 2 alternating components of a multiple schedule of reinforcement FR8 (food)/FR1 (food) + RR 50% (shocks randomly delivered with 50 ± 15% of the presses). The nonpunished and punished periods were signaled by one cue light above the right (safety signal) or left lever (punishment signal), respectively. On the test session (safety signal withdrawal), the safety signal was turned off at the end of the 1st nonpunished period, but the punishment signal was not presented (every press was food rewarded and no shocks were delivered). During this period, Ss exhibited a strong blockade of responding that lessened over time. This suppression was not caused by intervening events (i.e., novelty, temporal conditioning, schedule of food delivery). The behavioral blockade induced by withdrawal of the safety signal was reduced by benzodiazepines.

1562. Thiébot, Marie-Hélène; Soubrié, Philippe & Sanger, David. (1988). **Anxiogenic properties of beta-CCE and FG 7142: A review of promises and pitfalls.** ***Psychopharmacology,*** 94(4), 452–463.
Behavioral effects of the benzodiazepine-receptor partial inverse agonists beta-CCE and FG 7142 are reviewed, and the claim that these compounds possess anxiogenic properties is examined. Findings from human studies and global observations in animals, as well as those from experiments on aggression in animals or from studies of pentylenetetrazole discrimination are not considered to be conclusive. Contradictory findings are discussed from various theoretical perspectives: (1) the ability of the models to measure increased anxiety; (2) the possible ability of the drugs to reveal latent anxiety; and (3) anxiety produced by a pro- or preconvulsant state. Several hypotheses are considered to account for the behavioral effects of beta-CCE and FG 7142 without assuming anxiogenic properties. Available data are insufficient to strongly support the notion that FG 7142 and beta-CCE are the anxiogenic drugs "par excellence" they are often claimed to be.

1563. Thompson, Travis & Pickens, Roy. (1975). **An experimental analysis of behavioral factors in drug dependence.** ***Federation Proceedings,*** 34(9), 1759–1770.

Discusses the operation of drug reinforcers in the self-administration experimental situation. Variables influencing the drug functions in the acquisition, maintenance, and elimination phases of dependency are examined. Inducing and noninducing procedures used in studying acquisition of drug-reinforced behaviors are identified, 2 major variables in the maintenance of drug self-administration (injection dose and schedule of drug presentation) are considered, and 3 procedures for reducing or eliminating control of a reinforcing drug (weakening the reinforcing properties of the drug, changing stimulus control of drug self-administration, properties of the drug, changing stimulus control of drug self-administration, and increasing the probability of punishment associated with drug taking) are evaluated. Reasons why scientific analysis is relevant to the study of a complex problem such as drug dependence and implications for the study of drug use in humans are examined.

1564. Treit, Dallas. (1985). **Evidence that tolerance develops to the anxiolytic effect of diazepam in rats.** ***Pharmacology, Biochemistry & Behavior,*** 22(3), 383–387.
Three experiments examined the development of tolerance to the anxiolytic effect of diazepam (DZP [1 mg/kg, ip]) in 48 naive male Sprague-Dawley rats (Exps I and II) and 36 naive male hooded rats (Exp III). Suppression of defensive burying was used as an animal model of anxiolytic action. Although tolerance to the suppressive effect of DZP was not apparent after chronic administration when Ss were tested with a low-intensity shock, anxiolytic tolerance was detected under exactly the same drug regimen when Ss were tested with somewhat higher intensity shocks: Under the latter conditions, chronically treated Ss buried significantly more than acutely treated Ss. This tolerance effect did not appear to depend on the injection environment, control vehicle, or strain of the S; under each of these experimental variations Ss chronically treated with DZP buried significantly more than acutely treated Ss when they received a moderately high intensity shock. Results suggest that tolerance to the anxiolytic effects of benzodiazepines may be detectable when the stimuli eliciting anxiety are relatively intense.

1565. Treit, Dallas. (1987). **Ro 15-1788, CGS 8216, picrotoxin, and pentylenetetrazol: Do they antagonize anxiolytic drug effects through an anxiogenic action?** ***Brain Research Bulletin,*** 19(4), 401–405.
Examined anxiogenic behavioral effects of agents that inhibit GABAergic function, particularly at sites on the gamma-aminobutyric acid (GABA)/benzodiazepine receptor complex, using the defensive burying test with 252 rats. Putative blockers of the GABA-receptor coupled chloride channel, picrotoxin and pentylenetetrazol, and the benzodiazepine receptor antagonists Ro 15-1788 and CGS 8216 each blocked the anxiolytic effect of chlordiazepoxide. However, these compounds failed to exert significant anxiogenic effects in the burying test. Findings suggest that different animal models of anxiolytic drug effects are not equally sensitive to the possible anxiogenic effects of drugs that act at the GABA/benzodiazepine receptor complex.

1566. Treit, Dallas. (1985). **Animal models for the study of anti-anxiety agents: A review.** ***Neuroscience & Biobehavioral Reviews,*** 9(2), 203–222.
Reviews and evaluates animal models for the study of anxiolytic agents, according to pharmacological and behavioral criteria. Most early animal models, including models based on the effects of anxiolytics on unconditioned reactions and models based on the effects of anxiolytics in traditional learning paradigms, have not provided a reliable basis for identifying compounds with potential anxiolytic action or for delineating the mechanisms of anxiolytic drug action. The possibility that phylogenetically prepared forms of defensive learning might serve as a basis for the study of anxiolytic agents is introduced.

1567. Treit, Dallas & Fundytus, M. (1988). **A comparison of buspirone and chlordiazepoxide in the shock-probe/burying test for anxiolytics.** ***Pharmacology, Biochemistry & Behavior,*** 30(4), 1071–1075.
The effects of chlordiazepoxide (CDP) and buspirone were compared, using the shock-probe/burying test for anxiolytics. Both agents decreased rats' burying behavior toward the continuous shock probe and increased the number of probe-shocks received. The relative potency of buspirone was greater than that of CDP.

1568. Trullas, Ramon; Ginter, Hillary & Skolnick, Phil. (1987). **A benzodiazepine receptor inverse agonist inhibits stress-induced ulcer formation.** ***Pharmacology, Biochemistry & Behavior,*** 27(1), 35–39.
Studied the effects of a benzodiazepine receptor (BZPR) inverse agonist (FG-7142) on gastric ulcer formation in restrained male rats. FG-7142 (10–50 mg/kg) reduced in a dose-dependent fashion both the number and cumulative length of gastric ulcers elicited by restraint for 2 hrs at 4°C, but did not affect ulcer formation in unrestrained Ss maintained in this environment. FG-7142 also reduced ulcer formation in restrained Ss maintained at 22°C for 5 hrs. The ability of FG-7142 to reduce restraint-stress-induced ulcer formation was blocked by the BZPR antagonist ZK-93426 and the β-adrenoceptor antagonist propranolol. Findings suggest that FG-7142 produces a BZPR-mediated reduction in gastric ulcer formation, which may result from its ability to increase activity of the sympathetic nervous system.

1569. Trulson, Michael E.; Arasteh, Kamyar & Ray, Donald W. (1986). **Effects of elevated calcium on learned helplessness and brain serotonin metabolism in rats.** ***Pharmacology, Biochemistry & Behavior,*** 24(3), 445–448.
In a test of the effects of elevated calcium on learned helplessness in 16 male Sprague-Dawley rats, Ss maintained on high-calcium drinking water showed significantly longer escape latencies than their noncalcium counterparts after they were pretreated with inescapable electric shocks. Lower levels of 5-hydroxyindoleacetic acid were found in the forebrain and brain stem of Ss on high-calcium drinking water. Implications for the relationship between depressive states and calcium homeostatis are noted.

1570. Tulp, M.; Olivier, B.; Schipper, J.; Van der Poel, G. et al. (1991). **Serotonin reuptake blockers: Is there preclinical evidence for their efficacy in obsessive-compulsive disorder?** ***Human Psychopharmacology Clinical & Experimental,*** 6(Suppl), 63–71.
Explores preclinical evidence for possible animal models for obsessive-compulsive disorder (OCD). Different animal models for anxiety clearly differentiate between the serotonin uptake inhibitors and the classical anxiolytics such as the benzodiazepines, drugs that are not very effective in the treatment of OCD. One animal anxiety paradigm, ultrasonic distress calls in rat pups, appears as an attractive model to differentiate anxiolytic agents and should be considered as a tool to detect putative anti-OCD properties of drugs. The

neurochemical profile of the selective 5-hydroxytryptamine (5-HT) uptake inhibitors such as fluvoxamine and fluoxetine leaves little room for alternative mechanistic hypotheses about the mechanism of action of their clinical efficacy, not only in OCD but also in depression.

1571. Vaccheri, Alberto et al. (1984). **Antidepressant versus neuroleptic activities of sulpiride isomers on four animal models of depression.** ***Psychopharmacology,*** 83(1), 28–33.
The atypical neuroleptic sulpiride is also prescribed for depression because of its activating effect. However, such an effect does not necessarily imply an action identical to that of classical antidepressants. In the present study, a laboratory comparison of the neuroleptic and antidepressant activities of sulpiride was conducted to contribute to a better definition of its psychotherapeutic profile. Sulpiride isomers were studied in male Sprague-Dawley rats in 4 behavioral models of depression that are thought to be influenced by neuroleptics in different ways. Desipramine (imipramine) and haloperidol were employed in each test as a standard antidepressant and neuroleptic, respectively. The 4 tests were (1) prevention of apomorphine-induced sedation; (2) antagonism of apomorphine-induced hypothermia; (3) behavioral despair (swim test); and (4) learned helplessness (FR2 lever-pressing escape). Desipramine ameliorated behavior in all tests; haloperidol ameliorated the response to Test 1, influenced that to Test 2 in a neurolepticlike way, and worsened the responses to Tests 3 and 4. Levosulpiride worked in a similar way to haloperidol and in all tests. Dextrosulpiride significantly and dose dependently ameliorated the responses to Test 3 and was inactive in the others. The results of Tests 2–4 indicated a neuroleptic profile of levosulpiride and suggest a potential "antidepressant" activity of dextrosulpiride that merits further investigation.

1572. Vale, A. L. & Ratcliffe, F. (1987). **Effect of lithium administration on rat brain 5-hydroxyindole levels in a possible animal model for mania.** ***Psychopharmacology,*** 91(3), 352–355.
Examined the effect of short term administration of lithium (2 or 4 mEq/kg) on the hyperactivity induced by a mixture of dextroamphetamine (1.18 mg/kg) and chlordiazepoxide (12.5 mg/kg) in female Sprague-Dawley rats. An attempt has been made to relate this action to changes in brain concentrations of 5-hydroxytryptamine (5-HT) and 5-hydroxyindoleacetic acid.

1573. Vanderwolf, C. H.; Baker, G. B. & Dickson, C. (1990). **Serotonergic control of cerebral activity and behavior: Models of dementia.** ***Annals of the New York Academy of Sciences,*** 600, 366–383.
Discusses behavioral studies of atropine-resistant cerebral activation, particularly those involving intra-brainstem injections of 5,7-dihydroxytryptamine (5,7-DHT) and reserpine. Studies show that 5,7-DHT injections and systemic administration of p-chlorophenylalanine led to severe depletion of serotonin in the forebrain. Behavioral data suggest that central serotonergic blockade produces impairments in a broad spectrum of behaviors. The effects produced are, in many cases, modest. When central serotonergic blockade is combined with central cholinergic blockade, the behavioral deficits produced are much greater than those produced by cholinergic or serotonergic blockade alone.

1574. Vanderwolf, C. H.; Dickson, C. T. & Baker, G. B. (1990). **Effects of *p*-chlorophenylalanine and scopolamine on retention of habits in rats.** ***Pharmacology, Biochemistry & Behavior,*** 35(4), 847–853.
Rats were trained on a conventional maze test or on a swim-to-platform test. Retention of swim-to-platform performance 7 days later was severely impaired by posttraining treatment with a combination of *p*-chlorophenylalanine (PCPA) and scopolamine (SCO), although neither drug alone had any effect. Retention of the maze habit was moderately impaired by SCO alone and severely impaired by a combination of SCO and PCPA but was unaffected by PCPA alone. Polygraphic recordings confirmed previous reports that a combination of PCPA and SCO can abolish neocortical low voltage fast activity and hippocampal rhythmical slow activity. Combined blockade of central cholinergic and serotonergic neurotransmission in rats may provide a useful animal model of Alzheimer's disease.

1575. Van Hest, Annemieke & Van Haaren, Frans P. (1988). **Behavioral teratology.** ***Psycholoog,*** 23(4), 150–153.
Discusses the possible long-term adverse effects of early exposure to various chemicals on brain development and maladaptive behavior in adulthood. The difficulty of establishing a causal link if the relationship has not been studied in suitable animal models is emphasized. The need for preventive measures in occupational settings and for funding of research to develop animal models is considered. (English abstract).

1576. Van Thiel, David H.; Gavaler, Judith S. & Lester, Roger. (1977). **Ethanol: A gonadal toxin in the female.** ***Drug & Alcohol Dependence,*** 2(5–6), 373–380.
Tested an animal model of the pathophysiologic mechanisms responsible for the development of sexual abnormalities in alcoholic women. Findings from studies with 100 female white Wistar rats suggest that (a) ethanol is toxic for the female gonad, (b) sexual dysfunctions can occur in female alcoholics in the absence of irreversible histologic and biochemical liver disease, and (c) differences in nutrition alone cannot account for the gonadal dysfunctions occurring in female alcoholics.

1577. Velíšek, L.; Ortová, M.; Velíšková, J.; Kubová, H. et al. (1989). **Influence of clonazepam and valproate on kainate-induced model of psychomotor seizures.** ***Activitas Nervosa Superior,*** 31(1), 66–67.
Demonstrated that both clonazepam (CZP) and valproate (VPA) do not influence the automatisms elicited by kainic acid (KA) in rats aged 7, 12, 18, 25, and 90 days. The effects of VPA and CZP were very good on the 1st stage of seizure generalization. The same generator may function for clonic seizures induced by KA, metrazol, and other convulsants.

1578. Vezina, Paul; Giovino, Adrienne A.; Wise, Roy A. & Stewart, Jane. (1989). **Environment-specific cross-sensitization between the locomotor activating effects of morphine and amphetamine.** ***Pharmacology, Biochemistry & Behavior,*** 32(2), 581–584.
Pre-exposed 48 male rats to either 10 or 20 mg/kg morphine sulfate either in activity boxes (conditioning groups [COND]) or in their home cages (HOME), and on alternate days exposed them to saline in the other environment or to saline in both environments (controls [CONs]). On the day following morphine pre-exposure, all Ss were administered

amphetamine sulfate prior to being tested in the activity boxes. COND Ss exposed to 10 mg/kg morphine showed higher levels of activity than HOME Ss or CONs. COND Ss pre-exposed to 20 mg/kg morphine was significantly more active than HOME Ss but not more active than CONs. Results support the view that changes in the mesolimbic dopamine system are responsible for sensitization to the locomotor activating effects of morphine and amphetamine.

1579. Vilkov, G. A.; Minoranskaya, A. P. & Kolmakova, T. S. (1983). **Changes of catecholamine metabolism in rabbit brain under the action of anticerebral antibodies and lithium ions.** ***Zhurnal Nevropatologii i Psikhiatrii imeni S.S. Korsakova,*** 83(9), 1395–1397.
Examined changes in thyrosine hydroxylase and dopamine-beta-hydroxylase activity in rabbit brains following intracisternal injection of immunoglobulin G isolated from the blood of brain-sensitized dogs or the serum of human schizophrenics. Also examined were catecholamine changes after iv injection of LiCl following pretreatment with anticerebral antibodies. Findings suggest that the mechanisms underlying these disturbances are similar. A positive effect of Li on catecholamine metabolism was observed, indicating the value of this animal model for examining the efficacy and mode of action of different psychotropic drugs. (English abstract).

1580. Vogel, Gerald W.; Hartley, P.; Neill, D.; Hagler, M. et al. (1988). **Animal depression model by neonatal clomipramine: Reduction of shock induced aggression.** ***Pharmacology, Biochemistry & Behavior,*** 31(1), 103–106.
Clomipramine (CLM), administered to neonatal rats, has been reported to produce adult behavioral and REM sleep abnormalities. Since these abnormalities resemble some found in human endogenous depression, it is suggested that adult rats may represent an animal model of depression. The validity of the animal depression model was tested by determining in rats the effect of neonatal CLM on adult shock-induced fighting. Experimental Ss were treated neonatally with CLM, and controls were treated neonatally with saline. When they matured, compared with control Ss, experimental Ss had significantly fewer offensive fighting responses and significantly more defensive fighting responses. Findings support the validity of the animal depression model produced by neonatal CLM.

1581. Vogel, Gerald W.; Neill, D.; Hagler, M.; Kors, D. et al. (1990). **Decreased intracranial self-stimulation in a new animal model of endogenous depression.** ***Neuroscience & Biobehavioral Reviews,*** 14(1), 65–68.
Demonstrated that a major factor of depression in humans, the diminished capacity for pleasure, appeared to be present in male rats treated neonatally with clomipramine, an antidepressant. At age 7 mo, bar-press responding for rewarding hypothalamic stimulation was reduced across a range of intensities compared with control rats treated with saline. At age 4 or 5 mo this effect was not seen, although other behavioral abnormalities were present at the younger age. The delayed onset of diminished intracranial self-stimulation may relate to the gradual insidious onset of endogenous depression in humans.

1582. Vogel, Gerald W.; Neill, D.; Hagler, M. & Kors, D. (1990). **A new animal model of endogenous depression: A summary of present findings.** ***Neuroscience & Biobehavioral Reviews,*** 14(1), 85–91.
Summarizes tests of the validity of a new animal model of endogenous depression (EDP). After neonatally administered clomipramine (CLI), an antidepressant, adult male rats showed behavioral abnormalities of the human disorder: decreased sexual, aggressive, and intracranial self-stimulation activities, as well as motor hyperactivity in a stressful situation. Preliminary evidence suggests that behavioral abnormalities in rats briefly treated with antidepressant treatments (imipramine, REM sleep deprivation) began to normalize. Lastly, after neonatal CLI, adult rats showed REM sleep abnormalities of endogenous depression: low REM latency, frequent sleep onset REM periods, and abnormal temporal course of REM rebound after REM sleep deprivation. Results support the hypothesis that in rats neonatal CLI produced adult animals that modelled EDP.

1583. Vogel, Gerald W.; Neill, D.; Kors, D. & Hagler, M. (1990). **REM sleep abnormalities in a new animal model of endogenous depression.** ***Neuroscience & Biobehavioral Reviews,*** 14(1), 77–83.
Examined whether adult rats treated neonatally with the antidepressant clomipramine (CLI) will show REM sleep abnormalities characteristic of human endogenous depression. Neonatal CLI produced Ss that at age 6 mo (and to a lesser degree at age 11 mo) had shorter REM latency, more sleep onset REM periods than control Ss treated with saline, and (after REM sleep deprivation) had an abnormal temporal course of REM rebound in the presence of a normal total REM rebound. These REM sleep abnormalities support the validity of the animal model of endogenous depression.

1584. Vorhees, Charles V.; Fernandez, Kathleen; Dumas, Ruth M. & Haddad, Raef K. (1984). **Pervasive hyperactivity and long-term learning impairments in rats with induced micrencephaly from prenatal exposure to methylazoxymethanol.** ***Developmental Brain Research,*** 15(1), 1–10.
Pregnant Long-Evans rats were given a single injection of methylazoxymethanol acetate (MAM [30 mg/kg, ip]) or saline on Day 14 of gestation. Between 49 and 192 days of age, all offspring were examined on open-field, figure-8, and hole-board tests of activity, as well as passive avoidance and Biel water maze tests of learning. MAM-exposed Ss showed no increase in mortality but weighed less than controls, a difference that remained relatively constant throughout the experiment. At 204–215 days of age, the MAM-exposed Ss were confirmed to be micrencephalic, a known effect of this drug at this dose and exposure period. On all tests of activity, MAM-exposed Ss were markedly hyperactive. Females also exhibited a pronounced impairment of normal activity habituation patterns, and males showed a marked impairment of passive avoidance performance. At 2 and 6 mo of age, MAM-exposed Ss also showed a pronounced deficit in learning a water maze. The MAM-induced brain and behavioral abnormalities provide a potentially useful animal model of congenital micrencephaly and associated mental retardation.

1585. Vorobyeva, T. M. & Geiko, V. V. (1990). **Characteristics of neurobiological mechanisms of predisposition to the development of alcoholism in progeny of rats hereditarily aggravated by alcoholism.** ***Zhurnal Vysshei Nervnoi Deyatel'nosti,*** 40(2), 389–392.
Studied: (1) the neurophysiological features of the bioelectrogenesis of the limbic-neocortical regions of the brain and (2) the organization of the system of positive emotional reactions in intact Ss and offspring of Ss with familial

alcoholism under conditions of developing ethanol addiction. Animal subjects: 56 male noninbred white rats (aged 3–4 mo) (1st, 2nd, and 4th generations). A realistic model of inherited familial alcoholism was created. Emotional behavior was assessed based on general motor and orienting-exploratory activity and Ss' responses to alcohol. Electrical brain activity was recorded.

1586. Wachtel, Helmut & Löschmann, Peter-A. (1986). **Effects of forskolin and cyclic nucleotides in animal models predictive of antidepressant activity: Interactions with rolipram.** ***Psychopharmacology,*** 90(4), 430–435.
Forskolin, a direct activator of the catalytic subunit of adenylate cyclase (AC), and the cyclic nucleotide analogs dibutyryl cAMP (dBcAMP), 8-bromo cAMP (8-BrcAMP) and dibutyryl cGMP (dBcGMP) were tested for their ability to reverse the hypothermia or hypokinesia of male NMRI mice depleted of presynaptic endogenous monoamines by pretreatment with reserpine, alpha-methyl-p-tyrosine and para-chlorophenylalanine. Findings suggest that antidepressant activity is crucially linked to enhanced cAMP availability within brain effector cells. The successful treatment of endogenously depressed patients with rolipram supports this assumption.

1587. Waddington, John L. (1990). **Spontaneous orofacial movements induced in rodents by very long-term neuroleptic drug administration: Phenomenology, pathophysiology and putative relationship to tardive dyskinesia.** ***Psychopharmacology,*** 101(4), 431–447.
Suggests that animal models should first seek to reproduce the phenomenology of tardive dyskinesia, and then search for pathophysiological correlates of such behavior. Research on orofacial function in rats administered neuroleptic drugs for substantial proportions of their adult lifespan is reviewed. This literature reveals the emergence of late-onset orofacial movements in a number of studies, but early-onset movements or no effect in others. Potential explanations for these discrepancies are considered. The relationship of these orofacial phenomena to dopaminergic and nondopaminergic function, and to clinical tardive dyskinesia, is evaluated.

1588. Wahlström, Göran; Stenström, Anders; Tiger, Gunnar; O'Neill, Cora et al. (1988). **Influence of age on effects induced by intermittent ethanol treatment on the ethanol drinking pattern and related neurochemical changes in the rat.** ***Drug & Alcohol Dependence,*** 22(1–2), 117–128.
Developed an experimental animal model of alcoholism, using male rats who had both voluntary and forced exposure to ethanol. Voluntary intake during treatment varied as exposure over time and differed in groups of Ss who began ethanol injections at different ages.

1589. Walton, Nancy Y. & Deutsch, J. A. (1978). **Self-administration of diazepam by the rat.** ***Behavioral & Neural Biology,*** 24(4), 533–538.
20 male Sprague-Dawley rats were offered diazepam either orally or by counterinjection over a 6-day period, with drug concentrations increasing every other day. Although diazepam intoxication was observed at all drug concentrations and with increasing severity as concentration was increased, no decrease in intake ensued. This apparent lack of conditioned aversion formation was not due to diazepam-induced amnesia, since Ss poisoned with lithium chloride after diazepam consumption showed no attenuation of conditioned aversion when compared to Ss poisoned after drinking flavored water. Since diazepam is not unpalatable and does not produce a strong conditioned aversion, it may be a useful compound for an animal model of environmental factors leading to drug abuse.

1590. Wauquier, A.; Melis, W. & Janssen, P. A. (1989). **Long-term neurological assessment of the post-resuscitative effects of flunarizine, verapamil and nimodipine in a new model of global complete ischaemia.** ***Neuropharmacology,*** 28(8), 837–846.
Evaluated neurological recovery in 108 anesthetized male rats following treatment with 3 calcium antagonists (flunarizine, verapamil, and nimodipine). This noninvasive model of global ischemia proved valuable because of the high survival rate associated with the procedure and was useful in demonstrating the protective effects of the drugs studied.

1591. Weekley, Bruce & Harlow, Henry J. (1985). **Effects of pharmacological manipulation of the renin-angiotensin system on the hibernation cycle of the 13-lined ground squirrel (*Spermophilus tridecemlineatus*).** ***Physiology & Behavior,*** 34(1), 147–149.
74 adult 13-lined ground squirrels were monitored for body temperature and state of torpor or arousal on a 2:22 light–dark cycle after receiving ip injections of 0.9% saline vehicle or with the vehicle plus various test drugs. Data suggest that arousal from the reentry into torpor may be involved with renal function and the renin-angiotensin system. However, there appears to be temporal changes in the effect of angiotensin on torpor. Possible mechanisms, including the antagonistic effect of melatonin on angiotensin, are discussed.

1592. Weinstein, Debbie; See, Ronald E. & Ellison, Gaylord. (1989). **Delayed appearance of facial tics following chronic fluphenazine administration to guinea pigs.** ***Pharmacology, Biochemistry & Behavior,*** 32(4), 1057–1060.
21 female guinea pigs were administered chronic fluphenazine decanoate or vehicle for 11 mo, and oral movements (OMs) were periodically observed. Shortly after the initiation of neuroleptic treatment, increased OMs were seen in the drugged Ss, but these did not persist. After 7 mo of neuroleptics, twitchlike movements of the orofacial region were observed in the drugged Ss; these dyskinetic movements were enhanced by administration of d-amphetamine. These twitchlike movements appear to be a better model of tardive dyskinesia in the guinea pig than the initially observed and normal-appearing OMs.

1593. Weiss, Jay M.; Simson, Prudence G.; Hoffman, Laura J.; Ambrose, Monica J. et al. (1986). **Infusion of adrenergic receptor agonists and antagonists into the locus coeruleus and ventricular system of the brain: Effects on swim-motivated and spontaneous motor activity.** ***Neuropharmacology,*** 25(4), 367–384.
Examined how pharmacological stimulation and blockade of alpha receptors affect active motor behavior in male Holtzman Sprague-Dawley albino rats. In Exp I, varying doses of piperoxan, yohimbine, clonidine, and norepinephrine (NE) were infused into various locations in the ventricular system of the brain, including the locus coeruleus region. When infused into the locus coeruleus region, small doses of piperoxan and yohimbine depressed activity in the swim test, while infusion of clonidine and NE had the opposite effect of stimulating activity. In Exp II, isoproterenol and

phenylephrine were infused into the ventricular system. Both postsynaptic receptor agonists selectively decreased activity in the swim test with respect to dose and location of infusion in the ventricular system.

1594. Weiss, Susan R. & Hodos, William. (1988). **Defective mirror-image discrimination in pigeons: A possible animal model of dyslexia.** ***Neuropsychiatry, Neuropsychology, & Behavioral Neurology,*** 1(3), 161–170.
Pigeons were trained to discriminate between 2 simultaneously presented stimuli, which were either lateral mirror images, vertical mirror images, or unsymmetrical stimuli. Nine of 48 Ss showed markedly impaired acquisition of the 3 discriminations, with the greatest and most persistent deficits on the lateral mirror-image problem. Three of the poorly performing Ss were subsequently given monocular occlusion, a treatment reported to help visually dyslexic children; the performance of all 3 Ss improved. The pronounced and selective difficulties in mirror-image discrimination in this group of Ss and their response to monocular occlusion suggest that they may serve as a useful animal model of dyslexia.

1595. Wenger, Galen R.; McMillan, D. E. & Chang, Louis W. (1982). **Behavioral toxicology of acute trimethyltin exposure in the mouse.** ***Neurobehavioral Toxicology & Teratology,*** 4(2), 157–161.
Studied the effects of trimethyltin (TMT) on the gross behavior, lethality, spontaneous motor activity (SMA), and responding of BALB/c mice under a multiple FR 30/FI 600 sec schedule of reinforcement. Following doses of 4, 5, and 6 mg/kg, the cumulative 48-hr lethality was 10% at 4 mg/kg and 100% at 5 and 6 mg/kg. No deaths were observed during the 1st 48 hrs following 3 mg/kg TMT. This nonlethal dose produced whole body tremors. The SMA of Ss receiving 3 mg/kg was reduced to 70% during the 1st 24-hr period following TMT, and some recovery of total activity was observed during the 2nd 24-hr period. Reduction in SMA was accompanied by a change in the normal circadian cycle of activity. Responding under the multiple FR 30/FI 600 schedule was severely disrupted. Rate of responding in both components was decreased 3 hrs after TMT, and the decrease became progressively larger over the next 48 hrs. In addition to the rate of FI 600 responding being reduced, the normal pattern of FI responding was altered with an increase in responding observed in the early portions of the FI. Results suggest that the mouse is much more sensitive to the effects of TMT than the rat and may have potential as an animal model in the study of TMT neurotoxicity.

1596. Whitehouse, Wayne G.; Walker, Joseph; Margules, David L. & Bersh, Philip J. (1983). **Opiate antagonists overcome the learned helplessness effect but impair competent escape performance.** ***Physiology & Behavior,*** 30(5), 731–734.
In an experiment with 84 male Long-Evans rats, Ss exposed to inescapable shocks exhibited deficiencies in learning to escape shock in a novel situation 24 hrs later (learned helplessness). Opiate antagonists naloxone (10 mg/kg, sc) and naltrexone (1 or 10 mg/kg) blocked the learned helplessness effect, allowing efficient escape performance on the subsequent test. In contrast, these drugs impaired the performance of Ss pretrained with escapable shocks and Ss with no previous exposure to shock. Both effects increased substantially with higher doses. Results suggest a significant role for endogenous opiates in the induction of learned helplessness as well as in the acquisition of efficient escape behavior.

1597. Whitton, P. S.; Sarna, G. S.; Datla, K. P. & Curzon, G. (1991). **Effects of tianeptine on stress-induced behavioural deficits and 5-HT dependent behaviour.** ***Psychopharmacology,*** 104(1), 81–85.
The tricyclic antidepressant drug tianeptine had an antidepressant-like effect on a rat model of depression based on observed behavioral deficits in male rats in open field activity on the day after 2-hr restraint. This deficit was opposed when tianeptine was given 2 hrs after the end of the restraint to either untreated Ss or to Ss previously given 10 mg/kg of the drug per day for 13 days. Tianeptine, given acutely, moderately enhanced the 5-hydroxytryptamine$_{1C}$ (5-HT$_{1C}$) receptor-dependent hypolocomotor effect of *m*-chlorophenylpiperazine, but did not alter other 5-HT$_1$ receptor subtype-dependent behavior. Acute tianeptine also decreased 5-HT$_2$ receptor-dependent body shakes induced by 5-hydroxytryptophan (5-HTP). Shakes induced by the 5-HT$_2$ agonist 1-(2,5-dimethoxy-4-iodophenyl)-2 aminopropane were unaffected.

1598. Wideman, Cyrilla H. & Murphy, Helen M. (1991). **Effects of vasopressin replacement during food-restriction stress.** ***Peptides,*** 12(2), 285–288.
Examined the effects of sc injections of vasopressin in vasopressin-deficient (Brattleboro/diabetes insipidus [DI]) rats that were observed during nonstress (habituation) and stress (food-restricted [FRS]) conditions as compared with other rats. Four groups of Ss were used: (1) FRS Long-Evans Ss with no injections (controls), (2) FRS DI Ss with no injections, (3) DI Ss injected with vasopressin, and (4) DI Ss injected with peanut oil. Measures included body weight, food intake, water intake, and gastric ulcer formation. With respect to body weight, water intake, and ulcer formation, 2 sets of Ss emerged. Vasopressin-injected DI Ss and controls coped with the stress of being FRS, but peanut-oil-injected DI Ss and DI Ss with no injections could not cope with the stress of being FRS.

1599. Wigal, Tim; Greene, Paul L. & Amsel, Abram. (1988). **Effects on the partial reinforcement extinction effect and on physical and reflex development of short-term in utero exposure to ethanol at different periods of gestation.** ***Behavioral Neuroscience,*** 102(1), 51–53.
Ethanol was intubated into pregnant rats at gestational days 7–9 (G7–9) or 14–16 (G14–16). At both gestational ages, ethanol intubation affected reflex development but not physical development of resulting offspring. At G14–16, but not at G7–9, ethanol intubation resulted in the elimination of the partial reinforcement extinction effect when pups were tested at 15 days of age. These latter results confirm and extend previous ones in which ethanol was administered in a liquid diet throughout gestation.

1600. Willner, Paul. (1990). **The role of slow changes in catecholamine receptor function in the action of antidepressant drugs.** ***International Review of Psychiatry,*** 2(2), 141–156.
Reviews the evidence that antidepressants alter the functioning of noradrenaline and dopamine receptor systems in the brain and assesses the significance of these changes for the clinical action of antidepressants. The down regulation of beta-adrenoceptors is unlikely to represent the mechanism of

action but rather represent an unwanted effect of antidepressant drugs. The enhanced responsiveness of alpha$_1$- and D_2 receptors seems relevant to the action of antidepressant drugs in several animal models of depression, including the learned helplessness and chronic mild stress models. Similar changes should now be sought in people.

1601. Wilson, James R. (1985). **Development of an animal model of alcohol dependence.** Conference of the Union College and the Society for the Study of Social Biology: Genetics and the human encounter with alcohol (1984, Schenectady, New York). ***Social Biology,*** 32(3–4), 229–240.
Developed an animal model to study the mechanisms that underlie dependence. Genetic selection was undertaken for differential severity of the alcohol withdrawal syndrome in the laboratory mouse. Results are presented for 12 generations of replicate line, bidirectional, within-family artificial selection. Results to date are discussed with reference to natural selection, whose effects are adding unforeseen complications to the selection response.

1602. Winders, Suzan E. & Grunberg, Neil E. (1989). **Nicotine, tobacco smoke, and body weight: A review of the animal literature. *Annals of Behavioral Medicine*** , 11(4), 125–133.
Reviews the literature on the effects of exposure to nicotine (NIC) or tobacco smoke on energy intake and energy expenditure (EE) in animals. These studies indicate that NIC causes changes in body weight (BW) and that these effects are mediated in part by changes in food consumption. In addition, changes in EE, especially changes in fat utilization, seem to be involved in the effects of NIC on BW.

1603. Wisniewski, Henryk M.; Sturman, John A. & Shek, Judy W. (1982). **Chronic model of neurofibrillary changes induced in mature rabbits by metallic aluminum. *Neurobiology of Aging,*** 3(1), 11–22.
A slurry of aluminum powder injected into the brains of New Zealand rabbits produced neurofibrillary changes in neurons of the spinal cord and cerebrum. This chronic animal model of neurofibrillary changes, induced in a mature nervous system, will allow better investigations of alterations in biochemistry, pathology, behavior, and cognition. Results are discussed in terms of research suggesting that neurofibrillary changes impair some cognitive functions and possible implications for senile dementia of the Alzheimer type.

1604. Wolfarth, Stanislaw & Ossowska, Krystyna. (1989). **Can the supersensitivity of rodents to dopamine be regarded as a model of tardive dyskinesia? *Progress in Neuro-Psychopharmacology & Biological Psychiatry,*** 13(6), 799–840.
Presents arguments derived from clinical work and animal experiments for or against the hypothesis suggesting that tardive dyskinesia (TD) is caused by supersensitivity to dopamine. Data presented prove that chronic administration of neuroleptics to schizophrenic patients cannot be the only factor inducing TD; symptoms similar or identical to those of TD are also observed in the course of other disorders not connected with neuroleptics (e.g., aging, schizophrenia). Pharmacological and biochemical data show that chronic administration of neuroleptics to animals induces an increase in the density of dopamine D-2 receptors. It seems that this receptor-mediated supersensitivity may concern both the postsynaptic and presynaptic D-2 dopamine receptors. However, it is not clear enough whether a dopamine D-1 receptor-mediated supersensitivity might also be a causal factor of TD.

1605. Woods, Joycelyn S. & Leibowitz, Sarah F. (1985). **Hypothalamic sites sensitive to morphine and naloxone: Effects on feeding behavior. *Pharmacology, Biochemistry & Behavior,*** 23(3), 431–438.
Investigated the feeding response of 57 brain-cannulated male Sprague-Dawley rats to hypothalamic injections of norepinephrine (NE), the opiate agonist morphine sulfate (MO), and the opiate antagonist naloxone (NAL) in 3 experiments. MO elicited feeding in a dose-dependent manner when injected into the paraventricular nucleus (PVN) of satiated Ss at doses of 0.78–100 nmoles. NAL, at doses of 3.13–200 nmoles, produced a dose-dependent suppression of feeding. Ss with brain cannulas aimed at the PVN, the perifornical hypothalamus (PFH), the dorsomedial nuclei (DMN), and ventromedial nuclei (VMN) were compared for their sensitivity to the feeding stimulatory effects of NE and MO (except in the DMN) and the feeding suppressive effects of NAL. Consistent with earlier reports, the PVN-cannulated Ss exhibited a reliable increase in feeding after NE injections; the VMN cannula yielded a small feeding response. MO, in contrast, strongly stimulated eating after administration into the PFH, as well as the PVN, apparently dissociating the NE and MO eating responses. With regard to NAL's suppressive effect on feeding, the PVN and PFH, which were sensitive to MO, also exhibited responsiveness to opiate antagonism, suggesting the existence in these areas of opiate receptors that modulate feeding. These mapping studies suggest that the opiate receptors affecting eating behavior are most dense within the PVN and PFH but are either less dense or have variable responsiveness within the DMN and VMN.

1606. Woolverton, William L. & Kleven, Mark S. (1988). **Evidence for cocaine dependence in monkeys following a prolonged period of exposure. *Psychopharmacology,*** 94(2), 288–291.
The behavioral consequences of prolonged continuous exposure to cocaine were examined in monkeys. Operant behavior was sampled for 0.5 hr every 6 hrs, and cocaine was continuously infused through an iv catheter. Cocaine (4.0–32 mg/kg/day) initially caused reductions in the rate of responding for food, and tolerance developed to this effect. When infusion of cocaine was terminated following a period of exposure during which cocaine dose was escalated to 32 mg/kg/day, there was marked suppression of operant behavior lasting as long as 72 hrs as well as observable changes in behavior (e.g., hyporesponsiveness). This demonstration of behavioral disruptions following discontinuation of cocaine exposure suggested that the preparation may be a useful animal model for further examining the possibility that exposure to cocaine can induce dependence.

1607. Worms, Paul; Kan, Jean-Paul; Wermuth, Camille G.; Roncucci, Romeo et al. (1987). **SR 95191, a selective inhibitor of type A monoamine oxidase with dopaminergic properties: I. Psychopharmacological profile in rodents. *Journal of Pharmacology & Experimental Therapeutics,*** 240(1), 241–250.
Examined the activity of SR 95191 in rodent models that are predictive of antidepressant activity and then investigated, in a variety of animal models, the possible mechanisms through which SR 95191 could be exerting its antide

pressant activity. Based on data obtained, it is postulated that SR 95191 has a unique profile of activity combining the properties of a selective type A monoamine oxidase (MAO) inhibitor and those of an atypical dopaminergic drug.

1608. Wultz, Boaz; Sagvolden, Terje; Moser, Edvard I. & Moser, May-Britt. (1990). **The spontaneously hypertensive rat as an animal model of Attention-Deficit Hyperactivity Disorder: Effects of methylphenidate on exploratory behavior.** ***Behavioral & Neural Biology,*** 53(1), 88–102.
Used the spontaneously hypertensive rat (SHR) as an animal model of attention-deficit hyperactivity disorder (ADHD) by examining the effects of 1–24 mg/kg methylphenidate (MET) on the exploratory behavior of 7 SHRs and 6 normotensive Wistar-Kyoto (WKY) control rats. A 2-compartment free-exploration open field was used. SHRs showed significant hyperactivity both in the field and home cage. Rearing in SHRs was more sensitive to the actions of MET than ambulation. SHR behavior was generally less susceptible to MET than WKY behavior, but there was no paradoxical effect of MET on SHR hyperactivity.

1609. Wurster, Richard M.; Griffiths, Roland R.; Findley, Jack D. & Brady, Joseph V. (1977). **Reduction of heroin self-administration in baboons by manipulation of behavioral and pharmacological conditions.** ***Pharmacology, Biochemistry & Behavior,*** 7(6), 519–528.
Four male baboons responded on a discrete-trial choice task on which trials occurred every 3 hrs throughout the day. Trials involved choosing between several mutually exclusive options, one of which was always associated with iv infusion of a unit dose of heroin. Experiments were undertaken to reduce the selection of the heroin option. Exp I used methods analogous to clinical situations involving opioid maintenance and subsequent detoxification. During initial baseline conditions, Ss consistently preferred an option of heroin and food over an option of saline and food. Selection of heroin was almost entirely eliminated when there was a mutually exclusive choice between heroin and food and chronic noncontingent morphine (150 mg/day). Decreasing the dose of noncontingent morphine produced an increased selection of heroin. In Exp II, initial baseline conditions were similar to Exp I. Food availability was subsequently made contingent on selection of options involving progressively lower doses of contingent heroin. These manipulations reduced heroin intake to about 15% of baseline levels. The experiments demonstrate the utility of animal models for studying procedures for the reduction of opiate self-administration.

1610. Yadin, Elna; Friedman, Eitan & Bridger, Wagner H. (1991). **Spontaneous alternation behavior: An animal model for obsessive-compulsive disorder?** ***Pharmacology, Biochemistry & Behavior,*** 40(2), 311–315.
Food-deprived rats were run in a T-maze in which both a black and a white goal box were equally baited with chocolate milk. Each S was given trials during which it was placed in the start box and allowed to make a choice. The mean number of choices until an alternation occurred was recorded. After a stable baseline of spontaneous alternation was achieved, the effects of manipulating the serotonergic system were tested. 5-methoxy-N,N-dimethyl tryptamine (5-MeODMT), and 8-hydroxy-2-(di-n-propylamino)-tetralin hydrobromide (8-OH-DPAT) disrupted spontaneous alternation. Chronic treatment with fluoxetine had a protective effect on the 5-MeODMT-induced disruption of spontaneous alternation behavior. Serotonergic manipulations of spontaneous alternation may be a simple animal model for the perseverative symptoms or indecisiveness seen in people diagnosed with obsessive-compulsive disorder.

1611. Yaksh, Tony L. (1989). **Behavioral and autonomic correlates of the tactile evoked allodynia produced by spinal glycine inhibition: Effects of modulatory receptor systems and excitatory amino acid antagonists.** ***Pain,*** 37(1), 111–123.
Intrathecal administration of glycine (strychnine) or gamma-aminobutyric acid (GABA [bicuculline]) but not opioid (nalozone), adrenergic (phentolamine), or serotonin (methysergide) receptor antagonists resulted in a dose-dependent organized agitation response to light tactile stimulation in male rats. In contrast, intrathecal injections of glutamate receptor antagonists resulted in a dose-dependent depression of the strychnine evoked hyperesthesia. Results suggest the operation of discriminable processing systems, the characteristics of which resemble the clinical phenomenon observed in patients suffering from sensory dysesthesia following central and peripheral horn injury.

1612. Yamada, Kenji. (1988). **Effects of dexamethasone and prednisolone on corticosterone levels in forced swimming rats.** ***Research Communications in Psychology, Psychiatry & Behavior,*** 13(3), 231–236.
Suggests that 3-, 4-, and 5-day forced-swimming rats may be appropriate as an animal model for depression, based on investigation of an endocrinological approach.

1613. Yamazaki, Naoki; Kiyota, Yoshihiro; Take, Yomei; Miyamoto, Masaomi et al. (1989). **Effects of idebenone on memory impairment induced in ischemic and embolization models of cerebrovascular disturbance in rats.** ***Archives of Gerontology & Geriatrics,*** 8(3), Spec Issue, 213–224.
Used 2 types of rat models of cerebrovascular disturbance (a cerebral ischemia model produced by 4-vessel occlusion and a cerebral embolization model produced by injecting microspheres into the carotid arteries) to evaluate the effects of a new cerebral metabolic enhancer, idebenone. Psychopharmacologic findings with male rats suggest that idebenone may exert an ameliorating effect on the cognitive dysfunction in the aged or disordered central nervous system (CNS).

1614. Yamazaki, Naoki; Nomura, Masahiko; Nagaoka, Akinobu & Nagawa, Yuji. (1989). **Idebenone improves learning and memory impairment induced by cholinergic or serotonergic dysfunction in rats.** ***Archives of Gerontology & Geriatrics,*** 8(3), Spec Issue, 225–239.
Used 2 rat models with a decline in central cholinergic or serotonergic activity to examine the effects of idebenone (IB), a cerebral metabolic enhancer, on impaired learning and memory motivated with food. Results of 2 experiments demonstrated the beneficial effects of IB on scopolamine-induced impairment of short-term memory and tryptophan-deficient diet-induced impairment of operant discrimination learning in male rats. IB may exert its effect on a pathogenic condition related to hypofunction of the cerebral cholinergic system, which is observed in presenile dementia and senile dementia of the Alzheimer type.

1615. Yokel, Robert A. (1983). **Repeated systemic aluminum exposure effects on classical conditioning of the rabbit.** ***Neurobehavioral Toxicology & Teratology,*** 5(1), 41–52.

Excessive aluminum exposure and accumulation has been implicated as the cause of 2 disorders that involve learning deficits (dialysis encephalopathy and Alzheimer's disease). To develop an animal model, 61 female New Zealand White rabbits were given 20 A1 lactate injections (0, 25, 50, 100, 200, or 400 µmole/kg, sc) over 4 wks. Dose-dependent weight reductions were observed. When the baseline frequency of nictitating membrane extension (NME) were determined, 2 wks later, differential classical conditioning of the NME was conducted. No treatment group differences were observed in frequency of baseline NME, amplitude of the response to shock, or shock threshold to produce NME, suggesting no aluminum effects on the Ss' ability to perform the response. All Ss developed the discrimination. The 2 highest dose groups acquired the CR less well than controls, as shown by a lower percent of CRs in the 2nd half of the conditioning sessions (80 and 74% of controls) and a greater latency to onset of the CR (327 and 310 msec vs 261 msec for controls). Results indicate that chronic systemic exposure of adult rabbits to A1 results in learning deficits not due to sensory or motor impairment of the learned response.

1616. Yokel, Robert A. & Pickens, Roy. (1976). **Extinction responding following amphetamine self-administration: Determination of reinforcement magnitude.** ***Physiological Psychology,*** 4(1), 39–42.
Observed several measures of extinction responding in 16 male Holtzman rats following responding for 1 of several doses of either dextro- or levoamphetamine (0.25, 0.5, 0.75, or 1.0 mg/kg) or dextro- or levomethylamphetamine (1.0, 1.5, 2.0, and 2.5 mg/kg). After daily 6-hr sessions of iv self-administration of drug, the drug was replaced by saline and extinction behavior observed. The number of responses to extinction did not differ significantly between drugs and doses used, but extinction times were greater for the larger doses and the dextro isomers. Response rate during extinction appeared to be a function of response rate during drug access, suggesting that the rate of responding was conditioned during drug self-administration. It is concluded that response rate during extinction and time to complete extinction are not reliable indicators of reinforcement magnitude due to this conditioning effect.

1617. York, James L. & Regan, Susan G. (1988). **Aftereffects of acute alcohol intoxication.** ***Alcohol,*** 5(5), 403–407.
Female rats showed quantifiable behavioral impairment (measured by rotarod performance and free operant activity), compared with controls, after blood alcohol levels had declined following acute alcohol intoxication. Findings may serve as an animal model of postintoxication syndrome in humans.

1618. Young, Keith A.; Randall, Patrick K. & Wilcox, Richard E. (1991). **Dose and time response analysis of apomorphine's effect on prepulse inhibition of acoustic startle.** ***Behavioural Brain Research,*** 42(1), 43–48.
Explored the dose–response characteristics of the dopamine agonist apomorphine (APO) on prepulse inhibition of acoustic startle in male rats. A 73-dB tone was used as a prestimulus to 100-dB white noise (PP trials), and 100-dB white noise alone (WN trials) during a 5–30 min or 5–90 min session. Results show that ip injections of APO initially elevated PP trial means to levels statistically indistinguishable from WN trial means. Results suggest that APO can dose-dependently disrupt prepulse inhibition of the acoustic startle response. After the initial startle amplitude elevation 5–10 min postinjection in the 5–90 min protocol, startle response amplitude of the 3.2 mg/kg APO dose group decreased during successive testing periods, in contrast to the pattern of vehicle controls, which did not decrease during the session. APO-induced loss of prepulse inhibition in rats shows potential for modeling sensory gating abnormalities of schizophrenia.

1619. Yu, Sue & Ho, I. K. (1990). **Effects of acute barbiturate administration, tolerance and dependence on brain GABA system: Comparison to alcohol and benzodiazepines.** National Institute on Alcohol Abuse and Alcoholism Neuroscience and Behavioral Research Branch Workshop on the Neurochemical Bases of Alcohol-Related Behavior (1989, Bethesda, Maryland). ***Alcohol,*** 7(3), 261–272.
Summarizes the evidence available as to how the gamma-aminobutyric acid (GABA) system plays a part in the actions and the development of tolerance to, and physical dependence on, barbiturates. The comparisons of the effects of alcohol, barbiturates, and benzodiazepines at different steps of GABA synapse are also presented. Inconsistent results in the literature may be due to differences in animal models or brain regions used, protocols, or techniques.

1620. Yurek, David M. & Randall, Patrick K. (1985). **Simultaneous catalepsy and apomorphine-induced stereotypic behavior in mice.** ***Life Sciences,*** 37(18), 1665–1673.
Examined the effects of domperidone (DOM), haloperidol (HAL), and chlorpromazine (CPZ) on the stereotypic behavior elicited by systemic apomorphine (APO) in mice. Results indicate that the intraventricular (iv) administration of HAL or CPZ induced catalepsy and blocked APO-induced stereotypic behavior. Low iv doses of DOM, sulpiride, and spiperone, equally cataleptogenic as HAL or CPZ, augmented rather than diminished stereotypic behavior produced by subsequent APO treatment. The resultant stereotypic behavior continued even while Ss were in a rigid cataleptic posture and was marked by persistent gnawing and licking. Prior to the induction of catalepsy and after recovery from it, Ss displayed the entire range of typical APO-induced behavior, including sniffing, climbing, gnawing, and licking. It is suggested that this animal model may be related to the clinical observation of the coexistence of tardive dyskinesia and drug-induced parkinsonism in individual patients.

1621. Zeltser, R.; Seltzer, Ze'ev; Eisen, A.; Feigenbaum, J. J. et al. (1991). **Suppression of neuropathic pain behavior in rats by a non-psychotropic synthetic cannabinoid with NMDA receptor-blocking properties.** ***Pain,*** 47(1), 95–103.
Examined the effects of HU211, a synthetic cannabinoid on autotomy, a behavioral model of neuropathic pain. Autotomy was induced in male Sabra rats ($N = 145$) by cutting the sciatic and saphenous nerves. Injections of HU211 with cupric chloride every 2nd day markedly suppressed autotomy during the injection period by delaying its average onset day and reducing the incidence of severe autotomy. Suppression of autotomy was retained in the postinjection period (for at least 30 days) but only when the drug was injected ip. Cupric chloride or HU211 alone was ineffective. General behavior and open field motor activity indicated that the effects of HU211 with CU^{++} on autotomy were not due to sedation or ataxia but presumably to antinociception mediated by N-methyl-D-aspartate (NMDA) receptor blockade.

1622. Zimmerberg, Betty; Carr, Kathryn L.; Scott, Amy; Lee, Helen H. et al. (1991). **The effects of postnatal caffeine exposure on growth, activity and learning in rats.** ***Pharmacology, Biochemistry & Behavior,*** 39(4), 883–888.
Used an animal model equivalent to the human 3rd-trimester or premature infant exposure to caffeine. Rat pups that received either 1 or 9 mg/kg of caffeine during the 1st wk of life grew more slowly, were hypoactive at 2 wks of age, and were impaired on an operant spatial learning task as adults. Adding visual cues to the operant task did not improve their performance. The persistent behavioral deficits noted after postnatal caffeine exposure were similar to the effects of adenosine. These behavioral deficits may reflect an upregulation of developing adenosine receptors that persists into adulthood subsequent to early chronic postnatal caffeine exposure. Thus, use of caffeine to stimulate respiration in premature infants may have short-term benefits but long-term consequences.

1623. Zini, R.; Morin, D.; Martin, P.; Puech, A. J. et al. (1991). **Interactions of flerobuterol, an antidepressant drug candidate, with *beta* adrenoceptors in the rat brain.** ***Journal of Pharmacology & Experimental Therapeutics,*** 259(1), 414–422.
Interactions of *dl*-flerobuterol with central *beta* adrenoceptors were investigated in male rats through receptor binding assays, synaptosomal uptake studies, cyclic adenosine monophosphate studies, and a learned helplessness procedure. *dl*-Flerobuterol was behaviorally active in the learned helplessness paradigm, a test highly sensitive to antidepressant drugs, which is based on the fact that exposure to incontrollable stress produces performance deficits in subsequent learning. The behavioral changes produced by stimulation of *beta* adrenoceptors with flerobuterol were similar to those caused by other centrally acting *beta* adrenergic agonists and by tricyclic or atypical antidepressants. They were also antagonized by (±)propranolol, which highly suggests an involvement of central *beta* adrenoceptors in the mechanism of flerobuterol action.

1624. Zivkovic, B.; Morel, E.; Joly, Danielle; Perrault, Gh. et al. (1990). **Pharmacological and behavioral profile of alpidem as an anxiolytic.** ***Pharmacopsychiatry,*** 23(Suppl 3), 108–113.
Evaluated the neuropharmacological and behavioral effects of alpidem (AP), an imidazo(1,2-a) pyridine that has been developed as an anxiolytic agent. In mice, AP inhibited marble-burying behavior and enhanced feeding under stressful conditions, as did benzodiazepines; in contrast to these drugs, however, AP was inactive against shock-induced fighting and shock-suppressed exploration. In rats, AP exerted anticonflict activity in the punished drinking test, but failed to antagonize punishment-induced inhibition of operant behavior. AP decreased motor performance in the rotarod test and only produced a deficit in muscle strength at doses that were more than 20 times higher than the doses active in anxiolytic tests. The effects of AP were antagonized by flumazenil, indicating that central omega receptors are involved in the action of this drug. (German abstract).

1625. Zuardi, A. W.; Rodrigues, J. Antunes & Cunha, J. M. (1991). **Effects of cannabidiol in animal models predictive of antipsychotic activity.** ***Psychopharmacology,*** 104(2), 260–264.
Compared the effects of cannabidiol (CBD) to those produced by haloperidol (HAL) in male rats submitted to experimental models predictive of antipsychotic activity. Several doses of CBD and HAL were tested in each model. CBD increased the effective doses of apomorphine for induction of sniffing and biting stereotyped behaviors. In addition, both CBD and HAL reduced the occurrence of stereotyped biting induced by apomorphine, increased plasma prolactin levels, and produced palpebral ptosis compared with control solutions. However, CBD did not induce catalepsy, in contrast to HAL. CBD's pharmacological profile is compatible with that of an "atypical" antipsychotic agent, though the mechanism of action may not be identical to that of the dopamine antagonists.

Developmental Psychology

1626. Amsel, Abram. (1986). **Developmental psychobiology and behaviour theory: Reciprocating influences.** ***Canadian Journal of Psychology,*** 40(4), 311–342.
Integrates aspects of infant development, psychobiology, and learning theory. Based on a large body of normative data, the author discusses levels of functioning in ontogeny, demonstrating the ages of first appearance in infancy of a number of reward-schedule effects. Levels of functioning in animals and humans are also examined, with reference to neuropsychological theories of memory function for humans. Findings from the author's work on the effects of hippocampal lesions and exposure to alcohol in utero on single alternation patterning and on the partial reinforcement extinction effect are also discussed. A modification of frustration theory is suggested, based on recent work relating differential hippocampal granule-cell genesis to exposure to a number of reward-schedule effects in infancy. (French abstract).

1627. Ginsburg, Benson E. (1982). **Genetic factors in aggressive behavior.** ***Psychoanalytic Inquiry,*** 2(1), 53–75.
The author (1958, 1966, 1968, 1977) discusses his research and that of others to illustrate that genetic approaches to the study of aggression provide an additional tool at the developmental, neurophysiological level that permits the judicious use of animal models with the possibility for direct testing of homology of mechanism in humans. Findings suggest that biological subtypes exist with varying degrees of lability to similar developmental circumstances and that sensitive periods during which later behavior can be most easily altered vary among genealogies. Similar environments can produce very different results in individuals of diverse genetic constitution, and underlying physiological mechanisms leading to the same endpoint may differ according to genetic constitution. It is therefore necessary to consider these factors to avoid the homogenization of differing etiologies simply because they reach similar endpoints.

1628. Hollander, C. F. (1978). **Experimental gerontological research.** ***Nederlands Tijdschrift voor Gerontologie,*** 9(3), 125–128.
The main goal of experimental aging research is to understand the mechanisms of aging in order to rationally approach the health problems of the aged as well as to increase their quality of life. Animal models are needed to achieve this goal. The approach of biomedical research on aging

should be along 3 main lines: the study of fundamental aspects of aging, the relationship between aging and disease, and the influence of environmental factors on aging. (Dutch summary).

Cognitive & Perceptual Development

1629. Teller, Davida Y. (1983). **Measurement of visual acuity in human and monkey infants: The interface between laboratory and clinic.** ***Behavioural Brain Research,*** 10(1), 15–23.
Describes the development of visual acuity in human and monkey (*Macaca nemestrina*) infants as revealed by a combination of preferential looking and operant techniques. Infants of both species demonstrate acuity of about 1 cycle/degree (20/600 Snellen equivalent) near birth. Acuity develops gradually in both species, reaching adult levels at 3–5 yrs in humans and about 1 yr in monkeys. Amblyopiogenic conditions, such as strabismus, occur spontaneously in infant monkeys as they do in human infants, and normal monkeys reared under deprivation conditions that mimic amblyopiogenic conditions in humans also become amblyopic. Results establish the infant monkey as an animal model for human visual development. Difficulties and limitations involved in the use of these techniques in clinical settings are described.

Psychosocial & Personality Development

1630. Suomi, Stephen J.; Mineka, Susan & DeLizio, Roberta D. (1983). **Short- and long-term effects of repetitive mother–infant separations on social development in rhesus monkeys.** ***Developmental Psychology,*** 19(5), 770–786.
Rhesus monkey mother–infant dyads were each subjected to 16 4-day physical separations between the infants' 3rd and 9th mo of life. Infants displayed protest behavior following each separation but only minimal signs of despair. Their protest diminished somewhat over repeated separations. The mothers' separation reactions were considerably milder (and changed little) over repeated separations. The separations appeared to retard the development of normal mother–infant relationsips: Relative to nonseparated control dyads, separated infants displayed excessive levels of infantile behaviors, although their mothers did not differ from control mothers in levels of any behavior. Near the end of their 1st yr, all infants were permanently separated from their mothers and housed as peer groups. Over the next 30 wks during peer housing, few behavioral differences emerged between previously separated and control Ss. However, when exposed to their mothers during preference tests, previously separated Ss seemed to avoid their mothers in sharp contrast to the mother-seeking activity displayed by control infants.

Social Processes & Social Issues

Sexual Behavior & Sexual Orientation

1631. Gooren, Louis J. (1988). **Biomedical theories of homosexuality: A critique. (Trans B. Strauß).** ***Zeitschrift für Sexualforschung,*** 1(2), 132–145.
Presents a critical review of theories and research on the biological substrates of homosexuality. Topics include: (1) biomedicine's view that procreation and preservation of the species are the driving forces of sexuality, (2) the concept of sexual differentiation of the brain, (3) the absence of animal models of human homosexuality, (4) the lack of evidence for a correlation between sex hormones and sexual orientation, (5) research on neuroendocrinological determinants of sexual orientation, and (6) research on the relationship between homosexuality and the immune system and implications for acquired immune deficiency syndrome (AIDS). (English abstract).

1632. Gooren, Louis. (1990). **The endocrinology of transsexualism: A review and commentary.** ***Psychoneuroendocrinology,*** 15(1), 3–14.
Reviews studies on sexual differentiation (SXD) of the central nervous system (CNS) obtained predominantly from laboratory animals and examines whether this information is pertinent to transsexualism (TSX) in humans. The discrepancy in transsexuals between their chromosomal, gonadal, and genital sex and their gender identity has given rise to the presumption that in them a divergence has occurred between prenatal gonadal/genital SXD and subsequent sexual brain differentiation. Also discussed are animal models in the study of TSX, hormonally induced SXD of the brain in humans, and the role of prenatal stress on SXD.

Drug & Alcohol Usage (Legal)

1633. Doyle, Teresa F. & Samson, Herman H. (1988). **Adjunctive alcohol drinking in humans.** ***Physiology & Behavior,*** 44(6), 775–779.
In an attempt to validate the animal model of adjunctive ethanol drinking in humans, 29 21–31 yr olds were allowed access to beer while playing a game that delivered monetary reinforcements on an FI schedule. Ss exposed to a longer FI schedule drank significantly more than those exposed to a shorter schedule, confirming the prediction made by the animal model. A pattern of ingestion characteristic of adjunctive drinking was also observed in the longer FI condition, providing evidence that ethanol drinking in humans can be schedule-induced.

Personality Psychology

Personality Traits & Processes

1634. Garau i Florit, Adriana. (1985). **The importance of research with animal models in the field of individual differences.** ***Quaderns/Cuadernos de Psicología,*** 9(2), 5–17.
Discusses the utility of animal models in the study of interindividual variability. The relevance of animal models is discussed in relation to theories of personality, learned helplessness, and the heritability of personality traits. (English & Spanish abstracts).

Psychological & Physical Disorders

1635. Ackerman, Sigurd H. (1975). **Restraint ulceration as an experimental disease model.** ***Psychosomatic Medicine,*** 37(1), 4–7.

Notes that the investigation of psychosomatic diseases in humans has profited extensively from experimental models of these diseases in animals. The extent of the similarity between the experimental disease and its human analog is discussed, questioning whether the experimental model can have the characteristic gross and histologic morphology, pathogenesis, natural history, typical time of onset over life, and possible genetic or other factors predisposing to its occurrence. A study of acute gastric ulceration in the rat produced by physical restraint is elaborated which resulted in an experimental model showing (a) a high-risk group for acute gastric erosions at a young age, (b) a risk factor of early maternal separation, (c) a distinguishing age-specific feature of frequent hemorrhage, and (d) testable hypotheses about pathophysiology (e.g., failure of body temperature regulation under conditions of acute nutritional deprivation). Case histories of peptic ulcer disease in human infancy and childhood are cited which show striking similarities to the animal model. Possible areas of research derived from the animal model are suggested.

1636. Black, Maureen & Dubowitz, Howard. (1991). **Failure-to-thrive: Lessons from animal models and developing countries.** ***Journal of Developmental & Behavioral Pediatrics,*** 12(4), 259–267.
Explores associations between undernutrition and poverty in determining behavioral outcome for infants with failure-to-thrive by reviewing findings from research in undernutrition among animal models, among children in developing countries where rates of infant undernutrition are extremely high, and among children in industrialized countries. The associations among undernutrition, poverty, and family functioning persist in both animal and human research, whether manipulated in laboratory settings or observed in natural settings. Although environmental support and stimulation appear to ameliorate many negative consequences associated with undernutrition, infants with a history of nutritional deprivation are at increased risk for behavioral and emotional problems. Recommendations for prevention and intervention follow an ecological framework and include knowledge of nutritional requirements and feeding approaches support for parents and families.

1637. Demaret, A. (1971). **The manic-depressive psychosis considered in an ethological perspective.** ***Acta Psychiatrica Belgica,*** 71(6), 429–448.
Interprets certain behaviors of manic-depressive patients as phylogenetic regressions and as connected to territorial behavior of animals. The fantasies of omnipotence among manics are viewed as analogous to an animal's possession of a territory; likewise depressive ideas of ruin are comparable to an animal's being deprived of his territory. It is noted that ethological interpretations do not preclude the use of other theoretical models as well.

1638. Fowles, Don C. (1988). **Psychophysiology and psychopathology: A motivational approach.** ***Psychophysiology,*** 25(4), 373–391.
Argues that a psychobiological model, involving a theory of motivation derived from the animal learning literature, offers an attractive theoretical bridge between neurochemical influences and the phenotypic features of psychiatric disorders. This model involves separate but interactive appetitive and aversive motivational systems that control behavioral activation (appetitive) and inhibition (aversive). Ways in which these motivational constructs can be relevant to psychopathology are discussed. It is noted that a series of studies have shown that heart rate during performance of a continuous motor task does respond to appetitive motivation, and that nonspecific skin conductance fluctuations respond to aversive stimulation in other contexts. Findings suggest that under the right circumstances appetitive motivation can be assessed via heart rate and aversive motivation via skin conductance.

1639. Freed, Earl X. (1973). **Drug abuse by alcoholics: A review.** ***International Journal of the Addictions,*** 8(3), 451–473.
Discusses methodological problems in studies of alcoholism and drug abuse and hypothesizes similarities and differences between drug addiction and alcoholism. Evidence for and against specific and generalized addiction is discussed. A review of relevant studies shows that about 20% of addicted individuals use both alcohol and some other addicting drug. Data suggest that the conjoint use of both substances is more prevalent among younger than older people, that the tempo of research is increasing, and that possibly animal models can be used in understanding addiction.

1640. Jenner, F. (1971). **Relevance of biological rhythms to studies of psychoses.** ***Vestnik Akademii Meditsinskikh Nauk Sssr,*** 26(5), 74–75.
Reviews current knowledge of biological rhythms in psychiatry. An attempt is made to assess the relevance of the fundamental work on the timing mechanism to the problems of periodic psychoses. Efforts to produce animal models of periodic psychoses are described. It is suggested that these may offer the most promising approach to the problems.

1641. McKinney, William T. (1974). **Animal models in psychiatry.** ***Perspectives in Biology & Medicine,*** 17(4), 529–541.
Reviews the history of the use of animal models of psychopathologic states and discusses their intrinsic validity. Models of psychosis and depression, socially-induced and biologically-induced models, and rehabilitation models for primates and other organisms are considered.

1642. McKinney, William T., Jr. & Bunney, William E., Jr. (1969). **Animal model of depression: I. Review of evidence: Implications for research.** ***Archives of General Psychiatry,*** 21(2), 240–248.
Demonstrates a need for an experimental system in which social and biological variables thought to be important in depression can be systematically manipulated and their relationship to depression clarified. Evidence suggesting this possibility came from 2 main sources: animal separation experiments, and anecdotal case histories of animals who had developed depressive-like syndromes. Six categories of implications for research are discussed: (a) evaluation of base-line data, (b) methods for induction of depression, (c) sensitization to depression, (d) methods to evaluate change, (e) reversal of depression, and (f) minimal requirements for an animal model of depression. The limitations of animal experimentation in reference to confirming or negating current psychiatric theories are recognized. However, a comparative approach might identify some new significant problems for research and can provide a system for investigation of concepts by allowing them to be tested on an observational and experimental level in animals.

1643. Schleifer, Steven J.; Keller, S. E. & Stein, M. (1985). **Stress effects on immunity.** 26th Annual Meeting of the Group-Without-A-Name International Psychiatric Research Society (1984, White Plains, New York). ***Psychiatric Journal of the University of Ottawa,*** 10(3), 125–131.
Presents a brief review of the main components of the immune system and of some of the immune assays used in this area of research. Several experimental models are considered that have been employed to investigate behavior and brain effects on immunity. It is suggested that an extensive network of central nervous system (CNS) and endocrine system processes may be involved in the modulation of the immune system in response to stressors. In humans, conjugal bereavement and major depressive disorder have been associated with decreased lymphocyte proliferative responses. Animal models utilizing physical trauma or separation experiences as stressors have found similar effects and have permitted the investigation of some of the neuroendocrine mechanisms that may be involved. It is concluded that the processes linking stress and the immune system are highly complex, involving several mechanisms.

1644. Stampfl, Thomas G. (1988). **The relevance of laboratory animal research to theory and practice: One-trial learning and the neurotic paradox.** ***Behavior Therapist,*** 11(4), 75–79.
Argues that animal laboratory research does not typically demonstrate the correspondence between the known features of human psychopathological conditions (e.g., phobia) and animal learning. The present author suggests that the combining of variables to solve difficult problems relating human psychopathology to animal learning models is facilitated by the study of human maladaptive behavior as it is seen through the application of relatively simple principles derived originally in the animal laboratory. An illustrative experimental example is provided.

1645. Utena, H. & Machiyama, Yu. (1971). **Schizophrenia model in animals.** ***Vestnik Akademii Meditsinskikh Nauk SSSR,*** 26(5), 64–67.
Discusses the need for a suitable animal model for the biological study of schizophrenia. The basic requirements for such a model are: similarities in behavioral changes, a sustained course of the illness, liability to exacerbations, and the absence of gross morphological lesions in the brain. A long-term administration of methamphetamine produced peculiar behavior disorders in Japanese monkeys (i.e., reduced behavioral activity; a persistent stereotype of fingering, gazing, or staring; a tendency toward isolation from other members of the group). These behavioral changes, which appeared in the chronic stage, continued to exist in some measure for more than 3 mo after termination of the drug administration. The significance of this model for the study of schizophrenia is discussed.

Psychological Disorders

1646. Coyle, Joseph T.; Singer, Harvey; McKinney, Michael & Price, Don. (1984). **Neurotransmitter specific alterations in dementing disorders: Insights from animal models.** Symposium held at the Inst of Pharmacological Research "Maria Negri": Biological markers in mental disorders (1983, Milan, Italy). ***Journal of Psychiatric Research,*** 18(4), 501–512.
Reviews research on synaptic neurochemical mechanisms in animal models of dementing disorders, noting that analysis of animal models for human dementing conditions allows for a more detailed characterization of neurotransmitter-related processes than is possible in postmortem samples of human brain tissue. It is noted that there has been a considerable change in the conceptualization of the pathophysiology of the cognitive impairments in dementing disorders as a result of synaptic neurochemical analyses. The demonstration of cholingergic deficits in the cortex and hippocampus of patients dying from Alzheimer's disease has led to the delineation of a relatively selective lesion of the cholinergic neurons in the basal forebrain complex, based on an understanding of these projections gleaned from animal experiments. In cat models of GM_1 gangliosidosis, variable alterations in neurotransmitter-related processes that are located in synaptic membranes have been described, indicating that a degeneration of neuronal pathways is not required for the development of the profound behavioral and motoric impairments of progressive dementia. It is concluded that research aided by animal models must continue to move forward from a classic histopathologic viewpoint and reliance on cell number to explore the synaptic neurochemical alterations that are likely to underlie dementing disorders.

1647. Hatotani, Noboru; Nomura, Junichi; Yamaguchi, Takahisa & Kitayama, Isao. (1977). **Clinical and experimental studies on the pathogenesis of depression.** ***Psychoneuroendocrinology,*** 2(2), 115–130.
Serum thyroid-stimulating hormone (TSH) responses to thyrotropin-releasing hormone (TRH) showed abnormal patterns in terms of diminished, delayed, and exaggerated responses in more than one-third of the 51 15–56 yr old depressed patients studied. Findings suggest the pathogenetic importance of hypothalamo-pituitary dysfunction in depressed patients. Because of this, a number of patients, especially those with diminished and delayed TSH responses to TRH, are prone to develop latent hypothyroidism which might make them resistant to antidepressants. To elucidate the underlying mechanism of these clinical findings, changes in brain monoamines of depression-model female Wistar rats were examined by the histochemical fluorescence method. Fluorescence intensity in nerve cells of the ascending noradrenergic system (A1, A2, A5, A6, A7) was markedly increased, and fluorescence intensity in cell bodies (A12) and nerve terminals (external median eminence) of the tubero-infundibular dopaminergic system was decreased.

1648. Hawkins, James et al. (1980). **Emotionality and REMD: A rat swimming model.** ***Physiology & Behavior,*** 25(2), 167–171.
The drugs (tricyclic antidepressants and MAO inhibitors) used in treating humans who suffer depressive illness typically reduce REM sleep. The authors report a nonpharmacological validation of a screening test that is selectively sensitive to those drugs. This indicates that REM sleep deprivation, not other drug effects, is operative. A measure of emotionality is defined that is sensitive to REM status in that context. Emotional defecation increases during REM rebound providing a model outcome that parallels clinical findings.

1649. Janiri, L.; Rosini, E. & Tempesta, E. (1985). **Dopamine and schizophrenia.** ***Archivio di Psicologia, Neurologia e Psichiatria,*** 46(3–4), 435–450.

Reviews recent research on the dopamine hypothesis of schizophrenia. Direct and indirect evidence of the pathophysiological role of dopamine in the etiology of schizophrenia is examined with reference to the principal characteristics of schizophrenic syndromes. Animal models and criteria for verifying the dopaminergic hypothesis are also discussed. (English abstract).

1650. Jesberger, James A. & Richardson, J. Steven. (1985). **Animal models of depression: Parallels and correlates to severe depression in humans.** ***Biological Psychiatry,*** 20(7), 764–784.
Discusses animal models that have been suggested to be valid for research into the neurobiology of depression and the neurochemical mechanisms of the antidepressant drugs. Although a depression predisposing biochemical abnormality has not been identified, it may be related to the neurochemical mechanisms that regulate impulse traffic in neural systems and maintain the homeostatic balance of neural activity within the brain. Therefore, the appropriate animal model for severe depression should have some disruption of neural functioning that is returned to normal by the chronic administration of antidepressant drugs. The surgical removal of olfactory bulbs in the rat meets this requirement and creates a relatively permanent disruption of normal neural functioning that results in abnormal neurochemical, endocrine, and behavioral activities that are consistent with similar parameters observed in patients with severe depression.

1651. Khanna, Sumant. (1988). **Biological correlates of obsessive compulsive disorder.** ***Indian Journal of Psychological Medicine,*** 11(1), 59–66.
Discusses research that has noted biological abnormalities associated with obsessive compulsive disorder (OCD). It is argued that OCD is a heterogeneous syndrome. At the neuroanatomical level, cortical (frontal and temporal) and subcortical (caudate) structures are now known to be involved, perhaps linked by a connecting circuit. Although most evidence has implicated the serotonergic system in the genesis of OCD, other substances may also be involved. Evidence is reviewed from family genetics, EEG, evoked potential, computerized tomography (CT), biochemical, and animal model studies.

1652. Khanna, Sumant. (1988). **Obsessive-compulsive disorder: Is there a frontal lobe dysfunction?** ***Biological Psychiatry,*** 24(5), 602–613.
Reviews evidence from electrophysiological, neuropsychological, scan, lesion, and psychosurgical studies suggesting that there is a possible frontal lobe dysfunction involved in obsessive-compulsive disorder (OCD). Evidence from animal models of OCD is also considered. Implications of frontal lobe studies for research on the role of other sites (e.g., caudate nuclei, cingulum) in OCD are noted. It is concluded that evidence on brain dysfunctions in OCD implicates the frontal lobe, with a shift in laterality toward the left side of the brain.

1653. Mineka, Susan & Kihlstrom, John F. (1978). **Unpredictable and uncontrollable events: A new perspective on experimental neurosis.** ***Journal of Abnormal Psychology,*** 87(2), 256–271.
Recent work has shown that unpredictable and/or uncontrollable events can produce a variety of cognitive, affective, and somatic disturbances to the organism. These disturbances are compared to and found to be quite similar to the symptoms of the classic cases of experimental neurosis described by Pavlov, W. H. Gantt, H. S. Liddell, J. H. Masserman, and J. M. Wolpe. The hypothesis is then developed that the common element in the experimental neurosis literature is that important life events become unpredictable or uncontrollable, or both. This interpretation is contrasted with the earlier physiological, psychodynamic, and behavioral interpretations made by the investigators themselves. The implications of this analysis of experimental neurosis for various issues in the predictability–controllability literature are discussed—for example, the interaction between unpredictability and controllability, the "threshold" for response to lack of predictability or controllability, and the lack vs the loss of predictability and controllability. Finally, the possible clinical relevance of this new perspective on experimental neurosis is discussed.

1654. Radil, T. & Radilová, J. (1977). **Some methodological aspects of research concerning experimental neuroses.** ***Československá Psychologie,*** 21(3), 199–204.
Discusses methodological problems connected with inducing neuroses or their isomorphic models in order to study their mechanisms by objective mechanisms. Subhuman primates are the most suitable Ss for such studies, as experimental manipulation of their social relations is possible and analytic neurophysiological and neuroendocrinological methods can be used. It is assumed that it is valid to generalize the results of such studies to humans. (Russian & English summaries).

1655. Siever, Larry J. & Davis, Kenneth L. (1985). **Overview: Toward a dysregulation hypothesis of depression.** ***American Journal of Psychiatry,*** 142(9), 1017–1031.
Suggests that the activity of neurotransmitter systems in the affective disorders and related psychiatric syndromes may be better understood as a reflection of a relative failure in their regulation, rather than as simple increases or decreases in their activity. A model organized around the concept of dysregulation posits that persistent impairment in neurotransmitter homeostatic regulatory mechanisms confers a trait vulnerability to unstable or erratic neurotransmitter output. Diabetes is used as an example of a dysregulation syndrome. Six criteria by which to examine a dysregulated neurotransmitter system are proposed and discussed. Evidence from clinical and animal model studies for dysregulation of the noradrenergic system in depression is examined with respect to these criteria, and a specific configuration of noradrenergic dysregulation in some forms of depression is proposed.

1656. Sitaram, Natraj & Gershon, Samuel. (1983). **From animals models to clinical testing: Promises and pitfalls.** ***Progress in Neuro-Psychopharmacology & Biological Psychiatry,*** 7(2–3), 227–228.
Criticizes the overemphasis on animal models of depression. The literature indicates that most animal models are not useful in predicting effective antidepressants or in generating hypotheses of the pathophysiology of depression. Almost every advance in drug treatment has resulted from astute clinical observation or serendipity. There is a need for rigorous clinical validation of the recent neuronal receptor hypotheses of depression. Clinical investigations so far have identified several potential neurobiological and neuroendocrine abnormalities that may be markers of the depressed state.

1657. Suomi, Stephen J. (1983). **Models of depression in primates.** ***Psychological Medicine,*** 13(3), 465–468.
Reviews the history of the use of models of depression in primates, beginning with the Harlows' investigation of separation-induced depression in young macaques. Since that time, models have shown that some monkeys are at risk for depression while others are not. Under stable, normal conditions, the 2 types are indistinguishable. Those at high-risk differ from others in behavioral and physiological ways. The ability to identify those at high-risk allows various preventive strategies to be tested.

1658. Telner, J. I. & Singhal, R. L. (1984). **Psychiatric progress: The learned helplessness model of depression.** ***Journal of Psychiatric Research,*** 18(3), 207–215.
The animal model has emerged as an alternative to clinical studies in psychiatry because it is able to provide greater experimental control and allows the exercise of ethical discretion. Although numerous animal models of depression have been proposed in the literature, most, if not all, fail to mimic human depressive symptomatology; their main function is to act as selective screens for antidepressant drugs. The learned helplessness approach has been suggested as an animal analog of depression because of its similarities to the human depressive state in terms of provocation, manifestation, and treatment. The learned helplessness model, which was originally based on animal experimentation, has been shown to be reproducible in humans, a finding not observed with other animal models of depression. Although this model has been much criticized in the past, recent reformulation adds credence to it as a more valid analog of human depression, given the additional cognitive constructs in depressed human Ss.

1659. Willner, Paul. (1983). **Dopamine and depression: A review of recent evidence: II. Theoretical approaches.** ***Brain Research Reviews,*** 6(3), 225–236.
Reviews empirical evidence relating dopamine (DA) to depression in clinical settings. Three behavioral approaches to depression are discussed: learned helplessness, reward system dysfunction, and reduced responsiveness to the environment. The role of DA in the related animal models is reviewed. On the basis of the literature discussed, it is concluded that the mesolimbic and nigro-striatal DA systems appear to be related primarily to responsiveness to the environment. This conclusion is discussed in relation to a review of the clinical symptomatology of major depression, which found strong support for the concept that lack of reactivity and loss of interest are associated with autonomous depression.

1660. Willner, Paul. (1984). **The validity of animal models of depression.** ***Psychopharmacology,*** 83(1), 1–16.
Reviews 18 animal models of depression in relation to 3 sets of validating criteria: predictive, face, and construct validity. Of the 18 models, 5 could only be assessed for predictive validity, 7 could be assessed for predictive and face validity, and 6 could potentially have predictive, face, and construct validity. Some traditional models (e.g., reserpine reversal and amphetamine potentiation) are rejected as invalid. The models with the highest overall validity are the intracranial self-stimulation, chronic stress, and learned helplessness models in rats and the primate separation model. It is concluded that although the validity criteria may suffer from some empirical limitations, in the final analysis an animal model is a theory of interspecific homology and, like all theories, must be judged ultimately for its power to generate testable hypotheses about the human condition being modeled.

Affective Disorders

1661. Healy, David. (1987). **The comparative psychopathology of affective disorders in animals and humans.** ***Journal of Psychopharmacology,*** 1(3), 193–210.
Reviews the psychopathology of human and animal affective disorders in the light of current operational criteria for the diagnosis of major depressive disorders. It is argued that the psychopathological tradition stemming from K. Jaspers (1965) may be more appropriate to animal models of affective disorders than the psychopathological positions in psychoanalysis, behaviorism, or current cognitive psychologies. The adoption of such a perspective results in a shift of emphasis from abnormalities of psychological content to demonstrable neuropsychological deficits and a definition of affective disorders as psychosomatic illnesses, possibly involving a pathology of circadian rhythmicity.

1662. Kostowski, Wojciech; Plaznik, Adam & Archer, Trevor. (1989). **Possible implications of 5-HT function for the etiology and treatment of depression.** ***New Trends in Experimental & Clinical Psychiatry,*** 5(2), 91–116.
Clinical and preclinical evidence suggest a role of 5-hydroxytryptamine (5-HT) in depressive states and in animal models, behavioral bioassays, and drug screening. The therapeutic efficacy of 5-HT precursors, 5-HT uptake inhibitors, and 5-HT receptor agonists derives from the behavioral aspects of 5-HT neurotransmission and may be linked to the mechanisms of depressive illness. The involvement of 5-HT receptors in the activity of antidepressants implicate a catecholaminergic component of serotonergic drug action. The authors conclude that the antagonistic balance between 5-HT and noradrenaline systems in several brain areas suggests that depression is probably not related to contemporaneous impairments or elevations of the activity of these systems.

1663. Lechin, Fuad; Van der Dijs, Bertha; Amat, Jose & Lechin, Marcel. (1986). **Central neuronal pathways involved in depressive syndrome: Experimental findings.** ***Research Communications in Psychology, Psychiatry & Behavior,*** 11(2–3), 145–192.
Discusses whether neurochemical findings concerning depression correspond to evolutionary stages of the same syndrome or whether one depressive syndrome may be expressed in different clinical ways. Areas discussed include animal models of depression; antidepressant mechanisms; changes in sensitivity of central beta-adrenoceptors, alpha-adrenoceptors, and dopaminergic receptors, serotonergic receptors; and 2 hypothesized types of depressive syndromes differentiated in terms of monoaminergic functioning. Support for the existence of these syndromes is discussed in relation to findings concerning regional cerebral concentrations of 5-hydroxytryptamine (5-HT) and norepinephrine.

1664. Lloyd, Kenneth G.; Zivkovic, Branimir; Scatton, Bernard; Morselli, Paolo L. et al. (1989). **The GABAergic hypothesis of depression.** 11th Annual Meeting of the Canadian College of Neuropsychopharmacology (1988, Montreal, Canada). ***Progress in Neuro-Psychopharmacology & Biological Psychiatry,*** 13(3–4), 341–351.
Summarizes the present status of (gamma-aminobutyric acid) GABAergic mechanisms related to depressive disorders. Topics discussed include psychopharmacology of GABA synapses in affective disorders and their animal models, biochemistry of GABA synapses in depression and the action of antidepressant drugs, and correlation between behavioral and biochemical observations. The upregulation of $GABA_B$ receptors and the downregulation of β-adrenergic receptors appear to be consistent with an antidepressant action and linked in an integrative control of cyclic AMP production.

1665. Price, John S. & Sloman, Leon. (1987). **Depression as yielding behavior: An animal model based on Schjelderup-Ebbe's pecking order.** ***Ethology & Sociobiology,*** 8(3, Suppl), 85–98.
Proposes an evolutionary model of depressive mood change based on social competition that postulates that depression evolved as the yielding component, or yielding subroutine, of ritual agonistic behavior (RAB). Since RAB is ubiquitous among present-day vertebrates, it is likely to have been an early vertebrate behavioral adaptation, and the neurochemical structures and processes that underlie yielding behavior in existing species could well be similar both to each other and to the structures and processes underlying at least some forms of human depressive states. The hen behavioral model developed by T. Schjelderup-Ebbe (1935) was used to generate a mathematical model that describes the mutual pecking of 2 birds in both symmetrical and asymmetrical relationships.

1666. Rubin, Robert T. (1989). **Pharmacoendocrinology of major depression.** ***European Archives of Psychiatry & Neurological Sciences,*** 238(5–6), 259–267.
The use of neuroendocrine (NEC) abnormalities in depression to help elucidate neurotransmitter (NT) dysfunction (NTD) has been a frustrating endeavor. Major depression may not be indicative of the same central nervous system (CNS) NTD in every patient. There is no supreme diagnostic algorithm for determining endogenous depression. NEC regulation of each of the anterior and posterior pituitary hormones depends on an interplay among many CNS NT systems. Pharmacologic probes to perturb both affective state and hormone secretion (e.g., current antidepressants) have relatively nonspecific effects in the functioning patient, and reliable animal models of depression are still unavailable.

1667. Steiner, Meir. (1989). **The neurochemistry of mood.** Canadian Consensus Symposium on Depression (1988, Ottawa, Canada). ***Psychiatric Journal of the University of Ottawa,*** 14(2), 342–343.
Shows that catecholamine and receptor theories of affective disorders have been inconclusive, indicating only a suggestive link between serotonergic and noradrenergic systems mediating the clinical effects of selective antidepressant drugs. Potential areas of study for increasing the understanding of melancholia are discussed, including clinical neuroscience, molecular genetics, brain imaging, circadian rhythms, and unique animal models.

1668. Willner, Paul. (1991). **Animal models as research tools in depression.** Special Issue: Affective disorders in old age. ***International Journal of Geriatric Psychiatry,*** 6(6), 469–476.
Discusses the use of animal models of depression (DP) as simulations for investigating the psychobiology of DP. Animal models currently used include experimental procedures based on stress, separation, pharmacological, and other models. Much of the work relating to the psychological features of DP arises from models based on stress, in particular, learned helplessness. The availability of a range of models affords the opportunity to examine patterns of depressive symptoms across models and to determine whether different types of models relate to different aspects of DP. Animal models of DP are making their greatest contribution as tools for investigating the mechanisms of action of antidepressant drugs. There is an absence of animal models that directly address the problems of DP and aging.

1669. Zucker, Irving. (1988). **Seasonal affective disorders: Animal models *Non fingo*.** Special Issue: Seasonal affective disorder: Mechanisms, treatments, and models. ***Journal of Biological Rhythms,*** 3(2), 209–223.
Evaluates the utility of animal models in the study of seasonal affective disorder (SAD). It is concluded that while there are currently no adequate models of SAD, it is clear that animal research in chronobiology continues to play an important role in understanding normal and aberrant seasonal rhythms in humans. It is suggested that the development of pharmacological and photic interventions for treating SAD is likely to benefit from animal research.

Schizophrenia & Psychotic States

1670. Braff, David L. & Geyer, Mark A. (1990). **Sensorimotor gating and schizophrenia: Human and animal model studies.** ***Archives of General Psychiatry,*** 47(2), 181–188.
Reviews the sensorimotor gating literature to demonstrate the utility of using clearly defined specifiable measures of attention in human and animal models of schizophrenia. Discussion focuses on how (1) sensorimotor gating research permits understanding of the functional significance of neurotransmitter abnormalities and (2) specific patterns of monoaminergic overactivity may cause the clinically significant cognitive disturbances that are a hallmark of schizophrenia.

1671. Braff, David L. & Geyer, Mark A. (1991). **"Pitfalls in animal models": Reply.** ***Archives of General Psychiatry,*** 48(4), 380.
D. L. Braff and M. A. Geyer (1990) respond to the comments of M. R. Cohen (1991) and argue that they are dealing with a complex multineuron, multineurotransmitter circuit and it is quite possible that residual imbalances "upstream" or "downstream" from the dopaminergic synapse are responsible for gating abnormalities in schizophrenics treated with antipsychotic medications.

1672. Cohen, Martin R. (1991). **Pitfalls in animal models.** ***Archives of General Psychiatry,*** 48(4), 379.
Comments on D. L. Braff and M. A. Geyer's (1990) animal model of acute regional excess dopaminergic activity. It is argued that the model cannot be considered the bridge that Braff and Geyer propose "between hypotheses of mesolimbic dopamine overactivity and information processing/sensorimotor gating abnormalities in schizophrenia."

1673. Crow, T. J. (1978). **Clinical Research Centre, Division of Psychiatry 1974–1977.** ***Psychological Medicine,*** 8(3), 515–523.
Describes the main areas of research since the establishment of the Clinical Research Centre's (CRC) Division of Psychiatry in 1974. The CRC was established by the Medical Research Council as a national program for patient-oriented research in England. Although national in scope, the CRC is affiliated with a district general hospital. The main areas of the division's research have been directed toward (a) elucidating the mechanism of action of neuroleptic drugs in acute schizophrenia; (b) investigating the nature of the psychological deficit in chronic schizophrenia, and its correlates; (c) conducting postmortem studies of monoamine and other neurohumoural mechanisms in schizophrenia; and (d) carrying out animal studies of the functions of specific neurohumoural systems and developing animal models of functional psychosis.

1674. Ellenbroek, B. A. & Cools, A. R. (1990). **Animal models with construct validity for schizophrenia.** ***Behavioural Pharmacology,*** 1(6), 469–490.
Presents 3 animal models of schizophrenia (SCZ) which are based on the construct that schizophrenic Ss are disturbed in their information processing. The 1st 2 models are based on the construct that schizophrenic Ss are less able to differentiate between relevant and irrelevant stimuli. The 3rd model deals with the amphetamine-induced changes in the behavior of socially living monkeys. This model seems to be related to both positive and negative symptoms of SCZ and is based on the construct that the negative symptoms are due to a compensatory mechanism which protects the Ss from "sensory flooding."

1675. Feer, Hans. (1986). **Animal model and human psychosis: Results of dualistic thought.** ***Schweizer Archiv für Neurologie und Psychiatrie,*** 137(5), 171–176.
Discusses animal models of human psychoses, the theory of evolution as the basis of biological psychiatry, and monistic vs dualistic conceptions of the mind–body relationship. Correlations between brain (external self) and intelligence (internal self) are considered.

1676. Feldkircher, Kathleen M. et al. (1984). **Schizophreniform behavior in rats: Effects of L-dopa on various behavioral and physiological phenomena.** ***Physiological Psychology,*** 12(2), 156–158.
Investigated the dopamine hypothesis of schizophrenia and the validity of using rats as animal models for schizophrenia, by examining parameters of 19 male Sprague-Dawley rats prior to and after levodopa (75 mg/kg, ip). Body weight, wheel-running behavior, open-field exploratory behavior, grooming behavior, body posture, and urine ketone levels were examined following injections and were compared to those observed prior to injection. Results show a lack of exploratory behavior, an absence of grooming, a common occurrence of catatonic-like behavior in the home cage, and an increase in urine ketone levels. It is concluded that the rat is a good animal model for schizophrenia and that ip injections of levodopa produce symptoms suggestive of this disorder.

1677. Geyer, Mark A. & Braff, David L. (1987). **Startle habituation and sensorimotor gating in schizophrenia and related animal models.** ***Schizophrenia Bulletin,*** 13(4), 643–668.
Reviews studies in which measures of startle response were used to clarify the importance of habituation and central inhibition deficits in schizophrenia and discusses the development of animal models of habituation and sensory gating. Evidence from animal studies provide support for a schizophrenialike loss of sensory gating with nucleus accumbens dopamine activity and extend findings of LSD-induced habituation deficits similar to those observed in schizophrenics. The utility of operationally defined measures of preattentive processes in studying schizophrenia is noted.

1678. Geyer, Mark A.; Swerdlow, Neal R.; Mansbach, Robert S. & Braff, David L. (1990). **Startle response models of sensorimotor gating and habituation deficits in schizophrenia.** 19th Annual Meeting of the Society for Neuroscience: Neural basis of behavior: Animal models of human conditions (1989, Phoenix, Arizona). ***Brain Research Bulletin,*** 25(3), 485–498.
Studies of prepulse inhibition and habituation of startle responses elicited by intense stimuli provide some unusual opportunities for cross-species explorations of attentional deficits characteristic of schizophrenic patients. Schizophrenic patients exhibit deficits in both the prepulse inhibition of startle and the habituation of startle. The behavioral plasticity of startle responses and the comparability of the test paradigms used in rats and humans greatly facilitate the development of animal models of specifiable behavioral abnormalities in schizophrenic patients. Examples of parallel animal and human models are sensorimotor gating and behavioral habituation. There is evidence to support the involvement of mesolimbic dopaminergic systems in the modulation of prepulse inhibition or sensorimotor gating as well as the importance of central serotonergic systems in the habituation of startle.

1679. Iversen, Susan D. (1987). **Is it possible to model psychotic states in animals?** ***Journal of Psychopharmacology,*** 1(3), 154–176.
Reviews benefits and limitations of animal models of schizophrenia, particularly as they relate to the dopamine overactivity (DAO) hypothesis, the limbic dopaminergic receptor hypothesis, the detection of novel antipsychotic drugs with reduced risk for tardive dyskinesia, and the psychopathology of schizophrenia. It is suggested that the DAO model comes close to fulfilling 2 of the 4 proposed criteria for an animal model: similarity of inducing conditions, similarity of behavioral symptoms, common underlying neurobiological mechanisms, and reversal of abnormal behaviors by the same clinically effective treatment techniques. It is concluded that the neuropathology of schizophrenia is still poorly understood, and animal models will probably play only a minor role in research in this area. The focus should turn from the basal ganglia to the higher centers of the limbic system and cortex.

1680. Joseph, Michael H.; Frith, Christopher D. & Waddington, John L. (1979). **Dopaminergic mechanisms and cognitive deficit in schizophrenia.** ***Psychopharmacology,*** 63(3), 273–280.
Proposes the hypothesis that schizophrenic symptoms are due to a breakdown in a mechanism by which conscious attention is limited and directed. It is shown that this mechanism can be modeled in terms of a simple nerve network in which every channel inhibits all the others. Failure of this inhibition would cause the defect hypothesized to occur in schizophrenia. It is also shown that if dopamine is

given a central role as a transmitter in such a network, the various predictions about the biochemistry of schizophrenia that follow are not only consistent with the evidence for the dopamine theory of schizophrenia, but also with much of the evidence held to be contrary to that theory. While not purporting to be an experimentally validated description of schizophrenia, this model goes beyond the single amine theories of schizophrenia and links dysfunctions in amine systems with specific behavioral control mechanisms. Given the current state of knowledge, such models can make only limited predictions about the biochemistry of schizophrenia. However, an attempt to link behavioral and biochemical systems in this way will be crucial for the development of viable animal models of schizophrenia.

1681. Kokkinidis, Larry & Anisman, Hymie. (1980). **Amphetamine models of paranoid schizophrenia: An overview and elaboration of animal experimentation.** ***Psychological Bulletin,*** 88(3), 551–579.
A review of the literature suggests that although some behaviors may be mediated by dopamine (DA) and other behaviors are largely subserved by noradrenergic mechanisms, the elicitation of a given behavior by amphetamine may be influenced by the interaction among several transmitters. It is hypothesized that behaviors that involve norepinephrine (NE) activity are stimulus bound and are related to attentional processes, whereas behaviors primarily subserved by DA appear to be motorically based. Behavioral augmentations frequently observed following repeated drug treatment are postulated to reflect alterations in DA activity, whereas the reduction of the drug's effect is ascribed to changes in attentional processes owing to variations in NE activity. The acute and chronic effects of dextro-amphetamine on attentional and arousal processes are related to current views of drug-induced psychosis in humans and to endogenous disorders.

1682. Yamamotová, A. & Lát, J. (1982). **The theoretical significance of a bimodal distribution of individual activation rates in animals and psychotics.** ***Activitas Nervosa Superior,*** 24(3), 188–190.
Individual constants of acceleration and deceleration (A and D) were ascertained in 15 8-mo-old male rats and in 32 10–49 yr old schizophrenics and manic depressives. Findings show that the variability of the A and D constants was virtually the same in animal and human Ss. In both cases, a bimodal distribution was observed with peaks around the same values. Findings support (1) the existence of common, general mechanisms of activation dynamics; (2) the effectiveness of systems analysis; and (3) the use of animal models in psychiatric research.

Neuroses & Anxiety Disorders

1683. Ashcroft, G. W.; Palomo, T.; Salzen, E. A. & Waring, H. L. (1987). **Anxiety and depression: A psychobiological approach.** ***Psicopatología,*** 7(2), 155–162.
Suggests that the fundamental mechanism underlying mood change is a change in the level of exploratory behavior; a fall in exploration is labelled as depression and increased exploration as a rise in mood. This implies that manipulation of brain systems involved in exploration will provide models of mood disorders and that manipulation of exploratory activity by drugs and psychological mechanisms in humans can be used to treat mood disorders. It is postulated that separation anxiety results in inhibition of exploration and leads to depressive symptoms. The biological bases of exploratory behavior and separation anxiety are discussed, citing clinical and animal evidence.

1684. Bennett-Levy, Jamie & Marteau, Theresa. (1984). **Fear of animals: What is prepared?** ***British Journal of Psychology,*** 75(1), 37–42.
Examined, in 2 groups of Ss, the characteristics of animals that humans are prepared to fear. In Group 1, 64 Ss (mean age 35.5 yrs) rated how fearful they were of 29 small, harmless animals, while 49 Group 2 Ss (mean age 35.1 yrs) made ratings of the perceptual characteristics of these same animals. Fear ratings were significantly correlated with animal characteristics ratings. It is suggested that preparedness to fear certain animals (e.g., snakes) is not a function of the animals per se but of their fear-evoking perceptual properties and their discrepancy from the human form. Implications for the clinical treatment of animal phobias as well as a model in which fear responses are desensitized to specific perceptual characteristics are discussed.

1685. Brown, James H. (1989). **Psychosocial issues.** Canadian Consensus Symposium on Depression (1988, Ottawa, Canada). ***Psychiatric Journal of the University of Ottawa,*** 14(2), 426–429.
Briefly reviews animal models that have been studied in depression research, including separation, learned helplessness, and stress models, and asserts that the relationship of the findings to human depression is not conclusively established. From human epidemiological studies, the only finding that clearly emerges is that there seems to be an excess of life events preceding the onset of depression. Yet, there are no clear, generally accepted conclusions about the possible importance of other factors, such as early loss, social support, social class, separation and loss, or self-esteem. The need for continued examination of the psychosocial aspects of depression is emphasized.

1686. Harris, James C. (1989). **Experimental animal modeling of depression and anxiety.** ***Psychiatric Clinics of North America,*** 12(4), 815–836.
Reviews historical and contemporary perspectives on animal models and provides an ethologic view of depression and anxiety. Topics discussed include techniques to induce symptoms of these disorders through biochemical and surgical manipulation, environmental stress, learned helplessness, social separation, selection of naturally vulnerable animals, and breeding procedures to produce vulnerable animals. Social and biologic rehabilitation treatment methods are also addressed. With increased recognition of psychiatric disorders in children and adolescents, animal modeling of disorders that begin during development assumes importance.

1687. Kalin, Ned H. & Carnes, Molly. (1984). **Biological correlates of attachment bond disruption in humans and nonhuman primates.** ***Progress in Neuro-Psychopharmacology & Biological Psychiatry,*** 8(3), 459–469.
Reviews the biological alterations that occur in nonhuman primates undergoing separation and compares these with changes associated with separation in humans. The data reviewed demonstrate that separation in humans and nonhuman primates can be an event with profound behavioral and physiological sequelae. Numerous biological alterations occur, including changes in activity, sleep, heart rate, tempera-

ture, endocrine function, immune function, and monoamine systems. Because of the complex interrelationships among the different systems, how they interact to produce the observed physiological effects remains unknown, and it is not known which, if any, of these changes are primary in the response to separation. While the biological alterations induced by separation in nonhumans are not identical to those seen in humans, it is possible that the separated monkey model will allow investigators to distinguish the neurobiological mechanisms that mediate and the factors that modulate separation response in primates.

1688. Kandel, Eric R. (1983). **From metapsychology to molecular biology: Explorations into the nature of anxiety.** ***American Journal of Psychiatry,*** 140(10), 1277–1293.
Demonstrates how cognitive psychology merges with neurobiology in the study of 2 learned abnormalities of behavior: anticipatory anxiety and chronic anxiety. Through the use of animal models, specific forms of mentation can now be explored on the cellular and molecular levels. Chronic anxiety and anticipatory anxiety in humans are closely paralleled by 2 forms of learned fear in the sea snail *Aplysia*: sensitization and aversive classical conditioning. In *Aplysia's* simple nervous system it is possible to delineate how the 2 forms are acquired and maintained. Both rely on the mechanisms of presynaptic facilitation. An augmented form of presynaptic facilitation accounts for the associative component of conditioning. These findings suggest that a simple set of cellular and molecular mechanisms in various combinations may underlie a wide range of both adaptive and maladaptive behavioral modifications. They also suggest that mental processes that appear phenotypically unrelated may share a fundamental unity on the cellular and molecular levels.

1689. Lal, Harbans & Emmett-Oglesby, M. W. (1983). **Behavioral analogues of anxiety: Animal models.** ***Neuropharmacology,*** 22(12-B), 1423–1441.
Describes the available animal models of anxiety in view of how well they reproduce behavioral and pathological features of the anxiety syndrome, allow investigation of neurobiological mechanisms that are not easily amenable to study in humans, and permit reliable evaluation of anxiolytic drugs as well as help in identifying anxiogenic effects of drugs and toxins. Animal models of anxiety are based on the presentation of aversive stimuli, either exteroceptive or interoceptive in origin. Studies on the methodology of drug discrimination are reviewed, and this method is proposed as a powerful new approach to determining parallels between properties of discriminating stimuli and human anxiety.

1690. Pentony, P. (1983). **The neurotic paradox revisited.** ***Australian Psychologist,*** 18(2), 251–260.
Reexamines O. H. Mowrer's (1948) formulation of the neurotic paradox as the existence of behavior that is simultaneously self-defeating and self-perpetuating, rigidly maintained although consistently punished. Examples from animal learning studies in which such neurotic behavior was established are presented and accounted for by invoking a circular causal or cybernetic model instead of Mowrer's linear causal framework. Learning processes are considered from an aspect that has been neglected by behavioral science. By linking the circular causal model with a hierarchical view of the learning process, it can be seen why it is difficult to achieve the change in the neurotic paradox that is sought by psychotherapy. There are ways of disrupting neurotic behavior patterns that are generated through the capacity for secondary learning. Such patterns can be broken by providing feedback to challenge the secondary-learning assumption of a particular activity.

1691. Pitman, Roger K. (1989). **Post-traumatic stress disorder, hormones, and memory.** ***Biological Psychiatry,*** 26(3), 221–223.
Suggests that the pathogenesis of posttraumatic stress disorder (PTSD) involves overstimulation of endogenous stress-responsive hormones due to an extremely traumatic event. These substances mediate an overconsolidation of the event, leading to the formation of a deeply engraved traumatic memory that subsequently manifests itself in the intrusive recollections and conditioned emotional responses of PTSD. This model offers potential explanations for 2 puzzling aspects of PTSD: paradoxical amnesia with regard to the trauma and delayed onset.

1692. Rapoport, Judith L. (1991). **Recent advances in obsessive-compulsive disorder.** ***Neuropsychopharmacology,*** 5(1), 1–10.
A growing body of evidence from clinical phenomenology, including associated disorders, brain imaging, and neuropharmacologic studies, links obsessive-compulsive disorder (OCD) to basal ganglia dysfunction and to the serotonin (5-hydroxytryptamine [5-HT]) system. At present, OCD is the psychiatric syndrome for which a specific neuroleptic dysfunction is most strongly suggested and for which an animal model has been found. It is proposed that dysfunction of basal ganglia-thalamic frontal cortical loops produces positive symptoms of excessive grooming, checking, and doubt most common in OCD. Data from clinical trials indicate that a spectrum of other abnormal behaviors (e.g., canine acral lick dermatitis in dogs, nail biting in humans) resembling excessive grooming in both animals and humans may be related to OCD.

1693. Rosen, Jules & Fields, Robert. (1988). **The long-term effects of extraordinary trauma: A look beyond PTSD.** ***Journal of Anxiety Disorders,*** 2(2), 179–191.
Traces a theoretical link between the neurochemical changes in the brain that are induced by stress and the long-term medical morbidity in trauma victims. Several animal models are presented that illustrate the prolonged impact of environmental stressors on the catecholamine system of the brain. Although evidence of brain changes has never been documented in trauma survivors, there is evidence of increased reactivity of the autonomic nervous system to stressful stimuli in Ss with posttraumatic stress disorder (PTSD). Increased reactivity to stress is associated with cardiovascular morbidity in the general population and may be an important factor in the long-term health of people who have experienced extraordinary stress.

1694. Skolnick, P.; Crawley, Jacqueline N.; Glowa, John R. & Paul, S. M. (1984). **β-Carboline-induced anxiety states.** 14th Collegium Internationale Neuro-Psychopharmacologicum Congress (1984, Florence, Italy). ***Psychopathology,*** 17(Suppl 3), 52–60.
Notes that several C-3-substituted beta-carbolines (e.g., the ethyl ester of beta-carboline-3-carboxylic acid) have high affinities for benzodiazepine receptors and can antagonize the principal pharmacologic actions of benzodiazepines. These compounds' intrinsic actions, which are best described as pharmacologically opposite to the benzodiaze-

pines, have enabled researchers to develop a chemically induced model of extreme stress or anxiety. The actions of such compounds in currently used animal models of anxiety are reviewed, and the effects of these compounds in primates, including humans, are discussed.

1695. Thiebot, M.-H. (1983). **Behavioral models of anxiety in animals.** ***Encéphale,*** 9(4, Suppl 2), 167–176.
Reviews behavioral models aimed at creating and studying anxiety and contends that they emphasize procedures in which blockade (partial or total) of ongoing behavior is produced by aversive events such as punishment, nonreward (frustration), or novelty in the situation with which the animal is faced. An analysis of methodological characteristics that are particularly pertinent is included, and the respective limitations of each procedure are discussed. An interpretation of the effects of anxiolytic compounds (particularly the benzodiazepines) is included.

1696. Van der Kolk, Bessel A. (1988). **The trauma spectrum: The interaction of biological and social events in the genesis of the trauma response.** ***Journal of Traumatic Stress,*** 1(3), 273–290.
Based on studies of disruptions of attachment bonds in nonhuman primates, the animal model of inescapable shock, and numerous studies of traumatized children and adults, an understanding of the nature of the biological changes that underlie the psychological response to trauma is emerging. The author explores (1) the nature of the biological alterations in response to traumatization; (2) how these biological shifts depend on the maturation of the central nervous system (CNS) cognitive processes, and the social matrix in which they occur; and (3) how these alterations can influence psychopathological and interpersonal processes. Implications for the treatment of trauma are discussed.

1697. Van der Kolk, Bessel; Greenberg, Mark; Boyd, Helene & Krystal, John. (1985). **Inescapable shock, neurotransmitters, and addiction to trauma: Toward a psychobiology of post traumatic stress.** ***Biological Psychiatry,*** 20(3), 314–325.
Discusses chronic posttraumatic stress, which has been described as a mental disorder with both psychological and physiological components. The behavioral sequelae of inescapable shock in animals and of massive psychic trauma in people show parallels. Inescapable shock in animals has been found to lead to both transient catecholamine depletion and subsequent stress-induced analgesia. It is postulated that the numbing and catatenoid reactions following trauma in humans correspond to the central nervous system (CNS) catecholamine depletion that follows inescapable shock in animals. The evidence for a human equivalent of stress-induced analgesia in animals is also explored. Although reexposure to trauma may produce a sense of calm and control due to opioid release, a cessation of traumatic stimulation will be followed by symptoms of opioid withdrawal and physiological hyperreactivity mediated by CNS noradrenergic hypersensitivity. This hyperreactivity is temporarily modified by reexposure to trauma. It is concluded that this factor could account for voluntary reexposure to trauma in many traumatized individuals and would provide a complementary formulation to the psychodynamic concept of attempted mastery of the psychosocial meaning of the trauma.

1698. Willner, Paul. (1990). **Animal models for clinical psychopharmacology: Depression, anxiety, schizophrenia.** ***International Review of Psychiatry,*** 2(3–4), 253–276.
Presents an overview of animal models (AMs) in the areas of depression, anxiety, and schizophrenia, with emphasis on assessment of the validity of the available models and their contributions to the understanding of the disorders modeled. Included is a summary of aspects of the disorder relevant to modeling, including drug responsiveness and diagnostic subgroups. It is argued that the validity of AMs approaches the limits set by an incomplete understanding of the clinical disorders. At present, the major contribution of AMs is to elucidate the physiological mechanisms underlying abnormal behavior and the action of psychotherapeutic drugs.

Behavior Disorders & Antisocial Behavior

1699. Aggleton, J. P.; Nicol, R. M.; Huston, A. E. & Fairbairn, A. F. (1988). **The performance of amnesic subjects on tests of experimental amnesia in animals: Delayed matching-to-sample and concurrent learning.** ***Neuropsychologia,*** 26(2), 265–272.
Six adult amnesic Korsakoff patients and 6 adult alcoholic controls were trained on a test of visual recognition, delayed matching-to-sample with trial unique stimuli. This test was modeled on comparable tasks used in the development of animal models of human amnesia. It was found that the Korsakoff Ss were severely impaired when the task difficulty was increased by lengthening the retention delay beyond 10 sec or by increasing the number of items intervening between sample presentation and test. The amnesic Ss were also impaired on the acquisition of a set of concurrent visual discriminations. Results bear similarities to those obtained from experimental amnesic syndromes in monkeys.

1700. Archer, John. (1990). **Pain-induced aggression: An ethological perspective.** ***Current Psychology: Research & Reviews,*** 8(4), 298–306.
Examines the motivational basis of pain-induced attack (PIA) by considering it from an ethological perspective. Studies supporting the view that shock-induced attack is fear-motivated are briefly reviewed before consideration of evidence that does not support this view from naturally occurring examples of PIA, and from examples of sex and hormonal effects. A reconciliation between the conflicting views is suggested by distinguishing between the functional (competitive-protective) and the causal (offensive-defensive) dimensions . It is proposed that PIA is protective in function but can be offensive or defensive (or both) in its form. Implications of this view are discussed (1) for the use of shock-induced fighting as an animal model of aggression and (2) for studies of human pain-induced aggression.

1701. Brown, Gerald L. & Goodwin, Frederick K. (1986). **Human aggression and suicide.** Special Issue: Suicide and life-threatening behavior. ***Suicide & Life-Threatening Behavior,*** 16(2), 223–243.
Reviews animal models of self-destructive behaviors and aggressive behaviors, 2 studies of human aggressive/impulsive/suicidal behavior, and the literature concerning human aggression and suicide. It is suggested that there have been a substantial number of scientific reports indicating that central nervous system (CNS) 5-hydroxytryptamine (5-HT) may be altered in aggressive/impulsive and suicidal behaviors in humans. These reports are largely consistent

with the animal data, and they constitute one of the most highly replicated sets of findings in biological psychiatry. The fact that CNS 5-HT was associated with animal aggression prior to its hypothesized relationship to suicidal behavior in humans suggests that some suicidal behavior may be a special kind of aggressive behavior in humans, although perhaps not absolutely unique to humans.

1702. Coccaro, Emil F. (1989). **Central serotonin and impulsive aggression.** Symposium of the XVIth Collegium Internationale Neuro-psychopharmacologicum: Serotonin in behavioural disorders (1988, Munich, Federal Republic of Germany). ***British Journal of Psychiatry,*** 155(Suppl 8), 52–62.
A reduction in central 5-hydroxytryptamine (5-HT) function may be associated with suicidal and impulsive aggressive behavior in both animal and human clinical studies. Strong and consistent associations have been reported between indices reflecting reduced presynaptic 5-HT activity and aggression in animal models. Reduced neuroendocrine (i.e., prolactin) responses in humans to fenfluramine, a 5-HT uptake inhibitor/releaser that activates both pre- and postsynaptic sides of the 5-HT synapse, strongly suggest that overall central 5-HT activity is reduced in mood and/or personality disorder patients with a history of suicidal and/or impulsive aggressive behavior.

1703. Eichelman, Burr. (1985). **Aggressive behavior: Animal models.** ***International Journal of Family Psychiatry,*** 6(4), 375–387.
Describes several animal models of aggressive behavior, with illustrations of how these models may have relevance for humans. Focus is on human violent behavior that is episodic, usually dystonic, and may be seen in the context of the family. For psychiatric diagnostic purposes, such violent human behavior can be encompassed in the diagnostic categories of Antisocial Personality Disorder and Intermittent Explosive Disorder of the Diagnostic and Statistical Manual of Mental Disorders (DSM-III). Various concepts of human violent behavior that animal models can be used to illustrate include (1) how genetic endowment affects the propensity for aggressive behavior; (2) how aggressive behavior, even if socially maladaptive, may serve to reduce stress and function as a coping behavior; and (3) how critical periods in development can play a role in the genesis of aggressive behavior.

1704. MacRae, James R.; Scoles, Michael T. & Siegel, Shepard. (1987). **The contribution of Pavlovian conditioning to drug tolerance and dependence.** Special Issue: Psychology and addiction. ***British Journal of Addiction,*** 82(4), 371–380.
Describes the contribution of Pavlov's classical conditioning theory to the understanding of drug tolerance and dependence. While traditional views of tolerance have assumed that the underlying physiological changes result simply from past exposure to the drug per se, it has become increasingly apparent that environmental cues present at the time of drug administration contribute to tolerance. The role of such drug-predictive environmental cues has been incorporated into a Pavlovian conditioning model of tolerance. Research evidence for this conditioning model is reviewed, and implications for the management of tolerance in clinical settings and for understanding of drug withdrawal symptoms and of drug overdose are discussed.

1705. McClearn, Gerald E. (1985). **Genetics and the human encounter with alcohol.** ***Social Biology,*** 32(3–4), 143–145.
Provides an introduction to research studies concerning the role of genes in determining some pharmacological, neurochemical, behavioral, or pathological response to alcohol. The logic of the use of animal models in addressing problems of human alcohol use, alcohol abuse, and alcoholism is discussed.

1706. Newman, Joseph P.; Widom, Cathy S. & Nathan, Stuart. (1985). **Passive avoidance in syndromes of disinhibition: Psychopathy and extraversion.** ***Journal of Personality & Social Psychology,*** 48(5), 1316–1327.
According to the physiological animal model proposed by E. E. Gorenstein and J. P. Newman (1980), psychopaths and extraverts may be characterized by a common psychological diathesis related to behavioral inhibition. One aspect of this diathesis involves deficient passive avoidance learning, which has been central to explanations of unsocialized and antisocial behavior. In 3 experiments, a passive avoidance task was completed by 90 14–18 yr old males, 40 male university students, and 40 18–50 yr old men and women. Ss were also assessed on measures including the Eysenck Personality Questionnaire and MMPI. Results support the prediction that psychopaths and extraverts would exhibit deficient passive avoidance relative to nonpsychopaths and introverts, respectively. The passive avoidance deficit was particularly evident in tasks that required Ss to inhibit a rewarded response to avoid punishment. The latter finding may be important for explaining the inconsistent results regarding passive avoidance learning in psychopaths. Discussion focuses on the importance of reward in mediating the passive avoidance deficit of disinhibited individuals and on the existence of an indirect relationship between psychopathy and extraversion—one that is consistent with the observed experimental parallels as well as with the more ambiguous evidence regarding a direct correlation between measures of the 2 syndromes.

1707. Poulos, Constantine X.; Hinson, Riley E. & Siegel, Shepard. (1981). **The role of Pavlovian processes in drug tolerance and dependence: Implications for treatment.** ***Addictive Behaviors,*** 6(3), 205–211.
Presents evidence for the crucial role of Pavlovian conditional compensatory responses in tolerance to opiates and alcohol. An analysis of the motivational role of Pavlovian conditional compensatory responses to craving and relapse is also discussed, and supportive experimental and epidemiological evidence are presented. Given the role ascribed to Pavlovian processes in tolerance, craving, and relapse, it is proposed that extinction of cues that elicit conditional compensatory responses is an essential factor for treatment. It is suggested that by virtue of prior Pavlovian conditioning, stress and depression may serve as cues to elicit conditional compensatory responses and attendant craving and these cues can also be extinguished by Pavlovian procedures. Explication of this conditioning analysis to the patient may itself be an important cognitive adjunct to treatment.

1708. Siegel, Shepard. (1984). **Pavlovian conditioning and heroin overdose: Reports by overdose victims.** ***Bulletin of the Psychonomic Society,*** 22(5), 428–430.

Tested a Pavlovian conditioning model of tolerance, which emphasizes the contribution of an association between predrug cues (e.g., environment) and the systemic effects of the drug to tolerance, by interviewing 10 former heroin addicts (mean age 28 yrs) who had survived an overdose. It is noted that, in agreement with the model, a recent animal experiment by S. Siegel et al (1982) indicated that the lethal effect of heroin, in drug-experienced rats, was influenced by the cues associated with its administration. To assess the applicability of these findings to instances of heroin overdose in humans, Ss in the present study were questioned about the circumstances of their overdose to ascertain the role of drug-associated cues. Ss' reports were consistent with the animal data, showing that overdose was more likely in unusual circumstances related to the environment or to drug administration. Findings suggest that the conditioning model may be relevant to some instances of overdose death among heroin addicts.

1709. Siegel, Shepard; Krank, Marvin D. & Hinson, Riley E. (1987). **Anticipation of pharmacological and nonpharmacological events: Classical conditioning and addictive behavior. *Journal of Drug Issues,*** 17(1–2), 83–110.
Reviews research that suggests an alternative interpretation of pharmacological phenomena such as tolerance, sensitization, and dependence that have typically been viewed as resulting from the operation of feedback mechanisms through which pharmacologically disturbed homeostatic functioning is countered by compensatory responses that restore physiological equilibrium. It is argued that feedforward mechanisms (i.e., regulatory responses made in anticipation of a drug) also contribute to drug effects. Such mechanisms operate on the basis of Pavlovian conditioning principles. The role of physiological feedforward mechanisms is also discussed with respect to immunology, exercise physiology, and stress.

1710. Tabakoff, Boris. (1983). **Current trends in biologic research on alcoholism. *Drug & Alcohol Dependence,*** 11(1), 33–37.
Two factors in the progress in alcohol research in the last decade have been the development of animal models of alcoholism and the elucidation of the importance of genetic factors as determinants of the response to ethanol. A major advance in the understanding of the biochemical mechanism by which ethanol produces intoxication has come from studies that demonstrate that ethanol can disorder neuronal membranes.

1711. Valzelli, Luigi. (1984). **Reflections on experimental and human pathology of aggression. *Progress in Neuro-Psychopharmacology & Biological Psychiatry,*** 8(3), 311–325.
Hypothesizes, on the basis of the proposed distinction between *normal* and *pathological* aggression in laboratory animals, an integration of the experimental findings derived from a specific animal model of aggression with the available clinical information on human violent behavior. The importance of the role played by the inhibitory control of brain functions appears essential in the regulation of emotions and behavior and is of great relevance in explaining the behavioral changes that follow induced or spontaneous impairment of the serotonergic system of the brain. Research indicates that genetic predisposition and induced or acquired defects of serotonergic inhibitory control coincide to precipitate anomalous strong aggression. The cluster of symptoms presented by laboratory rats in consequence of the serotonergic discontrol has unexpected similarities with several pathological conditions of humans. This fact confers to laboratory experiments the value of a tool aimed at a better understanding of the biological mechanisms that underlie corresponding alterations of human conduct, with special reference to pathological aggression and violence.

Substance Abuse & Addiction

1712. Cummins, J. T.; Sack, M. & von Hungen, Kern. (1990). **The effect of chronic ethanol on glutamate binding in human and rat brain. *Life Sciences,*** 47(10), 877–882.
Examined changes in glutamate binding (GB) in male rats following a 5-day course of alcohol administration and in human brain postmortem tissues from male alcoholics. Highly significant decreases were found in GB to the N-methyl-D-aspartate receptor in the CA_1 region of the hippocampus in both the rats and the alcoholics. No significant effect of alcohol administration was shown on GB in the caudate, parietal cortex and the CA_3 region of the hippocampus. These results help validate the use of the gavage animal model for studies on alcoholism.

1713. Emmett-Oglesby, M. W.; Mathis, D. A.; Moon, R. T. & Lal, H. (1990). **Animal models of drug withdrawal symptoms. *Psychopharmacology,*** 101(3), 292–309.
Focuses on the use of animals to detect aspects of drug withdrawal (DW) that may be related to human subjective phenomena, including detection of withdrawal-related stimuli. Procedures using discrimination of the anxiogenic drug pentylenetetrazole have proven useful for detecting withdrawal in animals. Other animal models of anxiety also detect withdrawal from dependence on benzodiazepines and ethanol, which suggests that all of these experiments may be assessing the same phenomena. Other experiments clearly document the utility of schedule-controlled behavior for studying DW, especially for cocaine, nicotine, phencyclidine, tetrahydrocannabinol, and caffeine.

1714. Gregg, Edward & Rejeski, W. Jack. (1990). **Social psychobiologic dysfunction associated with anabolic steroid abuse: A review. *Sport Psychologist,*** 4(3), 275–284.
Reviews human and nonhuman primate research on the effects of anabolic/androgenic steroids (ASs). Descriptive research and anecdotal reports on athletes suggest that ASs have a variety of psychological and behavioral effects including psychotic episodes and increased aggression. Recent investigations with a nonhuman primate model confirm that the effects of ASs on psychological states and overt behavior range from active (e.g., mania and aggression) to more passive states (e.g., depression and social withdrawal). ASs also have profound physiological effects of a biobehavioral origin that pose a risk for cardiovascular disease. ASs' effects appear to be due to an interaction between its pharmacologic properties and the social milieu. Prevention and education strategies to curb AS abuse are discussed.

1715. Hartnoll, R. (1990). **Non-pharmacological factors in drug abuse. *Behavioural Pharmacology,*** 1(4), 375–384.
Argues that nonpharmacological factors (NPFs) are more important than pharmacological factors for understanding and responding to drug misuse. The relevance of laboratory models of drug abuse to drug taking in human societies is examined, and their significance relative to other explanatory frameworks is discussed. NPFs that may be of greater

importance than self-administration models of drug misuse derive from social perspectives (notably economic and market factors, socioeconomic conditions, and cultural and subcultural processes) and psychological approaches (individual, social learning, cognitive, and developmental).

1716. Henningfield, Jack E.; Cohen, Caroline & Heishman, Stephen J. (1991). **Drug self-administration methods in abuse liability evaluation.** Conference on Clinical Testing of Drug Abuse Liability: Consensus statement and recommendations (1990, Barcelona, Spain). ***British Journal of Addiction,*** 86(12), 1571–1577.
Discusses the human drug self-administration (HDSA) paradigm as an extension of the animal model developed in the 1960s and explores how the paradigm can be used to investigate the determinants and correlates of drug-seeking and drug-taking behavior. The basic components of the HDSA model are described. Some of its applications include assessment of the reinforcing effects of drugs, analysis of behavioral and pharmacological mechanisms of drug self-administration and measurement of the abuse liability, behavioral toxicity, and aversive effects of drugs. Some of the strengths and limitations of using the paradigm with human research Ss are also presented. It is concluded that the HDSA model should not replace other measures of abuse liability testing in humans, but should be incorporated into comprehensive programs of drug abuse assessment.

1717. Hutchings, Donald E. (1990). **Issues of risk assessment: Lessons from the use and abuse of drugs during pregnancy.** Special Issue: Qualitative and quantitative comparability of human and animal developmental neurotoxicity. ***Neurotoxicology & Teratology,*** 12(3), 183–189.
Discusses therapeutic uses and risks associated with methadone, cannabis, cocaine, and phencyclidine and examines limitations of the developmental animal and clinical pediatric literature with respect to problems of quantitative risk assessment. Human and animal developmental findings for methadone and cannabis are compared with respect to long-term behavioral effects, with emphasis on pharmacological and interpretive issues. The neonatal withdrawal or abstinence syndrome withdrawal phenomena and attempts to develop animal models are discussed. Methodological considerations are discussed in the context of adequacy of the Environmental Protection Agency's developmental neurotoxicology battery to characterize risk for abuse substances, as well as the National Institute on Drug Abuse medication compounds.

1718. Lewis, Michael J. (1990). **Alcohol: Mechanisms of addiction and reinforcement.** ***Advances in Alcohol & Substance Abuse,*** 9(1–2), 47–66.
Examines positive and negative reinforcement mechanisms that play a significant role in alcohol abuse and alcoholism. The role of euphoria and the anxiolytic effects of alcohol are considered as the basis of positive reinforcement, and physical dependence and aversive consequence of drinking as the basis of negative reinforcement. The motivational significance of each of these is discussed with respect to various animal models of addiction and clinical and human research. Brain neurochemistry, neuropharmacology, and genetic research data are evaluated from the perspective of reinforcement mechanisms involved with alcohol addiction.

1719. Pryor, Gordon T. (1990). **Persisting neurotoxic consequences of solvent abuse: A developing animal model for toluene-induced neurotoxicity.** ***National Institute on Drug Abuse: Research Monograph Series,*** 101, 156–166.
Toluene, a significant component of many abused solvent products, is the only pure solvent that has frequently been reported to be inhaled for its euphoric effects. Reports suggest that toluene causes moderate to severe neurologic dysfunction in heavy solvent abusers. A list of clinical symptoms commonly reported in patients who have abused solvents containing toluene was used to develop behavioral tests that might reflect similar signs and symptoms in rats chronically exposed to toluene. Several of the signs and symptoms form a cluster referred to as cerebellar ataxia. These experiments provide a partial animal model of the syndrome associated with human abuse of toluene-containing solvents.

1720. Schindler, Charles W.; Katz, Jonathan L. & Goldberg, Steven R. (1988). **The use of second-order schedules to study the influence of environmental stimuli on drug-seeking behavior.** ***National Institute on Drug Abuse: Research Monograph Series,*** 84, 180–195.
Presents a hypothetical case history to illustrate the effects of environmental factors on drug-seeking behavior (DSB), addiction, and relapse. It is argued that (1) the environment in which addiction develops may come to elicit the same feeling as the drug itself and its withdrawal symptoms, independent of the drug, through the process of Pavlovian conditioning; and (2) the sequence of behaviors that precede drug-taking and their associated stimuli may also become associated with the drug through the process of 2nd-order conditioning. Stimulus effects on the acquisition, maintenance, and extinction of DSB are examined. An animal model of human drug abuse supports the idea of Pavlovian and 2nd order conditioning of DSB. A review of a number of experiments using 2nd-order schedule indicates the importance of environmental stimuli in maintaining DSB.

1721. Stolerman, I. P. (1990). **What animal work may tell.** ***British Journal of Addiction,*** 85(10), 1251–1252.
Comments on G. Edwards's (1990) article on alcohol withdrawal syndrome and presents contributions from animal research on this topic. Animal work has progressed beyond the replication of objective signs of withdrawal to the extent that discriminative drug effects in animals are not merely models for the human phenomena but are homologous with them in terms of behavioral mechanisms.

1722. Sweeney, Donal F. (1989). **Alcohol versus Mnemosyne: Blackouts.** ***Journal of Substance Abuse Treatment,*** 6(3), 159–162.
Examines the alcohol–memory disturbance by reviewing data about blackouts and current thinking about the study of memory. From this data, a working definition of a blackout is proposed that may have significance in both the medical and medical-legal areas. The use of animal models to study human memory is questioned. Human experiments with alcohol and/or benzodiazepines might supply answers to questions concerning blackouts and other alcohol-related mental events.

1723. Wood, Ronald W. (1990). **Animal models of drug self-administration by smoking.** ***National Institute on Drug Abuse: Research Monograph Series,*** 99, 159–171.

Chemicals subject to abuse by inhalation may be divided into those that are volatile at room temperature (inhalants) and those that require heat for self-administration (smoking). These substance abuse practices are compared using examples of animal models. Forcible exposure to combustion products by chamber or tracheostomy is considered useful for some toxicologic purposes, but yields no information on self-administration and varies in the ability to provide adequate identification or characterization of the health consequences of smoking. Drug self-administration by smoking is criticized as being of restricted and limited value and lacking in control procedures. Substances used in these studies include tobacco, cannabis, dimethyltryptamine, and cocaine. The influence of aerosol phenomena on the design of suitable animal models is discussed.

Developmental Disorders & Autism

1724. Amsel, Abram. (1990). **Arousal, suppression, and persistence: Frustration theory, attention, and its disorders.** Special Issue: Development of relationships between emotion and cognition. ***Cognition & Emotion,*** 4(3), 239–268.
Summarizes frustration theory and a 4-stage theory of persistence as they relate to attentional mechanisms. Preliminary studies of reward-schedule effects on children with attention deficit hyperactivity disorder (ADHD) are considered. ADHD may be caused by the failure of ongoing behavior to be counter-conditioned to frustrative and disruptive stimuli. Additional developmental research is described, focusing on the effects of lesions and fetal alcohol on the emergence and development of reward-schedule phenomenon. Neuroanatomical and behavioral effects of fetal alcohol are discussed in relation to theories of attention and hyperactivity.

1725. Mailman, Richard B.; Lewis, Mark H. & Kilts, Clinton D. (1981). **Animal models related to developmental disorders: Theoretical and pharmacological analyses.** ***Applied Research in Mental Retardation,*** 2(1), 1–12.
Discusses 3 general approaches to the establishment of animal models of developmental disorders, with particular reference to hyperkinesis. It is concluded that, despite the employment of sophisticated physiological techniques, animal models suffer from a number of conceptual and empirical weaknesses and have yet to offer major pragmatic gains.

1726. Ruppenthal, Gerald C. et al. (1983). **Pigtailed macaques (*Macaca nemestrina*) with trisomy X manifest physical and mental retardation.** ***American Journal of Mental Deficiency,*** 87(5), 471–476.
Although no clear phenotype exists at birth, later speech and language problems and significantly lower full-scale IQs in trisomic females vs control groups have been reported. However, XXX aneuploidies have not been reported previously in nonhuman primates. In the present study, 3 pigtailed macaques from a nursery population were predicted to be genetically abnormal based on observations of anatomical and behavioral development. All 3 exhibited delays in skeletal, visual, intellectual, and social development, suggesting a chromosomal syndrome. Karyotypes showed that 2 Ss were XXX females, and the 3rd was a mosaic XX/XXX female.

Mental Retardation

1727. Davison, A. N. (1977). **The biochemistry of brain development and mental retardation: The Eleventh Blake Marsh Lecture delivered before the Royal College of Psychiatrists, 7 February, 1977.** ***British Journal of Psychiatry,*** 131, 565–574.
Considers that mental retardation may be associated with a number of environmental factors such as undernutrition, lead poisoning, or exposure to neuroactive drugs during a critical period of brain development. Possible biochemical mechanisms operating in these various conditions and in animal models are reviewed in relation to the vulnerable period hypothesis. Small brains are common in the mentally retarded, and it is suggested that this may be related to a developmental abnormality, particularly at the level of the synapse.

1728. Epstein, Charles J. (1984). **Mental Retardation Research Center, University of California, San Francisco.** ***American Journal of Mental Deficiency,*** 88(5), 590–593.
The Mental Retardation Research Center focuses on 3 elements of research: (1) a concern with the biology of interferon in trisomy 21 (Down's syndrome) cells and with the effects of interferon on the trisomic immune system; (2) the general problem of aneuploidy, particularly in the mechanisms by which chromosome imbalance results in deleterious effects on development and function; and (3) the development of an animal model for human trisomy 21.

1729. Epstein, Charles J.; Cox, David R. & Epstein, Lois B. (1985). **Mouse trisomy 16: An animal model of human trisomy 21 (Down syndrome).** Symposium of the National Down Syndrome Society: Molecular structure of the number 21 chromosome and Down syndrome (1984, New York, New York). ***Annals of the New York Academy of Sciences,*** 450, 157–168.
Discusses the use of the trisomy 16 mouse model developed when identified loci known to be present on human chromosome 21 were mapped on mouse chromosome 16. Trisomic mice can be produced and have been used to define general properties with regard to cell distributions. The model has been used in the form of fully trisomic embryos and fetuses and in a combination of trisomic cells with diploid ones in the form of viable trisomy 16 ^{7} 2n chimeras or mosaics. Studies of the phenotype of mouse trisomy 16 are indicative of retarded brain development that may lead to a reduced cortical surface and permanent deficiency of cortical neurons and to arrested development of the neurons within certain neurotransmitter systems. Defects in the heart and in the immunologic and hematologic systems have also been found. Although there are many interesting parallels, it cannot be said that Down's syndrome has been reproduced in the mouse. It is envisioned that the model may be used to investigate cell surface molecules, endocardial cushion defects, exogenous agents, and possible electrophysiologic changes.

1730. Hardy, John; Irving, Nick & Kessling, Anna. (1989). **Down on chromosome 21?** ***Trends in Neurosciences,*** 12(6), 209–211.
Argues that the relative importance of different genes on chromosome 21 in determining the Down's syndrome (DS) phenotype can be investigated by 1 of 2 main approaches. One is molecular and clinical investigation of individuals

who have unbalanced translocations of part of chromosome 21, and the other is the examination of animals that have extra copies of genes expressed on human chromosome 21. The part of the chromosome required for the DS phenotype is defined as the Down's obligate region. Issues are raised concerning DS cases with partial trisomy 21, definition of DS phenotype, animal models of DS such as the mouse with trisomy 16, the super-oxide dimutase gene, and Alzheimer's disease.

1731. Massa, E. (1986). **Atypical Marfan syndrome with mental retardation.** ***Psichiatria Generale e dell'Età Evolutiva,*** 24(3), 5–25.
Reports on 2 atypical cases of Marfan syndrome with mental retardation. The clinical picture of Marfan syndrome is compared with manifestations of experimental lathyrism. The author suggests that the human and the experimental illness may both be regarded as collagenoses (i.e., diseases of a macromolecular nature). Pathogenic theories on lathyrism are discussed; while the hypothesis of a direct lathyrogenesis in laboratory animals is accepted, an indirect lathyrogenlike sequence for the still-unresolved etiology of the Marfan syndrome is preferred. (English abstract).

Eating Disorders

1732. Van Vort, Walter B. (1988). **Is sham feeding an animal model of bulimia?** ***International Journal of Eating Disorders,*** 7(6), 797–806.
Explores sham-feeding (SF) of animals as a model of bulimia because SF resembles bulimia as 2 forms of hyperphagia; one associated with hunger and palatable, easy-to-ingest food and one with operational interference with gastric and postgastric safety mechanisms. Topics discussed include satiety mechanisms and reward processes for normal meal size, and bulimia and the W. T. McKinney (1974) criteria.

Physical & Somatoform & Psychogenic Disorders

1733. Abel, Ernest L. (1980). **Fetal alcohol syndrome: Behavioral teratology.** ***Psychological Bulletin,*** 87(1), 29–50.
The fetal alcohol syndrome (FAS) is a pattern of physical malformations observed in the offspring of women who drink alcohol during pregnancy. The most serious effect of in utero exposure to alcohol is mental retardation. Although the physical characteristics associated with the FAS have been attributed to the direct effects of alcohol, conditions secondary to alcohol intake (e.g., altered nutrition) cannot be eliminated as etiological factors in the impairment in cognitive function. Although animal models have been developed to study the question of direct vs indirect causation, there is little agreement in the results of these studies, and the methodology used leaves much to be desired in terms of adequate controls.

1734. Berga, Sarah L. & Girton, Lyda G. (1989). **The psychoneuroendocrinology of functional hypothalamic amenorrhea.** ***Psychiatric Clinics of North America,*** 12(1), 105–116.
Women with functional hypothalamic amenorrhea display multiple neuroendocrine aberrations suggestive of altered central neurotransmission. The role of antecedent stress as an explanation for both the neurochemical changes and the dysfunctional behavior of these women is examined by utilizing concepts provided by the animal model of inescapable shock and the human condition of posttraumatic stress disorder (PTSD).

1735. Campbell, Byron A.; Sananes, Catherine B. & Gaddy, James R. (1984). **Animal models of infantile amnesia, benign senescent forgetfulness, and senile dementia.** United States Environmental Protection Agency, Neurotoxicology Division, and the Johns Hopkins University Neurotoxicology Program Conference: Cross species extrapolation in neurotoxicology (1984, Raleigh, North Carolina). ***Neurobehavioral Toxicology & Teratology,*** 6(6), 467–471.
Discusses the application of J. H. Jackson's (1958) principle of hierarchical development and dissolution of function to infantile amnesia and memory loss in senescence. When the Jacksonian model is generalized to include life-span changes in memory, it predicts a last-in, first-out appearance and disappearance of memory processes. Those memory capacities that are the last to appear in ontogeny should be the first to be compromised in aging. To evaluate this proposition in a specific context, the rodent literature on long-term memory in infant, adult, and aged animals was surveyed. Three types of memorial processes that emerged sequentially in development were identified and examined in adult and aged rats. Although strong support of the Jacksonian principle did not emerge from this analysis, the data were sufficiently positive to suggest that the theory is viable and vigorous enough to guide future research on both normal and pathological processes of development and aging.

1736. Corraze, Jacques. (1976). **Neurotic excoriations: An ethological approach.** ***Evolution Psychiatrique,*** 41(2), 389–436.
Reviews psychoanalytic views that neurodermatitis reflects erotization of the skin and displacement of unconscious sexual conflicts. Examination of data from observations of animals, largely primates, suggests that behavior such as grooming and scratching can be viewed in terms of a displacement model which may contribute to the understanding of basic mechanisms in human excoriations.

1737. Fernández Teruel, A; Roca, M.; Ugarte, B. & Muntaner, C. (1988). **Psychosocial variables in asthma: Within reach of your relation: Beginning with psychoneuroimmunology.** ***Psiquis: Revista de Psiquiatría, Psicología y Psicosomática,*** 9(6–7), 30–36.
Reviews the clinical and experimental literature concerning the roles of psychosocial variables in the induction and treatment of asthma attacks. Topics discussed include: psychoneuroimmunology, aversive conditioning, classical conditioning, behavior therapy, systematic desensitization, hypnosis, and meditation. The need to develop an animal model of psychogenic asthma is emphasized. (English abstract).

1738. Goetsch, Virginia L. (1989). **Stress and blood glucose in diabetes mellitus: A review and methodological commentary.** ***Annals of Behavioral Medicine,*** 11(3), 102–107.
Reviews studies seeking to demonstrate a direct psychological stress–blood glucose relationship in diabetics. An overview of the pathophysiology and health risks of diabetes is provided, along with a review of stress investigations utilizing laboratory animal models and chemical infusion studies.

Data suggest that stress is important in understanding blood glucose fluctuations in at least some diabetics and that well-controlled laboratory stress procedures can produce a reliable, externally-valid index of the effects of stress.

1739. Jemmott, John B. & Locke, Steven E. (1984). **Psychosocial factors, immunologic mediation, and human susceptibility to infectious diseases: How much do we know?** ***Psychological Bulletin,*** 95(1), 78–108.
Recent evidence from animal models of stress suggests that stress can impair immunologic competence, rendering the host more vulnerable to infection and neoplasm. The present authors review studies on the relationship between psychosocial factors and human immunologic functioning, focusing on studies bearing on the relationship of psychosocial factors to altered susceptibility to infectious diseases and those bearing on the relationship of such factors to specific aspects of the human immune response. Findings indicate that a variety of psychosocial variables, disease states, and aspects of both humoral and cell-mediated immune responses have been investigated, and evidence favors the view that psychosocial variables may play a role in modulating the human immune response. More research is needed before it can be definitively concluded that the relationship between stress and human susceptibility to infectious diseases is a psychoimmunologic nexus. Relationships between the endocrine and immune systems are also discussed.

1740. Laudenslager, Mark L. & Reite, Martin L. (1984). **Losses and separations: Immunological consequences and health implications.** ***Review of Personality & Social Psychology,*** 5, 285–312.
Explores evidence for the hypothesis that psychosocial stressors, such as major separations and losses, disrupt the immune system and thereby contribute to a host of physical illnesses. Stressors are considered in terms of 3 dimensions: controllability, predictability, and chronicity. The growing evidence for 2 links in the hypothesized causal chain—the link between psychosocial stressors and illness and the link between stressors and disrupted immune functioning—are reviewed. Since much of the relevant evidence for humans is correlational, 2 animal models are examined: nonhuman primate mother–infant separation and learned helplessness in rodents.

1741. Mitler, Merrill M.; Nelson, Sara & Hajdukovic, Roza. (1987). **Narcolepsy: Diagnosis, treatment, and management.** ***Psychiatric Clinics of North America,*** 10(4), 593–606.
Characterizes narcolepsy as having 4 classic symptoms, including sudden sleep attacks, emotionally precipitated flaccid paralysis (cataplexy), hypnagogic hallucinations, and paralysis on awakening or falling asleep. Narcoleptic bouts are characterized by immediate REM sleep where non-REM sleep would be expected. Another associated symptom of narcolepsy is poor or disrupted nocturnal sleep. A strong association is noted between narcolepsy and the human leucocyte antigen-DR2. Animal models' and human narcoleptics' drug responses suggest a widespread release of dopamine and a brain-stem-specific hypersensitivity to acetylcholine. Cataplexy is suggested to be the most important diagnostic symptom.

1742. Money, John. (1985). **Pediatric sexology and hermaphroditism.** ***Journal of Sex & Marital Therapy,*** 11(3), 139–156.
Argues that a new 3-term paradigm (nature/critical period/nurture) is needed to explain the phenomenology of hermaphroditism and the differentiation of gender identity and role (GIR) in individual cases. Data from a study by the author and colleagues (1984) on the incidence of homoerotic imagery and practices in females with a history of the early-treated 46, XX congenital virilizing adrenal hyperplasia syndrome are compared with the findings of A. C. Kinsey et al (1953), who reported the incidence of those homoerotic experiences in females drawn from the general population. An animal model that demonstrates the existence of both male and female sexual schemas in the same brain is provided. Consideration of this human and animal evidence leads to the hypothesis that sexual dimorphism that is programmed into the brain under the influence of prenatal hormones is not sex-irreducible but sex-shared and threshold-dimorphic. It is concluded that a complete theory of the differentiation of all constituents of masculinity or femininity of GIR must be applicable to all of the syndromes of hermaphroditism and to the genesis of all GIR phenomena, including transvestism, transsexualism, and a heterosexual GIR.

1743. Rowland, Neil E. & Bellush, Linda L. (1989). **Diabetes mellitus: Stress, neurochemistry and behavior.** ***Neuroscience & Biobehavioral Reviews,*** 13(4), 199–206.
Compares neurochemical alterations in several rodent models of insulin-dependent diabetes and discusses their relevance to behavioral and physiological pathology in humans. In the majority of rodent models, reductions in metabolism of norepinephrine, dopamine, and serotonin (5-hydroxytryptamine [5HT]) in the central nervous system (CNS) have been reported. Insulin, the treatment for severe diabetes, has effects on monoamines opposite to that of chronic hyperglycemia. Both in rodent models and in humans, there is evidence of enhanced hormonal and behavioral responsiveness to stress. Findings in rodent models indicate that hormonal responses to stress are related to CNS monoamine activity. The mechanisms responsible for both hormonal and CNS alterations in diabetes, as well as their involvement in behavioral pathology, can best be investigated further using animal models.

1744. Silbergeld, Ellen K. (1985). **The relevance of animal models for neurotoxic disease states.** ***International Journal of Mental Health,*** 14(3), 26–43.
Discusses the use of animal models to develop preclinical markers of exposure and effects for neuropsychiatric disorders associated with exposure to environmental pollutants. Animal models can be used to provide the basic data for establishing exposure markers under conditions in which external or environmental exposure can be controlled and quantitated and various tissues or fluids can be sampled during or after exposure. Animal models, however, usually reflect human disease states in terms of exposure rather than effect. In the absence of agreement on the relationship between certain well-described human mental illnesses and behavioral states in animals, fundamental behavioral states in animals such as locomotor activity, active and passive avoidance, and simple maze learning are likely to be most useful for determining neurotoxic effects. The need to increase communication between clinical and experimental toxicologists is stressed.

1745. Surwit, Richard S. & Feinglos, Mark N. (1984). **Stress and diabetes.** ***Behavioral Medicine Update,*** 6(1), 8–11.

Environmental and physical stress increase hyperglycemia and glucose intolerance in diabetes mellitus. These effects are probably mediated through the central and autonomic nervous system as well as through the hypothalamic–pituitary axis. Recent experiments have shown that both immobilization and shaking stress increased plasma glucose in both lean and obese rats, with the increase being significantly greater in obese rats. Administration of epinephrine produced similar results, indicating that the environment plays a crucial role in the expression of the diabetic phenotype in this animal model of diabetes. Other studies, including those by the present authors (1983, 1984), have indicated the effectiveness of EMG-biofeedback-assisted relaxation training in improving glucose tolerance.

1746. Taylor, Eric A. (1986). **Childhood hyperactivity.** ***British Journal of Psychiatry,*** 149, 562–573.
Discusses the history, taxonomy, prevalence, etiology, developmental course, and treatment of childhood hyperactivity. Hyperactivity is distinguished from conduct disorder, and findings with regard to its origins in brain damage, perinatal trauma to the brain, minimal brain dysfunction, and neurophysiological unresponsiveness are described. Also presented are animal models, the possibility of genetic inheritance or of assault by chemical agents, and psychosocial factors. Treatments include drug, dietary, and psychological approaches. It is concluded that severe degrees of inattentive and restless behavior constitute a major problem for development and are different from (but overlap) conduct disorder. Family, school, and peer relationships are viewed as more important than the core problem in determining adult outcome.

1747. Weiner, Herbert. (1982). **The prospects for psychosomatic medicine: Selected topics.** ***Psychosomatic Medicine,*** 44(6), 491–517.
Discusses the social contexts and situations associated with a wide variety of illness and disease, the differences between health and disease, the "choice" of a disease, and the timing of disease onset. The traditionalist in medicine remains unconvinced by the answers provided to these issues: He or she seeks to know the mechanisms by which a person's inability to cope with events and experiences and their impact lead to disease. The author suggests that the understanding of these mechanisms will come about only by studying animal models. Benefits derived from the use of animal models are discussed: (1) the role of the brain in the immune response, (2) the participation of the brain in the regulation of bodily functions, (3) the role of the brain in the pathogenesis of disease, and (4) the role of experience in altering brain, behavior, and bodily function. Although little or nothing is known about the physiology of thought and emotion, more is being learned about the mechanisms by which behavior, biologically significant to the organism, and bodily function are temporarily correlated. It is concluded that psychosomatic medicine is making a major contribution to concepts of health, illness, and disease.

1748. Weinstock, Ruth S.; Wright, Herbert N.; Spiegel, Allen M.; Levine, Michael A. et al. (1986). **Olfactory dysfunction in humans with deficient guanine nucleotide-binding protein.** ***Nature,*** 322(6080), 635–636.
The guanine nucleotide-binding stimulatory protein (G_s) couples hormone–receptor interaction to the activation of adenylate cyclase and the generation of cyclic adenosine monophosphate. Because studies using frog neuroepithelium have indicated that the sense of smell is mediated by a G_s-adenylate cyclase system, the present study tested olfaction in the only known animal model of G_s deficiency: G_s-deficient (type 1a) pseudohypoparathyroidism (PHP), which occurs in humans. Results, based on data from 12 20–60 yr old patients with PHP, show that all G_s-deficient Ss had impaired olfaction when compared with PHP patients who had normal G_s activity. Findings indicate that human olfactory impairment can be related to G_s deficiency and suggest that G_s-deficient PHP patients may be resistant to cyclic adenosine monophosphate-mediated actions in other nonendocrine systems.

1749. Zola-Morgan, Stuart & Squire, Larry R. (1990). **The neuropsychology of memory: Parallel findings in humans and nonhuman primates.** Conference of the National Institute of Mental Health et al: The development and neural bases of higher cognitive functions (1989, Philadelphia, Pennsylvania). ***Annals of the New York Academy of Sciences,*** 608, 434–456.
Insights into human amnesia and new developments with animal models of amnesia in nonhuman primates have contributed to understanding the organization of memory in the brain. Two issues raised by these studies concern the type of memory affected by amnesia and the brain structures that must be damaged to cause amnesia. Memory impairment in humans and nonhuman primates is described, emphasizing that amnesia affects only declarative memory. The features of amnesia that have guided the development of a model in monkeys, including selectivity, are considered. Recent animal studies have related memory impairment to damage in the hippocampus and adjacent, neuroanatomically related cortical regions in the medial temporal lobe. The performance of amnesic patients on tests designed for monkeys is also discussed.

Cardiovascular Disorders

1750. Folkow, Björn. (1991). **Mental "stress" and hypertension: Evidence from animal and experimental studies.** Meeting of the American Institute of Stress (1991, Montreux, Switzerland). ***Integrative Physiological & Behavioral Science,*** 26(4), 305–308.
Three etiological components of the psychosocial environment's contribution to multifactorial primary hypertension in rats and humans are polygenetically linked predisposition, environmental factors, and structural cardiovascular adaptation. Animal studies show that the ordinary psychosocial environment for laboratory rats can reinforce primary hypertension if the right genetic predisposition is present; these models are suggested to be valid for humans because they are based on biological processes that are similar in nature and organization in rats and humans.

1751. Natelson, Benjamin H. (1983). **Stress, predisposition and the onset of serious disease: Implications about psychosomatic etiology.** ***Neuroscience & Biobehavioral Reviews,*** 7(4), 511–527.
Hypothesizes that lethal disease does not usually occur in healthy animals or people but does occur when covert or overt disease exists or when a predisposition for disease exists, based on a literature review and the present author's assessment of the human literature on sudden death. Further support for the hypotheses is presented from 2 animal models: stress-induced heart failure in the cardiomyopathic

hamster and stress-induced sensitization of digitalis-toxic ventricular arrhythmias. This analysis suggests a different view from the classical perspective on psychosomatic disease.

1752. Podrid, Philip J. (1984). **Role of higher nervous activity in ventricular arrhythmia and sudden cardiac death: Implications for alternative antiarrhythmic therapy.** Conference on the clinical pharmacology of cardiac antiarrhythmic agents: Classical and current concepts reevaluated (1983, New York, NY). ***Annals of the New York Academy of Sciences,*** 432, 296–313.
A review of animal models and clinical experience indicates that the CNS has important effects on cardiac function. Any type of stress has profound effects on the CNS and can affect hypothalamic function. When the hypothalamus is stimulated, the autonomic nervous system, especially the sympathetic limb, is activated. Neural traffic carried by the sympathetic nervous system can affect myocardial electrophysiologic properties and provoke arrhythmia. Evidence is presented that supports the view that the parasympathetic nervous system has an important effect on both the inotropic and chronotropic properties of the ventricle. There is also evidence that the parasympathetic nervous system can affect the electrical properties of the ventricular muscle. Animal and human studies concerning the psychological factor involved in ventricular arrhythmia are described. Antiarrhythmic drug therapy remains the cornerstone of a treatment program for suppressing ventricular arrhythmia and preventing sudden death.

1753. Zakharzhevsky, V. B. (1989). **Psychogenia, neurosis, and cortico-visceral conception.** ***Zhurnal Vysshei Nervnoi Deyatel'nosti,*** 39(6), 1003–1009.
Studied neurotic and psychosomatic aspects of psychogenic problems, particularly the significance of risk factors in the genesis of neurosis and the influence of cortical mechanisms in early diagnosis of neurosis and psychosomatic illnesses. Separate studies of "neurotized" dogs and 2 groups of human Ss with hypertension of various origins were conducted. The present author used his own experimental and clinical data to construct a binomial model of psychosomatic pathology that included acute and chronic visceral destabilization and "neurotizing" influences. (English abstract).

Neurological Disorders & Brain Damage

1754. Arendt, Thomas; Bigl, Volker & Schugens, Markus M. (1989). **Ethanol as a neurotoxin: A model of the syndrome of cholinergic deafferentation of the cortical mantle.** ***Brain Dysfunction,*** 2(4), 169–180.
Ibotenic acid was injected into rat brains or ethanol was administered to rats in drinking water. Ss were later assessed for behavior, neurons in the basal nucleus of the Meynert, and activity of choline acetyltransferase. The basal forebrain cholinergic projection system in humans who had died with Alzheimer's disease (AD) or Korsakoff's disease were compared to patients who had died without neurological or psychiatric illness and these results were compared to those of the rat study. Chronic intake of ethanol produced the same kind of cholinergic deafferentation of the cortical mantle, with comparable functional symptoms, as observed in AD and may provide a useful animal model of this and other chronic progressive neuropsychiatric disorders.

1755. Bartus, Raymond T.; Dean, Reginald L.; Pontecorvo, Michael J. & Flicker, Charles. (1985). **The cholinergic hypothesis: A historical overview, current perspective, and future directions.** ***Annals of the New York Academy of Sciences,*** 444, 332–358.
The authors discuss the cholinergic hypothesis of geriatric memory dysfunction, which proposes the following: (1) Significant, functional disturbances in cholinergic activity occur in the brains of aged and especially demented patients; (2) these disturbances play an important role in memory loss and related cognitive problems associated with old age and dementia; and (3) proper enhancement or restoration of cholinergic function may significantly reduce the severity of the cognitive loss. The initial empirical foundation for the cholinergic hypothesis can be traced to at least 4 distinct areas of study: biochemical determinations of human brain tissue, particularly from Alzheimer's patients; animal psychopharmacological studies; clinical pharmacological observations; and basic neuroscience research. The current status of treatment approaches with cholinomimetic agents is that there is hope for future drug development, but there is no immediate promise of effective therapeutic intervention. Future directions for research include the development of characteristic-specific cholinergic agents and reliable animal models of primary symptoms.

1756. Bruce, Moira E. (1984). **Scrapie and Alzheimer's disease.** ***Psychological Medicine,*** 14(3), 497–500.
Suggests that research on Alzheimer's disease (AD) has been limited by the lack of satisfactory experimental models and discusses the parallels between AD and scrapie, an infectious neurological disease found in sheep and goats. It is argued that, even though attempts to transmit AD to laboratory animals have either failed or have been inconclusive, scrapie still provides valuable models for this type of degenerative pathology.

1757. Dowson, Jonathan H. (1989). **Neuronal lipopigment: A marker for cognitive impairment and long-term effects of psychotropic drugs.** ***British Journal of Psychiatry,*** 155, 1–11.
The volume of neuronal lipopigment (NLP) has been positively correlated with advancing age, Alzheimer dementia, and the neuronal ceroidoses. Various changes in NLP have been reported in association with the chronic administration of some psychotropic drugs: dihydroergotoxine, ethanol, phenethylamine, centrophenoxine, and chlorpromazine. An increase in the volume of NLP may indicate increased functional activity of the cell, impaired removal of pigment, or anoxia. Chronic administration of agents which can be correlated with decreased NLP in animal models might protect neuronal function against any adverse effects associated with lipopigment accumulation in normal aging, anoxia, or certain degenerative diseases.

1758. Hardy, John. (1988). **Mouse models of human neurogenetic disorders.** ***Trends in Neurosciences,*** 11(3), 89–90.
Discusses advances and problems inherent in the construction, analysis, and use of mouse models of muscular dystrophy and other human neurogenetic disorders. It is suggested that an alternative to studying the effects of spontaneous mutation in mice is genetic engineering aimed at producing rodent homologues of the human disease.

1759. Jaworski, Martine & Edwards, Edmond. (1991). **Integrated genetic databases in the study of neuropsychiatric diseases: Inborn errors of cerebral metabolic pathways?** Special Issue: Perspectives in Canadian neuro-psychopharmacology: Proceedings of the 13th Annual Canadian College of Neuro-psychopharmacology. ***Progress in Neuro-Psychopharmacology & Biological Psychiatry,*** 15(2), 171–181.
Illustrates the application of information on the mapping and sequencing of the human genome and that of other model organisms to neuropsychiatric disease, using neuroendocrine and neuropharmacologic data and other genetic data bases. Results demonstrate that over 30 candidate loci for neuropsychiatric disease have been mapped in humans (spread over 14 chromosomes in the human genome), and that at least 6 homologous loci have been mapped in mice. The best current candidate gene locus for a subtype of schizophrenia located on chromosome 5q11-13 is in the serotonergic pathway. A subset of neuropsychiatric disorders may be viewed as inborn errors of cerebral metabolic pathways primarily affecting the biogenic amine pathways.

1760. Kesner, Raymond P.; Adelstein, Ted & Crutcher, Keith A. (1987). **Rats with nucleus basalis magnocellularis lesions mimic mnemonic symptomatology observed in patients with dementia of the Alzheimer's type.** ***Behavioral Neuroscience,*** 101(4), 451–456.
College students, healthy elderly subjects, patients diagnosed with mild or moderate dementia of the Alzheimer's type, as well as rats with small or large lesions of nucleus basalis magnocellularis (NBM) were tested on an order memory task for a 6- or 8-item list of varying spatial locations. Similar patterns of order memory deficits as a function of serial order position were observed in rats with small or large NBM lesions and patients with mild or moderate dementia of the Alzheimer's type. The results provide support for the possibility that rats with NBM lesions might mimic the mnemonic symptomatology of Alzheimer's disease.

1761. Kesner, Raymond P.; Adelstein, Ted B. & Crutcher, Keith A. (1989). **Equivalent spatial location memory deficits in rats with medial septum or hippocampal formation lesions and patients with dementia of the Alzheimer's type.** ***Brain & Cognition,*** 9(2), 289–300.
47 Ss including college students; healthy elderly adults; patients diagnosed with mild or moderate dementia of the Alzheimer's type; and rats with small or large lesions of the medial septum (MS), dorsal hippocampal formation (DHF), or nucleus basalis magnocellularis (NBM) were tested on an item memory task for a 5- or 6-item list of varying spatial locations. Equivalent patterns of item memory deficits as a function of serial order position were observed in rats with small or large MS or DHF lesions and patients with mild or moderate dementia of the Alzheimer's type. No deficits were found for NBM-lesioned rats. Results indicate that rats with MS and DHF lesions mimic the mnemonic symptomatology of patients with Alzheimer's disease.

1762. Lipkin, W. Ian; Carbone, Kathryn M.; Wilson, Michael C.; Duchala, Cynthia S. et al. (1988). **Neurotransmitter abnormalities in Borna disease.** ***Brain Research,*** 475(2), 366–370.
Studied the acute phase of Borna disease (BD) in 4–6 wk old inbred male rats, which was characterized by hyperactivity, aggression, in most cases ataxia, and inflammatory cell infiltrates in the brain. Results are discussed with regard to the pathogenesis of neurologic disturbances in BD and other inflammatory central nervous system (CNS) diseases.

1763. Mohs, Richard C. (1988). **Memory impairment in amnesia and dementia: Implications for the use of animal models.** Special Issue: Experimental models of age-related memory dysfunction and neurodegeneration. ***Neurobiology of Aging,*** 9(5–6), 465–468.
Notes that dementia, particularly that due to Alzheimer's disease, is more common than amnesia but differs from the amnestic syndromes in 3 ways: (1) a variety of cognitive functions including language and praxis are affected; (2) the condition is usually progressive with gradual loss of several neuronal populations; and (3) the etiology of dementia is not well understood at present, but clinical data suggest some role for genetic, viral, vascular, and toxic environmental factors. It is argued that animal models should give greater attention to these important features of the common clinical dementias.

1764. Olton, David S. (1985). **Strategies for the development of animal models of human memory impairments.** ***Annals of the New York Academy of Sciences,*** 444, 113–121.
Discusses strategies for evaluating the validity of different animal models of human memory impairments, arguing that the ultimate test of the validity of any model is its ability to predict effects in the system that is being modeled. Steps in developing an effective animal model include dissociations, sensitivity and selectivity, performance of normal animals, performance of animals with altered brain function, and interpretation of observed behavioral changes.

1765. Ostrosky-Solís, Feggy & Madrazo-Navarro, Ignacio. (1990). **Parkinson's disease: Symptomatology, pathogenesis and treatments.** ***Revista Mexicana de Psicología,*** 7(1–2), 81–95.
Reviews the symptomatology, neuroanatomy, and pathology of Parkinson's disease and recent research in animal models of Parkinson's disease. Emphasis is on experimental studies demonstrating that reduced dopamine levels and clinical manifestations of reduced dopamine levels are reversible by transplantation of dopamine containing cells, including cells from the substantia nigra or from the adrenal medulla. Results of clinical studies carried out in 50 patients with Parkinson's disease that received autografts of adrenal medullary tissue to the caudate nucleus and results of clinical studies of 7 patients that received fetal tissue grafts of mesencephalic or adrenal tissue are also reported. (English abstract).

1766. Pappas, Bruce A.; Ferguson, H. Bruce & Saari, Matti. (1976). **Minimal brain dysfunction: Dopamine depletion? III.** ***Science,*** 194(4263), 451–452.
The proposed animal model of hyperkinesis of minimal brain damage (MBD) proposed by B. L. Shaywitz et al (1976) is interesting because of the general utility of such models in determining neurochemical bases of this and other syndromes. It is argued, however, that their failure to use appropriate control groups which would allow them to relate their behavioral alterations only to the proposed critical pharmacological manipulation (depletion of brain dopamine) is questionable. Other methodological deficiencies are dis-

cussed, including the measurement of only 2 gross effects of the manipulations, a failure to recognize that 6-hydroxydopamine also causes behavioral alterations similar to those reported and the failure to report body weight data, a significant variable in MBD.

1767. Price, Donald L.; Kitt, Cheryl A.; Struble, Robert G.; Whitehouse, Peter J. et al. (1985). **Neurobiological studies of transmitter systems in aging and in Alzheimer-type dementia.** Institute for Child Development Research Conference: Hope for a new neurology (1984, New York, New York). ***Annals of the New York Academy of Sciences,*** 457, 35–51.
Discusses the clinical syndrome of Alzheimer-type dementia and reviews recent advances in the area of diagnostic studies. Focus is on the nature and distribution of structural and chemical pathologies (especially those involving specific neuronal systems); the development of nonhuman primate models of aging and disease; and prospects for using these models to design and test new therapeutic approaches. Dysfunction and death of specific neuronal systems are important processes occurring in aging and in Alzheimer's and in Parkinson's disease. The availability of animal models (including aged monkeys, macaques with cholinergic deficiencies, and monkeys with nigrostriatal pathology) allows the opportunity to assess the efficacies of new pharmacotherapies, neural grafts, and trophic factors.

1768. Prince, David A. & Wong, Robert K. (1981). **Human epileptic neurons studied in vitro.** ***Brain Research,*** 210(1–2), 323–333.
Used an in vitro neocortical brain slice technique to study electrophysiological properties of neurons from brain biopsies in 10 patients undergoing neurosurgical treatment for a variety of conditions, including focal epilepsy. Results suggest that intracellular events in human neurons involved in epileptogenesis are similar in appearance to those in various animal models. Neurons in chronic epileptogenic foci retained some of their abnormal properties within brain slices maintained in vitro.

1769. Roberts, Gareth W. (1987). **Herpes virus in Alzheimer's disease: A refutation.** ***Archives of Neurology,*** 44(1), 12.
Refutes the hypothesis that herpes simplex virus is involved in the etiology of Alzheimer's disease (AD) by noting that, although the herpes simplex virus has a predilection for the temporal lobe (the main site of pathology in AD) and that recovered encephalitis cases display amnesiac syndromes similar to those found in AD, immunocytochemical, neuropathological, and animal model studies have failed to detect active or latent viral infections in the majority of AD cases.

1770. Robertson, Harold A. & Cottrell, Georgia A. (1985). **Some observations on the kindling process.** Eighth Annual Meeting of the Canadian College of Neuropsychopharmacology: Perspectives in Canadian neuro-psychopharmacology (1985, London, Canada). ***Progress in Neuro-Psychopharmacology & Biological Psychiatry,*** 9(5–6), 539–544.
Discusses kindling as an animal model of epilepsy in which repeated administration of a subconvulsant stimulus, electrical or chemical, produces a gradually increasing electroencephalographic and behavioral response culminating in a behavioral seizure. Two factors involved in kindling are the neuroanatomical basis and neuronal systems necessary for the seizure to occur. Kindling is a permanent change in the sensitivity of the brain to a stimulus. Permanent changes in neurotransmitter levels or receptor parameters have not been conclusively demonstrated in kindling. Other possible permanent changes include changes in Ca^{++}-activated mechanisms that alter neuronal structures such as dendritic spines. Kindling requires an interstimulus interval of at least 1–2 hrs, suggesting that the biological process that leads to kindling occurs in this critical period. Recent experiments with cysteamine imply that the events in this period can be manipulated chemically.

1771. Robinson, Robert G. (1983). **Investigating mood disorders following brain injury: An integrative approach using clinical and laboratory studies.** ***Integrative Psychiatry,*** 1(2), 35–39.
Reports results from clinical and laboratory research conducted by the author and colleagues, which outline 2 distinct phenomena. One is a lateralized emotional response to stroke, with left-hemisphere-injured patients demonstrating depressive symptoms and right-hemisphere-injured patients showing undue cheerfulness associated with apathy and loss of interest. The 2nd phenomenon is an intrahemispheric difference in mood as evidenced by significantly greater depression in patients with left-anterior than in those with left-posterior hemisphere brain injury. Laboratory experiments have produced parallels in the animal model and suggested possible neurochemical mechanisms. Injury to one hemisphere in humans may produce an entirely different neurochemical response than an identical injury to the other hemisphere and represents a plausible explanation for why injury to one or the other hemisphere of the human brain can produce such different emotional states. Whether noradrenergic neurons specifically are involved in poststroke mood disorders is considered not as important as the laboratory demonstration that injury to the 2 sides of the brain can have different biochemical responses and corresponding behavioral consequences. It is concluded that the major point of these investigations is the importance of using an integrative approach to the analysis of the clinical phenomena in psychiatry.).

1772. Schultz, Wolfram. (1984). **Recent physiological and pathophysiological aspects of Parkinsonian movement disorders.** ***Life Sciences,*** 34(23), 2213–2223.
Reviews the literature concerning Parkinsonian movement disorders, focusing on the dopamine system as part of the basal ganglia and animal models for Parkinsonism. The pathophysiology of Parkinsonian motor symptoms is discussed in terms of hypokinesis, rigidity, and tremor, and the normal activity of neurons in the substantia nigra and associated motor areas in relation to movement is described. Data support the contention that Parkinsonian hypokinesia is due to a failure of basic behavioral activating mechanisms.

1773. Schwam, Elias; Gamzu, Elkan & Vincent, George. (1984). **New York Academy of Sciences' conference on memory dysfunctions.** ***Neurobiology of Aging,*** 5(3), 243–248.
Outlines the papers presented at a conference on memory dysfunctions sponsored by the New York Academy of Sciences and held on June 13–15, 1984. The conference was divided into 6 major topics: human neuropsychology, animal models of amnesic syndromes, the biochemistry of memory, neurobiological markers of memory loss, amnesia and aging, and pharmacological strategies in the treatment of memory loss.

1774. Schwarcz, Robert et al. (1984). **Excitotoxic models for neurodegenerative disorders.** ***Life Sciences,*** 35(1), 19–32.
Reviews recent research on the neurotoxic properties of excitatory amino acids and their possible relevance for the study of human neurodegenerative disorders (NDDs). The emergence of NDDs using intracerebral infusions of the excitotoxins kainic and ibotenic acids has led to the hypothesis that endogenous excitotoxins may exist that are linked to the pathogenesis of human diseases. Although a biochemical link between endogenous excitotoxins and human NDDs remains elusive, pharmacological blockade of excitotoxicity may constitute a novel therapeutic strategy for the treatment of NDDs.

1775. Shaywitz, Bennett A. (1976). **Minimal brain dysfunction: Dopamine depletion? IV.** ***Science,*** 194(4263), 452–453.
Replies to methodological criticisms of the B. A. Shaywitz et al (1976) study of dopamine deficiencies in minimal brain damage (MBD) or hyperkinetic syndromes by J. W. Kalat, J. H. McLean et al, and B. A. Pappas et al (all 1976). Further modifications of the animal model of MBD and certain criteria that would have to be fulfilled by it are specified.

1776. Smith, Gwenn. (1988). **Animal models of Alzheimer's disease: Experimental cholinergic denervation.** ***Brain Research Reviews,*** 13(2), 103–118.
Reviews the animal literature focusing on models of dementia and pharmacological reversal of induced deficits. Behavioral effects resulting from (1) lesions to the nucleus basalis of Meynert, (2) administration of a specific cholinotoxin (AF64A), and (3) scopolaine-induced dementia (SID) are discussed. Significant mnestic and cerebral metabolic deficits have been observed acutely after lesion, which are responsive to pharmacological reversal and recovery over time. Administration of AF64A produces similar deficits that are less responsive to pharmacological reversal. SID is pharmacologically reversible. It is the lack of pharmacological response and recovery of function that distinguishes Alzheimer's disease from the experimental animal models, as well as the presence in humans of noncholinergic neurochemical and cytoskeletal abnormalities.

1777. Solomon, Paul R.; Beal, M. Flint & Pendlebury, William W. (1988). **Age-related disruption of classical conditioning: A model systems approach to memory disorders.** Special Issue: Experimental models of age-related memory dysfunction and neurodegeneration. ***Neurobiology of Aging,*** 9(5–6), 535–546.
Suggests that, using a well characterized model system such as classical eyeblink conditioning, it should be possible to both characterize the changes in learning and memory that accompany aging and to investigate their neural substrates. The authors' strategy for using the conditioned eyeblink preparation for studying age-related memory deficits included investigating conditioning deficits in (1) humans across the life span, (2) rabbits across the life span, (3) Alzheimer's disease patients, and (4) rabbits with aluminum-induced neurofibrillary degeneration. Exemplary data from each of these lines of research are presented. It is argued that if similar deficits occur in each of these groups, it may be possible to begin to form hypotheses about the neurobiology of age-related memory disorders.

1778. Solomon, Paul R.; Groccia-Ellison, Maryellen; Levine, Elizabeth; Blanchard, Sharon et al. (1990). **Do temporal relationships in conditioning change across the life span? Perspectives from eyeblink conditioning in humans and rabbits.** Conference of the National Institute of Mental Health et al: The development and neural bases of higher cognitive functions (1989, Philadelphia, Pennsylvania). ***Annals of the New York Academy of Sciences,*** 608, 212–238.
Argues that the optimal temporal relationship between associated events may vary as a function of age and that this variation may be a factor in age-related memory disorders (ARMDs). A model systems approach, which advocates studying a well-characterized learned response in a relatively simple and well-controlled preparation, is applied to ARMDs. A strategy for using the classically conditioned eyeblink response to study ARMDs is presented. This strategy involves studying conditioning and modeling in rabbits and humans across the lifespan, and in Alzheimer's disease patients. Possible mechanisms and neurobiological substrates of age-related changes in temporal relationships are discussed.

1779. Squire, Larry R.; Zola-Morgan, Stuart & Chen, Karen S. (1988). **Human amnesia and animal models of amnesia: Performance of amnesic patients on tests designed for the monkey.** ***Behavioral Neuroscience,*** 102(2), 210–221.
The performance of amnesic patients was assessed on five tasks, which have figured prominently in the development of animal models of human amnesia in the monkey. The amnesic patients were impaired on four of these tasks (delayed nonmatching to sample, object-reward association, 8-pair concurrent discrimination learning, and an object discrimination task), in correspondence with previous findings for monkeys with bilateral medial temporal or diencephalic lesions. Moreover, performance of the amnesic patients correlated with the ability to verbalize the principle underlying the tasks and with the ability to describe and recognize the stimulus materials. These tasks therefore seem to be sensitive to the memory functions that are affected in human amnesia, and they can provide valid measures of memory impairment in studies with monkeys. For the fifth task (24-hour concurrent discrimination learning), the findings for the amnesic patients did not correspond to previous findings for operated monkeys. Whereas monkeys with medial temporal lesions reportedly learn this task at a normal rate, the amnesic patients were markedly impaired. Monkeys may learn this task differently than humans.

1780. Task Force on Alzheimer's Disease. (1985). **Department of Health and Human Services's Task Force on Alzheimer's disease: Report and recommendations.** ***Neurobiology of Aging,*** 6(1), 65–71.
Discusses 5 of the 9 areas addressed by the US Department of Health and Human Service's Task Force on Alzheimer's disease (AD) in a problem-oriented approach aimed at better defining research needs, options, and training, and service and policy issues relative to AD. The 5 areas addressed include research on epidemiology, etiology and pathogenesis, diagnosis, clinical course, and treatment. Future research directions include continuation of exploration of the cholinergic hypothesis in the development of new treatment agents for the primary cognitive impairment in AD; evaluation of the indications for, and efficacy of, psychopharmacologic agents in the treatment of secondary psychiatric symptoms and syndromes occurring concomitantly with AD; identifica-

tion of naturally occurring animal diseases and the development of new experimental animal models, based on actual neuropathological findings in human populations; and development of new and more specific psychometric models that can more accurately distinguish among memory, motivation, and attention effects over time.

1781. Zigmond, Michael J. & Stricker, Edward M. (1984). **Parkinson's disease: Studies with an animal model.** ***Life Sciences,*** 35(1), 5–18.
Reviews research on Parkinson's disease (PD), which has been associated with degeneration of dopaminergic (DA) neurons of the nigrostriatal bundle. Many neurological features of PD can be produced in rats by selective destruction of central DA neurons using the neurotoxin 6-hydroxydopamine (6-OHDA). Two aspects of PD that have been investigated in these rats are discussed. The near-total degeneration of nigrostriatal bundle neurons that is required before neurological symptoms emerge is considered. It appears that the loss of DA neurons is accompanied by an exponential increase in the ratio of tyrosine hydroxylase activity to dopamine content. Thus, after the brain lesions there may be a compensatory increase in the capacity of residual DA neurons to synthesize and release transmitter. The severe neurological deficits produced by stress in patients who are only mildly impaired otherwise are also considered. It appears that a variety of stressors produce an abrupt but transient increase in DA activity in the striatum of intact animals and that this increase is markedly attenuated by 6-OHDA.

Vision & Hearing & Sensory Disorders

1782. Kolata, Gina. (1985). **What causes nearsightedness?** ***Science,*** 229(4719), 1249–1250.
Discusses an animal model of myopia developed by E. Raviola and T. N. Wiesel (1985). The model suggests that myopia may be caused by abnormal influences of the nervous system on the developing eye. It is suggested that when visual perception is distorted, a regulatory molecule hypothesized to control the growth of the eye may be released in abnormal amounts.

Health & Mental Health Treatment & Prevention

1783. Azcarate, Carlos L. (1975). **Minor tranquilizers in the treatment of aggression.** ***Journal of Nervous & Mental Disease,*** 160(2), 100–107.
Discusses clinical trials designed to evaluate the efficacy of drugs in human aggression. The potential antiaggressive action of minor tranquilizers in humans has received little attention in spite of the claimed "taming effect" in some animal studies. A recent report examining the literature regarding the effects of benzodiazepines on animal models of aggressive behavior has pointed out the lack of consistency in such findings. Similar observations have been noted in humans where reduction in aggressive manifestations is contrasted with an increase in hostility in a few studies, as well as with the appearance of "paradoxical" rage reactions. Some variables that could account for such discrepancies are discussed. A study designed to test the efficacy of 2 benzodiazepines, at dosages higher than those usually recommended, being carried out in anxious, aggression-prone individuals with poor impulse control, is described.

1784. Blehar, Mary C. & Rosenthal, Norman E. (1989). **Seasonal affective disorders and phototherapy: Report of a National Institute of Mental Health-sponsored workshop.** ***Archives of General Psychiatry,*** 46(5), 469–474.
Discusses several aspects of phototherapy for seasonal affective disorder (SAD), including (1) diagnostic, clinical, and epidemiologic issues; (2) critical issues in research on the use of phototherapy; (2) biologic effects of light and mechanisms of action of phototherapy; and (4) the value of animal models in the study of SAD.

1785. Masserman, Jules H. (1987). **Origins of social behaviors.** 11th World Congress of Social Psychiatry (1986, Rio de Janeiro, Brazil). ***American Journal of Social Psychiatry,*** 7(3), 146–152.
Traces the evolution of normal and deviant behaviors from animals to corresponding patterns in humans as related to individual and social modes of treatment. Five biodynamic principles are inferred from the observation of animals: motivation, individuation, adaptability, neurotigenesis, and therapeutics. It is contended that humans operate on 3 more highly developed universal and ultimate conations: physical, social, and existential. These universal human needs suggest that social psychiatry best serves humanity by clarifying and employing basic principles of individuals and communal dynamics. It is concluded that the most productive clinical interviews occur when the patient regards the therapist not only as an expert in restoring somatic well-being but also as a sincere friend and empathetic mentor. Seven dynamics of therapy are discussed.

1786. Schroeder, Stephen R.; Bickel, Warren K. & Richmond, Glenn. (1986). **Primary and secondary prevention of self-injurious behaviors: A life-long problem.** ***Advances in Learning & Behavioral Disabilities,*** 5, 63–85.
Suggests that treatment goals for self-injurious behavior (SIB) in people with development disabilities be cast in terms of primary, secondary, and tertiary prevention, since the ultimate goal is to reduce the risk to the client. Primary prevention deals with genetic counseling, animal models of neuropathological function, and sensory deprivation due to impoverished environments. Secondary prevention deals with biological antecedents affecting SIB, sensory integration, neurochemical models, behavioral antecedents, and assessment of the controlling stimuli. Tertiary prevention refers to assistance to the client and the support system to prevent attrition and long-term disability.

1787. Snyder, Solomon H.; Taylor, Kenneth M.; Coyle, Joseph T. & Meyerhoff, James L. (1970). **The role of brain dopamine in behavioral regulation and the actions of psychotropic drugs.** ***American Journal of Psychiatry,*** 127(2), 199–207.
By comparing biochemical and behavioral actions of d- and l-isomers of amphetamine, it is shown that locomotor hyperactivity, an animal model for the central stimulant effects of amphetamine, is mediated by brain norepinephrine. By contrast, stereotyped, compulsive gnawing behavior in rats, which resembles symptoms of amphetamine psychosis, appears to be regulated by brain dopamine. Since haloperidol, a potent blocker of dopamine receptors, is uniquely efficacious in treating Gilles de la Tourette's disease, it is suggested that hyperactivity of dopamine systems in the brain may be a factor in the pathophysiology of this condition.

Psychotherapy & Psychotherapeutic Counseling

1788. Masserman, Jules H. (1989). **The dynamics of contemporary therapies.** ***Journal of Contemporary Psychotherapy,*** 19(4), 257–270.
Deviations of behavior induced in animals by motivational conflicts can be alleviated by pharmacologic, environmental, dyadic, or group retraining procedures analogous to those employed clinically. To be successful in humans, basic therapeutic modalities involving rapport, anamnestic review, cognitive reorientations, and guided rehabilitation should resonate with universal and urgent aspirations for physical vitality, social security, and existential serenity. Parameters of therapy are illustrated by 2 case examples (somatoanalysis with a 45-yr-old woman and family crisis therapy).

Behavior Therapy & Behavior Modification

1789. Baum, Morrie. (1988). **On the validity of the animal model for exposure therapy (flooding).** ***Behavioural Psychotherapy,*** 16(1), 38–44.
Reviews the generality of exposure effects in animal laboratory studies and assesses the validity of the animal model for exposure therapy, using N. Bond's (1984) criteria. The specific criticisms of the animal model raised by I. M. Marks (1977, 1981, 1982) are addressed. It is concluded that the animal model for exposure therapy is substantially valid.

1790. Becker, Johannes; Euler, Harald A. & Mielke, Rosemarie. (1976). **Contextual reorganization of behavior after induced behavioral change.** ***Zeitschrift für Klinische Psychologie,*** 5(2), 79–91.
Examined simple scalability as a model of the reorganization of response hierarchies after induced behavioral change. In 3 experiments with a multiple-response baseline procedure with preschool children and gerbils as Ss, the effects of withdrawal of behavioral options or of punishment on the individual response hierarchies were investigated. The observed reorganization effects contradicted the assumption of simple scalability. The induced behavioral changes resulted in changes of the hierarchies of the remaining responses. As an alternative model, the theory of elimination by aspects is discussed and also rejected.

1791. Fernández Castro, Jordi. (1984). **Avoidance behavior and flooding techniques: Theory and comparison of animal and human experiments.** ***Quaderns/Cuadernos de Psicología,*** 8(1), 131–151.
Formulates a theoretical model for avoidance behavior and reviews a set of experiments, conducted in the behavior laboratory at the Autonomous University of Barcelona, Spain, concerning flooding (blocking of response) and its effects on emotional and avoidance response extinction. The avoidance analogical model for anxious behavior in humans is revised. (English & Spanish abstracts).

1792. Wolpe, Joseph. (1979). **The experimental model and treatment of neurotic depression.** ***Behaviour Research & Therapy,*** 17(6), 555–565.
Identifies normal depression as an additional category of depression (i.e., besides endogenous and neurotic depression). The features of learned helplessness have little in common with clinical neurotic depression, which is shown to be similar to the depression in experimental neuroses: the latter is anxiety-based and reliably removed by deconditioning. 25 cases treated in this manner are presented—22 recovered from their depression, and recovery was maintained in 19 patients who were followed up for at least 6 mo.

Group & Family Therapy

1793. Buirski, Peter. (1975). **Some contributions of ethology to group therapy: Dominance and hierarchies.** ***International Journal of Group Psychotherapy,*** 25(2), 227–235.
Explores the extent to which the dynamics of the social grouping of animals in nature can serve as a model for the understanding of process in psychotherapy groups. Group dynamics necessary for the survival and perpetuation of animal groups serve the function of resistance in therapy groups. Clinical examples of the displacement of lower ranking by higher ranking members and the modification of hierarchical arrangements by alliances between members are given. Implications for the practice of cotherapy are noted.

Clinical Psychopharmacology

1794. Abernethy, Darrell R. (1987). **Development of memory-enhancing agents in the treatment of Alzheimer's disease.** ***Journal of the American Geriatrics Society,*** 35(10), 957–958.
Describes the drug development process for therapeutic agents for senile dementia of the Alzheimer's type (SDAT) prior to the general clinical availability. It is noted that lack of specific animal models of SDAT and a limited understanding of the clinical disease in humans make selection of a therapeutic approach difficult.

1795. Amaducci, L.; Angst, J.; Bech, P.; Benkert, O. et al. (1990). **Consensus Conference on the Methodology of Clinical Trials of "nootropics," Munich, June 1989: Report of the Consensus Committee.** ***Pharmacopsychiatry,*** 23(4), 171–175.
The 2nd Consensus Conference on the Methodology of Clinical Trials held in 1989 was devoted to drugs intended to improve attention, cognition, and memory in patients suffering from symptoms of dementia. These drugs are called cognition enhancing agents, antidementia drugs, or nootropic in the broad sense. The conference focused on the following points: terminology and classification of nootropic drugs, value of animal models and preclinical human studies, problems of design and patient characteristics, assessment instruments and statistical analysis, administrative regulations, and ethical problems.

1796. Ammar, Salomon & Martin, Patrick. (1991). **Models of dopaminergic-agonists efficacy in experimental and clinical psychopharmacology.** ***Psychologie Française,*** 36(3), 221–232.
Studied the effects of dopaminergic agonists on an experimental model of learned helplessness behavior in rats and in patients with depression or Parkinson's syndrome. Human Ss: 25 male and female French adults (Parkinson's syndrome). 10 male and female French adults (aged 20–49 yrs) (major depression). The effects of apomorphine, bromocriptine, and etrabamine on learned helplessness in rats were evaluated in a shock-avoidance task. In Parkinson's Ss neurological and psychopathological evaluations were carried out immediately, at 90 min, and 6–8 days after treatment with L-dopa (187.5 mg). Affective signs of depression were evaluated during 15 days of treatment with 5–15 mg/day

bromocriptine. Tests used: The Depression Rating Scale by M. Hamilton (1967) and the Rating Scale for Emotional Blunting by R. Abrams and M. A. Taylor (1978). An analysis of variance (ANOVA) and other statistical tests were used. (English abstract).

1797. Bartus, Raymond T. (1990). **Drugs to treat age-related neurodegenerative problems: The final frontier of medical science?** ***Journal of the American Geriatrics Society,*** 38(6), 680–695.
Examines the biomedical revolution in research into age-related neurodegeneration and 1st-, 2nd-, and 3rd-generation drug treatments. Social-political, medical, and scientific factors contributing to increased research into diseases such as Alzheimer's disease are reviewed. The 1st-generation treatment, which consists of nootropics (neural metabolic enhancers), is discussed in terms of definitions, animal models, and problems in clinical trials. Selective neurotransmitter modifiers comprise the 2nd-generation treatments and rest on the hypothesis that neurotransmitter systems are implicated in dementia. Arguments against such hypotheses are reviewed. Third-generation compounds are those that reduce or retard the pathogenesis and involve 3 approaches: neurotrophic factors, excitatory amino acid toxicity, and modulators of abnormal proteins.

1798. Bernstein, Ilene L. (1985). **Learning food aversions in the progression of cancer and its treatment.** ***Annals of the New York Academy of Sciences,*** 443, 365–380.
Discusses studies that examine the learned food aversions that arise as a result of antineoplastic drug therapy and of neoplastic disease. In studies with cancer patients receiving chemotherapy, it has been found that both adults and children can acquire learned taste aversions in a single trial and that the cognitive development of adults apparently does not override this conditioning. In studies of animal models of drug-induced aversions, findings suggest that the propensity to avoid proteins rather than carbohydrates has some adaptive value, particularly if potential sources of toxicosis in the natural diet of rats or humans are more likely to be proteins. A prevalence of aversions to familiar diet items has been found in pediatric cancer patients. Studies with animals suggest that tumor growth is associated with the development of strong aversions to the available diet. The aversions appear to play a causal role in the development of tumor-induced anorexia. It is suggested that physiological consequences of tumor growth may act as an unconditioned stimulus/stimuli (UCS) in taste aversion conditioning.

1799. Bollini, P. et al. (1984). **Drugs: Guide and caveats to explanatory and descriptive approaches: II. Drugs in psychiatric research.** Symposium held at the Inst of Pharmacological Research "Maria Negri": Biological markers in mental disorders (1983, Milan, Italy). ***Journal of Psychiatric Research,*** 18(4), 391–400.
Notes that drugs in psychiatric research, and antidepressants in particular, are acquiring a role that must be interpreted with caution in view of clinical evidence on drug responses. Over the last 20 yrs, animal models, mediators, and drugs have produced a mass of experimental data difficult to relate to the epidemiological and clinical side of psychiatric disorders. Studies of biochemical descriptors of drug actions as markers of disease and its outcome have drawn puzzling pictures, often contradictory and unstable in terms of the populations to whom they can be applied. Antidepressant drugs are used as a model to illustrate this situation. Controlled clinical trials with antidepressant drugs over the last 10 yrs have persisted in searching for short-term pharmacological effects rather than the medium- or long-term impact of medication in large populations. To establish a positive role for antidepressant drugs, they must be studied in a natural context in which depressed patients are treated with all necessary follow-up. Apart from challenges within the fields of pharmacology and behavioral and biochemical sciences, the existence of parallel lines of evidence (e.g., effective nonpharmacological therapeutic strategies) must be considered.

1800. Cole, W. & Lapierre, Y. D. (1986). **The use of tryptophan in normal-weight bulimia.** ***Canadian Journal of Psychiatry,*** 31(8), 755–756.
Reports on the successful diminution of bingeing behavior in a normal weight bulimic female (aged 21 yrs) who was administered 1 g/day oral tryptophan, a serotonin precursor, in a 6-wk trial. A cessation of both the urge to binge and the behavior itself was noted. Findings are consistent with animal models suggesting the regulation of satiety is dependent on the inhibitory action of serotonin. (French abstract).

1801. Cook, Leonard. (1982). **Animal psychopharmacological models: Use of conflict behavior in predicting clinical effects of anxiolytics and their mechanism of action.** ***Progress in Neuro-Psychopharmacology & Biological Psychiatry,*** 6(4–6), 579–583.
Discusses how animal models with a high predictability of clinical efficacy are useful in pharmacological studies of anxiety. It has been found that conflict behavior in rats and monkeys is useful in identifying pharmacological properties that are highly correlated with clinical antianxiety effects. Recent data show that one can also measure the anticonflict activity of diazepam directly in humans. Having established the correlation of such effects in animals with clinical antianxiety activity in patients, this model has been useful in exploring possible mechanisms of action of anxiolytics. A high correlation was found of the potency of benzodiazepine compounds that bind to the benzodiazepine receptor with their anticonflict effects. In addition, nonbenzodiazepine agents that bind to this receptor also have anticonflict effects. Studies with methysergide and cinanserin in the anticonflict model lend support to the hypothesis that such antianxiety effects may be related to an interaction with the serotonergic system. Studies of the role of GABA in antianxiety activity are discussed. Anticonflict activity of diazepam is antagonized by the recently described specific benzodiazepine antagonist compound RO 15-1788.

1802. Domino, Edward F. & Kovacic, Beverly. (1983). **Monkey models of tardive dyskinesia.** ***Modern Problems of Pharmacopsychiatry,*** 21, 21–33.
Developed a primate model for the study of the development of tardive dyskinesia (TD) in which the long-term treatment with neuroleptics (fluphenazine) of 5 *Cebus apella* and 13 *Macaca speciosa* monkeys resulted in the development of 2 distinct motor syndromes, one corresponding to the early appearing extrapyramidal symptoms of neuroleptic-treated patients, the other corresponding to TD.

1803. Dowson, Jonathan H. (1989). **Drug treatments for cognitive impairment due to ageing and disease: Current and future strategies.** ***International Journal of Geriatric Psychiatry,*** 4(6), 345–353.

Reviews research on drugs, including dihydroergotoxine, bufluomedil, paracetam, and centrophenoxine, that have been claimed to have a beneficial effect on cognitive impairment in the aged and in various diseases. Future strategies will be directed toward reducing the adverse effects of the aging process and neutralizing the pathogenic mechanisms related to genetic factors, infections, toxins, trauma, anoxia, and diet. Prophylactic administration of drugs with antioxidant activity can reduce adverse effects of the production of "free radicals" in animal models. Other drugs with therapeutic potential include antagonists of the excitatory effects of glutamate and aspartate.

1804. Dysken, Maurice. (1987). **A review of recent clinical trials in the treatment of Alzheimer's dementia.** ***Psychiatric Annals,*** 17(3), 178–191.
Reviews clinical trials that have been published over the past 5 yrs involving drugs tested for the treatment of Alzheimer's disease (AD). Knowledge about neurotransmitter deficits in the brains of AD patients and experience in the uses of pharmacological agents in animal models have suggested a variety of pharmacological treatment strategies. Cholinergic agents (e.g., lecithin) are used to restore deficient levels of brain acetylcholine in AD patients; physostigmine has been used with limited success to improve cognitive functioning. Bethanechol has resulted in some improvement in cognitive, social, and emotional functioning of AD patients. Other types of drugs have been used with varied success (e.g., opioid antagonists, ergoloid mesylates, vasopressin, lithium, and combinations of drugs).

1805. File, Sandra E. (1985). **Animal models for predicting clinical efficacy of anxiolytic drugs: Social behaviour.** ***Neuropsychobiology,*** 13(1–2), 55–62.
Describes the behavioral, physiological, and pharmacological validation of one animal test of anxiety, the social interaction test. The effects of anxiolytic and anxiogenic drugs and manipulations of catecholaminergic and serotonergic pathways are considered. Because the test is able to distinguish anxiolytic from sedative and antidepressant drug effects, it is argued that the test should prove useful as an animal screen for detecting novel putative anxiolytics and for research into neural mechanisms underlying anxiety.

1806. Gamzu, E. R. & Gracon, S. I. (1988). **Drug improvement of cognition: Hope and reality.** Symposium: Clinical and cognitive aspects of anxiety (1988, Paris, France). ***Psychiatrie & Psychobiologie,*** 3(Spec Issue B), 115–123.
Addresses the possibility of pharmacological enhancement of cognition. Recent research aimed at developing new drugs for diseases of cognition focuses mainly on Alzheimer's disease and emphasizes mechanistic/biochemical approaches. A series of compounds called nootropics protect animals against disruptions of learning and memory, offering hope that cognitive deficits may be treatable through drugs. However, clinical development is complicated by many factors, including the poor correlation between animal models of cognitive loss and clinical disease states, an exception being the amnesic effects of benzodiazepines. (French abstract).

1807. Gamzu, Elkan. (1985). **Animal behavioral models in the discovery of compounds to treat memory dysfunction.** ***Annals of the New York Academy of Sciences,*** 444, 370–393.
Describes a pragmatic approach to the discovery of new chemical agents to treat memory dysfunction that focuses on the use of animal behavioral tests to predict clinical utility. The approach is independent of any theoretical assumptions about disease states or specific pharmacological mechanisms of drug action and is not mutually exclusive of mechanistic approaches to the study of pharmacological alleviation of memory dysfunction.

1808. Greenblatt, David J.; Miller, Lawrence G. & Shader, Richard I. (1990). **Benzodiazepine discontinuation syndromes.** Symposium: Benzodiazepines: Therapeutic, biologic, and psychosocial issues (1988, Belmont, Massachusetts). ***Journal of Psychiatric Research,*** 24(Suppl 2), 73–79.
Discusses discontinuation syndromes, including recurrence, rebound, and withdrawal, as distinguished by the character, time course, and intensity of symptoms. Recurrence may occur since benzodiazepines suppress, reduce, or contain disorders but do not cure them. Symptom rebound appears within hours to days and is qualitatively similar to the disorder for which the drug was taken. Withdrawal is characterized by excess sensitivity to light and sound, tachycardia, mild systolic hypertension, tremulousness, sweating, and abdominal distress. Any combination of syndromes may occur; however, a systematic tapering schedule may obviate much of the risk. Animal models of the mechanism of benzodiazepine discontinuation syndrome are discussed, and the limitations on their application are noted.

1809. Guez, David. (1989). **Drug development in age-related memory disorders.** Symposium: Memory and aging (1988, Lausanne, Switzerland). ***Archives of Gerontology & Geriatrics,*** 1, 191–194.
Provides an overview of the major steps and obstacles to be overcome in the development of a drug for age-related memory disorders. Discussed are problems in drug development (pathophysiology, animal models, reliable markers, efforts toward clarification of the definition, clinical evaluation methods). Preclinical questions in drug development involve choice of a direction for the chemistry synthesis, study of the neurochemistry of age-related memory disorders, and evaluation involving prediction of promnestic activity, the use of 2 treatment groups, and identification of undesirable side effects. The importance of a multidisciplinary approach, of cooperation between researchers within and without the pharmaceutical industry, and of focusing projects emerge as priorities in resolving the difficulties.

1810. Hollister, Leo E. (1990). **Problems in the search for cognition enhancers.** Second International I.T.E.M.-Labo Symposium on Strategies in Psychopharmacology: Cognition enhancers—from animals to man (1988, Paris, France). ***Pharmacopsychiatry,*** 23(Suppl 2), 33–36.
Reviews the main problems that have limited progress in discovering clinically useful cognitive enhancers, including establishing suitable animal models for developing such drugs. The best animal models would employ learning paradigms that mimic the problems encountered by the disabled patient. Testing new compounds in the clinic also has problems. Ss should not be so deteriorated to require a miracle drug to show any effect. The characteristic variability in function of such patients must be taken into account to assure that any changes are not simply within the normal range of variability. A new kind of statistical approach to

analyzing data, based more on individual responses than on group means, may be needed. Reference is made to the problems of discovering and testing a drug to treat Alzheimer's disease. (German abstract).

1811. Itil, Turan M. (1983). **Nootropics: Status and perspectives.** ***Biological Psychiatry,*** 18(5), 521–523.
Nootropics, drugs that affect higher cerebral functions without producing immediate behavioral changes, have therapeutic indications for geriatric and organic brain syndromes, posttraumatic and posthypoxic events, and learning and speech disabilities in children. However, because of their selective effects on brain function and different focus of treatment, problems arise in determining pharmacological profiles using the traditional animal models and in determining clinical efficacy in humans.

1812. Janke, Wilhelm. (1985). **Clinical efficacy of drugs predicted from drug effects after short-term administration in animals, normal subjects and patients.** ***Neuropsychobiology,*** 13(1–2), 53–54.
Presents lists of the basic types of prediction problems, basic questions to be asked in the prediction of therapeutic drug efficacy, and types of criteria to be predicted. A taxonomy of predictors is also provided. It is argued that preclinical data used within adequate prediction models may prove to be important tools for the prediction of clinical long-term effects.

1813. Kelwala, Surendra; Stanley, Michael & Gershon, Samuel. (1983). **History of antidepressants: Successes and failures.** ***Journal of Clinical Psychiatry,*** 44(5, Sect 2), 40–48.
Discusses 11 compounds that had fulfilled laboratory animal pharmacologic screens to predict antidepressant activity but failed as antidepressants in clinical trials. Most were screened on animal models that measure enhancement of monoaminergic systems. Other agents are discussed because of their unconventional pharmacologic profiles. Many, like mianserin, trazodone, zimelidine, and salbutamol do not fulfill the criteria for an antidepressant in animal models but are reported to have clinical antidepressant activity. Many were selected for discussion because one or more of their pharmacologic properties are at variance with the current theories of depression. A summary of their actions on norepinephrine and 5-HT turnover, alpha$_2$- and beta-adrenergic receptors, 5-HT and 5-HT$_2$ receptors, and beta-receptor-mediated cyclic AMP activity is presented. The last 25 yrs of research with antidepressants have seen an increasing list of neuronal sites that are influenced by these drugs. There are many factors arguing against the theory that antidepressants have a unitary mechanism of action. Perhaps the mechanism of action of antidepressants is more complex than has hitherto been suspected. Far more basic pharmacologic work needs to be done before a comprehensive theory of depression can evolve.

1814. Langston, J. William. (1985). **MPTP neurotoxicity: An overview and characterization of phases of toxicity.** Meeting of the American Society for Pharmacology & Experimental Therapeutics: Pharmacological features of the dopaminergic neurotoxin MPTP (1984, Indianapolis, Indiana). ***Life Sciences,*** 36(3), 201–206.
Reviews the syndrome of 1-methyl-4-phenyl-1,2,3,6-tetrahydropyridine (MPTP) neurotoxicity. MPTP produces alloyed parkinsonism in humans when injected systemically and has been effective in producing an animal model of Parkinson's disease in nonhuman primates. A distinction between toxic and pharmacologic effects of MPTP is drawn, and basic criteria for animal models of human disease are reviewed, particularly as they relate to MPTP. A comparison between a unique population of humans ($N = 7$) affected by MPTP and nonhuman primates (15 squirrel monkeys) is presented. It is concluded that MPTP intoxication can be divided into 3 phases: acute, subacute, and chronic. The possibility of a 4th phase of late symptom appearance and/or progression in exposed but asymptomatic humans and animals is considered. Potential underlying mechanisms for the clinical and behavioral observations of each phase are discussed.

1815. Lapin, I. P. (1988). **Phenylethylamine as an endogenous anxiogenic substance and a common link in anxiety, mania, depression, and schizophrenia.** ***Trudy Leningradskogo Nauchno-Issledovatel'skogo Psikhonevrologicheskogo Instituta im V M Bekhtereva,*** 119, 92–101.
Discusses the role of phenethylamine (beta-phenylethylamine) as an endogenous anxiogenic amine, the role of beta-phenyl-gamma-aminobutyric acid (phenibut) and baclofen as antagonists of phenethylamine, the antagonism of tranquilizers for phenethylamine and anxiogenic substances, use of phenethylamine in animal models of anxiety, possibilities of using baclofen in psychiatry, and excretion of phenethylamine in patients with mania, depression, and schizophrenia. (English abstract).

1816. Leonard, B. E. (1988). **Lofepramine: Pharmacology and mode of action.** Symposium: Advances in the chemotherapy of depression: Lofepramine (1988, London, England). ***International Clinical Psychopharmacology,*** 3(Suppl 2), 25–38.
Discusses the pharmacokinetic aspects of lofepramine, as well as its effects on the cardiovascular and peripheral cholinergic system, cardiovascular effects, effects on central neurotransmission, and effects on the bulbectomized rat model of depression by S. Jancsar and B. E. Leonard (1983). As there is substantial evidence to show that lofepramine is an effective antidepressant with a major metabolite, desipramine, which is also therapeutically active, the question is raised as to whether lofepramine is active in its own right or merely a pro-drug for desipramine. A review of clinical and pharmacological evidence by Leonard (1987) reached the conclusion that lofepramine is a novel pharmacologically active tricyclic antidepressant. There are differences, however, between lofepramine and desipramine, the reasons for which are not yet apparent.

1817. Leonard, Brian E. (1984). **Pharmacology of new antidepressants.** ***Progress in Neuro-Psychopharmacology & Biological Psychiatry,*** 8(1), 97–108.
Attempts to show, through a literature review, how the tricyclic and nontricyclic ("second-generation") antidepressants, while differing in acute pharmacological profiles, have a similar effect on central neurotransmission following chronic administration. Animal models of depression and studies in depressed patients emphasize the importance of adrenergic receptor malfunction in the etiology of the disease. Experimental and clinical studies suggest that all chronically administered antidepressants, irrespective of their acute pharmacological profile, can normalize central noradrenergic receptor function. Such a hypothesis is seen as a way to explain the slow duration of onset of the antidepressant effect and similar therapeutic efficacies of all forms of treatment.

1818. Lloyd, K. G. et al. (1983). **The potential use of GABA agonists in psychiatric disorders: Evidence from studies with progabide in animal models and clinical trials.** ***Pharmacology, Biochemistry & Behavior,*** 18(6), 957–966.
Progabide (PG), a new antiepileptic GABA agonist of moderate affinity for GABA receptors, was studied in a number of psychiatric disorders and the results compared with the action in animal models. In an animal model for anxiety (the aversive response to periaqueductal grey stimulation in the rat), PG had a similar action to that of diazepam. However, in clinical trials to date, the effect of PG was inferior to that of benzodiazepines. Because PG diminished both nigrostriatal dopamine (DA) neuron activity and the effects of striatal DA receptor activation, a trial in schizophrenic patients was conducted. PG was devoid of evident antipsychotic action. However, a certain improvement in responsiveness to the environment and in social interactions was noticed in hebephrenic and schizoaffective syndromes. This lack of antipsychotic effect of PG may reflect the weak activity of GABA agonists on limbic DA neurons. In these clinical trials, a definite improvement of affect and mood was noted in those patients receiving PG. In clinical trials in depressed patients, PG produced significant reduction in depressive symptoms, an action similar to that of imipramine both for global clinical rating and Hamilton Rating Scale for Depression. This antidepressant activity was reflected by the action of PG in behavioral models of depression such as olfactory bulbectomy, learned helplessness, and the sleep–wake cycle.

1819. Palomo-Alvarez, Tomás. (1987). **Animal models of depression and therapeutic prospects. (Trans B. Shapiro).** ***Psicopatología,*** 7(3), 419–426.
Derives an animal model of depression from a clinical model of human depression for the purpose of biochemical research on depressive behavior. Animal models allow for controlled experimentation concerning neuronal mechanisms and new antidepressant therapies. The polydimensional model of depression (F. Alonso-Fernández, 1986) is suggested as being able to separate apparent neurobiological contradictions according to differential measures along 4 separate dimensions. Experimental confirmation of the model would provide criteria for the selection, design, and development of antidepressant drugs.

1820. Post, Robert M. (1990). **Sensitization and kindling perspectives for the course of affective illness: Toward a new treatment with the anticonvulsant carbamazepine.** ***Pharmacopsychiatry,*** 23(1), 3–17.
Discusses 2 preclinical models (behavioral sensitization to psychomotor stimulants and electrophysiological kindling) for conceptualizing mechanisms underlying the progressive and evolving aspects of manic-depressive illness and examines the acute and long-term effectiveness of the anticonvulsant carbamazepine. Peripheral-type and α-2 adrenergic benzodiazepine receptors and stabilization of type-2 sodium channels may be involved in the anticonvulsant effects of carbamazepine. Gamma-aminobutyric $acid_B$ ($GABA_B$) mechanisms are thought to be related to the antinociceptive, but not the anticonvulsant or psychotropic, effects of carbamazepine. Many neurotransmitters remain candidates for the psychotropic effects. An animal model requiring chronic administration of carbamazepine to show efficacy is reported. (German abstract).

1821. Rudin, Donald O. (1983). **The three pellagras.** ***Journal of Orthomolecular Psychiatry,*** 12(2), 91–110.
Indicates that there are substrate, vitamin, and modulator deficiency types of pellagraform disease that are vitamin-sensitive, vitamin-resistant, and vitamin-refractory. The history and chemistry of pellagra, animal models of the disease, and therapeutic considerations in humans are discussed. The possible role of substrate pellagra in mental disorders such as schizophrenia, manic-depressive psychosis, and phobic neurosis is considered, and therapeutic results obtained through treatment with essential fatty acids are outlined.

1822. Rupniak, N. M.; Jenner, P. & Marsden, C. D. (1986). **Acute dystonia induced by neuroleptic drugs.** ***Psychopharmacology,*** 88(4), 403–419.
Indicates that about 2.5% of patients treated with neuroleptic drugs develop acute dystonia within 48 hrs of commencing therapy. The symptoms remit on drug withdrawal or following anticholinergic therapy. Acute dystonia can also be reliably induced in many primate species by neuroleptic treatment with comparable time course, symptomatology, and pharmacological characteristics to those observed in man. It is suggested that some unknown, possibly species-specific or even genetic factors may determine an individual's susceptibility to develop dystonia. Use of a rodent model of dystonia might enable more detailed analysis of biochemical correlates of dystonic behavior.

1823. Siegel, Shepard & Ellsworth, Delbert W. (1986). **Pavlovian conditioning and death from apparent overdose of medically prescribed morphine: A case report.** ***Bulletin of the Psychonomic Society,*** 24(4), 278–280.
Suggests that a Pavlovian conditioning model of tolerance emphasizes the contribution of an association between predrug cues and the systemic effect of the drug to the display of tolerance. Consistent with the model, results of a recent animal experiment and retrospective reports by human overdose victims are discussed that indicate that the pernicious effects of heroin are influenced by environmental cues associated with the systemic effects of the drug. A case report describing death from apparent morphine overdose in a man receiving the drug for medical purposes is presented. The circumstances of this death are readily interpretable by the conditioning analysis of tolerance.

1824. Simpson, Dale M. & Foster, Dodi. (1986). **Improvement in organically disturbed behavior with trazodone treatment.** ***Journal of Clinical Psychiatry,*** 47(4), 191–193.
Four elderly male patients (aged 62–72 yrs) with severe organic brain syndromes and disturbed behavior were tested with trazodone after therapy with neuroleptics had been ineffective. The behavior of all 4 Ss improved substantially. In each case, the clinical presentation did not suggest a depressive illness. The improvement in behavior may be related to the taming effect of trazodone in animal models of aggression. Trazodone may have value as an alternative to neuroleptics for the treatment of organically disturbed behavior.

1825. Söderpalm, Bo. (1987). **Pharmacology of the benzodiazepines; with special emphasis on alprazolam.** Symposium: Panic disorder (1986, Gothenburg, Sweden). ***Acta Psychiatrica Scandinavica,*** 76(335, Suppl), 39–46.

Outlines aspects of the pharmacokinetic and the pharmacodynamic properties of the benzodiazepines (BDZs) as well as their suggested mode of action. A new group of BDZs, the triazolo-BDZs, is described. One of these compounds, alprazolam (APZ), exhibits a different clinical profile as compared to traditional BDZs. Effective in the treatment of generalized anxiety, it has been proven effective also in the treatment of panic disorder. A commonly applied animal model (the Vogel conflict model) for the study of anxiety-related mechanisms is described, and original animal data from experiments aiming at elucidating the mechanism behind the antipanic effect of APZ are presented.

1826. Soroko, Francis E. & Maxwell, Robert A. (1983). **The pharmacologic basis for therapeutic interest in bupropion.** ***Journal of Clinical Psychiatry,*** 44(5, Sect 2), 67–73.
In a study of its primary pharmacologic properties, bupropion (BP), a compound chemically dissimilar to tricyclic antidepressants and MAO inhibitors, was found to be active in animal models that are predictive of antidepressant activity in humans. BP was also found to be pharmacologically and biochemically distinct from tricyclics and MAO inhibitors. Furthermore, it lacked anticholinergic activity, was not sympathomimetic, and was at least 10-fold weaker as a cardiac depressant than the tricyclic antidepressants. It is concluded that BP would be better tolerated and safer in humans than standard therapies and that its pharmacologic and biochemical profile holds out the possibility of novel antidepressant actions.

1827. Stanley, Michael et al. (1980). **Metoclopramide: Antipsychotic efficacy of a drug lacking potency in receptor models.** ***Psychopharmacology,*** 71(3), 219–225.
Metoclopramide is a substituted benzamide derivative, structurally similar to procainanide and sulpiride. In behavioral, biochemical, and neuroendocrine tests it has neuroleptic dopamine (DA) antagonist properties, but it lacks potency in currently used DA receptor models. In previous clinical studies using low doses or dubious measures, it was not efficacious as an antipsychotic. The present findings from 8 24–56 yr old male psychiatric patients suggest that the drug has a clinical profile similar to known neuroleptics when used in a dose range predicted from animal models (i.e., 40–1,000 mg/day for 24 days). Findings question the validity and universality of several predictive models and of hypotheses purporting to explain molecular mechanisms of action of neuroleptic agents. The drug's inactivity in receptor models suggests that an as yet unidentified DA receptor subpopulation may be the mediator of many DA-dependent neurobiologic phenomena.

1828. Taylor, Duncan P. (1990). **Serotonin agents in anxiety.** ***Annals of the New York Academy of Sciences,*** 600, 545–557.
The side effects and unwanted or unnecessary ancillary pharmacological properties of benzodiazepine anxiolytic drugs resulted in a search for new agents with improved profiles of activity. Buspirone was one of the first novel drugs to emerge from this search. Investigations into its mechanism of action revealed a key role for serotonin in the pharmacotherapy of anxiety. A variety of serotonergic agents are now in preclinical and clinical development as anxiolytics, including 5-hydroxytryptamine$_{1A}$ (5-HT$_{1A}$) partial agonists, 5-HT$_2$ antagonists, and 5-HT$_3$ antagonists. The clinical efficacy of these drugs will prompt the development of new animal models of psychopathology.

1829. Taylor, Duncan P. et al. (1982). **Dopamine and antianxiety activity.** ***Pharmacology, Biochemistry & Behavior,*** 17(1), 25–35.
Clinical trials have indicated that buspirone is effective in the treatment of anxiety with efficacy and dosage comparable to diazepam. Buspirone appears to only interact with the dopaminergic system with reasonable potency and exhibits properties of both a dopamine (DA) agonist and antagonist. This suggests that DA is implicated in the etiology and expression of anxiety. A discussion of this implication is presented with a review of the clinical efficacy of nonbenzodiazepine drugs, especially DA agonists and antagonists, in the management of anxiety. In addition, neuropharmacological studies that have investigated the role of DA in animal models of anxiety are considered. The multiplicity of DA receptors and their regional localization in the brain are considered in the formulation of a hypothesis that features a role for dopaminergic agents in the pharmacotherapy of anxiety.

1830. Van der Kolk, Bessel A. (1987). **The drug treatment of post-traumatic stress disorder.** Special Issue: Drug treatment of anxiety disorders. ***Journal of Affective Disorders,*** 13(2), 203–213.
Examines etiological models of posttraumatic stress disorder (PTSD), physiological sequelae of traumatization, and PTSD and the endogenous opioid system. The psychopharmacological treatment of PTSD is discussed in terms of medications that affect the noradrenergic system, benzodiazepines, lithium and carbamazepine, and antidepressant drugs. It is contended that while many psychotropic agents have been proposed for the treatment of various symptoms of PTSD, controlled studies are needed to clarify the relative merits of particular psychotropic agents on the various PTSD symptoms. Impressions in open studies have utilized global ratings, rather than studied the effects on specific symptoms. The animal model of inescapable shock provides a good model for understanding the biological alterations produced by overwhelming trauma and suggests a variety of pharmacological treatment interventions.

1831. Van der Staay, Frans J. & Schuurman, Teunis. (1991). **Pharmacologic influences in the elderly.** ***Psycholoog,*** 26(11), 495–498.
Discusses the various strategies used to treat cognitive impairment connected with dementia (e.g., Alzheimer's disease) in the elderly, with emphasis on 2 therapies: (1) cholinergic precursor therapy, and (2) blocking of calcium ion channels. The importance of clinical testing based on adequate animal models is considered. (English abstract).

1832. Wetzel, Hermann; Wiedemann, Klaus; Holsboer, Florian & Benkert, Otto. (1991). **Savoxepine: Invalidation of an "atypical" neuroleptic response pattern predicted by animal models in an open clinical trial with schizophrenic patients.** ***Psychopharmacology,*** 103(2), 280–283.
12 schizophrenic inpatients (aged 20–67 yrs) were treated with either a stable dose of 0.5 mg per day of savoxepine or higher doses of up to 20 mg/day. Mean total Brief Psychiatric Rating Scale scores and subscores demonstrated a moderate improvement of mainly positive schizophrenic symptoms. Savoxepine in a broad dose range produced typical untoward extrapyramidal symptoms in the majority of Ss.

Savoxepine may not possess the expected "atypical" neuroleptic response pattern, and the predictive validity of the animal models in question used to separate antipsychotic effects from extrapyramidal reactions may be ill-founded.

Specialized Interventions

1833. Booth, David A. (1988). **Mechanisms from models: Actual effects from real life: The zero-calorie drink-break option.** Meetings of the North American Association for the Study of Obesity Workshop: The effect of sweeteners on food intake (1987, Boston, Massachusetts). ***Appetite,*** 11(Suppl 1), 94–102.
Discusses animal and human laboratory models showing that the suppression of appetite by a modest amount of readily assimilable energy, such as a caloric sweetener, is not likely to last longer than an hour. The transience of their satiating effect constitutes a mechanism whereby the sugars, starch, alcohol, and fats in drinks and the snack foods eaten with them could add to energy intake that is subsequently uncompensated and so contributes to weight gain. It is suggested that the effectiveness of a zero-calorie drink-break option can be tested by correlating separately reported real-life eating habits and weight changes across people whose circumstances are similar.

1834. Lerer, Bernard & Belmaker, Robert H. (1982). **Receptors and the mechanism of action of ECT. *Biological Psychiatry,*** 17(4), 497–511.
Reviews research in which the animal models used have employed repeated ECS and have led to persistent changes in dopamine, serotonergic, or noradrenergic receptors. Neuroendocrine studies in humans do not thus far support the animal data suggesting an increased sensitivity in these receptor systems following repeated ECS. Future research should focus on replicating animal data in humans and on comparing the physiological effects of ECT and lithium in the treatment of mania, depression, and affective disorders.

Drug & Alcohol Rehabilitation

1835. Kleber, Herbert D. & Gawin, Frank H. (1987). **"The physiology of cocaine craving and 'crashing'": In reply. *Archives of General Psychiatry,*** 44(3), 299–300.
Replies to comments by C. A. Dackis and M. S. Gold (1987) concerning the present authors' (1986) description of the cocaine "crashing" syndrome. It is concluded that although dopaminergic hypofunction after long-term stimulant use is firmly established in animal models, it is unlikely that a single adaptation, such as dopamine depletion, explains all of the brain's adaptive responses to a powerful stimulant.

1836. Naranjo, C. A.; Cappell, Howard & Sellers, E. M. (1981). **Pharmacological control of alcohol consumption: Tactics for the identification and testing of new drugs. *Addictive Behaviors,*** 6(3), 261–269.
Proposed that potentially useful agents can be detected if new drugs are systematically tested for effects on alcohol consumption in appropriate animal models. However, the efficacy and safety of these drugs must be assessed according to the classic 4 phases of human pharmacology, and claims of efficacy must be supported by randomized controlled trials. It is concluded that a better knowledge of the neurochemical basis of alcohol consumption should lead to the identification of more specific and active pharmacologic treatments.

1837. Sinclair, J. D. (1987). **The feasibility of effective psychopharmacological treatments for alcoholism. *British Journal of Addiction,*** 82(11), 1213–1223.
Discusses the effects of lithium, anxiolytics, dopaminergic compounds, 5-hydroxytryptamine (5-HT) uptake inhibitors, opiates, opiate antagonists, and alcohol-sensitizing drugs on the alcohol consumption of human alcoholics and laboratory animals. There is increasing evidence that each of these classes of drugs might be beneficial in reducing alcohol abuse, including recent controlled studies on alcoholics with lithium, 5-HT uptake blockers, and disulfiram. The fact that some drugs acting on the central nervous system (CNS) are able to promote abstinence supports the feasibility of psychopharmacological treatment for alcoholism. Since the known drugs that appear beneficial against human alcohol abuse also suppress voluntary alcohol selection by experimental animals, it is suggested that studies with animals could be used for screening other potentially effective drugs.

Section II. Citations to the Dissertation Literature on Animal Models of Human Pathology

This section contains selected citations to the dissertation literature, focusing on the behavioral and psychological aspects of animal models of human pathology. All citation references are covered in *Dissertation Abstracts International* (formerly *Dissertation Abstracts*). References were retrieved from the PsycINFO database and are sorted alphabetically by first author within the major and minor classification categories used by *Psychological Abstracts* and the PsycINFO database.

Animal Experimental & Comparative Psychology

1838. Adams, Nelson. (1983). **Development of social dominance in domestic Norway rats: Effects of captivity and social experience.** 43(9-B), 3068–3069.

1839. Kuntz, Judith A. (1978). **Mother–infant interactions in the Gunn rat: A model for the study of neonatal at-risk conditions.** 38(7-B), 3459.

1840. Lower, Jerold S. (1967). **Approach-avoidance conflict as a determinant of peptic ulceration in the rat.** 27(10-B), 3691.

1841. Maier, Steven F. (1969). **Failure to escape traumatic electric shock: Incompatible skeletal-motor responses or learned helplessness?** 30(3-B), 1383.

1842. Schneider, Mary L. (1988). **A rhesus monkey model of human infant individual differences.** 48(9-B), 2804.

1843. Seligman, Martin E. (1967). **The disruptive effects of unpredictable shock.** 28(4-B), 1714.

1844. Stone, William S. (1986). **Circadian behaviors and memory in animal models of aging and Alzheimer's disease.** 47(4-B), 1778.

1845. Wagner, Henry R. (1976). **An evaluation of learned helplessness as a model of depression in the rat.** 37(4-B), 1950–1951.

1846. Weiss, Jay M. (1968). **Effects of coping responses on stress.** 29(1-B), 400–401.

Learning & Motivation

1847. Baker, Stanley R. (1976). **A subhuman animal analogue of implosive therapy: Facilitated extinction of avoidance responding and learned fear following response prevention.** 37(3-B), 1424.

1848. Chatman, James E. (1979). **Learned helplessness in rats: Associative interference or learned inactivity.** 40(6-B), 2871–2872.

1849. Eldred, Nancy L. (1982). **Effects of preexposure to shock on autoshaping.** 43(5-B), 1642.

1850. Fallon, Joseph H. (1980). **The effects of inescapable noise and inescapable shock on subsequent response acquisition: A test of the helplessness hypothesis.** 40(7-B), 3465.

1851. Follick, Michael J. (1978). **Aggression during learned helplessness pretraining as a coping response.** 39(1-B), 377–378.

1852. Gade, Paul A. (1974). **The effects of varying the predictability, controllability and intensity of electric shocks on learned helplessness.** 35(5-B), 2462.

1853. Grant, Kathleen A. (1984). **An experimental analysis of oral ethanol self administration in the free feeding rat.** 45(2-B), 711.

1854. Hunsberger, Jack C. (1984). **The effects of stimulus generalization on escape in the rat.** 44(8-B), 2582.

1855. Ivnik, Robert J. (1976). **A test of the generality of learned helplessness in rats.** 36(12-B, Pt 1), 6415–6416.

1856. Jackson, Raymond L. (1980). **Learned helplessness, inactivity and associative deficits in rats.** 40(11-B), 5439–5440.

1857. Lowry, Michael A. (1985). **An analysis of stimulus functioning during acquisition of behavior chains: A comparison of forward and backward chaining.** 45(9-B), 3099.

1858. Martasian, Paula J. (1990). **Retention of distributed versus massed response prevention treatments in an animal model.** 51(6-B), 3167.

1859. Maxwell, Marianne E. (1978). **Changes in social dominance as a function of learned helplessness.** 38(9-B), 4470.

1860. Mineka, Susan. (1975). **The effects of irrelevant flooding on the extinction of avoidance responses.** 36(1-B), 477.

1861. Palese, Robert P. (1977). **Acquisition of a discrete-trial-operant appetitive response in rats after exposure to response-independent food.** 38(5-B), 2403–2404.

1862. Roca, Carmen S. (1977). **Stimulus generalization and immunization generalization of learned helplessness in the rat.** 37(9-B), 4701.

1863. Sutton, Betty R. (1979). **Shuttle-box escape-avoidance deficits and changes in monoamine levels following inescapable foot shock.** 39(8-B), 4092.

1864. Welker, Robert L. (1975). **Learned laziness: A replication and extension.** 35(8-B), 4236–4237.

Social & Instinctive Behavior

1865. Hamm, Thomas E. (1980). **A nonhuman primate model of the effect of sex and social interaction on coronary artery atherosclerosis.** 41(4-B), 1319.

Physiological Psychology & Neuroscience

1866. Anthony, Amelia H. (1990). **Somatic and behavioral characterization of a rodent model of congenital hypothyroidism.** 51(4-B), 2089.

1867. Boelkins, Richard C. (1973). **Mother–infant separation: Behavioral analysis of an animal model of depression.** 33(8-B), 3971–3972.

1868. Briner, Wayne E. (1988). **The ultrastructural correlates of aging and presbycusis in the C57 mouse anteroventral cochlear nucleus.** 48(10-B), 3144.

1869. Drugan, Robert C. (1985). **Anxiety and the benzodiazepine/GABA-chloride ionophore receptor complex: Implications for learned helplessness, coping and behavioral depression.** 45(9-A), 2802.

1870. Glavin, Gary B. (1976). **The pathogenesis and control of experimental gastric ulceration in the rat.** 37(6-B), 3110.

1871. Guile, Michael N. (1982). **Nociception and gastric lesions in reaction to signaled versus unsignaled aversive events.** 43(2-B), 550.

1872. Gurley, Kelly R. (1988). **Activity-stress ulcers: Effects of varied feeding regimes, a tryptophan-supplemented diet, and an extended habituation period to the activity wheel.** 49(5-B), 1984.

1873. Haenlein, Marianne. (1985). **Vigilance performance in rats: An animal model relevant to attention deficit disorder.** 45(12-B, Pt 1), 3987–3988.

1874. Henault, Mark A. (1986). **Nocturnal locomotor abnormalities in an animal model of Huntington's disease: The effects of homotopic transplantation of embryonic striatal tissue.** 47(5-B), 2218.

1875. Knoth, Russell L. (1989). **Multisensory discrimination and representational memory deficits in post-thiamine deficient rats: Testing an animal model of human Wernicke-Korsakoff's disease.** 50(4-B), 1681.

1876. Marais, James. (1982). **Ultrastructural and biochemical changes associated with the development of acute stress ulcers.** 43(5-B), 1329.

1877. McCaughran, James A. (1987). **Hyperthermia-induced convulsions in the developing rat: An animal model of human infantile febrile convulsions.** 48(1-B), 298.

1878. Meyers, David P. (1990). **Experiments and observations on the behaviour and brain of the aging female mouse with special reference to Dementia of the Alzheimer Type.** 51(2-B), 600.

1879. Parker, Vivienne M. (1989). **The socially isolated rat as a model of anxiety.** 49(10-B), 4597–4598.

1880. Pass, Harold L. (1974). **The effect of chlorpromazine and stimulation of the mesencephalic reticular formation on central arousal and on the visual evoked response.** 35(6-B), 2997.

1881. Shull, Ronald N. (1985). **Analysis of a possible animal model of the hyperactive syndrome involving damage to the hippocampal system.** 45(8-B), 2730.

1882. Squire, Jonathan M. (1986). **Voluntary exercise, chronic stress, and blood pressure and heart rate responses in the borderline hypertensive rat.** 47(3-B), 1322.

1883. Stewart, Karen T. (1988). **Circadian activity rhythms in an animal model of depression.** 48(8-B), 2495.

1884. Strupp, Barbara J. (1982). **Malnutrition and animal models of cognitive development.** 42(12-B, Pt 1), 4965.

1885. Visintainer, Madelon A. (1983). **Learned helplessness, stress, and tumor rejection.** 43(12-B), 4165.

1886. Warren, Donald A. (1987). **Learned helplessness, immunization, and contextual fear.** 47(8-B), 3575.

1887. Witt, Ellen D. (1980). **Neurological, neuropathological, and behavioral effects of thiamine deficiency in the rhesus monkey (*Macaca mulatta*).** 40(9-B), 4544–4545.

Genetics

1888. Curley, Michael D. (1978). **The gerbil as a model of human grand mal epilepsy.** 38(7-B), 3430.

Neuropsychology & Neurology

1889. Bharucha, Vandana A. (1990). **Focal cerebral ischemia in the rat: Membrane failure and behavioral deficits.** 51(3-B), 1539.

1890. Blau, Alan J. (1989). **A comparison of the effects of dorsal cortex and nucleus basalis lesions on the acquisition and reversal of pattern discriminations by turtles (*Chrysemys picta*): An animal model of Alzheimer's disease.** 49(9-B), 3994.

1891. Gorman, Linda K. (1990). **Behavioral characterization of memory deficits following traumatic brain injury.** 51(3-B), 1135.

1892. Hall, Robert S. (1990). **MRI, brain iron and experimental Parkinson's disease.** 50(9-B), 4270.

1893. Roland, Barbara L. (1991). **Gastric mucosal erosions induced by lateral hypothalamic damage: Neuronal and dopaminergic mechanisms.** 52(1-B), 563–564.

1894. Wieland, Douglas S. (1988). **A neuropharmacological analysis of learned helplessness in rat.** 48(12-B, Pt 1), 3722.

1895. Wozniak, David F. (1985). **The effect of brain lesions on gastric erosions in the rat.** 45(12-B, Pt 1), 3990.

Electrophysiology

1896. Brett, Claude W. (1977). **The generality of learned helplessness theory: Effect of electroconvulsive shock.** 38(4-B), 1871.

1897. Edmiston, Marilyn. (1975). **Conflict and the stress response in the laboratory rat.** 35(8-B), 4247.

1898. Terzano, Karen E. (1990). **The effect of kindling on juvenile rats and their acquisition of a passive avoidance task.** 51(3-B), 1542.

1899. Wepman, Barry J. (1974). **The effect of early shock with or without a warning signal on adult behavior and susceptibility to stress-induced ulcers in the rat.** 34(8-B), 4103.

Physiological Processes

1900. Butt, Gaye E. (1978). **Conditions affecting stress ulceration in ulcer-susceptible rats.** 38(10-B), 5055.

1901. Greenberg, Danielle. (1984). **Early separation and the development of impaired thermoregulation in rats: A risk factor in gastric ulcer susceptibility.** 45(1-B), 394.

1902. Gruber, Barry L. (1982). **Development of an animal model to study mechanisms mediating biofeedback learning.** 42(8-B), 3485.

1903. Kirby, Debra A. (1981). **Manipulating renal responses to noxious stimuli in the cynomolgus monkey: A renal/behavioral model of essential hypertension.** 42(5-B), 2119.

1904. McGinnis, Ralph. (1989). **Investigations of the brain-pituitary-adrenal axis in the hypercorticosteronemic, genetically obese (ob/ob) mouse, an animal model of Cushing's disease.** 50(4-B), 1684.

1905. West, David B. (1984). **Abnormal perinatal and infant nutrition as a cause of obesity: Development of an animal model.** 45(2-B), 517.

Psychophysiology

1906. Johnson, Elizabeth A. (1987). **Learned helplessness, stress-induced analgesia, and the immune response.** 47(9-B), 3958.

Psychopharmacology

1907. Abbott, Frances V. (1981). **Studies of morphine analgesia in an animal model of tonic pain.** 42(4-B), 1655.

1908. Alley, Michael C. (1978). **The influence of trace metals upon seizure activity in primate and rodent models of epilepsy: A pharmacological evaluation.** 38(10-B), 4755.

1909. Barcus, Robert A. (1979). **Food additives and hyperactivity in dogs: An animal model of the hyperactive child syndrome.** 39(10-B), 5054.

1910. Barnes, Janine M. (1990). **Neurochemical assessments of the actions of novel agents having nootropic and anxiolytic potential: The measurement in brain areas of *in vitro* neurotransmitter release, enzyme activity, and binding of a radioligand, to assess activity of agents improving cognition or reducing anxiety in animal models.** 50(10-B), 4463.

1911. Builione, Robert S. (1989). **The behavioral and genetic determinants of anxiety as measured by the effects of anti-anxiety agents in mice.** 50(5-B), 2147.

1912. Critchley, Martyn A. (1989). **Drug action on anxiety models with special reference to serotonergic mechanisms.** 49(8-B), 3133.

1913. Eison, Michael S. (1980). **The regional distribution of amphetamine and its effects upon local glucose utilization in discrete areas of rat brain during continuous amphetamine administration.** 41(1-B), 398–399.

1914. Fernandez, Kathleen. (1982). **Assessment of maternal–infant interactions in developing a rodent model of the Fetal Alcohol Syndrome.** 43(4-B), 1296–1297.

1915. Finlay, Janet M. (1990). **Intravenous self-administration of midazolam in the rat: Behavioral and neurochemical characterization.** 51(3-B), 1133–1134.

1916. Fryer, Katharine H. (1986). **Exploration of the rat model of the fetal alcohol syndrome.** 47(4-B), 1761.

1917. Gill, Kathryn J. (1990). **A critical evaluation of the use of animal models in alcohol research: An examination of voluntary alcohol consumption in rodents.** 50(12-B, Pt 1), 5921–5922.

1918. Hoffman, Laura J. (1986). **Clonidine withdrawal: A procedure to create long-lasting effects in a new animal model of depression.** 47(2-B), 789.

1919. King, Alan G. (1980). **A pharmacological investigation of an animal model of the petit mal seizure.** 41(6-B), 2141.

1920. Kochhar, Abha. (1986). **An animal model of ethanol-induced locomotor stimulation: Effects on the central cholinergic system.** 46(10-B), 3403.

1921. Koenigshofer, Kenneth A. (1979). **Dopaminergic and cholinergic factors controlling amphetamine aversiveness in the conditioned taste aversion experiment: An animal model for the detection of neuroleptic potency and extrapyramidal disturbance?** 40(2-B), 970.

1922. McLean, Maria S. (1987). **Neurobehavioral consequences of hyperphenylalaninemia in neonatal rats.** 48(1-B), 296.

1923. McMaster, Suzanne B. (1984). **An animal model for the study of exercise-induced changes in CNS drug sensitivity.** 45(2-B), 716.

1924. Miller, William C. (1989). **Examination of the motor abnormalities and changes in neuronal activity in the globus pallidus in the primate MPTP model of Parkinsonism.** 50(1-B), 73.

1925. Morrow, Nancy S. (1990). **Investigation of running and gastric ulceration in the rat during activity-stress.** 50(9-B), 4268.

1926. Patton, Jim H. (1979). **The behavioral and physiological effects of administering lead to neonatal rats: A test of the model of childhood hyperactivity.** 39(11-B), 5635.

1927. Phillips, Nona K. (1984). **A gestational monkey model: Effects of phenytoin vs. seizures on neonatal outcome.** 45(5-B), 1602–1603.

1928. Pignatiello, Michael F. (1989). **An investigation of the antidepressant and opiate antagonistic effects of MIF-1 in the unpredictable chronic stress animal model of depression.** 49(11-B), 5058.

1929. Rupich, Reta C. (1989). **Behavioral and physiological effects of estradiol cyclodextrin: An animal model.** 49(11-B), 5058.

1930. Sconzert, Alan C. (1982). **The role of GABAergic and cholinergic neurotransmitter systems in the development and maintenance of the ECC syndrome, an animal model of movement disorder.** 42(10-B), 4050–4051.

1931. See, Ronald E. (1990). **The effects of chronic neuroleptic administration on motor behavior and brain receptors in an animal model of tardive dyskinesia.** 50(8-B), 3752–3753.

1932. Soblosky, Joseph S. (1983). **Effects of tricyclic antidepressants and serotonergic manipulations in an animal model of depression utilizing chronic stress.** 44(2-B), 639.

1933. Spanos, Linda J. (1990). **An investigation of the analgesic and tolerance-inducing properties of intrathecally administered narcotics in an animal model of chronic pain.** 51(5-B), 2301.

1934. Vogel, Richard A. (1976). **Effects of carbon monoxide, hypoxic hypoxia, and drugs on animal models of complex learned behavior.** 37(5-B), 2565.

1935. Wasserman, Stephanie J. (1990). **Haloperidol-induced temporal licking pattern: An animal model of tardive dyskinesia.** 51(4-B), 2106–2107.

Psychological & Physical Disorders

1936. Stark, Larry G. (1969). **A neuropharmacological comparison of three models of epilepsy including the baboon, Papio papio.** 29(11-B), 4306.

Psychological Disorders

1937. Gorkin, Brett D. (1979). **Use of a rat model to explore the etiology of Korsakoff's psychosis in the chronic alcoholic.** 40(3-B), 1413.

1938. Johnson, Joel O. (1986). **Biochemical correlates in an animal model of depression.** 47(2-B), 518.

Behavior Disorders & Antisocial Behavior

1939. Hunter, Robert E. (1988). **Uric acid and adenosine binding in young rats: A potential model of hyperactivity.** 49(5-B), 1988.

Health & Mental Health Treatment & Prevention

Clinical Psychopharmacology

1940. Jessberger, James A. (1985). **An investigation of antidepressant mechanisms in an animal model of depression.** 46(5-B), 1523.

Section III. Selected References to Books on Animal Models of Human Pathology

This section contains citations of authored and edited books on animal models of human pathology. References are sorted alphabetically by first author. Citations from edited books include chapter information for relevant chapters only. Book and chapter citations have been indexed using descriptors from the *Thesaurus of Psychological Index Terms* (6th Edition, 1991). The section is not all inclusive; coverage of books in *Psychological Abstracts* was suspended between 1980 and 1987 while a new database was being devised.

1941. Ader, Robert; Felten, David L. & Cohen, Nicholas (Eds.). (1991). **Psychoneuroimmunology (2nd ed.).** San Diego, CA: Academic Press, Inc. xxvii, 1218 pp. ISBN 0-12-043782-1 (hardcover).

1942. The influence of conditioning on immune responses -*Robert Ader and Nicholas Cohen.* (pp. 611-646).

1943. Behavioral adaptations in autoimmune disease-susceptible mice -*Robert Ader, Lee J. Grota, Jan A. Moynihan and Nicholas Cohen.* (pp. 685-708).

1944. Autoimmunity and cognitive decline in aging and Alzheimer's disease -*Michael J. Forster and Harbans Lal.* (pp. 709-748).

1945. Behavioral consequences of virus infection -*Linda S. Crnic.* (pp. 749-769).

1946. Stress-induced changes in immune function in animals: Hypothalamo-pituitary-adrenal influences -*Steven E. Keller, Steven J. Schleifer and Melissa K. Demetrikopoulos.* (pp. 771-787).

1947. Behavioral sequelae of autoimmune disease -*Randolph B. Schiffer and Steven A. Hoffman.* (pp. 1037-1066).

1948. Ader, Robert; Weiner, Herbert & Baum, Andrew (Eds.). (1988). **Experimental foundations of behavioral medicine: Conditioning approaches.** Hillsdale, NJ: Lawrence Erlbaum Associates, Inc. xii, 222 pp. ISBN 0-8058-0139-1 (hardcover).

1949. Learned aspects of cardiovascular regulation -*Philip M. McCabe, Neil Schneiderman, Ray W. Winters, Christopher G. Gentile [and] Alan H. Teich.* (pp. 1-23).

1950. Pavlovian conditioning of endocrine responses -*Mark E. Stanton and Seymour Levine.* (pp. 25-46).

1951. The placebo effect as a conditioned response -*Robert Ader.* (pp. 47-66).

1952. The treatment of scoliosis by continuous automated postural feedback -*Barry Dworkin and Susan Dworkin.* (pp. 67-86).

1953. Relaxation training in essential hypertension: Prospects and problems -*W. Stewart Agras.* (pp. 87-110).

1954. The synthesis of medical and behavioral sciences with respect to bronchial asthma -*Thomas L. Creer.* (pp. 111-158).

1955. Stress, behavior, and glucose control in diabetes mellitus -*Richard S. Surwit.* (pp. 159-173).

1956. A proposal for a curriculum in behavioral biology and medicine in medical schools -*Herbert Weiner.* (pp. 175-184).

1957. Training of family physicians in behavioral medicine -*Hiram B. Curry.* (pp. 185-193).

1958. Teaching behavioral concepts in cardiovascular disease with remarks on challenges to medical education -*Alvin P. Shapiro.* (pp. 195-207).

1959. Adler, Martin W. & Cowan, Alan (Eds.). (1990). **Testing and evaluation of drugs of abuse. Modern methods in pharmacology, Vol. 6.** New York, NY: John Wiley & Sons. xi, 297 pp. ISBN 0-471-56743-4 (hardcover).

1960. Assessment of physical dependence techniques for the evaluation of abused drugs -*Mario D. Aceto.* (pp. 67-79).

1961. Drugs of abuse and the fetus and neonate: Testing and evaluation in animals -*Ian S. Zagon and Patricia J. McLaughlin.* (pp. 241-254).

1962. Altman, Harvey J. (Ed.). (1987). **Alzheimer's disease: Problems, prospects, and perspectives.** New York, NY: Plenum Press. xiii, 397 pp. ISBN 0-306-42662-5 (hardcover).

1963. On possible relationships between Alzheimer's disease, age-related memory loss and the development of animal models -*R. T. Bartus and R. L. Dean.* (pp. 129-139).

1964. Serotonin, Alzheimer's disease and learning and memory in animals -*H. J. Normile and H. J. Altman.* (pp. 141-156).

1965. Neural implants: A strategy for the treatment of Alzheimer's disease -*D. M. Gash.* (pp. 165-170).

1966. Appels, A. (Ed.). (1991). **Behavioral observations in cardiovascular research.** Amsterdam, , Netherlands [US Location: Rockland, MA]: Swets & Zeitlinger. 130 pp. ISBN 90-265-1036-5 (hardcover).

1967. Coping strategies and cardiovascular risk: A study of rats and mice -*J. M. Koolhaas and B. Bohus.* (pp. 45-58).

1968. Archer, John & Browne, Kevin (Eds.). (1989). **Human aggression: Naturalistic approaches.** London, England [US Location: New York, NY]: Routledge. xvi, 284 pp. ISBN 0-415-03036-6 (hardcover); 0-415-03037-4 (paperback).

1969. Experimental animal models of aggression: What do they say about human behaviour? -*D. Caroline Blanchard and Robert J. Blanchard.* (pp. 94-121).

1970. Archer, Trevor & Nilsson, Lars-Göran (Eds.). (1989). **Aversion, avoidance, and anxiety: Perspectives on aversively motivated behavior.** Hillsdale, NJ: Lawrence Erlbaum Associates, Inc. xvi, 491 pp. ISBN 0-8058-0132-4 (hardcover).

1971. Aversively motivated behavior: Which are the perspectives? *-Lars-Göran Nilsson and Trevor Archer.* (pp. 437-465).

1972. Avoli, Massimo; Gloor, Pierre; Kostopoulos, George & Naquet, Robert (Eds.). (1990). **Generalized epilepsy: Neurobiological approaches.** Boston, MA: Birkhäuser. xv, 481 pp. ISBN 0-8176-3445-2 (hardcover); 3-7643-3445-2 (paperback).

1973. Historical introduction *-H. H. Jasper.* (pp. 1-15).

1974. The relationship between sleep spindles and spike-and-wave bursts in human epilepsy *-P. Kellaway, J. D. Frost, Jr. and J. W. Crawley.* (pp. 36-48).

1975. In vitro electrophysiology of a genetic model of generalized epilepsy *-G. Kostopoulos and C. Psarropoulou.* (pp. 137-157).

1976. Spontaneous spike-and-wave discharges in Wistar rats: A model of genetic generalized nonconvulsive epilepsy *-M. Vergnes, Ch. Marescaux, A. Depaulis, G. Micheletti, and J. -M. Warter.* (pp. 238-253).

1977. Role of dopamine in generalized photosensitive epilepsy: Electroencephalographic and biochemical aspects *-L. F. Quesney and T. A. Reader.* (pp. 298-313).

1978. Animal models of generalized convulsive seizures: Some neuroanatomical differentiation of seizure types *-K. Gale.* (pp. 329-343).

1979. Mechanisms underlying generalized tonic-clonic seizures in the rat: Functional significance of calcium ions *-E.-J. Speckmann, J. Walden, D. Bingmann, A. Lehmenkühler and U. Altrup.* (pp. 344-354).

1980. Ballenger, James C. (Ed.). (1990). **Neurobiology of panic disorder. Frontiers of clinical neuroscience, Vol. 8.** New York, NY: Wiley-Liss. xvi, 391 pp. ISBN 0-471-56210-6 (hardcover).

1981. Animal models of anxiety *-Susan R. B. Weiss and Thomas W. Uhde.* (pp. 3-27).

1982. Preclinical studies of the mechanisms of anxiety and its treatment *-Sandra E. File.* (pp. 31-48).

1983. Bateson, Peter P. G. & Klopfer, Peter H. (Eds.). (1989). **Perspectives in ethology, Vol. 8: Whither ethology?** New York, NY: Plenum Press. xiv, 278 pp. ISBN 0-306-42948-9 (hardcover).

1984. Animal psychology: The tyranny of anthropocentrism *-J. E. R. Staddon.* (pp. 123-135).

1985. Berg, Joseph M. (Ed.). (1986). **Science and service in mental retardation.** London, England [US Location: New York, NY]: Methuen. xix, 475 pp. ISBN 0-416-40650-5 (hardcover).

1986. Primary prevention of neural tube defects in an animal model, the curly-tail mouse *-M. J. Seller.* (pp. 271-276).

1987. Mental defect due to iodine deficiency: A major international public health problem that can be eradicated *-B. S. Hetzel.* (pp. 297-306).

1988. Berkley, Mark A. & Stebbins, William C. (Eds.). (1990). **Comparative perception, Vol. 1: Basic mechanisms. Wiley series in neuroscience, Vol. 2.** New York, NY: John Wiley & Sons. xiv, 527 pp. ISBN 0-471-63167-1 (hardcover, Vol. 1); 0-471-52428-X (hardcover, set).

1989. Vision following loss of cortical directional selectivity *-Tatiana Pasternak.* (pp. 407-428).

1990. Experimentally induced and naturally occurring monkey models of human amblyopia *-Ronald G. Boothe.* (pp. 461-486).

1991. Bernardi, Giorgio; Carpenter, Malcolm B.; Di Chiara, Gaetano; Morelli, Micaela & Stanzione, Paolo (Eds.). (1991). **The basal ganglia III. Advances in behavioral biology, Vol. 39.** New York, NY: Plenum Press. xviii, 775 pp. ISBN 0-306-43720-1 (hardcover).

1992. L-dopa-induced chorea and dystonia in MPTP-treated squirrel monkeys *-N. M. J. Rupniak, S. Boyce, M. J. Steventon and S. D. Iversen.* (pp. 533-539).

1993. A primate model of Huntington's disease: Unilateral striatal lesions and neural grafting in the baboon (papio papio) *-D. Riche, P. Hantraye, O. Isacson and M. Maziere.* (pp. 561-572).

1994. Dopaminergic dysfunctions in neonatal hypothyroidism *-A. Vaccari and Z. L. Rossetti.* (pp. 597-606).

1995. Advances in the understanding of neural mechanisms in movement disorders *-I. J. Mitchell, J. M. Brotchie, W. C. Graham, R. D. Page, R. G. Robertson, M. A. Sambrook and A. R. Crossman.* (pp. 607-616).

1996. Bevan, Paul; Cools, Alexander R. & Archer, Trevor (Eds.). (1989). **Behavioural pharmacology of 5-HT.** Hillsdale, NJ: Lawrence Erlbaum Associates, Inc. 519 pp. ISBN 0-8058-0135-9 (hardcover).

1997. Mechanism of action of 8-OH-DPAT on a rat model for human depression *-G. A. Kennet and G. Curzon.* (pp. 225-229).

1998. Reversal of helpless behaviour in rats by serotonin uptake inhibitors *-P. Martin, A. M. Laporte, P. Soubrie, S. El Mestikawy and M. Hamon.* (pp. 231-233).

1999. 5-HT[subscript]2 antagonists increase tactile startle habituation in an animal model of a habituation deficit in schizophrenia *-M. A. Geyer.* (pp. 243-246).

2000. The relationship between various animal models of anxiety, fear-related psychiatric symptoms and response to serotonergic drugs *-C. L. Broekkamp and F. Jenck.* (pp. 321-335).

2001. Ultrasonic vocalizations by rat pups as an animal model for anxiolytic activity: Effects of serotonergic drugs *-J. Mos and B. Olivier.* (pp. 361-366).

2002. Antianxiety effect of various putative 5-HT[subscript]1 receptor agonists on the conditioned defensive burying paradigm *-A. Fernandez-Guasti and E. Hong.* (pp. 377-382).

2003. Behavioural effects of 5-HT[subscript]3 antagonists in animal models for aggression, anxiety and psychosis *-J. Mos, J. A. M. van der Heyden and B. Olivier.* (pp. 389-395).

2004. Birren, James E. & Schaie, K. Warner (Eds.). (1985). **Handbook of the psychology of aging (2nd ed.). The handbooks of aging.** New York, NY: Van Nostrand Reinhold Co, Inc. xvii, 931 pp. ISBN 0-442-21401-4 (hardcover).

2005. The neural basis of aging *-William Bondareff.* (pp. 95-112).

2006. Behavioral genetics *-Gerald McClearn and Terryl T. Foch.* (pp. 113-143).

2007. Birren, James E. & Schaie, K. Warner (Eds.). (1990). **Handbook of the psychology of aging (3rd ed.). The handbooks of aging.** San Diego, CA: Academic Press, Inc. xviii, 552 pp. ISBN 0-12-101280-8 (hardcover).

2008. Mammalian models of learning, memory, and aging *-Diana S. Woodruff-Pak.* (pp. 234-257).

2009. Bloom-Feshbach, Jonathan & Bloom-Feshbach, Sally (1987). **The psychology of separation and loss: Perspectives on development, life transitions, and clinical practice. The Jossey-Bass social and behavioral science series.** San Francisco, CA: Jossey-Bass Inc, Publishers. xxxiii, 587 pp. ISBN 1-55542-040-0 (hardcover).

2010. Maternal loss in nonhuman primates: Implications for human development *-Edward H. Plimpton and Leonard A. Rosenbaum.* (pp. 63-86).

2011. Bolwig, Tom G. & Trimble, Michael R. (Eds.). (1989). **The clinical relevance of kindling.** Chichester, England [US Location: New York, NY]: John Wiley & Sons. xi, 302 pp. ISBN 0-471-92449-0 (hardcover).

2012. Chemical kindling *-Claude G. Wasterlain, Anne M. Morin, Denson G. Fujikawa and Jeff M. Bronstein.* (pp. 35-53).

2013. Kindling and antiepileptic drugs *-M. Schmutz and K. Klebs.* (pp. 55-68).

2014. Kindling and neurotransmitter systems: Seizure suppression by intracerebral grafting of fetal neurons *-David I. Barry, Jørn Kragh and Benedikte Wanscher.* (pp. 75-85).

2015. Kindling and memory *-Jerzy Majkowski.* (pp. 87-102).

2016. Kindling, anxiety and personality *-Robert E. Adamec.* (pp. 117-135).

2017. Kindling, psychopathology and cerebral mechanisms in ethanol withdrawal *-Ralf Hemmingsen.* (pp. 137-145).

2018. Bond, Nigel W. & Siddle, David A. T. (Eds.). (1989). **Psychobiology: Issues and applications.** Amsterdam, Netherlands: North-Holland. xv, 658 pp. ISBN 0-444-88509-9 (hardcover, set); 0-444-88524-2 (hardcover).

2019. Part VI: Addiction: Theory and therapy . (pp. 575-651).

2020. Bornstein, Marc H. (Ed.). (1989). **Maternal responsiveness: Characteristics and consequences. New directions for child development, No. 43: The Jossey-Bass social and behavioral sciences series.** San Francisco, CA: Jossey-Bass Inc, Publishers. 112 pp. ISBN 1-55542-864-9 (paperback).

2021. Maternal responsiveness in human and animal mothers *-Alison S. Fleming.* (pp. 31-47).

2022. Boulton, Alan A.; Baker, Glen B. & Martin-Iverson, Mathew T. (Eds.). (1991). **Animal models in psychiatry, II. Neuromethods, 19.** Clifton, NJ: Humana Press, Inc. xx, 386 pp. ISBN 0-89603-198-5 (hardcover, Vol. 1); 0-89603-177-2 (hardcover, Vol. 2).

2023. Multisystem regulation of performance deficits induced by stressors: An animal model of depression *-Hymie Anisman, Steve Zalcman, Nola Shanks and Robert M. Zacharko.* (pp. 1-59).

2024. The olfactory bulbectomized rat as a model of major depressive disorder *-J. Steven Richardson.* (pp. 61-79).

2025. A cholinergic supersensitivity model of depression *-David H. Overstreet and David S. Janowsky.* (pp. 81-114).

2026. Methods of assessing circadian rhythms in animal models of affective disorders *-Naoto Yamada and Saburo Takahashi.* (pp. 115-146).

2027. Animal models of anxiety and the screening and development of novel anxiolytic drugs *-D. J. Sanger.* (pp. 147-198).

2028. Pharmacological evaluation of potential animal models for the study of antipanic and panicogenic treatment effects *-Randall L. Commissaris and David J. Fontana.* (pp. 199-232).

2029. Methods for drug studies in aggressive behavior *-Michael H. Sheard.* (pp. 233-248).

2030. Animal models of human aggression *-J. M. Koolhaas and B. Bohus.* (pp. 249-271).

2031. Animal models of mental retardation *-Trevor Archer, Ernest Hård and Stefan Hansen.* (pp. 273-314).

2032. Animal models of memory disorders *-David H. Overstreet and Roger W. Russell.* (pp. 315-368).

2033. Boulton, Alan A.; Baker, Glen B. & Martin-Iverson, Mathew Thomas (Eds.). (1991). **Animal models in psychiatry, I. Neuromethods, 18.** Clifton, NJ: Humana Press, Inc. xix, 411 pp. ISBN 0-89603-198-5 (hardcover).

2034. Methods for assessing the validity of animal models of human psychopathology *-Paul Willner.* (pp. 1-23).

2035. Animal models with parallels to schizophrenia - *Melvin Lyon*. (pp. 25-65).

2036. The hippocampal-lesion model of schizophrenia - *Nestor A. Schmajuk and Mabel Tyberg*. (pp. 67-102).

2037. An animal model of stimulant psychoses -*Mathew T. Martin-Iverson*. (pp. 103-149).

2038. Animal models of hallucinations: Continuous stimulants -*Gaylord D. Ellison*. (pp. 151-195).

2039. Animal models for the symptoms of mania - *Melvin Lyon*. (pp. 197-244).

2040. Animal models in tardive dyskinesia -*Helen Rosengarten, Jack W. Schweitzer and Arnold J. Friedhoff*. (pp. 245-266).

2041. Activity anorexia: An animal model and theory of human self-starvation -*W. David Pierce and W. Frank Epling*. (pp. 267-311).

2042. An animal model of attention deficit -*Joram Feldon and Ina Weiner*. (pp. 313-361).

2043. A computerized methodology for the study of neuroleptic-induced oral dyskinesias -*Gaylord D. Ellison and Ronald E. See*. (pp. 363-397).

2044. Bradley, Philip B. & Hirsch, Steven R. (Eds.). (1986). **The psychopharmacology and treatment of schizophrenia. British Association for Psychopharmacology monograph, No. 8 and Oxford medical publications' 2nd series.** Oxford, England [US Location: New York, NY]: Oxford University Press. 457 pp. ISBN 0-19-261260-3 (hardcover).

2045. Animal models of schizophrenia -*S. D. Iversen*. (pp. 71-102).

2046. Brain, Paul Frederic; Parmigiani, Stefano; Blanchard, Robert J. & Mainardi, Danilo (Eds.). (1990). **Fear and defence. Ettore Majorana international life sciences series, Vol. 8.** London, England [US Location: New York, NY]: Harwood Academic Publishers. x, 420 pp. ISBN 3-7186-5015-0 (hardcover).

2047. Anti-predator defense as models of animal fear and anxiety -*R. J. Blanchard and D. C. Blanchard*. (pp. 89-108).

2048. Fear pathways in the brain: Implications for a theory of the emotional brain -*J. E. LeDoux*. (pp. 163-177).

2049. Evolutionary psychiatric and biochemical aspects of emotional attachment -*D. Benton*. (pp. 289-308).

2050. Brauth, Steven E.; Hall, William S. & Dooling, Robert J. (Eds.). (1991). **Plasticity of development.** Cambridge, MA: MIT Press. viii, 182 pp. ISBN 0-262-02326-1 (hardcover).

2051. Uptight and laid-back monkeys: Individual differences in the response to social challanges -*Stephen J. Suomi*. (pp. 27-56).

2052. Bridge, T. Peter; Mirsky, Allan F. & Goodwin, Frederick K. (Eds.). (1988). **Psychological, neuropsychiatric, and substance abuse aspects of AIDS. Advances in biochemical psychopharmacology, Vol. 44.** New York, NY: Raven Press, Publishers. xvi, 261 pp. ISBN 0-88167-396-X (hardcover).

2053. Ethanol-associated immunosuppression -*Thomas R. Jerrells, Cheryl A. Marietta, George Bone, Forrest F. Weight and Michael J. Eckhardt*. (pp. 173-185).

2054. Brown, Serena-Lynn & van Praag, Herman M. (Eds.). (1991). **The role of serotonin in psychiatric disorders. Clinical and experimental psychiatry monograph, No. 4.** New York, NY: Brunner/Mazel, Inc. xiii, 349 pp. ISBN 0-87630-589-3 (hardcover).

2055. The role of serotonin in the regulation of anxiety -*Rene S. Kahn, Oren Kalus, Scott Wetzler and Herman M. van Praag*. (pp. 129-160).

2056. Capaldi, Elizabeth D. & Powley, Terry L. (Eds.). (1990). **Taste, experience, and feeding.** Washington, DC: American Psychological Association. xiii, 275 pp. ISBN 1-55798-091-8 (hardcover).

2057. Taste and food preferences in human obesity - *Adam Drewnowski*. (pp. 227-240).

2058. Cappell, H. D. & LeBlanc, A. E. (Eds.). (1975). **Biological and behavioural approaches to drug dependence.** Toronto, ON, Canada: Alcoholism & Drug Addiction Research Foundation. x, 179 pp.

2059. Carlson, John G. & Seifert, A. Ronald (Eds.). (1991). **International perspectives on self-regulation and health. Plenum series in behavioral psychophysiology and medicine.** New York, NY: Plenum Press. xviii, 291 pp. ISBN 0-306-43557-8 (hardcover).

2060. Opioid analgesia and descending systems of pain control -*G.F. Gebhart*. (pp. 207-221).

2061. Behavioral-anatomical studies of the central pathways subserving orofacial pain -*J. Peter Rosenfeld and James G. Broton*. (pp. 239-254).

2062. Cheren, Stanley (Ed.). (1989). **Psychosomatic medicine: Theory, physiology, and practice, Vols. 1 & 2. International Universities Press stress and health series, Monographs 1 & 2.** Madison, CT: International Universities Press, Inc. xi, 977 pp. ISBN 0-8236-5725-6 (hardcover, Vol. 1); 0-8236-5726-4 (hardcover, Vol. 2).

2063. The interaction between brain behavior and immunity -*Malcolm P. Rogers*. (pp. 279-330).

2064. Chrousos, George P.; Loriaux, D. Lynn & Gold, Philip W. (Eds.). (1988). **Mechanisms of physical and emotional stress. Advances in experimental medicine and biology, Vol. 245.** New York, NY: Plenum Press. xi, 530 pp. ISBN 0-306-43017-7 (hardcover).

2065. Implications of behavioral sensitization and kindling for stress-induced behavioral change -*R. M. Post, S. R. B. Weiss and A. Pert*. (pp. 441-463).

2066. Cicchetti, Dante & Toth, Sheree L. (Eds.). (1991). **Internalizing and externalizing expressions of dysfunction. Rochester Symposium on Developmental Psychopathology, Vol. 2.** Hillsdale, NJ: Lawrence Erlbaum Associates, Inc. v, 312 pp. ISBN 0-8058-0933-3 (hardcover).

2067. What can primate models of human developmental psychopathology model? *-Gene Sackett and Patricia Gould.* (pp. 265-292).

2068. Commons, Michael L.; Mazur, James E.; Nevin, John Anthony & Rachlin, Howard (Eds.). (1987). **Quantitative analyses of behavior, Vol. 5: The effect of delay and of intervening events on reinforcement value. Quantitative analyses of behavior.** Hillsdale, NJ: Lawrence Erlbaum Associates, Inc. xvii, 344 pp. ISBN 0-89859-800-1 (hardcover).

2069. Quantification of individual differences in self-control *-A. W. Logue, Monica L. Rodriguez, Telmo E. Peña-Correal and Benjamin C. Mauro.* (pp. 245-265).

2070. Effects of prior learning and current motivation on self-control *-Robert Eisenberger and Fred A. Masterson.* (pp. 267-282).

2071. Coyle, Joseph T. (Ed.). (1987). **Animal models of dementia: A synaptic neurochemical perspective. Neurology and neurobiology series, Vol. 33.** New York, NY: Alan R. Liss, Inc. xvi, 313 pp. ISBN 0-8451-2735-7 (hardcover).

2072. Parkinsonism: Insights from animal models utilizing neurotoxic agents *-Michael J. Zigmond, Edward M. Stricker and Theodore W. Berger.* (pp. 1-38).

2073. Excitotoxins and Huntington's disease *-Robert Schwarcz and Ira Shoulson.* (pp. 39-68).

2074. Animal models for age-related memory disturbances *-Raymond T. Bartus and Reginald L. Dean.* (pp. 69-79).

2075. Basal forebrain cholinergic neurons and Alzheimer's disease *-Gary L. Wenk and David S. Olton.* (pp. 81-101).

2076. Mood and cognitive disorders following stroke *-Rajesh M. Parikh and Robert G. Robinson.* (pp. 103-135).

2077. Microencephaly: Cortical hypoplasia induced by methylazoxymethanol *-Paul R. Sanberg, Timothy H. Moran and Joseph T. Coyle.* (pp. 253-278).

2078. Cristofalo, Vincent J. & Lawton, M. Powell (Eds.). (1991). **Annual review of gerontology and geriatrics, Vol. 10: Special focus on the biology of aging. Annual review of gerontology and geriatrics.** New York, NY: Springer Publishing Co, Inc. xiii, 225 pp. ISBN 0-8261-6492-7 (hardcover).

2079. Skeletal muscle weakness and fatigue in old age: Underlying mechanisms *-John A. Faulkner, Susan V. Brooks and Eileen Zerba.* (pp. 147-166).

2080. Dietary restriction as a probe of mechanisms of senescence *-Edward J. Masoro and Roger J. M. McCarter.* (pp. 183-197).

2081. Cullen, John H. (Ed.). (1974). **Experimental behaviour: A basis for the study of mental disturbance.** Dublin, Ireland: Irish U Press. 440 pp. ISBN 0-7165-2231-4 (hardcover).

2082. Effects of impoverished early environment: Maternal influences on adaptive function *-Seymour Levine.* (pp. 17-30).

2083. Syndromes resulting from object deprivation: Environmental restrictions on chickens *-Richard H. Porter.* (pp. 33-42).

2084. Syndromes resulting from object deprivation: Effects on development of rhesus monkeys *-Gene P. Sackett.* (pp. 43-69).

2085. Syndromes resulting from object deprivation: Analogues in human behaviour *-George D. Scott.* (pp. 70-80).

2086. Syndromes resulting from maternal deprivation: Maternal and peer affectional deprivation in primates *-Harry F. Harlow.* (pp. 85-98).

2087. Syndromes resulting from maternal deprivation: Analogues in human behaviour *-K. Feigenbaum.* (pp. 99-114).

2088. Factors in family relationships: Hormonal factors *-Richard Michael.* (pp. 126-130).

2089. Factors in family relationships: Adult male parental behaviour in feral- and isolation-reared monkeys (Macaca mulatta) *-William K. Redican, Jody Gomber and G. Mitchell.* (pp. 131-146).

2090. Factors in family relationships: Syndromes of paternal deprivation in man *-Henry B. Biller.* (pp. 147-171).

2091. Madness—An ethological perspective *-M. W. Fox.* (pp. 179-208).

2092. Syndromes resulting from social isolation: Lower animals *-Roger Ewbank.* (pp. 211-215).

2093. Syndromes resulting from social isolation: Primates *-G. Mitchell.* (pp. 216-223).

2094. Syndromes resulting from social isolation: Abnormalities of personal space in violent prisoners *-Augustus F. Kinzel.* (pp. 224-229).

2095. Syndromes resulting from social isolation: Need for experimental evaluation of the role of psychosocial stimuli in disease as a guide to rational health action *-Aubrey Kagan.* (pp. 230-239).

2096. Syndromes resulting from over-rigid control: Psychophysiological syndromes resulting from overly-rigid environmental control; concurrent and contingent animal models *-Joseph V. Brady.* (pp. 242-261).

2097. Syndromes resulting from over-rigid control: Man—Family *-Derek Russell Davis.* (pp. 262-268).

2098. Syndromes resulting from over-rigid control: Man–Society *-Eileen Kane.*

2099. Syndromes resulting from over-lax control: Causes and consequences of social contact in lower animals -*Bibb Latané*. (pp. 286-293).

2100. Syndromes resulting from over-lax control: Man—Family -*Jules H. Masserman*. (pp. 294-299).

2101. Syndromes resulting from over-lax control: Man–Society -*R. Carino*. (pp. 300-305).

2102. Behavioural engineering—Reconstruction of responses: The process and practice of participant modelling treatment -*Albert Bandura*. (pp. 312-333).

2103. Syndromes resulting from defects in communication: Lower animals -*Grant Noble*. (pp. 335-345).

2104. Syndromes resulting from defects in communication: Concordant preferences as a precondition for affective but not for symbolic communication (or how to do experimental anthropology) -*David Premack*. (pp. 346-361).

2105. Syndromes resulting from defects in communication: Man -*Colin Cherry*. (pp. 362-369).

2106. Syndromes resulting from defective satisfaction of physical appetites: Lower animals -*Alistair N. Worden and David E. Hathway*. (pp. 373-383).

2107. Syndromes resulting from defective satisfaction of physical appetites: Primates -*Alistair N. Worden, Ralph Heywood and David E. Hathway*. (pp. 384-394).

2108. Syndromes resulting from defective satisfaction of physical appetites: Death, survival and continuity of life -*Robert Jay Lifton*. (pp. 395-403).

2109. Syndromes resulting from defective satisfaction of physical appetites: Loss of sleep and mental illness -*Stuart A. Lewis*. (pp. 404-409).

2110. Cullen, John H. (Ed.). (1974). **Experimental behaviour: A basis for the study of mental disturbance.** New York, NY: John Wiley & Sons. 440 pp.

2111. Syndromes resulting from over-rigid control: I. Psychophysiological syndromes resulting from overly-rigid environmental control: Concurrent and contingent animal models -*Joseph V. Brady*. (pp. 242-261).

2112. Dachowski, Lawrence & Flaherty, Charles F. (Eds.). (1991). **Current topics in animal learning: Brain, emotion, and cognition.** Hillsdale, NJ: Lawrence Erlbaum Associates, Inc. xii, 437 pp. ISBN 0-8058-0441-2 (hardcover).

2113. Incentive contrast and selected animal models of anxiety -*Charles F. Flaherty*. (pp. 207-243).

2114. Animal models of Alzheimer's disease: Role of hippocampal cholinergic systems in working memory -*Thomas J. Walsh and James J. Chrobak*. (pp. 347-379).

2115. Davey, Graham (Ed.). (1987). **Cognitive processes and Pavlovian conditioning in humans.** Chichester, England [US Location: New York, NY]: John Wiley & Sons. x, 298 pp. ISBN 0-471-90791-X (hardcover).

2116. An integration of human and animal models of Pavlovian conditioning: Associations, cognitions, and attributions -*Graham C. L. Davey*. (pp. 83-114).

2117. Latent inhibition and human Pavlovian conditioning: Research and relevance -*David A. T. Siddle and Bob Remington*. (pp. 115-146).

2118. Denny, Maurice Ray (Ed.). (1991). **Fear, avoidance, and phobias: A fundamental analysis.** Hillsdale, NJ: Lawrence Erlbaum Associates, Inc. xii, 455 pp. ISBN 0-8058-0316-5 (hardcover); 0-8058-0317-3 (paperback).

2119. The psycho- and neurobiology of fear systems in the brain -*Jaak Panksepp, David S. Sacks, Loring J. Crepeau and Bruce B. Abbott*. (pp. 7-59).

2120. A clinician's plea for a return to the development of nonhuman models of psychopathology: New clinical observations in need of laboratory study -*Donald J. Levis*. (pp. 395-427).

2121. Diamant, Louis (Ed.). (1991). **Psychology of sports, exercise, and fitness: Social and personal issues.** New York, NY: Hemisphere Publishing Corp. xiv, 277 pp. ISBN 1-56032-170-9 (hardcover).

2122. Sports, exercise, and eating disorders -*Julie A. Pruitt, Ruth V. Kappius and Pamela S. Imm*. (pp. 139-151).

2123. Ellis, Norman R. (Ed.). (1975). **Aberrant development in infancy: Human and animal studies.** Hillsdale, NJ: Lawrence Erlbaum Associates, Inc. viii, 287 pp.

2124. An animal model for the small-for-gestational age infant: Some behavioral and morphological findings -*Victor H. Denenberg and Darlene DeSantis*. (pp. 77-88).

2125. Paradoxical effects of amphetamine on behavioral arousal in neonatal and adult rats: A possible animal model of the calming effect of amphetamine on hyperkinetic children -*Byron A. Campbell and Patric K. Randall*. (pp. 105-112).

2126. Engs, Ruth C. (Ed.) & Alcohol & Drug Problems Assn. (1990). **Women: Alcohol and other drugs.** Dubuque, IA: Kendall/Hunt Publishing Co. xiv, 173 pp. ISBN 0-8403-5571-8 (paperback).

2127. Physiological effects of cocaine, heroin and methadone -*Janet L. Mitchell and Gina Brown*. (pp. 53-60).

2128. Eysenck, H. J. (Ed.). (1973). **Handbook of abnormal psychology.** San Diego, CA: EdITS Publishers. xvi, 906 pp. ISBN 0-912736-13-5 (hardcover).

2129. Animal studies bearing on abnormal behaviour -*P. L. Broadhurst*. (pp. 721-754).

2130. Fisher, Seymour; Raskin, Allen & Uhlenhuth, E. H. (Eds.). (1987). **Cocaine: Clinical and biobehavioral aspects.** New York, NY: Oxford University Press. ix, 256 pp. ISBN 0-19-504068-6 (hardcover).

2131. Reinforcing and discriminative stimulus effects of cocaine: Analysis of pharmcological mechanisms - *James H. Woods, Gail D. Winger and Charles P. France.* (pp. 21-65).

2132. Frederickson, Robert C. A.; McGaugh, James L. & Felten, David L. (Eds.). (1991). **Peripheral signaling of the brain: Role in neural-immune interactions and learning and memory. Neuronal control of bodily function: Basic and clinical aspects, Vol. 6.** Lewiston, NY: Hogrefe & Huber Publishers. xvi, 528 pp. ISBN 0-88937-035-4 (hardcover, Toronto); 3-456-81863-7 (hardcover, Bern).

2133. Gut peptides as modulators of memory -*John E. Morley and James F. Flood.* (pp. 379-387).

2134. Vasopressin and habituation to spatial novelty in two genetic models of vasopressin-deficiency in rats -*Adolfo G. Sadile.* (pp. 485-490).

2135. Friedman, Howard S. (Ed.). (1992). **Hostility, coping, & health.** Washington, DC: American Psychological Association. xvi, 263 pp. ISBN 1-55798-138-8 (hardcover).

2136. Behavioral influences on coronary artery disease: A nonhuman primate model -*Stephen B. Manuck, Jay R. Kaplan, Thomas B. Clarkson, Michael R. Adams and Carol A. Shively.* (pp. 99-106).

2137. Galaburda, Albert M. (Ed.). (1989). **From reading to neurons. Issues in the biology of language and cognition.** Cambridge, MA: MIT Press. xxii, 545 pp. ISBN 0-262-07115-0 (hardcover).

2138. Animal models of developmental dyslexia: Brain lateralization and cortical pathology -*Gordon F. Sherman, Glenn D. Rosen and Albert M. Galaburda.* (pp. 389-404).

2139. Galanter, Marc; Begleiter, Henri; Deitrich, Richard; Gallant, Donald M.; Goodwin, Donald; Gottheil, Edward; Paredes, Alfonso; Rothschild, Marcus; Van Thiel, David H. & Cancellare, Denise (Eds.). (1991). **Recent developments in alcoholism, Vol. 9: Children of alcoholics.** New York, NY: Plenum Press. xxii, 382 pp. ISBN 0-306-43840-2 (hardcover).

2140. Basic animal research -*Stata Norton and Lois A. Kotkoskie.* (pp. 95-115).

2141. Galanter, Marc (Ed.). (1989). **Recent developments in alcoholism, Vol. 7: Treatment research.** New York, NY: Plenum Press. xxxi, 371 pp. ISBN 0-306-43042-8 (hardcover).

2142. Serotonin and ethanol preference -*William J. McBride, James M. Murphy, Lawrence Lumeng and Ting-Kai Li.* (pp. 187-209).

2143. Galanter, Marc (Ed.); American Medical Society on Alcoholism; Research Society on Alcoholism & National Council on Alcoholism. (1986). **Recent developments in alcoholism, Vol. 4.** New York, NY: Plenum Press. xxv, 453 pp. ISBN 0-306-42170-4 (hardcover).

2144. The alcohol withdrawal syndrome: A view from the laboratory -*Dora B. Goldstein.* (pp. 231-240).

2145. Gellatly, Angus; Rogers, Don & Sloboda, John A. (Eds.). (1989). **Cognition and social worlds. Keele cognition seminars, 2.** Oxford, England [US Location: New York, NY]: Clarendon Press/Oxford University Press. x, 257 pp. ISBN 0-19-852173-1 (hardcover).

2146. Machiavellian monkeys: Cognitive evolution and the social world of primates -*Andrew Whiten and Richard W. Byrne.* (pp. 22-36).

2147. Gibbons, Robert J. (Ed.). (1974). **Research advances in alcohol and drug problems.** New York, NY: John Wiley & Sons. xvii, 428 pp.

2148. The use of animal models for the study of drug abuse -*Charles R. Schuster and Chris E. Johanson.*

2149. Gray, Jeffrey A. (1982). **The neuropsychology of anxiety: An enquiry into the functions of the septo-hippocampal system. Oxford psychology series.** New York, NY: Clarendon Press/Oxford University Press. 548 pp. ISBN 0-19-852109-X (hardcover).
Preface * Introduction: The behavioral inhibition system * The behavioural effects of anti-anxiety drugs * The septo-hippocampal system: Anatomy * The septo-hippocampal system: Electrophysiology * Approaches to septo-hippocampal function * The behavioural effects of septal and hippocampal lesions * Hippocampal electrical activity and behaviour * The behavioural effects of stimulating the septo-hippocampal system * The hippocampus and memory * A theory of septo-hippocampal function * The role of the ascending projections to the septo-hippocampal system * Long-term effects of stress: The relation between anxiety and depression * The role of the prefrontal cortex * The symptoms of anxiety * The treatment of anxiety * The anxious personality * Résumé * References * Author index * Subject index

2150. Gray, Jeffrey A. (1971). **The psychology of fear and stress.** New York, NY: McGraw-Hill Book Company. 255 pp.
Introduction * Fears, innate and acquired * The expression of fear * The inheritance of fear * The physiology of the emotions: Fear and stress * An excursion into social biology: Fear and sex * The route from gene to behaviour: Sex differences and fear * The early environment and fearfulness * Punishment and conflict * Fear and frustration * The learning of active avoidance * A conceptual nervous system for avoidance behaviour * Fear and the central nervous system * Man: Neurosis, neuroticism, therapy * Notes * Bibliography * Acknowledgments * Index

2151. Green, Leonard & Kagel, John H. (Eds.). (1990). **Advances in behavioral economics, Vol. 2.** Norwood, NJ: Ablex Publishing Corp. xiv, 325 pp. ISBN 0-89391-449-5 (hardcover).

2152. The economics of leisure in psychological studies of choice -*Sandra M. Schrader and Leonard Green.* (pp. 226-252).

2153. Greenhill, Laurence L. & Osman, Betty B. (Eds.). (1991). **Ritalin: Theory and patient management.** New York, NY: Mary Ann Liebert, Inc, Publishers. 338 pp. ISBN 0-913113-53-0 (hardcover).

2154. Ritalin and brain metabolism -*Christine A. Redman and Alan J. Zametkin.* (pp. 301-308).

2155. Grunert, Klaus G. & Ölander, Folke (Eds.). (1989). **Understanding economic behaviour. Theory and decision library: Series A: Philosophy and methodology of the social sciences, Vol. 11.** Dordrecht, Netherlands [US Location: Boston, MA]: Kluwer Academic Publishers. vii, 440 pp. ISBN 0-7923-0482-9 (hardcover).

2156. The quantitative analysis of economic behavior with laboratory animals -*Steven R. Hursh, Thomas G. Raslear, Richard Bauman and Harold Black.* (pp. 393-407).

2157. Gunnar, Megan R. & Nelson, Charles A. (Eds.). (1992). **Developmental behavioral neuroscience. The Minnesota symposia on child psychology, Vol. 24.** Hillsdale, NJ: Lawrence Erlbaum Associates, Inc. xiii, 249 pp. ISBN 0-8058-0977-5 (hardcover).

2158. Basic mechanisms of human locomotor development -*Hans Forssberg, Virgil Stokes and Helga Hirschfeld.* (pp. 37-73).

2159. Hellhammer, Dirk H.; Florin, Irmela & Weiner, Herbert (Eds.). (1988). **Neurobiological approaches to human disease. Neuronal control of bodily function: Basic and clinical aspects, Vol. 2.** Stuttgart, Federal Republic of Germany [US Location: Lewiston, NY]: Hans Huber Publishers, Inc. x, 451 pp. ISBN 0-920887-27-9 (hardcover, Toronto); 3-456-81638-3 (hardcover, Bern).

2160. Transformation of emotion into motion: Role of mesolimbic noradrenaline and neostriatal dopamine -*Alexander R. Cools.* (pp. 15-28).

2161. Central neurochemical mechanisms in experimental stress ulcer -*Gary B. Glavin.* (pp. 86-104).

2162. The function of the autonomic system as interface between body and environment. Old and new concepts: W. B. Cannon and W. R. Hess revisited -*Wilfrid Jänig.* (pp. 143-173).

2163. Hypertension and the brain -*Herbert Weiner.* (pp. 197-207).

2164. Research directions in behavioral medicine -*Fritz A. Henn, Emmeline Edwards and Joel Johnson.* (pp. 215-224).

2165. Psychological determinants of when stressors stress -*J. Bruce Overmier.* (pp. 236-259).

2166. Limbic-midbrain mechanisms and behavioral physiology of interactions with CRF, ACTH, and adrenal hormones -*Béla Bohus.* (pp. 267-285).

2167. Differential control of ACTH-related peptides and the importance of the behavioral situation -*Peter G. Smelik.* (pp. 286-289).

2168. Stress and the immune response -*Rudy E. Ballieux and Cobi J. Heijnen.* (pp. 301-306).

2169. Expectancy and activation: An attempt to systematize stress theory -*Holger Ursin.* (pp. 313-334).

2170. Adrenocortical activity and disease, with reference to gastric pathology in animals -*Robert Murison and J. Bruce Overmier.* (pp. 335-343).

2171. Effects of benzodiazepam on salivary cortisol under experimental stress -*Dirk Hellhammer, Ingmar Gutberlet, Jürgen Konermann, Uta Müller and Ludger Rolf.* (pp. 413-416).

2172. Hertzberg, Leonard J.; Ostrum, Gene F. & Field, Joan Roberts (Eds.). (1990). **Violent behavior, Vol. 1: Assessment & intervention.** Costa Mesa, CA: PMA Publishing Corp. xii, 339 pp. ISBN 0-89335-220-9 (hardcover, Vol. 1).

2173. Aggression, violence, and violence prevention: An ethological perspective -*Slobodan B. Petrovich.* (pp. 3-26).

2174. Hindmarch, Ian & Stonier, P. D. (Eds.). (1989). **Human psychopharmacology: Measures and methods, Vol. 2. A Wiley medical publication.** Chichester, England [US Location: New York, NY]: John Wiley & Sons. xi, 296 pp. ISBN 0-471-91255-7 (hardcover, Vol. 2).

2175. From animals to man: Advantages, problems and pitfalls of animal models in psychopharmacology -*B. E. Leonard.* (pp. 23-66).

2176. Hrushesky, William J. M.; Langer, Robert & Theeuwes, Felix (Eds.). (1991). **Temporal control of drug delivery. Annals of the New York Academy of Sciences, Vol. 618.** New York, NY: New York Academy of Sciences. xvii, 641 pp. ISBN 0-89766-633-X (hardcover); 0-89766-634-8 (paperback).

2177. The multifrequency (circadian, fertility cycle, and season) balance between host and cancer -*William J. M. Hrushesky.* (pp. 228-256).

2178. The effect of surgical timing within the fertility cycle on breast cancer outcome -*Avrum Bluming and William J. M. Hrushesky.* (pp. 277-291).

2179. Hutchings, Donald E. (Ed.). (1989). **Prenatal abuse of licit and illicit drugs. Annals of the New York Academy of Sciences, Vol. 562.** New York, NY: New York Academy of Sciences. xii, 388 pp. ISBN 0-89766-521-X (hardcover); 0-89766-522-8 (paperback).

2180. Maternal-fetal pharmacokinetics and fetal dose-response relationships -*Hazel H. Szeto.* (pp. 42-55).

2181. Mothers who smoke and the lungs of their offspring -*Adrien C. Moessinger.* (pp. 101-104).

2182. Developmental neurotoxicity of nicotine, carbon monoxide, and other tobacco smoke constituents -*Charles F. Mactutus.* (pp. 105-122).

2183. The behavioral and neuroanatomical effects of prenatal alcohol exposure in animals -*Edward P. Riley and Susan Barron.* (pp. 173-177).

2184. Prenatal cocaine exposure to the fetus: A sheep model for cardiovascular evaluation -*James R. Woods, Jr., Mark A. Plessinger, Kimberly Scott and Richard K. Miller.* (pp. 267-279).

2185. Cocaine effects on the developing central nervous system: Behavioral, psychopharmacological, and neurochemical studies *-Linda Patia Spear, Cheryl L. Kirstein and Nancy A. Frambes.* (pp. 290-307).

2186. Prenatal amphetamine effects on behavior: Possible mediation by brain monoamines *-Lawrence D. Middaugh.* (pp. 308-318).

2187. Neurobehavioral effects of prenatal caffeine *-Thomas J. Sobotka.* (pp. 327-339).

2188. Ingle, David J. & Shein, Harvey M. (Eds.). (1975). **Model systems in biological psychiatry.** Cambridge, MA: MIT Press. xi, 196 pp. ISBN 0-262-09015-5 (hardcover).

2189. Animal models of schizophrenia *-Steven Mathysse and Suzanne Haber.* (pp. 4-25).

2190. Animal models and schizophrenia *-Conan Kornetsky and Robert Markowitz.* (pp. 26-50).

2191. Biological models in the study of false neurochemical synaptic transmitters *-Ross J. Baldessarini and Josef E. Fischer.* (pp. 51-79).

2192. Tissue- and cell-culture models in the study of neurotransmitter and synaptic function *-Harvey M. Shein.* (pp. 80-96).

2193. Central dopaminergic neurons: A model for predicting the efficacy of putative antipsychotic drugs? *-Benjamin S. Bunney and George K. Aghajanian.* (pp. 97-112).

2194. The frog's visual system as a model for the study of selective attention *-David J. Ingle.* (pp. 113-131).

2195. Processes controlling aggressive behavior in cichlid fish *-Walter Heiligenberg.* (pp. 132-148).

2196. A model system for neurophysiological investigations of behavioral plasticity *-Dennis L. Glanzman, Timothy J. Teyler and Richard F. Thompson.* (pp. 149-165).

2197. The goldfish as a model experimental animal for studies of biochemical correlates of the information-storage process *-Victor E. Shashoua.* (pp. 166-189).

2198. Ingle, David J. & Shein, Harvey M. (Eds.). (1975). **Model systems in biological psychiatry.** Cambridge, MA: MIT Press. xi, 196 pp.

2199. Janković, Branislav D.; Marković, Branislav M. & Spector, Novera Herbert (Eds.). (1987). **Neuroimmune interactions: Proceedings of the Second International Workshop on Neuroimmunomodulation. Annals of the New York Academy of Sciences, Vol. 496.** New York, NY: New York Academy of Sciences. 756 pp. ISBN 0-89766-387-X (hardcover); 0-89766-388-8 (paperback).

2200. Conditioning phenomena and immune function *-Robert Ader, Lee J. Grota and Nicholas Cohen.* (pp. 532-544).

2201. Jeannerod, Marc (Ed.). (1987). **Neurophysiological and neuropsychological aspects of spatial neglect. Advances in psychology, No. 45.** Amsterdam, Netherlands [US Location: New York, NY]: Elsevier Science Publishing Co, Inc. xiv, 346 pp. ISBN 0-444-70193-1 (hardcover).

2202. Animal models for the syndrome of spatial neglect *-A. David Milner.* (pp. 259-288).

2203. Keehn, J. D. (Ed.). (1979). **Psychopathology in animals: Research and clinical implications.** New York, NY: Academic Press, Inc. xiv, 334 pp. ISBN 0-12-403050-5 (hardcover).

2204. Psychopathology in animal and man *-J. D. Keehn.* (pp. 1-27).

2205. Natural animal addictions: An ethological perspective *-Ronald K. Siegel.* (pp. 29-60).

2206. Genetic analysis of deviant behavior *-John L. Fuller.* (pp. 61-79).

2207. Animal models of psychopathology: Studies in naturalistic colony environments *-Gaylord D. Ellison.* (pp. 81-101).

2208. Psychosis and drug-induced stereotypies *-Melvin Lyon and Erik Bardrum Nielsen.* (pp. 103-142).

2209. The role of conditioning in drug tolerance and addiction *-Shepard Siegel.* (pp. 143-168).

2210. The kindling effect: An experimental model of epilepsy? *-John Gaito.* (pp. 169-195).

2211. Behavioral anomalies in aversive situations *-Hank Davis.* (pp. 197-222).

2212. Experimental depression in animals *-V. A. Colotla.* (pp. 223-238).

2213. Fears in companion dogs: Characteristics and treatment *-David Hothersall and David S. Tuber.* (pp. 239-255).

2214. The infrahuman avoidance model of symptom maintenance and implosive therapy *-Donald J. Levis.* (pp. 257-277).

2215. Psychotherapy from the standpoint of a behaviorist *-C. B. Ferster.* (pp. 279-303).

2216. Ethics and animal experimentation: Personal views *-F. L. Marcuse and J. J. Pear.* (pp. 305-334).

2217. Kelly, Dennis D. (Ed.). (1986). **Stress-induced analgesia. Annals of the New York Academy of Sciences, Vol. 467.** New York, NY: New York Academy of Sciences. 449 pp. ISBN 0-89766-329-2 (hardcover); 0-89766-330-6 (paperback).

2218. Hyperalgesia induced by emotional stress in the rat: An experimental animal model of human anxiogenic hyperalgesia *-Catherine Vidal and Joseph Jacob.* (pp. 73-81).

2219. Kimmel, H. D. (Ed.). (1971). **Experimental psychopathology: Recent research and theory.** New York, NY: Academic Press, Inc. xiii, 264 pp. ISBN 0-12-407250-X (hardcover).

2220. The principle of uncertainty in neurotigenesis - *Jules H. Masserman.* (pp. 13-32).

2221. Experimental basis for neurotic behavior -*W. Horsley Gantt.* (pp. 33-48).

2222. Frustration, persistence, and regression -*Abram Amsel.* (pp. 51-69).

2223. Psychopathology: An analysis of response consequences -*Jack Sandler and Robert S. Davidson.* (pp. 71-93).

2224. Vicious circle behavior -*Kenneth B. Melvin.* (pp. 96-115).

2225. Experimental psychopathology and the psychophysiology of emotion -*Joseph V. Brady, Jack D. Findley and Alan Harris.* (pp. 119-146).

2226. Schedule-controlled modulation of arterial blood pressure in the squirrel monkey -*W. H. Morse, J. Alan Herd, R. T. Kelleher and Susan A. Grose.* (pp. 147-164).

2227. Pathological inhibition of emotional behavior -*H. D. Kimmel.* (pp. 165-181).

2228. Dominance and aggression -*Frank A. Logan.* (pp. 185-201).

2229. Psychopathology in monkeys -*Harry F. Harlow and Margaret K. Harlow.* (pp. 203-229).

2230. Psychopathological consequences of induced social helplessness during infancy -*Joseph B. Sidowski.* (pp. 231-248).

2231. Kitterle, Frederick L. (Ed.). (1991). **Cerebral laterality: Theory and research: The Toledo Symposium.** Hillsdale, NJ: Lawrence Erlbaum Associates, Inc. xiii, 234 pp. ISBN 0-8058-0471-4 (hardcover).

2232. Prosimians as animal models in the study of neural lateralization -*Jeannette P. Ward.* (pp. 1-17).

2233. Functional lateralization in monkeys -*Charles R. Hamilton and Betty A. Vermire.* (pp. 19-34).

2234. Klein, Stephen B. & Mowrer, Robert R. (Eds.). (1989). **Contemporary learning theories: Instrumental conditioning theory and the impact of biological constraints on learning.** Hillsdale, NJ: Lawrence Erlbaum Associates, Inc. xiii, 293 pp. ISBN 0-8058-0318-1 (hardcover); 0-8058-0319-X (hardcover, set).

2235. Schedule-induced polydipsia: Is the rat a small furry human? (An analysis of an animal model of human alcoholism) -*Anthony L. Riley and Cora Lee Wetherington.* (pp. 205-236).

2236. Kohnstamm, Geldolph A. (Ed.). (1986). **Temperament discussed: Temperament and development in infancy and childhood.** Lisse, Netherlands: Swets & Zeitlinger. 200 pp. ISBN 90-265-0783-6 (paperback).

2237. Stability does not mean stability -*Jan Strelau.* (pp. 59-62).

2238. Kohnstamm, Geldolph A.; Bates, John E. & Rothbart, Mary Klevjord (Eds.). (1989). **Temperament in childhood.** Chichester, England [US Location: New York, NY]: John Wiley & Sons. xvii, 641 pp. ISBN 0-471-91692-7 (hardcover).

2239. Temperamental reactivity in non-human primates -*J. D. Higley and S. J. Suomi.* (pp. 153-167).

2240. Kolb, Bryan & Tees, Richard C. (Eds.). (1990). **The cerebral cortex of the rat.** Cambridge, MA: MIT Press. xii, 645 pp. ISBN 0-262-11150-0 (hardcover); 0-262-61064-7 (paperback).

2241. The rat as a model of cortical function -*B. Kolb and R. C. Tees.* (pp. 3-17).

2242. Koob, George F.; Ehlers, Cindy L. & Kupfer, David J. (Eds.). (1989). **Animal models of depression.** Boston, MA: Birkhäuser. xiii, 295 pp. ISBN 0-8176-3407-X (hardcover, US); 3-7643-3407-X (hardcover, Switzerland).

2243. Basis of development of animal models in psychiatry: An overview -*William T. McKinney.* (pp. 3-17).

2244. Animal models: Promises and problems -*Conan Kornetsky.* (pp. 18-29).

2245. Non-homologous animal models of affective disorders: Clinical relevance of sensitization and kindling -*Robert M. Post and Susan R. B. Weiss.* (pp. 30-54).

2246. The HPA system and neuroendocrine models of depression -*Ned H. Kalin.* (pp. 57-73).

2247. The use of an animal model to study post-stroke depression -*Robert G. Robinson.* (pp. 74-98).

2248. Social zeitgebers: A peer separation model of depression in rats -*Cindy L. Ehlers, Tamara L. Wall, Stephen P. Wyss and R. Ian Chaplin.* (pp. 99-110).

2249. Electrophysiology of the locus coeruleus: Implications for stress-induced depression -*Jay M. Weiss and Peter E. Simson.* (pp. 111-134).

2250. Motor activity and antidepressant drug: A proposed approach to categorizing depression syndromes and their animal models -*Martin H. Teicher, Matacha I. Barber, Janet M. Lawrence and Ross J. Baldessarini.* (pp. 135-161).

2251. Anhedonia as an animal model of depression - *George F. Koob.* (pp. 162-183).

2252. Models of depression used in the pharmaceutical industry -*James L. Howard, Robert M. Ferris, Barrett R. Cooper, Francis E. Soroko, Ching M. Wang and Gerald T. Pollard.* (pp. 187-203).

2253. Pharmacological, biochemical, and behavioral analyses of depression: Animal models -*Robert M. Zacharko and Hymie Anisman.* (pp. 204-238).

2254. Pharmacologic probes in primate social behavior - *R. Francis Schlemmer, Jr., Jennifer E. Young and John M. Davis.* (pp. 239-260).

2255. The neuropharmacology of serotonin and sleep: An evaluation -*John D. Fernstrom and Ross H. Pastel.* (pp. 261-280).

2256. Krasnegor, Norman A. & Bridges, Robert S. (Eds.). (1990). **Mammalian parenting: Biochemical, neurobiological, and behavioral determinants.** New York, NY: Oxford University Press. xii, 502 pp. ISBN 0-19-505600-0 (hardcover).

2257. The biosocial context of parenting in human families -*Leon Eisenberg.* (pp. 9-24).

2258. Kupfer, David J.; Monk, Timothy H. & Barchas, Jack D. (Eds.). (1988). **Biological rhythms and mental disorders.** New York, NY: Guilford Press. xviii, 357 pp. ISBN 0-89862-746-X (hardcover).

2259. Neuroendocrine substrates of circannual rhythms -*Irving Zucker.* (pp. 219-251).

2260. Kurstak, Edouard (Ed.). (1991). **Psychiatry and biological factors.** New York, NY: Plenum Medical Book Co/ Plenum Press. xx, 311 pp. ISBN 0-306-43621-3 (hardcover).

2261. Herpes simplex virus type 1 transcription during latent infections of mouse and man: Implications for dementia -*Anne M. Deatly, Ashley T. Haase and Melvyn J. Ball.* (pp. 277-286).

2262. Kurstak, Edouard; Lipowski, Z. J. & Morozov, P. V. (Eds.). (1987). **Viruses, immunity, and mental disorders.** New York, NY: Plenum Medical Book Co/Plenum Press. xiv, 468 pp. ISBN 0-306-42337-5 (hardcover).

2263. Animal models in behavioral neurovirology -*Dennis McFarland and John Hotchin.* (pp. 189-198).

2264. Lande, Jeffrey S.; Scarr, Sandra Wood & Gunzenhauser, Nina (Eds.). (1989). **Caring for children: Challenge to America.** Hillsdale, NJ: Lawrence Erlbaum Associates, Inc. xi, 327 pp. ISBN 0-8058-0255-X (hardcover); 0-8058-0256-8 (paperback).

2265. Day care and the promotion of emotional development: Lessons from a monkey laboratory -*J. D. Higley, J. S. Lande and S. J. Suomi.* (pp. 77-91).

2266. Last, Cynthia G. (Ed.) & Hersen, Michel (1988). **Handbook of anxiety disorders. Pergamon general psychology series, Vol. 151.** Elmsford, NY: Pergamon Press, Inc. ix, 647 pp. ISBN 0-08-032766-4 (hardcover).

2267. Animal models -*William T. McKinney.* (pp. 171-180).

2268. Lewandowsky, Stephan; Dunn, John C. & Kirsner, Kim (Eds.). (1989). **Implicit memory: Theoretical issues.** Hillsdale, NJ: Lawrence Erlbaum Associates, Inc. x, 338 pp. ISBN 0-8058-0358-0 (hardcover).

2269. Episodically unique and generalized memories: Applications to human and animal amnesics -*Michael S. Humphreys, John D. Bain and Jennifer S. Burt.* (pp. 139-156).

2270. Lister, Richard G. & Weingartner, Herbert J. (Eds.). (1991). **Perspectives on cognitive neuroscience.** New York, NY: Oxford University Press. xvi, 508 pp. ISBN 0-19-506151-9 (hardcover).

2271. Approaches to the treatment of Alzheimer's disease -*Anthony C. Santucci, Vahram Haroutunian, Linda M. Bierer and Kenneth L. Davis.* (pp. 467-483).

2272. Ludwig, Frédéric C. (Ed.). (1991). **Life span extension: Consequences and open questions.** New York, NY: Springer Publishing Co, Inc. xi, 160 pp. ISBN 0-8261-7450-7 (hardcover).

2273. Life span extension by means other than control of disease -*Thomas C. Cesario and Daniel Hollander.* (pp. 43-54).

2274. Mann, J. John (Ed.). (1989). **Models of depressive disorders: Psychological, biological, and genetic perspectives. The depressive illness series.** New York, NY: Plenum Press. xii, 185 pp. ISBN 0-306-43277-3 (hardcover).

2275. Animal models -*Fritz A. Henn.* (pp. 93-107).

2276. Martin, Paul R. (Ed.). (1991). **Handbook of behavior therapy and psychological science: An integrative approach. Pergamon general psychology series, Vol. 164.** New York, NY: Pergamon Press, Inc. xii, 563 pp. ISBN 0-08-036129-3 (hardcover).

2277. Selective associations in the origins of phobic fears and their implications for behavior therapy -*Michael Cook and Susan Mineka.* (pp. 413-434).

2278. A contextual analysis of fear extinction -*Mark E. Bouton.* (pp. 435-453).

2279. Martinez, Joe L. Jr. & Kesner, Raymond P. (Eds.). (1991). **Learning and memory: A biological view (2nd ed.).** San Diego, CA: Academic Press, Inc. xvii, 563 pp. ISBN 0-12-474992-5 (hardcover); 0-12-474991-7 (paperback).

2280. Memory changes with age: Neurobiological correlates -*C. A. Barnes.* (pp. 259-296).

2281. Learning and memory: Vertebrate models -*Jeffrey P. Pascoe, William F. Supple and Bruce S. Kapp.* (pp. 359-407).

2282. Electrical brain stimulation used to study mechanisms and models of memory -*Robert F. Berman.* (pp. 409-438).

2283. Maser, Jack D. & Seligman, Martin E. P. (Eds.). (1977). **Psychopathology: Experimental models. A series of books in psychology.** San Francisco, CA: W. H. Freeman & Co, Publishers. 474 pp. ISBN 0-7167-0368-8 (hardcover); 0-7167-0367-X (paperback).

2284. Model 1: Obesity: Bidirectional influences of emotionality, stimulus responsivity, and metabolic events in obesity -*Judith Rodin.* (pp. 27-65).

2285. Model 2: Addiction: An opponent process theory of motivation. The affective dynamics of drug addiction -*Richard L. Solomon.* (pp. 66-103).

2286. Model 3: Depression: Learned helplessness and depression -*William R. Miller, Robert A. Rosellini and Martin E. P. Seligman.* (pp. 104-130).

2287. Model 4: Depression: Production and alleviation of depressive behaviors in monkeys *-Stephen J. Suomi and Harry F. Harlow.* (pp. 131-173).

2288. Model 5: Phobias and obsessions: Phobias and obsessions: Clinical phenomena in search of a laboratory model *-Isaac Marks.* (pp. 174-213).

2289. Model 6: Neurosis: Experimental neurosis: Neuropsychological analysis *-Earl Thomas and Louise DeWald.* (pp. 214-231).

2290. Model 7: Psychosomatic disorders: Ulcers *-Jay Weiss.* (pp. 232-269).

2291. Model 8: Psychosomatic disorders: Psychosomatic disorders and biofeedback: A psychobiological model of disregulation *-Gary E. Schwartz.* (pp. 270-307).

2292. Model 9: Minimal brain dysfunction: The neonatal split-brain kitten: A laboratory analogue of minimal brain dysfunction *-Jeri A. Sechzer.* (pp. 308-333).

2293. Model 10: Catatonia: Tonic immobility: Evolutionary underpinnings of human catalepsy and catatonia *-Gordon G. Gallup, Jr. and Jack D. Maser.* (pp. 334-357).

2294. Model 11: Schizophrenia: Movement, mood and madness: A biological model of schizophrenia *-Steven M. Paul.* (pp. 358-386).

2295. Model 12: Sex: Sexual diversity *-Ingeborg L. Ward.* (pp. 387-403).

2296. Masserman, Jules H. (Ed.). (1968). **Animal and human: Scientific proceedings of the American Academy of Psychoanalysis. Science and psychoanalysis, Vol. 12.** New York, NY: Grune & Stratton, Inc/W. B. Saunders Co. x, 277 pp.

2297. Genotypic factors in the ontogeny of behavior *-Benson E. Ginsburg.* (pp. 12-17).

2298. Inhibition of son–mother mating among free ranging rhesus monkeys *-Donald Stone Sade.* (pp. 18-38).

2299. Evolution of emotional responses: Evidence from recent research on nonhuman primates *-David A. Hamburg.* (pp. 39-54).

2300. Implications of primate research for understanding infant development *-Gordon D. Jensen and Ruth A. Bobbitt.* (pp. 55-81).

2301. Socialization of wolves *-Jerome H. Woolpy.* (pp. 82-94).

2302. Alliances and aggression among rhesus monkeys *-Jules H. Masserman, Stanley Wechkin and Marvin Woolf.* (pp. 95-100).

2303. Scope and potential of primate research *-William A. Mason.* (pp. 101-118).

2304. The evolution of human behavior *-Leon Moses.* (pp. 119-123).

2305. Fixed motor patterns in ethologic and psychoanalytic theory *-Sol Kramer.* (pp. 124-155).

2306. An analysis of the role of sucking in early infancy *-Wagner H. Bridger and Beverly M. Birns.* (pp. 156-165).

2307. The application of imprinting to psychodynamics *-Yasuhiko Taketomo.* (pp. 166-189).

2308. Primate parallels and biocultural models *-Mariam K. Slater.* (pp. 190-197).

2309. Ethology on the couch *-Colin G. Beer.* (pp. 198-213).

2310. An evolutionary-adaptational-ecologic view of human behavior *-Leon Moses.* (pp. 214-236).

2311. Cognition, thought and affect in the organization of experience *-Joseph Barnett.* (pp. 237-250).

2312. Comparative and experimental approaches to behavior *-Jules H. Masserman.* (pp. 251-261).

2313. The psychoanalyst's role in community mental health centers *-Robert C. Heath and William H. Shelton.* (pp. 262-264).

2314. McCubbin, James A.; Kaufmann, Peter G. & Nemeroff, Charles B. (Eds.). (1991). **Stress, neuropeptides, and systemic disease.** San Diego, CA: Academic Press, Inc. xvi, 485 pp. ISBN 0-12-482490-0 (hardcover).

2315. The role of endogenous opioids in chronic congestive heart failure *-Robert P. Frantz and Changseng Liang.* (pp. 429-444).

2316. McGaugh, James L.; Weinberger, Norman M. & Lynch, Gary (Eds.). (1990). **Brain organization and memory: Cells, systems, and circuits.** New York, NY: Oxford University Press. xvii, 409 pp. ISBN 0-19-505496-2 (hardcover).

2317. The dissection by Alzheimer's disease of cortical and limbic neural systems relevant to memory *-Gary W. Van Hoesen.* (pp. 234-261).

2318. McGuigan, Frank J. & Ban, Thomas A. (Eds.). (1987). **Critical issues in psychology, psychiatry, and physiology: A memorial to W. Horsley Gantt. Monographs in psychobiology: An integrated approach, Vol. 2.** New York, NY: Gordon and Breach Science Publishers. xvii, 379 pp. ISBN 2-88124-137-9 (hardcover).

2319. The foundations of behavior therapy and the contributions of W. Horsley Gantt *-Cyril M. Franks.* (pp. 133-148).

2320. Medin, Douglas L.; Roberts, William A. & Davis, Roger T. (Eds.). (1976). **Processes of animal memory.** Hillsdale, NJ: Lawrence Erlbaum Associates, Inc. xi, 267 pp.

2321. Animal models and memory models *-Douglas L. Medin.* (pp. 113-134).

2322. Milkman, Harvey B. & Sederer, Lloyd I. (Eds.). (1990). **Treatment choices for alcoholism and substance abuse.** Lexington, MA: Lexington Books/D. C. Heath and Company. xxxii, 395 pp. ISBN 0-669-20019-0 (hardcover).

2323. A review of genetic influences on psychoactive substance use and abuse *-Allan C. Collins and Christopher M. de Fiebre.* (pp. 7-23).

2324. Monat, Alan & Lazarus, Richard S. (Eds.). (1991). **Stress and coping: An anthology (3rd ed.).** New York, NY: Columbia University Press. xxi, 598 pp. ISBN 0-231-07456-5 (hardcover); 0-231-07457-3 (paperback).

2325. The concept of coping *-Richard S. Lazarus and Susan Folkman.* (pp. 189-206).

2326. Moore, Bert S. & Isen, Alice M. (Eds.). (1990). **Affect and social behavior. Studies in emotion and social interaction.** New York, NY: Cambridge University Press & Paris, France: Editions de la Maison des Sciences de l'Homme. ix, 277 pp. ISBN 0-521-32768-7 (hardcover); 2-7351-0311-0 (hardcover, France).

2327. Affect and aggression *-Paul A. Bell and Robert A. Baron.* (pp. 64-88).

2328. Nebes, Robert D. & Corkin, S. (Eds.). (1990). **Handbook of neuropsychology, Vol. 4. Handbook of neuropsychology.** Amsterdam, Netherlands [US Location: New York, NY]: Elsevier Science Publishing Co, Inc. xiv, 383 pp. ISBN 0-444-90492-1 (hardcover, series); 0-444-81234-2 (hardcover, vol. 4).

2329. Commissurotomy studies in animals *-G. Berlucchi.* (pp. 9-47).

2330. Animal models of age-related cognitive decline *-C. A. Barnes.* (pp. 169-196).

2331. Newman, John D. (Ed.). (1988). **The physiological control of mammalian vocalization.** New York, NY: Plenum Press. xv, 435 pp. ISBN 0-306-43003-7 (hardcover).

2332. Primate models for the management of separation anxiety *-J. C. Harris and J. D. Newman.* (pp. 321-330).

2333. Orban, Guy A.; Singer, Wolf & Bernsen, Niels Ole (Eds.). (1991). **Cognitive neuroscience: Research directions in cognitive science: European perspectives, Vol. 4.** Hove, England [US Location: Hillsdale, NJ]: Lawrence Erlbaum Associates, Inc. xviii, 170 pp. ISBN 0-86377-114-9 (hardcover).

2334. Neuroscience in the context of cognitive science and artificial intelligence *-Guy A. Orban and W. Singer.* (pp. 1-14).

2335. Osofsky, Joy Doniger (Ed.). (1987). **Handbook of infant development (2nd ed.). Wiley series on personality processes.** New York, NY: John Wiley & Sons. xix, 1391 pp. ISBN 0-471-88565-7 (hardcover).

2336. Behavioral teratogenesis: Long-term influences on behavior from early exposure to environmental agents *-Charles V. Vorhees and Elizabeth Mollnow.* (pp. 913-971).

2337. Parin, V. V. (Ed.). (1969). **Systemic organization of physiological functions.** Moscow, USSR: Medicina, Izdatel'stvo. 444 pp.

2338. Motor learning, visceral learning and homeostasis *-N. E. Miller.* (pp. 363-372).

2339. Perecman, Ellen (Ed.). (1989). **Integrating theory and practice in clinical neuropsychology.** Hillsdale, NJ: Lawrence Erlbaum Associates, Inc. xxviii, 438 pp. ISBN 0-8058-0285-1 (hardcover).

2340. The neuropsychology of attention: Elements of a complex behavior *-Allan F. Mirsky.* (pp. 75-91).

2341. Pfafflin, Sheila M.; Sechzer, Jeri A.; Fish, Jefferson M. & Thompson, Robert L. (Eds.). (1990). **Psychology: Perspectives and practice. Annals of the New York Academy of Sciences, Vol. 602.** New York, NY: New York Academy of Sciences. ix, 235 pp. ISBN 0-89766-601-1 (hardcover); 0-89766-602-X (paperback).

2342. Nutritional requirements for normative development of the brain and behavior *-Brian L. G. Morgan.* (pp. 127-132).

2343. Porter, Roger J.; Mattson, Richard H.; Cramer, Joyce A.; Diamond, Ivan & Schoenberg, Devera G. (Eds.). (1990). **Alcohol and seizures: Basic mechanisms and clinical concepts.** Philadelphia, PA: F. A. Davis. xviii, 342 pp. ISBN 0-8036-7008-7 (hardcover).

2344. Alcohol withdrawal seizures: Genetic animal models *-John Crabbe and Ann Kosobud.* (pp. 126-139).

2345. Ramsey, Christian N. Jr. (Ed.). (1989). **Family systems in medicine.** New York, NY: Guilford Press. xx, 615 pp. ISBN 0-89862-103-8 (hardcover).

2346. Psychoneuroimmunology *-Joan Borysenko.* (pp. 243-256).

2347. Family systems, stress, the endocrine system, and the heart *-Robert S. Eliot.* (pp. 283-293).

2348. Rifkin, Harold & Porte, Daniel Jr. (Eds.). (1990). **Ellenberg and Rifkin's diabetes mellitus: Theory and practice (4th ed.).** New York, NY: Elsevier Science Publishing Co, Inc. xvi, 972 pp. ISBN 0-444-01499-3 (hardcover).

2349. Diabetes in animals *-Eleazar Shafrir.* (pp. 299-340).

2350. Riley, Edward P. & Vorhees, Charles V. (Eds.). (1986). **Handbook of behavioral teratology.** New York, NY: Plenum Press. xx, 522 pp. ISBN 0-306-42246-8 (hardcover).

2351. Comparison and critique of government relations for behavioral teratology *-Charles V. Vorhees.* (pp. 49-66).

2352. Behavioral teratology of anticonvulsant and anti-anxiety medications *-Charles V. Vorhees.* (pp. 211-241).

2353. Prenatal phenobarbital: Effects on pregnancy and offspring *-Lawrence D. Middaugh.* (pp. 243-266).

2354. Animal models of behavioral effects of early lead exposure *-Nellie K. Laughlin.* (pp. 291-319).

2355. Risch, Samuel Craig (Ed.). (1991). **Central nervous system peptide mechanisms in stress and depression. Progress in psychiatry, No. 30.** Washington, DC: American Psychiatric Press, Inc. xvii, 128 pp. ISBN 0-88048-249-4 (hardcover).

2356. Animal studies implicating a role of corticotropin-releasing hormone in mediating behavior associated with psychopathology *-Ned H. Kalin and Lorey K. Takahashi.* (pp. 53-72).

2357. Rockstein, Morris & Sussman, Marvin L. (Eds.). (1976). **Nutrition, longevity, and aging.** New York, NY: Academic Press, Inc. 296 pp.

2358. Rohrbaugh, John W.; Parasuraman, Raja & Johnson, Ray Jr. (Eds.). (1990). **Event-related brain potentials: Basic issues and applications.** New York, NY: Oxford University Press. x, 384 pp. ISBN 0-19-504891-1 (hardcover).

2359. Animal models of cognitive event-related potentials *-Jennifer S. Buchwald.* (pp. 57-75).

2360. Cognitive constructs in animal and human studies *-Raymond Kesner.* (pp. 76-85).

2361. Rosenthal, Norman E. & Blehar, Mary C. (Eds.). (1989). **Seasonal affective disorders and phototherapy.** New York, NY: Guilford Press. ix, 386 pp. ISBN 0-89862-741-9 (hardcover).

2362. The photoperiodic phenomena: Seasonal modulation of the "day within" *-Colin S. Pittendrigh.* (pp. 87-104).

2363. Seasonal variations in body weight and metabolism in hamsters *-George N. Wade.* (pp. 105-126).

2364. Seasonal affective disorder, hibernation, and annual cycles in animals: Chipmunks in the sky *-N. Mrosovsky.* (pp. 127-148).

2365. Seasonal affective disorders: Animal models non fingo *-Irving Zucker.* (pp. 149-164).

2366. Rovee-Collier, Carolyn & Lipsitt, Lewis P. (Eds.). (1990). **Advances in infancy research, Vol. 6.** Norwood, NJ: Ablex Publishing Corp. xxxvi, 294 pp. ISBN 0-89391-512-2 (hardcover).

2367. An animal model of retarded cognitive development *-Barbara J. Strupp and David A. Levitsky.* (pp. 149-220).

2368. Roy-Byrne, Peter P. & Cowley, Deborah S. (Eds.). (1991). **Benzodiazepines in clinical practice: Risks and benefits. Clinical practice, No. 17.** Washington, DC: American Psychiatric Press, Inc. xvi, 227 pp. ISBN 0-88048-453-5 (hardcover).

2369. Interactions of benzodiazepines with psychological and behavioral treatment *-Peter P. Roy-Byrne and Richard P. Swinson.* (pp. 203-211).

2370. Russell, Roger W.; Flattau, Pamela Ebert & Pope, Andrew M. (Eds.). (1990). **Behavioral measures of neurotoxicity: Report of a symposium.** Washington, DC: National Academy Press. xiii, 432 pp. ISBN 0-309-04047-7 (hardcover).

2371. Exposure to neurotoxins throughout the life span: Animal models for linking neurochemical effects to behavioral consequences *-Hanna Michalek and Annita Pintor.* (pp. 101-123).

2372. Animal models of dementia: Their relevance to neurobehavioral toxicology testing *-David H. Overstreet and Elaine L. Bailey.* (pp. 124-136).

2373. Bridging experimental animal and human behavioral toxicology studies *-Deborah A. Cory-Slechta.* (pp. 137-158).

2374. Methods and issues in evaluating the neurotoxic effects of organic solvents *-Beverly M. Kulig.* (pp. 159-183).

2375. Animal models: What has worked and what is needed *-Robert C. MacPhail.* (pp. 184-188).

2376. Sandler, Merton; Coppen, Alec & Harnett, Sara (1991). **5-hydroxytryptamine in psychiatry: A spectrum of ideas.** Oxford, England [US Location: New York, NY]: Oxford University Press. xiii, 349 pp. ISBN 0-19-262011-8 (hardcover).

2377. Stimulation of rat pineal melatonin synthesis by a single electroconvulsive shock: Chronobiological effect of antidepressant therapy? *-G. F. Oxenkrug, P. J. Requintina, I. M. McIntyre and R. Davis.* (pp. 110-115).

2378. Serotonin in anxiety: Evidence from animal models *-M. Briley and P. Chopin.* (pp. 177-197).

2379. Anxiogenic effect of the 5-HT[subscript] 1[subscript]C agonist m-chlorophenylpiperazine *-G. Curzon, G. A. Kennett and P. Whitton.* (pp. 198-206).

2380. Does 5-HT have a role in anxiety and the action of anxiolytics? *-M. Palfreyman and J. H. Kehne.* (pp. 207-227).

2381. d-Fenfluramine and animal models of eating disorders *-B. Guardiola-Lemaitre.* (pp. 303-308).

2382. 5-HT[subscript]3 receptor antagonists and their potential in psychiatric disorders *-M. G. Palfreyman, S. M. Sorenson, A. A. Carr, H. C. Cheng and M. W. Dudley.* (pp. 324-330).

2383. Atypical antipsychotic drugs: The 5-HT[subscript]2/DA[subscript]2 ratio *-H. Y. Meltzer.* (pp. 331-335).

2384. Scheibel, Arnold B. & Wechsler, Adam F. (Eds.). (1990). **Neurobiology of higher cognitive function. UCLA forum in medical sciences, No. 29.** New York, NY: Guilford Press. xiv, 370 pp. ISBN 0-89862-425-8 (hardcover).

2385. The ontogeny of anatomic asymmetry: Constraints derived from basic mechanisms *-Glenn D. Rosen, Albert M. Galaburda and Gordon F. Sherman.* (pp. 215-238).

2386. Schneider, Linda H.; Cooper, Steven J. & Halmi, Katherine A. (Eds.). (1989). **The psychobiology of human eating disorders: Preclinical and clinical perspectives. Annals of the New York Academy of Sciences, Vol. 575.** New York, NY: New York Academy of Sciences. xii, 626 pp. ISBN 0-89766-541-4 (hardcover); 0-89766-542-2 (paperback).

2387. Animal models of human eating disorders *-Gerard P. Smith.* (pp. 63-74).

2388. Metabolic rate and feeding behavior -*Stylianos Nicolaidis and Patrick Even.* (pp. 86-105).

2389. Schopler, Eric & Mesibov, Gary B. (Eds.). (1987). **Neurobiological issues in autism. Current issues in autism.** New York, NY: Plenum Press. xxii, 418 pp. ISBN 0-306-42451-7 (hardcover).

2390. The neurochemical basis of symptoms in the Lesch-Nyhan Syndrome: Relationship to central symptoms in other developmental disorders -*George R. Breese, Robert A. Mueller and Stephen R. Schroeder.* (pp. 145-160).

2391. Possible brain opioid involvement in disrupted social intent and language development of autism -*Jaak Panksepp and Tony L. Sahley.* (pp. 357-372).

2392. Schulz, S. Charles & Tamminga, Carol A. (Eds.). (1989). **Schizophrenia: Scientific progress.** New York, NY: Oxford University Press. xv, 418 pp. ISBN 0-19-505527-6 (hardcover).

2393. Animal models of schizophrenic disorders -*W. T. McKinney.* (pp. 141-154).

2394. Limbic system: Localization of PCP drug action in rat and schizophrenic manifestations in humans -*C. A. Tamminga, G. K. Thaker, L. D. Alphs and T. N. Chase.* (pp. 163-172).

2395. A possible animal model of defect state schizophrenia -*R. J. Wyatt, R. Fawcett and D. Kirch.* (pp. 184-189).

2396. Seligman, Martin E. P. (1975). **Helplessness: On depression, development, and death. A series of books in psychology.** San Francisco, CA: W. H. Freeman & Co, Publishers. xv, 250 pp. ISBN 0-7167-0752-7 (hardcover); 0-7167-0751-9 (paperback).
Preface * Introduction * Controllability * Response independence and response contingency * Experimental studies * Helplessness saps the motivation to initiate responses * Helplessness disrupts the ability to learn * Helplessness produces emotional disturbance * Theory: Cure and immunization * Physiological approaches to helplessness * Depression * Anxiety and unpredictability * Systematic desensitization and uncontrollability * Emotional development and education * Maternal deprivation * Predictability and controllability in childhood and adolescence * Death * Death from helplessness in animals * Death from helplessness in humans * Notes * Bibliography * Name index * Subject index

2397. Serban, George & Kling, Arthur (Eds.). (1976). **Animal models in human psychobiology.** New York, NY: Plenum Press. xiv, 297 pp. ISBN 0-306-30864-9 (hardcover).

2398. New perspectives in psychiatry: Relevance of the psychopathological animal model to the human -*George Serban, Pierre Pichot, Alfred F. Freedman and Sol Kittay.* (pp. 1-6).

2399. Factors affecting responses to social separation in rhesus monkeys -*Stephen J. Suomi.* (pp. 9-26).

2400. Human personality development in an ethological light -*John Bowlby.* (pp. 27-36).

2401. Phylogenetic and cultural adaptation in human behavior -*Irenäus Eibl-Eibesfeldt.* (pp. 77-98).

2402. Unpredictability in the etiology of behavioral deviations -*Jules H. Masserman.* (pp. 99-110).

2403. Animal models of violence and hyperkinesis: Interaction of psychopharmacologic and psychosocial therapy in behavior modification -*Samuel S. Corson, E. O'Leary Corson, L. Eugene Arnold and Walter Knopp.* (pp. 111-139).

2404. Coping behavior and neurochemical changes in rats: An alternative explanation for the original "learned helplessness" experiments -*Jay Weiss, H. I. Glazer and L. A. Pohorecky.* (pp. 141-173).

2405. The use of differences and similarities in comparative psychpathology -*R. A. Hinde.* (pp. 187-202).

2406. Animal models for brain research -*José M. R. Delgado.* (pp. 203-218).

2407. Drug effects on foot-shock-induced agitation in mice -*Samuel Irwin, Roberta G. Kinohi and Elaine M. Carlson.* (pp. 219-237).

2408. Indole hallucinogens as animal models of schizophrenia -*Edward F. Domino.* (pp. 239-259).

2409. Animal models for human psychopathology: Observations from the vantage point of clinical psychopharmacology -*Dennis L. Murphy.* (pp. 265-271).

2410. The significance of ethology for psychiatry -*G. Serban.* (pp. 279-289).

2411. Shapiro, Alvin P. & Baum, Andrew (Eds.). (1991). **Behavioral aspects of cardiovascular disease. Perspectives in behavioral medicine.** Hillsdale, NJ: Lawrence Erlbaum Associates, Inc. xv, 370 pp. ISBN 0-8058-0771-3 (hardcover).

2412. Nonhuman primates as a model for evaluating behavioral influences on atherosclerosis, and cardiac structure and function -*J. R. Kaplan, S. B. Manuck, M. R. Adams, J. K. Williams, A. P. Selwyn and T. B. Clarkson.* (pp. 105-129).

2413. Simon, Pierre; Soubrié, Philippe & Widlocher, D. (Eds.). (1988). **An inquiry into schizophrenia and depression. Animal models of psychiatric disorders, Vol. 2.** Basel, Switzerland [US Location: New York, NY]: S. Karger, AG. vi, 212 pp. ISBN 3-8055-4757-9 (hardcover).

2414. A cross-species model of psychosis. Sensorimotor gating deficits in schizophrenic patients and dopaminergically activated rats -*N. R. Swerdlow, G. F. Koob, M. A. Geyer, R. Mansbach and D. L. Braff.* (pp. 1-18).

2415. Toward an animal model of schizophrenic attention disorder -*P. R. Solomon and A. Crider.* (pp. 21-42).

2416. Order and disorder of behaviour: The dopamine connection -*J. L. Evenden and C. N. Ryan.* (pp. 49-88).

2417. Speculations on the developmental neurobiology of protest and despair *-G. W. Kraemer.* (pp. 101-139).

2418. Neurochemical correlates of post-amphetamine depression and sensitization in animals. Implications for behavioral pathology *-L. Kokkinidis.* (pp. 148-173).

2419. The learned helplessness model of human depression *J. B. Overmier and D. H. Hellhammer.* (pp. 177-202).

2420. Simon, Pierre; Soubrié, Philippe & Widlocher, D. (Eds.). (1988). **Selected models of anxiety, depression and psychosis. Animal models of psychiatric disorders, Vol. 1.** Basel, Switzerland [US Location: New York, NY]: S. Karger, AG. vi, 198 pp. ISBN 3-8055-4667-X (hardcover).

2421. Animal models of human psychopathology *-J. B. Overmier and J. Patterson.* (pp. 1-35).

2422. The slow therapeutic action of antipsychotic drugs. A possible mechanism involving the role of dopamine in incentive learning *-R. J. Beninger.* (pp. 36-51).

2423. The potentiated startle response as a measure of conditioned fear and its relevance to the neurobiology of anxiety *-M. Davis.* (pp. 61-89).

2424. The olfactory bulbectomized rat model of depression *-B. E. Leonard.* (pp. 98-109).

2425. Animal models of aversion *-F. G. Graeff.* (pp. 115-141).

2426. How good is social interaction as a test of anxiety? *-S. E. File.* (pp. 151-166).

2427. Nootropic drugs in dementia: Preclinical prospects and clinical lessons *-R. J. Katz.* (pp. 175-188).

2428. Smotherman, William P. & Robinson, Scott R. (Eds.). (1988). **Behavior of the fetus.** Caldwell, NJ: Telford Press. ix, 231 pp. ISBN 0-936923-13-X (hardcover); 0-936923-14-8 (paperback).

2429. Dimensions of fetal investigation *-W. P. Smotherman & S. R. Robinson.* (pp. 19-34).

2430. Sonderegger, Theo B. (Ed.). (1985). **Nebraska Symposium on Motivation, 1984: Psychology and gender. Current theory and research in motivation, Vol. 32.** Lincoln, NE: University of Nebraska Press. xx, 326 pp. ISBN 0-8032-4152-6 (hardcover); 0-8032-9150-7 (paperback).

2431. Perinatal psychoactive drug use: Effects on gender, development, and function in offspring *-Joan C. Martin.* (pp. 227-266).

2432. Sonderegger, Theo B. (Ed.). (1992). **Perinatal substance abuse: Research findings and clinical implications. The Johns Hopkins series in environmental toxicology.** Baltimore, MD: Johns Hopkins University Press. x, 355 pp. ISBN 0-8018-4275-1 (hardcover).

2433. Methodological issues: Laboratory animal studies of perinatal exposure to alcohol or drugs and human studies of drug use during pregnancy *-Joanne Weinberg, Theo B. Sonderegger and Ira J. Chasnoff.* (pp. 13-50).

2434. Stahl, S. M.; Iversen, S. D. & Goodman, E. C. (Eds.). (1987). **Cognitive neurochemistry.** Oxford, England [US Location: New York, NY]: Oxford University Press. xiv, 395 pp. ISBN 0-19-854225-9 (hardcover).

2435. Amnesia, personal memory, and the hippocampus: Experimental neuropsychological studies in monkeys *-David Gaffan.* (pp. 46-56).

2436. Primate models of senile dementia *-N. M. J. Rupniak and S. D. Iversen.* (pp. 57-72).

2437. Do hippocampal lesions produce amnesia in animals? *-J. N. P. Rawlins.* (pp. 73-89).

2438. Problems in the cognitive neurochemistry of Alzheimer's disease *-Daniel Collerton.* (pp. 272-302).

2439. Startsev, V. G.; Schweinler, M. (Trans.) & Pahn, V. (Trans.). (1976). **Primate models of human neurogenic disorders.** Hillsdale, NJ: Lawrence Erlbaum Associates, Inc. x, 198 pp.

2440. Steinhauer, Stuart R.; Gruzelier, J. H. & Zubin, J. (Eds.). (1991). **Neuropsychology, psychophysiology, and information processing. Handbook of schizophrenia, Vol. 5.** Amsterdam, Netherlands [US Location: New York, NY]: Elsevier Science Publishing Co, Inc. xi, 687 pp. ISBN 0-444-90437-9 (hardcover, series); 0-444-81267-9 (hardcover, vol. 5).

2441. Possible animal models of some of the schizophrenias and their response to drug treatment *-S. M. Antelman.* (pp. 161-183).

2442. Strelau, Jan & Eysenck, Hans J. (Eds.). (1987). **Personality dimensions and arousal. Perspectives on individual differences.** New York, NY: Plenum Press. xviii, 325 pp. ISBN 0-306-42437-1 (hardcover).

2443. Individual characteristics of brain limbic structures interactions as the basis of Pavlovian/Eysenckian typology *-Pavel V. Simonov.* (pp. 121-131).

2444. Stricker, Edward M. (Ed.). (1990). **Neurobiology of food and fluid intake. Handbook of behavioral neurobiology, Vol. 10.** New York, NY: Plenum Press. xxii, 553 pp. ISBN 0-306-43458-X (hardcover).

2445. Comparative studies of feeding *-F. Reed Hainsworth and Larry L. Wolf.* (pp. 265-296).

2446. Food selection *-Paul N. Rozin and Jay Schulkin.* (pp. 297-328).

2447. Sudakov, Konstantin V.; Ganten, Detlev & Nikolov, Nicola A. (Eds.). (1989). **Perspectives on research in emotional stress. Systems research in physiology, Vol. 3.** London, England [US Location: New York, NY]: Gordon and Breach Science Publishers. xi, 400 pp. ISBN 2-88124-699-0 (hardcover).

2448. The role of stress in experimental neuroses *-W. H. Bridger and D. M. Stoff.* (pp. 217-224).

2449. Operant conditioning of the cardiovascular adjustments to exercise -*B. T. Engel and M. Talan.* (pp. 353-366).

2450. Tamminga, Carol A. & Schulz, S. Charles (Eds.). (1991). **Schizophrenia research. Advances in neuropsychiatry and psychopharmacology, Vol. 1.** New York, NY: Raven Press, Publishers. xviii, 373 pp. ISBN 0-88167-675-6 (hardcover).

2451. Gene expression: Implications for the study of schizophrenia and other neuropsychiatric disorders -*Edward I. Ginns.* (pp. 31-38).

2452. The limbic system in schizophrenia: Pharmacologic and metabolic evidence -*Carol A. Tamminga, Hiro Kaneda, Robert Buchanan, Brian Kirkpatrick, Gunvant K. Thaker, Mary Beth Yablonski and Henry H. Holcomb.* (pp. 99-109).

2453. An animal model for childhood autism: Memory loss and socioemotional disturbances following neonatal damage to the limbic system in monkeys -*Jocelyne Bachevalier.* (pp. 129-140).

2454. Tasman, Allan & Goldfinger, Stephen M. (Eds.). (1991). **American Psychiatric Press review of psychiatry, Vol. 10.** Washington, DC: American Psychiatric Press, Inc. xvii, 675 pp. ISBN 0-88048-436-5 (hardcover).

2455. Molecular biological and neurobiologic contributions to our understanding of Alzheimer's disease -*Joseph T. Coyle.* (pp. 515-527).

2456. Tuma, A. Hussain & Maser, Jack (Eds.). (1985). **Anxiety and the anxiety disorders.** Hillsdale, NJ: Lawrence Erlbaum Associates, Inc. xxxv, 1020 pp. ISBN 0-89859-532-0 (hardcover).

2457. Issues in the neuropsychology of anxiety -*Jeffrey A. Gray.* (pp. 5-25).

2458. Benzodiazepine-GABA interactions: A model to investigate the neurobiology of anxiety -*E. Costa.* (pp. 27-52).

2459. The neuroendocrinology of anxiety -*Peter Stokes.* (pp. 53-76).

2460. Animal models of anxiety-based disorders: Their usefulness and limitations -*Susan Mineka.* (pp. 199-244).

2461. The limitations of animal models in understanding anxiety -*Frederick H. Kanfer.* (pp. 245-259).

2462. Theoretical models relating animal experiments on fear to clinical phenomena -*Neal E. Miller.* (pp. 261-272).

2463. Ursin, Holger & Murison, Robert (Eds.). (1983). **Biological and psychological basis of psychosomatic disease. Advances in the biosciences, Vol. 42.** Oxford, England [US Location: New York, NY]: Pergamon Press, Inc. x, 286 pp. ISBN 0-08-029774-9 (hardcover).

2464. The stress concept -*Holger Ursin and Robert Murison.* (pp. 7-13).

2465. Coping: An overview -*Seymour Levine.* (pp. 15-26).

2466. Positive and negative expectancies: The rat's reward environment and pituitary-adrenal activity -*Gary D. Coover.* (pp. 45-60).

2467. Biological and psychological bases of gastric ulceration -*Robert Murison and Eva Isaksen.* (pp. 239-248).

2468. Sustained activation and psychiatric illness -*Gary D. Coover, Holger Ursin and Robert Murison.* (pp. 249-258).

2469. Sustained activation and disease -*Holger Ursin, Robert Murison and Stein Knardahl.* (pp. 269-277).

2470. van Praag, Herman M.; Plutchik, Robert & Apter, Alan (Eds.). (1990). **Violence and suicidality: Perspectives in clinical and psychobiological research. Clinical and experimental psychiatry, Vol. 3.** New York, NY: Brunner/Mazel, Inc. xviii, 332 pp. ISBN 0-87630-551-6 (hardcover).

2471. Serotonergic involvement in aggressive behavior in animals -*Berend Olivier, Jan Mos, Martin Tulp, Jacques Schipper, Sjaak den Daas and Geert van Oortmerssen.* (pp. 79-137).

2472. Parallels in aggression and serotonin: Consideration of development, rearing history, and sex differences -*J. D. Higley, Stephen J. Suomi and Markku Linnoila.* (pp. 245-256).

2473. Monoaminergic control of waiting capacity (impulsivity) in animals -*P. Soubrié and J. C. Bizot.* (pp. 257-272).

2474. Dopamine agonist-induced dyskinesias, including self-biting behavior, in monkeys with supersensitive dopamine receptors -*Menek Goldstein.* (pp. 316-323).

2475. Verny, Thomas R. (Ed.). (1987). **Pre- and perinatal psychology: An introduction.** New York, NY: Human Sciences Press, Inc. 296 pp. ISBN 0-89885-327-3 (hardcover).

2476. The infantile amnesia paradigm: Possible effects of stress associated with childbirth -*Janis W. Catano and Victor M. Catano.* (pp. 36-51).

2477. Wachs, Theodore D. & Plomin, Robert (Eds.). (1991). **Conceptualization and measurement of organism-environment interaction.** Washington, DC: American Psychological Association. xii, 191 pp. ISBN 1-55798-126-4 (hardcover).

2478. Toward a more temporal view of organism-environment interaction -*Gene P. Sackett.* (pp. 11-28).

2479. Wada, Juhn A. (Ed.). (1990). **Kindling 4. Advances in behavioral biology, Vol. 37.** New York, NY: Plenum Press. xiii, 477 pp. ISBN 0-306-43605-1 (hardcover).

2480. Basic mechanisms underlying seizure-prone and seizure-resistant sleep and awakening states in feline kindled and penicillin epilepsy -*M. N. Shouse, A. King, J. Langer, K. Wellesley, T. Vreeken, K. King, J. Siegel and R. Szymusiak.* (pp. 313-327).

2481. Kindling, anxiety and limbic epilepsy: Human and animal perspectives *-R. Adamec.* (pp. 329-341).

2482. Wagner, Hugh & Manstead, Antony (Eds.). (1989). **Handbook of social psychophysiology. Wiley handbooks of psychophysiology.** Chichester, England [US Location: New York, NY]: John Wiley & Sons. xvi, 447 pp. ISBN 0-471-91156-9 (hardcover).

2483. The neurobiology of emotions: Of animal brains and human feelings *-Jaak Panksepp.* (pp. 5-26).

2484. Walker, Clarence Eugene (Ed.). (1991). **Clinical psychology: Historical and research foundations. Applied clinical psychology.** New York, NY: Plenum Press. xii, 539 pp. ISBN 0-306-43757-0 (hardcover).

2485. Animal models of psychopathology *-Susan Mineka and Richard Zinbarg.* (pp. 51-86).

2486. Psychological research in depression and suicide: A historical perspective *-E. Edward Beckham.* (pp. 183-201).

2487. Walsh, Roger N. & Greenough, William T. (Eds.). (1976). **Environments as therapy for brain dysfunction.** New York, NY: Plenum Press. viii, 376 pp.

2488. Warburton, David M. (Ed.). (1990). **Addiction controversies.** London, England [US Location: New York, NY]: Harwood Academic Publishers. xiii, 386 pp. ISBN 3-7186-5045-2 (hardcover).

2489. Drug addiction as a psychobiological process *-Michael A. Bozarth.* (pp. 112-134).

2490. Weiner, Herbert; Florin, Irmela; Murison, Robert & Hellhammer, Dirk (Eds.). (1989). **Frontiers of stress research. Neuronal control of bodily function: Basic and clinical aspects, Vol. 3.** Stuttgart, Federal Republic of Germany [US Location: Lewiston, NY]: Hans Huber Publishers, Inc. xii, 458 pp. ISBN 0-920887-39-2 (hardcover, Toronto); 3-456-81701-0 (hardcover, Bern).

2491. Reproductive behavior and physiology in prenatally stressed males *-Ingeborg L. Ward and O. Byron Ward.* (pp. 9-20).

2492. Neurochemical basis of stress-induced depression *-Jay M. Weiss, Prudence G. Simpson and Peter E. Simson.* (pp. 37-50).

2493. Stress and the immune response *-Rudy E. Ballieux and Cobi J. Heijnen.* (pp. 51-55).

2494. Disease consequences of early maternal separation *-Sigurd H. Ackerman.* (pp. 85-93).

2495. Cardiovascular effects of emotional behavior in animals and humans *-Gianfranco Parati, Roberto Casadei and Giuseppe Mancia.* (pp. 100-110).

2496. Central pathways of emotional plasticity *-Joseph E. LeDoux.* (pp. 122-136).

2497. Stress and the immune response: Interactions of peptides, gonadal steroids and the immune system *-Charles J. Grossman.* (pp. 181-190).

2498. Behavioral stress enhances sensitivity of serotonergic system in an animal model of depression *-Joseph N. Hingtgen, John S. Gerometta, Aimee R. Mayeda, Jay R. Simon, John R. Hofstetter and Morris H. Aprison.* (pp. 376-378).

2499. West Virginia Dept of Mental Health, Div on Alcoholism & Drug Abuse & West Virginia U, Medical School, Dept of Behavioral Medicine & Psychiatry. (1973). **Ninth Annual West Virginia School on Alcohol and Drug Abuse Studies: June 17–22, 1973: Selected papers.** Morgantown, WV: West Virginia U. 165 pp.

2500. Experimental studies of alcohol dependence in animal models *-Fred W. Ellis and James R. Pick.*

2501. Willner, Paul (Ed.). (1991). **Behavioural models in psychopharmacology: Theoretical, industrial and clinical perspectives.** Cambridge, England [US Location: New York, NY]: Cambridge University Press. xiii, 540 pp. ISBN 0-521-39192-X (hardcover).

2502. Behavioural models in psychopharmacology *-Paul Willner.* (pp. 3-18).

2503. Animal models of anxiety *-Simon Green and Helen Hodges.* (pp. 21-49).

2504. Screening for anxiolytic drugs *-D. N. Stephens and John S. Andrews.* (pp. 50-75).

2505. Animal models of anxiety: A clinical perspective *-Malcolm Lader.* (pp. 76-88).

2506. Animal models of depression *-Paul Willner.* (pp. 91-125).

2507. Screening for new antidepressant compounds *-Wojciech Danysz, Trevor Archer and Christopher J. Fowler.* (pp. 126-156).

2508. The clinical relevance of animal models of depression *-J. F. W. Deakin.* (pp. 157-174).

2509. Animal models of eating disorders *-Anthony M. J. Montgomery.* (pp. 177-214).

2510. Screening methods for anorectic, anti-obesity and orectic agents *-Jerry Sepinwall and Ann C. Sullivan.* (pp. 215-236).

2511. Animal models of eating disorders: A clinical perspective *-Donald V. Coscina and Paul E. Garfinkel.* (pp. 237-250).

2512. Animal models of mania and schizophrenia *-Melvin Lyon.* (pp. 253-310).

2513. Pharmacological evaluation of new antipsychotic drugs *-Sven Ahlenius.* (pp. 311-330).

2514. Animal models of schizophrenia and mania: Clinical perspectives *-John Cookson.* (pp. 331-356).

2515. Animal models of Alzheimer's disease and dementia (with an emphasis on cortical cholinergic systems) *-Steven D. Dunnett and Timory M. Barth.* (pp. 359-418).

2516. Strategies for drug development in the treatment of dementia -*John D. Salamone*. (pp. 419-436).

2517. Dementia: The role of behavioral models -*Harvey J. Altman, Samuel Gershon and Howard J. Normile*. (pp. 437-450).

2518. Animal models of drug abuse and dependence - *Andrew J. Goudie*. (pp. 453-484).

2519. Screening for abuse and dependence liabilities - *David J. Sanger*. (pp. 485-502).

2520. The relevance of behavioural models of drug abuse and dependence liabilities to the understanding of drug misuse in humans -*Richard Hartnoll*. (pp. 503-519).

2521. Wolf, Marion E. & Mosnaim, Aron D. (Eds.). (1990). **Posttraumatic stress disorder: Etiology, phenomenology, and treatment.** Washington, DC: American Psychiatric Press, Inc. xvi, 270 pp. ISBN 0-88048-299-0 (hardcover).

2522. Interrelationships between biological mechanisms and pharmacotherapy of posttraumatic stress disorder -*Matthew J. Friedman*. (pp. 205-225).

2523. Wolpaw, Jonathan R.; Schmidt, John T. & Vaughan, Theresa M. (Eds.). (1991). **Activity-driven CNS changes in learning and development. Annals of the New York Academy of Sciences, Vol. 627.** New York, NY: New York Academy of Sciences. xi, 399 pp. ISBN 0-89766-637-2 (hardcover); 0-89766-638-0 (paperback).

2524. Morphological aspects of synaptic plasticity in Aplysia: An anatomical substrate for long-term memory -*Craig H. Bailey and Mary Chen*. (pp. 181-196).

Section IV. Author Index

This Author Index contains references to all of the records published in this bibliography. All authors whose works are cited are listed alphabetically by surname. This index is intended to be a name index only and not a person index. For example, a listing for "Barclay, A." will be listed separately from "Barclay, Allan," although the names may refer to the same person; two listings for "Barclay, A." may refer to two different authors. As many as four authors are listed for each record; if there are more than four authors, then the first author is listed, followed by "et al." Numbers cited refer to citation numbers in the bibliography.

AUTHOR INDEX

AUTHOR INDEX

AUTHOR INDEX

AUTHOR INDEX

AUTHOR INDEX

AUTHOR INDEX

AUTHOR INDEX

AUTHOR INDEX

AUTHOR INDEX

AUTHOR INDEX

AUTHOR INDEX

Section V. Subject Index

This is a subject index to entries in Sections I and II and III. Subject terms in this index are taken from the *Thesaurus of Psychological Index Terms* (6th Edition, 1991). The term animal models is not included in the index. *See* references are used to refer to preferred forms of entry or from conceptually broader terms to narrower ones. *See Also* references alert the reader to more specific terms. Each entry is indexed with the most specific terms that are appropriate for the contents of that entry. The numbers cited refer to the entry number in the bibliography and are listed in numerical order.

SUBJECT INDEX

SUBJECT INDEX

SUBJECT INDEX

SUBJECT INDEX

APA's Involvement in Animal Models Research

APA has actively participated in the development of animal models of human behavioral pathology through sustained efforts in support of the use of animals in research and through advocacy for increased funding for such studies. These APA efforts are carried out mainly through the Science Directorate and the Public Policy Office and include the following:

The Committee on Animal Research and Ethics (CARE) advises APA on policy issues concerning the use of animals in research. Over several years, the Committee submitted extensive comments on proposed changes in the federal Animal Welfare Regulations, which helped shape the final regulations. These regulations govern every research protocol using animals and affect the design and conduct of all experiments using animal models.

APA's "Guidelines for Ethical Conduct in the Care and Use of Animals" are an elaboration of the "Ethical Principles of Psychologists" that all APA members follow. The Guidelines are specific for the use of animal models in behavioral research and serve as a teaching guide in the education of psychology students on the ethical care and use of experimental animals.

The Science Directorate Conference Program has supported 22 conferences since 1988, several of which covered research issues based on animal models of human behavioral pathology and issues on the maintenance of animals essential to this research. The proceedings published by the APA in the volume, *Through the Looking Glass: Issues of Psychological Well-Being in Nonhuman Primates* (1991), illustrates some of the major concerns APA has grappled with in this area.

APA advocacy on behalf of appropriations for research has included recommendations for increased funding for the federal agencies that support studies on human behavioral pathology. These agencies include the National Institutes of Health (NIH), the National Science Foundation (NSF), the Department of Defense (DOD), and the National Aeronautics and Space Administration (NASA).

The Funding Research Bulletin, compiled and disseminated through BITNET, provides free bimonthly information on current funding opportunities to researchers on human behavioral pathology using animals models.

The Science Directorate, in cooperation with federal agencies, such as the National Institute of Mental Health (NIMH), compiles and disseminates current information on behavioral research methods with animals. Such information provides guidance to researchers on standard practices and methods, as well as to institutional review committees that must approve all behavioral research protocols using animal models.

Search Strategy Used to Retrieve References for the Bibliography

Jody Kerby

The following is the refined search strategy executed on the PsycINFO database (through DIALOG) to retrieve journal and dissertation records on animal models of human pathology. Explanatory comments are provided to explain the search statements.

S animal models/de (Searches all records with the descriptor term for animal models since term was added to the 1988 *Thesaurus of Psychological Index Terms).*

S1 613

S animal( )model? (Searches free text for records prior to 1988)

39527 ANIMAL
68892 MODEL?
S2 1312 ANIMAL( )MODEL?

S (sh=33 or sh=32 or sh=23) and model?

153440 SH=33
138457 SH=32
69323 SH=23
S3 28587 (SH=33 or SH=32 or SH=23) AND MODEL?

S S3/animal (Searches for records on animal research in content areas of mental and psychological disorders, treatment and prevention, and human experimental psychology that don't use the specific term "animal model")

S4 303 S3/animal

S s1 or s2 or s4

S5 1414

An examination of the search results indicated that some behavior animal models were missed in the above research. To remedy these omissions, pathologies and authors were searched. Examples follow:

Limitall/animal (Used to limit search to animal research and search pathologies or authors).

S learned( )helplessness/ti,id,de
or
Au=Harlow, H?

[Books were retrieved through a search of the PsycBOOKS database in-house and by searching authors in the Book Chapters & Books portion of **PsycLIT**. Efforts were made to obtain (and add to the database) important books and chapters published from 1980 to 1987, when coverage of books in *Psychological Abstracts* was suspended while a new database was being devised. The efforts were not entirely successful; all books desired were not obtained and entered.]

UPDATE YOUR SEARCH AS FOLLOWS:

animal model/de (searches for records specifically on animal models)

or use **limitall/animal** (restricts all searches to animal research) and area of interest (e.g., **learned helplessness**)

and **ud=9206:999**

UPDATE ON PSYCLIT:

animal model/de and **ud>9206**

or

pathology (e.g., **learned helplessness**) or author (e.g., **Harlow-H-F**) and **po=animal** and **ud>9206** (If the population is primarily human, records will be classified as po=human, and some records will be missed with this strategy.)